AN ABBREVIATED LIST OF LAPLACE TRANSFORM PAIRS

$f(t)(t > 0^-)$	TYPE	$F(s)$
$\delta(t)$	(impulse)	1
$u(t)$	(step)	$\dfrac{1}{s}$
t	(ramp)	$\dfrac{1}{s^2}$
e^{-at}	(exponential)	$\dfrac{1}{s + a}$
$\sin \omega t$	(sine)	$\dfrac{\omega}{s^2 + \omega^2}$
$\cos \omega t$	(cosine)	$\dfrac{s}{s^2 + \omega^2}$
te^{-at}	(damped ramp)	$\dfrac{1}{(s + a)^2}$
$e^{-at} \sin \omega t$	(damped sine)	$\dfrac{\omega}{(s + a)^2 + \omega^2}$
$e^{-at} \cos \omega t$	(damped cosine)	$\dfrac{s + a}{(s + a)^2 + \omega^2}$

AN ABBREVIATED LIST OF OPERATIONAL TRANSFORMS

$f(t)$	$F(s)$
$Kf(t)$	$KF(s)$
$f_1(t) + f_2(t) - f_3(t) + \cdots$	$F_1(s) + F_2(s) - F_3(s) + \cdots$
$\dfrac{df(t)}{dt}$	$sF(s) - f(0^-)$
$\dfrac{d^2f(t)}{dt^2}$	$s^2F(s) - sf(0^-) - \dfrac{df(0^-)}{dt}$
$\dfrac{d^nf(t)}{dt^n}$	$s^nF(s) - s^{n-1}f(0^-) - s^{n-2}\dfrac{df(0^-)}{dt}$ $-s^{n-3}\dfrac{d^2f(0^-)}{dt} - \cdots - \dfrac{d^{n-1}f(0^-)}{dt}$
$\displaystyle\int_0^t f(x)\, dx$	$\dfrac{F(s)}{s}$
$f(t - a)u(t - a),\ a > 0$	$e^{-as}F(s)$
$e^{-at}f(t)$	$F(s + a)$
$f(at),\ a > 0$	$\dfrac{1}{a}F\left(\dfrac{s}{a}\right)$
$tf(t)$	$-\dfrac{dF(s)}{ds}$
$t^nf(t)$	$(-1)^n\dfrac{d^nF(s)}{ds^n}$
$\dfrac{f(t)}{t}$	$\displaystyle\int_s^\infty F(u)\, du$

ELECTRIC CIRCUITS

FOURTH EDITION

JAMES W. NILSSON

PROFESSOR EMERITUS
IOWA STATE UNIVERSITY

ELECTRIC CIRCUITS

FOURTH EDITION

JAMES W. NILSSON

PROFESSOR EMERITUS
IOWA STATE UNIVERSITY

WITH CONTRIBUTIONS FROM

SUSAN A. RIEDEL
MARQUETTE UNIVERSITY

DAVID P. SHATTUCK
UNIVERSITY OF HOUSTON

NORMAN WITTELS
WORCESTER POLYTECHNIC INSTITUTE

 ADDISON-WESLEY PUBLISHING COMPANY

Reading, Massachusetts • Menlo Park, California
New York • Don Mills, Ontario • Wokingham, England
Amsterdam • Bonn • Sydney • Singapore
Tokyo • Madrid • San Juan • Milan • Paris

TO ANNA

This book is in the
Addison-Wesley Series in Electrical and Computer Engineering

Senior Sponsoring Editor, Eileen Bernadette Moran
Production Supervisor, David Dwyer
Production Coordinator, Genevra Hanke
Copy Editor, Jerrold A. Moore
Text Designer, Sally Bindari, Designworks, Inc.
Illustrators, Capricorn Design
Cover Designer, Peter M. Blaiwas
Senior Manufacturing Manager, Roy Logan
Compositor, Interactive Composition Corporation
Art Coordinator, Amanda G. Lewis, Designworks, Inc.

Library of Congress Cataloging-in-Publication Data

Nilsson, James William.
    Electric circuits / James W. Nilsson with contributions by Susan
A. Riedel   . . .   [et al.] — 4th ed.
        p.   cm.
    ISBN 0-201-54987-5
    1. Electric circuits.   II. Riedel, Susan A.   II. Title.
TK454.N54   1993
621.319′2—dc20   92-23491
                                                                    CIP

1 2 3 4 5 6 7 8 9 10—DO—95949392

PREFACE

INTRODUCTORY REMARKS

Nineteen ninety-three marks the tenth anniversary of the first edition of *Electric Circuits*. I developed the first edition of the text based on over 30 years of teaching experience, keenly aware that the educational approaches that had worked well for me might not meet the teaching needs of my colleagues at other institutions. We have, over the years, received considerable validation in the form of market success and much direct feedback that *Electric Circuits* is a highly useful educational tool. I feel most honored and fortunate that the text has been used by over 150,000 students and am gratified that the value of our effort has continued to find so much favor among educators at a wide variety of colleges and universities around the world.

While the fourth edition of *Electric Circuits* represents a major revision of previous editions of the text, the underlying teaching approaches and philosophies are unchanged. There are some specific aims that distinguish this text from the other editions.

To build an understanding of concepts and ideas explicitly in terms of previous learning. The learning challenges faced by students of engineering circuit analysis are prodigious, and each new concept is built on a foundation of many other concepts. In *Electric Circuits,* much attention is paid to helping students recognize how new concepts and ideas fit on top of those previously learned.

To emphasize the relationship between conceptual understanding and problem-solving approaches. Our research surveys of this course indicate that a high percentage of faculty view the practical issue of developing students' problem-solving skills as a central challenge. As do other circuit analysis texts, *Electric Circuits* makes considerable use of examples and simple drill exercises to demonstrate problem-solving approaches and to give students practice opportunities. We do so not with the

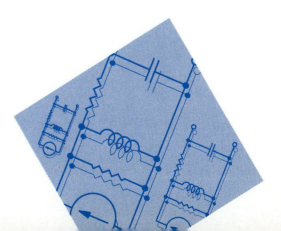

primary aim of giving students procedural models for solving problems; rather, we emphasize problem solving as a process in which one applies conceptual understanding to the solution of a practical problem. As such, in both the textual development and in the worked-out examples, we place great emphasis on problem-solving thought processes based on concepts rather than the use of rote procedures. Students are encouraged to think through problems before attacking them, and we often pause to consider the broader implications of a specific problem-solving situation. This approach is illustrated in Example 4.6, pp. 118–9.

To provide students with a strong foretaste of engineering practice. There are limited opportunities in a sophomore-year circuit analysis course to introduce students to real-world engineering experiences. I have worked hard to make the most of the opportunities that do exist, and this is reflected in a number of ways, including:

- *Homework solutions requiring insight.* Solution of a large percentage of the homework problems requires the kind of insight an engineer is expected to display when solving real-world problems. The worked-out examples, which emphasize engineering thought processes, also support the solution of real-world problems.

- *Realistic application problems.* I have included a large number of realistic application-type problems and exercises to help stimulate students' interest in engineering activity.

- *Use of realistic circuit parameters.* I have made a strong effort to develop problems and exercises that use realistic values and represent realizable physical situations.

What's New in the Fourth Edition

At first glance, the well-established practice of revising circuit analysis textbooks every three years may seem odd: after all, the principles of circuit analysis are well established, and even such recent innovations as the increased use of computers as problem-solving aids and the introduction of the operational amplifier in the course have become standardized. It may be surprising to hear, therefore, that we have come to regard each revision of *Electric Circuits* as a tremendous opportunity to make significant improvements in the book. The fourth edition has been a major undertaking and has involved far closer collaboration between myself and reviewers, contributors, and Addison-Wesley than has been the case in any previous edition.

In our work on the fourth edition of *Electric Circuits,* we were motivated by a strong desire to ensure that the text better supports the major learning styles of students in the 1990s. Compared to students of ten years ago, for example, today's undergraduates need more frequent visual support of learning to supplement verbal explanation, greater reinforcement through examples and drill exercises, and more opportunity to solve simple confidence-building homework exercises. This goal led us to incorporate significant changes in the design and pedagogy of the text. The major areas of change are outlined below.

NEW TEXT LAYOUT AND DESIGN

We have developed the interior design of the fourth edition to improve how students use this text to learn electric circuit analysis. With clarity and consistency as our overriding goals, we have adopted an interior design that creates an open, comfortable feel and achieves a highly efficient organization of information. The following information outlines some of our design improvements.

OVERALL ENHANCEMENT OF CLARITY AND CONSISTENCY OF FORMAT.
Too often, students are intimidated by the ocean of details and pedagogical "features" in modern circuits texts. Students might find themselves spending more time looking for and trying to sort through a dizzying array of information than studying the actual material. We wanted to support the learning process by creating a layout that establishes and sustains a higher level of comfort and accessibility than do previous editions. We have met these standards by:

Consistently using specific design elements. We have created a consistently applied visual language—horizontal rules to isolate topics and color to act as a visual signal in highlighting and identifying information—to present the material clearly and accurately. Every pedagogical element—chapter openers, example sections, drill exercises, chapter summaries, and end-of-chapter problems—is consistently formatted throughout the book.

Identifying different components by format for the reader. It is important that students acquire an ease and familiarity with different textual elements and learn to use each to its best advantage. This is achieved through the following design elements.

- *End-of-Chapter Problems* are cleanly displayed in an innovative design to enable students and professors to readily isolate particular problems and their respective figures. The layout of the end-of-chapter problems helps professors find problems for assignment more easily than the two-column problem sections typical of most engineering texts.

- *Drill Exercises* have been redesigned to allow more vertical space for scratch work and to distinguish the problem and solution parts within each exercise.

- *Example sections* are enclosed by horizontal rules to keep examples clearly separated from the text.

- *An abbreviated chapter table of contents* appears at the beginning of each chapter to quickly reference each chapter section.

- *Chapter summaries* are reformatted into color, bulleted lists that break down chapter information and remind students of key points.

"TIGHT" INTEGRATION OF TEXT AND ILLUSTRATIONS TO ENHANCE THE LEARNING PROCESS. Our close study of texts in this field (including previous editions of this text) led us to realize that students are often faced with the difficult task of searching around a page and even on other pages for illustrative material that accompanies text development. *Electric Circuits, 4/e* has been carefully designed so that:

Illustrations are uniformly placed on the page. In most textbooks, illustrations and their correlative text are often isolated from each other in a fashion that impedes learning. Illustrations in the fourth edition are now consistently located in the margins next to the corresponding text. This feature makes it convenient to look back and forth between the text and illustration.

Figures are repeated as needed. Illustrations that are referred to several times throughout the book are now often repeated to keep the student focused on the learning issue at hand and to prevent unnecessary page flipping.

EFFECTIVE PEDAGOGICAL USE OF COLOR. Use of a two-color format in a text is only warranted when it is used to enhance the learning process. The first edition of this book was one of the first electrical engineering textbooks to be published in a two-color format, and we feel that *Electric Circuits* has served as a positive standard for the functional use of color. In the fourth edition, the judicious use of a second color improves upon the text's pedagogical usefulness. We apply the second color exclusively to:

Operate as a visual cue. Color is used in figure caption numbers and footnote identifiers.

Enhance visualization in selected figures. In the figures, color is used only to differentiate information (such as varying plots on a graph) and to highlight key points (such as denoting current direction).

Add a fresh look to the chapter openers. We use color to stimulate the interest of students and to draw their attention into the chapters.

REVISED AND EXPANDED PROBLEM SETS

In addition to the design of the problem sets, a number of improvements have been made to the end-of-chapter problems. I have paid particular attention to expanding the range of homework problems, and have incorporated a larger number of confidence-building practice problems. The number of end-of-chapter problems has been increased. Most of the increase in problem selection occurs in the first 10 chapters. A majority of the problems are new or revised in this edition.

ADDITIONAL EXAMPLES AND DRILL EXERCISES

To provide students with additional learning support, I have increased the number of worked-out examples and drill exercises with most of the increase coming in the early chapters.

MINOR TOPICAL REORGANIZATION AND REWRITING

I have rewritten and reorganized some topics in response to the suggestions of reviewers and users of the third edition:

Chapter 1 has been rewritten to give the student more insight into electrical engineering as a career, thus making the connection that circuits serves as an introduction to this exciting profession.

Natural and step response of *RL* and *RC* circuits were topics previously covered in Chapters 8 and 9. Now, they have been combined to form a single chapter, *Chapter 8*. I rewrote this material to strengthen the tie between the natural and step response of a first-order circuit.

Natural and step response of *RLC* circuits, now located in *Chapter 9* (previously Chapter 10), has been rewritten with more emphasis placed on the thought process for finding the initial value of the unknown variable and its first derivative. This new approach highlights the circuit analysis aspects of finding the solution to the second-order circuit.

Introduction to the rms value of a signal has been introduced earlier (now in *Chapter 10*) to support the general discussion of the characteristics of a sinusoidal signal.

SOME EXPANSION AND CONTRACTION OF TOPICS

There are a small number of topical additions:

A section on Bartlett's bisection theorem has been added to *Chapter 6,* The Operational Amplifier, to reinforce the concept of common-mode and differential-mode signal components (Section 6.10).

The three-phase circuits chapter (*Chapter 12*) has been shortened by dropping the discussion of the two-wattmeter method for measuring power.

Additional changes have been made to strengthen the pedagogy of the book. Included among these changes are the following:

• The circuit model of a flashlight was rewritten as an example of constructing a circuit model based on the knowledge of the behavior of the system components.

• An example of constructing a circuit model based on terminal measurements was added in *Chapter 2* to expand the discussion of modeling.

• Several new examples have been introduced in *Chapter 2* to illustrate the interconnection of ideal sources, paying attention to both valid and invalid connections.

• The discussion of source transformations was expanded to include an ideal voltage source in parallel with a resistor and an ideal current source in series with a resistor (Section 4.9).

• An example of low-pass filter used in a telephone circuit has been added to the discussion of scaling (Example 14.10, pp. 583–6).

• The introduction to the decibel has been expanded to include decibels with respect to one milliwatt or dBm (Section 17.8).

EXAMPLES AND DRILL EXERCISES

Numerical examples are used extensively throughout the text to help readers understand how theory is applied to the analysis of a circuit structure (we have added roughly 20% new examples in the fourth edition). Because many students seem to value worked examples more than any other aspect of the text, these devices represent an important opportunity to influence the development of student's problem-solving behavior. The nature and format of the examples in *Electric Circuits* are a reflection of the overall teaching approach of the text. When presenting a solution, I place great emphasis on the importance of problem solving as a thought process that applies underlying concepts. By emphasizing this idea even in the solution of simple problems, I hope to communicate that problem-solving approaches based on conceptual understanding can help students handle the novelties and complexities that they will encounter later on. Some characteristics of the examples include:

- Encouraging the student to study the problem or the circuit and to make initial observations prior to diving into a solution pathway;

- Emphasizing the individual stages of the solution as part of solving the problem systematically, without suggesting that there are rote procedures for problem solving;

- Exploring decision making, that is, the idea that we are often faced with choosing among many different solution approaches;

- Challenging one's results, that is, emphasizing the importance of checking and testing answers based on one's knowledge of circuit theory and the real world.

Drill exercises are included in the text to give readers an opportunity to test their understanding of the material they have just read (we have added roughly 5% new drill exercises in the fourth edition). The text of the drill exercises is presented in a double-column format as a way of signaling readers to stop and solve the exercise before proceeding to the next section.

HOMEWORK PROBLEMS

The homework problems in *Electric Circuits* have been consistently rated by users of the text as one of the book's most attractive features. It is useful to get this feedback because it has served over the years to quiet my editors' complaints that it sometimes takes me several hours to develop a single problem. As with previous editions, I have spent a high percentage of my revision efforts on the development of the homework problem sets. The overall number of problems has been increased by 25% in the fourth edition, and 90% of the problems are new or revised. In fact, the fourth edition includes over 900 problems.

Given the wide range of functionality of the homework problems, I feel that instructors need to be aware of the different types of problems presented in this text and the goals I have adhered to in their development. Therefore, I have designed the problems around the following objectives:

- To give students practice in using the analytical techniques developed in the text;

- To show students that analytical techniques are tools, not objectives;

- To give students practice in choosing the analytical method to be used in obtaining a solution;

- To show students how the results from one solution can be used to find other information about a circuit's operation;

- To encourage students to challenge the solution either by using an alternate method or by testing the solution to see if it makes sense in terms of known circuit behavior;

- To introduce students to practical applications of circuit analysis;

- To introduce students to design oriented problems;

- To give students practice in deriving and manipulating equations where quantities of interest are expressed as functions of circuit variables such as R, L, C, ω, and so forth; this type of problem also supports the design process;

- To give students practice in using a computer, via PSpice®, to analyze circuit behavior; it is important at the introductory level to use the computer to solve problems whose solutions can be challenged by the beginning student;

- To challenge students with problems that will stimulate their interest in both electrical and computer engineering.

With the foregoing goals in mind I have designated ten categories of homework problems, which are described more fully in a supplement to this text, the *Instructor's Road Map*. The ten categories are: practice, analytic tool, open method, additional information, solution check, practical, design, derivation, hints, and PSpice. By categorizing the end-of-chapter problems, I hope to facilitate assignment planning and to give professors a greater arsenal of pedagogical tools to help students master concepts in this field.

PREREQUISITES

In writing the first fifteen chapters of the text, I have assumed that the reader has taken a course in elementary differential and integral calculus. I have also assumed that the reader has had an introductory physics course, at either the high school or university level, that introduces the concepts of energy, power, electric charge, electric current, electric potential, and electromagnetic fields. In writing Chapters 15–20, I have assumed the student has had, or is enrolled in, an introductory course in differential equations.

COURSE OPTIONS

The text has been designed for use in either a two-semester or a three-quarter sequence. Assuming three lectures per week, the first eleven chapters can be covered during the first semester, leaving *Chapters 12–20* for the second semester. If heavy use is made of the PSpice supplement, some topics may have to be omitted from the coverage. On an academic quarter schedule, the book can be subdivided into three parts: *Chapter 1–8, Chapters 9–15,* and *Chapters 16–20.*

The text can also be used in a single-semester introduction to circuit analysis. After covering *Chapters 1–4, Chapters 7–9* (omitting Sections 8.7 and 9.5), and *Chapters 10 and 11,* the instructor can choose from *Chapter 6* (operational amplifiers), *Chapter 12* (three-phase circuits), *Chapter 13* (mutual inductance), *Chapter 14* (resonance), and *Chapter 18* (Fourier series) to develop the desired emphasis.

The introduction of operational-amplifier circuits into the text has been presented so that this material can be omitted without interfering with the reading of the subsequent chapters. For example, if *Chapter 6* is omitted, the instructor simply omits Section 8.7, Section 9.5, and those problems and drill exercises in the chapters following *Chapter 6* that pertain to operational amplifiers.

There are four appendixes at the end of the book to help readers make effective use of their mathematical background. Appendix A reviews Cramer's method of solving simultaneous linear equations and simple matrix algebra; complex numbers are reviewed in Appendix B; Appendix C is devoted to an abbreviated table of trigonometric identities that are useful in circuit analysis; and an abbreviated table of useful integrals is given in

Appendix D. A fifth appendix, Appendix E, provides a comprehensive list of the Examples with titles and corresponding page numbers.

SUPPLEMENTS

In addition to the attention we have given to the revision of *Electric Circuits,* we have put care into the development of supplements that capitalize on the learning potential in the fourth edition. Students and professors are constantly challenged in terms of time and energy by the confines of the classroom and the importance of integrating new information and technologies into an electric circuits course. Through the following supplements, we believe we have succeeded in making some of these challenges more manageable.

INSTRUCTOR'S ROAD MAP

This time-saving supplement is the first of its kind in the introductory circuits field. The *Instructor's Road Map* is designed to enable professors to orient themselves quickly to this text and to easily find the kind of end-of-chapter problems they plan to assign. To this end, we have provided an overview of the table of contents of the fourth edition and the end-of-chapter problems. For quick reference, we have categorized these problems into ten groups (see the Homework Problems section of this preface for more details), and also created a list of titled examples with corresponding page numbers to facilitate course planning. Finally, a blank "Notes" page appears on the back of each problem and example list page in the Road Map to provide a space for the professor to record his assignments and to log information about the problems and examples. This supplement is available free to all adopters.

INTRODUCTION TO PSPICE®

Over the past few years, more and more instructors have begun to use circuit simulation software to supplement their basic course in electric circuits. To meet the needs of instructors who want instructional and reference materials to help them integrate software simulation tools into their electric circuits courses, a free manual accompanies this fourth edition of *Electric Circuits.* This manual is a substantially revised version of the manual, *In-*

troduction to SPICE, which accompanied the third edition of *Electric Circuits.*

The new manual, *Introduction to PSpice,* focuses on the PSpice circuit simulation package from MicroSim. Features distinct to PSpice are covered, including voltage- and current-controlled switches and the commands .STEP and .LIB. Also included are descriptions of the software package PROBE and a graphics post-processor bundled with PSpice, which can be used to produce high quality plots of PSpice output.

The topics covered by the manual have been re-ordered to correspond with the coverage of those same topics in the text. In addition, margin notes appearing in both the text and the manual will enable professors to integrate PSpice efficiently in their circuits course.

Four appendixes have been added. The first is a guide to using the menus for the PSpice Control Shell and the PROBE software. The second summarizes the options that control PSpice simulation. The third identifies the differences between PSpice and SPICE. The fourth is a quick reference to PSpice syntax.

Revised in content, the manual's format and style has been updated to follow the same design as *Electric Circuits.* The manual is again published as a separate booklet to facilitate its use at the computer. It uses many of the pedagogical features in the text, such as the two-color format and worked-out examples. The manual has a large number of problems designed to teach the student how to describe the circuit under simulation to PSpice, and to teach the importance of validating the simulation results to ensure that the computer-generated output makes sense in terms of known circuit behavior. *Introduction to PSpice* is included free with each copy of the text.

SOLUTIONS MANUAL, VOLUMES I AND II

The solutions manual contains complete worked solutions with supporting figures to all of the 900-plus end-of-chapter problems in the fourth edition. Volume I covers Chapters 1–10, and Volume II covers Chapters 11–20. These supplements, available free to all adopting faculty, were checked for accuracy by ten circuits professors. The manuals are not available for sale to students.

TRANSPARENCY MASTERS

This supplement consists of enlargements of selected figures from the fourth edition. Transparency masters will be free upon request to all adopters.

CircuitTutor® by TutorWare, Inc., Developed by Burks Oakley II for Macintosh® and Windows®

CircuitTutor is an interactive software tutorial for electric circuits designed to encourage the development of the analytical skills required to master circuits problems. When working through problems, the program offers students immediate feedback and reinforcement, as well as step-by-step guidance upon request. *CircuitTutor* serves as a supplement to course lectures and textual material. It is available for both Macintosh and IBM-PC® compatible computers.

ACKNOWLEDGMENTS

It is not possible, after over forty years of teaching, to mention all the people who have contributed to my development as an engineering educator. I do wish to continue to acknowledge those colleagues who generously offered their advice and counsel during the writing of the first three editions of this text. The support of both Iowa State University and the University of Notre Dame during the earlier development of this text is also greatly appreciated.

In developing the fourth edition, I would like to express my appreciation for the contributions of Professors Susan A. Riedel (Marquette University), David P. Shattuck (University of Houston), and Norman Wittels (Worcester Polytechnic Institute). Their wise counsel and drafts for rewriting Chapters 1, 8, 9, and 10, the chapter summaries, and the PSpice manual have improved the pedagogical character of the book. I would also like to express my appreciation to Iowa State University Professors Thomas Scott (who contributed the section on Bartlett's bisection theorem and the example of scaling a low-pass filter used in a telephone circuit) and Curran Swift (who reviewed the PSpice problems and the final manuscript) for their continued interest in the evolution of the text.

Special thanks are also due to Professor David P. Shattuck and the University of Houston for allowing me to return to the classroom during the development phase of the fourth edition.

I would also like to thank the following professors who undertook the task of checking the end-of-chapter problems for accuracy: James Gottling, Ohio State University; M. H. Hopkins Jr., Virginia Polytechnic Institute; Charles M. McKeough, Villanova University; Robert H. Miller, Virginia Polytechnic

Institute; Burks Oakley II, University of Illinois at Urbana-Champaign; Curran Swift, Iowa State University; John Taylor, Texas A & M University; Arthur T. Tiedemann, University of Wisconsin; Norman Wittels, Worcester Polytechnic Institute; and Xiao-Bang Xu, Clemson University.

There are many people behind the scenes at Addison-Wesley who have worked hard to produce the fourth edition. They have done a superb job in producing this book. A special thanks to David Dwyer, Production Supervisor, for his competence and Bette Aaronson, Managing Editor, for her continued interest. It is a pleasure for me to recognize the tremendous support and encouragement I have received from Eileen Bernadette Moran, Senior Engineering Editor, and the continued support of Don Fowley, Executive Editor. Their genuine interest in engineering education and quality textbooks has made my job that much more enjoyable.

BRIEF CONTENTS

CONTENTS

APPENDIX C   AN ABBREVIATED TABLE OF TRIGONOMETRIC IDENTITIES   899

APPENDIX D   AN ABBREVIATED TABLE OF INTEGRALS   901

APPENDIX E   LIST OF EXAMPLES   903

ANSWERS TO SELECTED PROBLEMS   909

INDEX   919

CIRCUIT VARIABLES

CHAPTER 1

Electrical engineering has been and continues to be an exciting and challenging profession for anyone who has a genuine interest in, and aptitude for, applied science and mathematics. Over the past century and a half, electrical engineers have played a dominant role in the development of systems that have changed the way people live and work. Satellite communication links, telephones, digital computers, televisions, diagnostic and surgical medical equipment, assembly-line robots, and electric power tools are representative components of systems that define a modern technological society. The need for continuous improvement and refinement of these existing systems, along with the desire to discover and develop new systems, continues to make electrical engineering an exciting area of study.

As you embark on the study of circuit analysis, you need to gain a feel for where electric circuits fits into the hierarchy of topics that comprise an introduction to electrical engineering. Hence we begin by presenting an overview of electrical engineering, some ideas about an engineering point of view as it relates to circuit analysis, and a review of the international system of units.

We then describe generally what circuit analysis entails. Next, we introduce the concepts of voltage and current. We follow those concepts with discussion of an ideal basic circuit element and the need for a polarity reference system. We conclude the chapter by describing how current and voltage relate to power and energy.

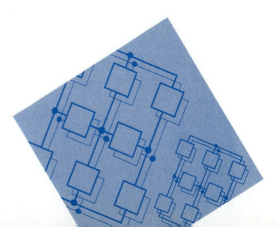

1.1 ELECTRICAL ENGINEERING: AN OVERVIEW

Electrical engineering is the profession concerned with systems that produce, transmit, and measure electric signals. Although closely associated with physics and mathematics, electrical engineering differs in that the emphasis is on applications of the physicist's models of natural phenomena combined with the mathematician's tools for manipulating those models to produce systems that meet practical needs. Electrical systems pervade our lives; they are found in homes, schools, workplaces, and transportation vehicles—everywhere. We begin by presenting a few examples from each of the five major classifications of electrical systems: (1) communication systems, (2) computer systems, (3) control systems, (4) power systems, and (5) signal-processing systems. Then we describe how electrical engineers analyze and design such systems.

Communication systems are electrical systems that generate, transmit, and distribute information. Well-known examples include television equipment, such as cameras, transmitters, receivers, and VCRs; radio telescopes, which are used to explore the universe; satellite systems, which return images of other planets and our own; radar systems used to coordinate plane flights; and telephone systems.

Figure 1.1 depicts the major components of a modern telephone system. Inside a telephone, a *microphone* turns sound waves into electric signals. They are carried to a switching station where they are combined with the signals from tens, hundreds, or thousands of other telephones. Depending on the distance to the other end of the circuit, the signals are sent through wires in underground coaxial cables, through air and space as radio waves, or through fiber-optic cables as pulses of light before they are routed to another phone where an *earphone* turns them back into sound. At each stage of the process, electrical circuits operate on the signals. Imagine the challenge involved in designing, building, and operating each circuit in a way that guarantees that all of the hundreds of thousands of simultaneous calls have high-quality connections.

Computer systems use electric signals to process information ranging from word processing to mathematical computations. Systems range in size and power from pocket calculators to personal computers to supercomputers that process weather data and model chemical interactions of complex organic molecules. These systems include networks of microcircuits, or *integrated circuits,* postage-stamp-sized assemblies of hundreds, thousands, or millions of electrical components which often operate

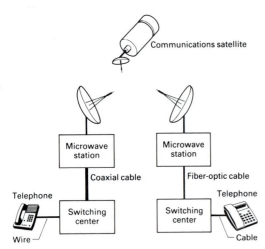

FIGURE 1.1 A telephone system.

at speeds and power levels close to fundamental physical limits—including the speed of light and the thermodynamic laws.

Control systems use electric signals to regulate processes such as temperatures, pressures, and flow rates in an oil refinery, or the temperature inside a steel reheat furnace, or the fuel–air mixture in a fuel-injected automobile engine. They also control mechanisms such as the motors, doors, and lights in elevators, the time and temperature in microwave ovens, the locks in the Panama Canal, and the autopilot and autolanding systems that fly and land airplanes.

Power systems generate and distribute electrical power. Electrical power, which is the foundation of our technology-based society, usually is generated in large quantities by thermal (coal- oil- or gas-fired), nuclear, and hydroelectric generators. Power is distributed by a grid of conductors that crisscross the country. A major challenge in designing and operating such a system is to provide sufficient redundancy and control so that failure of any piece of equipment cannot plunge a city, state, or region into darkness.

Signal processing is used to describe a wide range of systems that do not have communication, computation, control, or power as their primary function. A stereo sound system and a computer-assisted tomography (CAT) scan system used in medical diagnosis are examples of signal-processing systems. For example, the output of the CAT scan system is an image such as that shown in Fig. 1.2, which the physician uses to diagnose diseases and injuries.

Considerable interaction takes place among the engineering disciplines involved in designing and operating these five classes of systems. Thus communication engineers use digital computers to control the flow of information. Computers contain control systems, and control systems contain computers. Power systems require extensive communication systems to coordinate the operation of components, which may be spread across a continent, safely and reliably. A signal-processing system may involve a communication link, a computer, and a control system.

A good example of the interaction among systems is the antilock brake system used in automobiles. The schematic diagram shown in Fig. 1.3 depicts the major components of the system. The purpose of an antilock braking system is to prevent skids when the wheels "lock" (stop turning), although the car is still moving. In an emergency, most drivers instinctively step on the brake pedal, locking the wheels and causing the car to skid— usually making a bad situation worse. The antilock computer monitors the drivetrain speed (an indication of how fast the car is moving), the speed of each wheel (all are the same in normal op-

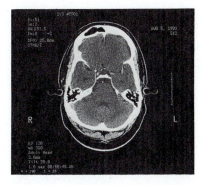

FIGURE 1.2 A CAT scan of an adult head.

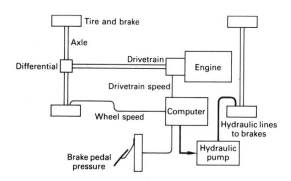

FIGURE 1.3 Top view of an automobile antilock braking system.

eration), and the brake pedal pressure (indicating how urgently the driver is trying to stop the car). If a wheel locks when the driver applies the brakes, the computer sends electric signals that modulate the output pressure of the hydraulic pump—part of the power-brake system—to unlock the wheel and provide the maximum stopping power to the car. Apart from the general problem of designing computer algorithms to correctly stop the car, the challenge is to design a computer that is safe, *even if the driver applies the brakes at the wrong time* or *even if components in the system fail*. In addition to the computer being *fail-safe*, it must be designed to operate over the extreme temperature range of −40°C to 120°C, and it must operate when installed near the engine spark plugs—a source of unwanted electrical signals known as *noise*.

Although an electrical engineer may be interested primarily in one area, he or she must also be knowledgeable in areas that interact with it. This interaction is part of what makes electrical engineering a challenging and exciting profession. The emphasis in engineering is to make things work so the engineer is free to acquire and use any technique, from any field, that helps to get the job done.

CIRCUIT THEORY

In a field as diverse as electrical engineering, you might well ask whether all of its branches have anything in common. The answer is: yes, electric circuits. This book contains material that all electrical engineers learn and use. Even those who are not actively involved in circuit analysis and design use the models, the mathematical techniques, and even the language of circuits as the intellectual framework for their work. A solid understanding of circuit theory gives you the foundation for learning—in your later courses and experiences as a practicing engineer—the details of how to design and operate systems such as those we have described.

Circuit theory is a special case of electromagnetic field theory—the study of static and moving electric charges. Although generalized field theory might seem to be an appropriate starting point for investigating electric signals, its application is not only cumbersome but also requires the use of advanced mathematics. Thus we make a few simplifying assumptions and use circuit theory instead. This approach yields the following benefits:

1. Circuit theory provides simple solutions (of sufficient accuracy) to problems that would otherwise become hopelessly complicated if we were to use electromagnetic field theory. We can analyze and build practical circuits with circuit theory.

2. The analysis and design of many useful electrical systems are less complicated if we divide them into subsystems, called *components*. We can then use the terminal behavior of each component to predict the behavior of the interconnection. The ability to derive circuit models of physical devices makes circuit theory an attractive approach.

3. Circuit analysis introduces a methodology for solving large networks of linked linear differential equations, which are prevalent throughout engineering and technology. The techniques and insights developed in this text for solving electrical circuits can be used to analyze and develop intuition about other engineering applications, including mechanical, structural, and hydraulic systems.

4. Circuit theory is an interesting area of study in its own right. Much of the remarkable development of human-made systems that depend on electrical phenomena can be attributed to the development of circuit theory as a separate discipline of study.

Even though circuit theory is a special case of field theory, understanding and using circuit theory prior to an in-depth study of fields is possible. Consequently, a course in electromagnetic field theory is not a prerequisite to understanding the material in this book. We do, however, assume that you have had an introductory physics course in which electrical and magnetic phenomena were discussed.

Three basic assumptions underlie circuit theory.

1. Charges in motion cause electrical signals. Signals propagate through a system at a finite velocity, usually at or near the speed of light. In circuit theory, we assume that the system is physically small enough that propagation effects can be ignored; that is, electrical effects happen instantaneously throughout the entire system. Ignoring spatial dimensions results in what we refer to as a *lumped-parameter* system.

2. The net charge on every component in the system is always zero. Thus no component can collect a net excess of charge, although some components, as you will learn later, can hold equal but opposite separated charges.

3. There is no magnetic coupling *between* the components in a system. As we demonstrate later, magnetic coupling can occur *within* a component.

That's it; there are no other assumptions. The benefits from using circuit theory are so great that engineers sometimes specifically design electrical systems to ensure that these assumptions are met. The importance of assumptions (2) and (3) become apparent after we introduce the basic circuit elements and the rules for analyzing interconnected elements.

However, we need to take a closer look at assumption (1). The question is: "How small does a system have to be to qualify as a lumped-parameter system?" We can get a quantitative handle on the question by noting that electrical effects propagate by wave phenomena. If the spatial wavelength of the electrical disturbance is large compared to the physical dimensions of the system, we have a lumped-parameter system. The wavelength λ is the velocity divided by the repetition rate or *frequency* of the signal; that is, $\lambda \approx c/f$. The frequency f is measured in hertz (formerly called cycles-per-second). For example, power systems in the United States operate at 60 Hz. If we use the speed of light ($c = 3 \times 10^8$ m/s) as the velocity of propagation, the spatial wavelength is 5×10^6 m or approximately 3100 mi. Radio signals are on the order of 10^9 Hz. At this frequency the wavelength is 0.3 m; thus physical dimensions on the order of centimeters are important. Whenever any of the pertinent dimensions of the system approach the wavelength, we must use electromagnetic field theory to analyze the system. Throughout this book we study lumped-parameter circuits.

COMMENTS FOR THE STUDENT

Engineering courses have two primary objectives that go hand-in-hand. One is to impart quantitative information about systems and components that reflects the current state of the art. The second is to develop techniques of analysis and synthesis that are applicable to a variety of specific problems. If you develop the habit of thinking in terms of realistic numbers that quantitatively describe systems already in existence, while at the same time focusing on the principles underlying the analysis of different systems and components, you will be in the best position to develop a successful career in a rapidly changing technology. As a practicing engineer, you will not be asked to solve problems that have already been solved. Whether you are trying to improve the performance of an existing system or creating a new system, you will be working on unsolved problems. As a student, you will devote most of your attention to the discussion of problems already solved. Discussion of how these problems were solved in the past will enable you to develop the skill to attack and solve problems in the future.

The following comments are offered to help you learn the techniques of circuit analysis and also to help you develop the skills necessary for a successful career in engineering.

1. Think through the solution to a problem before you begin calculating. Almost every problem that you will face as an engineer can be solved multiple ways, often with widely different levels of difficulty and probabilities of success. As part of your presolution strategy, try to think of several different solution methods and of a way to choose among them. If you get bogged down or become unsure of your answers, be prepared to back up and try a different approach. This is a useful skill, and with practice, you will become good at making intelligent selections. Working some of the problems several different ways to help you understand and compare various techniques also is a good idea.

2. Whenever possible, reduce complex circuits to simpler ones that are easier to solve. We present analytic techniques for replacing portions of circuits with simpler equivalents. Use these simplifying methods whenever you can.

3. Always check your answer. First, and foremost, *does it make sense*? If you ask yourself this question after working *each problem,* you will begin to develop an intuition that will give you confidence in your answers. As you develop more analytic methods, you can rework the problem using these alternative methods to verify that you get the same result. This approach is time-consuming but worthwhile, especially if the result is important; safety-critical designs are *always* checked by several independent means.

1.2 INTERNATIONAL SYSTEM OF UNITS

One distinguishing characteristic of engineering is a concern with quantitative measurements. Engineers compare theory to experiment and compare competing engineering designs by means of quantitative measures. Modern engineering is a multidisciplinary profession in which teams of engineers work together on projects, and they can communicate their results in a meaningful way only if they all use the same units. The International System of Units (abbreviated SI) is used by all the major engineering societies and most engineers throughout the world; hence we use it in this book.

The SI units are based on six *defined* quantities: (1) length, (2) mass, (3) time, (4) electric current, (5) thermodynamic temperature, and (6) luminous intensity. These quantities, along with the basic unit and symbol for each, are listed in Table 1.1. Although not strictly SI units, the familiar time units of minute (60 s), hour (3600 s), and so on, often are used in engineering

TABLE 1.1

THE INTERNATIONAL SYSTEM OF UNITS (SI)

QUANTITY	BASIC UNIT	SYMBOL
Length	Meter	m
Mass	Kilogram	kg
Time	Second	s
Electric current	Ampere	A
Thermodynamic temperature	Degree kelvin	K
Luminous intensity	Candela	cd

PSpice definitions of scaling factors for circuit element values: Table 1

calculations. In addition, defined quantities are combined to form *derived* units. Some, such as force, energy, power, and electric charge, you already know through previous physics courses. We introduce and relate derived units to electric circuits throughout this book.

In many cases the SI unit is either too small or too large to use conveniently. Standard prefixes corresponding to powers of 10, as listed in Table 1.2, are then applied to the basic unit. All of these prefixes are correct, but engineers often use only the ones for powers divisible by 3; thus centi, deci, deka, and hecto rarely are used. Also, engineers often select the prefix that places the base number in the range between 1 and 1000. Suppose that a time calculation yields a result of 10^{-5} s, that is, 0.00001 s. Most engineers would describe this quantity as 10 μs, that is, $10^{-5} = 10 \times 10^{-6}$ s, rather than as 0.01 ms or 10,000,000 ps.

Some non-SI units are still used in parts of the world. Table 1.3 gives the conversion factors for several of these quantities.

TABLE 1.2

STANDARDIZED PREFIXES TO SIGNIFY POWERS OF 10

PREFIX	SYMBOL	POWER
atto	a	10^{-18}
femto	f	10^{-15}
pico	p	10^{-12}
nano	n	10^{-9}
micro	μ	10^{-6}
milli	m	10^{-3}
centi	c	10^{-2}
deci	d	10^{-1}
deka	da	10
hecto	h	10^{2}
kilo	k	10^{3}
mega	M	10^{6}
giga	G	10^{9}
tera	Tj	10^{12}

TABLE 1.3

CONVERSION FACTORS

QUANTITY	MULTIPLY NUMBER OF	BY	TO OBTAIN NUMBER OF
Force	Pounds (lb)	4.448	Newtons (N)
Length	Inches (in.)	2.54×10^{-2}	Meters (m)
	Feet (ft)	0.3048	
	Miles (mi)	1609	
Area	Square inches (in^2)	6.452×10^{-4}	Square meters (m^2)
	Square feet (ft^2)	0.0929	
Volume	Cubic feet (ft^3)	28.3168	Liters (L)
	Cubic inches (in^3)	16.3871	Milliliters (mL)
	Gal (U.S. liquid) (gal)	3.7854	Liters (L)
	Gal (British) (gal)	4.546	Liters (L)
	Cubic Feet (ft^3)	0.02832	Cubic Meters (m^3)
Energy	British thermal units (Btu)	1.055×10^3	Joules (J)
	Kilowatt hours (kWh)	3.6×10^6	
Power	Horsepower (hp)	746	Watts (W)

DRILL EXERCISES

1.1 How many dollars per millisecond would the federal government have to collect to retire a deficit of $300 billion in one year?

ANSWER: 9.51/ms.

1.2 There are approximately 142 million passenger cars registered in the United States. Assume that the average passenger-car battery stores 440 watt-hours (Wh) of energy. Estimate (in gigawatt-hours) the total energy stored in U.S. passenger cars.

ANSWER: 62.48 GWh.

1.3 A high-resolution computer display monitor has 1280 × 1024 picture elements, or pixels. Each picture element contains 24 bits of information. If a byte is defined as eight bits, how many megabytes are required per display?

ANSWER: 3.93 Mbytes.

1.4 Some species of bamboo can grow 250 mm/day. Assuming the individual cells in the plant are 10 μm long, how long, on average, does it take a bamboo stalk to grow a 1 cell length?

ANSWER: 3.5 s.

1.5 One liter of paint covers approximately 10 m^2 of wall. How thick is the layer before it dries? (*Hint:* 1 L = 1×10^6 mm^3.)

ANSWER: 0.1 mm.

1.6 How long does it take for light to travel across a room which is 19′8¼″ wide?

ANSWER: 20 ns.

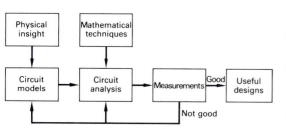

FIGURE 1.4 A conceptual model for the electrical engineering profession.

1.3 CIRCUIT ANALYSIS: AN OVERVIEW

Before becoming involved with the details of circuit analysis, we need to take a broad look at circuits. The purpose of this overview is to help you keep things in perspective as you study the various aspects of circuit analysis that make up the whole. Historically, progress in electrical engineering can be described by the conceptual model shown in Fig. 1.4. Insights about interesting physical phenomena are used to construct mathematical models for the behavior of electrical systems, called *circuits,* and the elements that make up the circuit, called *components.* The challenge is to develop models, called *ideal circuit elements,* that will predict the physical behavior of real components accurately and result in mathematical equations that can be solved. The element and circuit models are combined with mathematical techniques to analyze circuits of interest. This analysis is compared quantitatively with measurements on actual circuits built with real components. If the results are good (sufficiently accurate), they are used to make practical designs. If the results are not good, the circuit models and/or the mathematical analysis methods are adjusted and the process is tried again. Thus engineering is a never-ending process in which models, components, and mathematical techniques are continuously refined to produce highly accurate and useful designs. Practicing engineers try to use mature circuit models so that the designs will have the expected performance the first time. In this book, we use models that have been tested for between 20 and 100 years; you can assume that they are mature. The ability to model actual electrical systems with ideal circuit elements makes circuit theory extremely useful to engineers.

Saying that the interconnection of ideal circuit elements can be used to predict quantitatively the behavior of a system implies that we can describe the interconnection with mathematical equations. In order for the mathematical equations to be useful, we must write them in terms of measurable quantities. In the case of circuits, these quantities are current and voltage, which we discuss in Section 1.4. The study of circuit analysis involves (1) understanding the behavior of each ideal circuit element in terms of its current and voltage and (2) understanding the constraints imposed on the current and voltage as a result of interconnecting the ideal elements.

1.4 VOLTAGE AND CURRENT

The concept of electric charge is the basis for describing all electrical phenomena. First, the charge is *bipolar;* that is, electrical effects are described in terms of positive and negative charges. Second, the electric charge exists in *discrete quantities*. Specifically, all quantities of charge are integral multiples of the electronic charge, 1.6022×10^{-19} C (coulomb). Third, electrical effects are attributed to both the separation of charge and charges in motion. In circuit theory, the separation of charge creates an *electric force* (voltage), and the motion of charge creates an *electric fluid* (current).

The concepts of voltage and current are useful from an engineering point of view because they can be expressed quantitatively. Whenever positive and negative charges are separated, energy is expended. *Voltage is the energy per unit charge that is created by the separation.* We express this ratio in differential form, or

$$v = \frac{dw}{dq}, \tag{1.1}$$

where

$v =$ the voltage in volts,

$w =$ the energy in joules, and

$q =$ the charge in coulombs.

The electrical effects caused by charges in motion depend on the rate of charge flow. Thus this rate of charge flow became a significant variable in scientific work and is known as the *electric current*, or

$$i = \frac{dq}{dt}, \tag{1.2}$$

where

$i =$ the current in amperes,

$q =$ the charge in coulombs, and

$t =$ the time in seconds.

Equations (1.1) and (1.2) are definitions for the magnitude of voltage and current, respectively. The bipolar nature of electric charge requires that we assign polarity references to these variables. We will do so in Section 1.5.

In most electric circuits, the currents involve such enormous numbers of charge carriers that the "graininess" of the charge flow owing to the discrete nature of the charge becomes insignificant. In other words, we can treat i as a continuous variable even though we know that it is composed of a finite (but *very* large) number of discrete carriers. One advantage of using circuit models is that we can model a component strictly in terms of the voltage and current at its terminals. Thus two physically different components could have the same relationship between the terminal voltage and terminal current. If they do, for purposes of circuit analysis, they are identical. Once we know how a component behaves at its terminals, we can analyze its behavior in a circuit. However, when *developing* circuit models, we are interested in a component's internal behavior. We might want to know, for example, whether charge conduction is taking place because of free electrons moving through the crystal lattice structure of a metal or whether it is because of electrons moving within the covalent bonds of a semiconductor material. However, these concerns are beyond the realm of circuit theory. In this book we use circuit models that have already been developed and do not discuss how component models are developed.

1.5 THE IDEAL BASIC CIRCUIT ELEMENT

PSpice PSpice description of data statements, used to represent the ideal basic circuit element: Section 1.1

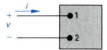

FIGURE 1.5 The basic circuit element.

PSpice PSpice discussion of polarity references for data statements: Section 1.1

At this point, we must define, at least in general terms, what we mean by an ideal basic circuit element. As we introduce each type of circuit element in subsequent chapters, we discuss its characteristics more extensively. For now, you need only to know that an ideal basic circuit element has three attributes: (1) it has only two terminals; (2) it is described mathematically in terms of the circuit variables of current and/or voltage; and (3) it cannot be subdivided into other elements. Figure 1.5 is a representation of an ideal basic circuit element. The box is blank because we are making no commitment at this time as to the type of circuit element it is. In Fig. 1.5, the voltage across the terminals of the box is denoted by v and the current in the circuit element is denoted i. The polarity reference for the voltage is indicated by the plus and minus signs, and the reference direction for the current is shown by the arrow placed alongside the current. The interpretation of these references can be summarized as follows. If the numerical value of v is a positive number, there is a drop in voltage from terminal 1 to terminal 2. If the numerical value of v is negative, there is a rise in voltage from terminal 1 to terminal 2. If the numerical value of i is positive, we can interpret the result as positive charge carriers flowing from terminal 1 to terminal 2 (that is, in the direction of the arrow) or as negative charge carriers flowing from terminal 2 to terminal 1 (that is, opposite the direction of the arrow). If the numerical value of i is negative, we know that positive charge carriers are flowing from terminal 2 to terminal 1 or that negative charge carriers are flowing from terminal 1 to terminal 2. Note that oppositely charged carriers flowing in opposite directions give the same algebraic sign to the current.

The assignments of the reference polarity for voltage and the reference direction for current are entirely arbitrary. However, once you have assigned the references, you must write all subsequent equations to agree with the chosen references. The most widely used sign convention applied to these references is called the *passive sign convention,* which we use throughout this book. The passive sign convention can be stated as follows:

> Whenever the reference direction for the current in an element is in the direction of the reference voltage drop across the element (as in Fig. 1.5), use a positive sign in the expression that relates the voltage to the current. Otherwise use a negative sign.

We apply this sign convention in all the analyses that follow. Our purpose for introducing it even before we have introduced the different types of basic circuit elements is to impress on you the fact that the selection of polarity references along with the adoption of the passive sign convention is *not* a function of the basic elements nor the type of interconnections made with the basic elements. We present the application and interpretation of the passive sign convention in power calculations in Section 1.6.

Before we discuss power (and energy), two more comments about the ideal basic circuit element are in order. We use the word *ideal* to imply that a basic circuit element does not exist as a realizable physical component. However, as we implied in Section 1.3, ideal elements can be connected to model actual devices and systems. The word ideal implies that we can formulate a precise mathematical relationship between the terminal voltage and current. We use the word *basic* to imply that the circuit element cannot be further reduced or subdivided into other elements. Thus the basic circuit elements form the building blocks for constructing circuit models, but they themselves cannot be modeled with any other type of element.

DRILL EXERCISES

1.7 The current at the terminals of the element in Fig. 1.5 is

$$i = 0, \quad t < 0;$$
$$i = 10e^{-2000t} \text{ A}, \quad t \geq 0.$$

Calculate the total charge (in microcoulombs) entering the element at its upper terminal.

ANSWER: 5000 μC.

1.8 The expression for the charge entering the upper terminal of Fig. 1.1 is

$$q = \frac{1}{\alpha^2} - \left(\frac{t}{\alpha} + \frac{1}{\alpha^2}\right)e^{-\alpha t} \text{ C}.$$

Find the maximum value of the current entering the terminal if $\alpha = 0.03679$ s^{-1}.

ANSWER: 10 A.

1.6 POWER AND ENERGY

Power and energy calculations also are important in circuit analysis. One reason is that although current and voltage are useful variables in the analysis and design of electrically based systems, the useful output of the system often is nonelectrical, and this output is conveniently expressed in terms of power or energy. Another reason is that all practical devices have limitations on the amount of power that they can handle. In the design process, therefore, voltage and current calculations by themselves are not sufficient.

We now relate power and energy to the circuit variables of voltage and current and at the same time discuss the power calculation in relation to the passive sign convention. We begin by first recalling from basic physics that power is the time rate of expending or absorbing energy. (A water pump rated 75 kW can deliver more liters per second than one rated 7.5 kW.) Mathematically, energy per unit time is expressed in the form of a derivative, or

$$p = \frac{dw}{dt},\tag{1.3}$$

where

$$p = \text{the power in watts,}$$
$$w = \text{the energy in joules, and}$$
$$t = \text{the time in seconds.}$$

Thus 1 W is equivalent to 1 J s^{-1}.

The power associated with the flow of charge follows directly from the definition of current and voltage, or

$$p = \frac{dw}{dt} = \left(\frac{dw}{dq}\right)\left(\frac{dq}{dt}\right) = vi,\tag{1.4}$$

where

$$p = \text{the power in watts,}$$
$$v = \text{the voltage in volts, and}$$
$$i = \text{the current in amperes.}$$

Equation (1.4) shows that the power associated with a basic circuit element is simply the product of the current in the element and the voltage across the element. Therefore power is a quantity associated with a pair of terminals, and we have to be able to tell from our calculation whether power is being delivered to the pair of terminals or extracted from them. This information comes from the correct application and interpretation of the passive sign convention.

If we use the passive sign convention, Eq. (1.4) is correct if the reference direction for the current is in the direction of the reference voltage drop across the terminals. Otherwise, Eq. (1.4) must be written with a minus sign. In other words, if the current reference is in the direction of a reference voltage rise across the terminals, the expression for the power is

$$p = -vi. \qquad (1.5)$$

The algebraic sign of power is based on charge movement through voltage drops and rises. As positive charges move through a drop in voltage, they lose energy, and as they move through a rise in voltage, they gain energy.

Figure 1.6 summarizes the relationship between the polarity references for voltage and current and the expression for power. Note that in all cases the algebraic sign assigned to the expression for power assumes that we are looking *toward* the terminals from the *outside* of the box.

We can now state the rule for interpreting the algebraic sign of power:

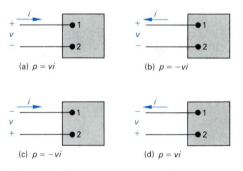

(a) $p = vi$ (b) $p = -vi$

(c) $p = -vi$ (d) $p = vi$

FIGURE 1.6 Polarity references and the expression of power.

> If the power is positive (that is, if $p > 0$), power is being delivered to the circuit inside the box. If the power is negative (that is, if $p < 0$), power is being extracted from the circuit inside the box.

For example, suppose that we have selected the polarity references shown in Fig. 1.6(b). Assume further that our calculations for the current and voltage yield the following numerical results:

$$i = 4 \text{ A} \qquad \text{and} \qquad v = -10 \text{ V}.$$

Then the power associated with the terminal pair 1, 2 is

$$p = -(-10)(4) = 40 \text{ W}.$$

Thus the circuit inside the box is absorbing 40 W.

The calculations for current can be interpreted as equivalent to positive charge carriers entering terminal 2 and leaving terminal 1 or, alternatively, as negative charge carriers entering terminal 1 and leaving terminal 2. The voltage result indicates a rise of 10 V from terminal 1 to terminal 2 or, alternatively, a drop of 10 V from terminal 2 to terminal 1.

To take this analysis one step further, assume that a colleague is solving the same problem but that she has chosen the reference polarities shown in Fig. 1.6(c). Her numerical values are

$$i = -4 \text{ A}, \qquad v = 10 \text{ V}, \qquad \text{and} \qquad p = 40 \text{ W}.$$

Note that interpreting these results in terms of this reference system gives the same conclusions that we obtained—namely, that the circuit inside the box is absorbing 40 W.

DRILL EXERCISES

1.9 Assume that a 15-V voltage drop occurs across an element from terminal 1 to terminal 2 and that a current of 5 A enters terminal 2.

    a) Specify the values of v and i for the polarity references shown in Fig. 1.6(a)–(d).

    b) State whether the circuit inside the box is absorbing or delivering power.

    c) How much power is the circuit absorbing?

ANSWER: a) Circuit 1.6(a): $v = $ 15 V, $i = -5$ A;
             circuit 1.6(b): $v = $ 15 V, $i = 5$ A;
             circuit 1.6(c): $v = -15$ V, $i = -5$ A;
             circuit 1.6(d): $v = -15$ V, $i = 5$ A;
        b) delivering;   c) -75 W.

1.10 Assume that the voltage at the terminals of the element in Fig. 1.5 corresponding to the current in Drill Exercise 1.7 is

$$v = 0, \quad t < 0;$$
$$v = 50e^{-2000t} \, \text{V}, \quad t \geq 0.$$

Calculate the total energy (in millijoules) delivered to the circuit element.

ANSWER: 125 mJ.

1.11 A high-voltage direct-current transmission line between Celilo, Oregon, and Sylmar, California, is operating at 800 kV and carrying 1800 A, as shown. Calculate the power (in megawatts) at the Oregon end of the line and state the direction of power flow.

ANSWER: 1440 MW, Celilo to Sylmar.

SUMMARY

The field of electrical engineering addresses five types of systems that depend on electrical phenomena: communication systems, computer systems, control systems, power systems, and signal-processing systems. Electromagnetic field theory is the discipline underlying the analysis, design, and operation of these systems. A special case of electromagnetic field theory, circuit analysis, is the subject of this book.

    Besides being an interesting area of study on its own, circuit analysis allows otherwise complicated field theory problems to be solved with sufficient accuracy for practical purposes. Circuit analysis also allows complicated systems to be analyzed in smaller, solvable pieces. In addition, its mathematical methods can be used to analyze nonelectrical systems.

The important concepts introduced were

- international system of units (SI), which enables engineers to communicate in a meaningful way about quantitative results;

- circuit analysis, which is based on the variables of current and voltage;

- voltage, which is the energy per unit charge created by separation and has the SI unit of volt;

- current, which is the rate of charge flow, and has the SI unit of ampere;

- ideal basic current element, which is a two-terminal component that cannot be subdivided and can be described mathematically in terms of its terminal voltage and current;

- passive sign convention, which requires the expression that relates the voltage and current at the terminals of an element to be positive when the reference arrow for the current points into the positive terminal of the reference voltage; and

- power, which is the energy per unit of time, equal to the product of the terminal voltage and current, and has the SI unit of watt.

PROBLEMS

1.1 A penny is approximately 1.5 mm thick. At what average velocity does a stack of pennies have to grow in order to accumulate 300 billion dollars in one year?

1.2 Assume a telephone signal travels through a cable at one half the speed of light. If it is approximately 5 Mm across the United States, how long does it take the signal to cross the country?

1.3 Estimate the time it takes to generate 1s of a film that uses computer-generated graphics if

  a) high-resolution film recorders have a resolution of 1200 × 1600 picture elements (pixels) per frame;

  b) each pixel requires 10 bits of data for each of the three primary colors—red, green, and blue;

  c) it takes 10 floating-point calculations to determine the value of each color per pixel;

  d) a motion picture runs 24 frames per second;

  e) a supercomputer can perform 225 million floating-point calculations per second.

1.4 In electronic circuits it is not unusual to encounter currents in the microampere range. Assume a 20-μA current is due to the flow of electrons.

   a) What is the average number of electrons per second that flow past a fixed reference cross section that is perpendicular to the direction of flow?

   b) Compare the size of this number to the number of micrometers between Miami and Seattle. You may assume the distance between Miami and Seattle is 3303 mi.

1.5 The current entering the upper terminal of Fig. 1.5 is 20 cos 5000t A. Assume the charge at the upper terminal is zero at the instant the current is passing through its maximum value. Find the expression for $q(t)$.

1.6 How much energy is imparted to an electron as it flows through a 24-V battery from the positive to the negative terminal? Express your answer in attojoules.

1.7 A current of 1600 A exists in a rectangular (0.4-by-16-cm) copper bus bar. The current is due to free electrons moving through the bus bar at an average velocity of v meters/second. If the concentration of free electrons is 10^{29} electrons per cubic meter and if they are uniformly dispersed throughout the bus bar, then what is the average velocity of an electron?

1.8 The line described in Drill Exercise 1.11 is 845 mi in length. The line contains four conductors. Each conductor weighs 2526 lb per 1000 ft. How many megatons of conductor are in the line?

1.9 The references for the voltage and current at the terminals of a circuit element are as shown in Fig. 1.6(b). The numerical values for v and i are −20 V and 5 A.

   a) Calculate the power at the terminals and state whether the power is being absorbed or delivered by the element in the box.

   b) Given that the current is due to electron flow, state whether the electrons are entering or leaving terminal 1.

   c) Do the electrons gain or lose energy as they pass through the element in the box?

1.10 Repeat Problem 1.9 with the current being −5 A.

1.11 Two electric circuits, represented by boxes A
and B, are connected as shown in Fig. P1.11.
The reference direction for the current i in the
interconnection and the reference polarity for
the voltage v across the interconnection are as
shown in the figure. For each of the following
sets of numerical values, calculate the power in
the interconnection and state whether the power
is flowing from A to B or vice versa.

a) $i = 10$ A, $v = 125$ V
b) $i = 5$ A, $v = -240$ V
c) $i = -12$ A, $v = 480$ V
d) $i = -25$ A, $v = -660$ V

FIGURE P1.11

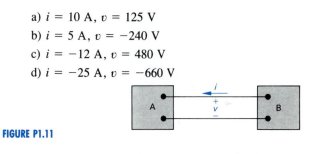

1.12 A 12-V battery supplies 100 mA to a radio.
How much energy does the battery supply in 4
hours?

1.13 The voltage and current at the terminals of an
automobile battery during a charge cycle are
shown in Fig. P1.13.

a) Calculate the total charge transferred to the
battery.

b) Calculate the total energy transferred to the
battery.

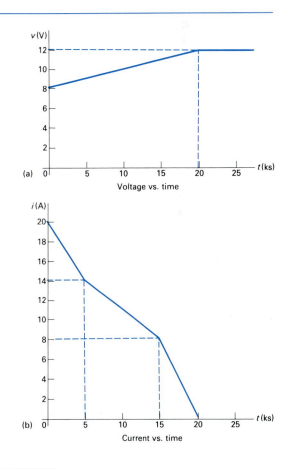

FIGURE P1.13

1.14 The voltage and current at the terminals of the circuit element in Fig. 1.5 are zero for $t < 0$. For $t \geq 0$ they are

$$v = 20e^{-20t} \text{ V}$$

$$i = 1.92e^{-20t} \text{ mA}.$$

Find the total energy delivered to the element.

1.15 The voltage and current at the terminals of the circuit element in Fig. 1.5 are zero for $t < 0$. For $t \geq 0$ they are

$$v = 50,000 \, te^{-4000t} \text{ V},$$

$$i = 32te^{-4000t} \text{ A},$$

a) Find the time (in microseconds) when the power delivered to the circuit element is maximum.

b) Find the maximum value of p in milliwatts.

c) Find the total energy delivered to the circuit element in microjoules.

1.16 The voltage and current at the terminals of the circuit element in Fig. 1.5 are zero for $t < 0$. For $t \geq 0$ they are

$$v = 50e^{-1600t} - 50e^{-400t} \text{ V},$$

$$i = 5e^{-1600t} - 5e^{-400t} \text{ mA}.$$

a) Find the power at $t = 625 \ \mu s$.

b) How much energy is delivered to the circuit element between 0 and 625 μs?

c) Find the total energy delivered to the element.

1.17 The voltage and current at the terminals of the circuit element in Fig. 1.5 are zero for $t < 0$. For $t \geq 0$ they are

$$v = (4 + 25t) \, 1000e^{-25t} - 4000e^{-20t} \text{ V},$$

$$i = (40 + 250t)e^{-25t} - 40e^{-20t} \text{ A}.$$

Calculate the power delivered to the element at $t = 4\text{ms}$.

1.18 The current and voltage at the terminals of the circuit element in Fig. 1.5 are shown in Fig. P1.18(a) and (b) respectively.

a) Sketch the power vs. t plot for $0 \leq t \leq 10s$.

b) Calculate the energy delivered to the circuit element at $t = 1$, 6, and 10 s.

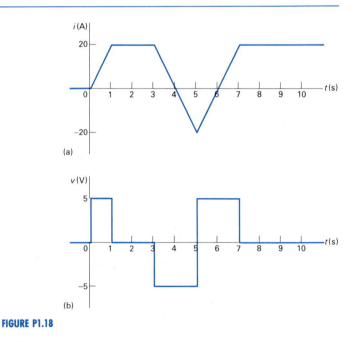

FIGURE P1.18

1.19 The voltage and current at the terminals of the circuit element in Fig. 1.5 are zero for $t < 0$ and $t > 40$s. In the interval between 0 and 40 seconds the expressions are

$$v = t(1 - 0.025t) \text{ V} \qquad 0 < t < 40s$$

and

$$i = 4 - 0.2t \text{ A} \qquad 0 < t < 40s$$

a) At what instant of time is the power being delivered to the circuit element maximum?

b) What is the power at the time found in part (a)?

c) At what instant of time is the power being extracted from the circuit element maximum?

d) What is the power at the time found in part (c)?

e) Calculate the net energy delivered to the circuit at 0, 10, 20, 30, and 40 s.

1.20 When a car has a dead battery it can often be started by connecting the battery from another car across its terminals. The positive terminals are connected together as are the negative terminals. The connection is illustrated in Fig. P1.20. Assume the current i in Fig. P1.20 is measured and found to be -40 A.

a) Which car has the dead battery?

b) If this connection is maintained for 1.5 minutes, how much energy is transferred to the dead battery?

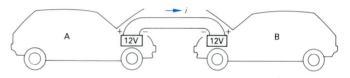

FIGURE P1.20

1.21 The voltage and current at the terminals of the element in Fig. 1.5 are

$$v = 250 \cos 800 \, \pi t \text{ V},$$

$$i = 8 \sin 800 \, \pi t \text{ A}.$$

a) Find the maximum value of the power being delivered to the element.

b) Find the maximum value of the power being extracted from the element.

c) Find the average value of p in the interval $0 \leq t \leq 2.5$ ms.

d) Find the average value of p in the interval $0 \leq t \leq 15.625$ ms.

1.22 The voltage and current at the terminals of the circuit element in Fig. 1.5 are zero for $t < 0$. For $t \geq 0$ they are

$$v = 400e^{-100t} \sin 200t \text{ V}$$

$$i = 5e^{-100t} \sin 200t \text{ A}.$$

a) Find the power absorbed by the element at $t = 10$ ms.

b) Find the total energy (in millijoules) absorbed by the element.

1.23 The manufacturer of a 1.5 V D-cell flashlight battery says that the battery will deliver 9 mA for 40 continuous hours. During that time the voltage will drop from 1.5 V to 1.0 V. Assume the drop in voltage is linear with time. How much energy does the battery deliver in this 40 hour interval?

1.24 a) In the circuit shown in Fig. P1.24, identify which elements have the voltage and current reference polarities defined in the passive convention.

b) The numerical values of the currents and voltages are,

$i_a = 2$ A, $\quad v_a = 5$ V, $\quad i_e = 5$ A, $\quad v_e = -20$ V,
$i_b = 3$ A, $\quad v_b = 1$ V, $\quad i_f = 2$ A, $\quad v_f = 20$ V,
$i_c = -2$ A, $\quad v_c = 7$ V, $\quad i_g = -2$ A, $\quad v_g = -3$ V,
$i_d = 1$ A, $\quad v_d = -9$ V, $\quad i_h = -3$ A, $\quad v_h = -12$ V.

How much total power is absorbed and how much is delivered in this circuit?

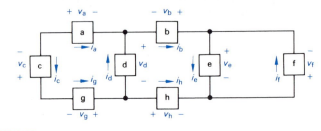

FIGURE P1.24

1.25 One method of checking calculations involving interconnected circuit elements is to see that the total power delivered equals the total power absorbed (conservation-of-energy principle). With this thought in mind, check the interconnection in Fig. P1.25 and state whether it satisfies this power check. The current and voltage values are

$i_a = -10$ A, $v_a = 160$ V, $i_b = 20$ A,
$v_b = -100$ V, $i_c = 6$ A, $v_c = 60$ V,
$i_d = 50$ A, $v_d = 800$ V, $i_e = -20$ A,
$v_e = 800$ V, $i_f = 14$ A, $v_f = -700$ V,
$i_g = 16$ A, and $v_g = 640$ V.

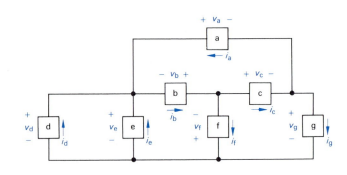

FIGURE P1.25

1.26 The numerical values of the voltages and currents in the interconnection seen in Fig. P1.26 are

$$v_a = 990 \text{ V}, \ i_a = -22.5 \text{ A};$$
$$v_b = 600 \text{ V}, \ i_b = -30 \text{ A};$$
$$v_c = 300 \text{ V}, \ i_c = 60 \text{ A};$$
$$v_d = 105 \text{ V}, \ i_d = 52.5 \text{ A};$$
$$v_e = -120 \text{ V}, \ i_e = 30 \text{ A};$$
$$v_f = 165 \text{ V}, \ i_f = 82.5 \text{ A};$$
$$v_g = 585 \text{ V}, \ i_g = 52.5 \text{ A};$$
$$v_h = -585 \text{ V, and } i_h = 82.5 \text{ A}.$$

Does the interconnection satisfy the power check?

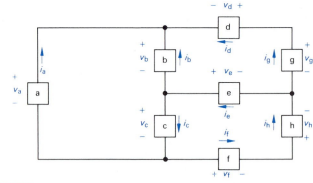

FIGURE P1.26

1.27 Assume you are an engineer in charge of a project and one of your subordinate engineers reports that the interconnection in Fig. P1.27 does not pass the power check. The data for the interconnection are given in Table P1.27.

a) Is the subordinate correct? Explain your answer.

b) If the subordinate is correct, can you find the error in the data?

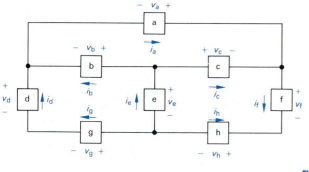

FIGURE P1.27

TABLE P1.27

ELEMENT	VOLTAGE (V)	CURRENT (A)
a	46.16	6
b	14.16	4.72
c	−32.0	−6.40
d	22.0	1.28
e	33.60	1.68
f	66.0	−0.40
g	2.56	1.28
h	−0.40	0.40

CIRCUIT ELEMENTS

CHAPTER 2

We begin the study of circuits with five ideal basic circuit elements:

1. voltage sources,

2. current sources,

3. resistors,

4. inductors, and

5. capacitors.

In this chapter we discuss the characteristics of voltage sources, current sources, and resistors. We begin with these three elements for several reasons. First, analysts use ideal sources in almost every circuit model of a practical electrical system. Second, an algebraic relationship between the terminal voltage and current describes the resistive element, and this mathematical simplicity makes it an attractive starting point. Third, modeling of many practical systems involves only sources and resistors. The second and third reasons combine to give a fourth reason for starting with sources and resistors—namely, that it enables you to learn the basic techniques of circuit analysis with only algebraic manipulations. Introduction of inductors and capacitors into circuit models requires that you solve integral and differential equations. However, even with inductors and capacitors, the basic circuit techniques stay the same. Thus by the time you start manipulating integral and differential equations, you will be familiar with methods of writing circuit equations. This approach means temporarily postponing discussion of inductors and capacitors.

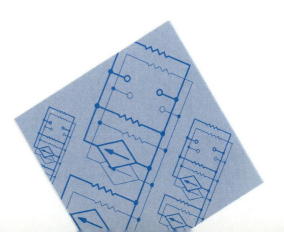

2.1 VOLTAGE AND CURRENT SOURCES

Before discussing ideal voltage and current sources, we need to consider the general nature of electrical sources. The term *source* means a device that is capable of converting nonelectric energy to electric energy and vice versa. A discharging battery converts chemical energy to electric energy, whereas a battery being charged converts electric energy to chemical energy. A dynamo is a machine that can convert mechanical energy to electric energy and vice versa. If operating in the mechanical-to-electrical mode, it is called a *generator*. If transforming from electrical to mechanical, it is referred to as a *motor*. The important thing to remember about these sources that are capable of reversible transformation is that they can either deliver or absorb electric power. These practical sources generally are *devices that tend to maintain either voltage or current*. This general behavior of practical devices led to the creation of the *ideal voltage source* and the *ideal current source* as basic circuit elements. The challenge is to model practical sources in terms of the ideal basic circuit elements.

Ideal voltage and current sources can be divided into two broad categories: independent sources and dependent sources. An *independent* source is independent of any other voltage or current that exists in the circuit to which the source is connected. A *dependent* source, however, depends on a voltage or current somewhere else in the circuit. These characteristics become more meaningful when we discuss some actual circuits. For now you need only be aware that both types of source are used in building circuit models of practical devices. We begin by describing independent sources.

IDEAL INDEPENDENT SOURCES

The *ideal independent voltage source* is a circuit element that maintains a prescribed voltage across its terminals regardless of the current in the device. Such a voltage source is capable of generating the prescribed terminal voltage whether the current in the source is zero or finite. Because the terminal voltage is not a function of the current, the ideal independent voltage source is defined completely by the prescribed voltage. The reference polarity for the prescribed voltage also must be given. Figure 2.1 shows the graphic, or circuit, symbol for the ideal independent voltage source. The prescribed voltage is given by the symbol v_s. The reference polarity of v_s is given by the plus and minus signs inside the circle.

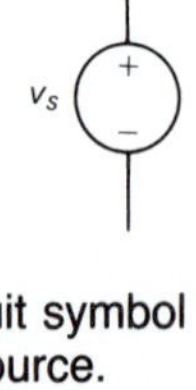

FIGURE 2.1 The circuit symbol for an ideal independent voltage source.

The *ideal independent current source* is a circuit element that maintains a prescribed current within its terminals regardless of the voltage across them. Such a current source is capable of generating the prescribed terminal current whether the terminal voltage is zero or finite. Because the current is not a function of the voltage across the terminals, the ideal independent current source is defined completely by the prescribed current and the reference direction. Figure 2.2 shows the circuit symbol for the ideal independent current source. The symbol i_s denotes the prescribed current, and the arrow inside the circle gives the reference direction.

IDEAL DEPENDENT SOURCES

An *ideal dependent*, or *controlled*, *voltage source* is a source in which either a voltage or a current at some other location in the circuit determines the voltage across its terminals. Thus a dependent voltage source can be either voltage controlled or current controlled. Figure 2.3 shows the circuit symbol for a dependent voltage source. The diamond-shaped source always indicates a dependent source. Either a voltage or a current somewhere else in the circuit controls the voltage source v_s. Letting v_x, or i_x, symbolize the controlling variable yields

$$v_s = \mu v_x \qquad \text{or} \qquad v_s = \rho i_x,$$

where μ and ρ are multiplying constants. Note that μ is dimensionless, whereas ρ carries the dimensions of volts per ampere.

With either an independent or a dependent voltage source, you cannot express the current in the source as a function of its terminal voltage. In other words, if you only know the terminal voltage of a voltage source, whether independent or dependent, you do not have enough information to determine the current the source may be carrying.

An *ideal dependent*, or *controlled*, *current source* is a source in which either a voltage or a current at some other location in the circuit determines the terminal current. Thus a dependent current source can be either voltage controlled or current controlled. Figure 2.4 shows the circuit symbol for a dependent current source. Either a voltage, v_x, or a current, i_x, somewhere else in the circuit controls the current source i_s. Hence

$$i_s = \alpha v_x \qquad \text{or} \qquad i_s = \beta i_x,$$

where α and β denote multiplying constants. Note that β is dimensionless, whereas α carries the dimension of amperes per volt.

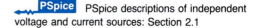
PSpice descriptions of independent voltage and current sources: Section 2.1

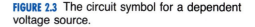

FIGURE 2.2 The circuit symbol for an ideal independent current source.

FIGURE 2.3 The circuit symbol for a dependent voltage source.

PSpice descriptions of dependent voltage and current sources: Section 2.2

FIGURE 2.4 The circuit symbol for a dependent current source.

With either an independent or a dependent current source, you cannot express the voltage across the source as a function of the terminal current. Thus if you only know the current in the source, whether independent or dependent, you do not have enough information to determine the terminal voltage of the source. Dependent, or controlled, sources are especially useful in building circuit models of electronic devices.

Because both independent and dependent sources are used to model devices that are capable of generating electric energy, they also are referred to as *active elements*. The other three basic circuit elements—resistors, inductors, and capacitors—are referred to as *passive elements*, because, when used without sources, they can model only devices that cannot generate electric energy.

Examples 2.1 and 2.2 illustrate how the characteristics of ideal independent and dependent sources limit the types of permissible interconnections of the sources.

E X A M P L E   2.1

Using the definitions of the ideal independent voltage and current sources, state which interconnections in Fig. 2.5 are permissible and which violate the constraints imposed by the ideal sources.

S O L U T I O N

Connection (a) is valid because the voltage across each source is identical. Connection (b) is valid because the current in each source is identical.

Connection (c) is not permissible because each ideal voltage source produces a different voltage. The connection is a contradiction as the voltage can't be both 10 V and 5 V.

Connection (d) is not permissible because each ideal current source produces a different current. The connection is a contradiction as the current can't be both 5 A and 2 A.

Connection (e) is valid because the voltage across the ideal voltage source is independent of its current and the current in the ideal current source is independent of its voltage.

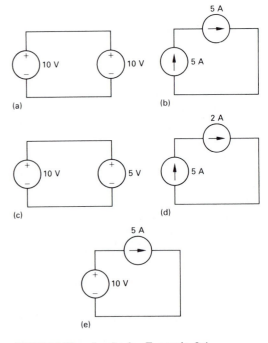

FIGURE 2.5 The circuits for Example 2.1.

EXAMPLE   2.2

Using the definitions of the ideal independent and dependent sources, state which interconnections in Fig. 2.6 are valid and which violate the constraints imposed by ideal sources.

SOLUTION

Connection (a) is invalid because the voltage across the independent voltage source is not identical to the voltage across the dependent voltage source. The voltage v_Δ cannot be both 5 V and 15 V.

Connection (b) is valid, because the voltage across the ideal voltage source is independent of its current. Similarly, the voltage across the dependent current source has no effect on its current.

Connection (c) is valid because the current in an ideal current source is independent of the voltage across its terminals. Similarly, the current in the dependent voltage source has no effect on its voltage.

Connection (d) is invalid because the current in the independent current source is not identical to the current in the dependent current source. The current i_Δ cannot be both 2 A and −6 A.

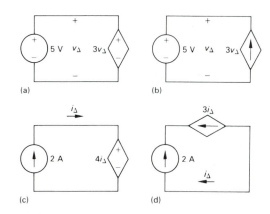

FIGURE 2.6 The circuit for Example 2.2.

2.2 ELECTRICAL RESISTANCE (OHM'S LAW)

Many useful electrical devices are designed to convert electric energy to thermal energy. Stoves, toasters, irons, and space heaters are examples of household appliances that rely on this conversion process. Such appliances take advantage of the thermal energy that arises whenever charge carriers flow through a metal. The larger the charge flow, the larger the amount of energy converted to heat. This characteristic behavior of metals such as copper and aluminum is referred to as the *resistance* of the material to the flow of electric charge. The circuit element used to model this behavior is the *resistor*. Figure 2.7 shows the circuit symbol for the resistor, with R denoting the resistance value of the resistor.

For purposes of circuit analysis, we must reference the current in the resistor to the terminal voltage. We can do so in two ways: either in the direction of the voltage drop across the resistor, as

 PSpice
Section 2.3

PSpice data statement for a resistor:

FIGURE 2.7 The circuit symbol for a resistor having a resistance R.

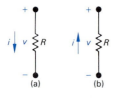

(a)   (b)

FIGURE 2.8 Two possible reference choices for the current and voltage at the terminals of a resistor: (a) current in the direction of the voltage drop across the resistor; (b) current in the direction of the voltage rise across the resistor.

FIGURE 2.9 The circuit symbol for an 8-Ω resistor.

shown in Fig. 2.8(a), or in the direction of the voltage rise across the resistor, as shown in Fig. 2.8(b). If we choose the former, the relationship between the voltage and current is

$$v = iR, \tag{2.1}$$

where

$$v = \text{the voltage in volts,}$$
$$i = \text{the current in amperes, and}$$
$$R = \text{the resistance in ohms.}$$

If we choose the second method, we must write

$$v = -iR, \tag{2.2}$$

where v, i, and R are, as before, measured in volts, amperes, and ohms, respectively. The algebraic signs used in Eqs. (2.1) and (2.2) are a direct consequence of the passive sign convention, which we introduced in Chapter 1. In both Eqs. (2.1) and (2.2), we assume that the resistance parameter itself is positive. Occasionally, a negative resistance appears in the circuit model of a device, and this implies that the device is a source of electric energy. The exact physical interpretation is best discussed when the detailed behavior of the device itself is being studied. For now, we assume that R is a positive constant.

Equations (2.1) and (2.2) are known as *Ohm's law,* after George Simon Ohm, a German physicist who established its validity early in the nineteenth century. Ohm's law is the algebraic relationship to which we referred in the introduction to this chapter. In the International System of Units, resistance is measured in ohms. The Greek letter omega (Ω) is the standard symbol for an ohm. Thus a resistor having a resistance of 8 Ω would appear in a circuit diagram as shown in Fig. 2.9.

Ohm's law expresses the voltage as a function of the current. However, expressing the current as a function of the voltage also is convenient. Thus, from Eq. (2.1),

$$i = \frac{v}{R}, \tag{2.3}$$

or, from Eq. (2.2),

$$i = -\frac{v}{R}. \tag{2.4}$$

The reciprocal of the resistance is referred to as *conductance,* is symbolized by the letter G, and is measured in siemens (S). Thus

$$G = \frac{1}{R}\text{S}. \tag{2.5}$$

An 8-Ω resistor has a conductance value of 0.125 S. In much of the professional literature, the unit used for conductance is the mho (ohm spelled backward), which is symbolized by an inverted omega ($\mho$). Therefore we may also describe an 8-Ω resistor as having a conductance of 0.125 mho, or 0.125 $\mho$.

The ideal resistor has several important properties. First, the resistance is *constant*. It is not a function of the current in the resistor nor the voltage across the resistor. This ideal resistor is referred to as a *linear time-invariant* resistor. Thus the use of the ideal resistor to model an actual device implies that a linear model is a sufficiently accurate representation of the device. Second, the resistor is *bilateral*. That is, if the polarity of the voltage reverses, the direction of the current reverses, and vice versa. Not all electrical devices are bilateral, so keep in mind that the ideal passive circuit elements (resistors, inductors, and capacitors) are bilateral elements. Third, the resistor is a *lumped element*, which means that the circuit element carries no information about spatial dimensions.

We may calculate the power at the terminals of a resistor in several ways. The first approach is to use the defining equation and simply calculate the product of the terminal voltage and current. For the reference system shown in Fig. 2.8(a) we write

$$p = vi. \tag{2.6}$$

For the reference system shown in Fig. 2.6(b) we must write

$$p = -vi. \tag{2.7}$$

We may also express the power at the terminals of a resistor by writing the power in terms of the current and the resistance. Regardless of the reference system,

$$p = i^2R. \tag{2.8}$$

Thus starting with Eq. (2.6), we get

$$p = vi = (iR)i = i^2R,$$

and starting with Eq. (2.7), we obtain

$$p = -vi = -(-iR)i = i^2R.$$

Equation (2.8) shows clearly that the power at the terminals of a positive resistor is always positive. Therefore a positive resistor always absorbs power from the circuit.

A third method of expressing the power at the terminals of a resistor is in terms of the voltage and resistance. The expression is independent of the polarity references, so

$$p = \frac{v^2}{R}. \tag{2.9}$$

Finally, note that we may also write Eqs. (2.8) and (2.9) in terms of the conductance, or

$$p = \frac{i^2}{G} \tag{2.10}$$

and

$$p = v^2 G. \tag{2.11}$$

Example 2.3 illustrates the application of Ohm's law in conjunction with an ideal source and a resistor. Power calculations at the terminals of a resistor also are illustrated.

E X A M P L E   2.3

In each circuit in Fig. 2.10, either the value of v or i is not known.

a) Calculate the values of v and i.

b) Determine the power dissipated in each resistor.

S O L U T I O N

a) The voltage v_a in Fig. 2.10(a) is a drop in the direction of the current in the resistor. Therefore

$$v_a = (1)(8) = 8 \text{ V}.$$

The current i_b in the 5-Ω resistor in Fig. 2.10(b) is in the direction of the voltage drop across the resistor. Thus

$$i_b = \frac{50}{5} = 10 \text{ A}.$$

The voltage v_c in Fig. 2.10(c) is a rise in the direction of the current in the resistor. Hence

$$v_c = -(1)(20) = -20 \text{ V}.$$

The current i_d in the 25-Ω resistor in Fig. 2.10(d) is in the direction of the voltage rise across the resistor. Therefore

$$i_d = \frac{-50}{25} = -2 \text{ A}.$$

b) The power dissipated in each of the four resistors is

$$p_{8\Omega} = \frac{(8)^2}{8} = (1)^2(8) = 8 \text{ W};$$

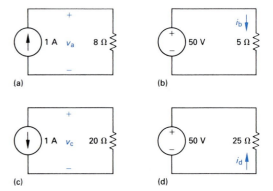

FIGURE 2.10 The circuits for Example 2.3.

$$p_{5\Omega} = \frac{(50)^2}{5} = (10)^2 5 = 500 \text{ W};$$

$$p_{20} = \frac{(-20)^2}{20} = (1)^2(20) = 20 \text{ W};$$

$$p_{25} = \frac{(50)^2}{25} = (-2)^2(25) = 100 \text{ W}.$$

Having introduced the general characteristics of ideal sources and the resistor, we next show how to use these elements to build the circuit model of a practical system.

2.3 CONSTRUCTION OF A CIRCUIT MODEL

We have already stated that one reason for an interest in the basic circuit elements is that they can be used to construct circuit models of practical systems. The skill required to develop a circuit model of a device or system is as demanding as the skill required to solve the derived circuit. Although we emphasize the skills required to solve circuits in this text, you should recognize that you also will need other skills in the practice of electrical engineering.

We illustrate development of a circuit model with two examples. In Example 2.4 we construct a circuit model based on knowledge of the behavior of the system's components and how the components are interconnected. In Example 2.5 we illustrate how a circuit model is created by measuring the terminal behavior of a device.

EXAMPLE 2.4

Construct a circuit model of a flashlight.

SOLUTION

We chose the flashlight to illustrate a practical system because its components are so familiar. Figure 2.11(a) shows a photograph of a widely available flashlight. Figure 2.11(b) shows the disassembled flashlight's components.

When a flashlight is regarded as an electrical system, the components of primary interest are (1) the batteries, (2) the lamp, (3) the connector, (4) the case, and (5) the switch. We now consider the circuit model for each component.

A dry-cell battery maintains a reasonably constant terminal voltage if the current demand is not excessive. Thus if the dry-cell battery is operating within its intended limits, we can model it with an ideal voltage source. The prescribed voltage then is constant and equal to the sum of two dry-cell values.

The ultimate output of the lamp is light energy, which is achieved by heating the filament in the lamp to a temperature high enough to cause radiation in the visible range. We can model the filament with an ideal resistor. Note in this case that although the resistor accounts for the amount of electric energy converted to thermal energy, it does not predict how much of the thermal energy is converted to light energy. The resistor used to represent the lamp does predict the steady current drain on the batteries, a characteristic of the system that also is of interest. In this model R_l symbolizes the lamp resistance.

The connector used in the flashlight serves a dual role. First, it provides an electrical conductive path between the dry cells and the case. Second, it is formed into a springy coil so that it also can apply mechanical pressure to the contacts between the batteries and the lamp. The purpose of this mechanical pressure is to minimize the contact resistance between the two dry cells and between the dry cells and the lamp. Hence, in choosing the wire for the connectors, we may find that the mechanical properties of the wire determine the choice of material and size of wire. Electrically, we can model the connector with an ideal resistor. A resistor labeled R_1 models the coiled connector.

The case also serves two purposes: one electrical and one mechanical. If the flashlight case is metal, it conducts current. That is, the case is one link in the electrical path between the batteries and the lamp. Because it is a metal conductor, we can model its electrical behavior with an ideal resistor, which we denote R_c. If the flashlight has a plastic case, a metal strip inside the case connects the coiled connector to the switch. This strip is necessary because the plastic case is not an electrical conductor. An ideal resistor also models the metal strip.

The final component is the switch. Electrically, the switch is a two-state device. It is either ON or OFF. An ideal switch offers no resistance to the current when it is in the ON state, but does offer infinite resistance to current when it is in the OFF state. These two states represent the limiting values of a resistor—that is, the ON state corresponds to a resistor with a numerical value of zero and the OFF state corresponds to a resistor with a numerical value

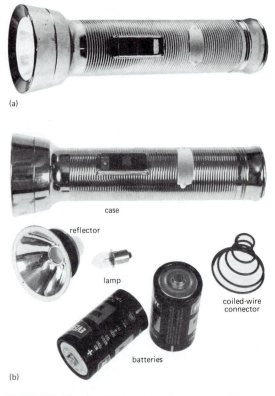

FIGURE 2.11 The flashlight viewed as an electrical system: (a) the flashlight; (b) the disassembled flashlight.

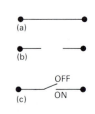

FIGURE 2.12 The circuit symbol for: (a) a short circuit; (b) an open circuit; (c) a switch.

of infinity. The two extreme values have the descriptive names *short circuit* ($R = 0$) and *open circuit* ($R = \infty$). Figure 2.12(a) and (b) show the graphic representation of a short circuit and an open circuit, respectively. The symbol shown in Fig. 2.12(c) represents the fact that a switch can be either a short circuit or an open circuit, depending on the position of its contacts.

We now construct the circuit model of the flashlight. Note that the flashlight's components are connected in tandem, or series. That is, starting with the dry-cell batteries, the positive terminal of one cell is connected to the negative terminal of the second cell, as shown in Fig. 2.13. The positive terminal of the second cell is connected to one terminal of the lamp. The other terminal of the lamp makes contact with one side of the switch, and the other side of the switch is connected to the metal case. The metal case is then connected to the negative terminal of the first dry cell by means of the metal spring. Note that the elements form a closed path or circuit. In Fig. 2.13 the dashed line depicts this closed path. Figure 2.14 shows the circuit model for the flashlight.

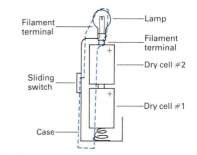

FIGURE 2.13 Arrangement of flashlight components.

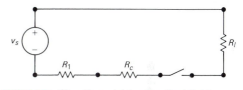

FIGURE 2.14 Circuit model for the flashlight.

Some general observations about the flashlight model are in order. First, note that we used the ideal resistor to model a lamp, a metal case, and a piece of coiled wire that provides mechanical pressure as well as an electrical connection. The choice of a resistor to model such diverse physical components demonstrates that selection of a circuit element must focus on the electrical phenomenon that the element is representing. Here, we used the resistor to model the flow of electric charge through a metal. Second, the resistance of the lamp filament also serves a useful function in the system: It generates the heat that produces the flashlight's light output. However, the resistance of the flashlight case and the coiled connector yield unwanted or parasitic effects. That is, the heat dissipated in the case and connector produces no useful output and at the same time represents a drain on the dry cells. In building circuit models of devices, you must always be alert to these unwanted parasitic effects; otherwise the models may not adequately represent the system. Third, note that building a circuit model for even this simple system requires approximations. We assumed an ideal switch, but in practical switches, contact resistance may be high enough to interfere with proper operation of the system. Our model does not predict this behavior. We also assumed that the coiled connector exerts enough pressure to eliminate any contact resistance between the dry cells. Our model does not predict the possible deleterious effect of inadequate pressure. Our use of an ideal voltage source ignores any internal dissipation of energy in the dry cells, which we could account for by adding an ideal resistor in series with the source. Our model assumes the internal loss to be negligible.

EXAMPLE 2.5

The voltage and current are measured at the terminals of the device illustrated in Fig. 2.15(a) and the values of v_t and i_t are tabulated in Fig. 2.15(b). Construct a circuit model of the device inside the box.

SOLUTION

Plotting the voltage as a function of the current yields the graph shown in Fig. 2.16. This figure illustrates that the terminal voltage is directly proportional to the terminal current, $v = 4i$. In terms of Ohm's law, the device inside the box behaves like a 4-Ω resistor. Therefore the circuit model for the device inside the box is a 4-Ω resistor.

We further illustrate this technique that utilizes terminal characteristics to construct a circuit model after introducing Kirchhoff's laws and circuit analysis. (See Drill Exercises 2.4–2.6 and Problems 2.9–2.12.

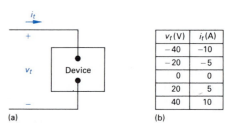

v_t(V)	i_t(A)
−40	−10
−20	−5
0	0
20	5
40	10

(a)   (b)

FIGURE 2.15 The device (a) and data (b) for Example 2.5.

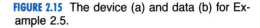

FIGURE 2.16 v_t versus i_t for the device in Fig. 2.15.

2.4 KIRCHHOFF'S LAWS

We now turn to circuit analysis, using the circuit model of the flashlight as the starting point. We redrew the circuit as Fig. 2.17, showing the switch in an ON state. We also assigned the terminal voltage and current variable for each resistor element. For convenience, we attached the same subscript to the voltage and current as previously assigned to the resistor. Note that in assigning the circuit variables, we also specified their reference polarities.

A circuit is said to be solved when the voltage across and the current in every element have been determined. For the circuit shown in Fig. 2.17, we can identify seven unknowns: i_s, i_1, i_c, i_l,

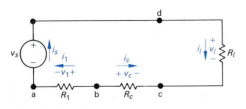

FIGURE 2.17 Circuit model of the flashlight with assigned voltage and current variables.

v_1, v_c, and v_l. Recall that v_s is a known voltage, as it represents the sum of the terminal voltages of the two dry cells—a constant voltage of 3 V. The problem is to find these seven unknowns. From algebra, you know that to find n unknown quantities you must solve n simultaneous independent equations. From the discussion of Ohm's law in Section 2.2, you learned that three of the necessary equations are

$$v_1 = i_1 R_1; \tag{2.12}$$

$$v_c = i_c R_c; \tag{2.13}$$

$$v_l = i_l R_l. \tag{2.14}$$

What about the other four equations?

The interconnection of circuit elements imposes constraints on the relationships between the terminal voltages and currents. These constraints are referred to as *Kirchhoff's laws,* after Gustav Kirchhoff, who first stated them in a paper published in 1848. The two laws that state the constraints in mathematical form are known as *Kirchhoff's current law* and *Kirchhoff's voltage law*.

Before we can state Kirchhoff's current law, we must first define *node*. A node is simply a point in a circuit at which two or more circuit elements join. For example, the nodes in the circuit of Fig. 2.17 are labeled a, b, c, and d. Node d stretches all the way across the top of the diagram (for convenience). The reason is that *any uninterrupted line segment in a circuit diagram is always interpreted as a connection having zero resistance*. Note that this configuration is consistent with the circuit representation of a short circuit. We can now state

> *Kirchhoff's current law:* The algebraic sum of all the currents at any node in a circuit equals zero.

An algebraic sign must be assigned to every current at the node. Assigning a positive sign to a current leaving a node requires assigning a negative sign to a current entering a node. Conversely, giving a negative sign to a current leaving a node requires giving a positive sign to a current entering a node.

Applying Kirchhoff's current law to the four nodes in the circuit shown in Fig. 2.17—using the convention that currents leaving a node are considered positive—yields four equations:

$$\text{Node a} \qquad i_s - i_1 = 0; \tag{2.15}$$

$$\text{Node b} \qquad i_1 + i_c = 0; \tag{2.16}$$

$$\text{Node c} \qquad -i_c - i_l = 0; \tag{2.17}$$

$$\text{Node d} \qquad i_l - i_s = 0. \tag{2.18}$$

Caution: Equations (2.15)–(2.18) are not an independent set because any one of the four can be derived from the other three. Therefore, in any circuit with n nodes, $n - 1$ independent current equations can be derived from Kirchhoff's current law.[†] Let's disregard Eq. (2.18) so that we have six independent equations, namely, Eqs. (2.12)–(2.17). We need one more, which we can derive from Kirchhoff's voltage law.

Before we can state Kirchhoff's voltage law, we must define *closed path*, or *loop*. Starting at an arbitrarily selected node, we trace a closed path in a circuit through selected basic circuit elements and return to the original node without passing through any intermediate node more than once. The circuit shown in Fig. 2.17 has only one closed path or loop. For example, choosing node d as the starting point and tracing the circuit clockwise, we form the closed path by moving through nodes c, b, a, and back to d. We can now state

> *Kirchhoff's voltage law:* The algebraic sum of all the voltages around any closed path in a circuit equals zero.

The phrase *algebraic sum* implies that we must assign an algebraic sign to each voltage in the loop. As we trace a closed path, a voltage will appear either as a rise or a drop in the tracing direction. Assigning a positive sign to a voltage rise requires assigning a negative sign to a voltage drop. Conversely, giving a negative sign to a voltage rise requires giving a positive sign to a voltage drop.

We now apply Kirchhoff's voltage law to the circuit shown in Fig. 2.17. We elect to trace the closed path clockwise and at the same time assign a positive algebraic sign to voltage drops. Starting at node d leads to the expression

$$v_l - v_c + v_1 - v_s = 0, \qquad (2.19)$$

which represents the seventh independent equation needed to find the seven unknown circuit variables mentioned earlier.

The thought of having to solve seven simultaneous equations to find the current delivered by a pair of dry cells to a flashlight lamp is not very appealing. Thus in the coming chapters we introduce you to analytic techniques that will enable you to solve a simple one-loop circuit by writing a single equation. However, before moving on to a discussion of these circuit techniques, we need to make several observations about the detailed analysis of the flashlight circuit. In general they are true and therefore are important to the discussions in subsequent chapters. They also support the contention that the flashlight circuit can be solved by defining a single unknown.

PSpice   PSpice control statements for dc analysis and output statements for formulating results permit PSpice solution of circuit problems: Section 2.3 and Chapter 3

[†] We say more about this observation in Chapter 4.

First, note that if you know the current in a resistor, you also know the voltage across a resistor, because current and voltage are directly related through Ohm's law. Thus you can associate one unknown variable with each resistor, either the current or the voltage. Choose, say, the current as the unknown variable. Then, once you solve for the unknown current in the resistor, you can find the voltage across the resistor. In general, if you know the current in a passive element, you can find the voltage across the passive element. We discuss the relationship between the current and voltage in inductors and capacitors in Chapter 7. The significance of this viewpoint is that it greatly reduces the number of simultaneous equations to be solved. For example, in the flashlight circuit we eliminate the voltages v_c, v_l, and v_1 as unknowns. Thus at the outset we reduce the analytic task to solving four simultaneous equations rather than seven.

The second general observation relates to the consequences of connecting only two elements to form a node. From Kirchhoff's current law, when only two elements connect to a node, if you know the current in one of the elements, you also know it in the second element. In other words, you need define only one unknown current for the two elements. When just two elements connect to form a node, the elements are said to be *in series*. (We say much more about the series connection in Chapter 3.) The importance of this second observation is obvious when you note that each node in the circuit shown in Fig. 2.17 involves only two elements. Thus you need to define only one unknown current. The reason is that Eqs. (2.15)–(2.17) lead directly to

$$i_s = i_1 = -i_c = i_l, \qquad (2.20)$$

which states that if you know any one of the element currents you know them all. For example, choosing to use i_s as the unknown eliminates i_1, i_c, and i_l. The problem is reduced to determining one unknown, namely, i_s.

Examples 2.6 and 2.7 illustrate how to write circuit equations based on Kirchhoff's laws. Example 2.8 illustrates how to use Kirchhoff's laws and Ohm's law to find an unknown current.

E X A M P L E   2.6

Sum the currents at each node in the circuit shown in Fig. 2.18. Note that there is no connection dot (●) in the center of the diagram, where the 4-Ω branch crosses the branch containing the ideal current source i_a.

SOLUTION

In writing the equations we use a positive sign for a current leaving a node. The four equations are

Node a	$i_1 + i_4 - i_2 - i_5 = 0;$
Node b	$i_5 + i_a + i_c = 0;$
Node c	$i_b - i_3 - i_4 - i_c = 0;$
Node d	$i_2 + i_3 - i_1 - i_b - i_a = 0.$

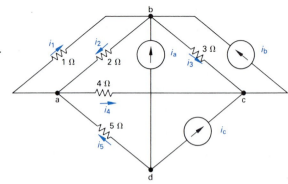

FIGURE 2.18 The circuit for Example 2.6.

EXAMPLE   2.7

Sum the voltages around each designated path in the circuit shown in Fig. 2.19.

SOLUTION

In writing the equations we use a positive sign for a voltage drop. The four equations are

Path a	$-v_1 + v_2 + v_4 - v_b - v_3 = 0;$
Path b	$-v_a + v_3 + v_5 = 0;$
Path c	$v_b - v_4 - v_c - v_6 - v_5 = 0;$
Path d	$-v_a - v_1 + v_2 - v_c + v_7 - v_d = 0.$

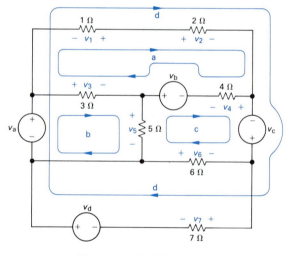

FIGURE 2.19 The circuit for Example 2.7.

EXAMPLE   2.8

a) Use Kirchhoff's laws and Ohm's law to find i_o in the circuit shown in Fig. 2.20.

b) Test the solution for i_o by verifying that the total power generated equals the total power dissipated.

S O L U T I O N

a) We begin by redrawing the circuit and assigning an unknown current to the 50-Ω resistor and unknown voltages across the 10-Ω and 50-Ω resistors. Figure 2.21 shows the circuit. The nodes are labeled a, b, and c to aid the discussion.

Because i_o also is the current in the 120-V source, we have two unknown currents and therefore must derive two simultaneous equations involving i_o and i_1. We obtain one of the equations by applying Kirchhoff's current law to either node b or c. Summing the currents at node b and assigning a positive sign to currents leaving the node gives

$$i_1 - i_o - 6 = 0.$$

We obtain the second equation from Kirchhoff's voltage law in combination with Ohm's law. Noting from Ohm's law that v_o is $10i_o$ and v_1 is $50i_1$, we sum the voltages around the closed path abca to obtain

$$-120 + 10i_o + 50i_1 = 0.$$

In writing this equation we assigned a positive sign to voltage drops in the direction of the trace. Solving these two equations for i_o and i_1 yields

$$i_o = -3 \text{ A} \qquad \text{and} \qquad i_1 = 3 \text{ A}.$$

b) The power dissipated in the 50-Ω resistor is

$$p_{50} = (3)^2 50 = 450 \text{ W}.$$

The power dissipated in the 10-Ω resistor is

$$p_{10} = (-3)^2(10) = 90 \text{ W}.$$

The power delivered to the 120-V source is

$$p_{120V} = -120i_o = -120(-3) = 360 \text{ W}.$$

The power delivered to the 6-A source is

$$p_{6A} = -v_1(6), \qquad \text{but} \qquad v_1 = 50i_1 = 150 \text{ V}.$$

Therefore

$$p_{6A} = -150(6) = -900 \text{ W}.$$

The 6-A source is developing 900 W and the 120-V source is absorbing 360 W. The total power absorbed is $360 + 450 + 90 = 900$ W. Therefore the solution verifies that the power developed equals the power absorbed.

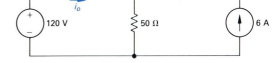

FIGURE 2.20 The circuit for Example 2.8.

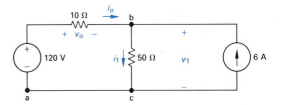

FIGURE 2.21 The circuit shown in Fig. 2.20, with the unknown i_1 defined.

DRILL EXERCISES

2.1 a) Show that Eq. (2.19) reduces to

$$i_s R_l + i_s R_c + i_s R_1 - v_s = 0.$$

b) Write the explicit expression for i_s in terms of v_s, R_1, R_c, and R_l.

ANSWER: $i_s = v_s/(R_l + R_c + R_1)$.

2.2 For the circuit shown, calculate (a) i_5; (b) v_1; (c) v_2; (d) v_5; and (e) the power delivered by the 24-V source.

ANSWER: (a) $i_5 = 3$ A; (b) $v_1 = -3$ V; (c) $v_2 = 6$ V; (d) $v_5 = 15$ V; (e) 72 W.

2.3 Use Ohm's law and Kirchhoff's laws to find the value of R in the circuit shown.

ANSWER: $R = 2\ \Omega$.

2.4 a) The terminal voltage and terminal current were measured on the device shown. The values of v_t and i_t are $v_t = 30$ V, $i_t = 0$ A; $v_t = 15$ V, $i_t = 3$ A; and $v_t = 0$ V, $i_t = 6$ A. Construct a circuit model for the device using an ideal voltage source and a resistor.

b) Use the model constructed in (a) to predict the power that the device will deliver to a 10-Ω resistor.

ANSWER: (a) A 30-V source in series with a 5-Ω resistor; (b) 40 W.

2.5 Repeat Drill Exercise 2.4 for an ideal current source and a resistor.

ANSWER: (a) A 6-A current source in parallel with a 5-Ω resistor; (b) 40 W.

2.6 In Section 2.2 we mentioned that a negative resistance can appear in the circuit model of a device. With this thought in mind, verify that the circuit inside the dashed box can be modeled with a negative resistance. (*Hint*: Find the ratio v_t/i_t.)

ANSWER: $R = -8\ \Omega$.

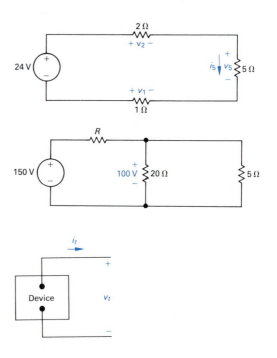

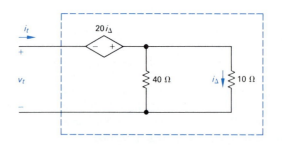

2.5 ANALYSIS OF A CIRCUIT CONTAINING A DEPENDENT SOURCE

We conclude this introduction to elementary circuit analysis with a discussion of a circuit that contains a dependent source, as depicted in Fig. 2.22. This type of circuit is of interest because it represents a structure encountered in the analysis and design of transistor amplifiers. The circuit elements inside the shaded box model the transistor, but we're not concerned with development of this transistor circuit model at this time. Here, our interest is in analyzing the circuit containing the model, that is, determining the current in each element of the circuit. We assume that the values of all the circuit elements—R_1, R_2, R_C, R_E, V_{CC}, V_0, and β—are known. Recall that once the unknown currents are found, we can find any voltage or power of interest. In Fig. 2.22 there are six unknown currents, designated i_1, i_2, i_B, i_C, i_E, and i_{CC}. In defining these six unknown currents, we took advantage of the resistor R_C being in series with the dependent current source βi_B. The problem is to derive six independent simultaneous equations involving these six unknowns. We can derive three of these equations by applying Kirchhoff's current law to any *three* of nodes a, b, c, and d. Let's use a, b, and c and label the currents away from a node as positive. Then,

$$i_1 + i_C - i_{CC} = 0; \qquad (2.21)$$

$$i_B + i_2 - i_1 = 0; \qquad (2.22)$$

$$i_E - i_B - i_C = 0. \qquad (2.23)$$

We obtain a fourth independent equation by using the constraint imposed by the dependent current source. That is, as R_C is in series with the dependent current source,

$$i_C = \beta i_B. \qquad (2.24)$$

We derive the remaining two equations by using Kirchhoff's voltage law.

In applying Kirchhoff's voltage law to the circuit shown in Fig. 2.22, we note that the voltage across the dependent current source is unknown and that it cannot be expressed as a function of the source current, βi_B. Therefore, in selecting two closed paths, we deliberately avoid any path including the dependent current source. Thus we use the paths bcdb and badb. Choosing voltage drops to be positive, we obtain

$$V_0 + i_E R_E - i_2 R_2 = 0 \qquad (2.25)$$

and

$$-i_1 R_1 + V_{CC} - i_2 R_2 = 0. \qquad (2.26)$$

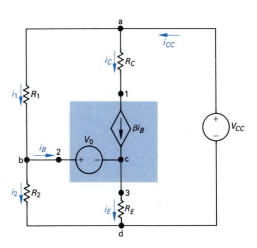

FIGURE 2.22 A circuit with a dependent current source.

We do not discuss the algebraic manipulations involved in solving these six simultaneous equations because our goal was to derive the equations. However, the solution for i_B is

$$i_B = \frac{(V_{CC}R_2)/(R_1 + R_2) - V_0}{(R_1R_2)/(R_1 + R_2) + (1 + \beta)R_E}. \qquad (2.27)$$

Problem 2.19 calls for you to verify Eq. (2.27). Note that once we know i_B, we can easily obtain the remaining currents. Problem 2.20 gives you the opportunity to analyze the circuit shown in Fig. 2.22 when numerical values are assigned to R_1, R_2, R_C, R_E, V_{CC}, V_0, and β.

Note that we defined the number of unknown variables in terms of the number of unknown currents. As with the flashlight circuit, we acknowledged that once we know the currents, we can easily calculate any unknown voltages or powers of interest. We use this circuit to illustrate some of the more powerful techniques of circuit analysis discussed in subsequent chapters. You will then be able to derive Eq. (2.27) in a single step.

Example 2.9 illustrates the application of Kirchhoff's laws and Ohm's law to a circuit containing a dependent source when the numerical values of the circuit components are known.

PSpice PSpice analysis of a circuit containing a dependent source: Example 1

E X A M P L E   2.9

a) Use Kirchhoff's laws and Ohm's law to find v_o in the circuit in Fig. 2.23.

b) Show that the solution for v_o is consistent with the constraint that the total power developed in the circuit equals the total power dissipated.

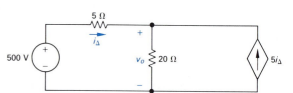

FIGURE 2.23 The circuit for Example 2.9.

S O L U T I O N

a) The circuit shown in Fig. 2.23 reveals that

1. once we know i_Δ, we also know $5i_\Delta$,

2. once we know the current in the 20-Ω resistor we can calculate v_o, and

3. the current in the 500-V source is i_Δ.

Therefore we have two unknown currents and hence must derive two simultaneous equations involving i_Δ and the current in the 20-Ω resistor. To aid the discussion we redrew the circuit as Fig. 2.24, adding the unknown current i_o and labeling the nodes. Summing the currents away from node b yields

$$i_o = 6i_\Delta.$$

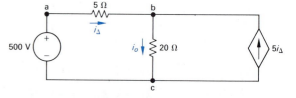

FIGURE 2.24 The circuit shown in Fig. 2.23, with the current in the 20-Ω resistor and node labels added.

Summing the voltages around the closed path abca generates

$$500 = 5i_\Delta + 20i_o$$
$$= 5i_\Delta + 20(6i_\Delta) = 125i_\Delta.$$

Hence $i_\Delta = 4$ A, $i_o = 24$ A, and $v_o = 20i_o = 480$ V.

b) The power delivered to the 500-V source is

$$p_{500V} = -500i_\Delta = -2000 \text{ W}.$$

The power delivered to the dependent current source is

$$p_{5i_\Delta} = -480(5i_\Delta) = -9600 \text{ W}.$$

Thus both sources are developing power, and the total developed power is 11,600 W.

The power dissipated in the 5-Ω resistor is

$$p_{5\Omega} = i_\Delta^2(5) = 80 \text{ W}.$$

and the power dissipated in the 20-Ω resistor is

$$p_{20\Omega} = \frac{v_o^2}{20} = \frac{480^2}{20} = 11{,}520 \text{ W}.$$

The total power dissipated in the circuit is 11,600 W.

DRILL EXERCISES

2.7 For the circuit shown find (a) the current i_1 in microamperes and (b) the voltage v in volts.

ANSWER: (a) $i_1 = 50 \ \mu\text{A}$; (b) $v = 4.175$ V.

2.8 The current i_ϕ in the circuit shown is 5 A. Calculate:

a) v_s;

b) the power absorbed by the independent voltage source;

c) the power delivered by the independent current source;

d) the power delivered by the controlled current source;

e) the total power dissipated in the two resistors.

ANSWER: (a) 50 V; (b) 500 W; (c) −250 W; (d) 3000 W; (e) 2250 W.

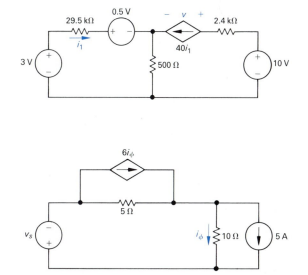

SUMMARY

We introduced linear circuit models for three basic circuit elements: voltage sources, current sources, and resistors. We also showed how to model a flashlight by using a voltage source and resistors. The development of the flashlight's equivalent circuit illustrated some challenges associated with modeling real systems and described the process an engineer uses to address these challenges. The key points of our analytical approach are the following:

- Circuit models are composed of interconnected basic circuit elements, and the circuit variables are the voltages and currents of each circuit element. Each element's power and energy are computed from its voltage and current.

- The circuit models introduced were

  voltage source, which constrains its own voltage, but its current is established by interconnected components;

  current source, which constrains its own current, but its voltage is established by interconnected components; and

  resistor, which constrains its voltage and current to be proportional to each other (*Ohm's Law*), but the actual values are established by interconnected components. The constant of proportionality is the *resistance* (voltage/current) or its inverse, the *conductance* (current/voltage).

- Circuits are described by

  nodes, which are points where circuit components join; and

  closed loops, which are paths traced through connected elements, starting and ending at the same node and encountering intermediate nodes only once.

- The voltages and currents of interconnected circuit elements obey Kirchhoff's laws.

  Kirchhoff's current law: The sum of the currents into (or out of) any node is zero.

  Kirchhoff's voltage law: The total voltage drop around any closed path is zero.

Circuit analysis is derived from these two simple laws, which are based on the physical principles of conservation of charge and energy respectively.

PROBLEMS

2.1 A pair of automotive headlamps are connected to a 12-V battery via the arrangement shown in Fig. P2.1. In the figure, the triangular symbol ▼ is used to indicate that the terminal is connected directly to the metal frame of the car.

   a) Construct a circuit model using resistors and an independent voltage source.

   b) Identify the correspondence between the ideal circuit element and the system component that it represents.

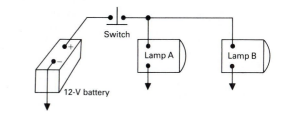

FIGURE P2.1

2.2 A simplified circuit model for a residential wiring system is shown in Fig. P2.2.

   a) How many basic circuit elements are there in this model?

   b) How many nodes are there in the circuit?

   c) How many of the nodes connect three or more basic elements?

   d) Identify the circuit elements that form a series pair.

   e) What is the minimum number of unknown currents?

   f) Describe seven closed paths in the circuit.

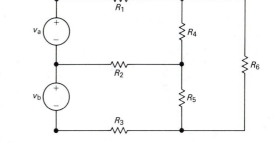

FIGURE P2.2

2.3 The voltage v_o in the circuit shown in Fig. P2.3 is 100 V. Find (a) i_a; (b) i_g; and (c) the power delivered by the independent current source.

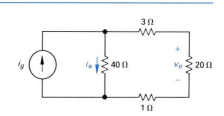

FIGURE P2.3

2.4 a) Is the interconnection of ideal sources in the circuit in Fig. P2.4 valid? Explain.

b) Identify which sources are developing power and which sources are absorbing power.

c) Verify that the total power developed in the circuit equals the total power absorbed.

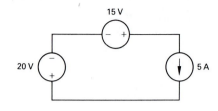

FIGURE P2.4

2.5 Is the interconnection in Fig. P2.5 valid? Explain.

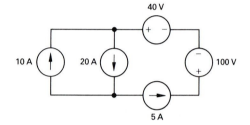

FIGURE P2.5

2.6 If the interconnection in Fig. P2.6 is valid, find the total power developed in the circuit. If the interconnection is not valid, explain why.

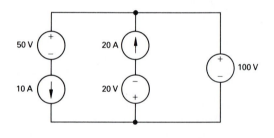

FIGURE P2.6

2.7 If the interconnection in Fig. P2.7 is valid, find the total power developed in the circuit. If the interconnection is not valid, explain why.

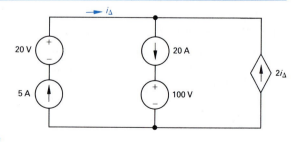

FIGURE P2.7

2.8 Given the circuit shown in Fig. P2.8, find each of the following:

    a) the value of i_a;

    b) the value of i_b;

    c) the value of v_o;

    d) the power dissipated in each resistor; and

    e) the power delivered by the 200-V source.

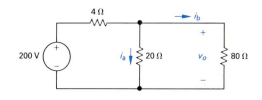

FIGURE P2.8

2.9 The current i_o in the circuit in Fig. P2.9 is 1 A.

    a) Find i_1.

    b) Find the power dissipated in each resistor.

    c) Verify that the total power dissipated in the circuit equals the power developed by the 150-V source.

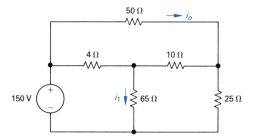

FIGURE P2.9

2.10 The currents i_a and i_b in the circuit in Fig. P2.10 are 4 A and -2 A, respectively.

    a) Find i_g.

    b) Find the power dissipated in each resistor.

    c) Find v_g.

    d) Show that the power delivered by the current source is equal to the power absorbed by all the other circuit elements.

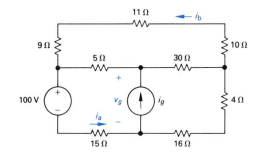

FIGURE P2.10

2.11 a) Find the currents i_a and i_b in the circuit in Fig. P2.11.

    b) Find the voltage v_g.

    c) Verify that the total power developed equals the total power dissipated.

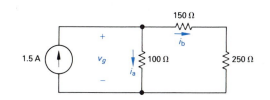

FIGURE P2.11

2.12 The currents i_1 and i_2 in the circuit in Fig. P2.12 are 21 and 14 A, respectively.

   a) Find the power supplied by each voltage source.

   b) Show that the total power supplied equals the total power dissipated in the resistors.

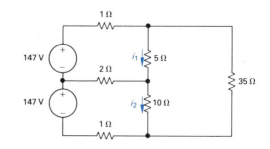

FIGURE P2.12

2.13 The voltage and current were measured at the terminals of the device shown in Fig. P2.13(a). The results are tabulated in Fig. P2.13(b).

   a) Construct a circuit model for this device using an ideal voltage source and a resistor.

   b) Use the model to predict the value of $i(t)$ when v_t is zero.

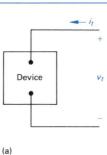

v_t (V)	i_t (A)
50	0
66	2
82	4
98	6
114	8
130	10

(a)   (b)

FIGURE P2.13

2.14 The voltage and current were measured at the terminals of the device shown in Fig. P2.14(a). The results are tabulated in Fig. P2.14(b).

   a) Construct a circuit model for this device using an ideal current source and a resistor.

   b) Use the model to predict the amount of power the device will deliver to a 20-Ω resistor.

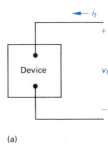
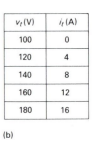

v_t (V)	i_t (A)
100	0
120	4
140	8
160	12
180	16

(a)   (b)

FIGURE P2.14

2.15 The table in Fig. P2.15(a) gives the relationship between the terminal voltage and current of the practical constant voltage source shown in Fig. P2.15(b).

   a) Plot v_s vs. i_s.

   b) Construct a circuit model of the practical source that is valid for $0 \leq i_s \leq 24$ A. (Use an ideal voltage source in series with an ideal resistor.)

   c) Use your circuit model to predict the current delivered to a 1-Ω resistor connected to the terminals of the practical source.

d) Use your circuit model to predict the current delivered to a short circuit connected to the terminals of the practical source.

e) What is the actual short-circuit current?

f) Explain why the answers to parts (d) and (e) are not the same.

v_s (volts)	i_s (amperes)
24	0
22	8
20	16
18	24
15	32
10	40
0	48

(a)

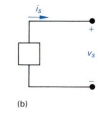

(b)

FIGURE P2.15

2.16 The table in Fig. P2.16(a) gives the relationship between the terminal current and voltage of the practical constant current source shown in Fig. P2.16(b).

a) Plot i_s vs. v_s.

b) Construct a circuit model of this current source that is valid for $0 \leq v_s \leq 75$ V. (Use an ideal current source in parallel with an ideal resistor.)

c) Use your circuit model to predict the current delivered to a 2.5-kΩ resistor.

d) Use your circuit model to predict the open-circuit voltage of the current source.

e) What is the actual open-circuit voltage?

f) Explain why the answers to parts (d) and (e) are not the same.

i_s(mA)	v_s(V)
20	0
17.5	25
15	50
12.5	75
9	100
4	125
0	140

(a)

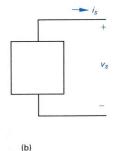

(b)

FIGURE P2.16

2.17 The voltage across the 16-Ω resistor in the circuit in Fig. P2.17 is 80 V, positive at the upper terminal.

a) Find the power dissipated in each resistor.

b) Find the power supplied by the 125-V ideal voltage source.

c) Verify that the power supplied equals the total power dissipated.

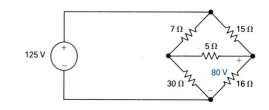

FIGURE P2.17

2.18 For the circuit shown in Fig. P2.18, find (a) R and (b) the power supplied by the 240-V source.

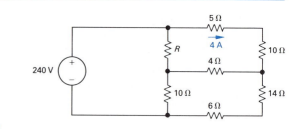

FIGURE P2.18

2.19 Derive Eq. (2.27). [*Hint:* Use Eqs. (2.23) and (2.24) to express i_E as a function of i_B. Solve Eq. (2.22) for i_2 and substitute the result into both Eqs. (2.25) and (2.26). Solve the "new" Eq. (2.26) for i_1 and substitute this result into the "new" Eq. (2.25). Replace i_E in the "new" Eq. (2.25) and solve for i_B.] Note that since i_{CC} appears only in Eq. (2.21), the solution for i_B involves the manipulation of only five equations.

2.20 For the circuit shown in Fig. 2.22, $R_1 = 40$ kΩ, $R_2 = 60$ kΩ, $R_C = 750$ Ω, $R_E = 120$ Ω, $V_{CC} = 10$ V, $V_0 = 600$ mV, and $\beta = 49$. Calculate i_B, i_C, i_E, v_{3d}, v_{bd}, i_2, i_1, v_{ab}, i_{CC}, and v_{13}. (*Note:* In the double subscript notation on voltage variables, the first subscript is positive with respect to the second subscript. See Fig. P2.20.)

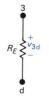

FIGURE P2.20

2.21 Find (a) i_1, (b) i_o, and (c) i_2 in the circuit in Fig. P2.21.

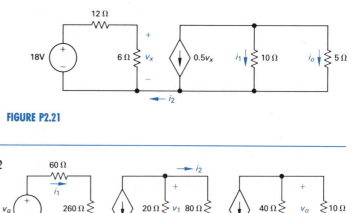

FIGURE P2.21

2.22 Find v_1 and v_g in the circuit shown in Fig. P2.22 when v_o equals 5 V. (*Hint:* Start at the right end of the circuit and work back toward v_g).

FIGURE P2.22

2.23 a) Find the voltage v_y in the circuit in Fig. P2.23.

b) Show that the total power generated in the circuit equals the total power absorbed.

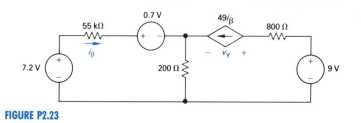

FIGURE P2.23

2.24 For the circuit shown in Fig. P2.24, calculate (a) v_1 and v_2 and (b) show that the power developed equals the power absorbed.

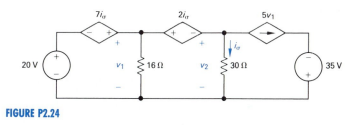

FIGURE P2.24

2.25 It is often desirable in designing an electric wiring system to be able to control a single appliance from two or more locations, for example, to control a lighting fixture from both the top and bottom of a stairwell. In home wiring systems, this type of control is implemented with three-way and four-way switches. A three-way switch is a three-terminal, two-position switch, and a four-way switch is a four-terminal, two-position switch. The switches are shown schematically in Fig. P2.25(a), which illustrates a three-way switch, and P2.25(b), which illustrates a four-way switch.

a) Show how two three-way switches can be connected between a and b in the circuit in

Fig. P2.25(c) so that the lamp l can be turned ON or OFF from two locations.

b) If the lamp (appliance) is to be controlled from more than two locations, four-way switches are used in conjunction with two three-way switches. One four-way switch is required for each location in excess of two. Show how one four-way switch plus two three-way switches can be connected between a and b in Fig. P2.25(c) to control the lamp from three locations. (*Hint:* The four-way switch is placed between the three-way switches.)

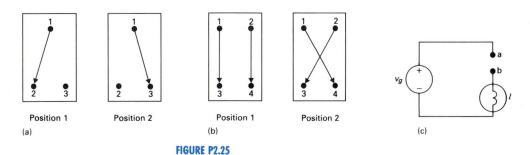

FIGURE P2.25

SIMPLE RESISTIVE CIRCUITS

CHAPTER 3

In this chapter we concentrate on solving simple resistive circuits. By simple we mean a circuit contains a small number of resistors and sources that are interconnected in such a way that the circuit can be easily analyzed by a direct application of Kirchhoff's laws in conjunction with Ohm's law. We solve these simple circuits before introducing more elegant techniques of circuit analysis for two reasons: (1) it gives us a chance to acquaint you thoroughly with the laws underlying the more sophisticated methods; and (2) it allows us to introduce you to some circuits that have important engineering applications. The sources in the circuits discussed in this chapter are limited to voltage and current sources that generate either constant voltages or currents, that is, voltages and currents that are invariant with time. We refer to sources of this type as *direct-current,* or *dc,* sources. Historically, a direct current was defined as a current produced by a constant voltage. Therefore a constant voltage became known as a direct current, or dc, voltage. You might think that if a constant current is called a direct current, a constant voltage is called a direct voltage. However, the term *direct current voltage,* or *dc voltage,* is universally used in science and engineering.

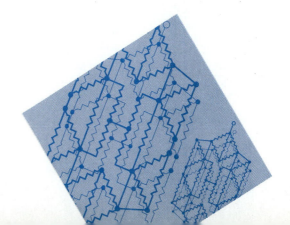

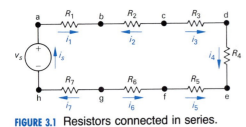

FIGURE 3.1 Resistors connected in series.

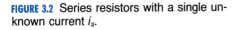

FIGURE 3.2 Series resistors with a single unknown current i_s.

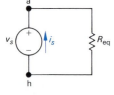

FIGURE 3.3 A simplified version of the circuit shown in Fig. 3.2.

3.1 RESISTORS IN SERIES

In the discussion of the circuit model of a flashlight we noted that circuit elements may be constrained to carry the same current. *Circuit elements that are connected in series carry the same current.* The resistors in the circuit shown in Fig. 3.1 are connected in series.

We can show that these resistors carry the same current by applying Kirchhoff's current law to each node in the circuit. The series interconnection in Fig. 3.1 requires that

$$i_s = i_1 = -i_2 = i_3 = i_4 = -i_5 = -i_6 = i_7, \quad \textbf{(3.1)}$$

which states that if we know any one of the seven currents, we know them all. Thus we can redraw Fig. 3.1 as shown in Fig. 3.2, retaining the identity of the single current i_s. To find i_s, we apply Kirchhoff's voltage law around the single closed loop. Defining the voltage across each resistor as a drop in the direction of i_s gives

$$-v_s + i_s R_1 + i_s R_2 + i_s R_3 + i_s R_4 + i_s R_5 + i_s R_6 + i_s R_7 = 0, \quad \textbf{(3.2)}$$

or

$$v_s = i_s(R_1 + R_2 + R_3 + R_4 + R_5 + R_6 + R_7). \quad \textbf{(3.3)}$$

The significance of Eq. (3.3) insofar as calculating i_s is concerned is that the seven resistors can be replaced by a single resistor whose numerical value is the sum of the individual resistors—that is,

$$R_{eq} = R_1 + R_2 + R_3 + R_4 + R_5 + R_6 + R_7 \quad \textbf{(3.4)}$$

and

$$v_s = i_s R_{eq}. \quad \textbf{(3.5)}$$

Thus we can redraw Fig. 3.2 as shown in Fig. 3.3.

In general, if k resistors are connected in series, the equivalent single resistor has a resistance equal to the sum of the k resistances, or

$$R_{eq} = \sum_{i=1}^{k} R_i = R_1 + R_2 + \cdots + R_k. \quad \textbf{(3.6)}$$

Note that the resistance of the equivalent resistor always is *larger* than the largest resistor in the series connection.

Another way to think about this concept of an equivalent resistance is to visualize the string of resistors as being inside a "black box." (*Note:* Electrical engineers use the term *black box* to imply an opaque container; that is, the contents are hidden from view. The engineer is then challenged to model the contents of the box by studying the relationship between the voltage and current at its terminals.) In terms of the relationship between the voltage and current at the box's terminals, determining whether the box contains k resistors or a single equivalent resistor is impossible. Figure 3.4 illustrates this method of studying the circuit shown in Fig. 3.2.

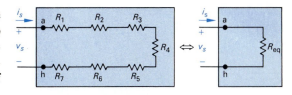

FIGURE 3.4 The "black box" equivalent of the circuit shown in Fig. 3.2.

3.2 RESISTORS IN PARALLEL

Circuit elements that are connected in parallel have the same voltage across their terminals. The circuit shown in Fig. 3.5 illustrates resistors connected in parallel. We can replace the parallel resistors with a single equivalent resistor, which is related to the individual resistors by

$$\frac{1}{R_{eq}} = \frac{1}{R_1} + \frac{1}{R_2} + \frac{1}{R_3} + \frac{1}{R_4}. \tag{3.7}$$

We derive Eq. (3.7) by directly applying Ohm's law and Kirchhoff's current law. In the circuit shown in Fig. 3.5, we let the currents i_1, i_2, i_3, and i_4 be the currents in the resistors R_1 through R_4, respectively. We also let the positive reference direction for each resistor current be down through the resistor, that is, from node a to node b. From Kirchhoff's current law,

$$i_s = i_1 + i_2 + i_3 + i_4. \tag{3.8}$$

The parallel connection of the resistors means that the voltage across each resistor must be the same. Hence, from Ohm's law,

$$i_1 R_1 = i_2 R_2 = i_3 R_3 = i_4 R_4 = v_s. \tag{3.9}$$

Therefore

$$i_1 = \frac{v_s}{R_1}, \qquad i_2 = \frac{v_s}{R_2}, \qquad i_3 = \frac{v_s}{R_3}, \qquad \text{and} \qquad i_4 = \frac{v_s}{R_4}. \tag{3.10}$$

Substituting Eq. (3.10) into Eq. (3.8) yields

$$i_s = v_s\left(\frac{1}{R_1} + \frac{1}{R_2} + \frac{1}{R_3} + \frac{1}{R_4}\right), \tag{3.11}$$

from which,

$$\frac{i_s}{v_s} = \frac{1}{R_{eq}} = \frac{1}{R_1} + \frac{1}{R_2} + \frac{1}{R_3} + \frac{1}{R_4}. \tag{3.12}$$

Equation (3.12) is what we set out to show: that the four resistors in the circuit shown in Fig. 3.5 can be replaced by a single

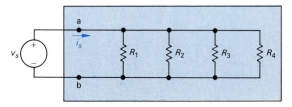

FIGURE 3.5 Resistors in parallel.

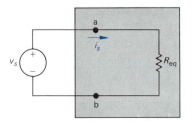

FIGURE 3.6 Replacing the four parallel resistors shown in Fig. 3.5 with a single equivalent resistor.

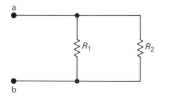

FIGURE 3.7 Two resistors connected in parallel.

equivalent resistor. The circuit shown in Fig. 3.6 illustrates the substitution. For k resistors connected in parallel, Eq. (3.7) becomes

$$\frac{1}{R_{eq}} = \sum_{i=1}^{k} \frac{1}{R_i} = \frac{1}{R_1} + \frac{1}{R_2} + \cdots + \frac{1}{R_k}. \qquad (3.13)$$

Note that the resistance of the equivalent resistor always is *smaller* than the resistance of the smallest resistor in the parallel connection. Sometimes, using conductance when dealing with resistors connected in parallel is more convenient. In that case, Eq. (3.13) becomes

$$G_{eq} = \sum_{i=1}^{k} G_i = G_1 + G_2 + G_3 + \cdots + G_k. \qquad (3.14)$$

Many times only two resistors are connected in parallel. Figure 3.7 illustrates this special case. We calculate the equivalent resistance from Eq. (3.13):

$$\frac{1}{R_{eq}} = \frac{1}{R_1} + \frac{1}{R_2} = \frac{R_2 + R_1}{R_1 R_2}, \qquad (3.15)$$

or

$$R_{eq} = \frac{R_1 R_2}{R_1 + R_2}. \qquad (3.16)$$

Thus for *just two* resistors in parallel, the equivalent resistance equals the product of the resistances divided by the sum of the resistances. Caution: The "product" divided by the "sum" applies *only* to two resistors in parallel. Example 3.1 illustrates the usefulness of these results.

E X A M P L E   3.1

Find i_s, i_1, and i_2 in the circuit shown in Fig. 3.8.

S O L U T I O N

Here, we show how to find the three specified currents by using series–parallel simplifications of the circuit. We begin by noting that the 3-Ω resistor is in series with the 6-Ω resistor. We therefore replace this series combination with a 9-Ω resistor, reducing the circuit to that shown in Fig. 3.9(a). We now can replace the parallel combination of the 9-Ω and 18-Ω resistors with a single resistance of $(18 \times 9)/(18 + 9)$, or 6 Ω. Figure 3.9(b) shows this further reduction of the circuit. The nodes x and y marked

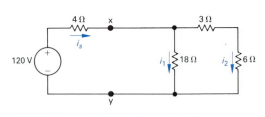

FIGURE 3.8 The circuit for Example 3.1.

on all diagrams facilitate tracing through the reduction of the circuit.

From Fig. 3.9(b) you can verify that i_s equals 120/10, or 12 A. Figure 3.10 shows the result at this point in the analysis. We added the voltage v_1 to help clarify the subsequent discussion. Using Ohm's law we compute the value of v_1:

$$v_1 = (12)(6) = 72 \text{ V}. \qquad (3.17)$$

But v_1 is the voltage drop from node x to node y, so we can return to the circuit shown in Fig. 3.9(a) and again use Ohm's law to calculate i_1 and i_2. Thus

$$i_1 = \frac{v_1}{18} = \frac{72}{18} = 4 \text{ A} \qquad (3.18)$$

and

$$i_2 = \frac{v_1}{9} = \frac{72}{9} = 8 \text{ A}. \qquad (3.19)$$

We have found the three specified currents by using series–parallel reductions in combination with Ohm's law.

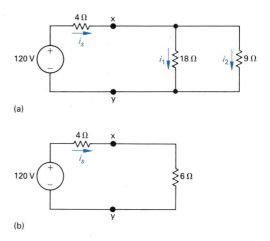

(a)

(b)

FIGURE 3.9 A simplification of the circuit shown in Fig. 3.8.

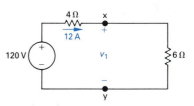

FIGURE 3.10 The circuit of Fig. 3.9(b) showing the numerical value of i_s.

Before leaving Example 3.1, we suggest that you take the time to show that the solution satisfies Kirchhoff's current law at every node and Kirchhoff's voltage law around every closed path. (Note that there are three closed paths that can be tested.) Showing that the power delivered by the voltage source equals the total power dissipated in the resistors also is informative. (See Problems 3.1 and 3.2.)

DRILL EXERCISES

3.1 For the circuit shown, find (a) the voltage v; (b) the power delivered to the circuit by the current source; and (c) the power dissipated in the 10-Ω resistor.

ANSWER: (a) 60 V; (b) 300 W; (c) 57.6 W.

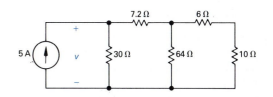

3.3 THE VOLTAGE-DIVIDER CIRCUIT

At times—especially in electronic circuits—developing more than one voltage level from a single voltage supply is necessary. We do so by using the circuit shown in Fig. 3.11, called a *voltage-divider circuit*.

We analyze this circuit by directly applying Ohm's law and Kirchhoff's laws. To aid the analysis, we introduce the currents i and i_o, as shown in Fig. 3.11(b). We begin by assuming that the load current i_o is zero. Thus, from Kirchhoff's current law, R_1 and R_2 carry the same current. Applying Kirchhoff's voltage law around the closed loop yields

$$v_s = iR_1 + iR_2 \tag{3.20}$$

or

$$i = \frac{v_s}{R_1 + R_2}. \tag{3.21}$$

Now we can use Ohm's law to calculate v_o:

$$v_o = iR_2 = v_s\frac{R_2}{R_1 + R_2}. \tag{3.22}$$

Equation (3.22) shows that v_o is a fraction of v_s, the fraction being the ratio of R_2 to $R_1 + R_2$. Obviously this ratio is always less than 1.0; thus the output voltage v_o is less than the source voltage v_s.

If v_o and v_s are specified, infinitely many combinations of R_1 and R_2 yield the proper ratio. For example, suppose that v_s equals 15 V and v_o is to be 5 V. Then $v_o/v_s = \frac{1}{3}$ and, from Eq. (3.22), we find that this ratio is satisfied whenever $R_2 = \frac{1}{2}R_1$. Other factors that enter into the selection of R_1, and hence R_2, are (1) the power loss in the voltage divider and (2) the value of the load resistor that will parallel R_2.

If the load on the voltage divider circuit is denoted by R_L, as shown in Fig. 3.12, the expression for the output voltage becomes

$$v_o = \frac{R_{eq}}{R_1 + R_{eq}}v_s, \tag{3.23}$$

where

$$R_{eq} = \frac{R_2 R_L}{R_2 + R_L}. \tag{3.24}$$

Substituting Eq. (3.24) into Eq. (3.23) yields

$$v_o = \frac{R_2}{R_1[1 + (R_2/R_L)] + R_2}v_s. \tag{3.25}$$

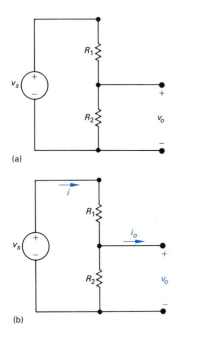

(a)

(b)

FIGURE 3.11 (a) The voltage-divider circuit; and (b) the divider circuit currents i and i_o.

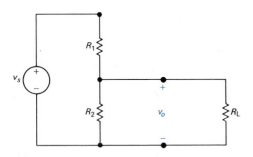

FIGURE 3.12 A voltage divider connected to a load R_L.

Note that Eq. (3.25) reduces to Eq. (3.22) as $R_L \to \infty$, as it should. Equation (3.25) shows that as long as $R_L \gg R_2$, the voltage ratio v_o/v_s essentially is undisturbed by the addition of the load on the divider.

Another characteristic of the voltage divider circuit of interest is the sensitivity of the divider to the tolerances of the resistors, as Example 3.2 illustrates.

EXAMPLE 3.2

The resistors used in the voltage divider circuit shown in Fig. 3.13 have a tolerance of $\pm 10\%$. Find the maximum and minimum value of v_o.

SOLUTION

From Eq. (3.22), the maximum value of v_o occurs when R_2 is 10 percent high and R_1 is 10 percent low and the minimum value of v_o will occur when R_2 is 10 percent low and R_1 is 10 percent high. Therefore

$$v_o(\text{max}) = \frac{(100)(110)}{110 + 22.5} = 83.02 \text{ V}$$

and

$$v_o(\text{min}) = \frac{100(90)}{90 + 27.5} = 76.60 \text{ V}.$$

Thus in making the decision to use 10 percent resistors in this voltage divider, we recognize that the no-load output voltage will lie between 76.60 and 83.02 V.

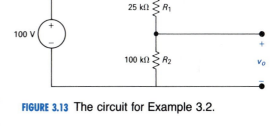

FIGURE 3.13 The circuit for Example 3.2.

DRILL EXERCISES

3.2 a) Find the no-load value of v_o in the circuit shown.

b) Find v_o when R_L is 450 kΩ.

c) How much power is dissipated in the 30-kΩ resistor if the load terminals are accidentally short-circuited?

d) What is the maximum power dissipated in the 50-kΩ resistor?

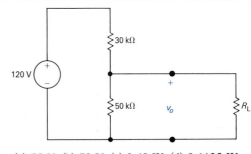

ANSWER: (a) 75 V; (b) 72 V; (c) 0.48 W; (d) 0.1125 W.

3.4 THE CURRENT-DIVIDER CIRCUIT

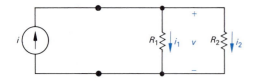

FIGURE 3.14 The current-divider circuit.

The *current-divider circuit,* shown in Fig. 3.14, consists of two resistors connected in parallel across a current source. The current divider is designed to divide the current i between R_1 and R_2. We find the relationship between the current i and the current in each resistor (that is, i_1 and i_2) by directly applying Ohm's law and Kirchhoff's current law. The voltage across the parallel resistors is

$$v = i_1 R_1 = i_2 R_2 = \frac{i R_1 R_2}{R_1 + R_2}. \quad (3.26)$$

From Eq. (3.26),

$$i_1 = \frac{i R_2}{R_1 + R_2}; \quad (3.27)$$

$$i_2 = \frac{i R_1}{R_1 + R_2}. \quad (3.28)$$

Equations (3.27) and (3.28) show that the current divides between two resistors in parallel such that the current in either resistor equals the current entering the parallel pair multiplied by the resistance of the other branch and divided by the sum of the resistors. Example 3.3 illustrates the use of the current-divider equation.

E X A M P L E   3.3

Find the power dissipated in the 6-Ω resistor shown in Fig. 3.15.

S O L U T I O N

First, we must find the current in the resistor by simplifying the circuit with series–parallel reductions. Thus the circuit shown in Fig. 3.15 reduces to that shown in Fig. 3.16. We find the current i_o by using the formula for current division:

$$i_o = \frac{(10)(16)}{16 + 4} = 8 \text{ A}.$$

Note that i_o is the current in the 1.6-Ω resistor in Fig. 3.15. We now can further divide i_o between the 6-Ω and 4-Ω resistors. The current in the 6-Ω resistor is

$$i_6 = \frac{(8)(4)}{10} = 3.2 \text{ A},$$

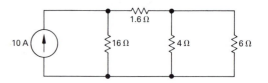

FIGURE 3.15 The circuit for Example 3.3.

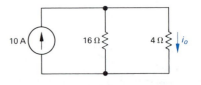

FIGURE 3.16 A simplification of the circuit shown in Fig. 3.15.

and the power dissipated in the 6-Ω resistor is

$$p = (3.2)^2(6) = 61.44 \text{ W.}$$

3.5 THE D'ARSONVAL METER MOVEMENT

Many instruments for electrical engineering measurements use a *d'Arsonval meter movement*, which is best described with the aid of Fig. 3.17. The movement consists of a movable coil placed in the field of a permanent magnet. The current in the coil creates a torque on the coil, which then rotates until this torque is exactly balanced by a restoring spring. As the coil rotates, it moves a pointer across a calibrated scale. The movement is designed so that the *deflection of the pointer is directly proportional to the current in the movable coil.* From a circuit point of view, the coil is described in terms of a voltage and current rating. For example, one commercially available meter movement is rated at 50 mV and 1 mA. When the coil is carrying its rated current, the voltage drop across the coil is the rated coil voltage, and the pointer is deflected to its full-scale position. The current and voltage ratings of the coil also specify the resistance of the coil. Thus a 50-mV, 1-mA movement has a resistance of 50 Ω.

The important thing to remember about instruments that use the d'Arsonval movement as a readout mechanism is that the pointer deflection is governed by the current in the coil. The significance of the deflection depends on what the coil current represents. In the next several sections, we show how the d'Arsonval movement is used as an ammeter, a voltmeter, an ohmmeter, and a null detector.

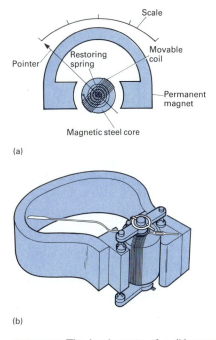

(a)

(b)

FIGURE 3.17 The basic parts of a d'Arsonval meter: (a) a schematic diagram; (b) a pictorial diagram.

3.6 THE AMMETER CIRCUIT

An ammeter is an instrument designed to measure current. Therefore the ammeter terminals are inserted in series with the current to be measured. A direct-current ammeter that uses a d'Arsonval movement consists of the movement in parallel with a resistor, as shown in Fig. 3.18. The purpose of the shunting resistor R_A is to control the amount of current that passes through the meter movement. The shunting resistor R_A and the meter movement form a current divider. Thus for a given

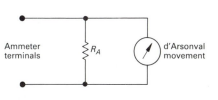

FIGURE 3.18 A direct-current ammeter circuit.

PSpice　　　PSpice illustrations of a zero-value
voltage source used as an ammeter to measure current:
Example 5
d'Arsonval movement, R_A determines the full-scale reading of
the ammeter. Example 3.4 illustrates the calculations involved in
determining R_A.

E X A M P L E   3.4

a) A 50-mV, 1-mA d'Arsonval movement is to be used in an
ammeter with a full-scale reading of 10 mA. Determine R_A.

b) Repeat (a) for a full-scale reading of 1 A.

c) How much resistance is added to the circuit when the 10-mA
ammeter is inserted to measure current?

d) Repeat (c) for the 1-A ammeter.

S O L U T I O N

a) From the statement of the problem, we know that when the
current at the terminals of the ammeter is 10 mA, then 9 mA
must be diverted through R_A. At the same time, we know that
when the movement carries 1 mA, the drop across its terminals is 50 mV. Ohm's law requires that

$$9 \times 10^{-3} R_A = 50 \times 10^{-3},$$

or

$$R_A = 50/9 = 5.555 \ \Omega.$$

b) When the full-scale deflection of the ammeter is 1 A, R_A must
carry 999 mA when the movement carries 1 mA. In this case
then,

$$999 \times 10^{-3} R_A = 50 \times 10^{-3},$$

$$R_A = 50/999 \cong 50.05 \ \text{m}\Omega.$$

c) Let R_m represent the equivalent resistance of the ammeter.
For the 10-mA ammeter

$$R_m = \frac{50 \ \text{mV}}{10 \ \text{mA}} = 5 \ \Omega,$$

or, alternatively,

$$R_m = \frac{(50)(50/9)}{50 + (50/9)} = 5 \ \Omega.$$

d) For the 1-A ammeter

$$R_m = \frac{50 \times 10^{-3}}{1} = 0.050 \ \Omega,$$

or, alternatively,

$$R_m = \frac{(50)(50/999)}{50 + (50/999)} = 0.050 \ \Omega.$$

DRILL EXERCISES

3.3 a) Find the current in the circuit shown.

b) If the milliammeter in Example 3.4(a) is used to measure the current, what will it read?

ANSWER: (a) 6.25 mA; (b) 5.88 mA.

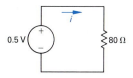

3.7 THE VOLTMETER CIRCUIT

A voltmeter is an instrument designed to measure voltage. Therefore the voltmeter terminals are placed in parallel with the voltage to be measured. A direct-current voltmeter that uses a d'Arsonval movement consists of the movement in series with a resistor, as shown in Fig. 3.19. The purpose of the series resistor R_v is to limit the voltage applied to the meter movement. The series resistor R_v and the d'Arsonval movement form a voltage divider. Thus for a given d'Arsonval movement, R_v determines the full-scale reading of the voltmeter. Example 3.5 illustrates the calculations involved to determine R_v.

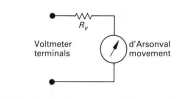

FIGURE 3.19 A direct-current voltmeter circuit.

E X A M P L E   3 . 5

a) A 50-mV, 1-mA d'Arsonval movement is to be used in a voltmeter where the full-scale reading is 150 V. Determine R_v.

b) Repeat (a) for a full-scale reading of 5 V.

c) How much resistance does the 150-V meter insert into the circuit?

d) Repeat (c) for the 5-V meter.

S O L U T I O N

a) Full-scale deflection requires 50 mV across the meter movement, and the movement has a resistance of 50 Ω. Therefore we apply Eq. (3.22) with $R_1 = R_v$, $R_2 = 50$, $v_s = 150$, and $v_o = 50$ mV:

$$50 \times 10^{-3} = \frac{150(50)}{R_v + 50}.$$

Solving for R_v gives

$$R_v = 149{,}950 \ \Omega.$$

b) For a full-scale reading of 5 V,

$$50 \times 10^{-3} = \frac{5(50)}{R_v + 50}, \quad \text{or} \quad R_v = 4950 \ \Omega.$$

c) If we let R_m represent the equivalent resistance of the meter,

$$R_m = \frac{150}{10^{-3}} = 150{,}000 \ \Omega,$$

or, alternatively,

$$R_m = 149{,}950 + 50 = 150{,}000 \ \Omega.$$

d) Then,

$$R_m = \frac{5}{10^{-3}} = 5000 \ \Omega,$$

or, alternatively,

$$R_m = 4950 + 50 = 5000 \ \Omega.$$

Insertion of either an ammeter or a voltmeter into a circuit disturbs the circuit in which the measurement is being made. An ammeter adds resistance in the branch in which the current is being measured. A voltmeter adds resistance across the terminals, where the voltage is being measured. *How much the meters disturb the circuit in which the measurements are being made depends on the resistance of the meters in comparison with the resistances of the circuit.* If the resistance of the branch without the ammeter is large compared to the meter resistance, insertion of the ammeter has a negligible effect. However, if the resistance of the branch is of the same order of magnitude as the ammeter resistance, insertion of the meter could significantly affect the current in the branch. In this case, the current measured by the ammeter would not be the same as the current in the branch without the ammeter.

The loading effect of a voltmeter depends on the resistance of the voltmeter compared to the resistance that the voltmeter shunts in the circuit. The higher the total resistance of the voltmeter circuit is, the smaller the loading effect is. Some commercial voltmeters carry a sensitivity rating in ohms per volt so that the user may quickly determine the total resistance that the voltmeter adds to the circuit. For example, the 150-V voltmeter circuit discussed in Example 3.5 would carry a sensitivity rating of 1000 Ω/V because the total resistance of the voltmeter is 150,000 Ω and the full-scale rating of the meter is 150 V. Direct-current voltmeters that use the d'Arsonval meter movement may have sensitivity ratings ranging from 100 to 20,000 Ω/V.

Problems 3.29 and 3.33 illustrate the importance of being aware of possible meter-loading effects, which are not peculiar to d'Arsonval meter movements. Any instrument used to make physical measurements extracts energy from the system while making measurements. The more energy extracted in relation to the amount of energy available in the system, the more severely the measurement is disturbed. Therefore you must always be conscious of the burden that the measuring device imposes on the system being measured.

DRILL EXERCISES

3.4 a) Find the voltage v across the 75-kΩ resistor in the circuit shown.

   b) If the 150-V voltmeter of Example 3.5(a) is used to measure the voltage, what will be the reading?

ANSWER: (a) 37.5 V; (b) 36.36 V.

3.8 THE OHMMETER CIRCUIT

An ohmmeter is an instrument designed to measure resistance. The resistance is deactivated at the time it is placed across the terminals of an ohmmeter. In other words, the resistance is not an energized component of a circuit at the time the measurement is made.

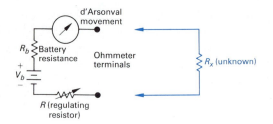

FIGURE 3.20 A d'Arsonval ohmmeter circuit.

The d'Arsonval *ohmmeter circuit*, shown in Fig. 3.20, consists of a meter movement in series with a battery and a regulating resistor. The ohmmeter circuit operates as follows. The ohmmeter terminals are short-circuited, and the regulating resistor is adjusted to give full-scale deflection of the meter (labeled zero resistance on the meter scale). When an unknown resistance is connected to the ohmmeter, the deflection is less than full scale. A calibrated scale (reading from right to left) may be constructed by connecting a series of known resistors across the ohmmeter and noting the deflection for each resistor. When an unknown resistance is connected to the ohmmeter, its value can be read from the calibrated scale. One of the disadvantages of the d'Arsonval ohmmeter circuit is the inherently nonuniform resistance scale; that is, the resistance scale will be cramped at the high-resistance end of the scale.

Successful operation of the d'Arsonval ohmmeter depends on a stable dc supply. The regulating resistor is used to compensate for changes in the internal resistance of the battery; that is, the regulating resistor holds $R + R_b$ constant. Then, as long as V_b is constant, the ohmmeter remains calibrated.

Although this ohmmeter circuit is not a precision instrument, it is extremely useful because it is so simple to use. This type of ohmmeter often is used to check the continuity of a circuit (that is, to determine whether $R_x < \infty$).

DRILL EXERCISES

3.5 In the ohmmeter circuit shown, the voltage drop across the 50-μA ammeter and the internal resistance of the 1.5-V battery are negligible. The variable resistor R is adjusted to give full-scale deflection of the microammeter when R_x is zero.

a) What is the numerical value of R?

b) What is the midscale reading of the ohmmeter in ohms?

c) If the variable resistor R can be reduced to zero, how low can the battery voltage drop before the ohmmeter cannot be adjusted to read 0 Ω when R_x is zero?

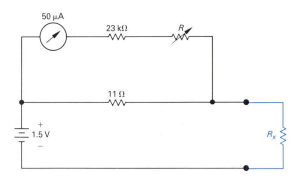

ANSWER: (a) 7 kΩ; (b) 11 Ω; (c) 1.15 V.

3.9 THE WHEATSTONE BRIDGE

The Wheatstone bridge circuit is a method used to measure precisely resistances of medium values, that is, resistances in the range of 1 Ω to 1 MΩ. In commercial models of the Wheatstone bridge, accuracies on the order of ±0.1 percent are possible. The bridge circuit consists of four resistance branches, a dc voltage source (usually a battery), and a detector. The detector is generally a d'Arsonval movement in the microamp range and is called a galvanometer. Figure 3.21 shows the circuit arrangement of the resistances, battery, and detector where R_1, R_2, and R_3 are known resistors and R_x is the unknown resistor.

To find the value of R_x, we adjust the variable resistor R_3 until there is no current in the microammeter branch of the bridge. We calculate the unknown resistor from the simple expression

$$R_x = \frac{R_2}{R_1} \cdot R_3. \tag{3.29}$$

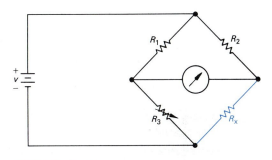

FIGURE 3.21 The Wheatstone bridge circuit.

The derivation of Eq. (3.29) follows directly from application of Kirchhoff's circuit laws to the bridge circuit. We redraw the bridge circuit as Fig. 3.22 to show the branch currents appropriate to the derivation of Eq. (3.29). When i_g is zero, that is, when the bridge is balanced, Kirchhoff's current law requires that

$$i_1 = i_3 \tag{3.30}$$

and

$$i_2 = i_x. \tag{3.31}$$

Now, because i_g is zero, there is no voltage drop across the detector, and therefore points a and b are at the same potential. Thus when the bridge is balanced, Kirchhoff's voltage law requires that

$$i_3 R_3 = i_x R_x \tag{3.32}$$

and

$$i_1 R_1 = i_2 R_2. \tag{3.33}$$

Combining Eqs. (3.30) and (3.31) with Eq. (3.32) gives

$$i_1 R_3 = i_2 R_x. \tag{3.34}$$

We obtain Eq. (3.29) by first dividing Eq. (3.34) by Eq. (3.33) and then solving the resulting expression for R_x:

$$\frac{R_3}{R_1} = \frac{R_x}{R_2}, \tag{3.35}$$

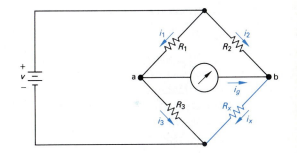

FIGURE 3.22 A balanced Wheatstone bridge ($i_g = 0$).

from which

$$R_x = \frac{R_2}{R_1} \cdot R_3. \tag{3.36}$$

Now that we have verified the validity of Eq. (3.29), several comments about the result are in order. First, note that if the ratio R_2/R_1 is unity, the unknown resistor R_x equals R_3. In this case, the bridge resistor R_3 must vary over a range that includes the value R_x. For example, if the unknown resistance were 1000 Ω and R_3 could be varied from 0 to 100 Ω, the bridge could never be balanced. Thus in order to cover a wide range of unknown resistors, we must be able to vary the ratio R_2/R_1. In a commercial Wheatstone bridge, R_1 and R_2 consist of decimal values of resistances that can be switched into the bridge circuit. Normally, the decimal values are 1, 10, 100, and 1000 Ω so that the ratio R_2/R_1 can be varied from 0.001 to 1000 in decimal steps. The variable resistor R_3 is usually adjustable in integral values of resistance from 1 to 11,000 Ω.

Although Eq. (3.29) implies that R_x can vary from zero to infinity, second-order effects enter into the operation of the bridge and limit the practical range of R_x to approximately 1 Ω to 1 MΩ. Low resistances are difficult to measure on the standard Wheatstone bridge because of thermoelectric voltages generated at the junctions of dissimilar metals and because of thermal heating effects, that is, i^2R effects. High resistances are difficult to measure accurately because of leakage currents. In other words, if R_x is large, the current leakage in the electrical insulation may be comparable to the current in the branches of the bridge circuit.

DRILL EXERCISES

3.6 The bridge circuit shown is balanced when $R_1 = 100$ Ω, $R_2 = 1000$ Ω, and $R_3 = 150$ Ω. The bridge is energized from a 5-V dc source.

a) What is the value of R_x?

b) If each bridge resistor is capable of dissipating 250 mW, will the bridge be damaged if left in the balanced state?

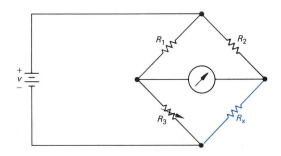

ANSWER: (a) 1500 Ω; (b) no; the total power delivered to the bridge is 110 mW.

3.10 DELTA-TO-WYE (OR PI-TO-TEE) EQUIVALENT CIRCUITS

The bridge configuration in Fig. 3.21 introduces an interconnection of resistances that warrants further discussion. If we replace the galvanometer with its equivalent resistance R_m, we can draw the circuit shown in Fig. 3.23. We cannot reduce the interconnected resistors of this circuit to a single equivalent resistance across the terminals of the battery if we are restricted to the simple series or parallel equivalent circuits introduced earlier in this chapter. The interconnected resistors can be reduced to a single equivalent resistor by means of a delta-to-wye (Δ-to-Y) or pi-to-tee (π-to-T) equivalent circuit.[†]

The resistors R_1, R_2, and R_m or R_3, R_m, and R_x in the circuit shown in Fig. 3.23 are referred to as a *delta* (Δ) *interconnection,* because the interconnection looks like the Greek letter Δ. It also is referred to as a *pi* (π) *interconnection* because the Δ can be shaped into a π without disturbing the electrical equivalence of the two configurations. The electrical equivalence between the Δ and π interconnections is apparent in Fig. 3.24.

The Δ-to-Y equivalent circuit means that the Δ configuration can be replaced with a Y configuration to make the terminal behavior of the two configurations identical. Figure 3.25 illustrates the Δ-to-Y transformation. The two circuits are equivalent at the terminals a, b, and c, provided that

$$R_1 = \frac{R_b R_c}{R_a + R_b + R_c}; \qquad (3.37)$$

$$R_2 = \frac{R_c R_a}{R_a + R_b + R_c}; \qquad (3.38)$$

$$R_3 = \frac{R_a R_b}{R_a + R_b + R_c}. \qquad (3.39)$$

The Y configuration also is referred to as the *tee* (T) *configuration* because the Y structure can be replaced with the T structure without disturbing the electrical equivalence of the two structures. The electrical equivalence of the Y and the T configurations is apparent from Fig. 3.26.

Equations (3.37)–(3.39) give the Y-connected resistors as functions of the Δ-connected resistors. Reversing the Δ-to-Y (π-to-T) transformation also is possible; that is, we can start with the Y structure and replace it with an equivalent Δ structure.

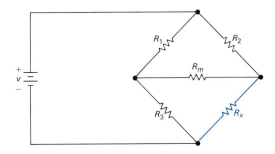

FIGURE 3.23 A resistive network generated by a Wheatstone bridge circuit.

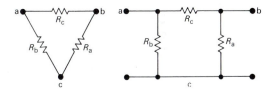

FIGURE 3.24 A Δ configuration viewed as a π configuration.

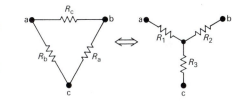

FIGURE 3.25 The Δ-to-Y transformation.

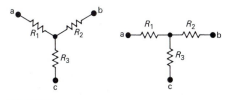

FIGURE 3.26 A Y structure viewed as a T structure.

[†] Δ and Y structures are present in a variety of useful circuits (not just resistive networks). Hence, the Δ-to-Y transformation is a helpful tool in circuit analysis.

The expressions for the three Δ resistors R_a, R_b, and R_c as functions of the three Y resistors R_1, R_2, and R_3 are

$$R_a = \frac{R_1 R_2 + R_2 R_3 + R_3 R_1}{R_1}; \qquad (3.40)$$

$$R_b = \frac{R_1 R_2 + R_2 R_3 + R_3 R_1}{R_2}; \qquad (3.41)$$

$$R_c = \frac{R_1 R_2 + R_2 R_3 + R_3 R_1}{R_3}. \qquad (3.42)$$

We can derive Eqs. (3.37)–(3.42) by noting that the two circuits are by definition equivalent with respect to their terminal behaviors. Thus if each circuit is placed in a black box, we can't tell by external measurements whether the box contains a set of Δ-connected resistors or a set of Y-connected resistors. This condition is true only if the resistance between corresponding terminal pairs is the same for each box. For example, the resistance between terminals a and b must be the same whether we use the Δ-connected set or the Y-connected set. Then,

$$R_{ab} = \frac{R_c(R_a + R_b)}{R_a + R_b + R_c} = R_1 + R_2; \qquad (3.43)$$

$$R_{bc} = \frac{R_a(R_b + R_c)}{R_a + R_b + R_c} = R_2 + R_3; \qquad (3.44)$$

$$R_{ca} = \frac{R_b(R_c + R_a)}{R_a + R_b + R_c} = R_1 + R_3. \qquad (3.45)$$

We obtain Eqs. (3.37)–(3.42) by straightforward algebraic manipulation of Eqs. (3.43)–(3.45). (Problem 3.54 gives you hints on how to start the manipulations.)

Example 3.6 illustrates the use of a Δ-to-Y transformation to simplify the analysis of a circuit.

E X A M P L E   3.6

Find the current and power supplied by the 40-V source in the circuit shown in Fig. 3.27.

S O L U T I O N

We are interested only in the current and power drain on the 40-V source, so the problem has been solved once we obtain the equivalent resistance across the terminals of the source. We can find this equivalent resistance easily after replacing either the

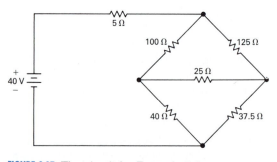

FIGURE 3.27 The circuit for Example 3.6.

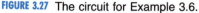

upper Δ (100, 125, 25 Ω) or the lower Δ (40, 25, 37.5 Ω) with its equivalent Y. We choose to replace the upper Δ. We compute the three Y resistances defined in Fig. 3.28 from Eqs. (3.37)–(3.39). Thus

$$R_1 = \frac{100 \times 125}{250} = 50 \ \Omega;$$

$$R_2 = \frac{125 \times 25}{250} = 12.5 \ \Omega;$$

$$R_3 = \frac{100 \times 25}{250} = 10\Omega.$$

Substituting the Y-resistors into the circuit shown in Fig. 3.27 produces the circuit shown in Fig. 3.29. From Fig. 3.29, we can easily calculate the resistance across the terminals of the 40-V source by series–parallel simplifications:

$$R_{eq} = 55 + \frac{(50)(50)}{100} = 80 \ \Omega.$$

The final step is to note that the circuit reduces to an 80-Ω resistor across a 40-V source, as shown in Fig. 3.30, from which it is apparent that the 40-V source delivers 0.5 A and 20 W to the circuit.

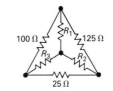

FIGURE 3.28 The equivalent Y resistors.

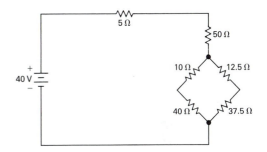

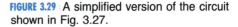

FIGURE 3.29 A simplified version of the circuit shown in Fig. 3.27.

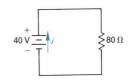

FIGURE 3.30 The final step in the simplification of the circuit shown in Fig. 3.27.

DRILL EXERCISES

3.7 a) Use a Δ-to-Y transformation to find the current i in the circuit shown.

b) Find v_1 and v_2. (*Hint:* Use the circuit that exists after the Δ-to-Y transformation.)

ANSWER: (a) 1 A; (b) $v_1 = 23.2$ V, $v_2 = 21$ V.

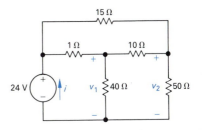

SUMMARY

We showed how to analyze simple resistive circuits by applying Ohm's law and Kirchhoff's current and voltage laws. Using these laws we developed methods for simplifying combinations of resistors. They are fundamental tools that engineers use to analyze and design circuits. One of the purposes of this chapter was to help you develop the skill of deciding when and how to use these techniques.

- *Series resistors:* Combine into a single resistor by adding their resistances.

- *Parallel resistors:* Combine into a single resistor by adding their conductances.

- Δ-to-Y or Y-to-Δ transformation: Replace three resistors in a delta with three resistors in a wye, or vice versa.

- π-to-T or T-to-π transformation: The pi–tee (or tee–pi) transformation is another name for the delta–wye (or wye–delta) transformation.

- *Voltage divider:* In series resistors, the ratio of a resistor's voltage to the total voltage across the resistors is the same as the ratio of its resistance to the total resistance.

- *Current divider:* In parallel resistors, the ratio of a resistor's current to the total current into the resistors is the same as the ratio of its conductance to the total conductance.

The d'Arsonval meter movement is an electromechanical device that looks electrically like a resistor and has a mechanical deflection directly proportional to its coil current. Specifically, we showed how to use the d'Arsonval movement to construct the

- *ammeter,* which uses a parallel resistor to determine the meter's current range;

- *voltmeter,* which uses a series resistor to determine the meter's voltage range;

- *ohmmeter,* which uses a series voltage source and a calibrating resistor to measure the current through an unknown resistor, with the unknown resistance determining the current, converting the ammeter scale to a resistance scale; and

- *bridge circuit,* which uses the d'Arsonval movement to detect voltage balance between two voltage dividers, with only one of the four bridge resistors unknown so that when the bridge is balanced its value is determined from the other three known resistors.

PROBLEMS

3.1 a) Show that the solution of the circuit in Fig. 3.8 (see Example 3.1) satisfies Kirchhoff's current law at junctions x and y.

b) Show that the solution of the circuit in Fig. 3.8 satisfies Kirchhoff's voltage law around every closed loop.

3.2 a) Find the power dissipated in each resistor in the circuit shown in Fig. 3.8.

b) Find the power delivered by the 120-V source.

c) Show that the power delivered equals the power dissipated.

3.3 Find the equivalent resistance R_{ab} for each of the circuits in Fig. P3.3.

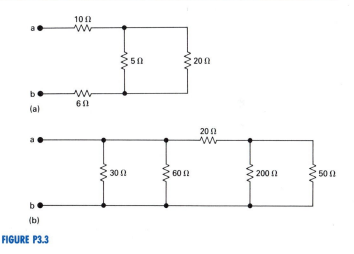

FIGURE P3.3

3.4 a) In the circuits in Fig. P3.4(a), (b), and (c) find the equivalent resistance R_{ab}.

b) For each circuit, find the power delivered by the source.

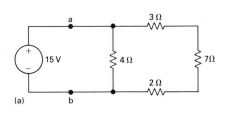

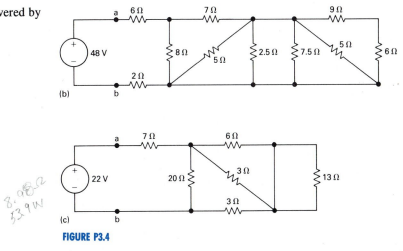

FIGURE P3.4

3.5 Find the power dissipated in the 12-Ω resistor in the circuit in Fig. P3.5.

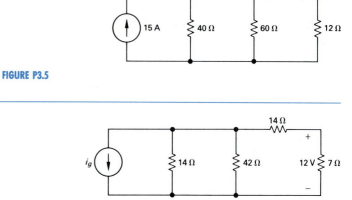

FIGURE P3.5

3.6 Find the value of i_g in the circuit in Fig. P3.6.

FIGURE P3.6

3.7 Find the equivalent resistance R_{ab} for each of the circuits in Fig. P3.7.

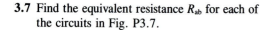

14.6 Ω

(b)

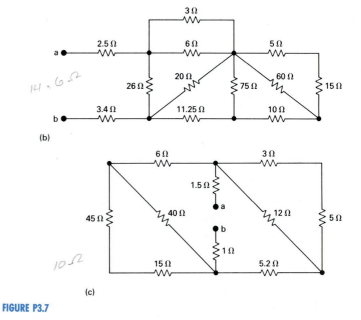

18 Ω

a 5Ω 9 Ω

30 Ω 20 Ω 5 Ω

b 3 Ω 10 Ω

(a)

20 Ω

10 Ω

(c)

FIGURE P3.7

3.8 For the circuit in Fig. P3.8 calculate

a) v_o,   5.76

b) the power dissipated in the 20-Ω resistor, and   5

c) the power developed by the current source.   41.2

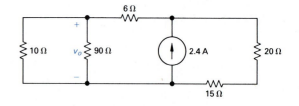

FIGURE P3.8

3.9 Find i_o and i_g in the circuit in Fig. P3.9.

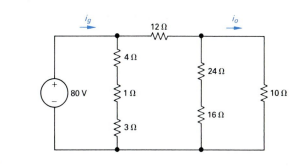

FIGURE P3.9

3.10 Find v_o and v_g in the circuit in Fig. P3.10.

FIGURE P3.10

3.11 For the circuit in Fig. P3.11 calculate (a) i_g and
(b) the power dissipated in the 30-Ω resistor.

20 A
1080 w

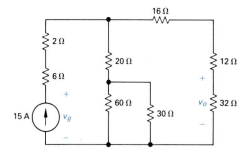

FIGURE P3.11

3.12 a) Find the voltage v_x in the circuit in Fig.
P3.12.

b) Replace the 18-V source with a general
voltage source equal to V_s. Assume V_s is positive at the upper terminal. Find v_x as a function of V_s.

FIGURE P3.12

3.13 Find v_o in the circuit in Fig. P3.13.

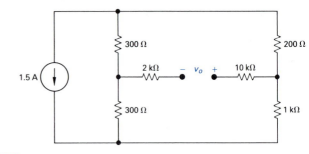

FIGURE P3.13

3.14 The current in the 12-Ω resistor in the circuit in Fig. P3.14 is 1 A, as shown.

a) Find v_g.

b) Find the power dissipated in the 20-Ω resistor.

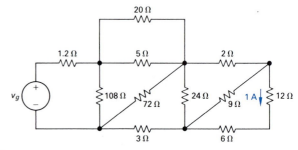

FIGURE P3.14

3.15 The high-voltage direct-current transmission line introduced in Drill Exercise 1.11 is 845 mi long. Each side of the circuit consists of two conductors in parallel. The resistance of each conductor is 0.0397 Ω/mi. The arrangement is illustrated in Fig. P3.15.

a) The voltage at the Oregon terminal of the line is 800 kV. Each conductor is carrying 1000 A as shown in the figure. Calculate the power received at the California end of the line and the efficiency of the power transmission from Oregon to California.

b) Repeat part (a) with the voltage at the Oregon terminal raised to 1000 kV and the current remaining at 1000 A/conductor.

c) Repeat part (b) with a third conductor added to each side of the circuit and the current remaining at 1000 A/conductor.

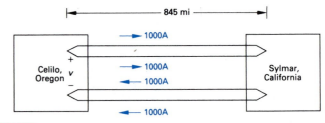

FIGURE P3.15

3.16 a) Calculate the no-load voltage v_o for the voltage-divider circuit shown in Fig. P3.16.

b) Calculate the power dissipated in R_1 and R_2.

c) Assume that only 1-W resistors are available. The no-load voltage is to be the same as in part (a). Specify the ohmic values of R_1 and R_2.

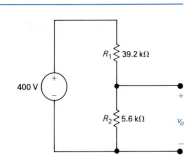

FIGURE P3.16

3.17 The no-load voltage in the voltage-divider circuit shown in Fig. P3.17 is 20 V. The smallest load resistor that is ever connected to the divider is 37.8 kΩ. When the divider is loaded, v_o is not to drop below 18 V.

a) Specify the numerical value of R_1 and R_2.

b) What is the maximum power dissipated in R_1?

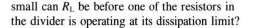

FIGURE P3.17

3.18 Assume the voltage divider in Fig. P3.17 has been constructed from 0.25-W resistors. How small can R_L be before one of the resistors in the divider is operating at its dissipation limit?

3.19 There is often a need to produce more than one voltage using a voltage divider. For example, the memory components of many personal computers require voltages of -12 V, 5 V, and $+12$ V all with respect to a common reference terminal.

Select the values of R_1, R_2, and R_3 in the circuit in Fig. P3.19 to meet the following design requirements:

1) The total power supplied to the divider circuit by the 24-V source is 60 W when the divider is unloaded; and

2) The three voltages, all measured with respect to the common reference terminal are: $v_1 = 12$ V, $v_2 = 5$ V, and $v_3 = -12$ V.

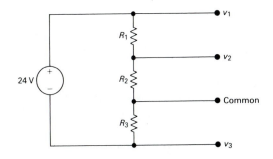

FIGURE P3.19

3.20 a) The voltage divider in Fig. P3.20(a) is loaded with the voltage divider shown in Fig. P3.20(b), that is, a is connected to a′ and b is connected to b′. Find v_o.

b) Now assume the voltage divider in Fig. P3.20(b) is connected to the voltage divider in Fig. P3.20(a) by means of a current-controlled voltage source as shown in Fig. P3.20(c). Find v_o.

c) What effect does adding the dependent-voltage source have on the operation of the voltage divider that is connected to the 380-V source?

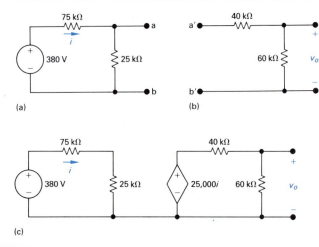

(a)   (b)

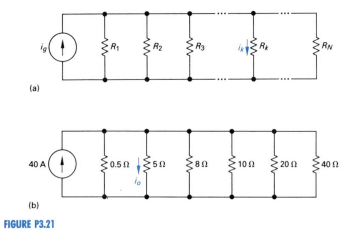

(c)

FIGURE P3.20

3.21 a) Show that the current in the kth branch of the circuit in Fig. P3.21(a) is equal to the source current i_g times the conductance of the kth branch divided by the sum of the conductances, that is

$$i_k = \frac{i_g G_k}{[G_1 + G_2 + G_3 + \ldots + G_k + \ldots + G_N]}$$

b) Use the result derived in part (a) to calculate the current in the 5-Ω resistor in the circuit in Fig. P3.21(b).

(a)

(b)

FIGURE P3.21

3.22 Specify the resistors in the circuit in Fig. P3.22 to meet the following design criteria:

$i_g = 1$ mA; $v_g = 1$ V; $i_1 = 2i_2$; $i_2 = 2i_3$; and $i_3 = 2i_4$.

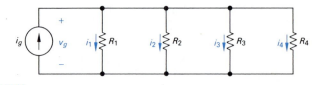

FIGURE P3.22

3.23 In the circuit in Fig. P3.23(a) the device labeled D represents a component that has the equivalent circuit shown in Fig. P3.23(b). The labels on the terminals of D show how the device is connected to the circuit. Find v_x and the power absorbed by the device.

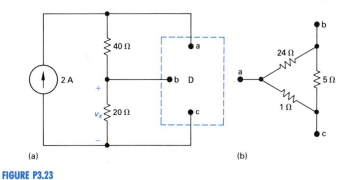

(a)   (b)

FIGURE P3.23

3.24 In Fig. P3.24(a) the box represents a field-effect transistor known as a FET. A simplified circuit model for the FET is shown in Fig. P3.24(b). Find v_o and v_{DS}.

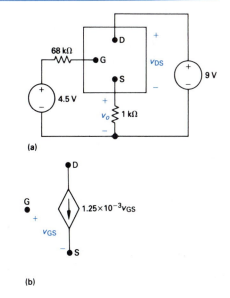

(a)

(b)

FIGURE P3.24

3.25 A d'Arsonval ammeter is shown in Fig. P3.25. Find the value of the shunt resistor R_A for each of the following full-scale readings: (a) 10 A, (b) 1 A, (c) 50 mA, and (d) 2 mA.

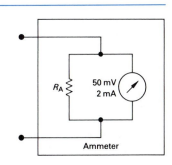

FIGURE P3.25

3.26 A shunt resistor and a 50-mV, 1-mA d'Arsonval movement are used to build a 10-A ammeter. A resistance of 0.01 Ω is placed across the terminals of the ammeter. What is the new full-scale range of the ammeter?

3.27 Two d'Arsonval ammeters are connected in parallel. Ammeter 1 uses a 2-μA, 500 μV movement and has a full-scale reading of 10 μA. Ammeter 2 uses a 1 μA, 1 mV movement and has a full-scale reading of 5 μA. What is the largest current these parallel-connected ammeters can read?

3.28 a) Show for the ammeter circuit in Fig. P3.28 that the current in the d'Arsonval movement is always one-fiftieth of the current being measured.

b) What would the fraction be if the 50-mV, 1-mA movement were used in a 1-A ammeter?

c) Would you expect a uniform scale on a dc d'Arsonval ammeter?

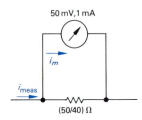

FIGURE P3.28

3.29 The ammeter in the circuit in Fig. P3.29 has a resistance of 1 Ω. What is the percent error in the reading of this ammeter if

$$\% \text{ error} = \left(\frac{\text{measured value}}{\text{true value}} - 1 \right) \times 100?$$

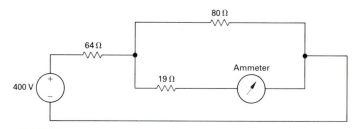

FIGURE P3.29

3.30 The paralleled ammeters described in Problem 3.27 are used to measure the current i_o in the circuit in Fig. P3.30. What is the percent error in the measured value?

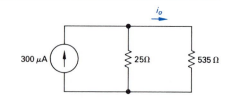

FIGURE P3.30

3.31 A d'Arsonval movement is rated at 1 mA and 20 mV. Assume 0.25-W precision resistors are available to use as shunts.

What is the largest full-scale-reading ammeter that can be constructed? Explain.

3.32 A d'Arsonval voltmeter is shown in Fig. P3.32. Find the value of R_V for each of the following full-scale readings: (a) 100 V, (b) 1 V, (c) 200 mV, and (d) 20 mV.

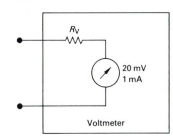

FIGURE P3.32

3.33 The voltage-divider circuit shown in Fig. P3.33 is designed so that the no-load output voltage is seventy-five hundredths of the input voltage. A d'Arsonval voltmeter having a sensitivity of 600 Ω/V and a full-scale rating of 400 V is used to check the operation of the circuit.

a) What will the voltmeter read if it is placed across the 320-V source?

b) What will the voltmeter read if it is placed across the 48-kΩ resistor?

c) What will the voltmeter read if it is placed across the 16-kΩ resistor?

d) Will the voltmeter readings obtained in parts (b) and (c) add to the reading recorded in part (a)? Explain why.

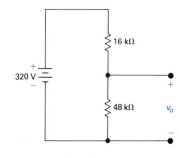

FIGURE P3.33

3.34 A multirange voltmeter consisting of a 50-mV, 1-mA meter movement and three resistances, R_1, R_2, and R_3, is shown in Fig. P3.34. The desired voltage ranges are 30, 150, and 300 V. The 30-, 150-, and 300-V terminals are as shown in Fig. P3.34.

a) Determine R_1, R_2, and R_3.

b) Assume that a 600-kΩ resistor is connected between the 150-V terminal and the common terminal. The voltmeter is then connected to an unknown voltage using the common terminal and the 300-V terminal. The voltmeter reads 240 V. What is the unknown voltage?

c) What is the maximum voltage the voltmeter in part (b) can measure?

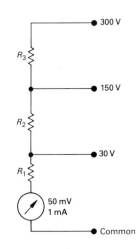

FIGURE P3.34

3.35 You have been told that the dc voltage of a power supply is about 400 V. When you go to the instrument room to get a dc voltmeter to measure the power supply voltage, you find that there are only two dc voltmeters available. One voltmeter is rated 300 V full scale and has a sensitivity of 900 Ω/V. The second voltmeter is rated 150 V full scale and has a sensitivity of 1200 Ω/V.

a) How can you use the two voltmeters to check the power supply voltage?

b) What is the maximum voltage that can be measured?

c) If the power supply voltage is 320 V, what will each voltmeter read?

3.36 Assume that in addition to the two voltmeters described in Problem 3.35 a 50-kΩ precision resistor is also available. The 50-kΩ resistor is connected in series with the series-connected voltmeters. This circuit is then connected across the terminals of the power supply. The reading on the 300-V voltmeter is 205.2 V and the reading on the 150-V voltmeter is 136.8 V. What is the voltage of the power supply?

3.37 A 200-kΩ resistor is connected from the 200-V terminal to the common terminal of a dual-scale voltmeter as shown in Fig. P3.37(a). This modified voltmeter is then used to measure the voltage across the 125-kΩ resistor in the circuit in Fig. P3.37(b).

a) What is the reading on the 600-V scale of the meter?

b) What is the percent error in the measured voltage?

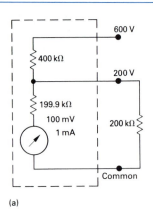

(a)

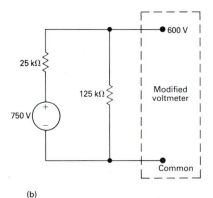

(b)

FIGURE P3.37

3.38 The voltmeter shown in Fig. P3.38(a) has a full-scale reading of 750 V. The meter movement is rated 75 mV and 1.5 mA. What is the percent error in the meter reading if it is used to measure the voltage v in the circuit of Fig. P3.38(b)?

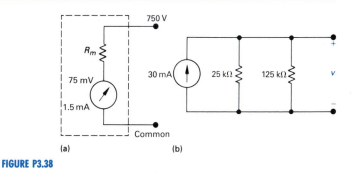

(a)   (b)

FIGURE P3.38

3.39 The elements in the circuit in Fig. 2.22 have the following values: $R_1 = 20$ kΩ, $R_2 = 80$ kΩ, $R_C = 0.82$ kΩ, $R_E = 0.20$ kΩ, $V_{CC} = 7.5$ V, $V_o = 0.6$ V, and $\beta = 39$.

a) Calculate the value of i_B in microamperes.

b) Assume a digital multimeter, when used as a dc ammeter, has a resistance of 1.0 kΩ. If the meter is inserted between terminals b and 2 to measure the current i_B, what will the meter read?

c) Using the calculated value of i_B in part (a) as the correct value, what is the percent error in the measurement?

3.40 The circuit model of a dc voltage source is shown in Fig. P3.40. The following voltage measurements are made at the terminals of the source: (1) With the terminals of the source open, the voltage is measured at 67.2 mV, and (2) with a 4-MΩ resistor connected to the terminals, the voltage is measured at 60 mV. All measurements are made with a digital voltmeter that has a meter resistance of 12 MΩ.

a) What is the internal voltage of the source (V_s) in millivolts?

b) What is the internal resistance of the source (R_s) in kilohms?

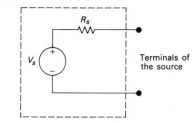

FIGURE P3.40

3.41 Assume in designing the multirange voltmeter shown in Fig. P3.41 you ignore the resistance of the meter movement.

a) Specify the values of R_1, R_2, and R_3.

b) For each of the three ranges calculate the percentage error that this design strategy produces.

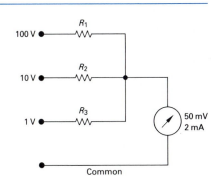

FIGURE P3.41

3.42 The components in the ohmmeter circuit shown in Fig. P3.42 are as follows: A is a 250-μA ammeter; B is a battery with negligible internal resistance and a terminal voltage of 9 V; R_1 is a fixed resistor of 12 kΩ; and R is an adjustable resistor.

a) Determine the setting of R so that the microammeter will deflect to its full-scale position when the ohmmeter terminals a, b are short-circuited.

b) Specify the resistance that the ohmmeter should indicate at 100% of full scale; 80% of full scale; 50% of full scale; 20% of full scale; and 0% of full scale.

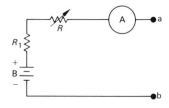

FIGURE P3.42

3.43 Find the current and voltage ratings of the d'Arsonval movement so that the ohmmeter in Fig. P3.43 reads full scale if shorted and half scale when connected to a 2000-Ω resistor.

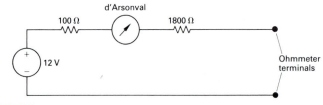

FIGURE P3.43

3.44 In the ohmmeter circuit in Fig. P3.44 the voltage drop across the 1-mA ammeter is negligible.

a) What value of R_o will produce a full-scale reading (1 mA) when R_x is zero?

b) Use the value of R_o found in part (a) to plot a graph of the meter current in mA vs R_x in kΩ for $0 \le R_x \le 50$ kΩ.

c) Explain why the scale on the d'Arsonval ohmmeter is nonlinear.

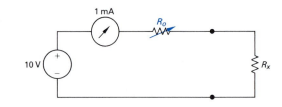

FIGURE P3.44

3.45 a) The ohmmeter circuit in Fig. P3.45 has been calibrated assuming $v_g = 9$ V, $R_g = 1$ Ω, and that the voltage drop across the 10-mA ammeter is negligible. What is the value of R_o?

b) Assume v_g drops to 8 V and R_o is adjusted

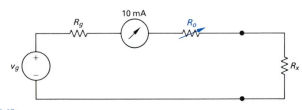

FIGURE P3.45

so that the ohmmeter reads zero when R_x is zero. What is the new value of R_o?

c) Assume that after R_o is adjusted to compensate

for the drop in v_g from 9 V to 8 V a resistance of 900 Ω is connected to the ohmmeter, i.e. $R_x = 900$ Ω. What will the ohmmeter read?

3.46 A multirange ohmmeter circuit is shown in Fig. P3.46. Note that the circuit uses a shunting resistor in parallel with the microammeter branch as well as resistors in series with the microammeter. The resistors R_3 through R_{10} have the following numerical values:

$R_3 = 20$ Ω, $\qquad R_7 = 990$ Ω,
$R_4 = 200$ Ω, $\qquad R_8 = 54$ Ω,
$R_5 = 2950$ Ω, $\qquad R_9 = 110$ Ω,
$R_6 = 29{,}500$ Ω, $\qquad R_{10} = 10$ Ω.

a) Determine the center-scale resistance reading of the ohmmeter for the four positions of the selector switch, that is, positions 1, 2, 3, and 4. Assume that the adjustable resistor is set to 500 Ω for all positions of the selector switch.

b) If the center-scale calibration of positions 1, 2, 3, and 4 on the multirange ohmmeter read as 30, 300, 3000, and 30,000, respectively, will the ohmmeter indicate the correct resistance to within $\pm 10\%$?

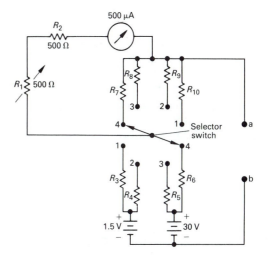

FIGURE P3.46

3.47 The variable resistor R_o is adjusted to give full-scale deflection when a short circuit is placed across the terminals a, b in the ohmmeter circuit in Fig. P3.47. When an unknown resistor (R_x) is placed across a, b, the meter deflection is 20% of full scale. The d'Arsonval meter movement is rated 100 mV and 10 μA. What is the value of R_x?

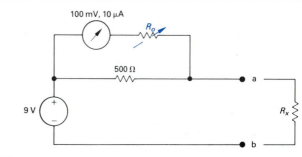

FIGURE P3.47

3.48 The bridge circuit shown in Fig. 3.21 is energized from a 24-V dc source. The bridge is balanced when $R_1 = 500$ Ω, $R_2 = 1000$ Ω, and $R_3 = 750$ Ω.

a) What is the value of R_x?

b) How much current (in milliamperes) does the dc source supply?

(*continued*)

c) Which resistor in the circuit absorbs the most power? How much power does it absorb?

d) Which resistor absorbs the least power? How much power does it absorb?

3.49 Find the detector current i_d in the unbalanced bridge in Fig. P3.49 if the voltage drop across the detector is negligible.

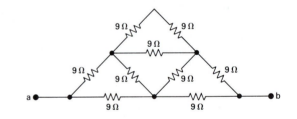

FIGURE P3.49

3.50 Assume the ideal voltage source in Fig. 3.22 is replaced by an ideal current source. Show that Eq. (3.36) is still valid.

3.51 Find R_{ab} in the circuit in Fig. P3.51.

FIGURE P3.51

3.52 a) Find the equivalent resistance R_{ab} in the circuit in Fig. P3.52 by using a Δ to Y transformation involving the resistors R_2, R_3, and R_4.

b) Repeat part (a) using a Y to Δ transformation involving resistors R_2, R_4, and R_5.

c) Give two additional Δ to Y or Y to Δ transformations that could be used to find R_{ab}.

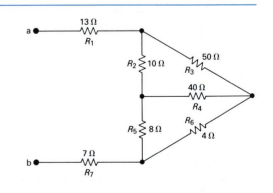

FIGURE P3.52

3.53 In the Wheatstone bridge circuit shown in Fig. 3.21, the ratio R_2/R_1 can be set to the following values: 0.001, 0.01, 0.1, 1.0, 10, 100, and 1000. The resistor R_3 can be varied from 1 to 11,110 Ω in increments of 1 Ω. An unknown resistor is known to lie between 4 and 5 Ω. What should be the setting of the R_2/R_1 ratio so that the unknown resistor can be measured to four significant figures?

3.54 Derive Eqs. (3.37)–(3.42) from Eqs. (3.43)–(3.45). The following two hints should help you get started in the right direction.

i) To find R_1 as a function of R_a, R_b, and R_c, first subtract Eq. (3.44) from Eq. (3.45) and then add this result to Eq. (3.43). Use similar manipulations to find R_2 and R_3 as functions of R_a, R_b, and R_c.

ii) To find R_b as a function of R_1, R_2, and R_3, take advantage of the derivations obtained by hint (i), namely, Eqs. (3.37)–(3.39). Note that these equations can be divided to

obtain

$$\frac{R_2}{R_3} = \frac{R_c}{R_b} \quad \text{or} \quad R_c = \frac{R_2}{R_3}R_b$$

and

$$\frac{R_1}{R_2} = \frac{R_b}{R_a} \quad \text{or} \quad R_a = \frac{R_2}{R_1}R_b.$$

Now use these ratios in Eq. (3.45) to eliminate R_a and R_c. Use similar manipulations to find R_a and R_c as functions of R_1, R_2, and R_3.

3.55 Show that the expressions for Δ-conductances as functions of the three Y-conductances are

$$G_a = \frac{G_2 G_3}{G_1 + G_2 + G_3},$$

$$G_b = \frac{G_1 G_3}{G_1 + G_2 + G_3},$$

and

$$G_c = \frac{G_1 G_2}{G_1 + G_2 + G_3},$$

where

$$G_a = \frac{1}{R_a}, \qquad G_1 = \frac{1}{R_1}, \text{ etc.}$$

3.56 For the circuit shown in Fig. P3.56, find (a) i_1, (b) i_2, (c) v, and (d) the power supplied by the voltage source.

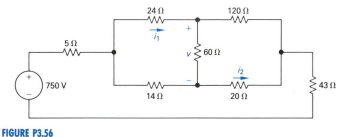

FIGURE P3.56

3.57 Find the equivalent resistance R_{ab} in the circuit in Fig. P3.57.

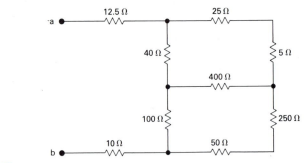

FIGURE P3.57

3.58 Use a wye-delta transformation to find: (a) i_o; (b) i_1; (c) i_2; and (d) the power delivered by the ideal current source in the circuit in Fig. P3.58.

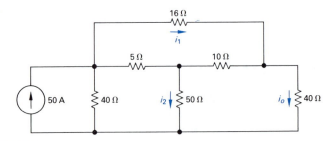

FIGURE P3.58

3.59 a) Find the resistance seen by the ideal voltage source in the circuit in Fig. P3.59.

b) If v_{ab} equals 400 V, how much power is dissipated in the 31-Ω resistor?

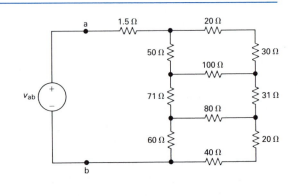

FIGURE P3.59

3.60 A voltage divider like that in Fig. 3.12 is to be designed so that $v_o = kv_s$ at no load ($R_1 = \infty$) and $v_o = \alpha v_s$ at full load ($R_L = R_o$). Note that by definition $\alpha < k < 1$.

a) Show that

$$R_1 = \frac{k - \alpha}{\alpha k} R_o \quad \text{and} \quad R_2 = \frac{k - \alpha}{\alpha(1 - k)} R_o.$$

b) Specify the numerical values of R_1 and R_2 if $k = 0.90$, $\alpha = 0.75$, and $R_o = 36$ kΩ.

c) If $v_s = 120$ V, specify the maximum power that will be dissipated in R_1 and R_2.

d) Assume the load resistor is accidentally short-circuited. How much power is dissipated in R_1 and R_2?

3.61 Resistor networks are sometimes used as volume-control circuits. In this application, they are referred to as "resistance attenuators" or "pads." A typical fixed-attenuator pad is shown in Fig. P3.61. In designing an attenuation pad, the circuit designer will select the values of R_1 and R_2 so that the ratio of v_o/v_i has a specified value and the resistance seen by the input voltage source R_{ab} has a specified value.

a) Show that if $R_{ab} = R_L$, then

$$R_L^2 = 4R_1(R_1 + R_2) \quad \text{and} \quad \frac{v_o}{v_i} = \frac{R_2}{2R_1 + R_2 + R_L}.$$

b) Select the values of R_1 and R_2 so that $R_{ab} = R_L = 600\ \Omega$ and $v_o/v_i = 0.6$.

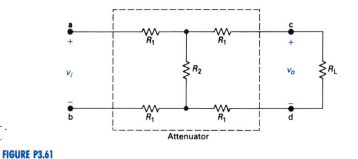

FIGURE P3.61

3.62 a) The fixed-attenuator pad shown in Fig. P3.62 is called a *bridged tee*. Use a Y-to-Δ transformation to show that $R_{ab} = R_L$ if $R = R_L$.

b) Show that when $R = R_L$ the voltage ratio v_o/v_i equals 0.50.

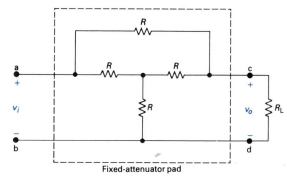

Fixed-attenuator pad

FIGURE P3.62

3.63 The design equations for the bridged-tee attenuator circuit in Fig. P3.63 are $R_{ab} = R_L$ when

$$R_2 = \frac{2RR_L^2}{3R^2 - R_L^2} \quad \text{and} \quad \frac{v_o}{v_i} = \frac{3R - R_L}{3R + R_L}$$

when R_2 has the value given above.

a) Design a fixed attenuator so that $v_i = 3.5v_o$ when $R_L = 300\ \Omega$.

b) Assume the voltage applied to the input of the pad designed in part (a) is 42 V. Which resistor in the pad dissipates the most power?

c) How much power is dissipated in the resistor in part (b)?

d) Which resistor dissipates the least power?

e) How much power is dissipated in the resistor in part (d)?

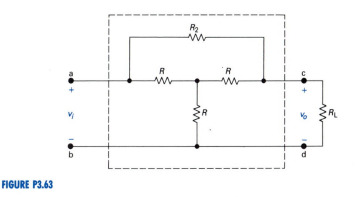

FIGURE P3.63

TECHNIQUES OF CIRCUIT ANALYSIS

CHAPTER 4

So far, we have analyzed relatively simple resistive circuits by applying Kirchhoff's laws in combination with Ohm's law. We can use this approach for all circuits, but as they become structurally more complicated and involve more and more elements, this direct method soon becomes cumbersome. In this chapter we introduce two powerful techniques of circuit analysis that aid the analysis of complex circuit structures: the node-voltage method and the mesh-current method. In addition to these two general analytic methods, we also discuss some additional techniques for simplifying circuits. We have already demonstrated how to use series–parallel reductions and Δ-to-Y transformations to simplify a structure. We now add source transformations and Thévenin and Norton equivalent circuits to those simplification techniques.

Before beginning the discussion of the node-voltage and mesh-current methods of circuit analysis, we need to reflect on the groundwork laid. In Chapter 1 we introduced current and voltage as the two variables used to describe the behavior of the basic circuit elements. In Chapter 2 we discussed voltage and current sources along with the circuit parameter of resistance. We began with these three types of basic circuit elements because all the basic techniques of circuit analysis can be explored using interconnections of just these elements. In both Chapters 2 and 3 we introduced circuit analysis through direct application of Ohm's law and Kirchhoff's laws. Ohm's law is crucial because it describes the relationship between the current and voltage at the terminals of a resistor. Kirchhoff's laws are important because they describe the constraints imposed on currents and voltages by interconnections of the basic elements. We also noted in Section 2.5 in discussing both the flashlight circuit and the dependent-source circuit that we can simplify the analysis problem by first concentrating on finding the element currents.

We now introduce the node-voltage and mesh-current methods of circuit analysis. Keep in mind that these two methods give us two *systematic* methods of describing circuits with the minimum number of simultaneous equations.

One additional comment: In this chapter we emphasize the *mechanics* of implementing the node-voltage and mesh-current methods. In Chapter 5 we show why either method leads to a set of independent simultaneous equations.

4.1 TERMINOLOGY

In order to discuss the more involved methods of circuit analysis, we must define a few basic terms necessary for a clear, concise description of important circuit features. So far, all the circuits presented have been *planar circuits,* that is, those circuits that can be drawn on a plane with no crossing branches. A circuit that can be drawn with crossing branches still is considered planar if it can be redrawn with no crossover branches. For example, the circuit shown in Fig. 4.1(a) is a planar circuit because it can be redrawn as shown in Fig. 4.1(b). Figure 4.2 shows an example of a nonplanar circuit.

The node-voltage method is applicable to both planar and nonplanar circuits, whereas the mesh-current method is limited to planar circuits. For nonplanar circuits, a technique known as the loop-current method replaces the mesh-current method. We discuss the loop-current method in Chapter 5.

DESCRIBING A CIRCUIT—THE VOCABULARY

In Section 1.6 we defined an ideal basic circuit element. When basic circuit elements are interconnected to form a circuit, the resulting interconnection is described in terms of nodes, paths, branches, loops, and meshes. We defined both a node and a closed path, or loop, in Section 2.4. Here, we restate those definitions and then define the terms *path, branch,* and *mesh.*

A *node* is a point in a circuit where two or more circuit elements join. Creating a *closed path,* or *loop,* involves starting at a selected node, tracing a set of connected basic circuit elements, and returning to the original starting node without passing through any intermediate node more than once.

A *path* is formed whenever a set of adjoining basic circuit elements is traced, in order, without passing through a connecting node more than once. A *branch* is a path that connects two nodes. A *mesh* is a special type of loop; that is, it does not contain any other loops within it.

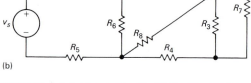

(a)

(b)

FIGURE 4.1 (a) A planar circuit; (b) the same circuit redrawn to verify that it is planar.

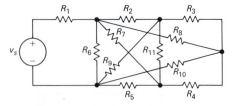

FIGURE 4.2 A nonplanar circuit.

The circuit shown in Fig. 4.3 illustrates these circuit characteristics. Note that the circuit has

1. seven nodes (a, b, c, d, e, f, and g);

2. ten branches (v_1, R_1, R_2, R_3, v_2, R_4, R_5, R_6, R_7, and I); and

3. four meshes (v_1–R_1–R_5–R_3–R_2, v_2–R_2–R_3–R_6–R_4, R_5–R_7–R_6, and R_7–I).

Note also that several loops are not meshes. For example, v_1–R_1–R_5–R_6–R_4–v_2 forms a closed loop, but it is not a mesh because other closed loops exist within it. A path that is neither closed nor a branch is the trace v_2–R_2–R_3–R_5–R_7.

In the work that follows, identifying only those nodes in the circuit that join *three or more* elements often is convenient. We call such nodes *essential nodes*. Also convenient is identifying only those paths that connect essential nodes *without passing through an essential node*. We call such paths *essential branches*. The circuit shown in Fig. 4.3 contains four essential nodes (b, c, e, and g) and seven essential branches (v_1–R_1; R_2–R_3; v_2–R_4; R_5; R_6; R_7; and I). Note that in general the number of essential nodes is less than or equal to the number of nodes and that the number of essential branches is less than or equal to the number of branches.

SIMULTANEOUS EQUATIONS—HOW MANY?

The numbers of nodes, branches, and meshes in a circuit determine the number of simultaneous equations that we must derive in order to solve the circuit. The reason is that the number of unknown currents in the circuit equals the number of branches, b, *where the current is not known*. For example, the circuit shown in Fig. 4.3 has nine branches in which the current is unknown. Recall that we must have b independent equations in order to solve a circuit with b unknown currents. If we let n represent the number of nodes in the circuit, we can derive $n - 1$ independent equations by applying Kirchhoff's current law to any set of $n - 1$ nodes. (Application of the current law to the nth node does not generate an independent equation because this equation can be derived from the previous $n - 1$ equations. See Drill Exercise 4.2.) Because we need b equations to describe a given circuit and because we can obtain $n - 1$ of these equations from Kirchhoff's current law, we must apply Kirchhoff's voltage law to independent loops or meshes to obtain the remaining $b - (n - 1)$ equations.

Thus by counting branches, nodes, and meshes we have established a systematic method for writing the necessary number of equations to solve a circuit. Specifically, we apply Kirchhoff's

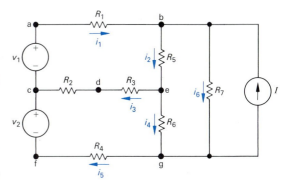

FIGURE 4.3 A circuit illustrating nodes, branches, meshes, paths, and loops.

current law to $n - 1$ junctions and Kirchhoff's voltage law to $b - (n - 1)$ independent loops (or meshes). These observations also are valid in terms of essential nodes and essential branches. Thus if we let n_e represent the number of essential nodes and b_e the number of essential branches *where the current is unknown,* we can apply Kirchhoff's current law at $n_e - 1$ nodes and Kirchhoff's voltage law around $b_e - (n_e - 1)$ loops or meshes.

A circuit may consist of disconnected parts. Hence you must recognize that the statements pertaining to the number of equations that can be derived from Kirchhoff's current law $(n - 1)$ and voltage law $[b - (n - 1)]$ apply to *connected circuits.* If a circuit has n nodes and b branches and is made up of s parts, the current law can be applied $n - s$ times and the voltage law $b - n + s$ times. Any two separate parts can be connected by a *single* conductor. This connection always causes two nodes to form one node. Moreover, no current exists in the single conductor, so any circuit made up of s disconnected parts can always be reduced to a connected circuit.

The Systematic Approach—An Illustration

We illustrate this systematic approach to deriving the simultaneous equations that describe a connected circuit in terms of its unknown currents. We use the circuit shown in Fig. 4.4 and write the equations on the basis of essential nodes and branches. The circuit has four essential nodes and six essential branches, denoted i_1–i_6, for which the current is unknown.

We derive three of the six simultaneous equations needed to describe this circuit by applying Kirchhoff's current law to any three of the four essential nodes. We use the nodes b, c, and e to get

$$-i_1 + i_2 + i_6 - I = 0;$$
$$i_1 - i_3 - i_5 = 0; \qquad \textbf{(4.1)}$$
$$i_3 + i_4 - i_2 = 0.$$

We derive the remaining three equations by applying Kirchhoff's voltage law around three meshes. Because the circuit has four meshes, we need to dismiss one mesh. We choose $R_7 - I$ because we don't know the voltage across I.[†] Using the other three meshes gives

$$R_1 i_1 + R_5 i_2 + i_3(R_2 + R_3) - v_1 = 0;$$
$$-i_3(R_2 + R_3) + i_4 R_6 + i_5 R_4 - v_2 = 0; \qquad \textbf{(4.2)}$$
$$-i_2 R_5 + i_6 R_7 - i_4 R_6 = 0.$$

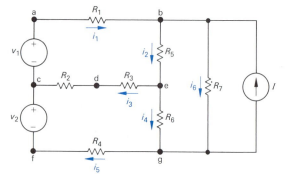

FIGURE 4.4 The circuit shown in Fig. 4.3 with six unknown branch currents defined.

[†] We say more about this decision in Section 4.7.

Rearranging Eqs. (4.1) and (4.2) to facilitate their solution yields the set

$$-i_1 + i_2 + 0i_3 + 0i_4 + 0i_5 + i_6 = I;$$

$$i_1 + 0i_2 - i_3 + 0i_4 - i_5 + 0i_6 = 0;$$

$$0i_1 - i_2 + i_3 + i_4 + 0i_5 + 0i_6 = 0;$$

$$R_1 i_1 + R_5 i_2 + (R_2 + R_3)i_3 + 0i_4 + 0i_5 + 0i_6 = v_1;$$

$$0i_1 + 0i_2 - (R_2 + R_3)i_3 + R_6 i_4 + R_4 i_5 + 0i_6 = v_2;$$

$$0i_1 + R_5 i_2 + 0i_3 - R_6 i_4 + 0i_5 + R_7 i_6 = 0.$$

(4.3)

Note that summing the current at the nth node (g in this example) gives

$$i_5 - i_4 - i_6 + I = 0. \tag{4.4}$$

Equation (4.4) is not independent because we can derive it by summing Eqs. (4.1) and then multiplying the sum by -1. Thus Eq. (4.4) is a linear combination of Eqs. (4.1) and therefore is not independent of them.

We now carry the procedure one step further. By introducing new variables we can describe a circuit with just $n - 1$ equations or just $b - (n - 1)$ equations. Therefore these new variables allow us to obtain a solution by manipulating fewer equations, a desirable goal even if a computer is to be used to obtain a numerical solution. The new variables are known as *node voltages* and *mesh currents*. The node-voltage method enables us to describe a circuit in terms of $n - 1$ (or $n_e - 1$) equations, whereas the mesh-current method enables us to describe a circuit in terms of $b - (n - 1)$ [or $b_e - (n_e - 1)$] equations. We begin with the node-voltage method.

DRILL EXERCISES

4.1 For the circuit shown, state the numerical value of the number of (a) branches, (b) branches where the current is unknown, (c) essential branches, (d) essential branches where the current is unknown, (e) nodes, (f) essential nodes, and (g) meshes.

ANSWER: (a) 11; (b) 9; (c) 9; (d) 7; (e) 6; (f) 4; (g) 6.

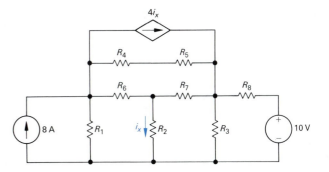

4.2 A current leaving a node is defined as positive.

a) Sum the currents at each node in the circuit shown.

b) Show that any one of the equations in (a) can be derived from the remaining two equations.

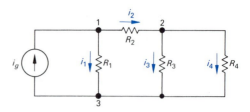

ANSWER: (a) 1: $i_1 - i_g + i_2 = 0$; 2: $i_3 + i_4 - i_2 = 0$; 3: $i_g - i_1 - i_3 - i_4 = 0$. (b) To derive any one equation from the other two equations, simply add the two equations and then multiply the resulting sum by -1.

4.3 a) If only the essential nodes and branches are identified in the circuit of Drill Exercise 4.1, how many simultaneous equations are needed to describe the circuit?

b) How many of these equations can be derived using Kirchhoff's current law?

c) How many must be derived using Kirchhoff's voltage law?

d) What two meshes should be avoided in applying the voltage law?

ANSWER: (a) 7; (b) 3; (c) 4; (d) R_4–R_5–$4i_x$ and 8 A–R_1.

4.4 a) How many separate parts does the circuit shown have?

b) How many nodes?

c) How many independent current equations can be written?

d) How many branches are there?

e) How many branches are there where the current is unknown?

f) How many equations must be written using the voltage law?

g) Assume that the lower node in each part of the circuit is joined by a single conductor. Repeat the calculations in (a)–(g).

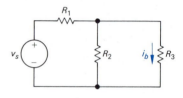

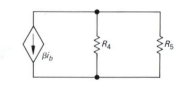

ANSWER: (a) 2; (b) 5; (c) 3; (d) 7; (e) 6; (f) 3; (g) 1, 4, 3, 7, 6, 3.

4.2 INTRODUCTION TO THE NODE-VOLTAGE METHOD

PSpice   Use the PSpice commands .OP or .DC to perform nodal analysis: Sections 3.1 and 3.2

We introduce the node-voltage method by using the essential nodes of the circuit. The first step is to make a neat layout of the circuit so that no branches cross over and to mark clearly the essential nodes on the circuit diagram, as in Fig. 4.5. This circuit has three essential nodes $(n_e = 3)$; therefore we need two $(n_e - 1)$ node-voltage equations to describe the circuit. The next step is to select one of the three essential nodes as a reference node. Although theoretically the choice is arbitrary, practically the choice for the reference node often is obvious. For example, the node with the most branches is usually a good choice. The optimum choice of the reference node (if one exists) will become apparent after you have gained some experience using this method. Because the lower node connects the most branches in the circuit shown in Fig. 4.5, we use it as the reference node. We flag the chosen reference node with the symbol ↓, as in Fig. 4.6.

After selecting the reference node, we define the node voltages on the circuit diagram. *A node voltage is defined as the voltage rise from the reference node to a nonreference node.* For this circuit, we must define two node voltages, which are denoted v_1 and v_2 in Fig. 4.6.

We are now ready to generate the node-voltage equations. We do so by writing the current leaving each branch connected to a nonreference node as a function of the node voltages and then summing these currents to zero in accordance with Kirchhoff's current law. Hence, with reference to the circuit shown in Fig. 4.6, the current away from node 1 through the 1-Ω resistor is the voltage drop across the resistor divided by the resistance (Ohm's law). The voltage drop across the resistor, in the direction of the current away from the node, is $v_1 - 10$. Therefore the current in the 1-Ω resistor is $(v_1 - 10)/1$. Figure 4.7 readily confirms these observations. It shows the 10-V–1-Ω branch, with the appropriate voltages and current. Note that summing the voltages around the closed path in accordance with Kirchhoff's voltage law verifies that the voltage drop across the 1-Ω resistor in the direction of i is $(v_1 - 10)$ V.

This same reasoning yields the current in every branch where the current is unknown. Thus the current away from node 1 through the 5-Ω resistor is $v_1/5$, and the current away from node 1 through the 2-Ω resistor is $(v_1 - v_2)/2$. The sum of the three currents leaving node 1 must equal zero; therefore the node-voltage equation derived at node 1 is

$$\frac{v_1 - 10}{1} + \frac{v_1}{5} + \frac{v_1 - v_2}{2} = 0. \qquad (4.5)$$

The node-voltage equation derived at node 2 is

$$\frac{v_2 - v_1}{2} + \frac{v_2}{10} - 2 = 0. \qquad (4.6)$$

Note that the first term in Eq. (4.6) is the current away from node 2 through the 2-Ω resistor, the second term is the current away from node 2 through the 10-Ω resistor, and the third term is the current away from node 2 through the current source.

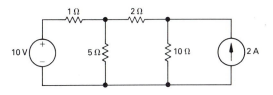

FIGURE 4.5 A circuit used to illustrate the node-voltage method of circuit analysis.

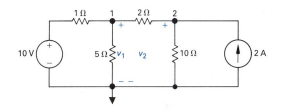

FIGURE 4.6 The circuit shown in Fig. 4.5 with reference node and the node voltages.

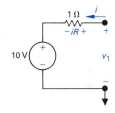

FIGURE 4.7 Computation of the branch current i.

Equations (4.5) and (4.6) are the two simultaneous equations that describe the circuit shown in Fig. 4.6 in terms of the node voltages v_1 and v_2. Solving for v_1 and v_2 yields

$$v_1 = \frac{100}{11} = 9.09 \text{ V} \quad \text{and} \quad v_2 = \frac{120}{11} = 10.91 \text{ V}.$$

Once the node voltages are known, all the branch currents can be calculated. Once the branch currents are known, the branch voltages and powers can be calculated. Example 4.1 illustrates the use of the node-voltage method to analyze a circuit.

EXAMPLE 4.1

a) Use the node-voltage method of circuit analysis to find the branch currents i_a, i_b, and i_c in the circuit shown in Fig. 4.8.

b) Find the power associated with each source, and state whether the source is delivering or absorbing power.

SOLUTION

a) We begin by noting that the circuit has two essential nodes; thus we need to write a single node-voltage expression. We select the lower node as the reference node and define the unknown node voltage as v_1. Figure 4.9 illustrates these decisions. Summing the currents away from node 1 generates the node-voltage equation:

$$\frac{v_1 - 50}{5} + \frac{v_1}{10} + \frac{v_1}{40} - 3 = 0.$$

Solving for v_1 gives

$$v_1 = 40 \text{ V}.$$

Hence

$$i_a = \frac{50 - 40}{5} = 2 \text{ A};$$

$$i_b = \frac{40}{10} = 4 \text{ A};$$

$$i_c = \frac{40}{40} = 1 \text{ A}.$$

b) The power associated with the 50-V source is

$$p_{50\text{V}} = -50 i_a = -100 \text{ W (delivering)}.$$

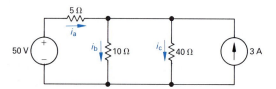

FIGURE 4.8 The circuit for Example 4.1.

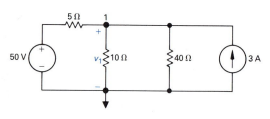

FIGURE 4.9 The circuit shown in Fig. 4.8 with reference node and the unknown node voltage v_1.

The power associated with the 3-A source is

$$p_{3A} = -3v_1 = -3(40) = -120 \text{ W (delivering)}.$$

We check these calculations by noting that the total delivered power is 220 W. The total power absorbed by the three resistors is $4(5) + 16(10) + 1(40)$, or 220 W, as we calculated and as it must be.

DRILL EXERCISES

4.5 a) For the circuit shown, use the node-voltage method to find v_1, v_2, and i_1.

b) How much power is delivered to the circuit by the 12-A source?

c) Repeat (a) and (b) for the 5-A source.

ANSWER: (a) 48 V, 64 V, -8 A; (b) 768 W; (c) -240 W.

4.6 Use the node-voltage method to find v in the circuit shown.

ANSWER: 15 V.

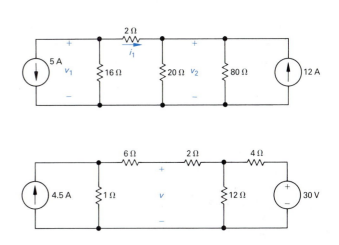

4.3 THE NODE-VOLTAGE METHOD AND DEPENDENT SOURCES

If the circuit contains dependent sources, the node-voltage equations must be supplemented with the constraint equations imposed by the presence of the dependent sources. Example 4.2 illustrates application of the node-voltage method to a circuit containing a dependent source.

PSpice Use the PSpice command .DC to perform nodal analysis on a circuit with dependent sources: Chapter 4

EXAMPLE 4.2

Use the node-voltage method to find the power dissipated in the 5-Ω resistor in the circuit shown in Fig. 4.10.

S O L U T I O N

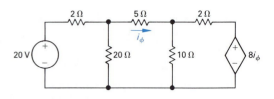

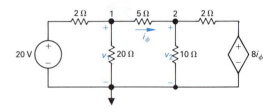

FIGURE 4.10 The circuit for Example 4.2.

We begin by noting that the circuit has three essential nodes. Hence, we need two node-voltage equations to describe the circuit. Four branches terminate on the lower node, so we select it as the reference node. The two unknown node voltages are defined on the circuit shown in Fig. 4.11. Summing the currents away from node 1 generates the equation

$$\frac{v_1 - 20}{2} + \frac{v_1}{20} + \frac{v_1 - v_2}{5} = 0.$$

Summing the currents away from node 2 yields

$$\frac{v_2 - v_1}{5} + \frac{v_2}{10} + \frac{v_2 - 8i_\phi}{2} = 0.$$

As written, these two node-voltage equations contain three unknowns, namely, v_1, v_2, and i_ϕ. To eliminate i_ϕ we must express this controlling current in terms of the node voltages, or

$$i_\phi = \frac{v_1 - v_2}{5}.$$

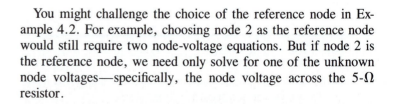

FIGURE 4.11 The circuit shown in Fig. 4.10 with reference node and node voltages.

Substituting this relationship into the node 2 equation simplifies the two node-voltage equations to

$$0.75v_1 - 0.2v_2 = 10;$$
$$-v_1 + 1.6v_2 = 0.$$

Solving for v_1 and v_2 gives

$$v_1 = 16 \text{ V} \qquad \text{and} \qquad v_2 = 10 \text{ V}.$$

Then,

$$i_\phi = \frac{16 - 10}{5} = 1.2 \text{ A}$$

and

$$p_{5\Omega} = (1.44)(5) = 7.2 \text{ W}.$$

You might challenge the choice of the reference node in Example 4.2. For example, choosing node 2 as the reference node would still require two node-voltage equations. But if node 2 is the reference node, we need only solve for one of the unknown node voltages—specifically, the node voltage across the 5-Ω resistor.

DRILL EXERCISES

4.7 a) Use the node-voltage method to find the power associated with each source in the circuit shown.

b) State whether the source is delivering power to the circuit or extracting power from the circuit.

ANSWER: (a) $p_{1.5A} = 15$ W, $p_{6i_2} = 1200$ W; $p_{80V} = 320$ W; (b) All sources are delivering power to the circuit.

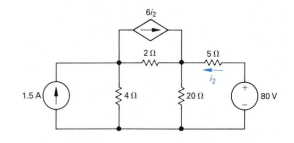

4.4 THE NODE-VOLTAGE METHOD: SOME SPECIAL CASES

When a voltage source is the only element between two essential nodes, the node-voltage method requires some additional manipulations. The circuit, with reference node and node voltages, shown in Fig. 4.12 depicts the nature of the problem. It has to do with writing the expression for the current leaving node 1 through the independent voltage source, because there is no resistance in series with the 100-V source. At first glance, the current in this branch may appear to be infinite, as implied by the expression $(v_1 - 100)/0$. However, closer inspection shows that v_1 must be 100 V, resulting in the indeterminate form 0/0. The observation that $v_1 = 100$ V allows us to apply the node-voltage method with no further difficulty. That is, recognizing that $v_1 = 100$ V, we see that there is only one unknown node voltage (v_2) and that solution of this particular circuit involves a single node-voltage equation. At node 2,

$$\frac{v_2 - v_1}{10} + \frac{v_2}{50} - 5 = 0. \tag{4.7}$$

But $v_1 = 100$ V, so Eq. (4.7) can be solved for v_2:

$$v_2 = 125 \text{ V}. \tag{4.8}$$

Knowing v_2, we can calculate the current in every branch. We leave to you verification that the current into node 1 in the branch containing the independent voltage source is 1.5 A.

Note that in the node-voltage method any voltage sources that are connected directly between essential nodes reduce the num-

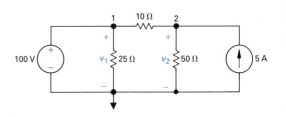

FIGURE 4.12 A circuit with a known node voltage.

ber of unknown node voltages. The reason is that whenever a voltage source connects two essential nodes, it constrains the difference between the node voltages at the essential nodes to equal the voltage of the source.

Suppose that the circuit shown in Fig. 4.13 is to be analyzed using the node-voltage method. The circuit contains four essential nodes, so we anticipate writing three node-voltage equations. However, two essential nodes are connected by an independent voltage source, and two other essential nodes are tied together with a current-controlled dependent voltage source. Hence, there actually is only one unknown node voltage. For example, note that if the voltage across the 50-Ω resistor is known, the voltage across the 100-Ω resistor is also known because of the presence of the dependent voltage source. Choosing which node to use as the reference node involves several possibilities. Either node on each side of the dependent voltage source looks attractive because, if chosen, one of the node voltages would be known to be either $+10i_\phi$ (left node is the reference) or $-10i_\phi$ (right node is the reference). The lower node looks even more attractive because, if chosen, one node voltage is immediately known (50 V) and five branches terminate there. We therefore opt for the lower node as the reference.

Figure 4.14 shows the redrawn circuit. In addition to flagging the reference node and defining the node voltages, we introduce the current i, which is needed to support the discussion that follows.

In writing the appropriate node-voltage equation at either node 2 or 3 we cannot express the current in the dependent voltage source branch as a function of the node voltages v_2 and v_3. To solve this dilemma, we introduce the unknown current i and then promptly eliminate it from the equations. Thus at node 2,

$$\frac{v_2 - v_1}{5} + \frac{v_2}{50} + i = 0, \qquad (4.9)$$

and at node 3,

$$\frac{v_3}{100} - i - 4 = 0. \qquad (4.10)$$

We eliminate i simply by adding Eqs. (4.9) and (4.10) to get

$$\frac{v_2 - v_1}{5} + \frac{v_2}{50} + \frac{v_3}{100} - 4 = 0. \qquad (4.11)$$

Equation (4.11) may be written directly, without resorting to the intermediate step represented by Eqs. (4.9) and (4.10). To do so, we consider nodes 2 and 3 to be a single node and simply

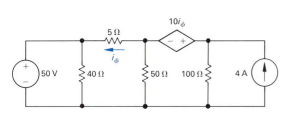

FIGURE 4.13 A circuit with a dependent voltage source connected between nodes.

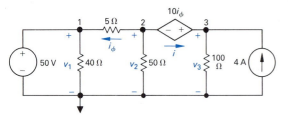

FIGURE 4.14 The circuit shown in Fig. 4.13 with the selected node voltages defined.

sum the currents away from the node in terms of the node voltages v_2 and v_3. Figure 4.15 illustrates this approach.

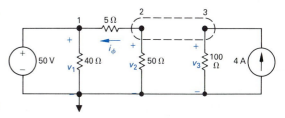

FIGURE 4.15 Considering nodes 2 and 3 to be a *supernode*.

THE CONCEPT OF A SUPERNODE

Combining nodes 2 and 3 is sometimes referred to as forming a *supernode*. Obviously, Kirchhoff's current law must hold for the supernode. Starting with the 5-Ω branch and moving counterclockwise around the supernode, we generate the equation

$$\frac{v_2 - v_1}{5} + \frac{v_2}{50} + \frac{v_3}{100} - 4 = 0, \qquad (4.12)$$

which is identical to Eq. (4.11). Therefore the benefit from creating the supernode is apparent. The supernode concept can be used whenever two essential nodes are connected by a voltage source element.

After Eq. (4.11) (or Eq. 4.12) has been derived, the next step is to reduce the expression to a single unknown node voltage. First, we eliminate v_1 from the equation because we know that $v_1 = 50$ V. Next we express v_3 as a function of v_2:

$$v_3 = v_2 + 10i_\phi. \qquad (4.13)$$

We now express the current controlling the dependent voltage source as a function of the node voltages:

$$i_\phi = \frac{v_2 - 50}{5}. \qquad (4.14)$$

Using Eqs. (4.13) and (4.14) and $v_1 = 50$ V reduces Eq. (4.11) to

$$v_2\left(\frac{1}{50} + \frac{1}{5} + \frac{1}{100} + \frac{10}{500}\right) = 10 + 4 + 1$$

$$v_2(0.25) = 15$$

$$v_2 = 60 \text{ V}.$$

From Eqs. (4.13) and (4.14),

$$i_\phi = \frac{60 - 50}{5} = 2 \text{ A} \qquad \text{and} \qquad v_3 = 60 + 20 = 80 \text{ V}.$$

ANOTHER ILLUSTRATION

Let's use the node-voltage method to analyze the circuit first introduced in Section 2.5 and shown again in Fig. 4.16. When we

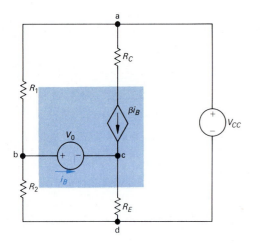

FIGURE 4.16 The transistor amplifier circuit shown in Fig. 2.22.

used the branch-current method of analysis in Section 2.5, we faced the task of writing and solving six simultaneous equations. We now view this circuit in terms of nodal analysis. The circuit has four essential nodes: Nodes a and d are connected by an independent voltage source, as are nodes b and c. Therefore the problem reduces to finding a single unknown node voltage $[(n_e - 1) - 2]$. Using d as the reference node and combining nodes b and c into a supernode, we obtain

$$\frac{v_b}{R_2} + \frac{v_b - V_{CC}}{R_1} + \frac{v_c}{R_E} - \beta i_B = 0, \qquad (4.15)$$

where v_b and v_c denote the voltage rise from the reference node to nodes b and c, respectively, and V_{CC} represents the voltage rise from the reference node to node a. We now eliminate both v_c and i_B from Eq. (4.15) by noting that

$$v_c = (i_B + \beta i_B)R_E \qquad (4.16)$$

and

$$v_c = v_b - V_0. \qquad (4.17)$$

Substituting Eqs. (4.16) and (4.17) into Eq. (4.15) yields

$$v_b\left[\frac{1}{R_1} + \frac{1}{R_2} + \frac{1}{(1 + \beta)R_E}\right] = \frac{V_{CC}}{R_1} + \frac{V_0}{(1 + \beta)R_E}. \qquad (4.18)$$

Solving Eq. (4.18) for v_b yields

$$v_b = \frac{V_{CC}R_2(1 + \beta)R_E + V_0R_1R_2}{R_1R_2 + (1 + \beta)R_E(R_1 + R_2)}. \qquad (4.19)$$

Using the node-voltage method to analyze the circuit in Fig. 4.15 reduced the problem from manipulating six simultaneous equations to manipulating three simultaneous equations. (See Problem 2.19.) We leave to you verification that, when Eq. (4.19) is combined with Eqs. (4.16) and (4.17), the solution for i_B is identical to Eq. (2.27). (See Problem 4.23.)

D R I L L   E X E R C I S E S

4.8 Use the node-voltage method to find v in the circuit shown.

ANSWER: 8 V.

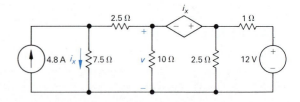

4.9 Use the node-voltage method to find v_1 in the circuit shown.

ANSWER: 120 V.

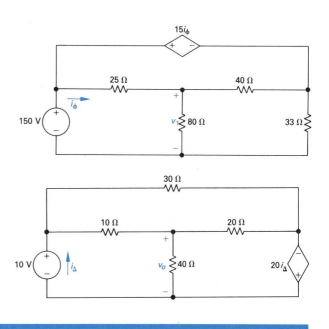

4.10 Use the node-voltage method to find v_o in the circuit shown.

ANSWER: 24 V.

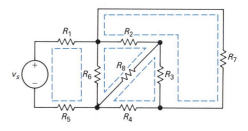

4.5 INTRODUCTION TO MESH CURRENTS

As stated in Section 4.1, the mesh-current method of circuit analysis enables us to describe a circuit in terms of $b - (n - 1)$ or $b_e - (n_e - 1)$ equations. For planar networks, the meshes in the network are identical to the "windows" formed when the network is drawn with no crossing branches. The circuit in Fig. 4.1(b) is shown again in Fig. 4.17, where the "windows" are identified by the closed dashed paths. Figure 4.17 contains seven essential branches where the current is unknown. Because the circuit contains four essential nodes, we must write four $[7 - (4 - 1)]$ mesh-current equations.

A *mesh current* is the current that exists only in the perimeter of a mesh. On a circuit diagram it appears as either a closed solid line or an almost-closed solid line that follows the perimeter of the appropriate mesh. An arrowhead on the solid line indicates the reference direction for the mesh current. Figure 4.18 shows four mesh currents used to describe the circuit in Fig. 4.17. Note that by definition mesh currents automatically satisfy Kirchhoff's current law, that is, at any node in the circuit, a given mesh current both enters and leaves the node.

Figure 4.18 also shows that identifying a mesh current in terms of a branch current is not always possible. For example,

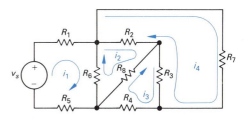

FIGURE 4.17 The circuit shown in Fig. 4.1(b) indicating how to use the "windows" in a planar circuit to identify circuit meshes.

FIGURE 4.18 The circuit shown in Fig. 4.17 with the mesh currents defined.

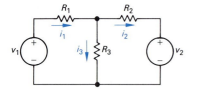

FIGURE 4.19 A circuit used to illustrate development of the mesh-current method of circuit analysis.

the mesh current i_2 is not equal to any branch current, whereas mesh currents i_1, i_3, and i_4 can be identified with branch currents. Thus measuring a mesh current is not always possible. For example, in the circuit shown in Fig. 4.18, there is no place where an ammeter can be inserted to measure the mesh current i_2. The fact that a mesh current can be a fictitious quantity doesn't mean that it is a useless concept. On the contrary, it is useful in circuit analysis.

The mesh-current method of circuit analysis evolves quite naturally from the branch-current equations. We can use the circuit shown in Fig. 4.19 to show the evolution of the mesh-current technique. We begin by using the branch currents (i_1, i_2, and i_3) to formulate the set of independent equations. For this circuit, $b_e = 3$ and $n_e = 2$. We can write only one independent current equation, so we need two independent voltage equations. Applying Kirchhoff's current law to the upper node and Kirchhoff's voltage law around the two meshes generates the set of equations:

$$i_1 = i_2 + i_3; \tag{4.20}$$

$$v_1 = i_1 R_1 + i_3 R_3; \tag{4.21}$$

$$-v_2 = i_2 R_2 - i_3 R_3. \tag{4.22}$$

We reduce this set of three equations to a set of two equations by solving Eq. (4.20) for i_3 and then substituting this expression for i_3 into Eqs. (4.21) and (4.22):

$$v_1 = i_1(R_1 + R_3) - i_2 R_3; \tag{4.23}$$

$$-v_2 = -i_1 R_3 + i_2(R_2 + R_3). \tag{4.24}$$

We can solve Eqs. (4.23) and (4.24) for i_1 and i_2 to replace the solution of three simultaneous equations with the solution of two simultaneous equations. We derived Eqs. (4.23) and (4.24) by substituting the $n_e - 1$ current equations into the $b_e - (n_e - 1)$ voltage equations. The value of the mesh-current method is that by defining mesh currents, we *automatically* eliminate the $n_e - 1$ current equations. Thus the mesh-current method is equivalent to a systematic substitution of the $n_e - 1$ current equations into the $b_e - (n_e - 1)$ voltage equations. The mesh currents for the circuit shown in Fig. 4.19 that are equivalent to eliminating the branch current i_3 from Eqs. (4.21) and (4.22) are shown in Fig. 4.20.

We now apply Kirchhoff's voltage law around the two meshes, expressing all voltages across resistors in terms of the mesh currents, to get the equations

$$v_1 = i_a R_1 + (i_a - i_b)R_3; \tag{4.25}$$

$$-v_2 = (i_b - i_a)R_3 + i_b R_2. \tag{4.26}$$

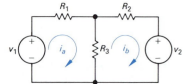

FIGURE 4.20 Mesh currents i_a and i_b.

Collecting the coefficients of i_a and i_b in Eqs. (4.25) and (4.26) gives

$$v_1 = i_a(R_1 + R_3) - i_b R_3; \qquad \textbf{(4.27)}$$

$$-v_2 = -i_a R_3 + i_b(R_2 + R_3). \qquad \textbf{(4.28)}$$

Note that Eqs. (4.27) and (4.28) and Eqs. (4.23) and (4.24) are identical in form, with the mesh currents i_a and i_b replacing the branch currents i_1 and i_2. Note also that the branch currents shown in Fig. 4.19 can be expressed in terms of the mesh currents shown in Fig. 4.20 by inspection, or

$$i_1 = i_a; \qquad \textbf{(4.29)}$$

$$i_2 = i_b; \qquad \textbf{(4.30)}$$

$$i_3 = i_a - i_b. \qquad \textbf{(4.31)}$$

The ability to write Eqs. (4.29)–(4.31) by inspection is crucial to the mesh-current method of circuit analysis. Once you know the mesh currents, you also know the branch currents. And once you know the branch currents, you can compute any voltages or powers of interest.

Because we have defined meshes as the "windows" of a planar circuit, we guarantee that the set of mesh-current equations that describe the circuit comprise an independent set. In Chapter 5 we prove that the mesh-current equations form an independent set. Example 4.3 illustrates how the mesh-current method is used to find source powers and a branch voltage.

E X A M P L E   4.3

a) Use the mesh-current method to determine the power associated with each voltage source in the circuit shown in Fig. 4.21.

b) Calculate the voltage v_o across the 8-Ω resistor.

S O L U T I O N

a) To calculate the power associated with each source, we need to know the current in each source. The circuit indicates that these source currents will be identical to mesh currents. Also, the circuit has seven branches where the current is unknown and five nodes. Therefore we need three $[b - (n - 1) = 7 - (5 - 1)]$ mesh-current equations to describe the circuit. Figure 4.22 shows the three mesh currents used to describe the circuit in Fig. 4.21.

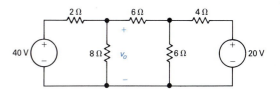

FIGURE 4.21 The circuit for Example 4.3.

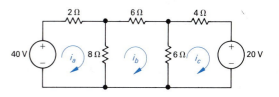

FIGURE 4.22 Three mesh currents used to analyze the circuit shown in Fig. 4.21.

If we assume that the voltage drops are positive, the three mesh equations are

$$-40 + 2i_a + 8(i_a - i_b) = 0;$$
$$8(i_b - i_a) + 6i_b + 6(i_b - i_c) = 0; \qquad \textbf{(4.32)}$$
$$6(i_c - i_b) + 4i_c + 20 = 0.$$

Reorganizing Eqs. (4.32) in anticipation of using Cramer's method for solving simultaneous equations gives

$$10i_a - 8i_b + 0i_c = 40;$$
$$-8i_a + 20i_b - 6i_c = 0; \qquad \textbf{(4.33)}$$
$$0i_a - 6i_b + 10i_c = -20.$$

The characteristic determinant is

$$\Delta = \begin{vmatrix} 10 & -8 & 0 \\ -8 & 20 & -6 \\ 0 & -6 & 10 \end{vmatrix}$$

$$= 10(200 - 36) + 8(-80)$$

$$= 1640 - 640 = 1000.$$

The three mesh currents are

$$i_a = \frac{\begin{vmatrix} 40 & -8 & 0 \\ 0 & 20 & -6 \\ -20 & -6 & 10 \end{vmatrix}}{1000} \qquad i_b = \frac{\begin{vmatrix} 10 & 40 & 0 \\ -8 & 0 & -6 \\ 0 & -20 & 10 \end{vmatrix}}{1000}$$

$$= \frac{40(200 - 36) - 20(48)}{1000} \qquad = \frac{10(-120) + 8(400)}{1000}$$

$$= 5.6 \text{ A}; \qquad = 2.0 \text{ A};$$

$$i_c = \frac{\begin{vmatrix} 10 & -8 & 40 \\ -8 & 20 & 0 \\ 0 & -6 & -20 \end{vmatrix}}{1000}$$

$$= \frac{10(-400) + 8(160 + 240)}{1000} = -0.80 \text{ A}.$$

The mesh current i_a is identical with the branch current in the 40-V source, so the power associated with this source is

$$p_{40\text{V}} = -40i_a = -224 \text{ W}.$$

The minus sign means that this source is delivering power to the network. The current in the 20-V source is identical to the mesh current i_c; therefore

$$p_{20\text{V}} = 20i_c = -16 \text{ W}.$$

The 20-V source also is delivering power to the network.

b) The branch current in the 8-Ω resistor in the direction of the voltage drop v_o is $i_a - i_b$. Therefore

$$v_o = 8(i_a - i_b) = 8(3.6) = 28.8 \text{ V}.$$

DRILL EXERCISES

4.11 Use the mesh-current method to find (a) the power delivered to the circuit shown by the 100-V source and (b) the power dissipated in the 15-Ω resistor.

ANSWER: (a) 600 W; (b) 240 W.

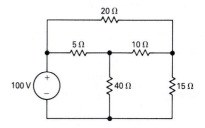

4.6 THE MESH-CURRENT METHOD AND DEPENDENT SOURCES

If the circuit contains dependent sources, the mesh-current equations must be supplemented by the appropriate constraint equations imposed by the presence of the dependent source or sources. Example 4.4 illustrates the application of the mesh-current method when the circuit includes a dependent source.

EXAMPLE 4.4

Use the mesh-current method of circuit analysis to determine the power dissipated in the 4-Ω resistor in the circuit shown in Fig. 4.23.

SOLUTION

This circuit has six branches where the current is unknown and four nodes. Therefore we need three mesh currents to describe the circuit. They are defined on the circuit shown in Fig. 4.24.

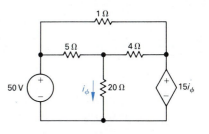

FIGURE 4.23 The circuit for Example 4.4.

The three mesh-current equations are

$$50 = 5(i_1 - i_2) + 20(i_1 - i_3);$$
$$0 = 5(i_2 - i_1) + 1i_2 + 4(i_2 - i_3); \qquad \textbf{(4.34)}$$
$$0 = 20(i_3 - i_1) + 4(i_3 - i_2) + 15i_\phi.$$

We now express the branch current controlling the dependent voltage source in terms of the mesh currents as

$$i_\phi = i_1 - i_3, \qquad \textbf{(4.35)}$$

which is the supplemental equation imposed by the presence of the dependent source. Substituting Eq. (4.35) into Eqs. (4.34) and collecting the coefficients of i_1, i_2, and i_3 generates

$$50 = 25i_1 - 5i_2 - 20i_3;$$
$$0 = -5i_1 + 10i_2 - 4i_3;$$
$$0 = -5i_1 - 4i_2 + 9i_3.$$

The characteristic determinant is

$$\Delta = \begin{vmatrix} 25 & -5 & -20 \\ -5 & 10 & -4 \\ -5 & -4 & 9 \end{vmatrix}$$

Expanding the characteristic determinant by the first column gives

$$\Delta = 25(90 - 16) + 5(-45 - 80) - 5(20 + 200) = 125.$$

Because we are calculating the power dissipated in the 4-Ω resistor, we compute the mesh currents i_2 and i_3:

$$i_2 = \frac{\begin{vmatrix} 25 & 50 & -20 \\ -5 & 0 & -4 \\ -5 & 0 & 9 \end{vmatrix}}{125} \qquad i_3 = \frac{\begin{vmatrix} 25 & -5 & 50 \\ -5 & 10 & 0 \\ -5 & -4 & 0 \end{vmatrix}}{125}$$

$$= \frac{-50(-45 - 20)}{125} = 26 \text{ A}; \qquad = \frac{50(20 + 50)}{125} = 28 \text{ A}.$$

The current in the 4-Ω resistor oriented from left to right is $i_3 - i_2$, or 2 A. Therefore the power dissipated is

$$p_{4\Omega} = (i_3 - i_2)^2(4) = (2)^2(4) = 16 \text{ W}.$$

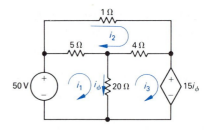

FIGURE 4.24 The circuit shown in Fig. 4.23 with the three mesh currents.

What if you were not told to use the mesh-current method? Would you have chosen the node-voltage method? It reduces the problem to finding one unknown node voltage because of the presence of two voltage sources between essential nodes. More about making such choices later.

DRILL EXERCISES

4.12 a) Use the expression $b - (n - 1)$ to determine the number of mesh-current equations needed to solve the circuit shown.

   b) Repeat (a) using $b_e - (n_e - 1)$.

   c) Use the mesh-current method to find how much power is being delivered to the dependent voltage source.

ANSWER: (a) 3; (b) 3; (c) −36 W.

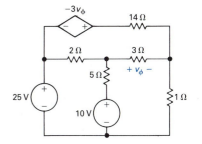

4.13 Use the mesh-current method to find v_o in the circuit shown.

ANSWER: 20 V.

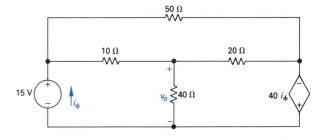

4.7 THE MESH-CURRENT METHOD: SOME SPECIAL CASES

When a branch includes a current source, the mesh-current method requires some additional manipulations. The circuit shown in Fig. 4.25 depicts the nature of the problem. We defined the mesh currents i_a, i_b, and i_c, as well as the voltage across the 5-A current source, to aid the discussion. Note that the circuit contains five essential branches where the current is unknown and four essential nodes. Hence, we need to write two $[5 - (4 - 1)]$ mesh-current equations in order to solve the circuit. Using the "windows" to define the meshes reduces the three unknown mesh currents to two unknown mesh currents. The reason is that the current source coupling meshes a and c limits the differences between i_c and i_a to equal 5 A. However, when we attempt to sum the voltages around either mesh a or mesh c, we must introduce the unknown voltage across the 5-A current source into the equations. We can eliminate this unknown voltage simply by introducing it into both mesh equations and then adding them. Thus for mesh a,

$$100 = 3(i_a - i_b) + v + 6i_a \qquad (4.36)$$

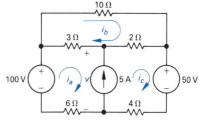

FIGURE 4.25 A circuit illustrating mesh analysis when a branch contains an independent current source.

and for mesh c,

$$-50 = 4i_c - v + 2(i_c - i_b). \qquad (4.37)$$

We now add Eqs. (4.36) and (4.37) to obtain

$$50 = 9i_a - 5i_b + 6i_c. \qquad (4.38)$$

Summing voltages around mesh b gives

$$0 = 3(i_b - i_a) + 10i_b + 2(i_b - i_c). \qquad (4.39)$$

We reduce Eqs. (4.38) and (4.39) to two equations and two unknowns by using the constraint that

$$i_c - i_a = 5. \qquad (4.40)$$

We leave to you verification that when Eq. (4.40) is combined with Eqs. (4.38) and (4.39), the solutions for the three mesh currents are

$$i_a = 1.75 \text{ A}, \qquad i_b = 1.25 \text{ A}, \qquad \text{and} \qquad i_c = 6.75 \text{ A}.$$

THE CONCEPT OF A SUPERMESH

We can derive Eq. (4.38) without introducing the unknown voltage v by using the concept of a *supermesh*. To create a supermesh, we mentally remove the current source from the circuit by simply avoiding this branch when writing the mesh-current equations. We express the voltages around the supermesh in terms of the mesh currents defined by the original "windows" of the circuit. Figure 4.26 illustrates the supermesh concept. When we sum the voltages around the supermesh denoted by the dashed line, we obtain the equation

$$-100 + 3(i_a - i_b) + 2(i_c - i_b) + 50 + 4i_c + 6i_a = 0, \qquad (4.41)$$

which reduces to

$$50 = 9i_a - 5i_b + 6i_c. \qquad (4.42)$$

Comparison of Eqs. (4.42) and (4.38) shows that they are identical. Thus the supermesh has eliminated the need for introducing the unknown voltage across the current source into these equations.

MESH-CURRENT ANALYSIS OF A FAMILIAR CIRCUIT

We can use the circuit first introduced in Section 2.5 (Fig. 2.22) to illustrate the mesh-current method when a branch contains a dependent current source. Figure 4.27 shows that circuit, with the three mesh currents denoted i_a, i_b, and i_c. This circuit has four essential nodes and five essential branches where the current is unknown. Therefore we know that the circuit can be ana-

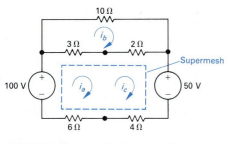

FIGURE 4.26 The circuit shown in Fig. 4.25 illustrating the concept of the supermesh.

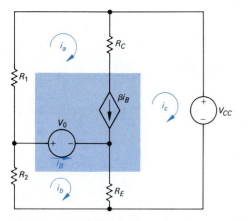

FIGURE 4.27 The circuit shown in Fig. 2.22 with the mesh currents i_a, i_b, and i_c.

·lyzed in terms of two $[5 - (4 - 1)]$ mesh-current equations. Although we defined three mesh currents in Fig. 4.27, the dependent current source forces a constraint between mesh currents i_a and i_c so that we have only two unknown mesh currents. Using the concept of the supermesh, we redrew the circuit as shown in Fig. 4.28.

We now sum the voltages around the supermesh in terms of the mesh currents i_a, i_b, and i_c to obtain

$$R_1 i_a + V_{CC} + R_E(i_c - i_b) - V_0 = 0. \qquad \textbf{(4.43)}$$

The mesh b equation is

$$R_2 i_b + V_0 + R_E(i_b - i_c) = 0. \qquad \textbf{(4.44)}$$

The constraint imposed by the dependent current source is

$$\beta i_B = i_a - i_c. \qquad \textbf{(4.45)}$$

The branch current controlling the dependent current source, expressed as a function of the mesh currents, is

$$i_B = i_b - i_a. \qquad \textbf{(4.46)}$$

From Eqs. (4.45) and (4.46),

$$i_c = (1 + \beta)i_a - \beta i_b. \qquad \textbf{(4.47)}$$

We now use Eq. (4.47) to eliminate i_c from Eqs. (4.43) and (4.44):

$$[R_1 (1 + \beta)R_E]i_a - (1 + \beta)R_E i_b = V_0 - V_{CC}; \qquad \textbf{(4.48)}$$

$$-(1 + \beta)R_E i_a + [R_2 + (1 + \beta)R_E]i_b = -V_0. \qquad \textbf{(4.49)}$$

We leave to you verification that the solution of Eqs. (4.48) and (4.49) for i_a and i_b gives

$$i_a = \frac{V_0 R_2 - V_{CC} R_2 - V_{CC}(1 + \beta)R_E}{R_1 R_2 + (1 + \beta)R_E(R_1 + R_2)}; \qquad \textbf{(4.50)}$$

$$i_b = \frac{-V_0 R_1 - (1 + \beta)R_E V_{CC}}{R_1 R_2 + (1 + \beta)R_E(R_1 + R_2)}. \qquad \textbf{(4.51)}$$

We also leave to you verification that when Eqs. (4.50) and (4.51) are used to find i_B, the result is the same as that given by Eq. (2.27).

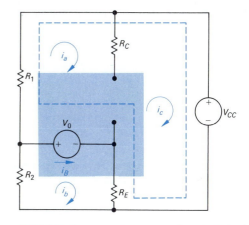

FIGURE 4.28 The circuit shown in Fig. 4.27 depicting the supermesh created by the presence of the dependent current source.

DRILL EXERCISES

4.14 Use the mesh-current method to find the power dissipated in the 2-Ω resistor in the circuit shown.

ANSWER: 72 W.

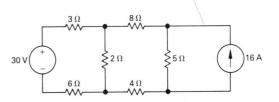

4.15 Use the mesh-current method to find the mesh current i_a in the circuit shown.

ANSWER: 10 A.

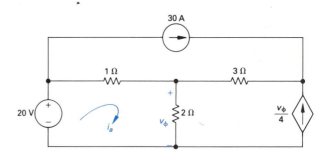

4.8 THE NODE-VOLTAGE METHOD VERSUS THE MESH-CURRENT METHOD

The greatest advantage of both the node-voltage and mesh-current methods is that they reduce the number of simultaneous equations that must be manipulated. They also require the analyst to be quite systematic in terms of organizing and writing the required simultaneous equations. It is natural to ask then, "When is the node-voltage method preferred to the mesh-current method, and vice versa?" As you might suspect, there is no clear-cut answer. One possible approach is to compare the number of simultaneous equations required for each method and then to select the one requiring the fewest. A second is to analyze the presence and location of voltage and current sources within the circuit. Voltage sources may require extra effort in formulating node-voltage equations, whereas current sources may require extra effort in formulating mesh-current equations.

Another point to consider when you are choosing between the two methods is what information about the circuit being analyzed is of primary interest. In other words, a complete solution of a circuit may not be needed, and therefore the particular piece of information of interest may influence the method you use. For example, in the circuit shown in Fig. 2.22, if only i_{CC} is of interest, the mesh-current method may be preferable, whereas if only the voltage across R_2 is of interest, the node-voltage method may be favored.

Perhaps the most important observation that we can make regarding these two methods of circuit analysis is that for any situation, some time spent thinking about the problem in relation to the various analytical approaches available is time well spent. Examples 4.5 and 4.6 illustrate the process of deciding between the node-voltage and mesh-current approaches.

EXAMPLE 4.5

Find the power dissipated in the 300-Ω resistor in the circuit shown in Fig. 4.29.

SOLUTION

To find the power dissipated in the 300-Ω resistor we need to find either the current in the resistor or the voltage across it. The mesh-current method yields the current in the 300-Ω resistor directly. For this circuit the mesh-current approach requires solving five simultaneous mesh equations, as depicted in Fig. 4.30. In writing the five mesh equations we must include the constraint $i_\Delta = -i_b$.

Before jumping in and writing the five mesh-current equations, let's look at the circuit in terms of the node-voltage method. Note that once we know the node voltages, we can calculate either the current in the 300-Ω resistor or the voltage across it. The circuit has four essential nodes, and therefore only three node voltage equations are required to describe the circuit. Because of the dependent voltage source between two essential nodes, we have to sum the currents at only two nodes. Hence, the problem is reduced to writing two node voltage equations and a constraint equation. Because the node-voltage method requires the manipulation of only three simultaneous equations, it is the more attractive approach.

Once the decision to use the node-voltage method has been made, the next step is to select a reference node. Two essential nodes in the circuit in Fig. 4.29 merit consideration. The first is the node where the 150-Ω, 300-Ω, 100-Ω, and 200-Ω resistors connect, as shown in Fig 4.31. If this node is selected, one of the unknown node voltages is the voltage across the 300-Ω resistor, namely, v_2 in Fig. 4.31. Once we know this voltage we calculate the power in the 300-Ω resistor by using the expression $p_{300} = v_2^2/300$. Note that in addition to the reference node, we defined the three node voltages v_1, v_2, and v_3 and indicated that nodes 1 and 3 form a supernode because they are connected by a dependent voltage source. In assigning the node voltages for the circuit in Fig. 4.31 we have placed each voltage next to its node. It is understood that a node voltage is a rise from the reference node; therefore it is not necessary to place the node voltage polarity references on the circuit diagram.

The second node that merits consideration as the reference node is the lower node in the circuit, as shown in Fig. 4.32. It is attractive as a reference node because it has the most branches connected to it and the node-voltage equations are easier to

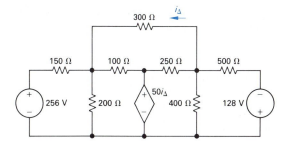

FIGURE 4.29 The circuit for example 4.5.

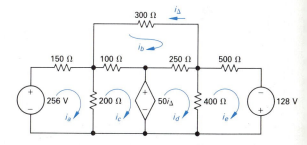

FIGURE 4.30 The circuit shown in Fig. 4.29 with the five mesh currents.

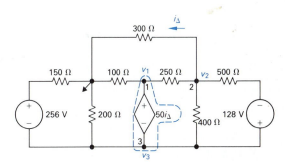

FIGURE 4.31 The circuit shown in Fig. 4.29 with a reference node.

write. However to find either the current in the 300-Ω resistor or the voltage across it requires an additional calculation once we know the node voltages v_a and v_c. For example, the current in the 300-Ω resistor is $(v_c - v_a)/300$, whereas the voltage across the resistor is $v_c - v_a$).

We compare these two choices of a reference node by means of the following sets of equations. The first set pertains to the circuit shown in Fig. 4.31, and the second set is based on the circuit shown in Fig. 4.32.

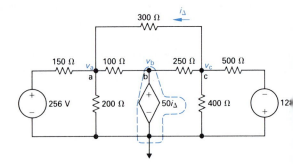

FIGURE 4.32 The circuit shown in Fig. 4.29 with an alternative reference node.

- Set 1

$$\frac{v_1}{100} + \frac{v_1 - v_2}{250} + \frac{v_3}{200} + \frac{v_3 - v_2}{400} + \frac{v_3 - (v_2 + 128)}{500}$$

$$+ \frac{v_3 + 256}{150} = 0;$$

$$\frac{v_2}{300} + \frac{v_2 - v_1}{250} + \frac{v_2 - v_3}{400} + \frac{v_2 + 128 - v_3}{500} = 0;$$

$$v_3 = v_1 - 50i_\Delta = v_1 - \frac{v_2}{6}.$$

- Set 2

$$\frac{v_a}{200} + \frac{v_a - 256}{150} + \frac{v_a - v_b}{100} + \frac{v_a - v_c}{300} = 0;$$

$$\frac{v_c}{400} + \frac{v_c + 128}{500} + \frac{v_c - v_b}{250} + \frac{v_c - v_a}{300} = 0;$$

$$v_b + 50i_\Delta = 50(v_c - v_a)300 = \frac{v_c - v_a}{6}.$$

We leave to you verification that the solution of either set leads to a power calculation of 16.57 W dissipated in the 300-Ω resistor.

E X A M P L E   4.6

Find the voltage v_o in the circuit in Fig. 4.33.

S O L U T I O N

At first glance the node-voltage method looks appealing because we may define the unknown voltage as a node voltage by choos-

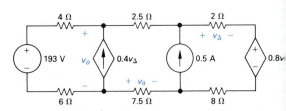

FIGURE 4.33 The circuit for Example 4.6.

ing the lower terminal of the dependent current source as the reference node. Further consideration of the node-voltage method reveals four essential junctions and two voltage controlled dependent sources. Therefore the node-voltage method requires manipulation of three node voltage equations and two constraint equations.

Let's now turn to the mesh-current method for finding v_o. The circuit contains three meshes, and when we determine the mesh current in the left "window," we can use it to calculate v_o. If we let i_a denote the clockwise left-window mesh current, then $v_o = 193 - 10i_a$. The presence of the two current sources reduces the problem to manipulating a single supermesh equation and two constraint equations. Hence, the mesh-current method is the more attractive technique here.

To help you compare the two approaches we summarize both methods. The mesh-current equations are based on the circuit shown in Fig. 4.34, and the node-voltage equations are based on the circuit shown in Fig. 4.35. The supermesh equation is

$$193 = 10i_a + 10i_b + 10i_c + 0.8v_\theta,$$

and the constraint equations are

$$i_b - i_a = 0.4v_\Delta = 0.8i_c; \qquad v_\theta = -7.5i_b; \qquad i_c - i_b = 0.5.$$

We use the constraint equations to write the supermesh equation in terms of i_a:

$$160 = 80i_a \qquad \text{or} \qquad i_a = 2\text{A};$$

$$v_o = 193 - 20 = 173 \text{ V}.$$

The node voltage equations are

$$\frac{v_o - 193}{10} - 0.4v_\Delta + \frac{v_o - v_a}{2.5} = 0;$$

$$\frac{v_a - v_o}{2.5} - 0.5 + \frac{v_a - (v_b + 0.8v_\theta)}{10} = 0;$$

$$\frac{v_b}{7.5} + 0.5 + \frac{v_b + 0.8v_\theta - v_a}{10} = 0.$$

The constraint equations are

$$v_\theta = v_b \qquad \text{and} \qquad v_\Delta = \left[\frac{v_a - (v_b + 0.8v_\theta)}{10}\right]2.$$

We use the constraint equations to reduce the node-voltage equations to three simultaneous equations involving v_o, v_a, and v_b. We leave to you verification that the node-voltage approach also gives $v_o = 173$ V.

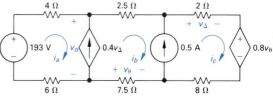

FIGURE 4.34 The circuit shown in Fig. 4.33 with the three mesh currents.

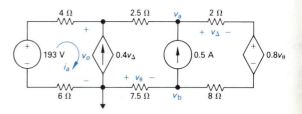

FIGURE 4.35 The circuit shown in Fig. 4.33 with node voltages.

DRILL EXERCISES

4.16 Find the power delivered by the 2-A current source in the circuit shown.

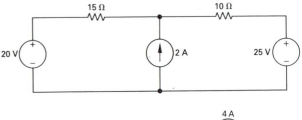

ANSWER: 70 W.

4.17 Find the power delivered by the 4-A current source in the circuit shown.

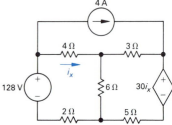

ANSWER: 40 W.

4.9 SOURCE TRANSFORMATIONS

Even though the node-voltage and mesh-current methods are powerful techniques for solving circuits, we are still interested in methods that can be used to simplify circuits. We begin expanding our list of simplifying techniques with source transformations. A *source transformation,* shown in Fig. 4.36, allows us to replace a voltage source in series with a resistor by a current source in parallel with the same resistor, or vice versa. The double-headed arrow emphasizes that a source transformation is bilateral; that is, we can start with either configuration and derive the other. The two configurations shown in Fig. 4.36 are equivalent with respect to the terminals a, b provided that

$$i_s = \frac{v_s}{R_s};$$ (4.52)

$$R_s = R_p.$$ (4.53)

We can verify Eqs. (4.52) and (4.53) with the following arguments. If the two circuits are equivalent with respect to the terminals a, b, they must be equivalent for *all* external values of R connected across a, b. Two extreme values of R that are easy to test are zero and infinity. For 0 Ω, or a short circuit, the voltage

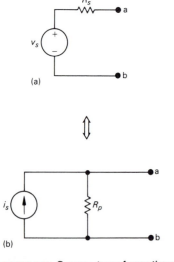

FIGURE 4.36 Source transformations.

source delivers a short-circuit current of v_s/R_s amperes, oriented from terminal a to b. The short-circuit current delivered by the current source is i_s, also oriented from terminal a to b. These two short-circuit currents are identical by virtue of Eq. (4.52).

For infinite external resistance, the source arrangement of Fig. 4.36(a) predicts that the voltage from a to b would be v_s, with terminal a positive. The voltage across a, b in the circuit shown in Fig. 4.36(b) is $i_s R_p$, which equals v_s by virtue of Eqs. (4.52) and (4.53). Terminal a also is positive, as it must be in order for the two source arrangements to be equivalent.

If the polarity of v_s is reversed, the orientation of i_s must be reversed in order to maintain equivalence.

Example 4.7 illustrates the usefulness of making source transformations in order to simplify a circuit-analysis problem.

EXAMPLE 4.7

a) For the circuit shown in Fig. 4.37, find the power associated with the 6-V source.

b) State whether the 6-V source is absorbing or delivering the power calculated in (a).

SOLUTION

a) If we study the circuit shown in Fig. 4.37 knowing that the power associated with the 6-V source is of interest, several approaches come to mind. The circuit has four essential nodes and six essential branches where the current is unknown. Thus we can find the current in the branch containing the 6-V source by solving either three $[6 - (4 - 1)]$ mesh-current equations or three $(4 - 1)$ node-voltage equations. Choosing the mesh-current approach involves solving the three mesh-current equations for the mesh current that corresponds to the branch current in the 6-V source. Choosing the node-voltage approach involves solving the three node-voltage equations for the voltage across the 30-Ω resistor, from which the branch current in the 6-V source can be calculated. However, by focusing on just one branch current, we can first simplify the circuit by using source transformations. We must reduce the circuit in a way that preserves the identity of the branch containing the 6-V source. We have no reason to preserve the identity of the branch containing the 40-V source. Beginning with this branch, we can transform the 40-V source in series with the 5-Ω resistor to an 8-A current source in parallel with a 5-Ω resistor, as shown in Fig.

FIGURE 4.37 The circuit for Example 4.7.

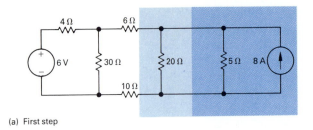

(a) First step

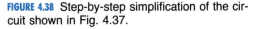

(b) Second step

4.38(a). Next, we can replace the parallel combination of the 20-Ω and 5-Ω resistors with a 4-Ω resistor. This 4-Ω resistor shunts the 8-A source and therefore can be replaced with a 32-V source in series with a 4-Ω resistor, as shown in Fig. 4.38(b). The 32-V source is in series with 20 Ω of resistance and hence, can be replaced by a current source of 1.6 A in parallel with 20 Ω, as shown in Fig. 4.38(c). The parallel combination of the 1.6-A current source and the 12-Ω resistor transforms to a voltage source of 19.2 V in series with 12 Ω. Figure 4.38(d) shows the result of this last transformation. The current in the direction of the voltage drop across the 6-V source is $(19.2 - 6)/16$, or 0.825 A. Therefore the power associated with the 6-V source is

$$p_{6V} = (0.825)(6) = 4.95 \text{ W.}$$

b) The voltage source is absorbing power.

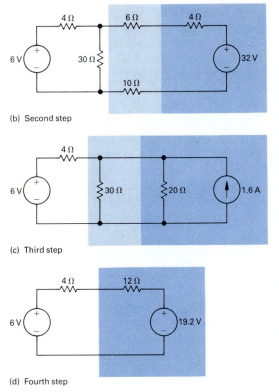

(c) Third step

(d) Fourth step

FIGURE 4.38 Step-by-step simplification of the circuit shown in Fig. 4.37.

A question that arises from use of the source transformation depicted in Fig. 4.36 is, "What happens if there is a resistance, R, in parallel with the voltage source or in series with the current source?" In both cases the resistance R has no effect on the equivalent circuit that predicts behavior with respect to the terminals a, b. Figure 4.39 summarizes this observation.

The two circuits depicted in Fig. 4.39(a) are equivalent with respect to the terminals a, b because they produce the same

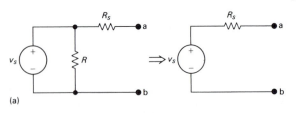

(a)

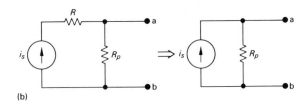

(b)

FIGURE 4.39 Equivalent circuits containing a resistance in parallel with a voltage source or in series with a current source.

open-circuit voltage and short-circuit current. The same can be said for the circuits in Fig. 4.39(b). Example 4.8 illustrates an application of the equivalent circuits depicted in Fig. 4.39.

EXAMPLE 4.8

a) Use source transformations to find the voltage v_o in the circuit shown in Fig. 4.40.

b) Find the power developed by the 250-V voltage source.

c) Find the power developed by the 8-A current source.

SOLUTION

a) We begin by removing the 125-Ω and 10-Ω resistors because the 125-Ω resistor is connected across the 250-V voltage source and the 10-Ω resistor is connected in series with the 8-A current source. We also combine the series-connected resistors into a single resistance of 20 Ω. Figure 4.41 shows the simplified circuit.

We now use a source transformation to replace the 250-V source and 25-Ω resistor with a 10-A source in parallel with the 25-Ω resistor, as shown in Fig. 4.42. We can now simplify the circuit shown in Fig. 4.42 by combining the parallel current sources into a single source and the parallel resistors into a single resistor. Figure 4.43 shows the result. Hence, $v_o = 20$ V.

b) The current supplied by the 250-V source equals the current in the 125-Ω resistor plus the current in the 25-Ω resistor. Thus

$$i_s = \frac{250}{125} + \frac{250 - 20}{25} = 11.2 \text{ A}.$$

Therefore the power developed by the voltage source is

$$p_{250\text{V}} \ (developed) = (250)(11.2) = 2800 \text{ W}.$$

c) To find the power developed by the 8-A current source, we first find the voltage across the source. If we let v_s represent the voltage across the source—positive at the upper terminal of the source—we obtain

$$v_s + 8(10) = v_o = 20 \quad \text{or} \quad v_s = -60 \text{ V},$$

and the power developed by the 8-A source is 480 W. Note that the 125-Ω and 10-Ω resistors do not affect the value of v_o but do affect the power calculations.

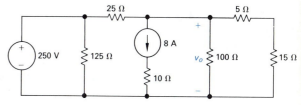

FIGURE 4.40 The circuit for Example 4.8.

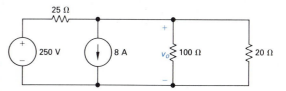

FIGURE 4.41 Simplified version of the circuit shown in Fig. 4.40.

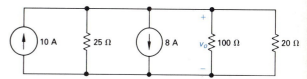

FIGURE 4.42 The circuit shown in Fig. 4.41 after a source transformation.

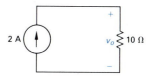

FIGURE 4.43 The circuit shown in Fig. 4.42 after combining sources and resistors.

DRILL EXERCISES

4.18 a) Use a series of source transformations to find the voltage v in the circuit shown.

b) How much power does the 120-V source deliver to the circuit?

ANSWER: (a) 48 V; (b) 374.4 W.

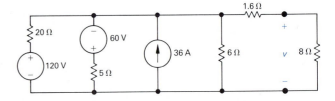

4.10 THÉVENIN AND NORTON EQUIVALENTS

At times in circuit analysis we want to concentrate on what happens at a specific pair of terminals in a circuit. For example, when we plug a toaster into an outlet, we are interested primarily in the voltage and current at the terminals of the toaster. We have little or no interest in the effect that connecting the toaster has on voltages or currents elsewhere in the circuit supplying the outlet. We can expand this interest in terminal behavior to the case of a set of appliances, each requiring a different amount of power. We then are interested in how the voltage and current delivered at the outlet changes as we change appliances. In other words, we want to focus on the behavior of the circuit supplying the outlet but only at the outlet terminals. Because interest in circuit behavior so often focuses on a pair of terminals, the Thévenin and Norton equivalent circuits are extremely valuable aids in analysis. Although here we discuss these equivalent circuits as they pertain to resistive circuits, you should be aware that Thévenin and Norton equivalent circuits may be used to represent any circuit made up of linear elements.

We can best describe the significance of the Thévenin equivalent circuit by reference to Fig. 4.44. Figure 4.44(a) represents any circuit made up of sources (both independent and dependent) and resistors. The letters a and b denote the pair of terminals of interest. Figure 4.44(b) shows the Thévenin equivalent. The circuit shown in Fig. 4.44(b) implies that the original interconnection of sources and resistors can be replaced by an independent voltage source V_{Th} in series with a resistor R_{Th}. Moreover, this series combination of V_{Th} and R_{Th} is equivalent to the original circuit in the sense that if we connect the same load across the terminals a, b of each circuit, we get the same voltage

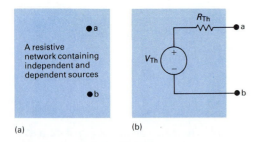

(a)   (b)

FIGURE 4.44 (a) A general circuit; (b) the Thévenin equivalent circuit.

and current at the terminals of the load. This equivalence holds for *all possible values of load resistance*.

In order to represent the original circuit by its Thévenin equivalent, we must be able to determine the Thévenin voltage V_{Th} and the Thévenin resistance R_{Th}. We find these two parameters of the Thévenin equivalent as follows. First, we note that if the load resistance is infinitely large, we have an open-circuit condition. The open-circuit voltage at the terminals a, b in the circuit shown in Fig. 4.44(b) is V_{Th}. By hypothesis, this must be the same as the open-circuit voltage at the terminals a, b in the original circuit. Therefore to calculate the Thévenin voltage V_{Th}, we simply calculate the open-circuit voltage in the original circuit.

Reducing the load resistance to zero gives us a short-circuit condition. If we place a short circuit across the terminals a, b of the Thévenin equivalent circuit, the short-circuit current directed from a to b is

$$i_{\text{sc}} = \frac{V_{\text{Th}}}{R_{\text{Th}}}. \tag{4.54}$$

By hypothesis, this short-circuit current must be identical to the short-circuit current that exists in a short circuit placed across the terminals a, b of the original network. From Eq. (4.54),

$$R_{\text{Th}} = \frac{V_{\text{Th}}}{i_{\text{sc}}}. \tag{4.55}$$

Thus the Thévenin resistance is the ratio of the open-circuit voltage to the short-circuit current.

FINDING A THÉVENIN EQUIVALENT

To find the Thévenin equivalent circuit of the circuit shown in Fig. 4.45, we first calculate the open-circuit voltage v_{ab}. Note that when the terminals a, b are open, there is no current in the 4-Ω resistor. Therefore the open-circuit voltage v_{ab} is identical to the voltage across the 3-A current source, labeled v_o on the circuit shown in Fig. 4.45. We find the voltage v_o by solving a single node-voltage equation. Choosing the lower node as the reference node, we get

$$\frac{v_o - 25}{5} + \frac{v_o}{20} - 3 = 0. \tag{4.56}$$

Solving for v_o yields

$$v_o = 32 \text{ V}. \tag{4.57}$$

Hence, the Thévenin voltage for the circuit is 32 V.

The next step in deriving the Thévenin equivalent circuit with respect to the terminals a, b is to place a short circuit across the

PSpice Use the PSpice command .TF to compute the Thévenin voltage and Thévenin resistance: Section 3.3

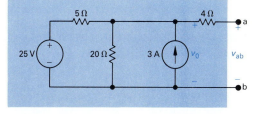

FIGURE 4.45 A circuit used to illustrate a Thévenin equivalent.

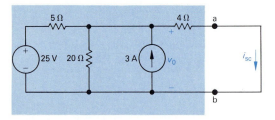

FIGURE 4.46 The circuit shown in Fig. 4.45 with terminals a and b short-circuited.

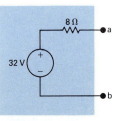

FIGURE 4.47 The Thévenin equivalent of the circuit shown in Fig. 4.45.

terminals and calculate the resulting short-circuit current. Figure 4.46 shows the circuit with the short in place. Note that the short-circuit current is in the direction of the open-circuit voltage drop across the terminals a, b. [If the short-circuit current is in the direction of the open-circuit voltage rise across the terminals a, b, a minus sign must be inserted in Eq. (4.55).]

The short-circuit current (i_{sc}) is easily found once v_o is known. Therefore the problem reduces to finding v_o with the short in place. Again, if we use the lower node as the reference node, the equation for v_o becomes

$$\frac{v_o - 25}{5} + \frac{v_o}{20} - 3 + \frac{v_o}{4} = 0. \qquad (4.58)$$

Solving Eq. (4.58) for v_o gives

$$v_o = 16 \text{ V}. \qquad (4.59)$$

Hence, the short-circuit current is

$$i_{sc} = \frac{16}{4} = 4 \text{ A}. \qquad (4.60)$$

We now find the Thévenin resistance by substituting the numerical results from Eqs. (4.57) and (4.60) into Eq. (4.55):

$$R_{Th} = \frac{V_{Th}}{i_{sc}} = \frac{32}{4} = 8 \text{ } \Omega. \qquad (4.61)$$

Figure 4.47 shows the Thévenin equivalent circuit for the circuit shown in Fig. 4.45.

We leave to you verification that if a 24-Ω resistor is connected across the terminals a, b in the circuit shown in Fig. 4.45, the voltage across the resistor will be 24 V and the current in the resistor will be 1 A. You can see by inspection that the Thévenin circuit in Fig. 4.47 predicts the same voltage and current if a 24-Ω resistor is connected across the terminals a, b.

THE NORTON EQUIVALENT

The Norton equivalent circuit consists of an independent current source in parallel with the Norton equivalent resistance. We can derive it from the Thévenin equivalent circuit simply by making a source transformation. Thus the Norton current equals the short-circuit current at the terminals of interest, and the Norton resistance is identical to the Thévenin resistance.

USING SOURCE TRANSFORMATIONS

Sometimes we can make effective use of source transformations to derive a Thévenin or Norton equivalent circuit. For example,

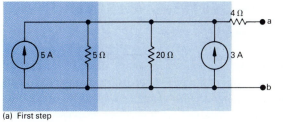

(a) First step

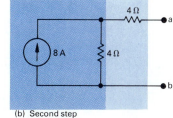

(b) Second step

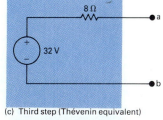

(c) Third step (Thévenin equivalent)

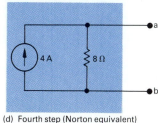

(d) Fourth step (Norton equivalent)

we can derive the Thévenin and Norton equivalent circuits of the circuit shown in Fig. 4.45 by making the series of source transformations shown in Fig. 4.48. This technique is most useful when the network contains only independent sources. The presence of dependent sources requires retaining the identity of the controlling voltages and/or currents, and this constraint usually prohibits continued reduction of the circuit by source transformations. We next discuss the problem of finding the Thévenin equivalent when the circuit contains dependent sources.

FIGURE 4.48 Step-by-step derivation of the Thévenin and Norton equivalent circuits of the circuit shown in Fig. 4.45.

DRILL EXERCISES

4.19 Find the Thévenin equivalent circuit with respect to the terminals a, b for the circuit shown.

ANSWER: $V_{ab} = V_{Th} = 64.8$ V, $R_{Th} = 6$ Ω.

4.20 Find the Norton equivalent circuit with respect to the terminals a, b for the circuit shown.

ANSWER: $I_N = 6$ A (directed toward a), $R_N = 6$ Ω.

4.21 A voltmeter with an internal resistance of 100 kΩ is used to measure the voltage v_{AB} in the circuit shown. What is the voltmeter reading?

ANSWER: 120 V.

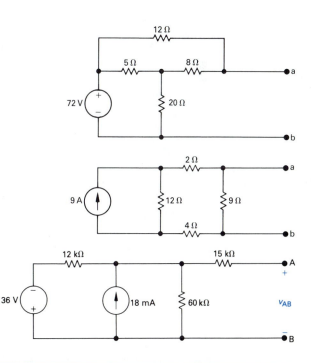

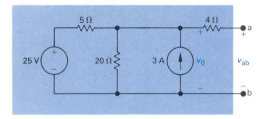

FIGURE 4.49 A circuit used to illustrate a Thévenin equivalent.

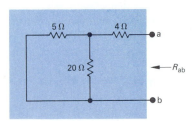

FIGURE 4.50 The circuit shown in Fig. 4.49 after deactivation of the independent sources.

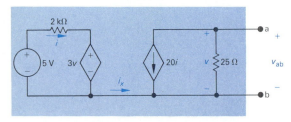

FIGURE 4.51 A circuit used to illustrate a Thévenin equivalent circuit when the circuit contains dependent sources.

4.11 MORE ON DERIVING A THÉVENIN EQUIVALENT

The technique for determining R_{Th} that we discussed and illustrated in Section 4.10 is not always the easiest method available. Two other methods generally are simpler to use. The first is useful if the network contains only independent sources. To calculate R_{Th} for such a network, we first deactivate all independent sources and then calculate the resistance seen looking into the network at the designated terminal pair. *A voltage source is deactivated by replacing it with a short circuit. A current source is deactivated by replacing it with an open circuit.* For example, consider the circuit shown in Fig. 4.49. Deactivating the independent sources simplifies the circuit to that shown in Fig. 4.50. The resistance seen looking into the terminals a, b is denoted R_{ab}, which consists of the 4-Ω resistor in series with the parallel combinations of the 5- and 20-Ω resistors. Thus

$$R_{ab} = R_{Th} = 4 + \frac{5 \times 20}{25} = 8 \ \Omega. \qquad \textbf{(4.62)}$$

Note that derivation of R_{Th} with Eq. (4.62) is much simpler than the derivation of R_{Th} with Eq. (4.61).

If the circuit, or network, contains dependent sources, an alternative procedure for finding the Thévenin resistance R_{Th} is as follows (see Fig. 4.51). We first deactivate all *independent* sources and then apply either a test voltage source or a test current source to the Thévenin terminals a, b. The Thévenin resistance equals the ratio of the voltage across the test source to the current delivered by the test source. Note that the circuit shown in Fig. 4.51 contains an independent 5-V source, a voltage-controlled voltage source, and a current-controlled current source. Also note the controlling signals. The voltage across the 25-Ω resistor controls the dependent voltage source. The current in the 2-kΩ resistor controls the dependent current source. We first derive the Thévenin equivalent circuit with respect to the terminals a, b using the open-circuit voltage and short-circuit current calculations. We then illustrate the alternative method for finding the Thévenin resistance.

DEPENDENT SOURCES—A NUMERICAL EXAMPLE

The first step in analyzing the circuit in Fig. 4.51 is to recognize that the current labeled i_x must be zero. (Note the absence of a return path for i_x to enter the left-hand portion of the circuit.) The open-circuit, or Thévenin, voltage will be the voltage across the 25-Ω resistor. With $i_x = 0$,

$$V_{Th} = v_{ab} = (-20i)(25) = -500i. \qquad \textbf{(4.63)}$$

The current i is:

$$i = \frac{5 - 3v}{2000} = \frac{5 - 3V_{Th}}{2000}. \qquad \textbf{(4.64)}$$

In writing Eq. (4.64), we recognize that the Thévenin voltage is identical to the control voltage. When we substitute Eq. (4.64) into Eq. (4.63), we obtain

$$V_{Th} = -5 \text{ V}. \qquad \textbf{(4.65)}$$

To calculate the short-circuit current, we place a short circuit across a, b. When the terminals a, b are shorted together, the control voltage v is reduced to zero. Therefore, with the short in place, the circuit shown in Fig. 4.51 becomes that shown in Fig. 4.52. With the short circuit shunting the 25-Ω resistor, all the current from the dependent current source appears in the short, so

$$i_{sc} = -20i. \qquad \textbf{(4.66)}$$

As the voltage controlling the dependent voltage source has been reduced to zero, the current controlling the dependent current source is

$$i = \frac{5}{2000} = 2.5 \text{ mA}. \qquad \textbf{(4.67)}$$

Substituting Eq. (4.67) into Eq. (4.66) yields a short-circuit current of

$$i_{sc} = -20(2.5) = -50 \text{ mA}. \qquad \textbf{(4.68)}$$

From Eqs. (4.65) and (4.68),

$$R_{Th} = \frac{V_{Th}}{i_{sc}} = \frac{-5}{-50} \times 10^3 = 100 \ \Omega. \qquad \textbf{(4.69)}$$

Figure 4.53 illustrates the Thévenin equivalent circuit for the circuit shown in Fig. 4.51. Note that the reference polarity marks on the Thévenin voltage source in Fig. 4.53 agree with Eq. (4.65).

DEPENDENT SOURCES—THE THÉVENIN RESISTANCE BY AN ALTERNATIVE METHOD

We now consider the alternative technique for finding the Thévenin resistance R_{Th}. We first deactivate the independent voltage source from the circuit and then excite the circuit from

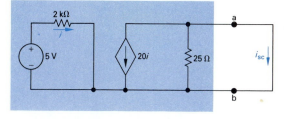

FIGURE 4.52 The circuit shown in Fig. 4.51 with terminals a and b short-circuited.

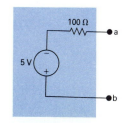

FIGURE 4.53 The Thévenin equivalent for the circuit shown in Fig. 4.51.

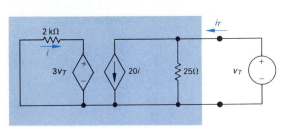

FIGURE 4.54 An alternative method for computing the Thévenin resistance.

the terminals a, b with either a test voltage source or a test current source. If we apply a test voltage source we will know the voltage of the dependent voltage source and hence the controlling current i. Therefore we opt for the test voltage source. Figure 4.54 shows the circuit for computing the Thévenin resistance. The externally applied test voltage source is denoted v_T and the current that it delivers to the circuit is labeled i_T. To find the Thévenin resistance, we simply solve the circuit shown in Fig. 4.54 for the ratio of the voltage to the current at the test source; that is, $R_{Th} = v_T/i_T$. From Fig. 4.54,

$$i_T = \frac{v_T}{25} + 20i; \qquad (4.70)$$

$$i = \frac{-3v_T}{2}\,\text{mA}. \qquad (4.71)$$

We then substitute Eq. (4.71) into Eq. (4.70) and solve the resulting equation for the ratio v_T/i_T:

$$
\begin{aligned}
i_T &= \frac{v_T}{25} - \frac{60v_T}{2000}; \\[2mm]
\frac{i_T}{v_T} &= \frac{1}{25} - \frac{6}{200} = \frac{50}{5000} = \frac{1}{100}.
\end{aligned}
\qquad (4.72)
$$

From Eqs. (4.72),

$$R_{Th} = \frac{v_T}{i_T} = 100\ \Omega. \qquad (4.73)$$

In general, these computations are easier than those involved in computing the short-circuit current. Moreover, in a network containing only resistors and dependent sources, you must use the alternative method because the ratio of the Thévenin voltage to the short-circuit current is indeterminate; that is, it is the ratio 0/0. (See Problems 4.66 and 4.67.)

DRILL EXERCISES

4.22 Find the Thévenin equivalent circuit with respect to the terminals a, b for the circuit shown.

ANSWER: $V_{Th} = v_{ab} = 20$ V, $R_{Th} = 0.625\ \Omega$.

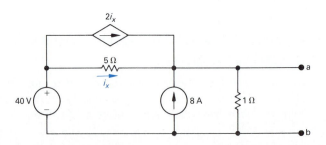

4.23 Find the Thévenin equivalent circuit with respect to the terminals a, b for the circuit shown.

ANSWER: $V_{Th} = v_{ab} = 30$ V, $R_{Th} = 10$ Ω.

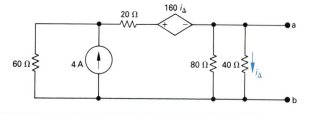

ILLUSTRATION OF A USEFUL APPLICATION

At times we can use a Thévenin equivalent to reduce one portion of a larger circuit in order to greatly simplify analysis of the larger network. Let's return to the circuit first introduced in Section 2.5 and subsequently analyzed in Sections 4.4 and 4.7. To aid our discussion of using Thévenin's theorem to analyze this circuit, we redrew the circuit and identified the branch currents of interest, as shown in Fig. 4.55. Once we know i_B, we can easily obtain the other branch currents. We arrive at that conclusion as follows. The current i_E is simply $(1 + \beta)i_B$. When we know i_E, we know the voltages v_{cd} and hence v_{bd}, because $v_{bd} = v_{cd} + V_0$. When we know the voltage v_{bd}, we can quickly compute the branch currents i_1 and i_2. Thus $i_2 = v_{bd}/R_2$ and $i_1 = i_2 + i_B$. Realizing that i_B is the key to finding the other branch currents, we redrew the circuit as shown in Fig. 4.56. You should be able to determine that this modification has no effect on the branch currents i_1, i_2, i_B, and i_E.

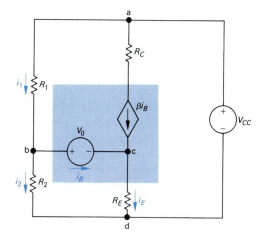

FIGURE 4.55 Application of a Thévenin equivalent in circuit analysis.

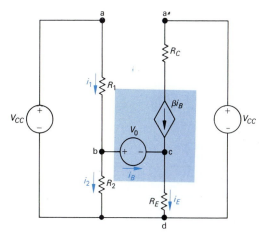

FIGURE 4.56 A modified version of the circuit shown in Fig. 4.55.

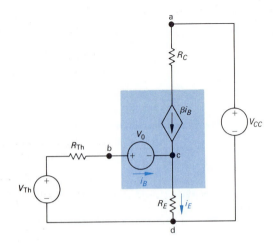

FIGURE 4.57 The circuit shown in Fig. 4.56 modified by a Thévenin equivalent.

Now we replace the circuit made up of V_{CC}, R_1, and R_2 with a Thévenin equivalent, with respect to the terminals b, d. The Thévenin voltage and resistance are

$$V_{Th} = \frac{V_{CC}R_2}{R_1 + R_2}; \qquad (4.74)$$

$$R_{Th} = \frac{R_1 R_2}{R_1 + R_2}. \qquad (4.75)$$

With the Thévenin equivalent, the circuit in Fig. 4.56 becomes that shown in Fig. 4.57.

We now derive an equation for i_B simply by summing the voltages around the left mesh. In writing this mesh equation, we recognize that $i_E = (1 + \beta)i_B$. Thus

$$V_{Th} = R_{Th}i_B + V_0 + R_E(1 + \beta)i_B, \qquad (4.76)$$

from which

$$i_B = \frac{V_{Th} - V_0}{R_{Th} + (1 + \beta)R_E}. \qquad (4.77)$$

When we substitute Eqs. (4.74) and (4.75) into Eq. (4.77), we get the same expression obtained in Eq. (2.27). Note that when we have incorporated the Thévenin equivalent into the original circuit, we get the solution for i_B by writing a single equation.

4.12 MAXIMUM POWER TRANSFER

Circuit analysis plays an important role in the analysis of systems designed to transfer power from a source to a load. We discuss power transfer in terms of two basic types of systems. One emphasizes the efficiency of the power transfer, and the other emphasizes the amount of the power transfer. Power utility systems are a good example of the first type because they are concerned with the generation, transmission, and distribution of large quantities of electric power. Communication and instrumentation systems are good examples of the second type because they are designed to transmit information via electric signals. In the transmission of information, or data, via electric signals, the power available at the transmitter or detector is limited. Thus transmitting as much of this power as possible to the receiver, or load, becomes desirable. In such applications the amount of power being transferred is small, so the efficiency of transfer is not a primary concern. We now consider maximum power transfer in systems that can be modeled by a purely resistive circuit.

Maximum power transfer can best be described with the aid of the circuit shown in Fig. 4.58. We assume a resistive network

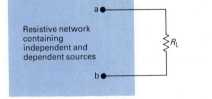

FIGURE 4.58 A circuit describing maximum power transfer.

containing independent and dependent sources and a designated pair of terminals, a, b, to which a load, R_L, is to be connected. The problem is to determine the value of R_L that permits maximum power delivery to R_L. The first step in finding the critical value of R_L is to recognize that a resistive network can always be replaced by its Thévenin equivalent. Therefore we redraw the circuit shown in Fig. 4.58 as that shown in Fig. 4.59. Replacing the original network by its Thévenin equivalent greatly simplifies the task of finding R_L. Derivation of R_L requires expressing the power dissipated in R_L as a function of the three circuit parameters V_{Th}, R_{Th}, and R_L. Thus

$$p = i^2 R_L = \left(\frac{V_{Th}}{R_{Th} + R_L}\right)^2 R_L. \qquad (4.78)$$

Next, we recognize that for a given circuit, V_{Th} and R_{Th} will be fixed. Therefore the power dissipated is a function of the single variable R_L. To find the value of R_L that maximizes the power, we use elementary calculus; that is, we solve for the value of R_L then dp/dR_L equals zero:

$$\frac{dp}{dR_L} = V_{Th}^2 \left[\frac{(R_{Th} + R_L)^2 - R_L \cdot 2(R_{Th} + R_L)}{(R_{Th} + R_L)^4}\right], \qquad (4.79)$$

and the derivative is zero when

$$(R_{Th} + R_L)^2 = 2R_L(R_{Th} + R_L). \qquad (4.80)$$

Solving Eq. (4.80) yields

$$R_L = R_{Th}. \qquad (4.81)$$

Thus maximum power transfer occurs when the load resistance R_L equals the Thévenin resistance R_{Th}. To find the maximum power delivered to R_L, we simply substitute Eq. (4.81) into Eq. (4.78):

$$p_{max} = \frac{V_{Th}^2 R_L}{(2R_L)^2} = \frac{V_{Th}^2}{4R_L}. \qquad (4.82)$$

The analysis of a circuit when the load resistor is adjusted for maximum power transfer is illustrated in Example 4.9.

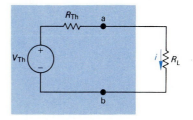

FIGURE 4.59 A circuit used to determine the value of R_L for maximum power transfer.

EXAMPLE 4.9

a) For the circuit shown in Fig. 4.60, find the value of R_L that results in maximum power being transferred to R_L.

b) Calculate the maximum power that can be delivered to R_L.

c) When R_L is adjusted for maximum power transfer, what percentage of power delivered by the 360-V source reaches R_L?

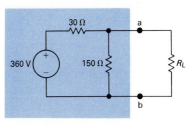

FIGURE 4.60 The circuit for Example 4.9.

SOLUTION

a) The Thévenin voltage for the circuit to the left of the terminals a, b is

$$V_{Th} = \frac{360}{180} \times 150$$
$$= 300 \text{ V}.$$

The Thévenin resistance is

$$R_{Th} = \frac{(150)(30)}{180}$$
$$= 25 \; \Omega.$$

Replacing the circuit to the left of the terminals a, b with its Thévenin equivalent gives the circuit shown in Fig. 4.61, which indicates that R_L must equal 25 Ω for maximum power transfer.

b) The maximum power that can be delivered to R_L is

$$p_{max} = \left(\frac{300}{50}\right)^2 (25)$$
$$= 900 \text{ W}.$$

c) When R_L equals 25 Ω, the voltage v_{ab} is

$$v_{ab} = \left(\frac{300}{50}\right)(25)$$
$$= 150 \text{ V}.$$

From Fig. 4.47, when v_{ab} equals 150 V, the current in the voltage source in the direction of the voltage rise across the source is

$$i_s = \frac{360 - 150}{30} = \frac{210}{30}$$
$$= 7 \text{ A}.$$

Therefore the source is delivering 2520 W to the circuit, or

$$p_s = -i_s(360)$$
$$= -2520 \text{ W}.$$

The percentage of the source power delivered to the load is

$$\frac{900}{2520} \times 100 = 35.71\%.$$

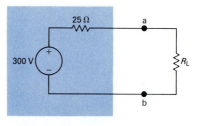

FIGURE 4.61 Reduction of the circuit shown in Fig. 4.60 by means of a Thévenin equivalent.

DRILL EXERCISES

4.24 a) Find the value of R that enables the circuit shown to deliver maximum power to the terminals a, b.

b) Find the maximum power delivered to R.

ANSWER: (a) 3 Ω; (b) 1.2 kW.

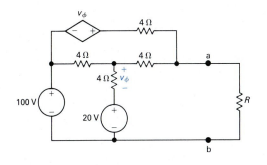

4.25 Assume that the circuit of Drill Exercise 4.24 is delivering maximum power to the load resistor R.

a) How much power is the 100-V source delivering to the network?

b) Repeat (a) for the dependent voltage source.

c) What percentage of the total power generated by these two sources is delivered to the load resistor R?

ANSWER: (a) 3000 W; (b) 800 W; (c) 31.58%.

4.13 SUPERPOSITION

The most distinguishing characteristic of a linear system is the principle of *superposition,* which states that, whenever a linear system is excited, or driven, by more than one independent source of energy, the total response is the sum of the individual responses. An individual response is the result of an independent source acting alone. Because we are dealing with circuits made up of interconnected linear-circuit elements, we can apply the principle of superposition directly to the analysis of such circuits when they are driven by more than one independent energy source. At present, we restrict the discussion to simple resistive networks; however, the principle is applicable to circuits containing inductance and capacitance as well as resistance. In fact, it is applicable to any linear system.

We demonstrate the use of the superposition principle by using it to find the branch currents in the circuit shown in Fig. 4.62. We begin by finding the branch currents resulting from the 120-V voltage source. We denote with a prime the component of the branch current resulting from the voltage source. Opening

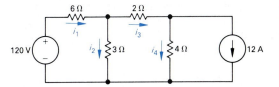

FIGURE 4.62 A circuit used to illustrate superposition.

FIGURE 4.63 The circuit shown in Fig. 4.62 with the current source deactivated.

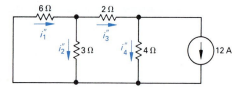

FIGURE 4.64 The circuit shown in Fig. 4.62 with the voltage source deactivated.

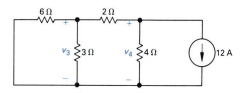

FIGURE 4.65 The circuit shown in Fig. 4.64 showing the node voltages v_3 and v_4.

the branch in which it appears deactivates the ideal current source. Figure 4.63 shows the circuit with the current source removed. The prime notation applied to the branch currents indicates that the currents in the circuit are the result of only the voltage source.

We can easily find the branch currents in the circuit in Fig. 4.63 once we know the node voltage across the 3-Ω resistor. Denoting this voltage v_1, we write

$$\frac{v_1 - 120}{6} + \frac{v_1}{3} + \frac{v_1}{2 + 4} = 0, \qquad (4.83)$$

from which,

$$v_1 = 30 \text{ V}. \qquad (4.84)$$

Now we can write the expressions for the branch currents i_1' through i_4' directly:

$$i_1' = \frac{120 - 30}{6} = 15 \text{ A}; \qquad (4.85)$$

$$i_2' = \frac{30}{3} = 10 \text{ A}; \qquad (4.86)$$

$$i_3' = i_4' = \frac{30}{6} = 5 \text{ A}. \qquad (4.87)$$

Replacing it with a short circuit deactivates the ideal voltage source in Fig. 4.62. Thus to find the component of the branch currents resulting from the current source, we must solve the circuit shown in Fig. 4.64. The double-prime notation for the currents indicates that these currents are the components of the total current resulting from the ideal current source.

We determine the branch currents in the circuit shown in Fig. 4.64 by first solving for the node voltages across the 3- and 4-Ω resistors, respectively. Figure 4.65 shows the two node voltages. The two node-voltage equations that describe the circuit are

$$\frac{v_3}{3} + \frac{v_3}{6} + \frac{v_3 - v_4}{2} = 0; \qquad (4.88)$$

$$\frac{v_4 - v_3}{2} + \frac{v_4}{4} + 12 = 0. \qquad (4.89)$$

Solving Eqs. (4.88) and (4.89) for v_3 and v_4, we get

$$v_3 = -12 \text{ V}; \qquad (4.90)$$

$$v_4 = -24 \text{ V}. \qquad (4.91)$$

Now we can write the branch currents i_1'' through i_4'' directly in terms of the node voltages v_3 and v_4:

$$i_1'' = \frac{-v_3}{6} = \frac{12}{6} = 2 \text{ A}; \qquad (4.92)$$

$$i_2'' = \frac{v_3}{3} = \frac{-12}{3} = -4 \text{ A}; \qquad (4.93)$$

$$i_3'' = \frac{v_3 - v_4}{2} = \frac{-12 + 24}{2} = 6 \text{ A}; \qquad (4.94)$$

$$i_4'' = \frac{v_4}{4} = \frac{-24}{4} = -6 \text{ A}. \qquad (4.95)$$

To find the branch currents in the original circuit—that is, the currents i_1, i_2, i_3, and i_4 in the circuit shown in Fig. 4.62—we simply add the currents given by Eqs. (4.92)–(4.95) to the currents given by Eqs. (4.85)–(4.87):

$$i_1 = i_1' + i_1'' = 15 + 2 = 17 \text{ A}; \qquad (4.96)$$

$$i_2 = i_2' + i_2'' = 10 - 4 = 6 \text{ A}; \qquad (4.97)$$

$$i_3 = i_3' + i_3'' = 5 + 6 = 11 \text{ A}; \qquad (4.98)$$

$$i_4 = i_4' + i_4'' = 5 - 6 = -1 \text{ A}. \qquad (4.99)$$

We leave to you verification that the currents given by Eqs. (4.96)–(4.99) are the correct values for the branch currents in the circuit shown in Fig. 4.62.

When applying the principle of superposition to linear circuits containing both independent and dependent sources, you must recognize that the dependent sources are never deactivated. Example 4.10 illustrates the application of superposition when the circuit contains both dependent and independent sources.

E X A M P L E   4.10

Use the principle of superposition to find v_o in the circuit shown in Fig. 4.66.

S O L U T I O N

We begin by finding the component of v_o resulting from the 10-V source. Figure 4.67 shows the circuit. With the 5-A source deactivated, v_Δ' must equal $(0.4v_\Delta')(10)$. Hence, v_Δ' must be zero, the branch containing the two dependent sources is open, and

$$v_o' = \frac{10}{25}(20) = 8 \text{ V}.$$

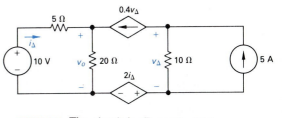

FIGURE 4.66 The circuit for Example 4.10.

When the 10-V source is deactivated, the current reduces to that shown in Fig. 4.68. We added a reference node and the node designations a, b, and c to aid the discussion.

Summing the currents away from node a yields

$$\frac{v_o''}{20} + \frac{v_o''}{5} - 0.4v_\Delta'' = 0, \quad \text{or} \quad 5v_o'' - 8v_\Delta'' = 0.$$

Summing the currents away from node b gives

$$0.4v_\Delta'' + \frac{v_b - 2i_\Delta''}{10} - 5 = 0, \quad \text{or} \quad 4v_\Delta'' + v_b - 2i_\Delta'' = 50.$$

We now use

$$v_b = 2i_\Delta'' + v_\Delta''$$

to find the value for v_Δ''. Thus

$$5v_\Delta'' = 50, \quad \text{or} \quad v_\Delta'' = 10 \text{ V}.$$

From the node a equation,

$$5v_o'' = 80, \quad \text{or} \quad v_o'' = 16 \text{ V}.$$

The value of v_o is the sum of v_o' and v_o'', or 24 V.

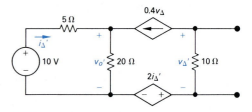

FIGURE 4.67 The circuit shown in Fig. 4.66 with the 5-A source deactivated.

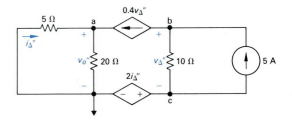

FIGURE 4.68 The circuit shown in Fig. 4.66 with the 10-V source deactivated.

DRILL EXERCISES

4.26 a) Use the principle of superposition to find the voltage v in the circuit shown.

   b) Find the power dissipated in the 40-Ω resistor.

ANSWER: (a) 40 V; (b) 40 W.

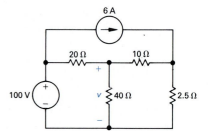

4.27 Use the principle of superposition to find the voltage v in the circuit shown.

ANSWER: 30 V.

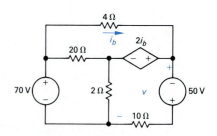

SUMMARY

Two extremely useful circuit analysis techniques are the node-voltage and the mesh-current methods. They enable us to minimize the number of circuit equations to be solved simultaneously in analyzing a circuit. They also encourage us to take a systematic approach to writing circuit equations. Two goals of this book are to help you develop the habits of thinking strategically before writing equations and of always checking your answers—habits that are marks of good engineers. In this chapter we showed you how to do both.

The key concepts and techniques introduced related to sources and resistors, but we show—beginning in Chapter 7—that they have much greater general application.

- *Terminology* plays an important role in describing circuits. Key definitions are:

  node, which is a point in the circuit where two or more elements join;

  path, which is an ordered trace through adjoining elements, passing through connecting nodes only once;

  closed path, which is an ordered trace through adjoining elements that starts and stops at the same node, passing through connecting nodes only once;

  branch, which is a path between any two nodes;

  mesh, which is a "window" in a planar circuit;

  essential node, which connects three or more branches; and

  essential branch, which connects essential nodes.

- *Planar circuit* is a circuit that can be drawn on a plane without branches crossing.

- *Node-voltage method* works with both planar and nonplanar circuits. Voltage variables are assigned at each essential node and Kirchhoff's current law is used to write one equation per essential node. The number of equations is one less than the number of essential nodes. The node voltages are used to find the branch currents.

- *Mesh-current method* works only with planar circuits. Mesh currents are assigned to each mesh and Kirchhoff's voltage law is used to write one equation per mesh. The number of equations equals one more than the difference between the number of essential branches and nodes. The mesh currents are used to find the branch currents.

• *Loop-current method* is similar in concept to the mesh-current method, except that it works with both planar and nonplanar circuits. We discuss this method in Chapter 5.

• *Source transformation* involves replacing, *insofar as its terminal voltage and currents are concerned,* a voltage source and a series resistor with a current source and a parallel resistor (or vice versa). The values of the voltage source, the current source, and the resistor are such that the equivalents have the same open circuit voltage and short circuit current.

• *Thévenin and Norton equivalents* allow any circuit made up of sources and resistors to be replaced, *insofar as its terminal voltage and current are concerned,* by an equivalent circuit containing a voltage source with a series resistor (Thévenin) or by a current source with a parallel resistor (Norton).

• *Superposition* is a method of analyzing a circuit containing multiple *independent* sources by activating one source at a time and summing the resulting voltages and currents to determine the voltages and currents that exist when all the independent sources are active. (*Note:* Dependent sources are not deactivated when applying superposition.)

PROBLEMS

4.1 Assume the current i_g in the circuit in Fig. P4.1 is known. The resistors R_1 through R_5 are also known.

a) How many unknown currents are there?

b) How many independent equations can be written using Kirchhoff's current law?

c) Write an independent set of KCL equations.

d) How many independent equations must be derived from Kirchhoff's voltage law?

e) Write a set of independent KVL equations.

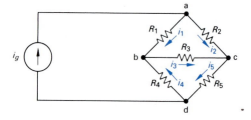

FIGURE P4.1

4.2 Use the node-voltage method to find v_o in the circuit in Fig. P4.2.

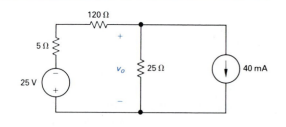

FIGURE P4.2

4.3 Use the node-voltage method to find v_1 and v_2 in the circuit shown in Fig. P4.3.

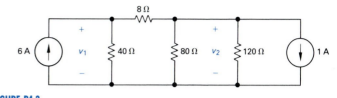

FIGURE P4.3

4.4 Use the node-voltage method to find how much power the 45-V source extracts from the circuit in Fig. P4.4.

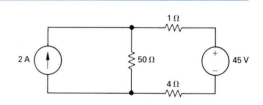

FIGURE P4.4

4.5 Use the node-voltage method to find v_1 and v_2 in the circuit in Fig. P4.5.

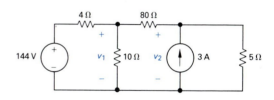

FIGURE P4.5

4.6 a) Use the node-voltage method to find v_1, v_2, and v_3 in the circuit in Fig. P4.6.

b) How much power does the 28-A current source deliver to the circuit?

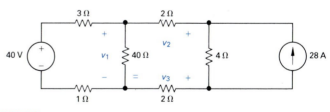

FIGURE P4.6

4.7 a) Use the node-voltage method to find the branch currents i_a through i_e in the circuit shown in Fig. P4.7.

b) Find the total power developed in the circuit.

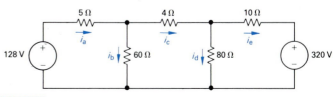

FIGURE P4.7

4.8 Use the node-voltage method to find the total power dissipated in the circuit in Fig. P4.8.

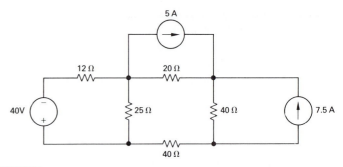

FIGURE P4.8

4.9 a) Use the node-voltage method to find the branch currents i_1 through i_6 in the circuit shown in Fig. P4.9.

b) Test your solution for the branch currents by showing that the total power dissipated equals the total power developed.

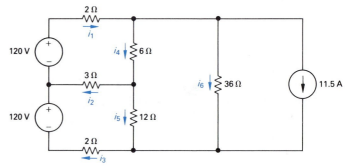

FIGURE P4.9

4.10 Use the node-voltage method to find v_1 and the power delivered by the 2-A current source in the circuit in Fig. P4.10.

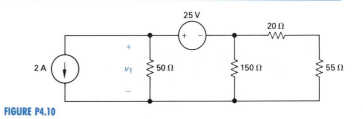

FIGURE P4.10

4.11 a) Use the node-voltage method to find the branch currents i_1, i_2, and i_3 in the circuit in Fig. P4.11.

b) Check your solution for i_1, i_2, and i_3 by showing that the power dissipated in the circuit equals the power developed.

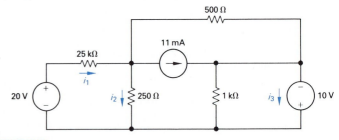

FIGURE P4.11

4.12 a) Find the power developed by the 40-mA current source in the circuit in Fig. P4.2.

b) Find the power developed by the 25-V voltage source in the circuit in Fig. P4.2.

c) Verify that the total power developed equals the total power dissipated.

4.13 A 100-Ω resistor is connected in series with the 40-mA current source in the circuit in Fig. P4.2.

a) Find v_o.

b) Find the power developed by the 40-mA current source.

c) Find the power developed by the 25-V voltage source.

d) Verify that the total power developed equals the total power dissipated.

e) What effect will any finite resistance connected in series with the 40-mA current source have on the value of v_o?

4.14 Use the node-voltage method to find the value of v_o in the circuit in Fig. P4.14.

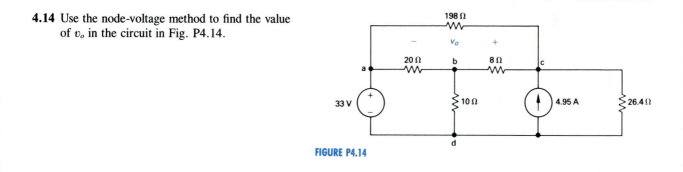

FIGURE P4.14

4.15 Check the solution for v_o in Problem 4.14 by first using a Y-to-Δ transformation to eliminate node b.

4.16 a) Use the node-voltage method to find the power dissipated in the 2-Ω resistor in the circuit in Fig. P4.16.

b) Find the power supplied by the 230-V source.

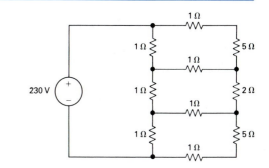

FIGURE P4.16

4.17 Use the node-voltage method to find v_o in the circuit in Fig. P4.17.

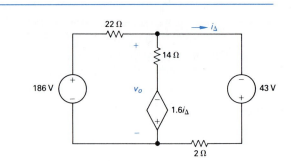

FIGURE P4.17

4.18 Use the node-voltage method to calculate the power delivered by the dependent voltage source in the circuit in Fig. P4.18.

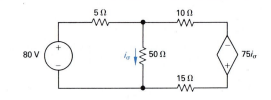

FIGURE P4.18

4.19 Use the node-voltage method to determine how much power the dependent voltage source in Fig. P4.19 delivers to the circuit.

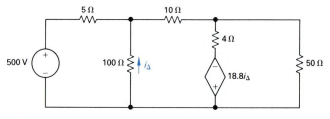

FIGURE P4.19

4.20 Use the node-voltage method to find v_Δ in the circuit in Fig. P4.20.

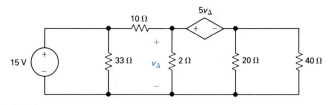

FIGURE P4.20

4.21 a) Find the node voltages v_1, v_2, and v_3 in the circuit in Fig. P4.21.

b) Find the total power dissipated in the circuit.

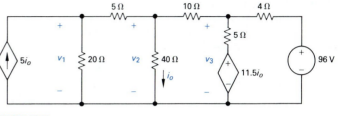

FIGURE P4.21

4.22 Assume you are a project engineer and one of your staff is assigned to analyze the circuit shown in Fig. P4.22. The reference node and node numbers given on the figure were assigned by the analyst. Her solution gives the values of v_3 and v_4 as 108 V and 81.60 V respectively.

Test these values by checking the total power developed in the circuit against the total power dissipated. Do you agree with the solution submitted by the analyst?

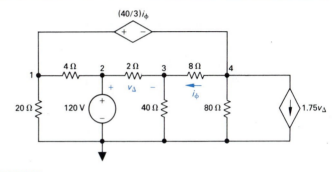

FIGURE P4.22

4.23 Show that when Eqs. (4.16), (4.17), and (4.19) are solved for i_B, the result is identical to Eq. (2.27).

4.24 Use the node-voltage method to find the power developed by the 20-V source in the circuit in Fig. P4.24.

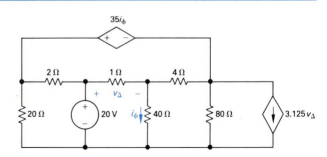

FIGURE P4.24

4.25 a) Use the node-voltage method to show that the output voltage v_o in the circuit in Fig. P4.25 is equal to the average value of the source voltages.

b) Find v_o if $v_1 = 100$ V, $v_2 = 80$ V, and $v_3 = -60$ V.

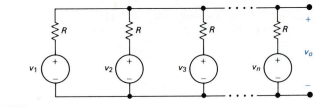

FIGURE P4.25

4.26 a) Use the mesh-current method to find the branch currents i_a, i_b, and i_c in the circuit in Fig. P4.26.

b) Repeat part (a) if the polarity of the 64-V source is reversed.

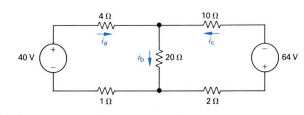

FIGURE P4.26

4.27 a) Use the mesh-current method to find the total power developed in the circuit in Fig. P4.27.

b) Check your answer by showing that the total power developed equals the total power dissipated.

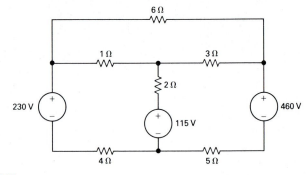

FIGURE P4.27

4.28 a) Use the mesh-current method to find how much power the 4-A current source delivers to the circuit in Fig. P4.28.

b) Find the total power delivered to the circuit.

c) Check your calculations by showing that the total power developed in the circuit equals the total power dissipated.

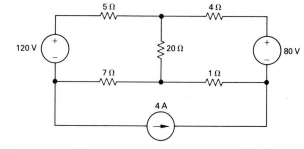

FIGURE P4.28

4.29 Use the mesh-current method to find the total power dissipated in the circuit in Fig. P4.29.

don't solve each mesh current independent of the other one

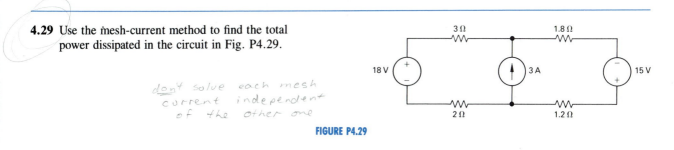

FIGURE P4.29

4.30 Assume the 18-V source in the circuit in Fig. P4.29 is reduced to 6 V. Find the total power dissipated in the circuit.

4.31 a) Assume the 18-V source in the circuit in Fig. P4.29 is reduced to 10 V. Find the total power dissipated in the circuit.

  b) Repeat part (a) if the 3-A current source is replaced by a short circuit.

  c) Explain why the answers to parts (a) and (b) are the same.

4.32 a) Use the mesh-current method to find the branch currents i_a through i_e in the circuit in Fig. P4.32.

  b) Check your solution by showing the total power developed in the circuit equals the total power dissipated.

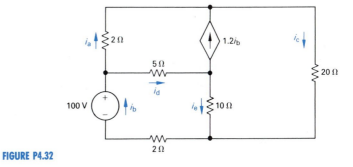

FIGURE P4.32

4.33 Use the mesh-current method to find the power delivered by the dependent voltage source in the circuit seen in Fig. P4.33.

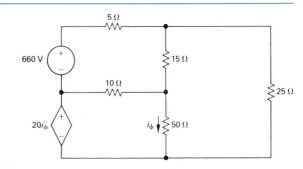

FIGURE P4.33

4.34 a) Use the mesh-current method to find v_o in the circuit in Fig. P4.34.

b) Find the power delivered by the dependent source.

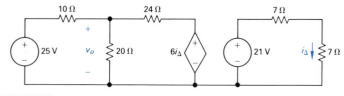

FIGURE P4.34

4.35 a) Use the mesh-current method to solve for i_Δ in the circuit in Fig. P4.35.

b) Find the power delivered by the independent current source.

c) Find the power delivered by the dependent voltage source.

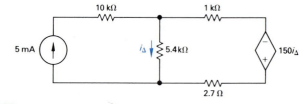

FIGURE P4.35

4.36 a) Use the mesh-current method to determine which sources in the circuit in Fig. P4.36 are generating power.

b) Find the total power dissipated in the circuit.

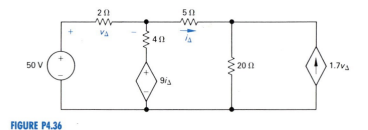

FIGURE P4.36

4.37 Use the mesh-current method to find the total power developed in the circuit in Fig. P4.37.

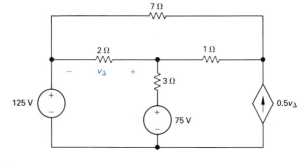

FIGURE P4.37

4.38 Use the mesh-current method to find the power dissipated in the 20-Ω resistor in the circuit in Fig. P4.38.

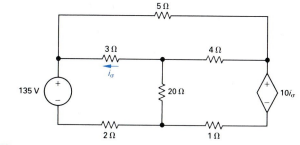

FIGURE P4.38

4.39 a) Use the mesh-current method to find the power delivered to the 25-Ω resistor in the circuit in Fig. P4.39.

b) What percent of the total power developed in the circuit is delivered to the 25-Ω resistor?

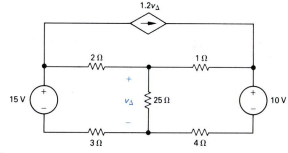

FIGURE P4.39

4.40 Use the mesh-current method to find the power developed in the dependent voltage source in the circuit in Fig. P4.40.

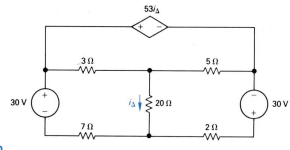

FIGURE P4.40

4.41 Assume you have been asked to find the power dissipated in the 5-Ω resistor in the circuit in Fig. P4.41.

a) Which method of circuit analysis would you recommend? Explain why.

b) Use your recommended method of analysis

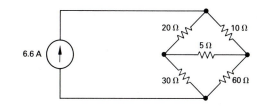

FIGURE P4.41

(*continued*)

to find the power dissipated in the 5-Ω resistor.

c) Would you change your recommendation if the problem had been to find the power de-

veloped by the 6.6-A current source? Explain.

d) Find the power developed by the 6.6-A current source.

4.42 A 160-Ω resistor is placed in parallel with the 6.6-A current source in the circuit in Fig. P4.41. Assume you have been asked to calculate the power developed by the current source.

a) Which method of circuit analysis, node-

voltage or mesh-current, would you recommend? Explain why.

b) Find the power developed by the current source.

4.43 a) Would you use the node-voltage or mesh-current method to find the power absorbed by the 5-V source in the circuit in Fig. P4.43? Explain your choice.

b) Use the method you selected in part (a) to find the power.

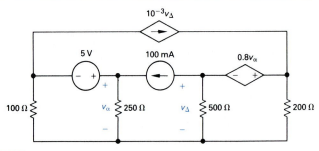

FIGURE P4.43

4.44 a) Find the branch currents i_a through i_e for the circuit shown in Fig. P4.44.

b) Check your answers by showing that the total power generated equals the total power dissipated.

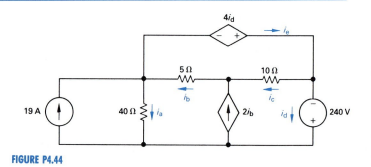

FIGURE P4.44

4.45 The circuit in Fig. P4.45 is a direct-current version of a typical three-wire distribution system. The resistors R_a, R_b, and R_c represent the resistances of the three conductors that connect the three loads R_1, R_2, and R_3 to the 125/250-V voltage supply. The resistors R_1 and R_2 represent loads connected to the 125-V circuits, and

R_3 represents a load connected to the 250-V circuit.

a) Calculate v_1, v_2, and v_3.

b) Calculate the power delivered to R_1, R_2, and R_3.

c) What percentage of the total power developed by the sources is delivered to the loads?

d) The R_b branch represents the neutral conductor in the distribution circuit. What adverse effect occurs if the neutral conductor is opened? (*Hint:* Calculate v_1 and v_2 and note that appliances or loads designed for use in this circuit would have a nominal voltage rating of 125 V.)

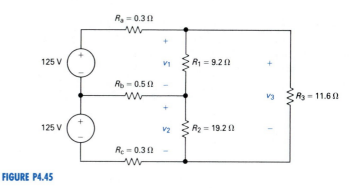

FIGURE P4.45

4.46 Show that whenever $R_1 = R_2$ in the circuit in Fig. P4.45 the current in the neutral conductor is zero. (*Hint:* Solve for the neutral conductor current as a function of R_1 and R_2.)

4.47 The variable dc-voltage source in the circuit in Fig. P4.47 is adjusted so that the power developed by the 5-A current source is zero. Find the value of the dc voltage.

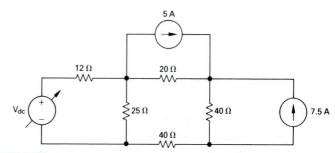

FIGURE P4.47

4.48 The variable dc-voltage source in the circuit in Fig. P4.48 is adjusted so that i_o is zero. Find the value of V_{dc}.

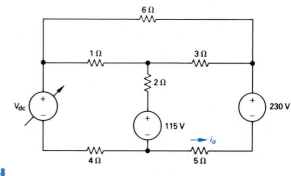

FIGURE P4.48

4.49 The variable resistor (R_o) in the circuit in Fig. P4.49 is adjusted until the power dissipated in the resistor (R_o) is 250 W.
Find the values of R_o which satisfy this condition.

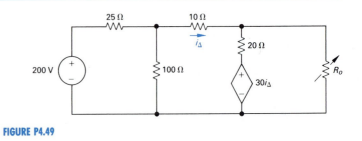

FIGURE P4.49

4.50 a) Use a series of source transformations to find the current i_o in the circuit in Fig. P4.50.

b) Verify your solution by using the node-voltage method to find i_o.

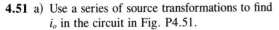

FIGURE P4.50

4.51 a) Use a series of source transformations to find i_o in the circuit in Fig. P4.51.

b) Verify your solution by using the mesh-current method to find i_o.

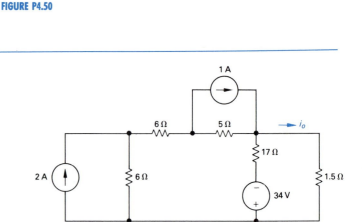

FIGURE P4.51

4.52 a) Find the current in the 5-kΩ resistor in the circuit in Fig. P4.52 by making a succession of appropriate source transformations.

b) Using the result obtained in part (a) work back through the circuit to find the power developed by the 120-V source.

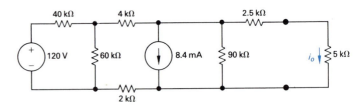

FIGURE P4.52

4.53 a) Use source transformations to find v_o in the circuit in Fig. P4.53.

b) Find the power developed by the 520-V source.

c) Find the power developed by the 1-A current source.

d) Verify that the total power developed equals the total power dissipated.

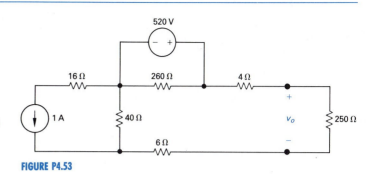

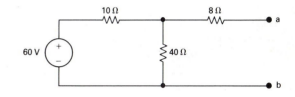

FIGURE P4.53

4.54 Find the Thévenin equivalent with respect to the terminals a,b for the circuit in Fig. P4.54.

FIGURE P4.54

4.55 Find the Thévenin equivalent with respect to the terminals a,b for the circuit in Fig. P4.55.

FIGURE P4.55

4.56 Find the Thévenin equivalent with respect to the terminals a,b for the circuit in Fig. P4.56.

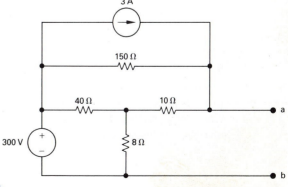

FIGURE P4.56

4.57 a) Find the Thévenin equivalent with respect to the terminals a,b for the circuit in Fig. P4.57 by finding the open-circuit voltage and the short-circuit current.

b) Solve for the Thévenin resistance by removing the independent sources. Compare your result to the Thévenin resistance found in part (a).

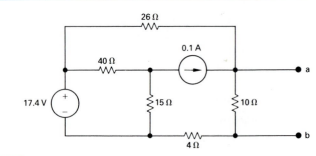

FIGURE P4.57

4.58 Find the Norton equivalent with respect to the terminals a,b in the circuit in Fig. P4.58.

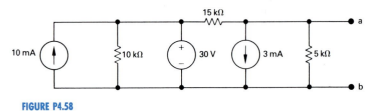

FIGURE P4.58

4.59 Determine i_o and v_o in the circuit shown in Fig. P4.59 when R_o is 0, 1, 3, 5, 10, 15, 25, 40, 55, 70, 85, and 95-Ω.

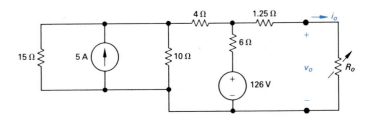

FIGURE P4.59

4.60 A voltmeter with a resistance of 85.5 kΩ is used to measure the voltage v_{ab} in the circuit in Fig. P4.60.

a) What is the voltmeter reading?

b) What is the percent error in the voltmeter reading if percent error is defined as [(measured − actual)/actual] × 100?

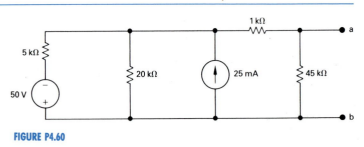

FIGURE P4.60

4.61 The Wheatstone bridge in the circuit shown in Fig. P4.61 is balanced when R_2 equals 3000 Ω. If the galvanometer has a resistance of 50 Ω, how much current will the galvanometer detect when the bridge is unbalanced by setting R_2 to 3003 Ω? (*Hint:* Find the Thévenin equivalent with respect to the galvanometer terminals when $R_2 = 3003$ Ω.) Note that once we have found the Thévenin equivalent with respect to the galvanometer terminals, it is easy to find the amount of unbalanced current in the galvanometer branch for different galvanometer movements.

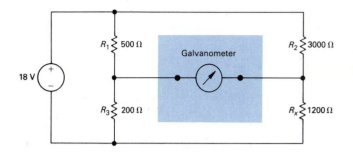

FIGURE P4.61

4.62 Determine the Thévenin equivalent with respect to the terminals a,b for the circuit shown in Fig. P4.62.

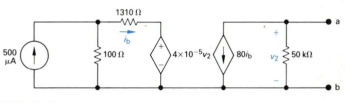

FIGURE P4.62

4.63 Find the Thévenin equivalent with respect to the terminals a,b for the circuit shown in Fig. P4.63.

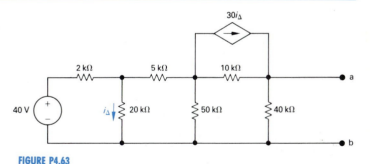

FIGURE P4.63

4.64 When a voltmeter is used to measure the voltage v_e in Fig. P4.64 it reads 5.5 V.

a) What is the resistance of the voltmeter?

b) What is the percent error in the voltage measurement?

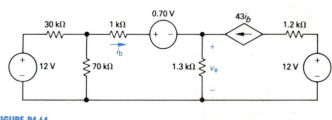

FIGURE P4.64

4.65 When an ammeter is used to measure the current i_ϕ in the circuit shown in Fig. P4.65 it reads 6 A.

a) What is the resistance of the ammeter?

b) What is the percent error in the current measurement?

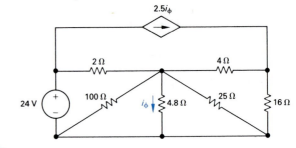

FIGURE P4.65

4.66 Find the Thévenin equivalent with respect to the terminals a,b for the circuit shown in Fig. P4.66.

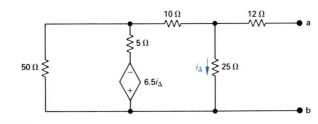

FIGURE P4.66

4.67 Find the Thévenin equivalent with respect to the terminals a,b in the circuit in Fig. P4.67.

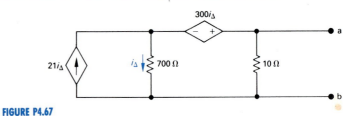

FIGURE P4.67

4.68 A Thévenin equivalent can also be determined from measurements made at the pair of terminals of interest. Assume the following measurements were made at the terminals a,b in the circuit in Fig. P4.68.

When a 20-Ω resistor is connected to the terminals a,b the voltage v_{ab} is measured and found to be 100 V.

When a 50-Ω resistor is connected to the terminals a,b the voltage is measured and found to be 200 V.

Find the Thévenin equivalent of the network with respect to the terminals a,b.

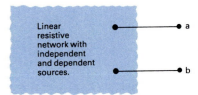

FIGURE P4.68

4.69 An automobile battery, when connected to a car radio, provides 12.5 V to the radio. When connected to a set of headlights it provides 11.8 V to the headlights. Assume the radio can be modeled as a 6-Ω resistor and the headlights can be modeled as a 0.75-Ω resistor. What are the Thévenin and Norton equivalents for the battery?

4.70
a) Calculate the power delivered to each resistor in Problem 4.59.
b) Plot the power delivered versus the resistance.
c) At what value of R is the power maximum?

4.71 The variable resistor in the circuit in Fig. P4.71 is adjusted for maximum power transfer to R_o.
a) Find the value of R_o.
b) Find the maximum power that can be delivered to R_o.

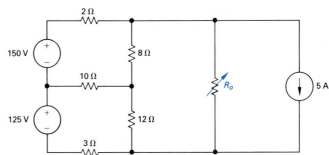

FIGURE P4.71

4.72 The variable resistor in the circuit in Fig. P4.72 is adjusted for maximum power transfer to R_o. What percentage of the total power developed in the circuit is delivered to R_o?

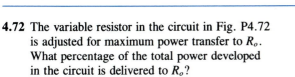

FIGURE P4.72

4.73 The variable resistor (R_L) in the circuit in Fig. P4.73 is adjusted for maximum power transfer to R_L.
a) Find the numerical value of R_L.
b) Find the maximum power transfered to R_L.

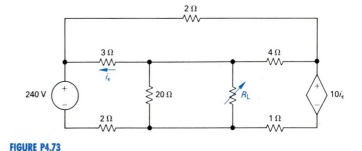

FIGURE P4.73

4.74 The variable resistor (R_o) in the circuit in Fig. P4.74 is adjusted for maximum power transfer to R_o.

a) Find the value of R_o.

b) Find the maximum power that can be delivered to R_o.

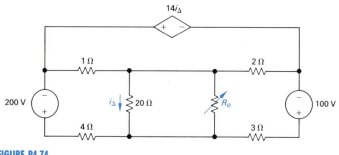

FIGURE P4.74

4.75 What percentage of the total power developed in the circuit in Fig. P4.74 is delivered to R_o?

4.76 The variable resistor (R_o) in the circuit in Fig. P4.76 is adjusted until it absorbs maximum power from the circuit. Find:

a) the value of R_o;

b) the maximum power; and

c) the percent of the total power developed in the circuit that is delivered to R_o.

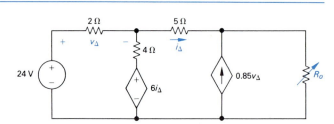

FIGURE P4.76

4.77 The variable resistor (R_o) in the circuit in Fig. P4.77 is adjusted for maximum power transfer to R_o.
What percent of the total power developed in the circuit is delivered to R_o?

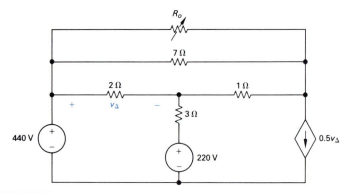

FIGURE P4.77

4.78 The 40-Ω resistor in the circuit in Fig. P4.24 is replaced with a variable resistor R_o. R_o is adjusted for maximum power transfer to R_o.

a) Find the numerical value of R_o.

b) Find the maximum power delivered to R_o.

c) How much power does the 20-V source deliver to the circuit when R_o is adjusted to the value found in part (a)?

4.79 A variable resistor R_o is connected across the terminals a,b in the circuit in Fig. P4.63. The variable resistor is adjusted until maximum power is transferred to R_o. Find:

a) the value of R_o;

b) the maximum power delivered to R_o; and

c) the percentage of the total power developed in the circuit that is delivered to R_o.

4.80 a) Find the value of the variable resistor R_o in the circuit in Fig. P4.80 that will result in maximum power dissipation in the 8-Ω resistor.

b) What is the maximum power that can be delivered to the 8-Ω resistor.

(*Hint:* Hasty conclusions could be hazardous to your career.)

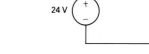

FIGURE P4.80

4.81 Use superposition to solve for i_o and v_o in the circuit in Fig. P4.81.

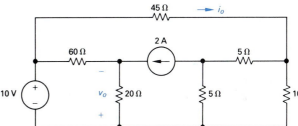

FIGURE P4.81

4.82 Use the principle of superposition to find the current i_o in the circuit shown in Fig. P4.82.

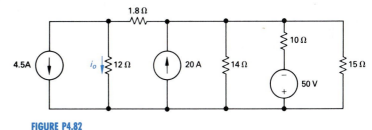

FIGURE P4.82

4.83 Use the principle of superposition to find v_o in the circuit in Fig. P4.83.

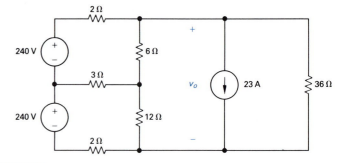

FIGURE P4.83

4.84 Use the principle of superposition to find the voltage v_o in the circuit in Fig. P4.84.

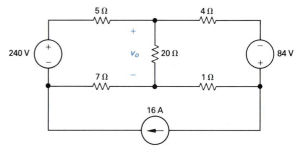

FIGURE P4.84

4.85 a) In the circuit in Fig. P4.85, before the 5-mA current source is attached to the terminals a,b the current i_o was calculated and found to be 3.5 mA. Use superposition to find the value of i_o after the current source is attached.

    b) Verify your solution by finding i_o when all three sources are acting simultaneously.

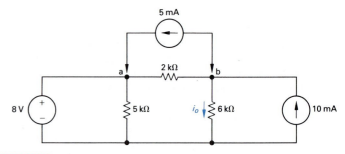

FIGURE P4.85

4.86 Use the principle of superposition to find v_o in the circuit in Fig. P4.86.

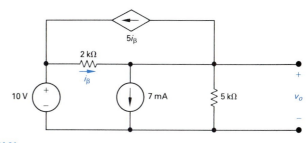

FIGURE P4.86

4.87 Use the principle of superposition to find the current entering the positive terminal of the 75-V source in the circuit in Fig. P4.37.

4.88 Use the principle of superposition to find the current i_Δ in the circuit in Fig. P4.40.

4.89 Use the principle of superposition to find v_Δ in the circuit in Fig. P4.39.

4.90 Use the principle of superposition to find the voltage across the dependent current source in the circuit in Fig. P4.44. Use the upper terminal of the dependent current source as the positive reference for the voltage.

4.91 Laboratory measurements on a dc voltage source yield a terminal voltage of 75 V with no load connected to the source and 60 V when loaded with a 20-Ω resistor.

a) What is the Thévenin equivalent with respect to the terminals of the dc voltage source?

b) Show that the Thévenin resistance of the

source is given by the expression

$$R_{\text{Th}} = \left(\frac{V_{\text{Th}}}{V_o} - 1 \right) R_{\text{L}},$$

where

V_{Th} = the Thévenin voltage,

V_o = the terminal voltage corresponding to the load resistance R_{L}.

4.92 Two ideal dc voltage sources are connected by electrical conductors that have a resistance of r ohms/meter as shown in Fig. P4.92. A load having a resistance of R ohms moves between the two voltage sources. Let x equal the distance between the load and the source V_1 and L equal the distance between the sources.

a) Show that

$$v = \frac{V_1 RL + R(V_2 - V_1)x}{RL + 2rlx - 2rx^2}.$$

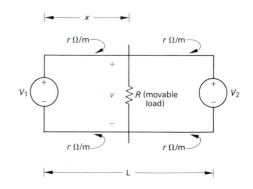

FIGURE P4.92

(*continued*)

b) Show that the voltage v will be minimum when

$$x = \frac{L}{V_2 - V_1}\left[-V_1 \pm \sqrt{V_1 V_2 - \frac{R}{2rL}(V_1 - V_2)^2}\right].$$

c) Find x when $L = 16$ km, $V_1 = 1000$ V, $V_2 = 1200$ V, $R = 3.9 \ \Omega$ and $r = 5 \times 10^{-5} \ \Omega/\text{m}$.

d) What is the minimum value of v for the circuit of part (c)?

4.93 Assume your supervisor has asked you to determine the power developed by the 1-V source in the circuit in Fig. P4.93. Before calculating the power developed by the 1-V source the supervisor asks you to submit a proposal describing how you plan to attack the problem. Furthermore he asks you to explain why you have chosen your proposed method of solution.

a) Describe your plan of attack and at the same time explain your reasoning.

b) Use the method you have outlined in (a) to find the power developed by the 1-V source.

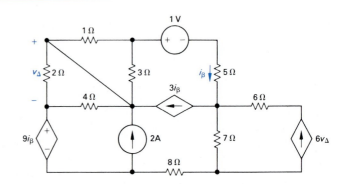

FIGURE P4.93

4.94 Find the power absorbed by the 5-A current source in the circuit in Fig. P4.94.

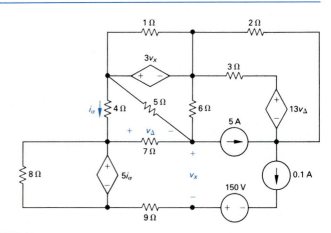

FIGURE P4.94

4.95 Find v_1, v_2, and v_3 in the circuit in Fig. P4.95.

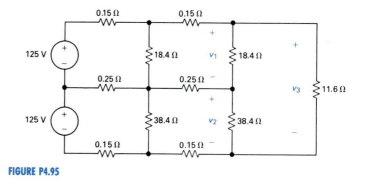

FIGURE P4.95

4.96 Find i_1 in the circuit in Fig. P4.96.

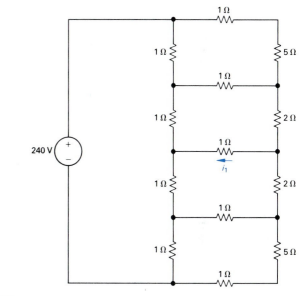

FIGURE P4.96

TOPOLOGY IN CIRCUIT ANALYSIS

CHAPTER 5

Now that you have gained some experience and some confidence in using both the node-voltage method and the mesh-current method, we introduce you to the topological properties of a circuit that allow the validity of these two powerful analytical techniques to be proved. This introduction and discussion of topological concepts also adds some depth of understanding to circuit analysis in general.

As noted earlier, circuit analysis revolves around two sets of constraints. The first is the set of constraints imposed on the circuit variables by the elements themselves, for example, Ohm's law. The second is the set of constraints forced on the variables by the interconnections of the elements, that is, Kirchhoff's laws. By focusing on the constraints imposed by interconnections circuit or network topology leads to an efficient means of writing the minimum number of independent equations that describe the circuit.

The first step in studying the topological properties of a network is to suppress the nature of the circuit elements that make up the network. We do so by constructing a *graph* of the circuit. The graph consists of redrawing the circuit, with a line representing each branch of the network. Obviously, the line conceals the nature of the element in a branch. Figure 5.1 shows a circuit and its graph.

The graph of a circuit is a visual aid that contains the pertinent information about the interconnections of nodes and branches. Its graph specifies completely the circuit's topological character.

Before we show how to use topology to derive node-voltage or mesh-current equations, a brief explanation of this important branch of mathematics is in order. The word *topology* refers to *the science of place*. In mathematics, topology is a branch of geometry in which figures are considered to be perfectly elastic. Their flexible nature allows elastic, as opposed to rigid, motions

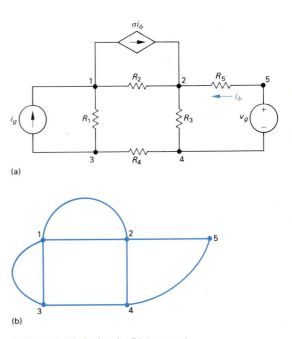

(a)

(b)

FIGURE 5.1 (a) A circuit; (b) its graph.

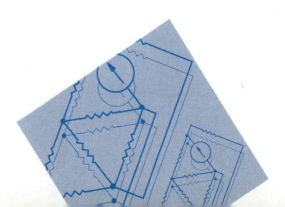

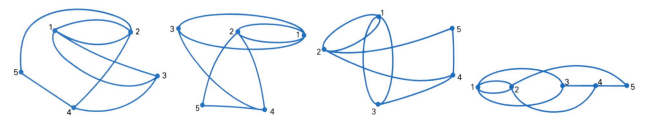

FIGURE 5.2 Four graphs that are topologically equivalent to the graph shown in Fig. 5.1(b).

to take place. Thus figures can be stretched, twisted, squeezed, pulled, and bent. Two figures are topologically equivalent only if one figure can be made to coincide with the other by an elastic deformation. Thus a sphere and a cube are topologically equivalent, as are a circle and a square. The topological properties of a figure are those that are invariant under elastic deformation. The topological properties of a circuit are those that are invariant with stretching, squeezing, bending, or twisting of the circuit graph. For example, all the graphs in Fig. 5.2 are topologically equivalent to the graph in Fig. 5.1(b) because they all depict the same interconnection information between the five nodes and eight branches.

Two networks also are topologically equivalent if they differ only in the elements that make up their branches. Thus the circuit shown in Fig. 5.3 is the same, topologically, as the circuit shown in Fig. 5.1(a).

To show how to use topology to derive the node-voltage and mesh-current methods of analysis, we need to expand on the vocabulary introduced in Chapter 4. Thus in addition to the terms node, branch, path, loop, mesh, planar, and nonplanar, we must add the concepts of *tree, cotree, link, cut set, fundamental cut set,* and *fundamental loop.*

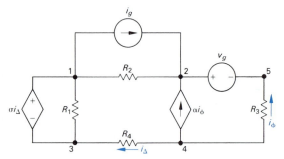

FIGURE 5.3 A circuit that is topologically the same as the circuit shown in Fig. 5.1(a).

5.1 TOPOLOGICAL CONCEPTS

A *tree* is defined as *any* set of connecting branches that connects every node to every other node *without forming any closed paths or loops.* Consider, for example, the graph shown in Fig. 5.4, where we labeled the branches and nodes for purposes of discussion. One of the trees in this graph consists of the branches a, d, and e; these three connected branches connect all four nodes of the graph without forming any loops. Figure 5.5 shows this tree of the graph, with the tree branches highlighted by heavy lines.

In general, a graph contains multiple trees. The graph shown in Fig. 5.4 has eight trees. The seven, in addition to the one

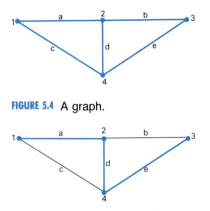

FIGURE 5.4 A graph.

FIGURE 5.5 A tree of the graph shown in Fig. 5.4.

shown in Fig. 5.5, are b,d,c; c,a,b; e,b,a; a,c,e; b,e,c; a,b,d; and c,d,e.

DRILL EXERCISES

5.1 Identify the eight possible trees for the graph shown.

ANSWER: a,c; a,d; a,e; b,c; b,d; b,e; e,d; and e,c.

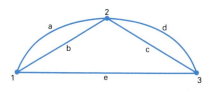

The number of branches in any selected tree is always one less than the number of nodes, that is, $n - 1$. This condition follows directly from the definition of a tree. Once a tree for a graph has been defined, the remaining branches are referred to as the *links*, or *chords*. The collection of links is called a *complementary tree*, or *cotree*. Thus the cotree of the tree shown in Fig. 5.5 consists of the branches b and c. The branches of a cotree may or may not be connected, whereas the branches of a tree are always connected. Also, cotrees may have loops, whereas, by definition, a tree contains no loops.

DRILL EXERCISES

5.2 Identify the cotrees for the graph of Drill Exercise 5.1.

ANSWER: b,d,e; b,c,e; b,c,d; a,d,e; a,c,e; a,c,d; a,b,c; and a,b,d.

The remaining topological property of a graph that is germane to our discussion is the *cut set*. A cut set is a *minimal* set of branches that, when cut, divides the graph into two groups of nodes. The adjective *minimal* means that a cut set of a graph cannot itself contain a cut set that would divide the original graph into the same two groups of nodes. For example, in the graph shown in Fig. 5.6, the branches a, c, f, and h are a cut set that divides the graph into the two groups of nodes shown in Fig. 5.7. Cutting the branches a, c, f, h, and e also would divide

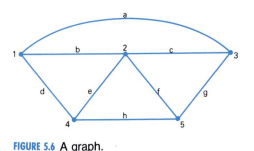

FIGURE 5.6 A graph.

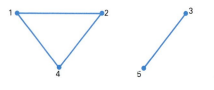

FIGURE 5.7 The two subgraphs of the graph shown in Fig. 5.6, created by the cut set a,c,f,h.

the original graph into two graphs, one containing nodes 1, 2, and 4, and the other nodes 3 and 5. However, this configuration is not a cut set because it contains the cut set a,c,f,h. Thus a cut set is the smallest number of branches needed to cut the graph into two specified groups of nodes.[†]

When a graph is cut into two subgraphs,[†] an isolated node may form one subgraph. Therefore in Fig. 5.6 d,e,h; a,b,d; b,c,e,f; a,c,g; and f,g,h all are cut sets. Just as there are many trees in a given graph, there also are many cut sets.

DRILL EXERCISES

5.3 Identify six cut sets for the graph shown in Fig. 5.4.   **ANSWER:** a,c; a,b,d; b,e; c,d,e; b,c,d; and a,d,e.

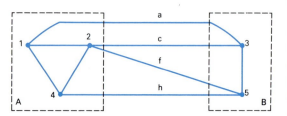

FIGURE 5.8 An elastic deformation of the graph shown in Fig. 5.6.

An important characteristic of a cut set that relates to circuit analysis is that the algebraic sum of the currents in the cut-set branches is zero. This characteristic becomes apparent when you view the cut set as the branches tying one portion of a circuit to another. For example, consider stretching the graph shown in Fig. 5.6 so that the cut-set branches a, c, f, and h clearly form the tie between the nodes 1,2,4 and 3,5. Figure 5.8 shows the resulting graph. Kirchhoff's current law requires that the algebraic sum of the currents in the cut-set branches a, c, f, and h equal zero; otherwise charge would accumulate in either the A portion or the B portion of the circuit. Although a graph may contain many loops and many cut sets, in order to derive an independent set of circuit equations we need to focus on a subset of these loops and cut sets, called *fundamental loops* and *fundamental cut sets*.

A fundamental loop is a loop that contains one and only one link. Returning to the graph in Fig. 5.6, we define the tree as the branches d, e, f, and g. Then the four fundamental loops are d,b̲,e; f,c̲,g; e,f,h̲; and d,a̲,g,f,e, with the underlined letter denoting the link in each loop. The set of fundamental loops in a circuit is not unique because the tree of a circuit is not unique. The number of fundamental loops is unique and equals the number of links; that is, $b' - n + 1$, where b' is the total number of branches in the graph.

[†] A graph *g* is said to be a subgraph of a graph *G* if all the nodes and all the branches in *g* are in *G* and each branch in *g* has the same end nodes in *g* as in *G*.

DRILL EXERCISES

5.4 Define the tree in the graph shown in Fig. 5.6 as the branches a, c, f, and h. Enumerate the fundamental loops in the graph.

ANSWER: a,c,b̲; a,c,f,h,d̲; h,f,e̲; and c,f,g̲.

A fundamental cut set is a cut set that contains one and only one tree branch. To find a fundamental cut set associated with a tree follow these steps: (1) select a tree branch; (2) divide the graph into two subgraphs, one containing all the tree branches connected to one node of the selected tree branch and the other containing all the tree branches connected to the other node of the selected tree branch; (3) separate these two subgraphs by stretching the selected tree branch; and (4) draw in just those links that span the two subgraphs. These links and the selected tree branch form a fundamental cut set.

To illustrate this procedure, we return to the graph shown in Fig. 5.6 and as before, select the tree as the set of branches d, e, f, and g. Next, we find the fundamental cut set associated with the tree branch labeled e. In accordance with step 2, we construct the graph shown in Fig. 5.9(a). Figure 5.9(b) shows the result of stretching the graph shown in Fig. 5.9(a), and Fig. 5.9(c) shows the construction of those links that span the two subgraphs. From Fig. 5.9(c), we identify the fundamental cut set associated with tree branch e as the branches a, b, h, and $\boxed{e}$, with the boxed letter denoting the tree branch. The three

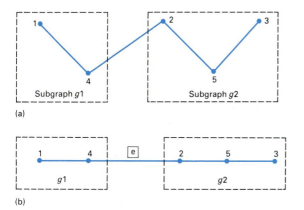

(a)

(b)

FIGURE 5.9 A graphic technique for finding the fundamental cut set associated with tree branch e: (a) the subgraphs associated with branch e; (b) separating the subgraphs by stretching branch e; (c) constructing the links that span the two subgraphs.

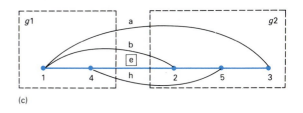

(c)

remaining fundamental cut sets are a,b,$\boxed{d}$; a,c,h,$\boxed{f}$; and a,c,$\boxed{g}$. The graphs associated with these cut sets appear in Fig. 5.10(a), (b), and (c), respectively.

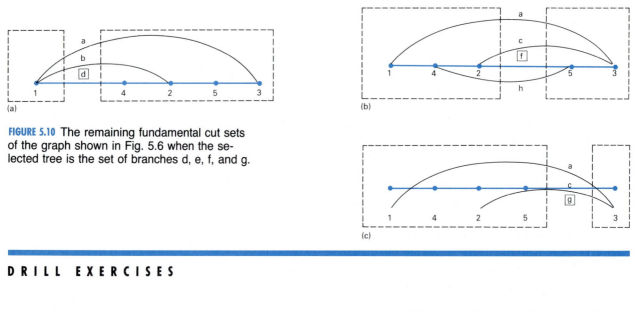

FIGURE 5.10 The remaining fundamental cut sets of the graph shown in Fig. 5.6 when the selected tree is the set of branches d, e, f, and g.

DRILL EXERCISES

5.5 Define the tree in the graph shown in Fig. 5.6 as branches b, e, h, and g. Identify the fundamental cut sets in the graph.

ANSWER: a,d,$\boxed{b}$; a,c,f,d,$\boxed{e}$; a,c,f,$\boxed{h}$; and a,c,$\boxed{g}$.

We now show how to use the topological properties of trees, links, fundamental loops, and fundamental cut sets to prove the validity of the node-voltage, loop-current, and mesh-current methods of analysis. In Section 5.2 we validate the node-voltage method, and in Section 5.3 we justify the loop-current method. In Section 5.4 we show how the mesh-current method can be used to replace the loop-current method in planar circuits.

5.2 A TOPOLOGICAL APPROACH TO THE NODE-VOLTAGE METHOD

Recall from the introduction to circuit analysis in Chapters 1 and 2 that current and voltage are the variables of primary interest. Therefore, in using the topological properties of a network to derive a set of independent equations that describe the network, we start by assigning a voltage and a current to each branch of the network graph. For convenience, we always assign the direction of the branch current in the direction of the voltage drop across the branch.

Derivation of the node-voltage method is rooted in two fundamental characteristics. First, once the tree-branch voltages are known, the link voltages also are known. Second, an independent set of equations involving the tree-branch voltages can be derived by summing the currents in every fundamental cut set. The cut-set currents must add to zero in accordance with Kirchhoff's current law. These equations are guaranteed to be an independent set because each fundamental cut set introduces a tree branch that is not in any other fundamental cut set. The circuit shown in Fig. 5.11 illustrates these characteristics.

The first step in the topological approach to analysis is to construct the circuit graph and assign a voltage and a current to each branch in the graph. Figure 5.12 presents the graph with the assigned variables for the circuit shown in Fig. 5.11.

The second step is to select a tree. Although there are in general many trees to choose from, we can facilitate circuit analysis by adhering, whenever possible, to the following guidelines.

1. Place branches in a tree if they contain voltage sources or if the branch voltage controls dependent sources.

2. Place branches in a cotree if they contain current sources or if the branch current controls dependent sources.

Hence, we select a tree consisting of the branches d, e, c, and g, as shown in Fig. 5.13. By defining the tree, we also define the fundamental cut sets. Here, the four fundamental cut sets are (1) a,b,$\boxed{d}$; (2) a,b,$\boxed{e}$,h; (3) a,$\boxed{c}$,f,h; and (4) $\boxed{g}$,f,h. For this circuit we know the tree-branch voltage v_d and therefore have only three unknown tree-branch voltages—namely, v_e, v_c, and v_g. Thus, we only have to sum the currents in cut sets (2)–(4). Figure 5.14 depicts these three cut sets.

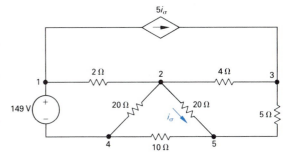

FIGURE 5.11 A circuit illustrating a topological approach to analysis.

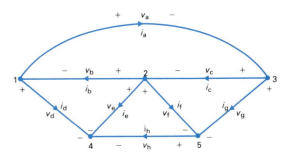

FIGURE 5.12 A graph, with assigned branch variables, for the circuit shown in Fig. 5.11.

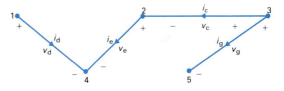
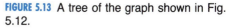

FIGURE 5.13 A tree of the graph shown in Fig. 5.12.

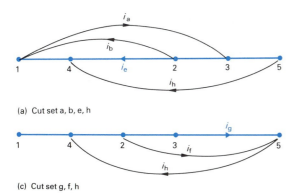

(a) Cut set a, b, e, h

(c) Cut set g, f, h

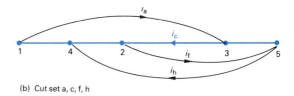

(b) Cut set a, c, f, h

FIGURE 5.14 A graphic representation of the pertinent cut sets.

If we choose the cut-set currents oriented toward the right as positive, then those oriented toward the left are negative. Then, the three cut-set current equations are

$$i_a - i_b - i_e - i_h = 0;$$
$$i_a - i_c + i_f - i_h = 0; \qquad (5.1)$$
$$i_g + i_f - i_h = 0.$$

To find the tree-branch voltages, we simply express the cut-set currents as functions of these voltages. Reference to both the circuit diagram and its graph yields the set of equations:

$$i_a = 5i_\sigma = 5i_f = \frac{5(v_g - v_c)}{20}, \qquad i_b = \frac{v_e - 149}{2},$$

$$i_c = \frac{v_c}{4}, \qquad i_e = \frac{v_e}{20}, \qquad i_f = \frac{v_g - v_c}{20}, \qquad (5.2)$$

$$i_g = \frac{v_g}{5}, \quad and \quad i_h = \frac{v_e + v_c - v_g}{10}.$$

Substituting Eqs. (5.2) into Eqs. (5.1) and multiplying the resulting equations by 20, generates the set of independent equations:

$$7v_g - 7v_c - 13v_e = -1490;$$
$$8v_g - 13v_c - 2v_e = 0; \qquad (5.3)$$
$$7v_g - 3v_c - 2v_e = 0.$$

The solutions for the tree-branch voltages are

$$v_g = 40 \text{ V};$$
$$v_c = 4 \text{ V}; \qquad (5.4)$$
$$v_e = 134 \text{ V}.$$

We leave to you verification that once the tree-branch voltages are known, the current, voltage, and power associated with any branch in the circuit can be found. (See Drill Exercise 5.7.)

DRILL EXERCISES

5.6 a) Derive Eqs. (5.3).

    b) Use Eqs. (5.3) to derive the results given in Eqs. (5.4).

ANSWER: (a) Derivation; (b) see Eqs. (5.4).

5.7 a) Find the power associated with each branch of the circuit shown in Fig. 5.11.

b) Verify that the power generated is equal to the power absorbed.

ANSWER: (a) $P_a = 99$ W; $P_b = 112.50$ W; $P_c = 4.0$ W; $P_d = -2458.50$ W; $P_e = 897.80$ W; $P_f = 64.80$ W; $P_g = 320$ W; $P_h = 960.40$; (b) $P_{gen} = P_{abs} = 2458.50$ W.

The transition from tree-branch voltages to node voltages is based on the fact that, if one node is selected as a reference, the resulting $n - 1$ node voltages always are expressible in terms of $n - 1$ tree-branch voltages. The reason is that, by definition, every tree consists of a set of branches that connects all the nodes without forming loops. We have already shown that the $n - 1$ tree-branch voltages constitute an independent set of variables that can be used to describe a circuit; thus the node voltages also form an independent set of $n - 1$ variables. The reason the node voltages are chosen more often than the tree-branch voltages is that they can be identified without specifying a tree.

For example, in the circuit shown in Fig. 5.11, we arbitrarily chose node 4 as the reference and let v_1, v_2, v_3, and v_5 denote the resulting node voltages. If we now select the tree depicted in Fig. 5.13, the node voltages as functions of the tree-branch voltages are

$$v_1 = v_d, \qquad v_2 = v_e,$$
$$v_3 = v_e + v_c, \qquad \text{and} \qquad v_5 = v_e + v_c - v_g. \tag{5.5}$$

Because v_c, v_d, v_e, and v_g are a set of independent variables, so are v_1, v_2, v_3, and v_5.

Now suppose that we retain the same reference node but change the tree to consist of the branches d, h, f, and c. Then, from the graph shown in Fig. 5.12,

$$v_1 = v_d, \qquad v_2 = v_h + v_f,$$
$$v_3 = v_h + v_f + v_c, \qquad \text{and} \qquad v_5 = v_h. \tag{5.6}$$

But v_c, v_d, v_f, and v_h are independent variables because they are a set of tree-branch voltages. Thus, as before, we conclude that v_1, v_2, v_3, and v_5 form a set of independent variables.

We conclude this introductory discussion of the topological approach to the node-voltage method with the following summary.

1. Once the tree-branch voltages are determined, the link voltages can be calculated, and therefore every branch voltage is known.

2. The tree-branch voltages form a set of independent variables that can be used to describe the circuit.

3. A set of independent equations involving the tree-branch voltages can be derived by summing the branch currents in the $n - 1$ fundamental cut sets in accordance with Kirchhoff's current law.

4. Once a reference node has been selected, the $n - 1$ node voltages can always be expressed as combinations of $n - 1$ tree-branch voltages, and therefore the node voltages can also be used as a set of independent variables to describe a circuit.

5. Node voltages are used more often than tree-branch voltages because they can be defined without reference to a specific tree.

DRILL EXERCISES

5.8 a) Use node 4 as the reference node in the circuit shown and calculate the numerical values of the node voltages v_1, v_2, v_3, and v_5 by using the node-voltage method.

   b) Check the answers to (a) by using the numerical values of the tree-branch voltages v_c, v_d, v_e, and v_g obtained in Drill Exercise 5.6(b).

ANSWER: (a) $v_1 = 149$ V; $v_2 = 134$ V; $v_3 = 138$ V; $v_5 = 98$ V; (b) $v_1 = v_d$, $v_2 = v_e$, $v_3 = v_e + v_c$, $v_5 = v_e + v_c - v_g$.

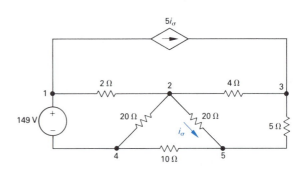

5.3 A TOPOLOGICAL APPROACH TO THE LOOP-CURRENT METHOD

The loop-current method is valid for both planar and nonplanar circuits. After introducing this method, we illustrate with a numerical example how to use it to analyze a nonplanar circuit. Then in Section 5.4 we show how the mesh-current method evolves from the loop-current method when the circuit is planar.

The key to the loop-current method is to recognize that, if we know the $b' - n + 1$ link currents, we can calculate the $n - 1$ tree-branch currents from the link currents simply by summing the currents in the $n - 1$ fundamental cut sets. Thus the $b' - n + 1$ link currents form a set of independent variables that we can use to describe a circuit. We derive a set of independent equations, using the link currents as variables, by defining a loop current for each of the $b' - n + 1$ fundamental loops in the circuit and then summing the voltages around each loop in accordance with Kirchhoff's voltage law. Note that, as each fundamental loop contains only one link, each loop current is identical to a link current.

We use the nonplanar circuit in Fig. 5.15 to illustrate the loop-current method. The circuit contains five nodes and ten branches, so we have to derive and solve six $(10 - 5 + 1)$ simultaneous equations.

Figure 5.16 shows a graph of the circuit, with the nodes and branches labeled to aid the discussion. Next, we select a tree and at the same time assign reference directions for all the branch currents. Figure 5.17 shows the results of these decisions. Note that the tree branches are g, h, j, and k and that the six link currents are i_a–i_f.

Now that we have selected the tree we can identify the six fundamental loops. Recall that each loop current coincides with a link current. The six fundamental loops are:

Loop	Branches
a	a, h, g
b	b, k, j, h, g
c	c, k, j
d	d, h, j
e	e, h, j, k
f	f, j, h, g

If we attempted to draw the six loop currents on the graph shown in Fig. 5.17, the resulting figure would be difficult to decipher. However, the information of primary interest at this point is the identification and orientation of the loop currents in each tree branch. Hence, we depict only this information in Fig. 5.18, where each tree branch is isolated and the loop currents, along with their directions, are shown next to the branch.

We derived the following set of loop-current equations by summing the voltages around each fundamental loop. Each equation starts with the voltage across the link and then proceeds around the fundamental loop containing that link. The tracing direction corresponds to the direction of the link current. The

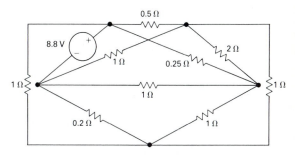

FIGURE 5.15 A nonplanar circuit illustrating the loop-current method.

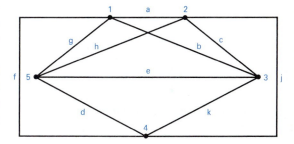

FIGURE 5.16 A labeled graph of the circuit shown in Fig. 5.15.

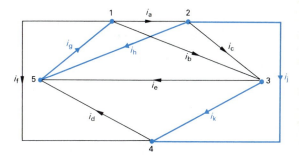
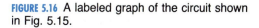

FIGURE 5.17 The graph shown in Fig. 5.16 with the selected tree and branch current references.

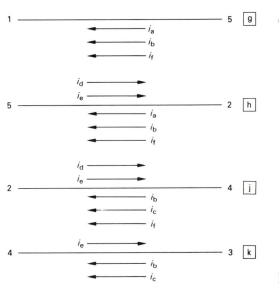

FIGURE 5.18 The identification and orientation of the loop currents in each tree branch.

equations start with loop a and proceed in alphabetical order:

$$0 = 0.5i_a + 1(i_a + i_b + i_f - i_d - i_e) - 8.8;$$

$$0 = 0.25i_b + 1(i_b + i_c - i_e) + 1(i_b + i_c + i_f - i_d - i_e)$$
$$+ 1(i_a + i_b + i_f - i_d - i_e) - 8.8;$$

$$0 = 2i_c + 1(i_b + i_c - i_e) + 1(i_b + i_c + i_f - i_d - i_e);$$

$$0 = 0.2i_d + 1(i_d + i_e - i_a - i_b - i_f)$$
$$+ 1(i_d + i_e - i_b - i_c - i_f); \qquad (5.7)$$

$$0 = 1i_e + 1(i_d + i_e - i_a - i_b - i_f)$$
$$+ 1(i_d + i_e - i_b - i_c - i_f) + 1(i_e - i_b - i_c);$$

$$0 = 1i_f + 1(i_b + i_c + i_f - i_d - i_e)$$
$$+ 1(i_a + i_b + i_f - i_d - i_e) - 8.8.$$

Rearranging Eqs. (5.7) to facilitate solution by a computer gives

$$8.8 = 1.5i_a + i_b + 0i_c - i_d - i_e + i_f;$$

$$8.8 = i_a + 3.25i_b + 2i_c - 2i_d - 3i_e + 2i_f;$$

$$0 = 0i_a + 2i_b + 4i_c - i_d - 2i_e + i_f;$$

$$0 = i_a - 2i_b - i_c + 2.2i_d + 2i_e - 2i_f; \qquad (5.8)$$

$$0 = -i_a - 3i_b - 2i_c + 2i_d + 4i_e - 2i_f;$$

$$8.8 = i_a + 2i_b + i_c - 2i_d - 2i_e + 3i_f.$$

Solving for the six loop (link) currents yields

$$i_a = 7.28 \text{ A};$$

$$i_b = 10.4 \text{ A};$$

$$i_c = -0.52 \text{ A};$$

$$i_d = 12.6 \text{ A}; \qquad (5.9)$$

$$i_e = 6.20 \text{ A};$$

$$i_f = 6.28 \text{ A}.$$

To find the tree-branch currents, we sum the currents in each fundamental cut set. The four fundamental cut sets are (1) $\boxed{g}$, b, a, f; (2) $\boxed{h}$, a, b, e, d, f; (3) $\boxed{j}$, c, b, e, d, f; and (4) $\boxed{k}$, e, b, c. The cut-set equations then are

$$i_g = i_a + i_b + i_f;$$

$$i_h = i_a + i_b + i_f - i_d - i_e;$$

$$i_j = i_d + i_e - i_b - i_c - i_f; \qquad (5.10)$$

$$i_k = i_b + i_c - i_e.$$

The numerical values of the tree-branch currents are

$$i_g = 23.96 \text{ A};$$

$$i_h = 5.16 \text{ A};$$

$$i_j = 2.64 \text{ A};$$

$$i_k = 3.68 \text{ A}.$$

(5.11)

Now that we know all the branch currents, we have enough information to calculate any branch voltage or power that may be of interest.

DRILL EXERCISES

5.9 Test the validity of the solution for the circuit shown in Fig. 5.15 by showing that the total power dissipated in the circuit equals the total power generated by the voltage source.

ANSWER: $P_{dis} = P_{gen} = 210.848$ W.

5.10 In the circuit shown in Fig. 5.15, v_g is increased to 10 V, and all the resistors are set equal to 1 Ω. Use the loop-current method to find

a) the branch currents and

b) the total power dissipated in the circuit.

ANSWER: a) All the link currents equal 5 A except i_c, which is zero, and the tree-branch currents are $i_g = 15$ A, $i_h = 5$ A, $i_j = i_k = 0$; b) 150 W.

5.4 THE MESH-CURRENT METHOD (PLANAR CIRCUITS)

Having demonstrated the application of the loop-current method of circuit analysis, we now show that in planar circuits the mesh-current method can be used to replace the loop-current method. For planar circuits where every "window" forms part of the boundary[†] of the graph, the mesh-current method is equivalent to the loop-current method because we can always construct a tree so that the cotree is the set of perimeter links. Then the fundamental loops and meshes are identical.

For example, consider the graph shown in Fig. 5.19 where, as before, the heavy lines represent the tree branches. Note that the

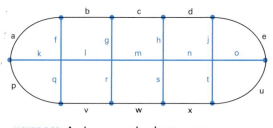

FIGURE 5.19 A planar graph where every "window" forms part of the boundary.

[†] The *boundary* of a planar graph is made up of the branches that form the perimeter of the graph. For example, in Fig. 5.12, the boundary consists of the branches a, g, h, and d.

meshes are identical to the fundamental loops. Because the number of fundamental loops is $b' - n + 1$, the number of meshes also equals $b' - n + 1$.

DRILL EXERCISES

5.11 Refer to the circuit graph shown in Fig. 5.19.

a) State how many nodes, branches, links, tree branches, fundamental cut sets, fundamental loops, and meshes it contains.

b) Identify the perimeter branches, tree branches, fundamental cut sets, fundamental loops, and meshes.

ANSWER: (a) 14 nodes, 23 branches, 10 links, 13 tree branches, 13 fundamental cut sets, 10 fundamental loops,

and 10 meshes; (b) a, b, c, d, e, u, x, w, v, and p; f, g, h, j, k, l, m, n, o, q, r, s, and t;

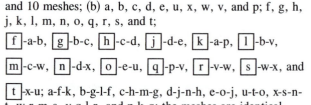

$\boxed{f}$-a-b, $\boxed{g}$-b-c, $\boxed{h}$-c-d, $\boxed{j}$-d-e, $\boxed{k}$-a-p, $\boxed{l}$-b-v,

$\boxed{m}$-c-w, $\boxed{n}$-d-x, $\boxed{o}$-e-u, $\boxed{q}$-p-v, $\boxed{r}$-v-w, $\boxed{s}$-w-x, and

$\boxed{t}$-x-u; a-f-k, b-g-l-f, c-h-m-g, d-j-n-h, e-o-j, u-t-o, x-s-n-t, w-r-m-s, v-q-l-r, and p-k-q; the meshes are identical with the fundamental loops.

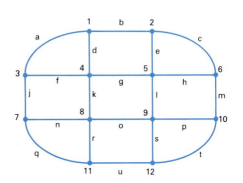

FIGURE 5.20 A planar graph with one interior "window".

What happens if the circuit graph has one or more interior "windows"? We show (1) that the number of meshes still equals $b' - n + 1$ and (2) that the mesh equations form an independent set. To aid the discussion of item (1), we use the circuit graph shown in Fig. 5.20, which has one interior "window". The graph has 20 branches and 12 nodes, so we expect to count $20 - 12 + 1$, or 9, meshes. The graph does indeed contain nine meshes.

We deduce that the number of meshes always equals the number of links by reasoning as follows. First, we select a tree and then add the links to the graph one at a time. Each time we add a link, it either forms a new loop or divides an existing loop in two. Hence, the addition of each link corresponds to adding a new loop to the graph. After the last link has been added, the loops correspond to the windows or meshes of the graph. That is, the number of meshes equals the number of links. Figure 5.21 shows the result of adding the first five links in this process for the graph shown in Fig. 5.20. Here, we selected branches j, a, d, k, r, u, s, l, e, c, and m to form the tree.

Now that we have established that the number of meshes in a planar circuit equals $b' - n + 1$, the remaining task is to show that the system of mesh equations is independent. Recall from algebra that a set of linear equations is *independent* if no equation in the set can be derived from a linear combination of the remaining equations. If one of the equations in the set can be derived from a linear combination of the remaining, we say

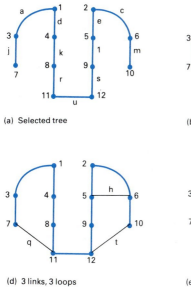

(a) Selected tree

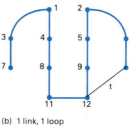

(b) 1 link, 1 loop

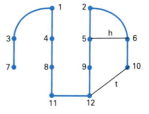

(c) 2 links, 2 loops

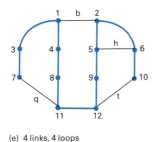

(d) 3 links, 3 loops

(e) 4 links, 4 loops

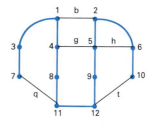

(f) 5 links, 5 loops

FIGURE 5.21 An illustration showing how the addition of a link adds a loop to the graph.

the equations are *dependent*. We can state this criterion for dependence in a set of mesh equations as follows. Let m_1, $m_2, \ldots, (m'_b - n + 1)$ represent each mesh equation in the set of equations. Let $c_1, c_2, \ldots, (c'_b - n + 1)$ be a set of constants, not all of which are zero. Then, if the equations are *dependent*, we can find values for the c's so that

$$c_1 m_1 + c_2 m_2 + \cdots + (c'_b - n + 1)(m'_b - n + 1) = 0. \quad (5.12)$$

If the sum in Eq. (5.12) cannot be made equal to zero, the equations are *independent*.

To show that the mesh equations form an independent set, we assume that the equations satisfy Eq. (5.12) and then proceed to show that we always get a contradiction. Before proceeding to the general case, let's consider a circuit with no interior meshes, because in this case verifying that the mesh equations form an independent set is easy. In the circuit graph shown in Fig. 5.22, we selected a tree so that each mesh contains a perimeter branch. The branch voltages with their reference polarities also are given.

If we sum clockwise around each mesh and assign a plus sign to a voltage drop, we get four mesh equations:

$$m_1: \quad -v_a + v_f - v_e = 0;$$

$$m_2: \quad v_b + v_g - v_f = 0;$$

$$m_3: \quad v_c - v_h - v_g = 0; \quad (5.13)$$

$$m_4: \quad -v_d + v_e + v_h = 0.$$

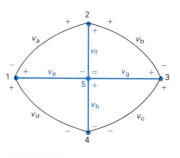

FIGURE 5.22 A circuit graph.

Note that each equation contains a voltage that does not appear in any other equation. The reason is that each mesh is identical to a fundamental loop and hence, contains only one link voltage. Hence, we can never find a set of constants c_1, c_2, c_3, and c_4, of which some are not zero, to satisfy Eq. (5.12). Therefore the mesh equations in Eqs. (5.13) are independent.

We now return to the general case where the circuit may contain one or more interior meshes. First, we acknowledge that some, but not all, of the constants c_1 to $(c_b' - n + 1)$ may be zero. For purposes of discussion, we number the meshes so that the first k constants—that is, c_1 to c_k—are not zero. Then Eq. (5.12) reduces to

$$c_1 m_1 + c_2 m_2 + \cdots + c_k m_k = 0. \tag{5.14}$$

We now show that Eq. (5.14) always leads to contradiction, reasoning as follows. By hypothesis, we can eliminate summing the voltages around the meshes from $k + 1$ to $b' - n + 1$. In this group of meshes, at least one mesh always has a branch common with at least one mesh in the group 1 to k. If we denote the voltage across this common branch v_x, in the 1 to k mesh equations only one equation will contain v_x. Hence, finding a set of nonzero coefficients (c_1 to c_k) to eliminate this voltage is impossible. Therefore Eq. (5.14) cannot be satisfied, and the set of $b' - n + 1$ mesh equations must be independent.

For example, consider the circuit graph shown in Fig. 5.23 and assume that c_1–c_9 are nonzero constants and that $c_{10} = 0$. Therefore we do not sum around mesh 10. Hence, v_x appears only in equation m_6, making impossible either

$$\sum_{k=1}^{9} c_k m_k = 0 \tag{5.15}$$

or

$$\sum_{k=1}^{10} c_k m_k = 0. \tag{5.16}$$

Therefore the 10 mesh equations are independent.

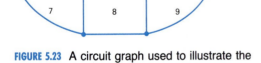

FIGURE 5.23 A circuit graph used to illustrate the independence of the mesh equations.

DRILL EXERCISES

5.12 Are the following three equations independent?   **ANSWER:** No.

$$6v_1 - 3v_2 + 4v_3 = 15. \qquad 2v_1 + 8v_2 - 10v_3 = 0.$$

$$-4v_1 + 11v_2 - 14v_3 = -15.$$

We summarize the topological approach to the loop-current and mesh-current methods of analysis as follows.

1. The loop-current method is based on summing the voltages around each fundamental loop of the circuit in accordance with Kirchhoff's voltage law.

2. The loop-current method generates a set of independent equations because each fundamental loop contains a link-branch voltage that appears in no other fundamental loop.

3. The loop-current method can be used for either planar or nonplanar circuits.

4. If a circuit is planar, the loop-current method can be replaced with the simpler mesh-current method.

5. The mesh-current method is simpler because in general identifying the meshes in a circuit is easier than identifying the fundamental loops.

6. The number of meshes in a circuit equals the number of links, that is, $b' - n + 1$.

7. The $b' - n + 1$ mesh equations are also independent.

SUMMARY

Topological circuit properties allow verification of the independence of the node-voltage and mesh-current equations. The key topological concepts introduced are:

- *tree,* which is a set of branches connecting all nodes without forming any closed loops, but the set is not unique because a circuit may have many trees;

- *link* (or chord), which is a branch not in a tree;

- *cotree,* which is the set of links;

- *cut set,* which is the *smallest* set of branches that, if removed from a graph, cuts the graph into two subgraphs (a graph can have many cut sets);

- *fundamental cut set,* which is a cut set containing a single tree branch; and

- *fundamental loop,* which is a closed loop containing exactly one link.

The topological properties of a circuit were used to justify both the node-voltage and mesh-current methods.

- The topological approach to the node-voltage method is summarized at the end of Section 5.2.

• The topological approach to the loop-current and mesh-current methods is summarized at the end of Section 5.4.

Three key points to remember about the topological approach are:

• node voltages are used in lieu of tree-branch voltages, because they can be defined without reference to a specific tree;

• mesh currents can be defined only in planar circuits; and

• in planar circuits mesh currents are used in lieu of link currents, because they can be defined without reference to a specific tree.

PROBLEMS

5.1 Identify three trees in the graph of Fig. P5.1.

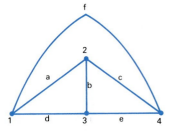

FIGURE P5.1

5.2 Identify eight trees in the graph of Fig. P5.2.

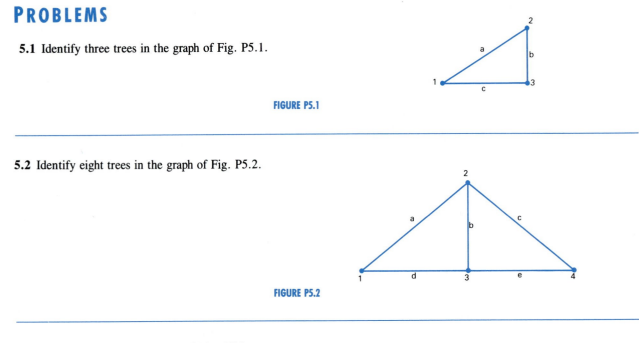

FIGURE P5.2

5.3 Specify 16 trees in the graph of Fig. P5.3.

FIGURE P5.3

5.4 Assume v_g, i_g, β, μ, α, ρ, and all the resistance values are known in the circuit in Fig. P5.4.

a) Construct a graph for the circuit.

b) Using the guidelines enumerated in Section 5.2 draw 10 different trees you could use to analyze the circuit.

c) How many unknown tree-branch voltages are there?

d) How many unknown link-currents are there?

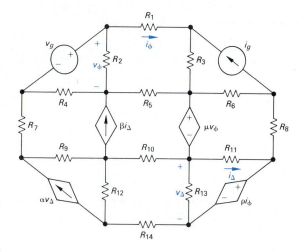

FIGURE P5.4

5.5 a) Construct a graph for the circuit shown in Fig. P5.5.

b) If you follow the guidelines enumerated in Sec. 5.2 how many different trees could you choose from to analyze the circuit?

c) How many fundamental cut-set equations must be solved to find the unknown tree-branch voltages?

d) Use the tree-branch voltage method to find the power absorbed by the 45-V source.

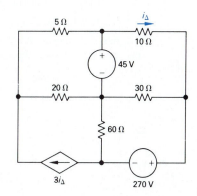

FIGURE P5.5

5.6 Use the tree-branch-voltage method to find v_1 and v_2 in the circuit in Fig. P5.6.

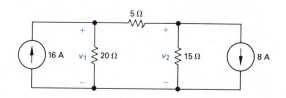

FIGURE P5.6

5.7 A graph for the circuit in Fig. P5.7(a) is shown in Fig. P5.7(b). The branch currents and voltages are defined on the graph.

a) Select a tree.

b) Using the tree selected in part (a) write the fundamental cut-set equations in terms of the branch currents.

c) Find the tree-branch voltages.

d) Find the total power developed in the circuit.

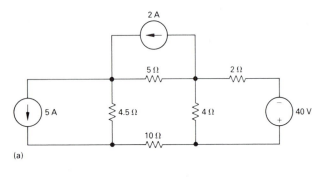

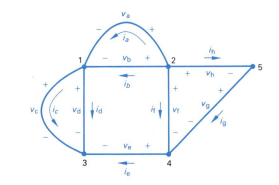

(a)   (b)

FIGURE P5.7

5.8 a) Draw a graph of the circuit in Fig. P5.8.

b) Select a tree for the graph.

c) Use the tree-branch-voltage method to find the branch currents i_1 through i_6.

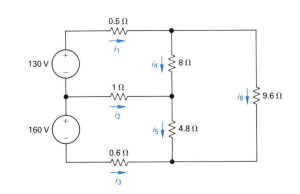

FIGURE P5.8

5.9 a) Use the tree-branch-voltage method of analysis to find v_o in the circuit in Fig. P5.9.

b) How much power does the dependent current source deliver to the circuit?

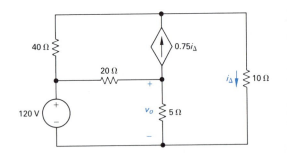

FIGURE P5.9

5.10 Use the tree-branch-voltage method to find i_Δ in the circuit in Fig. P5.10.

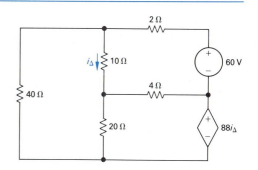

FIGURE P5.10

FIGURE P5.10

5.11 Use the tree-branch-voltage method to find v_Δ in the circuit in Fig. P5.11.

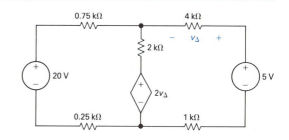

FIGURE P5.11

5.12 Use the tree-branch-voltage method to find the total power developed in the circuit in Fig. P5.12.

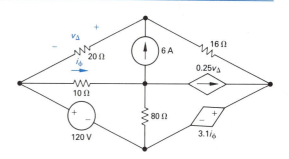

FIGURE P5.12

5.13 Use the loop-current method to find the branch currents i_1 through i_6 in the circuit in Fig. P5.8.

5.14 a) Draw a graph for the circuit in Fig. P5.14 and then select a tree using the guidelines given in Sec. 5.2.

   b) Use the loop-current method to find the voltage v_a.

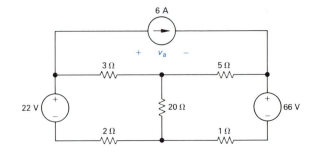

FIGURE 5.14

5.15 a) Construct a graph for the circuit in Fig. P5.15.

   b) How many different trees are there if you use the guidelines set forth in Sec. 5.2?

   c) Select one of the trees noted in part (b). Use the loop-current method to find the current i_ϕ.

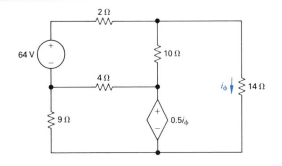

FIGURE P5.15

5.16 Use the loop-current method to find the total power developed in the circuit in Fig. P5.16.

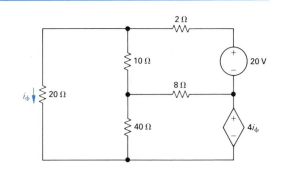

FIGURE P5.16

5.17 Use the tree-branch-voltage method to find the total power developed in the circuit in Fig. P5.16.

5.18 Use the loop-current method to find the total power developed in the circuit in Fig. P5.18.

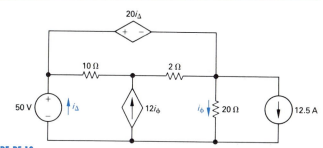

FIGURE P5.18

5.19 Use the loop-current method to find the power delivered by the independent voltage source in the circuit in Fig. P5.19 when v_g is 193.5 V.

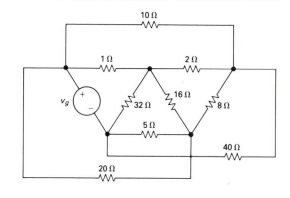

FIGURE P5.19

5.20 In Problem 5.19 you were told to use the loop-current method to find the power delivered by the independent voltage source. Now choose your own method of analysis to find the power developed by the independent voltage source.

5.21 a) Using the recommendations enumerated in Section 5.2 of the text, select a tree for the circuit shown in Fig. P5.21.

   b) Write the loop-current equations corresponding to the tree selected in part (a).

   c) Use the solution of the equations in part (b) to find the total power dissipated in the circuit.

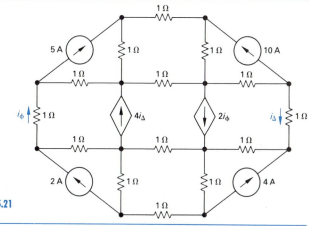

FIGURE P5.21

5.22 a) Use the mesh-current method to find the total power dissipated in the circuit shown in Fig. P5.21.

   b) Comment on the advantages the mesh-current method has over the loop-current method when analyzing planar circuits.

THE OPERATIONAL AMPLIFIER

CHAPTER 6

The electronic circuit known as an operational amplifier has become increasingly important. However, a detailed analysis of this circuit requires an understanding of electronic devices such as diodes and transistors. You may wonder, then, why we are introducing the circuit before discussing the circuit's electronic components. There are several reasons. First, circuit theory is important because it enables you to design and analyze electrical systems that have important applications and hence commercial value. Second, you can develop an appreciation for how the operational amplifier can be used as a circuit building block by focusing on its terminal behavior. Thus at an introductory level you need not fully understand the operation of the electronic components that govern terminal behavior. Third, the circuit model of the operational amplifier requires the use of a dependent source. Thus you have a chance to use a dependent source in a practical circuit rather than as an abstract circuit component. Fourth, you can combine the operational amplifier with resistors to perform some very useful functions, such as scaling, summing, sign changing, and subtracting. Fifth, after introducing inductors and capacitors in Chapter 7 we show you how to use the operational amplifier to design integrating and differentiating circuits. Finally, the operational amplifier plays an important role in the design of the digital multimeter. In order to get some feel for this important laboratory instrument, you must understand the terminal behavior of the operational amplifier.

Our focus on the terminal behavior of the operational amplifier implies taking a "black-box" approach to its operation; that is, we are not interested in the internal structure of the amplifier nor the currents and voltages that exist in this internal structure. The important thing to remember is that the internal behavior of the

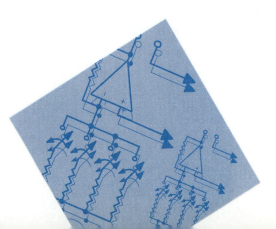

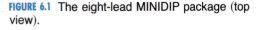 PSpice provides several options for analysis of circuits containing op amps: Chapter 5

amplifier accounts for the voltage and current constraints imposed at the terminals. (For now, we ask that you accept these constraints on faith.)

The operational-amplifier circuit first came into existence as a basic building block in the design of analog computers. It was referred to as *operational* because it was used to implement the mathematical operations of integration, differentiation, addition, sign changing, and scaling. In recent years, the range of application has broadened beyond implementing mathematical operations; however, the original name for the circuit persists. Engineers and technicians have a penchant for creating technical jargon; hence the operational amplifier is widely known as the *op amp*.

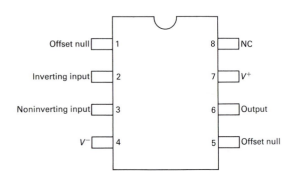 PSpice circuit element libraries contain models for actual components, such as the μA741, which can be included in PSpice circuit analysis: Section 5.3

6.1 OPERATIONAL-AMPLIFIER TERMINALS

Because we are stressing the terminal behavior of the operational amplifier, we begin by discussing the terminals on a commercially available amplifier. In 1968, Fairchild Semiconductor introduced an operational amplifier that has found widespread acceptance: the μA741. (The μA prefix is used by Fairchild to indicate a microcircuit fabrication of the amplifier.) The amplifier is available in several different packages. For our discussion, we assume an eight-lead MINIDIP[†] package. Figure 6.1 shows a top view of the package, with the terminal designations given alongside the terminals. The terminals of primary interest are

1. the inverting input,
2. the noninverting input,
3. the output,
4. the positive power supply (V^+), and
5. the negative power supply (V^-).

The remaining three terminals are of little or no concern. The offset null terminals may be used in an auxiliary circuit to compensate for a degradation in amplifier performance because of aging and imperfections that occur in the circuit during its fabri-

FIGURE 6.1 The eight-lead MINIDIP package (top view).

[†] DIP is an abbreviation for *d*ual *i*n-line *p*ackage. This means that the terminals on each side of the package are in line, and at the same time, terminals on opposite sides of the package also line up.

cation. However, the degradation of performance in most cases is negligible, so the offset terminals often are unused and play a secondary role in circuit analysis. We assume that operational amplifiers require no trimming. Terminal 8 is of no interest simply because it is an unused terminal: NC stands for no connection, which means that the terminal is not connected to the amplifier circuit. (When the μA741 is packaged in a 14-lead DIP package, there are seven unused terminals.)

Figure 6.2 shows a widely used circuit symbol for the operational amplifier that contains the five terminals of primary interest. Using word labels for the terminals is inconvenient when the operational amplifier is embedded in a circuit, so we simplify the terminal designations in the following way. The noninverting input terminal is labeled plus ($+$), and the inverting terminal is labeled minus ($-$). The power supply terminals, which are always drawn outside the triangle, are marked V^+ and V^-. The terminal at the apex of the triangular box always is understood to be the output terminal. Figure 6.3 summarizes these simplified designations.

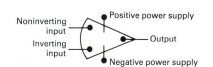

FIGURE 6.2 The circuit symbol for an operational amplifier.

FIGURE 6.3 A simplified circuit symbol for the op amp.

6.2 **TERMINAL VOLTAGES AND CURRENTS**

We are now ready to introduce the terminal voltages and currents used to describe the behavior of the operational amplifier, beginning with the voltage variables. Those used to describe the behavior of an operational amplifier are measured from a common reference node.[†] Figure 6.4 shows the voltage variables with their *reference* polarities. All voltages are considered as voltage rises from the common node. This convention is the same as that used in the node-voltage method of analysis. A positive supply voltage (V_{CC}) is connected between V^+ and the common node. A negative supply voltage ($-V_{CC}$) is connected between V^- and the common node. The voltage between the inverting input terminal and the common node is denoted v_1. The voltage between the noninverting input terminal and the common node is designated as v_2. The voltage between the output terminal and the common node is denoted by v_o.

Figure 6.5 shows the current variables with their *reference* directions. Note that all the current reference directions are into

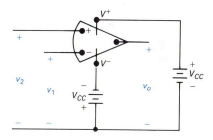

FIGURE 6.4 Terminal voltage variables.

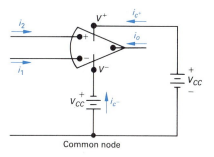

FIGURE 6.5 Terminal current variables.

[†] The reference node is external to the operational amplifier. It is the reference terminal of the circuit in which the operational amplifier is embedded.

the terminals of the operational amplifier: i_1 is the current into the inverting input terminal; i_2 is the current into the noninverting input terminal; i_o is the current into the output terminal; i_{c^+} is the current into the positive power supply terminal; and i_{c^-} is the current into the negative power supply terminal.

In order to predict the behavior of an operational amplifier when circuit elements are externally connected to its terminals, we must recognize the constraints imposed on the terminal voltages and currents by the amplifier itself. Those imposed on the terminal voltages are

$$v_o = A(v_2 - v_1) \qquad (6.1)$$

and

$$-V_{CC} \leq v_o \leq V_{CC}. \qquad (6.2)$$

Equation (6.1) states that the output voltage is proportional to the difference between v_2 and v_1. The proportionality constant A is known as the *open-loop voltage gain*. (We explain the significance of the term *open loop* later.) Equation (6.2) states that the output voltage is bounded. In particular, v_o must lie between $\pm V_{CC}$, the power supply voltages. If v_o is at either limiting value, we say that the operational amplifier is saturated. The amplifier is operating in its linear range so long as $v_o < |V_{CC}|$. The graph shown in Fig. 6.6 summarizes the significance of Eqs. (6.1) and (6.2). Note that the x-axis variable is $v_2 - v_1$, that is, the difference between the two input voltages.

The importance of the voltage constraints lies in knowing the typical numerical values of V_{CC} and A. The dc power supply voltages seldom exceed 20 V and A is rarely less than 10,000, or 10^4. Figure 6.6 indicates that in the linear range of operation the magnitude of $v_2 - v_1$ must be less than $20/10^4$, or 2 mV. Hence, when the operational amplifier is operating in its linear region,

$$v_1 \approx v_2. \qquad (6.3)$$

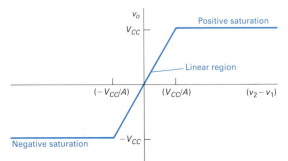

FIGURE 6.6 The voltage transfer characteristic of an op amp.

We soon show that Eq. (6.3) is valuable in predicting the performance of linear op-amp circuits.

The result expressed in Eq. (6.3) is referred to as the *virtual short* condition at the input of the amplifier. The question naturally arises as to how the virtual short is maintained across the input terminals when the operational amplifier is embedded in a circuit. The answer is that a signal is fed back from the output terminal to the *inverting* input terminal. This arrangement is known as *negative* feedback because the signal fed back subtracts from the input signal. Feedback is an important concept that is investigated in depth in introductory electronics and systems analysis courses. Here, we simply point out that, in all circuits where the operational amplifier is not operating open loop, there is a circuit connection between the output and inverting input terminal.

From Kirchhoff's current law we know that the sum of the currents entering the operational amplifier is zero, or

$$i_1 + i_2 + i_o + i_{c^+} + i_{c^-} = 0. \tag{6.4}$$

The constraint imposed on the currents by the operational amplifier is that i_1 and i_2 must be extremely small in comparison to the other terminal currents. In an ideal operational amplifier the input terminal currents are zero. Thus

$$i_1 = i_2 \approx 0, \tag{6.5}$$

which indicates that the input resistance to an operational amplifier is large. Typical values range from hundreds of kilohms to thousands of megohms. Equation (6.5) is useful in analyzing circuits containing operational amplifiers.

Substituting the constraint given by Eq. (6.5) into Eq. (6.4) gives

$$i_o = -(i_{c^+} + i_{c^-}). \tag{6.6}$$

The significance of Eq. (6.6) is that, even though the current at the input terminals is negligible, there may still be appreciable current at the output terminal.

Before we start analyzing circuits containing operational amplifiers, let's further simplify the circuit symbol. The first step is to remove the dc power supplies. We can do so simply by indicating the power supply voltage next to the appropriate terminal. Figure 6.7 shows the simplified symbol.

We can simplify further when we know that the amplifier is operating within its linear range. In this situation, the dc voltages $\pm V_{CC}$ do not enter into the circuit equations, with the implicit understanding that $v_o < |V_{CC}|$. For linear operation of the op amp, we can remove the power supply terminals from the symbol, as shown in Fig. 6.8. A word of caution: Because the power supply terminals have been omitted, there is a danger of inferring from the symbol that $i_1 + i_2 + i_o = 0$. We have already noted that such is not the case; that is, $i_1 + i_2 + i_o + i_{c^+} + i_{c^-} = 0$. In other words, the constraint that $i_1 = i_2 \cong 0$ does not imply that $i_o \cong 0$.

A couple of additional comments are in order. The positive and negative power supply voltages do not have to be equal in magnitude. In the linear operating range, v_o must lie between the two supply voltages. For example, if $V^+ = 15$ V and $V^- = -10$ V, then -10 V $\leq v_o \leq 15$ V. The open-loop voltage gain, A, is not constant under all operating conditions. For now, however, we assume that it is. A discussion of how and why the open-loop gain A can change must be delayed until after you

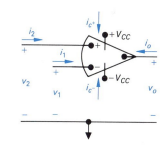

FIGURE 6.7 The op-amp symbol with power supplies removed.

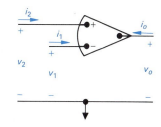

FIGURE 6.8 The op-amp symbol with the power supply terminals removed.

have studied the electronic devices and components that are used to fabricate the amplifier.

Before starting to analyze circuits that contain operational amplifiers, we must address the question, "How do we know whether the operational amplifier is operating in its linear region?" The answer is, "We don't!" We get around this dilemma by assuming linear operation and then, based on our results, go back and check the circuit to see if there is a contradiction. For example, if we assume linear operation and calculate v_o to equal 10 V and then check and see that V_{CC} is 6 V, we know immediately that the assumption of linear operation is wrong. Thus we conclude that the operational amplifier is saturated at 6 V and that the output can't be 10 V, as predicted by the linear model.

PSpice PSpice can be used to model an ideal op amp in a circuit: Section 5.1

Example 6.1 illustrates analysis of a circuit containing an operational amplifier by judicious application of Eqs. (6.3) and (6.5). When we use these equations to predict the behavior of a circuit containing an operational amplifier, in effect we are using an idealized model of the amplifier.

E X A M P L E   6.1

The operational amplifier in the circuit shown in Fig. 6.9 is ideal.

a) Calculate v_o if $v_a = 1$ V and $v_b = 0$ V.

b) Repeat (a) for $v_a = 1$ V and $v_b = 2$ V.

c) If $v_a = 1.5$ V, specify the range of v_b that avoids amplifier saturation.

S O L U T I O N

a) Because $v_2 = v_b$ by virtue of the external connection and $v_1 = v_2$ because of the constraint imposed by the operational amplifier, we have $v_1 = 0$. Therefore the current in the 25-kΩ resistor from left to right is $\frac{1}{25}$ mA. The current in the 100-kΩ resistor from right to left will be $v_o/100$ mA. If we apply Kirchhoff's current law to the inverting input junction, recalling that the current into the operational amplifier is negligible, we obtain

$$\frac{1}{25} + \frac{v_o}{100} = 0.$$

Hence, v_o is -4 V. Note that as v_o lies between ±10 V, the operational amplifier is in its linear range of operation.

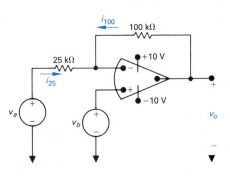

FIGURE 6.9 The circuit for Example 6.1.

b) Using the same thought process as in (a), we get

$$v_2 = v_b = v_1 = 2 \text{ V};$$

$$i_{25} = \frac{v_a - v_1}{25} = \frac{1 - 2}{25} = -\frac{1}{25}\text{mA};$$

$$i_{100} = \frac{v_o - v_1}{100} = \frac{v_o - 2}{100}\text{mA};$$

$$i_{25} = -i_{100}.$$

Therefore $v_o = 6$ V. Again, v_o lies within ± 10 V.

c) As before, $v_1 = v_2 = v_b$ and $i_{25} = -i_{100}$. Because $v_a = 1.5$ V,

$$\frac{1.5 - v_b}{25} = -\frac{v_o - v_b}{100}.$$

Solving for v_b as a function of v_o gives

$$v_b = \frac{1}{5}(6 + v_o).$$

Now, if the amplifier is to be within the linear range of operation, $-10 \le v_o \le 10$ V. Substituting these limits on v_o into the expression for v_b, we see that v_b is limited to

$$-0.8 \text{ V} \le v_b \le 3.2 \text{ V}.$$

We are now ready to discuss the operation of some important operational-amplifier circuits, using Eqs. (6.3) and (6.5) to idealize the behavior of the amplifier itself.

DRILL EXERCISES

6.1 Assume that the operational amplifier in the circuit shown is ideal.

a) Calculate v_o for the following values of v_s: 0.4, 0.72, 2.0, -0.6, -0.8, and -2.0 V.

b) Specify the range of v_s required to avoid amplifier saturation.

ANSWER: (a) -5, -9, -10, 7.5, 10, and 15 V; (b) $-1.2 \text{ V} \le v_s \le 0.8$ V.

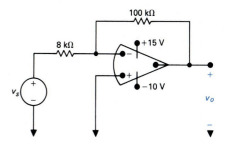

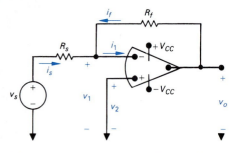

FIGURE 6.10 An inverting-amplifier circuit.

6.3 THE INVERTING-AMPLIFIER CIRCUIT

Figure 6.10 shows the inverting-amplifier circuit. We assume that the operational amplifier is operating in its linear range. Note that the circuit external to the operational amplifier consists of two resistors, R_f and R_s, a voltage signal source, v_s, and a short circuit connected between the noninverting input terminal and the common node.

We now analyze this circuit, assuming an ideal op amp. The goal is to obtain an expression for the output voltage, v_o, as a function of the source voltage, v_s. First, we note that $v_2 = 0$ *because of the external short circuit*. We have already discovered through Eq. (6.3) that the op amp forces v_1 to be nearly equal to v_2 (within a few millivolts if A is at least 10^4). Therefore, for all practical purposes, $v_1 = 0$, so

$$i_s = \frac{v_s}{R_s} \tag{6.7}$$

and

$$i_f = \frac{v_o}{R_f}. \tag{6.8}$$

Now we invoke the constraint stated in Eq. (6.5), namely, that the terminal current i_1 is negligible. Then,

$$i_f = -i_s. \tag{6.9}$$

Substituting Eqs. (6.7) and (6.8) into Eq. (6.9) yields the sought-after result:

$$v_o = \frac{-R_f}{R_s} v_s. \tag{6.10}$$

Note that the output voltage is an inverted (sign reversal), scaled replica of the input. The scaling factor is the ratio R_f/R_s. The minus sign in Eq. (6.10) is, of course, the reason for referring to the circuit as an inverting amplifier.

The result given by Eq. (6.10) is valid only if the op amp shown in the circuit in Fig. 6.10 is ideal, that is, if the open-loop gain, A, is infinite and the input current, i_1, is zero. For a practical op amp, Eq. (6.10) is an approximation. In most cases, the approximation is a good one. (We say more about this later.) Equation (6.10) is important because it tells us that if the open-loop gain is large, we can control the gain of the inverting amplifier with the external resistors R_f and R_s. The upper limit on the gain, R_f/R_s, is determined by the power supply voltage and the value of the signal voltage v_s. If we assume equal power supply voltages, that is, $V^+ = -V^- = V_{CC}$, we get

$$|v_o| < V_{CC};$$

$$\left| \frac{R_f}{R_s} v_s \right| < V_{CC}; \qquad \frac{R_f}{R_s} < \left| \frac{V_{CC}}{v_s} \right|.$$

For example, if $V_{CC} = 15$ V and $v_s = 10$ mV, the ratio R_f/R_s must be less than 1500.

In the inverting amplifier circuit shown in Fig. 6.10, the resistor R_f provides the negative feedback connection. That is, it connects the output terminal to the inverting input terminal. If R_f is removed, the feedback path is opened and the amplifier is said to be operating open loop. Figure 6.11 shows the open-loop operation.

Opening the feedback path drastically changes the behavior of the circuit. First, the output voltage is now

$$v_o = -Av_1, \qquad \textbf{(6.11)}$$

assuming as before that $V^+ = -V^- = V_{CC}$; then $|v_1| < V_{CC}/A$ for linear operation. Because the inverting input current is zero for an ideal op amp, the voltage drop across R_s is zero, and the inverting input voltage equals the signal voltage v_s; that is, $v_1 \approx v_s$. Hence, the amplifier can operate open loop in the linear mode only if $|v_s| < V_{CC}/A$. If $|v_s| > V_{CC}/A$, the op amp simply saturates. In particular, if $v_s < -V_{CC}/A$ the op amp saturates at $+V_{CC}$, and if $v_s > V_{CC}/A$ the op amp saturates at $-V_{CC}$.

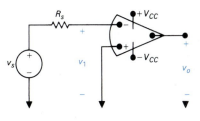

FIGURE 6.11 An inverting amplifier operating open loop.

DRILL EXERCISES

6.2 The source voltage v_s in the circuit in Drill Exercise 6.1 is −640 mV. The 100-kΩ feedback resistor is replaced by a variable resistor R_x. What range of R_x allows the inverting amplifier to operate in its linear range?

ANSWER: $0 \le R_x \le 187.5$ kΩ.

6.4 THE SUMMING-AMPLIFIER CIRCUIT

The output voltage of a summing amplifier is an inverted, scaled sum of the voltages applied to the input of the amplifier. Figure 6.12 shows a summing amplifier with three input voltages.

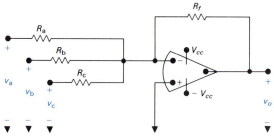

FIGURE 6.12 A summing amplifier.

We obtain the relationship between the output voltage v_o and the three input voltages, v_a, v_b, and v_c, by summing the currents away from the inverting input terminal:

$$\frac{v_1 - v_a}{R_a} + \frac{v_1 - v_b}{R_b} + \frac{v_1 - v_c}{R_c} + \frac{v_1 - v_o}{R_f} + i_1 = 0. \tag{6.12}$$

Assuming an ideal operational amplifier, we can use the constraints that $v_1 = v_2 = 0$ and $i_1 = 0$ to reduce Eq. (6.12) to

$$v_o = -\left(\frac{R_f}{R_a} v_a + \frac{R_f}{R_b} v_b + \frac{R_f}{R_c} v_c\right). \tag{6.13}$$

Equation (6.13) states that the output voltage is an inverted, scaled sum of the three input voltages.

We get a simple scaled sum, with inversion, by making $R_a = R_b = R_c = R_s$. Then Eq. (6.13) reduces to

$$v_o = -\frac{R_f}{R_s}(v_a + v_b + v_c). \tag{6.14}$$

Finally, if we make $R_f = R_s$, the output voltage is just the inverted sum of the input voltages; that is,

$$v_o = -(v_a + v_b + v_c). \tag{6.15}$$

Although we illustrated the summing amplifier with just three input signals, the number of input voltages can be increased as needed for an application. As in the case of the inverting-amplifier circuit, the scaling factors in the summing-amplifier circuit are determined by the external resistors R_f, R_a, R_b, R_c, . . . , R_n.

DRILL EXERCISES

6.3 a) Find v_o in the circuit shown if $v_a = 0.1$ V and $v_b = 0.25$ V.

b) If $v_b = 0.25$ V, how large can v_a be before the op amp saturates?

c) If $v_a = 0.10$ V, how large can v_b be before the op amp saturates?

d) Repeat (a), (b), and (c) with the polarity of v_b reversed.

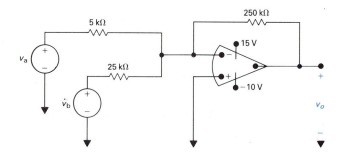

ANSWER: (a) -7.5 V; (b) 0.15 V; (c) 0.5 V;
(d) -2.5, 0.25, and 2 V.

6.5 THE NONINVERTING-AMPLIFIER CIRCUIT

Figure 6.13 depicts the noninverting-amplifier circuit. The signal source is represented by v_g in series with the resistor R_g. In deriving the expression for the output voltage as a function of the source voltage, we assume an ideal operational amplifier operating within its linear range. Thus, as before, we use Eqs. (6.3) and (6.5) as the basis for the derivation. Because the op-amp input current is zero, we can write $v_2 = v_g$, and, from Eq. (6.3), v_1 also equals v_g. Now, because the input current ($i_1 = i_2 = 0$) is zero, the resistors R_f and R_s form an unloaded voltage divider across v_o. Therefore

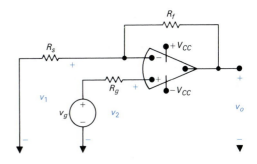

FIGURE 6.13 A noninverting amplifier.

$$v_1 = v_g = \frac{v_o R_s}{R_s + R_f}. \tag{6.16}$$

Solving Eq. (6.16) for v_o gives us the sought-after expression:

$$v_o = \frac{R_s + R_f}{R_s} v_g. \tag{6.17}$$

Operation in the linear range requires that

$$\frac{R_s + R_f}{R_s} < \left| \frac{V_{CC}}{v_g} \right|.$$

Note again that because of the ideal operational amplifier, we can express the output voltage as a function of the input voltage and the external resistors—in this case, R_s and R_f. Problems 6.16 and 6.17 illustrate a simple application of the noninverting-amplifier circuit.

DRILL EXERCISES

6.4 Assume that the operational amplifier in the circuit shown is ideal.

   a) Find the output voltage when the variable resistor is set to 80 kΩ.

   b) How large can R_x be before the amplifier saturates?

ANSWER: (a) 9 V; (b) 160 kΩ.

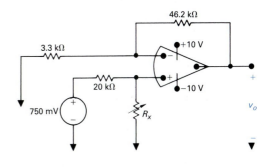

6.6 THE DIFFERENCE-AMPLIFIER CIRCUIT

The output voltage of a difference amplifier is proportional to the difference of the two input voltages. We analyze the difference-amplifier circuit shown in Fig. 6.14, assuming an ideal operational amplifier operating in its linear region. We derive the relationship between v_o and the two input voltages v_a and v_b by summing the currents away from the inverting input node:

$$\frac{v_1 - v_a}{R_a} + \frac{v_1 - v_o}{R_b} + i_1 = 0. \tag{6.18}$$

Because the op amp is ideal, we use the constraints

$$i_1 = i_2 = 0 \tag{6.19}$$

and

$$v_1 = v_2 = \frac{R_d}{R_c + R_d} v_b. \tag{6.20}$$

Combining these constraints with Eq. (6.18) gives the desired relationship:

$$v_o = \frac{R_d (R_a + R_b)}{R_a (R_c + R_d)} v_b - \frac{R_b}{R_a} v_a. \tag{6.21}$$

Equation (6.21) shows that the output voltage is proportional to the difference between a scaled replica of v_b and a scaled replica of v_a. In general, the scaling factor applied to v_b is not the same as the scaling factor applied to v_a. However, the scaling factor applied to each input voltage can be made equal by setting

$$\frac{R_a}{R_b} = \frac{R_c}{R_d}. \tag{6.22}$$

When the ratios given in Eq. (6.22) are equal, the expression for the output voltage reduces to

$$v_o = \frac{R_b}{R_a} (v_b - v_a). \tag{6.23}$$

Equation (6.23) indicates that the output voltage can be made a scaled replica of the difference between the input voltages v_b and v_a.

As in the previous ideal-amplifier circuits, the scaling is controlled by the external resistors. Furthermore, the relationship between the output voltage and the input voltages is not affected by connecting a nonzero load resistance across the output of the amplifier.

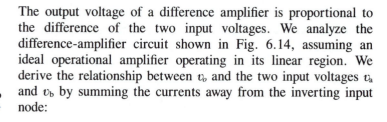

FIGURE 6.14 A difference amplifier.

DRILL EXERCISES

6.5 a) Use the principle of superposition to derive Eq. (6.21).

b) Derive Eqs. (6.22) and (6.23).

ANSWER: (a) $v'_o = \dfrac{-R_b}{R_a} v_a$, $v''_o = \dfrac{R_d(R_a + R_b)}{R_a(R_c + R_d)} v_b$, $v_o = v'_o + v''_o$; (b) derivation.

6.6 a) In the difference amplifier shown, $v_b = 6.0$ V. What range of values for v_a will result in linear operation?

b) Repeat (a) with the 20-kΩ resistor decreased to 5 kΩ.

ANSWER: (a) 3.5 V $\leq v_a \leq$ 8.5 V; (b) 1.25 V $\leq v_a \leq$ 6.25 V.

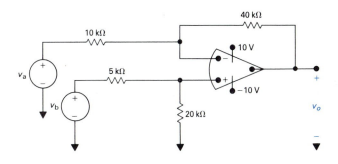

6.7 AN EQUIVALENT CIRCUIT FOR THE OPERATIONAL AMPLIFIER

We now consider a more realistic model that predicts the performance of the operational amplifier in its *linear region* of operation. In building a more realistic circuit model, we modify three assumptions used in characterizing an ideal op amp: (1) a finite input resistance, R_i; (2) a finite open-loop gain, A; and (3) a nonzero output resistance, R_o. The circuit diagram shown in Fig. 6.15 illustrates the more realistic model.

Whenever we use the equivalent circuit shown in Fig. 6.15, we disregard the assumptions that $v_1 = v_2$ (Eq. 6.3) and $i_1 = i_2 = 0$ (Eq. 6.5). Furthermore, Eq. (6.1) is no longer valid because of the presence of the nonzero output resistance, R_o.

Another way to think about the circuit shown in Fig. 6.15 is to reverse the thought process. That is, we can say that the circuit shown in Fig. 6.15 reduces to the ideal model of the operational amplifier when $R_i \rightarrow \infty$, $A \rightarrow \infty$, and $R_o \rightarrow 0$. For the μA741 op amp, the typical values of R_i, A, and R_o are 2 MΩ, 10^5, and 75 Ω, respectively.

Although the presence of R_i and R_o makes the analysis of circuits containing op amps more cumbersome, the analysis remains straightforward. To illustrate, we analyze both the invert-

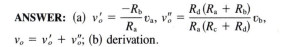 PSpice can be used to model a more realistic op amp in a circuit: Section 5.1

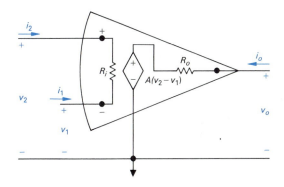

FIGURE 6.15 An equivalent circuit for an operational amplifier.

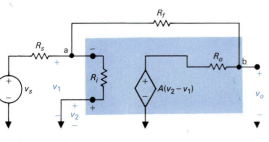

FIGURE 6.16 An inverting-amplifier circuit.

ing and noninverting amplifiers using the equivalent circuit shown in Fig. 6.15. We begin with the inverting amplifier.

ANALYSIS OF THE INVERTING OP-AMP CIRCUIT USING THE MORE REALISTIC MODEL

If we use the op-amp circuit shown in Fig. 6.15, the circuit for the inverting amplifier is that depicted in Fig. 6.16. As before, our goal is to express the output voltage, v_o, as a function of the source voltage, v_s. We obtain the desired expression by writing the two node-voltage equations that describe the circuit and then solving the resulting set of equations for v_o. The two nodes are labeled a and b in Fig. 6.16. Also note that $v_2 = 0$ by virtue of the external short-circuit connection at the noninverting input terminal. The two node-voltage equations are

$$\text{Node a:} \quad \frac{v_1 - v_s}{R_s} + \frac{v_1}{R_i} + \frac{v_1 - v_o}{R_f} = 0; \quad \textbf{(6.24)}$$

$$\text{Node b:} \quad \frac{v_o - v_1}{R_f} + \frac{v_o - A(-v_1)}{R_o} = 0. \quad \textbf{(6.25)}$$

We rearrange Eqs. (6.24) and (6.25) so that the solution for v_o by Cramer's method becomes apparent:

$$\left(\frac{1}{R_s} + \frac{1}{R_i} + \frac{1}{R_f}\right) v_1 - \frac{1}{R_f} v_o = \frac{1}{R_s} v_s; \quad \textbf{(6.26)}$$

$$\left(\frac{A}{R_o} - \frac{1}{R_f}\right) v_1 + \left(\frac{1}{R_f} + \frac{1}{R_o}\right) v_o = 0. \quad \textbf{(6.27)}$$

Solving for v_o yields

$$v_o = \frac{-A + (R_o/R_f)}{\dfrac{R_s}{R_f}\left(1 + A + \dfrac{R_o}{R_i}\right) + \left(\dfrac{R_s}{R_i} + 1\right) + \dfrac{R_o}{R_f}} v_s. \quad \textbf{(6.28)}$$

Note that Eq. (6.28) reduces to Eq. (6.10) as $R_o \rightarrow 0$, $R_i \rightarrow \infty$, and $A \rightarrow \infty$.

If the inverting amplifier shown in Fig. 6.16 were loaded at its output terminals with a load resistance of R_L ohms, the relationship between v_o and v_s would become

$$v_o = \frac{-A + (R_o/R_f)}{\dfrac{R_s}{R_f}\left(1 + A + \dfrac{R_o}{R_i} + \dfrac{R_o}{R_L}\right) + \left(1 + \dfrac{R_o}{R_L}\right)\left(1 + \dfrac{R_s}{R_i}\right) + \dfrac{R_o}{R_f}} v_s. \quad \textbf{(6.29)}$$

Problems 6.30, 6.31, and 6.34 will familiarize you with numerical calculations involving Eqs. (6.28) and (6.29).

ANALYSIS OF THE NONINVERTING OP-AMP CIRCUIT USING THE MORE REALISTIC MODEL

When we use the equivalent circuit shown in Fig. 6.15 to analyze the noninverting amplifier, we obtain the circuit depicted in Fig. 6.17. Here, the voltage source v_g in series with the resistance R_g represents the signal source. The resistor R_L denotes the load on the amplifier. The analysis of the amplifier consists of deriving an expression for v_o as a function of v_g. We do so by writing the node-voltage equations at nodes a and b. At node a,

$$\frac{v_1}{R_s} + \frac{v_1 - v_g}{R_g + R_i} + \frac{v_1 - v_o}{R_f} = 0, \qquad (6.30)$$

and at node b,

$$\frac{v_o - v_1}{R_f} + \frac{v_o}{R_L} + \frac{v_o - A(v_2 - v_1)}{R_o} = 0. \qquad (6.31)$$

Because the current in R_g is the same current as in R_i, we have

$$\frac{v_2 - v_g}{R_g} = \frac{v_1 - v_g}{R_i + R_g}. \qquad (6.32)$$

We use Eq. (6.32) to eliminate v_2 from Eq. (6.31), giving a pair of equations involving the unknown voltages v_1 and v_o. This algebraic manipulation leads to

$$v_1 \left(\frac{1}{R_s} + \frac{1}{R_g + R_i} + \frac{1}{R_f} \right) - v_o \left(\frac{1}{R_f} \right) = v_g \left(\frac{1}{R_g + R_i} \right); \qquad (6.33)$$

$$v_1 \left[\frac{AR_i}{R_o(R_i + R_g)} - \frac{1}{R_f} \right]$$

$$+ v_o \left(\frac{1}{R_f} + \frac{1}{R_o} + \frac{1}{R_L} \right) = v_g \left[\frac{AR_i}{R_o(R_1 + R_g)} \right]. \qquad (6.34)$$

Solving for v_o yields

$$v_o = \frac{[(R_f + R_s) + (R_s R_o / AR_i)] v_g}{R_s + \dfrac{R_o}{A}(1 + K_r) + \dfrac{R_f R_s + (R_f + R_s)(R_i + R_g)}{AR_i}}, \qquad (6.35)$$

where

$$K_r = \frac{R_s + R_g}{R_i} + \frac{R_f + R_s}{R_L} + \frac{R_f R_s + R_f R_g + R_g R_s}{R_i R_L}.$$

Note that Eq. (6.35) reduces to Eq. (6.17) when $R_o \to 0$, $A \to \infty$, and $R_i \to \infty$. For the unloaded ($R_L = \infty$) noninverting

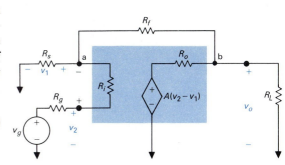

FIGURE 6.17 A noninverting-amplifier circuit.

amplifier, Eq. (6.35) simplifies to

$$v_o = \frac{[(R_f + R_s) + R_s R_o / A R_i]\, v_g}{R_s + \dfrac{R_o}{A}\left(1 + \dfrac{R_s + R_g}{R_i}\right) + \dfrac{1}{A R_i}[R_f R_s + (R_f + R_s)(R_i + R_g)]}.$$

(6.36)

Note that, in the derivation of Eq. (6.36) from Eq. (6.35), K_r reduces to $(R_s + R_g)/R_i$. Problem 6.32 illustrates the effect of R_i, A, and R_o on the performance of a noninverting amplifier.

6.8 THE DIFFERENTIAL MODE

The operational amplifier is also useful as a linear differential amplifier. In the differential mode, the amplifier is meant to produce an amplified replica of the difference between v_1 and v_2. We touched on this possibility in Section 6.3 when we briefly discussed the possibility of the inverting amplifier operating open loop with $v_2 = 0$. We now introduce the differential mode when neither v_1 nor v_2 is set equal to zero. From the previous discussion, the difference between v_1 and v_2 obviously must be extremely small; otherwise, the high-gain amplifier will simply saturate. Thus the operational amplifier is useful as a linear differential amplifier in applications where v_1 and v_2 are approximately equal. The bridge circuit shown in Fig. 6.18 is a good example of such an application. The purpose of the bridge structure is to generate a signal $v_1 - v_2$ when the bridge arm resistor R_x changes from its balance value of $R_1 R_2 / R_4$. That is, when $\epsilon = 0$, the bridge is balanced and $v_1 = v_2$ or $v_1 - v_2 = 0$. When R_x changes (because of some phenomenon such as temperature or strain), the voltage $v_1 - v_2$ is nearly proportional to ϵ if ϵ is relatively small. We leave to you (see Problem 6.36) to show that, for an ideal operational amplifier with infinite input resistance,

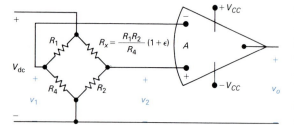

FIGURE 6.18 Using an operational amplifier in a differential input mode.

$$v_1 - v_2 = \frac{V_{dc} R_1 R_4}{(R_1 + R_4)^2} \frac{\epsilon}{1 + [R_1/(R_1 + R_4)]\epsilon}. \qquad \textbf{(6.37)}$$

We also are assuming the bridge is energized from a regulated power supply, so the bridge voltage V_{dc} is constant for all values of R_x.

For small values of ϵ, Eq. (6.37) simplifies to

$$v_1 - v_2 \approx \frac{V_{dc} R_1 R_4}{(R_1 + R_4)^2}\epsilon. \qquad \textbf{(6.38)}$$

As R_1, R_4, and V_{dc} all are fixed quantities, the input to the differential amplifier is proportional to ϵ. The output of an ideal operational amplifier is $-A(v_1 - v_2)$; therefore, in the circuit shown in Fig. 6.18, v_o is

$$v_o = \frac{-V_{dc} R_1 R_4 A\epsilon}{(R_1 + R_4)^2} \quad \text{for} \quad |\epsilon| \ll 1. \qquad \textbf{(6.39)}$$

If the amplifier is to operate in its linear region, from Eq. (6.39),

$$\frac{V_{dc} R_1 R_4 A\epsilon}{(R_1 + R_4)^2} \leq V_{CC}$$

or

$$|\epsilon| \leq \frac{V_{CC}(R_1 + R_4)^2}{V_{dc} R_1 R_4 A}. \qquad \textbf{(6.40)}$$

The inequality of Eq. (6.40) indicates that a large open-loop gain, A, means that the amplifier can tolerate only a small difference in v_1 and v_2 before saturating. Typically, $v_1 - v_2$ is limited to the μV range.

When ϵ is zero, v_o should be zero because $v_1 = v_2$. However, even with $v_1 = v_2$, there will be some output from the amplifier, because, in practice, getting perfect balance between the two channels in the amplifier is not possible. The fact that the differential amplifier yields an infinitesimal output when the same signal is applied to the two input terminals (common mode) represents a flaw in the operation of the amplifier. How well an operational amplifier can reject a common-mode signal is referred to as the *common-mode rejection* property of the amplifier. The ability of a differential amplifier to reject a signal common to both input terminals is expressed quantitatively by the *common-mode rejection ratio* (CMRR), which we discuss in Section 6.9.

DRILL EXERCISES

6.7 In the bridge circuit shown in Fig. 6.18, the circuit parameters are $V_{dc} = 1.5$ V, $V_{CC} = 20$ V, $R_1 = 10$ kΩ, $R_2 = 2$ kΩ, $R_4 = 1$ kΩ, and $A = 2 \times 10^6$.

a) Find $v_1 - v_2$ in microvolts when R_x is larger than its balance value by 1 Ω.

b) Specify the range of R_x within which the operational amplifier does not saturate.

ANSWER: (a) 6.20 μV; (b) 19,998.387 $\Omega \leq R \leq$ 20,001.613 Ω.

6.9 THE COMMON-MODE REJECTION RATIO

When we use the operational amplifier in the differential mode, one measure of its effectiveness is its ability to produce zero output if the voltages v_1 and v_2 are identical. That is, as

$$v_o = A(v_2 - v_1), \tag{6.41}$$

we would expect that $v_o = 0$ when $v_2 = v_1$. In practical operational amplifiers, $v_o \neq 0$ when $v_1 = v_2$. This inability to produce a zero output when v_1 and v_2 are the same is caused by unavoidable imperfections in the electronic components used in fabricating the amplifier. These imperfections make identical amplifying channels impossible. The nature of the problem can be best described in terms of the circuit shown in Fig. 6.19.

If both channels through the operational amplifier were identical, the output would be the same, except for sign inversion, regardless of the channel used. For example, if v_2 were made zero by grounding the noninverting input terminal and v_1 were set equal to v_s,

$$v_o = A_1 v_s, \tag{6.42}$$

where A_1 represents the gain of the amplifier via the inverting channel.

We now reverse the process, grounding the inverting input terminal and setting v_2 equal to v_s. Then,

$$v_o = A_2 v_s, \tag{6.43}$$

where A_2 represents the gain of the amplifier via the noninverting channel. For an ideal operational amplifier we would expect

$$A_1 v_s = -A_2 v_s. \tag{6.44}$$

In terms of the open-loop gain previously used in the analysis, the ideal op amp requires that

$$A = -A_1 = A_2. \tag{6.45}$$

In order to quantitatively discuss the difference between channels, we rewrite the expression for the output voltage as a function of the two input voltages:

$$v_o = A_1 v_1 + A_2 v_2. \tag{6.46}$$

Note that Eq. (6.46) reduces to the more familiar form of Eq. (6.41) if we idealize the operational amplifier by assuming that Eq. (6.45) is valid. Also, Eq. (6.46) is a direct application of the principle of superposition. Thus Eq. (6.46) reinforces the fact that we are assuming operation within the linear range of the operational amplifier, or $v_o \leq |V_{CC}|$.

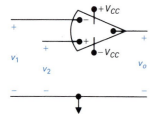

FIGURE 6.19 A circuit used to illustrate the concept of the common-mode rejection ratio.

COMMON-MODE AND DIFFERENTIAL-MODE COMPONENTS

The next step in development of the common-mode rejection ratio is to regard all signal voltages as being made up of two components: a common-mode component and a differential-mode component. We justify the need for such a viewpoint by considering the following two sets of signals. We assume that the two voltage sets are described in terms of dc voltages. The first set of voltages is $v_1 = +10 \ \mu V$ and $v_2 = -10 \ \mu V$, and the second set is $v_1 = 50 \ \mu V$ and $v_2 = 30 \ \mu V$. Note that in both sets the difference between v_1 and v_2 is 20 μV; the sum in the first set is 0, whereas in the second set it is 80 μV. Because the two sets have the same difference, we would expect each set to generate the same output from an operational amplifier operating in the differential mode. However, we have already noted that the output will not be the same because of differences in the two channels. We now identify the outputs as different because the two sets have different common-mode components. The *common-mode component* of the two signals is defined as the average value of the two signals, or

$$v_c = \tfrac{1}{2}(v_1 + v_2). \tag{6.47}$$

The *differential-mode component* is defined as the difference between v_1 and v_2, namely,

$$v_d = v_1 - v_2. \tag{6.48}$$

When we solve Eqs. (6.47) and (6.48) for v_1 and v_2, we get the desired expressions for the signal voltages as functions of common-mode and differential-mode components:

$$v_1 = v_c + \tfrac{1}{2}v_d \tag{6.49}$$

and

$$v_2 = v_c - \tfrac{1}{2}v_d. \tag{6.50}$$

In terms of the two sets of signals mentioned, the first set has a common-mode component of 0 V and a differential-mode component of 20 μV. The second set has a common-mode component of 40 μV and a differential-mode component of 20 μV. That is,

Set 1: $\quad v_c = \tfrac{1}{2}(v_1 + v_2) = \tfrac{1}{2}(+10 - 10) = 0; \tag{6.51}$

$\qquad\quad v_d = v_1 - v_2 = 10 - (-10) = 20 \ \mu V. \tag{6.52}$

Set 2: $\quad v_c = \tfrac{1}{2}(v_1 + v_2) = \tfrac{1}{2}(50 + 30) = 40 \ \mu V; \tag{6.53}$

$\qquad\quad v_d = 50 - 30 = 20 \ \mu V. \tag{6.54}$

Using the common-mode and differential-mode components, we

can describe the two sets of signals as

Set 1: $v_1 = v_c + \frac{1}{2}v_d = 0 + \frac{1}{2}(20) = +10 \ \mu V;$ **(6.55)**

$v_2 = v_c - \frac{1}{2}v_d = 0 - \frac{1}{2}(20) = -10 \ \mu V;$ **(6.56)**

Set 2: $v_1 = 40 + \frac{1}{2}(20) = 50 \ \mu V;$ **(6.57)**

$v_2 = 40 - \frac{1}{2}(20) = 30 \ \mu V.$ **(6.58)**

OUTPUT IN TERMS OF V_c AND V_D

Now that we have introduced the concept of signals being described in terms of common-mode and differential-mode components, we can describe quantitatively the operational amplifier's ability to reject the common-mode component. We begin by returning to Eq. (6.46) and writing v_1 and v_2 in terms of their common-mode and differential-mode components:

$$v_o = A_1(v_c + \tfrac{1}{2}v_d) + A_2(v_c - \tfrac{1}{2}v_d).$$ **(6.59)**

Collecting coefficients of v_c and v_d, we rewrite Eq. (6.59) as

$$v_o = (A_1 + A_2)v_c + \tfrac{1}{2}(A_1 - A_2)v_d,$$ **(6.60)**

where $A_1 + A_2$ is the gain for the common-mode signal and $\frac{1}{2}(A_1 - A_2)$ is the gain for the differential-mode signal. We emphasize this result by letting

$$A_c = A_1 + A_2$$ **(6.61)**

and

$$A_d = \tfrac{1}{2}(A_1 - A_2).$$ **(6.62)**

Substituting Eqs. (6.61) and (6.62) into Eq. (6.60) yields

$$v_o = A_c v_c + A_d v_d.$$ **(6.63)**

Note that for an ideal op amp where $A_2 = -A_1 = A$, we have $A_c = 0$ and $A_d = -A$. Thus an operational amplifier that exhibits a small value for A_c and a large value for A_d approaches the characteristics of an ideal op amp. The magnitude of the ratio of A_d/A_c is the common-mode rejection ratio, CMRR:

$$\text{CMRR} \triangleq \left| \frac{A_d}{A_c} \right|.$$ **(6.64)**

This ratio usually is expressed in decibels. Use of the decibel as a unit for measuring ratios is introduced in Chapter 17. For now we simply note that, when the common-mode rejection ratio is expressed in decibels, we write

$$\text{CMRR} = 20 \log_{10} \left| \frac{A_d}{A_c} \right|.$$ **(6.65)**

Typical values of the CMRR range from 60 to 120 dB. Equation (6.65) indicates that this range translates into ratios ranging from 10^3 to 10^6.

We can measure the common-mode gain by making $v_1 = v_2$ and measuring the corresponding output. If $v_1 = v_2$, then $v_c = v_1$ and $v_d = 0$. Thus, from Eq. (6.63), $v_o = A_c v_1$, where A_c is simply the ratio of the measured output to the input. Figure 6.20 summarizes the measurement of A_c.

We can measure the differential-mode gain by making $v_1 = -v_2$. In this case, $v_c = 0$ and $v_d = 2v_1$. We measure A_d with the circuit shown in Fig. 6.21, where

$$v_1 = \frac{v_s}{2} = -v_2. \qquad (6.66)$$

Because $v_c = 0$ and $v_d = 2v_1$,

$$v_o = A_d v_d = A_d \left(\frac{2v_s}{2}\right) = A_d v_s, \qquad (6.67)$$

which gives

$$A_d = \frac{v_o}{v_s}. \qquad (6.68)$$

After measuring A_c and A_d, we calculate the common-mode rejection ratio from Eq. (6.64). We calculate the rejection in decibels from Eq. (6.65).

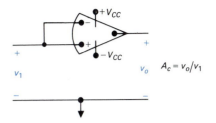

FIGURE 6.20 A circuit for the measurement of A_c.

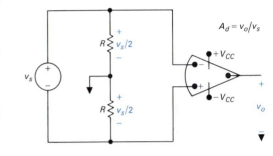

FIGURE 6.21 A circuit for the measurement of A_d.

DRILL EXERCISES

6.8 The common-mode and differential-mode gains of the amplifier shown are 10^3 and 10^6, respectively. The input voltages to the amplifier are $v_1 = 15.6\ \mu V$ and $v_2 = 4.4\ \mu V$. Assume that the amplifier is operating in its linear region. Compute (a) the common-mode and differential-mode components of v_1 and v_2; (b) the CMRR in decibels; (c) the channel gains A_1 and A_2; and (d) the output voltage v_o. (e) Repeat (d) if $v_1 = v_2 = 15.6\ \mu V$ and if $v_1 = v_2 = 4.4\ \mu V$.

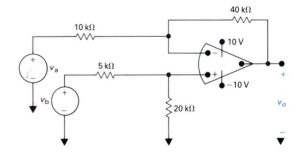

ANSWER: (a) $v_c = 10\ \mu V$, $v_d = 11.2\ \mu V$; (b) 60 dB; (c) $A_2 = -999{,}500$, $A_1 = 1{,}000{,}500$; (d) 11.21 V; (e) 15.6 mV, 4.4 mV.

6.10 BARTLETT'S BISECTION THEOREM

In Section 6.9, we introduced the concept that two signals can be thought of as being made up of a common-mode component and a differential-mode component. We now show how common-mode and differential-mode voltages can be used to analyze a symmetric circuit or network.

A network is called *symmetric* if the right half is a mirror image of the left half. Electronic amplifier circuits often are symmetric, and the use of differential-mode and common-mode voltages, together with superposition, greatly simplifies their analysis. Only half the circuit then requires analysis.

Consider a symmetric network split into mirror-image halves joined by connecting wires (without crossovers) as shown in Fig. 6.22. If $v_1 = v_2$ (common-mode excitation) there is no current in any wire connecting the network halves. That is, if v_1 produced a current directed to the right, v_2 would produce an equal current directed to the left, making the total current zero. Consequently, all the connecting wires may be cut (network bisected) without altering any current or voltage inside either half of the network.

If $v_1 = -v_2$ (differential-mode excitation) there is no potential difference between any pair of wires connecting the network halves. If v_1 causes a potential v_x between any pair, then v_2 produces an equal and opposite potential. Consequently, all connecting wires may be shorted without affecting any current or voltage in either half of the network.

These two observations on symmetric networks with common-mode or differential-mode excitation are known as *Bartlett's theorem*. We demonstrate their use by solving for currents i_1 and i_2 in the circuit shown in Fig. 6.23.

The common-mode input voltage is

$$v_c = \frac{6 + 4}{2} = 5 \text{ V}. \tag{6.69}$$

Figure 6.24(a) shows this voltage applied to the half-circuit with the connections open. Note that the horizontal 12-Ω resistor is split into two 6-Ω resistors in series, whereas the vertical 2-Ω resistor becomes two 4-Ω resistors in parallel. By inspection, the common-mode current is

$$i_c = \frac{5}{10} = 0.5 \text{ A}. \tag{6.70}$$

The differential-mode voltage is

$$v_1 - v_2 = 6 - 4 = 2 \text{ V}. \tag{6.71}$$

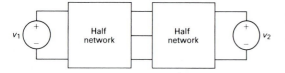

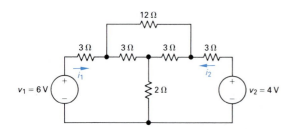

FIGURE 6.22 Symmetric network divided into connected halves.

FIGURE 6.23 Symmetric network with unsymmetric excitation.

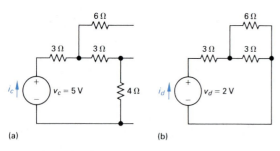

(a)   (b)

FIGURE 6.24 (a) Common-mode half-circuit; (b) differential-mode half-circuit.

Figure 6.24(b) shows this voltage applied to the half-circuit with the connections shorted. (The 4-Ω resistor is shorted and hence removed.) The differential-mode current is

$$i_d = \frac{2}{5} = 0.4 \text{ A}. \qquad \textbf{(6.72)}$$

By analogy with Eqs. (6.49) and (6.50),

$$i_1 = i_c + \frac{1}{2}i_d = 0.5 + 0.2 = 0.7 \text{ A}; \qquad \textbf{(6.73)}$$

$$i_2 = i_c - \frac{1}{2}i_d = 0.5 - 0.2 = 0.3 \text{ A}. \qquad \textbf{(6.74)}$$

Bartlett's theorem applies to symmetric linear networks that contain controlled sources. Linear models of differential amplifier circuits are of this form.

SUMMARY

This chapter introduced the operational amplifier, a versatile device that can be used to form a large number of practical circuits. Although all the circuits in this chapter contained only op amps in conjunction with sources and resistors, in later chapters we show how the same techniques can be applied to circuits that also include inductors and capacitors. We used the op amp to illustrate the general principle that different models of the same device are used for different purposes. Thus we considered both an ideal model with infinite input resistance, zero output resistance, and infinite gain and a more realistic model in which the input resistance, output resistance, and gain all were finite. We pointed out that a more detailed analysis of the operational amplifier as a circuit component must await presentation of electronic circuits and devices.

We discussed the following circuits containing op amps.

* *Inverting amplifier:* The output voltage is a negative multiple of the input voltage.

* *Summing amplifier:* In this variation of the inverting amplifier, the output voltage is a negative multiple of the sum of several input voltages.

* *Non-inverting amplifier:* The output voltage is a multiple of the input voltage.

* *Difference amplifier:* The output voltage is a multiple of the difference between two input voltages.

• *Differential operation:* In this open-loop operation, the high gain of the amplifier is used to amplify a small difference in the input voltages.

We also discussed the

• *common-mode signal,* which is the average value of two signals (either voltage or current);

• *differential-mode signal,* which is the difference between two signals (either voltage or current);

• *common-mode rejection ratio (CMRR),* which is a measure of how effective an operational amplifier is in rejecting the common-mode component of a signal; and

• *Bartlett's bisection theorem,* which is a technique for analyzing symmetric circuits by combining common-mode and differential-mode signal components with the principle of superposition.

PROBLEMS

6.1 The operational amplifier in the circuit in Fig. P6.1 is ideal.

a) Calculate v_o if $v_a = 1.5$ V and $v_b = 0$ V.

b) Calculate v_o if $v_a = 3.0$ V and $v_b = 0$ V.

c) Calculate v_o if $v_a = 1.0$ V and $v_b = 2$ V.

d) Calculate v_o if $v_a = 4.0$ V and $v_b = 2$ V.

e) Calculate v_o if $v_a = 6.0$ V and $v_b = 8$ V.

f) If $v_a = 4.5$ V, specify the range of v_b such that the amplifier does not saturate.

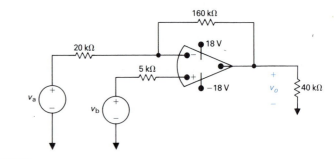

FIGURE P6.1

6.2 The operational amplifier in the circuit in Fig. P6.2 is ideal.

a) Calculate v_o.

b) Calculate i_o.

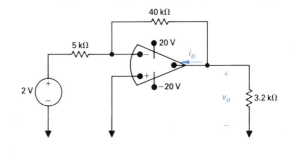

FIGURE P6.2

6.3 A voltmeter with a full-scale reading of 10 V is used to measure the output voltage in the circuit in Fig. P6.3. What is the reading of the voltmeter? Assume the operational amplifier is ideal.

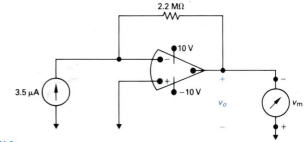

FIGURE P6.3

6.4 Find i_b in the circuit in Fig. P6.4 if the operational amplifier is ideal.

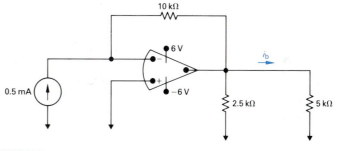

FIGURE P6.4

6.5 A circuit designer claims that the circuit in Fig. P6.5 will produce an output voltage that will vary between ±5 as v_g varies between 0 and 5 V. Assume the operational amplifier is ideal.

a) Draw a graph of the output voltage v_o as a function of the input voltage v_g for $0 \le v_g \le$ 5 V.

b) Do you agree with the designer's claim?

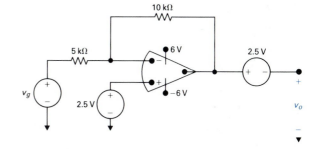

FIGURE P6.5

6.6 The operational amplifier in the circuit in Fig. P6.6 is ideal. Calculate:

a) v_1;

b) v_o;

c) i_2; and

d) i_o when $v_g = 150$ mV.

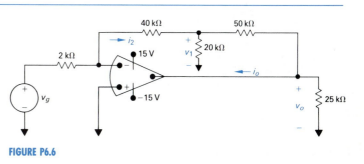

FIGURE P6.6

6.7 a) The operational amplifier in the circuit shown in Fig. P6.7 is ideal. The adjustable resistor R_Δ has a maximum value of 120 kΩ, and α is restricted to the range of $0.25 \leq \alpha \leq 0.8$. Calculate the range of v_o if $v_g = 40$ mV.

b) If α is not restricted, at what value of α will the operational amplifier saturate?

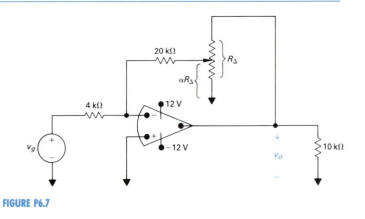

FIGURE P6.7

6.8 The operational amplifier in the circuit in Fig. P6.8 is ideal.

a) Find the range of values for σ for which the operational amplifier does not saturate.

b) Find i_o (in μA) when $\sigma = 0.272$.

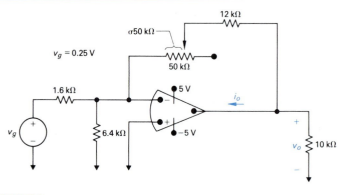

FIGURE P6.8

6.9 The operational amplifier in Fig. P6.9 is ideal.

a) Find v_o if $v_a = 1$ V, $v_b = 1.5$ V and $v_c = -4$ V.

b) The voltages v_a and v_c remain at 1 V and -4 V respectively. What are the limits on v_b if the operational amplifier operates within its linear region?

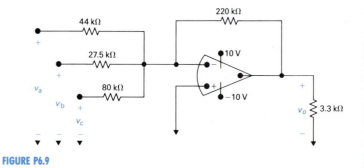

FIGURE P6.9

6.10 a) The operational amplifier in Fig. P6.10 is ideal. Find v_o if $v_a = 15$ V, $v_b = 10$ V, $v_c = 8$ V, and $v_d = 12$ V.

b) Assume v_a, v_c, and v_d retain their values as given in part (a). Specify the range of v_b such that the operational amplifier operates within its linear range.

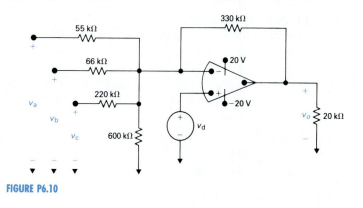

FIGURE P6.10

6.11 The 330-kΩ feedback resistor in the circuit in Fig. P6.10 is replaced by a variable resistor R_f. The voltages v_a through v_d have the same values as given in Problem 6.10(a).

a) What value of R_f will cause the operational amplifier to saturate? Note that $0 \le R_f \le \infty$.

b) When R_f has the value found in part (a) what is the current (in microamperes) into the output terminal of the operational amplifier?

6.12 The operational amplifiers in the circuit in Fig. P6.12 are ideal. Find i_a.

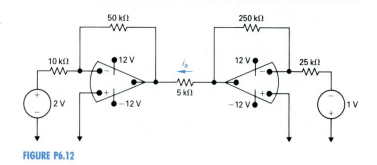

FIGURE P6.12

6.13 The variable resistor R_o in the circuit in Fig. P6.13 is adjusted until the source current i_g is zero. The operational amplifiers are ideal and $0 \le v_g \le 1.2$ V.

a) What is the value of R_o?

b) If $v_g = 1.0$ V, how much power (in μW) is dissipated in R_o?

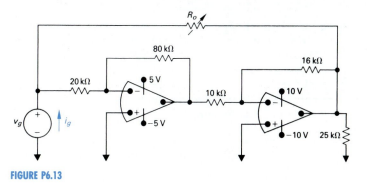

FIGURE P6.13

6.14 The circuit inside the shaded area in Fig. P6.14 is a constant current source for a limited range of values of R_L.

a) Find the value of i_L for $R_L = 4$ kΩ.

b) Find the maximum value for R_L for which i_L will have the value of part (a).

c) Assume that $R_L = 7$ kΩ. Explain the operation of the circuit. You can assume that $i_1 = i_2 \approx 0$ under all operating conditions.

d) Sketch i_L versus R_L for $0 \leq R_L \leq 7$ kΩ.

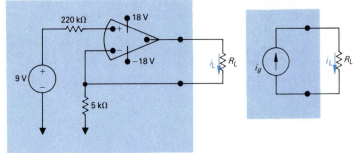

FIGURE P6.14

6.15 a) Find i_a in the circuit in Fig. P6.15 assuming the operational amplifier is ideal and is operating in its linear range.

b) How large can R be before the operational amplifier saturates?

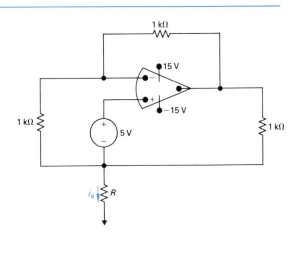

FIGURE P6.15

6.16 Refer to the circuit in Fig. 6.12, where the operational amplifier is assumed to be ideal. Given that $R_a = 3$ kΩ, $R_b = 5$ kΩ, $R_c = 25$ kΩ, $v_a = 150$ mV, $v_b = 100$ mV, $v_c = 250$ mV, and $V_{cc} = \pm 6$ V, specify the range of R_f for which the operational amplifier operates within its linear region.

6.17 The output voltage of a summing amplifier similar to that shown in Fig. 6.12 is to be the inverted weighted sum of the four input signals. Specifically,

$$v_o = -(2v_a + 4v_b + 6v_c + 8v_d).$$

If $R_f = 48$ kΩ, draw a circuit diagram of the amplifier and specify the values of R_a, R_b, R_c, and R_d.

6.18 The operational amplifier in the circuit shown in Fig. P6.18 is ideal.

a) Calculate v_o when v_g equals 4 V.

b) Specify the range of values of v_g so that the operational amplifier operates in a linear mode.

c) Assume that v_g equals 2 V and that the 63-kΩ resistor is replaced with a variable resistor. What value of the variable resistor will cause the operational amplifier to saturate?

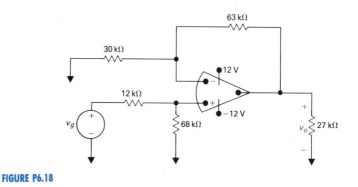

FIGURE P6.18

6.19 Assume that the ideal op amp in the circuit seen in Fig. P6.19 is operating in its linear region.

a) Show that $v_o = [(R_1 + R_2)/R_1]v_s$.

b) What happens if $R_1 \to \infty$ and $R_2 \to 0$?

c) Explain why this circuit is referred to as a voltage follower when $R_1 = \infty$ and $R_2 = 0$.

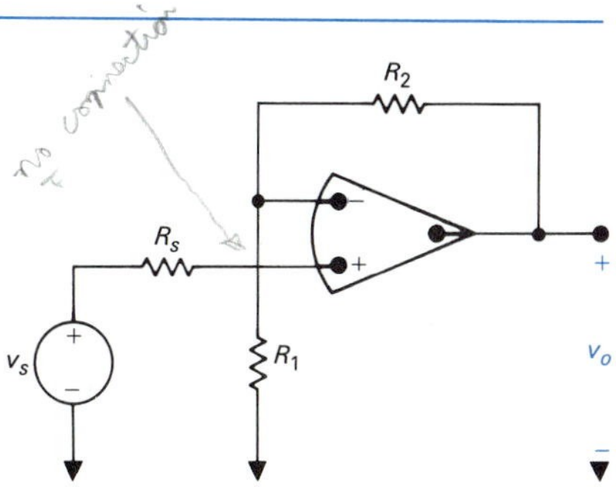

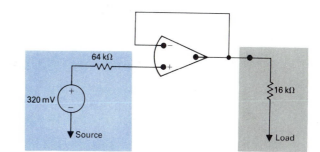

FIGURE P6.19

6.20 Assume that the ideal op amp in the circuit in Fig. P6.20 is operating in its linear region.

a) Calculate the power delivered to the 16-kΩ resistor.

b) Repeat part (a) with the operational amplifier removed from the circuit, that is, with the 16-kΩ resistor connected in the series with the voltage source and the 64-kΩ resistor.

c) Find the ratio of the power found in part (a) to that found in part (b).

d) Does the insertion of the op amp between the source and the load serve a useful purpose? Explain.

FIGURE P6.20

6.21 The operational amplifier in the noninverting amplifier shown in Fig. P6.21 is ideal. The signal voltages v_a and v_b are 800 mV and 400 mV respectively.

a) Calculate v_o in V.

b) Find i_a and i_b in μA.

c) What are the weighting factors associated with v_a and v_b?

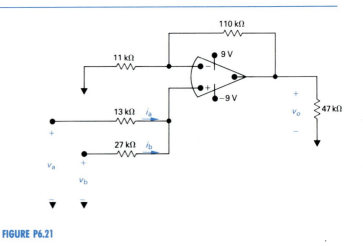

FIGURE P6.21

6.22 The circuit in Fig. P6.22 is a noninverting summing amplifier. The operational amplifier is ideal.

a) Specify the numerical values of R_a and R_c so that

$$v_o = v_a + 2v_b + 3v_c.$$

b) Calculate (in μA) i_a, i_b, and i_c when $v_a = 0.7$ V, $v_b = 0.4$ V, and $v_c = 1.1$ V.

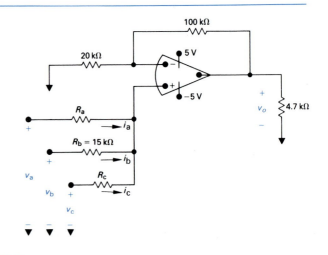

FIGURE P6.22

6.23 The operational amplifier in the noninverting summing amplifier of Fig. P6.23 is ideal.

a) Find the value of R_g so that

$$v_o = 1.8\, v_a + 7.2\, v_b + 14.4\, v_c$$

b) Find (in μA) i_a, i_b, i_c, i_g, and i_h when $v_a = 0.50$ V, $v_b = 0.25$ V, and $v_c = 0.15$ V.

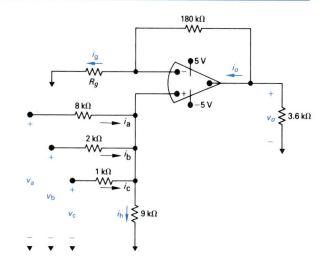

FIGURE P6.23

6.24 The operational amplifier in the circuit of Fig. P6.24 is ideal. Plot v_o versus α when $R_f = 4R_1$ and $v_g = 10$ V. Use increments of 0.1 and note by hypothesis that $0 \leq \alpha \leq 1.0$.

FIGURE P6.24

6.25 Select values of R_a, R_b, and R_f for the amplifier circuit of Fig. P6.25 such that $v_o = 10(v_b - v_a)$ and the voltage source v_b sees an input resistance of 220 kΩ. Use the ideal model for the operational amplifier.

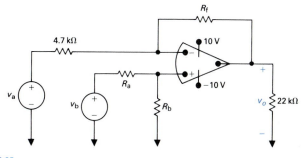

FIGURE P6.25

6.26 The resistors in the difference amplifier shown in Fig. 6.14 are $R_a = 24$ kΩ, $R_b = 75$ kΩ, $R_c = 130$ kΩ, and $R_d = 120$ kΩ. The signal voltages v_a and v_b are 8 and 5 V, respectively, and $V_{CC} = \pm 20$ V.

a) Find v_o.

b) What is the resistance seen by the signal source v_a?

c) What is the resistance seen by the signal source v_b?

6.27 Specify R_a, R_b, R_c, and R_d in the difference amplifier of Fig. 6.14 to meet the following criteria: $v_o = 4.2v_b - 6v_a$; the resistance seen by the signal source v_b is 450 kΩ; and the resistance seen by the signal source v_a is 21 kΩ when the output voltage v_o is zero.

6.28 Select the values of R_1 and R_f in the circuit in Fig. P6.28 so that

$$v_o = 5000(i_b - i_a).$$

The operational amplifier is ideal.

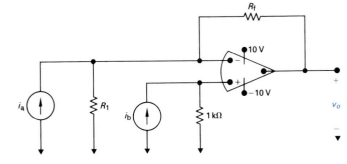

FIGURE P6.28

6.29 The operational amplifier in the adder-subtracter circuit shown in Fig. P6.29 is ideal.

a) Find v_o when $v_a = 1$ V, $v_b = 2$ V, $v_c = 3$ V, and $v_d = 4$ V.

b) If v_a, v_b, and v_d are held constant, what values of v_c will saturate the op amp?

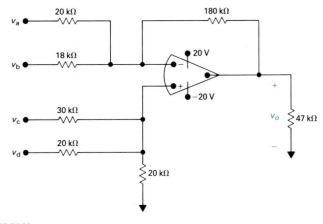

FIGURE P6.29

6.30 The inverting amplifier in the circuit in Fig. P6.30 has an input resistance of 500 kΩ, an output resistance of 5 kΩ, and an open-loop gain of 250,000. Assume that the amplifier is operating in its linear region. Calculate the following:

a) the voltage gain (v_o/v_g) of the amplifier;

b) the value of v_1 in microvolts when $v_g = 100$ mV;

c) the resistance seen by the signal source (v_g).

d) Repeat parts (a), (b), and (c) using the ideal model for the op amp.

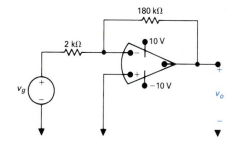

FIGURE P6.30

6.31 Repeat Problem 6.30 given that the inverting amplifier is loaded with a 1600-Ω resistor.

6.32 The operational amplifier in the noninverting amplifier circuit of Fig. P6.32 has an input resistance of 560 kΩ, an output resistance of 8 kΩ, and an open-loop gain of 50,000. Assume that the op amp is operating in its linear region. Calculate the following:

a) the voltage gain (v_o/v_g);

b) the inverting and noninverting input voltages v_1 and v_2 (in mV) if $v_g = 1$ V;

c) the difference $(v_2 - v_1)$ in microvolts when $v_g = 1$ V;

d) the current drain in picoamperes on the signal source v_g when $v_g = 1$ V.

e) Repeat parts (a) through (d) assuming an ideal op amp.

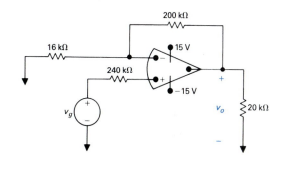

FIGURE P6.32

6.33 Assume the input resistance of the operational amplifier in Fig. P6.33 is infinite and its output resistance is zero.

a) Find v_o as a function of v_g and the open-loop gain A.

b) What is the value of v_o if $v_g = 1$ V and A = 194?

c) What is the value of v_o if $v_g = 1$ V and A = ∞?

d) How large does A have to be so that v_o is 99% of its value in part (c)?

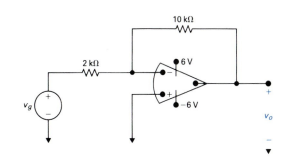

FIGURE P6.33

6.34 a) Find the Thévenin equivalent circuit with respect to the output terminals a,b for the inverting amplifier of Fig. P6.34. The dc signal source has a value of 880 mV. The operational amplifier has an input resistance of 500 kΩ, an output resistance of 2 kΩ, and an open loop gain of 100,000.

b) What is the output resistance of the inverting amplifier?

c) What is the resistance (in ohms) seen by the signal source v_g when the load at the terminals a,b is 330 Ω?

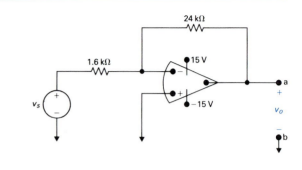

FIGURE P6.34

6.35 Find the Thévenin equivalent with respect to the terminals a,b in the circuit in Fig. P6.35 if the operational amplifier is ideal.

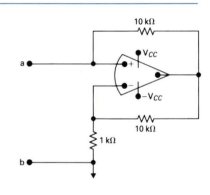

FIGURE P6.35

6.36 a) Derive Eq. (6.37). Note that Eq. (6.37) assumes an ideal operational amplifier and a constant bridge voltage of V_{dc} volts.

b) Verify the approximation given by Eq. (6.38) by using long division to expand

$$\frac{\epsilon}{1 + [R_1/(R_1 + R_4)]\epsilon}$$

into a power series. Simplify the algebra by letting $R_1/(R_1 + R_4) = \alpha$. Note that $\alpha < 1$ by definition. Comment on the significance of higher-order terms for small ϵ.

6.37 The two operational amplifiers in the circuit in Fig. P6.37 are ideal. Calculate v_{o1} and v_{o2}.

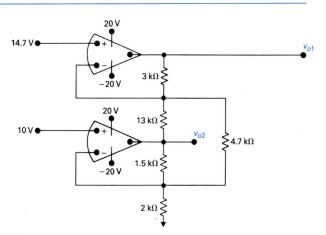

FIGURE P6.37

6.38 The resistor R_f in the circuit seen in Fig. P6.38 is adjusted until the ideal operational amplifier saturates. Specify R_f in kilohms.

FIGURE P6.38

6.39 The operational amplifiers in the circuit of Fig. P6.39 are ideal. Find v_x, i_a, and i_o.

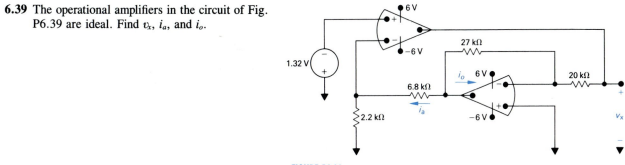

FIGURE P6.39

6.40 Find v_o and i_o in the circuit shown in Fig. P6.40 if the operational amplifiers are ideal.

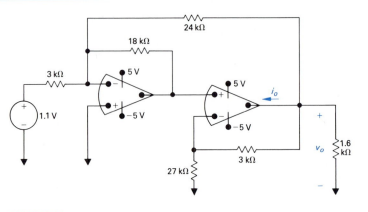

FIGURE P6.40

6.41 The operational amplifiers in the circuit shown in Fig. P6.41 are ideal.

a) Find v_o as a function of α, σ, v_{g1}, and v_{g2} when the op amps operate within their linear range.

b) Describe the behavior of the circuit when $\alpha = \sigma = 1.0$.

c) Describe the behavior of the circuit when $\alpha = \sigma = 0$.

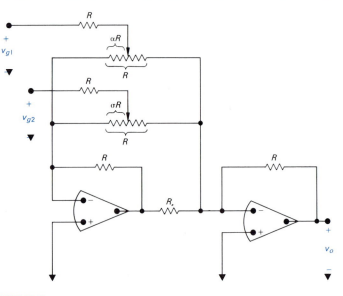

FIGURE P6.41

6.42 The voltage v_g shown in Fig. P6.42(a) is applied to the inverting amplifier shown in Fig. P6.42(b). Sketch v_o vs. t assuming the operational amplifier is ideal.

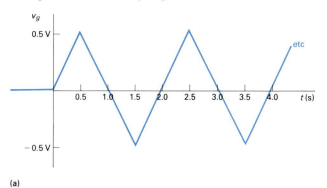

(a)

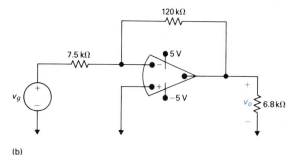

(b)

FIGURE P6.42

6.43 The signal voltage v_g in the circuit shown in Fig. P6.43 is described by the following equations:

$$v_g = 0 \qquad t \leq 0$$

$$v_g = 4 \sin (5\pi/3)t \text{ V} \qquad 0 \leq t \leq \infty.$$

Sketch v_o vs. t assuming the operational amplifier is ideal.

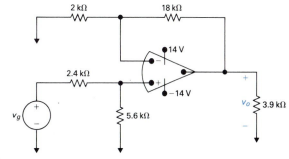

FIGURE P6.43

6.44 Use Bartlett's bisection theorem to find i_a through i_e in the circuit in Fig. P6.44.

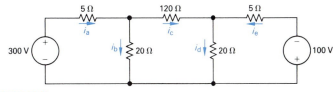

FIGURE P6.44

6.45 Use Bartlett's bisection theorem to find i_1 and i_2 in the circuit in Fig. P6.45. (*Hint:* Observe that $v_2 = 0$).

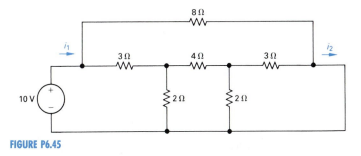

FIGURE P6.45

Inductance and Capacitance

CHAPTER 7

We now introduce the last two ideal circuit elements mentioned in Chapter 2, namely, inductors and capacitors. First, we assure you that the circuit analysis techniques introduced in Chapters 3, 4, and 5 apply to circuits containing inductors and capacitors. Therefore, once you understand the terminal behavior of inductors and capacitors in terms of current and voltage, you can use Kirchhoff's laws to describe any interconnections with the other basic elements.

Recall from Chapter 1 that circuit theory has the advantage of providing a relatively simple description of practical components that would be unnecessarily complicated if given in terms of electromagnetic field theory. Inductors and capacitors are circuit elements that are easier to describe in terms of circuit variables than field variables. However, before we focus on the circuit descriptions, a brief review of the field concepts underlying these basic elements is in order.

Inductors are circuit elements based on phenomena associated with magnetic fields. The source of the magnetic field is charge in motion, or current. If the current is varying with time, the magnetic field is varying with time. A time-varying magnetic field induces a voltage in any conductor linked by the field. The circuit parameter of inductance relates the induced voltage to the current. We discuss this quantitative relationship in Section 7.1.

Capacitors are circuit elements based on phenomena associated with electric fields. The source of the electric field is separation of charge, or voltage. If the voltage is varying with time, the electric field is varying with time. A time-varying electric field produces a displacement current in the space occupied by the field. The circuit parameter of capacitance relates the displacement current to the voltage. The displacement current is equal to the conduction current at the terminals of the capacitor;

therefore capacitance relates the circuit current to the voltage. We discuss this quantitative relationship in Section 7.2.

Energy can be stored in both magnetic and electric fields. Hence you should not be too surprised to learn that inductors and capacitors are capable of storing energy. For example, energy can be stored in an inductor and then released to "fire" a spark plug. Energy can be stored in a capacitor and then released to "fire" a flashbulb. Here, we are talking about energy storage, not energy generation. Therefore in ideal inductors and capacitors, only as much energy can be extracted as has been stored. Because inductors and capacitors cannot generate energy, they are classified as *passive elements*. Energy storage is not unique to electrical systems. Two of the most common examples of mechanical devices used to store energy are springs and flywheels.

We now describe the behavior of inductors and capacitors in terms of current and voltage.

7.1 THE INDUCTOR

Resistance is the circuit parameter used to describe a resistor, and inductance is the circuit parameter used to describe an inductor. Inductance is symbolized by the letter L, is measured in henrys (H), and is represented graphically as a coiled wire—a reminder that inductance is a consequence of a conductor linking a magnetic field. Figure 7.1(a) shows the inductor. Assigning the reference direction of the current in the inductor in the direction of the voltage drop across the terminals of the inductor, as shown in Fig. 7.1(b), yields

$$v = L\frac{di}{dt}, \tag{7.1}$$

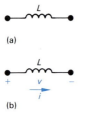

FIGURE 7.1 The graphic symbol for an inductor with an inductance of L henrys.

where v is measured in volts, L in henrys, i in amperes, and t in seconds. If the current is in the direction of the voltage rise across the inductor, Eq. (7.1) is written with a minus sign.

Note from Eq. (7.1) that the voltage across the terminals of an inductor is proportional to the time rate of change of current in the inductor. We can make two important observations at this time. First, if the current is constant, the voltage across the ideal inductor is zero. Thus the inductor appears as a short circuit to a constant, or dc, current. Second, current cannot change instantaneously in an inductor; that is, the current cannot change by a finite amount in zero time. Equation (7.1) tells us that this change would require an infinite voltage, and infinite voltages are not possible. For example, when someone opens the switch on an inductive circuit in a physically realizable system, the switch arcs over, preventing the current from dropping to zero instantaneously. (Switching inductive circuits is an important engineering problem because the arcing and voltage surges that can occur must be controlled to prevent equipment damage. The first step to understanding the nature of this problem is to master the introductory material presented in this chapter and in the two that follow.) Example 7.1 illustrates application of (Eq. 7.1) to a simple circuit.

PSpice PSpice data statement for an inductor: Chapter 6

E X A M P L E   7.1

The independent current source in the circuit shown in Fig. 7.2 generates zero current for $t < 0$ and a pulse $10te^{-5t}$ for $t > 0$.

a) Sketch the current waveform.

b) At what instant of time is the current maximum?

c) Express the voltage across the terminals of the 100-mH inductor as a function of time.

d) Sketch the voltage waveform.

e) Is the voltage maximum when the current is maximum?

f) At what instant of time does the voltage change polarity?

g) Is there ever an instantaneous change in voltage across the inductor? If so, at what time?

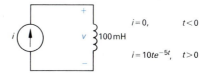

$$i = 0, \qquad t < 0$$

$$i = 10te^{-5t}, \quad t > 0$$

FIGURE 7.2 The circuit for Example 7.1.

PSpice Use PSpice transient analysis (.TRAN) to examine a circuit's behavior as a function of time: Section 7.2

S O L U T I O N

a) Figure 7.3 shows the current waveform.

b) $\dfrac{di}{dt} = 10(-5te^{-5t} + e^{-5t}) = 10e^{-5t}(1 - 5t);$

$\dfrac{di}{dt} = 0$ when $t = \frac{1}{5}$ s. (See Fig. 7.3.)

c) $v = L\dfrac{di}{dt} = (0.1)10e^{-5t}(1 - 5t) = e^{-5t}(1 - 5t)$ V, $t > 0;$

$v = 0, \qquad t < 0.$

d) Figure 7.4 shows the voltage waveform.

e) No; the voltage is proportional to di/dt, not i.

f) At 0.2 s, which corresponds to the moment when di/dt is passing through zero and changing sign.

g) Yes, at $t = 0$. Note that the voltage can change instantaneously across the terminals of an inductor.

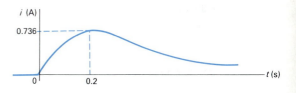

FIGURE 7.3 Current waveform for Example 7.1.

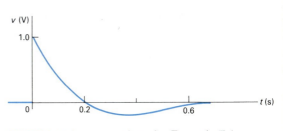

FIGURE 7.4 Voltage waveform for Example 7.1.

CURRENT IN AN INDUCTOR IN TERMS OF THE VOLTAGE ACROSS THE INDUCTOR

Equation (7.1) expresses the voltage across the terminals of the inductor as a function of the current in the inductor. Also desirable is the ability to express the current in the inductor as a function of the voltage. To find i as a function of v, we start by multiplying both sides of Eq. (7.1) by a differential time dt:

$$v\, dt = L\left(\frac{di}{dt}\right) dt. \tag{7.2}$$

Recognizing that multiplying the rate at which i varies with t by a differential change in time generates a differential change in i, we write Eq. (7.2) as

$$v\, dt = L\, di. \tag{7.3}$$

To find i as a function of v, we simply integrate both sides of Eq. (7.3). For convenience, we interchange the two sides of the equation and write

$$L\int_{i(t_0)}^{i(t)} dx = \int_{t_0}^{t} v\, d\tau, \tag{7.4}$$

in which x and τ are variables of integration. Then from Eq. (7.4),

$$i(t) = \frac{1}{L}\int_{t_0}^{t} v\, d\tau + i(t_0), \tag{7.5}$$

where $i(t)$ is the current corresponding to t and $i(t_0)$ is the value of the inductor current when we initiate the integration, namely, t_0. In many practical applications, t_0 is zero and Eq. (7.5) becomes

$$i(t) = \frac{1}{L}\int_{0}^{t} v\, d\tau + i(0). \tag{7.6}$$

Equations (7.1) and (7.5) both give the relationship between the voltage and current at the terminals of an inductor. Equation (7.1) expresses the voltage as a function of current, whereas Eq. (7.5) expresses the current as a function of voltage. In both equations the reference direction for the current is in the direction of the voltage drop across the terminals. Note that $i(t_0)$ carries its own algebraic sign. If the initial current is in the same direction as the reference direction for i, it is a positive quantity. If the initial current is in the opposite direction from the reference direction for i, it is a negative quantity. Example 7.2 illustrates the application of Eq. (7.5).

E X A M P L E   7.2

The voltage pulse applied to the 100-mH inductor shown in Fig. 7.5 is 0 for $t < 0$ and is given by the expression

$$v(t) = 20te^{-10t}$$

for $t > 0$. Also assume $i = 0$ for $t \leq 0$.

a) Sketch the voltage as a function of time.

b) Find the inductor current as a function of time.

c) Sketch the current as a function of time.

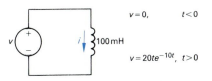

FIGURE 7.5 The circuit for Example 7.2.

S O L U T I O N

a) The voltage as a function of time is shown in Fig. 7.6.

b) The current in the inductor is 0 at $t = 0$. Therefore the current for $t > 0$ is

$$i = \frac{1}{0.1} \int_0^t 20\tau e^{-10\tau} \, d\tau + 0$$

$$= 200 \left[\frac{-e^{-10\tau}}{100} (10\tau + 1) \right] \Big|_0^t$$

$$= 2(1 - 10te^{-10t} - e^{-10t}) \, \text{A}, \qquad t > 0.$$

c) Figure 7.7 shows the current as a function of time.

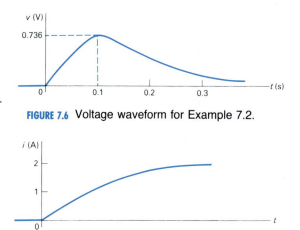

FIGURE 7.6 Voltage waveform for Example 7.2.

FIGURE 7.7 Current waveform for Example 7.2.

Note in Example 7.2 that i approaches a constant value of 2 A as t increases. We say more about this result after discussing the energy stored in an inductor.

POWER AND ENERGY IN THE INDUCTOR

The power and energy relationships for the inductor can be derived directly from the current and voltage relationships. If the current reference is in the direction of the voltage drop across the terminals of the inductor, the power is

$$p = vi. \qquad (7.7)$$

Remember that power is in watts for voltage in volts and current in amperes. If we express the inductor voltage as a function of the inductor current, Eq. (7.7) becomes

$$p = Li\frac{di}{dt}. \qquad (7.8)$$

We can also express the current in terms of the voltage:

$$p = v \left[\frac{1}{L} \int_{t_0}^{t} v \, d\tau + i(t_0) \right]. \tag{7.9}$$

Equation (7.8) is most useful in expressing the energy stored in the inductor. Power is the time rate of expending energy, so

$$p = \frac{dw}{dt} = Li \frac{di}{dt}. \tag{7.10}$$

Multiplying both sides of Eq. (7.10) by a differential time gives the differential relationship

$$dw = Li \, di. \tag{7.11}$$

Both sides of Eq. (7.11) are integrated with the understanding that the reference for zero energy corresponds to zero current in the inductor. Thus

$$\int_{0}^{w} dx = L \int_{0}^{i} y \, dy;$$
$$w = \tfrac{1}{2} Li^2. \tag{7.12}$$

As before, we utilize different symbols of integration to avoid confusion with the limits placed on the integrals. In Eq. (7.12), the energy is in joules when inductance is in henrys and current in amperes. To illustrate the application of Eqs. (7.7) and (7.12), we return to Examples 7.1 and 7.2 by means of Example 7.3.

EXAMPLE   7.3

a) For Example 7.1, plot i, v, p, and w versus time. Line up the plots vertically to allow easy assessment of each variable's behavior.

b) In what time interval is energy being stored in the inductor?

c) In what time interval is energy being extracted from the inductor?

d) What is the maximum energy stored in the inductor?

e) Evaluate the integrals

$$\int_{0}^{0.2} p \, dt \qquad \text{and} \qquad \int_{0.2}^{\infty} p \, dt$$

and comment on their significance.

f) Repeat parts (a)–(d) for Example 7.2.

g) In Example 7.2, why is there a sustained current in the inductor as the voltage approaches zero?

S O L U T I O N

a) The plots of i, v, p, and w follow directly from the expressions for i and v obtained in Example 7.1 and are shown in Fig. 7.8. In particular, $p = vi$ and $w = (\frac{1}{2})Li^2$.

b) Energy is being stored in the time interval 0 to 0.2 s, that is, in the interval when $p > 0$.

c) Energy is being extracted in the time interval 0.2 s to ∞, that is, in the interval when $p < 0$.

d) $w_{max} = 27.07$ mJ

e) From Example 7.1,

$$i = 10te^{-5t} \text{ A} \qquad \text{and} \qquad v = e^{-5t}(1 - 5t) \text{ V.}$$

Therefore

$$p = vi = 10te^{-10t} - 50t^2e^{-10t} \text{ W.}$$

Thus

$$\int_0^{0.2} p\,dt = 10\left[\frac{e^{-10t}}{100}(-10t - 1)\right]_0^{0.2}$$

$$- 50\left\{\frac{t^2e^{-10t}}{-10} + \frac{2}{10}\left[\frac{e^{-10t}}{100}(-10t - 1)\right]\right\}_0^{0.2}$$

$$= 0.2e^{-2} = 27.07 \text{ mJ;}$$

$$\int_{0.2}^{\infty} p\,dt = 10\left[\frac{e^{-10t}}{100}(-10t - 1)\right]_{0.2}^{\infty}$$

$$- 50\left\{\frac{t^2e^{-10t}}{-10} + \frac{2}{10}\left[\frac{e^{-10t}}{100}(-10t - 1)\right]\right\}_{0.2}^{\infty}$$

$$= -0.2e^{-2} = -27.07 \text{ mJ.}$$

Based on the definition of p, the area under the plot of p versus t represents the energy expended over the interval of integration. Hence the integration of the power between 0 and 0.2 s represents the energy stored in the inductor during this time interval. The integral of p over the interval 0.2 s to ∞ is the energy extracted from the inductor. Note that in this time interval all the energy originally stored is removed. That is, after the current pulse has passed, no energy is stored in the inductor.

f) The plots of v, i, p, and w follow directly from the expressions for v and i given in Example 7.2 and are shown in Fig. 7.9. Note that in this case the power always is positive, and hence energy always is being stored in the inductor during the duration of the voltage pulse.

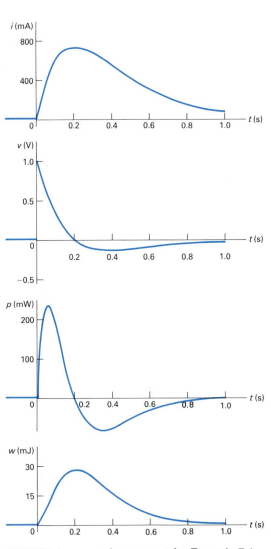

FIGURE 7.8 i, v, p, and w versus t for Example 7.1.

PSpice   Use PSpice plots (.PLOT) or the graphical post-processor PROBE to examine plots of circuit variable values: Section 7.3

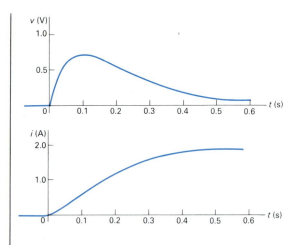

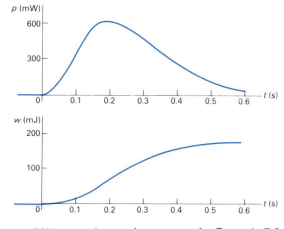

FIGURE 7.9 v, i, p, and w versus t for Example 7.2.

g) The application of the voltage pulse stores energy in the inductor. Because the inductor is ideal, this energy cannot dissipate after the voltage subsides to zero. Therefore a sustained current circulates in the circuit. A lossless inductor obviously is an ideal circuit element. Practical inductors require the insertion of a resistor in the circuit model. (More about this later.)

Examples 7.1, 7.2, and 7.3 illustrated the application of the basic equations that relate voltage and current at the terminals of an inductor and the subsequent calculations of power and energy. Next we consider the corresponding relationships for the capacitor.

DRILL EXERCISES

7.1 The current source in the circuit shown generates the current pulse

$$i_g(t) = 0, \qquad t < 0;$$
$$i_g(t) = 5e^{-200t} - 5e^{-800t} \text{ A}, \qquad t \geq 0.$$

Find (a) $v(0)$; (b) the instant of time, greater than zero, when the voltage v passes through zero; (c) the expression for the power delivered to the inductor; (d) the instant when the power delivered to the induc-

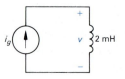

tor is maximum; (e) the maximum power; (f) the maximum energy stored in the inductor; and (g) the instant of time when the stored energy is maximum.

ANSWER: (a) 6 V; (b) 2.31 ms; (c) $50e^{-1000t} - 10e^{-400t} - 40e^{-1600t}$ W; (d) 616.58 μs; (e) 4.26 W; (f) 5.58 mJ; and (g) 2.31 ms.

7.2 THE CAPACITOR

The circuit parameter of capacitance is represented by the letter C, is measured in farads (F), and is symbolized by two short parallel conductive plates, as shown in Fig. 7.10(a). Because the farad is an extremely large quantity of capacitance, practical capacitors are based on submultiples of the farad. The most frequently encountered values lie in the picofarad (pF) to microfarad (μF) range. The graphic symbol for the capacitor is a reminder that capacitance occurs whenever electrical conductors are separated by a dielectric, or insulating, material. This condition implies that electric charge is not transported through the capacitor. Although the application of a voltage to the terminals of the capacitor cannot move a charge through the dielectric, it can displace a charge within the dielectric. As the voltage varies with time, the displacement of charge within the dielectric varies with time, causing what is known as the *displacement current*. At the terminals, the displacement current is indistinguishable from a conduction current. The current at the capacitor's terminals is proportional to the rate at which the voltage across the capacitor varies with time, or mathematically,

$$i = C\frac{dv}{dt}, \tag{7.13}$$

where i is measured in amperes, C in farads, v in volts, and t in seconds.

Equation (7.13) reflects the passive sign convention shown in Fig. 7.10(b); that is, the current reference is in the direction of the voltage drop across the capacitor. If the current reference is in the direction of the voltage rise, Eq. (7.13) is written with a minus sign.

Two important observations follow from Eq. (7.13). First, voltage cannot change instantaneously across the terminals of a capacitor. Equation (7.13) indicates that such a change would produce infinite current, a physical impossibility. Second, if the voltage across the terminals is constant, the capacitor current is zero. The reason is that a conduction current cannot be estab-

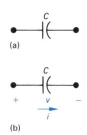

(a)

(b)

FIGURE 7.10 The circuit symbol for a capacitor.

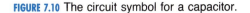

PSpice PSpice data statement for a capacitor: Chapter 6

lished in the dielectric material of the capacitor. Only a time-varying voltage can produce a displacement current. Thus a capacitor appears as an open circuit to a constant voltage.

Equation (7.13) gives the capacitor current as a function of the capacitor voltage. Expressing the voltage as a function of the current also is useful. To do so we multiply both sides of Eq. (7.13) by a differential time dt and then integrate the resulting differentials:

$$i\,dt = C\,dv \qquad \text{or} \qquad \int_{v(t_0)}^{v(t)} dx = \frac{1}{C}\int_{t_0}^{t} i\,d\tau.$$

Carrying out the integration of the left-hand side of the second equation gives

$$v(t) = \frac{1}{C}\int_{t_0}^{t} i\,d\tau + v(t_0). \qquad (7.14)$$

In many practical applications of Eq. (7.14), the initial time is zero; that is, $t_0 = 0$. Thus Eq. (7.14) becomes

$$v(t) = \frac{1}{C}\int_{0}^{t} i\,d\tau + v(0). \qquad (7.15)$$

We can easily derive the power and energy relationships for the capacitor. From the definition of power,

$$p = vi = Cv\frac{dv}{dt}, \qquad (7.16)$$

or

$$p = i\left[\frac{1}{C}\int_{t_0}^{t} i\,d\tau + v(t_0)\right]. \qquad (7.17)$$

Combining the definition of energy with Eq. (7.16) yields

$$dw = Cv\,dv,$$

from which

$$\int_{0}^{w} dx = C\int_{0}^{v} y\,dy,$$

or

$$w = \tfrac{1}{2}Cv^2. \qquad (7.18)$$

In the derivation of Eq. (7.18), the reference for zero energy corresponds to zero voltage.

Examples 7.4 and 7.5 illustrate application of the current, voltage, power, and energy relationships for the capacitor.

EXAMPLE 7.4

The voltage pulse described by the following equations is impressed across the terminals of a $0.5\text{-}\mu\text{F}$ capacitor:

$$v(t) = 0, \qquad\qquad t \le 0;$$
$$v(t) = 4t \text{ V}, \qquad\quad 0 \le t \le 1;$$
$$v(t) = 4e^{-(t-1)} \text{ V}, \qquad 1 \le t \le \infty.$$

a) Derive the expressions for the capacitor current, power, and energy.

b) Sketch the voltage, current, power, and energy as functions of time. Follow the instructions given in Example 7.3.

c) Specify the interval of time when energy is being stored in the capacitor.

d) Specify the interval of time when energy is being delivered by the capacitor.

e) Evaluate the integrals

$$\int_0^1 p\, dt \qquad \text{and} \qquad \int_1^\infty p\, dt$$

and comment on their significance.

SOLUTION

a) From Eq. (7.13),

$$i = (0.5 \times 10^{-6})(0) = 0, \qquad\qquad t < 0;$$
$$i = (0.5 \times 10^{-6})(4) = 2\ \mu\text{A}, \qquad\qquad 0 < t < 1;$$
$$i = (0.5 \times 10^{-6})(-4e^{-(t-1)}) = -2e^{-(t-1)}\ \mu\text{A}, \quad 1 < t < \infty.$$

The expression for the power is derived from Eq. (7.16):

$$p = 0, \qquad\qquad\qquad\qquad t < 0;$$
$$p = (4t)(2) = 8t\ \mu\text{W}, \qquad\qquad 0 \le t < 1;$$
$$p = (4e^{-(t-1)})(-2e^{-(t-1)}) = -8e^{-2(t-1)}\ \mu\text{W}, \qquad 1 < t \le \infty.$$

The energy expression follows directly from Eq. (7.18):

$$w = 0, \qquad\qquad\qquad\qquad t < 0,$$
$$w = \tfrac{1}{2}(0.5)16t^2 = 4t^2\ \mu\text{J}, \qquad\qquad 0 \le t \le 1,$$
$$w = \tfrac{1}{2}(0.5)16e^{-2(t-1)} = 4e^{-2(t-1)}\ \mu\text{J}, \qquad 1 \le t \le \infty.$$

b) Figure 7.11 shows the voltage, current, power, and energy as functions of time.

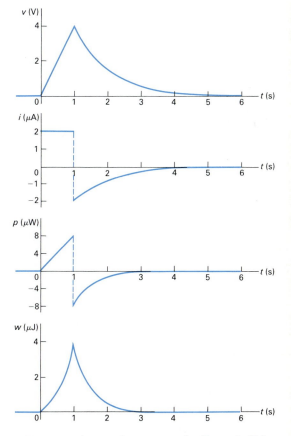

FIGURE 7.11 v, i, p, and w versus t for Example 7.4.

c) Energy is being stored in the capacitor whenever the power is positive. Hence energy is being stored in the interval 0 to 1 s.

d) Energy is being delivered by the capacitor whenever the power is negative. Thus energy is being delivered by the capacitor for all t greater than 1 s.

e) The integral of $p\,dt$ is the energy associated with the time interval corresponding to the limits on the integral. Thus the first integral represents the energy stored in the capacitor between 0 and 1 s, whereas the second integral represents the energy returned, or delivered, by the capacitor in the interval 1 s to ∞:

$$\int_0^1 p\,dt = \int_0^1 8t\,dt = 4t^2 \bigg|_0^1 = 4\ \mu\text{J};$$

$$\int_1^\infty p\,dt = \int_1^\infty (-8e^{-2(t-1)})\,dt = (-8)\frac{e^{-2(t-1)}}{-2}\bigg|_1^\infty = -4\ \mu\text{J}.$$

The voltage applied to the capacitor returns to zero as time increases without limit, so the energy returned by this ideal capacitor must equal the energy stored.

EXAMPLE 7.5

An uncharged 0.2-μF capacitor is driven by a triangular current pulse. The current pulse is described by

$$\begin{aligned}
i(t) &= 0, & t &\le 0; \\
i(t) &= 5000t\ \text{A}, & 0 &\le t \le 20\ \mu\text{s}; \\
i(t) &= 0.2 - 5000t\ \text{A}, & 20 &\le t \le 40\ \mu\text{s}; \\
i(t) &= 0, & t &\ge 40\ \mu\text{s}.
\end{aligned}$$

a) Derive the expressions for the capacitor voltage, power, and energy for each of the four time intervals needed to describe the current.

b) Plot i, v, p, and w versus t. Align the plots as specified in the previous examples.

c) Why does a voltage remain on the capacitor after the current returns to zero?

SOLUTION

a) For $t \leq 0$, v, p, and w all are zero. For $0 \leq t \leq 20$ μs,

$$v = 5 \times 10^6 \int_0^t 5000\tau \, d\tau + 0 = 12.5 \times 10^9 t^2 \text{ V};$$

$$p = vi = 62.5 \times 10^{12} t^3 \text{ W};$$

$$w = \tfrac{1}{2} C v^2 = 15.625 \times 10^{12} t^4 \text{ J}.$$

For 20 $\mu s \leq t \leq 40$ μs,

$$v = 5 \times 10^6 \int_{20 \, \mu s}^t (0.2 - 5000\tau) \, d\tau + 5.$$

(Note that 5 V is the voltage on the capacitor at the end of the preceding interval.) Then,

$$v = (10^6 t - 12.5 \times 10^9 t^2 - 10) \text{ V},$$

$$p = vi$$

$$= (62.5 \times 10^{12} t^3 - 7.5 \times 10^9 t^2 + 2.5 \times 10^5 t - 2) \text{ W};$$

$$w = \tfrac{1}{2} C v^2$$

$$= (15.625 \times 10^{12} t^4 - 2.5 \times 10^9 t^3 + 0.125 \times 10^6 t^2$$

$$- 2.0 \times 10^{-6} t + 10^{-5}) \text{ J}.$$

For 40 $\mu s \leq t$,

$$v = 10 \text{ V};$$

$$p = vi = 0;$$

$$w = \tfrac{1}{2} C v^2 = 10 \text{ } \mu\text{J}.$$

b) The excitation current and the resulting voltage, power, and energy are plotted in Fig. 7.12.

c) Note that the power is always positive during the duration of the current pulse, which means that energy is continuously being stored in the capacitor by the current pulse. When the current returns to zero, the energy stored in the capacitor is trapped because the ideal capacitor offers no means for dissipating energy. Thus a voltage remains on the capacitor after i returns to zero.

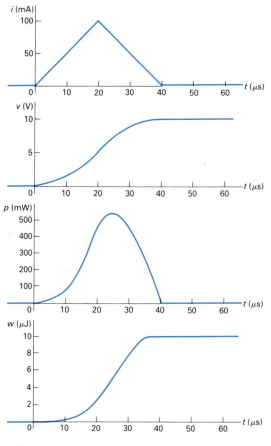

FIGURE 7.12 i, v, p, and w versus t for Example 7.5.

Problem 7.8 further pursues the solution to this example.

7.2 The voltage at the terminals of the 0.5-μF capacitor is 0 for $t < 0$ and $100e^{-20,000t} \sin 40,000t$ V for $t > 0$. Find (a) $i(0)$; (b) the power delivered to the capacitor at $t = \pi/80$ ms; and (c) the energy stored in the capacitor at $t = \pi/80$ ms.

0.5 μF

ANSWER: (a) 2 A; (b) -20.79 W; (c) 519.20 μJ.

7.3 The current in the capacitor of Drill Exercise 7.2 is 0 for $t < 0$ and $2 \cos 50,000t$ A for $t \geq 0$. Find (a) $v(t)$; (b) the maximum power delivered to the capacitor at any one instant of time; and (c) the maximum energy stored in the capacitor at any one instant of time.

ANSWER: (a) $80 \sin 50,000t$ V; (b) 80 W; (c) 1.6 mJ.

7.3 SERIES–PARALLEL COMBINATIONS OF INDUCTANCE AND CAPACITANCE

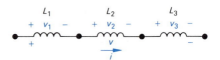

FIGURE 7.13 Inductors in series.

Just as series–parallel combinations of resistors can be reduced to a single equivalent resistor, series–parallel combinations of inductors or capacitors can be reduced to a single inductor or capacitor. Figure 7.13 shows inductors in series. In the series connection, the inductors are forced to carry the same current; thus we define only one current for the series combination. The voltage drops across the individual inductors are

$$v_1 = L_1 \frac{di}{dt}, \qquad v_2 = L_2 \frac{di}{dt}, \qquad \text{and} \qquad v_3 = L_3 \frac{di}{dt}.$$

The voltage across the series connection is

$$v = v_1 + v_2 + v_3 = (L_1 + L_2 + L_3)\frac{di}{dt},$$

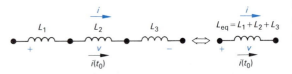

FIGURE 7.14 An equivalent circuit for inductors in series carrying an initial current $i(t_0)$.

from which it should be apparent that the equivalent inductance of series-connected inductors is the sum of the individual inductances. For n inductors in series,

$$L_{eq} = L_1 + L_2 + L_3 + \cdots + L_n. \qquad \textbf{(7.19)}$$

If the original inductors carry an initial current of $i(t_0)$, the equivalent inductor carries the same initial current. Figure 7.14

shows the equivalent circuit for series inductors carrying an initial current.

Inductors in parallel have the same terminal voltage. In the equivalent circuit the current in each inductor is a function of the terminal voltage and the initial current in the inductor. For the three inductors in parallel shown in Fig. 7.15, the currents for the individual inductors are

$$i_1 = \frac{1}{L_1} \int_{t_0}^{t} v \, d\tau + i_1(t_0);$$

$$i_2 = \frac{1}{L_2} \int_{t_0}^{t} v \, d\tau + i_2(t_0);$$

$$i_3 = \frac{1}{L_3} \int_{t_0}^{t} v \, d\tau + i_3(t_0). \tag{7.20}$$

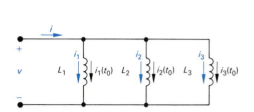

FIGURE 7.15 Three inductors in parallel.

The current at the terminals of the three parallel inductors is the sum of the inductor currents:

$$i = i_1 + i_2 + i_3. \tag{7.21}$$

Substituting Eq. (7.20) into Eq. (7.21) yields

$$i = \left(\frac{1}{L_1} + \frac{1}{L_2} + \frac{1}{L_3}\right) \int_{t_0}^{t} v \, d\tau + i_1(t_0) + i_2(t_0) + i_3(t_0). \tag{7.22}$$

Now we can interpret Eq. (7.22) in terms of a single inductor; that is,

$$i = \frac{1}{L_{eq}} \int_{t_0}^{t} v \, d\tau + i(t_0). \tag{7.23}$$

Comparing Eq. (7.23) with (7.22) yields

$$\frac{1}{L_{eq}} = \frac{1}{L_1} + \frac{1}{L_2} + \frac{1}{L_3} \tag{7.24}$$

and

$$i(t_0) = i_1(t_0) + i_2(t_0) + i_3(t_0). \tag{7.25}$$

Figure 7.16 shows the equivalent circuit for the three parallel inductors in Fig. 7.15.

The results expressed in Eqs. (7.24) and (7.25) can be extended to n inductors in parallel:

$$\frac{1}{L_{eq}} = \frac{1}{L_1} + \frac{1}{L_2} + \cdots + \frac{1}{L_n} \tag{7.26}$$

and

$$i(t_0) = i_1(t_0) + i_2(t_0) + \cdots + i_n(t_0). \tag{7.27}$$

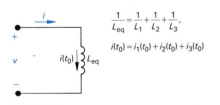

FIGURE 7.16 An equivalent circuit for three inductors in parallel.

Capacitors connected in series can be reduced to a single equivalent capacitor. The reciprocal of the equivalent capacitance is equal to the sum of the reciprocals of the individual capacitances. If each capacitor carries its own initial voltage, the initial voltage on the equivalent capacitor is the algebraic sum of the initial voltages on the individual capacitors. Figure 7.17 and the following equations summarize these observations:

$$\frac{1}{C_{\text{eq}}} = \frac{1}{C_1} + \frac{1}{C_2} + \cdots + \frac{1}{C_n}; \tag{7.28}$$

$$v(t_0) = v_1(t_0) + v_2(t_0) + \cdots + v_n(t_0). \tag{7.29}$$

We leave derivation of the equivalent circuit for series-connected capacitors as an exercise. (See Problem 7.24.)

The equivalent capacitance of capacitors connected in parallel is simply the sum of the capacitances of the individual capacitors, as Fig. 7.18 and the following equation show:

$$C_{\text{eq}} = C_1 + C_2 + \cdots + C_n. \tag{7.30}$$

Capacitors connected in parallel must carry the same voltage. Therefore, if there is an initial voltage across the original parallel capacitors, this same initial voltage appears across the equivalent capacitance C_{eq}. The derivation of the equivalent circuit for capacitors connected in parallel is left as an exercise. (See Problem 7.25.)

We say more about series–parallel equivalent circuits of inductors or capacitors in Chapter 8, where we interpret results based on their use.

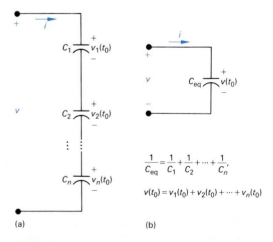

(a)   (b)

FIGURE 7.17 An equivalent circuit of capacitors connected in series: (a) series capacitors; (b) the equivalent circuit.

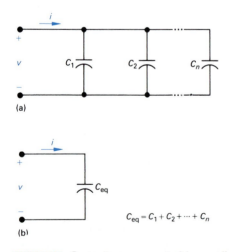

(a)

(b)

FIGURE 7.18 Capacitors connected in parallel: (a) capacitors in parallel; (b) the equivalent circuit.

DRILL EXERCISES

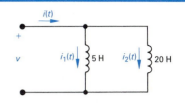

7.4 The initial values of i_1 and i_2 in the circuit shown are -2 and $+4$ A, respectively. The voltage at the terminals of the paralleled inductors for $t \geq 0$ is $-40e^{-5t}$ V.

a) If the parallel inductors are replaced by a single inductor, what is its inductance?

b) What is the initial current and its reference direction in the equivalent inductor?

c) Use the equivalent inductor to find $i(t)$.

d) Find $i_1(t)$ and $i_2(t)$. Verify that the solutions for $i_1(t)$, $i_2(t)$, and $i(t)$ satisfy Kirchhoff's current law.

ANSWER: (a) 4H; (b) 2 A, down; (c) $2e^{-5t}$ A; (d) $i_1(t) = 1.6e^{-5t} - 3.6$ A, $i_2(t) = 0.4e^{-5t} + 3.6$ A.

7.5 The current at the terminals of the two capacitors shown is $240e^{-10t}$ μA for $t \geq 0$. The initial values of v_1 and v_2 are -10 and -5 V, respectively. Calculate the total energy trapped in the capacitors as $t \to \infty$.

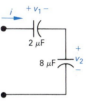

ANSWER: 20 μJ.

SUMMARY

We introduced two ideal circuit elements: the inductor and the capacitor. We stressed the following important points.

- Inductance is a linear circuit parameter that relates the voltage induced by a time-varying magnetic field to the current producing the time-varying magnetic field.

- Capacitance is a linear circuit parameter that relates the current induced by a time-varying electric field to the voltage producing the time-varying electric field.

- Both the inductor and the capacitor are linear, passive elements that can store and release energy. However, neither element can generate or dissipate energy.

- The instantaneous power at the terminals of an inductor or capacitor can be positive or negative, depending on whether energy is being delivered to or extracted from the element.

- An inductor *does not* permit an instantaneous change in its terminal current.

- A capacitor *does not* permit an instantaneous change in its terminal voltage.

- An inductor *does* permit an instantaneous change in its terminal voltage.

- A capacitor *does* permit an instantaneous change in its terminal current.

- An inductor appears as a short circuit in relation to a constant terminal current.

- A capacitor appears as an open circuit in relation to a constant terminal voltage.

Table 7.1 summarizes the mathematical equations, based on the passive sign convention, that describe the terminal behavior of the inductor and capacitor.

TABLE 7.1

TERMINAL EQUATIONS FOR IDEAL INDUCTORS AND CAPACITORS*

INDUCTORS		CAPACITORS	
1. $v = L\dfrac{di}{dt}$	(V)	1. $v = \dfrac{1}{C}\displaystyle\int_{t_0}^{t} i\, d\tau + v(t_0)$	(V)
2. $i = \dfrac{1}{L}\displaystyle\int_{t_0}^{t} v\, d\tau + i(t_0)$	(A)	2. $i = C\dfrac{dv}{dt}$	(A)
3. $p = vi = Li\dfrac{di}{dt}$	(W)	3. $p = vi = Cv\dfrac{dv}{dt}$	(W)
4. $w = \frac{1}{2}Li^2$	(J)	4. $w = \frac{1}{2}Cv^2$	(J)

*The equations in this table are based on the passive sign convention.

PROBLEMS

7.1 Evaluate the integral

$$\int_0^{\infty} p\, dt$$

for Example 7.2. Comment on the significance of the result.

7.2 The triangular current pulse shown in Fig. P7.2 is applied to a 25-mH inductor.

a) Write the expressions that describe $i(t)$ in the four intervals $t < 0$, $0 \le t \le 5$ ms, 5 ms $\le t \le 10$ ms, and $t > 10$ ms.

b) Derive the expressions for the inductor voltage, power, and energy. Use the passive sign convention.

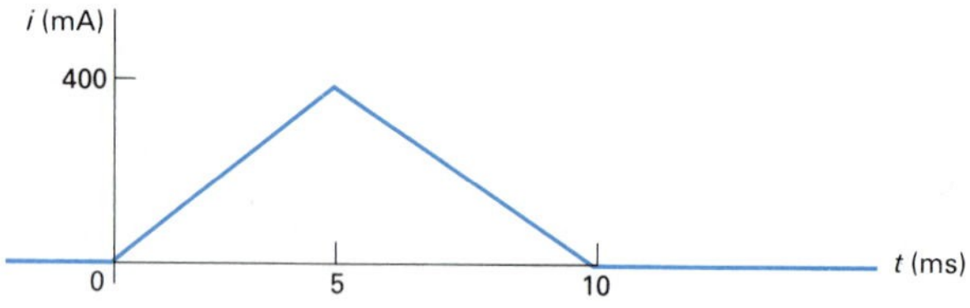

FIGURE P7.2

7.3 The voltage at the terminals of the 200-μH inductor in Fig. P7.3(a) is shown in Fig. P7.3(b). The inductor current i is known to be zero for $t \leq 0$.

a) Derive the expressions for i for $t \geq 0$.

b) Sketch i vs. t for $0 \leq t \leq \infty$.

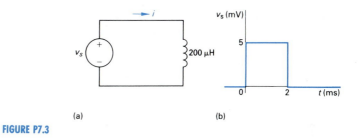

(a)   (b)

FIGURE P7.3

7.4 The current in the 2.5-mH inductor in Fig. P7.4 is known to be 1 A for $t \leq 0$. The inductor voltage for $t \geq 0^+$ is given by the expressions

$$v_L(t) = 3\,e^{-4t}\ \text{mV}, \quad 0^+ \leq t < 2\text{s}$$

$$v_L(t) = -3\,e^{-4(t-2)}\ \text{mV}, \quad 2^+\text{s} \leq t \leq \infty.$$

Sketch $v_L(t)$ and $i_L(t)$ for $0 \leq t \leq \infty$.

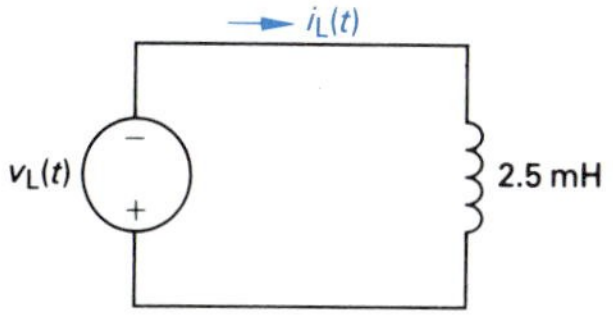

FIGURE P7.4

7.5 a) Find the inductor current in the circuit in Fig. P7.5 if $v = -50 \sin 250t$ V, $L = 20$ mH, and $i(4\pi\ \text{ms}) = -10$ A.

b) Sketch v, i, p, and w, versus time. In making these sketches, use the format used in Fig. 7.8. Plot over one complete cycle of the voltage waveform.

c) Describe the subintervals in the time interval between 0 and 4π ms when power is being

absorbed by the inductor. Repeat for the subintervals when power is being delivered by the inductor.

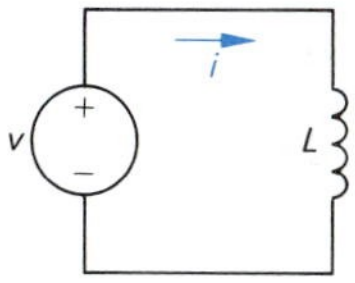

FIGURE P7.5

7.6 The current in a 25-mH inductor is known to be -10 A for $t \leq 0$ and $i = -[10 \cos 400t + 5 \sin 400t]e^{-200t}$ A for $t \geq 0$. Assume the passive sign convention.

a) At what instant of time is the voltage across the inductor maximum?

b) What is the maximum voltage?

7.7 The current in a 50-μH inductor is known to be

$$i_L = 18t\,e^{-10t}\ \text{A for } t \geq 0.$$

a) Find the voltage across the inductor for $t > 0$. (Assume the passive sign convention.)

b) Find the power (in μW) at the terminals of the inductor when $t = 200$ ms.

c) Is the inductor absorbing or delivering power at 200 ms?

d) Find the energy (in μJ) stored in the inductor at 200 ms.

e) Find the maximum energy (in μJ) stored in the inductor, and the time (in ms) when it occurs.

7.8 The current in and the voltage across a 5-H inductor are known to be zero for $t \leq 0$. The voltage across the inductor is given by the graph in Fig. P7.8 for $t \geq 0$.

a) Derive the expressions for the current as a function of time in the intervals $0 \leq t \leq 1s$, $1s \leq t \leq 3s$, $3s \leq t \leq 5s$, $5s \leq t \leq 6s$, and $6s \leq t \leq \infty$.

b) For $t > 0$ what is the current in the inductor when the voltage is zero?

c) Sketch i vs t for $0 \leq t \leq \infty$.

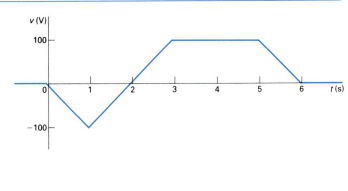

FIGURE P7.8

7.9 The current in a 4-H inductor is

$$i = 10 \text{ A}, \ t \leq 0;$$

$$i = (B_1 \cos 4t + B_2 \sin 4t)e^{-0.5t} \text{ A}, \ t \geq 0.$$

The voltage across the inductor (passive-sign convention) is 60 V at $t = 0$.
Calculate the power at the terminals of the inductor at $t = 1s$. State whether the inductor is absorbing or delivering power.

7.10 The current in a 20-mH inductor is known to be

$$i = 40 \text{ mA}, \ t \leq 0$$

$$i = A_1 e^{-10,000t} + A_2 e^{-40,000t} \text{ A}, \ t \geq 0.$$

The voltage across the inductor (passive-sign convention) is 28 V at $t = 0$.

a) Find the expression for the voltage across the inductor for $t > 0$.

b) Find the time, greater than zero, when the power at the terminals of the inductor is zero.

7.11 Assume in Problem 7.10 that the value of the voltage across the inductor at $t = 0$ is -68 V instead of 28 V.

a) Find the numerical expressions for i and v for $t \geq 0$.

b) Specify the time intervals when the inductor

is storing energy and the time intervals when the inductor is delivering energy.

c) Show that the total energy extracted from the inductor is equal to the total energy stored.

7.12 Initially there was no energy stored in the 5-H inductor in the circuit in Fig. P7.12 when it was placed across the terminals of the voltmeter. At $t = 0$ the inductor was switched instantaneously to position b where it remained for 1.6 s before returning instantaneously to position a. The d'Arsonval voltmeter has a full scale reading of

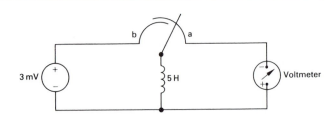

FIGURE P7.12

20 V and a sensitivity of 1000 Ω/volt. What will the reading of the voltmeter be at the in-

stant the switch returns to position a if the inertia of the d'Arsonval movement is negligible?

7.13 The voltage across the terminals of a 0.20-μF capacitor is

$$v = 150 \text{ V}, t \leq 0$$

$$v = A_1 t e^{-5000t} + A_2 e^{-5000t} \text{ V}, t \geq 0.$$

The initial current in the capacitor is 250 mA. Assume the passive sign convention.

a) What is the initial energy stored in the capacitor?

b) Evaluate the coefficients A_1 and A_2.

c) What is the expression for the capacitor current?

7.14 The voltage at the terminals of the capacitor in Fig. 7.10 is known to be

$$v = -20 \text{ V}, t \leq 0$$

$$v = 100 - 40e^{-2000t}(3 \cos 1000t + \sin 1000t) \text{ V},$$
$$t \geq 0.$$

If $C = 0.4$ μF:

a) Find the current in the capacitor for $t < 0$.

b) Find the current in the capacitor for $t > 0$.

c) Is there an instantaneous change in the voltage across the capacitor at $t = 0$?

d) Is there an instantaneous change in the current in the capacitor at $t = 0$?

e) How much energy (in μJ) is stored in the capacitor at $t = \infty$?

7.15 The rectangular-shaped current pulse shown in Fig. P7.15 is applied to a 0.1-μF capacitor. The initial voltage on the capacitor is a 15-V drop in the reference direction of the current. Assume the passive sign convention. Derive the expression for the capacitor voltage for each of the following time intervals:

a) $0 \leq t \leq 10$ μs

b) 10 μs $\leq t \leq 20$ μs

c) 20 μs $\leq t \leq 40$ μs

d) 40 μs $\leq t \leq \infty$.

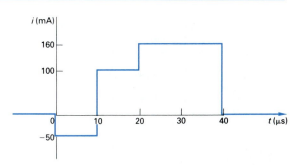

7.16 The current pulse shown in Fig. P7.16 is applied to a 0.25-μF capacitor. The initial voltage on the capacitor is zero.

a) Find the charge on the capacitor at $t = 15$ μs.

b) Find the voltage on the capacitor at $t = 30$ μs.

c) How much energy is stored in the capacitor by the current pulse?

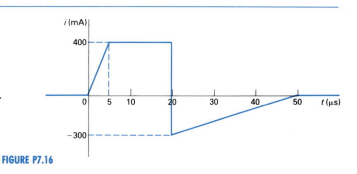

7.17 The initial voltage on the 5-μF capacitor shown in Fig. P7.17(a) is 4 V. The capacitor current has the waveform shown in Fig. P7.17(b).

a) How much energy, in microjoules, is stored in the capacitor at $t = 400\ \mu$s?

b) Repeat part (a) for $t = 600\ \mu$s.

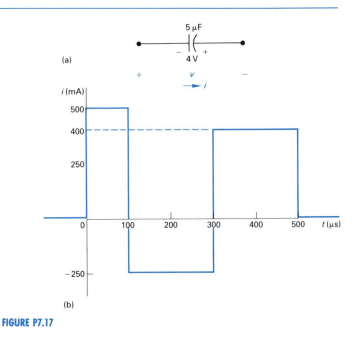

FIGURE P7.17

7.18 A 20-μF capacitor is subjected to a voltage pulse having a duration of 1 sec. The pulse is described by the following equations:

$$v_c(t) = 30t^2\ \text{V} \qquad 0 \le t \le 0.5\ \text{s}$$
$$v_c(t) = 30(t - 1)^2\ \text{V} \qquad 0.5\ \text{s} \le t \le 1.0\ \text{s}$$
$$v_c(t) = 0 \qquad \text{elsewhere.}$$

Sketch the current pulse that exists in the capacitor during the 1-s interval.

7.19 The expressions for voltage, power, and energy derived in Example 7.5 involved both integration and manipulation of algebraic expressions. As an engineer, you cannot accept such results on faith alone. That is, you should develop the habit of asking yourself, "Do these results make sense in terms of the known behavior of the circuit they purport to describe?" With these thoughts in mind, test the expressions of Example 7.5 by performing the following checks:

a) Check the expressions to see whether the voltage is continuous in passing from one time interval to the next.

b) Check the power expression in each interval by selecting a time within the interval and see whether it gives the same result as the corresponding product of v and i. For example, test at 10 and 30 μs.

c) Check the energy expression within each interval by selecting a time within the interval and see whether the energy equation gives the same result as $\frac{1}{2}Cv^2$. Use 10 and 30μs as test points.

7.20 Assume that the initial energy stored in the inductors of Fig. P7.20 is zero. Find the equivalent inductance with respect to the terminals a, b.

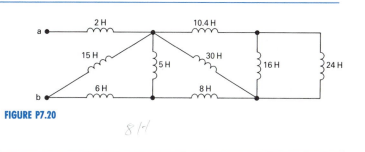

FIGURE P7.20

7.21 The two parallel inductors in Fig. P7.21 are connected across the terminals of a black box at $t = 0$. The resulting voltage v for $t \geq 0$ is known to be $64e^{-4t}$ V. It is also known that $i_1(0) = -10$A and $i_2(0) = 5$A.

a) Replace the original inductors with an equivalent inductor and find $i(t)$ for $t \geq 0$.

b) Find $i_1(t)$ for $t \geq 0$.

c) Find $i_2(t)$ for $t \geq 0$.

d) How much energy is delivered to the black box in the time interval $0 \leq t \leq \infty$?

e) How much energy was initially stored in the parallel inductors?

f) How much energy is trapped in the ideal inductors?

g) Do your solutions for i_1 and i_2 agree with the answer obtained in part (f)?

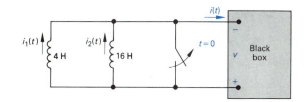

FIGURE P7.21

7.22 The three inductors in the circuit in Fig. P7.22 are connected across the terminals of a black box at $t = 0$. The resulting voltage for $t \geq 0$ is known to be

$$v_b = 2000 \, e^{-100t} \text{ V.}$$

If $i_1(0) = -6$A and $i_2(0) = 1$ A, find

a) $i_0(0)$;

b) $i_0(t)$, $t \geq 0$;

c) $i_1(t)$, $t \geq 0$;

d) $i_2(t)$, $t \geq 0$;

e) the initial energy stored in the three inductors;

f) the total energy delivered to the black box; and

g) the energy trapped in the ideal inductors.

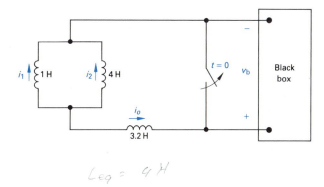

FIGURE P7.22

7.23 For the circuit shown in Fig. P7.22 how many milliseconds after the switch is opened is the energy delivered to the black box 80% of the total amount delivered?

7.24 Derive the equivalent circuit for a series connection of ideal capacitors. Assume that each capacitor has its own initial voltage. Denote these initial voltages as $v_1(t_0)$, $v_2(t_0)$, ..., etc. (*Hint:* Sum the voltages across the string of capacitors, recognizing that the series connection forces the current in each capacitor to be the same.)

7.25 Derive the equivalent circuit for a parallel connection of ideal capacitors. Assume that the initial voltage across the paralleled capacitors is $v(t_0)$. (*Hint:* Sum the currents into the string of capacitors, recognizing that the parallel connection forces the voltage across each capacitor to be the same.)

7.26 Find the equivalent capacitance with respect to the terminals a, b for the circuit shown in Fig. P7.26.

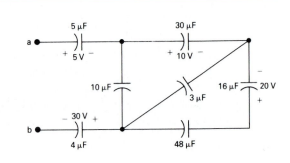

FIGURE P7.26

7.27 The four capacitors in the circuit in Fig. P7.27 are connected across the terminals of a black box at $t = 0$. The resulting current i_b for $t \geq 0$ is known to be

$$i_b = -5e^{-50t} \text{ mA}.$$

If $v_a(0) = -20$ V, $v_c(0) = -30$ V, and $v_d(0) = 250$ V, find for $t \geq 0$, (a) $v_b(t)$, (b) $v_a(t)$, (c) $v_c(t)$, (d) $v_d(t)$, (e) $i_1(t)$, and (f) $i_2(t)$.

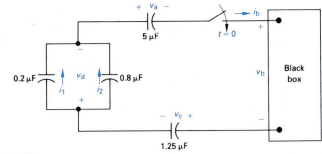

FIGURE P7.27

7.28 For the circuit in Fig. P7.27 calculate:

a) the initial energy stored in the capacitors;

b) the final energy stored in the capacitors;

c) the total energy delivered to the black box;

d) the percentage of the initial energy stored that is delivered to the black box; and

e) the time, in milliseconds, it takes to deliver 7500 μJ to the black box.

7.29 The two series-connected capacitors in Fig. P7.29 are connected to the terminals of a black box at $t = 0$. The resulting current $i(t)$ for $t \geq 0$ is known to be $800e^{-25t}$ μA.

a) Replace the original capacitors with an equivalent capacitor and find $v_o(t)$ for $t \geq 0$.

b) Find $v_1(t)$ for $t \geq 0$.

c) Find $v_2(t)$ for $t \geq 0$.

d) How much energy is delivered to the black box in the time interval $0 \leq t \leq \infty$?

e) How much energy was initially stored in the series capacitors?

f) How much energy is trapped in the ideal capacitors?

g) Do the solutions for v_1 and v_2 agree with the answer obtained in part (f)?

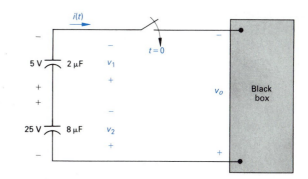

FIGURE P7.29

7.30 At $t = 0$ a series-connected capacitor and inductor are placed across the terminals of a black box, as shown in Fig. P7.30. For $t \geq 0$ it is known that

$$i_o = e^{-80t} \sin 60t \text{ A}.$$

If $v_c(0) = -300$ V find v_o for $t \geq 0$.

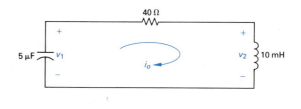

FIGURE P7.30

7.31 The current in the circuit in Fig. P7.31 is known to be

$$i_0 = 5 \, e^{-2000t}[2 \cos 4000t + \sin 4000t] \text{ A}$$

for $t \geq 0^+$.

Find $v_1(0^+)$ and $v_2(0^+)$.

FIGURE P7.31

RESPONSE OF FIRST-ORDER
RL AND *RC* CIRCUITS

CHAPTER 8

In Chapter 7 we discussed the terminal behavior of ideal inductors and capacitors, noting that an important attribute of these circuit elements is their ability to store energy. We now determine the currents and voltages that arise when the energy stored in either an inductor or a capacitor is released or acquired, with the analysis divided into three phases.

In the first phase of the analysis we discuss the currents and voltages that arise when stored energy in an inductor or capacitor is released to a resistive network. In this first phase the resistive network may contain dependent sources, but there are no independent sources in the circuit when the energy is released. Thus, when determining voltages and currents, we reduce the structure of the circuit to one of the two equivalent forms shown in Fig. 8.1. The currents and voltages are referred to as the *natural response* of the circuit. This designation emphasizes that the nature of the circuit itself, not external sources of excitation, determine the currents and voltages.

In the second phase of the analysis we discuss the problem of finding currents and voltages generated in *RL* and *RC* circuits when either dc voltage or current sources are suddenly applied to the circuit. The response of a circuit to the sudden application of a constant voltage or current source is referred to as the *step response* of the circuit.

In the third phase of the analysis we develop a general method that can be used to find the response of *RL* and *RC* circuits to any abrupt change in a dc voltage or current source. In doing so we demonstrate that the thought process for finding the natural response is the same as that for finding the response to step changes in dc voltage or current sources.

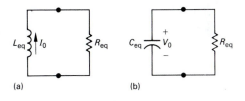

(a)   (b)

FIGURE 8.1 The two forms of circuits for natural response: (a) *RL* Circuit; (b) *RC* Circuit.

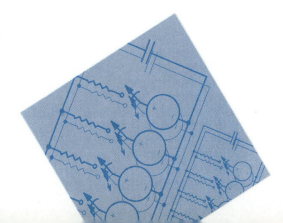

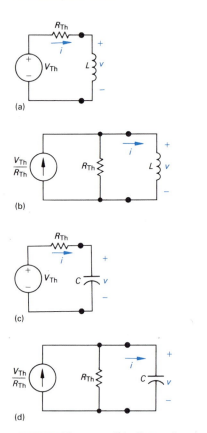

FIGURE 8.2 Four possible first-order circuits: (a) Inductor connected to a Thévenin equivalent; (b) inductor connected to a Norton equivalent; (c) capacitor connected to a Thévenin equivalent; (d) capacitor connected to a Norton equivalent.

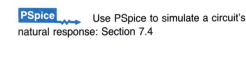

Use PSpice to simulate a circuit's natural response: Section 7.4

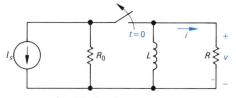

FIGURE 8.3 An *RL* circuit.

We limit the analysis in this chapter to single time-constant or first-order circuits.[†] We say more about these descriptive terms later, but for now we can define the type of circuit we are analyzing in terms of the equivalent circuits introduced earlier. A first-order circuit occurs whenever a circuit can be reduced to a Thévenin or Norton equivalent connected to the terminals of an equivalent inductor or capacitor. Figure 8.2 shows the four possibilities. A first-order circuit is made up of sources, resistors, and either inductors or capacitors. Both inductors and capacitors are not present in a first-order circuit. Furthermore, if more than one inductor—or capacitor—exists in the circuit, they must be interconnected so that they can be replaced by a single equivalent element. Note that when there are no independent sources in the circuit, the Thévenin voltage or Norton current is zero. Thus when finding the natural response, we reduce the four circuits shown in Fig. 8.2 to the two circuits shown in Fig. 8.1.

After introducing the techniques for analyzing the natural and step response of first-order circuits, we discuss some special cases of interest. The first is that of *sequential switching*, involving circuits in which switching can take place at two or more instants in time. Next, we extend the discussion to the case of the *unbounded response*. Finally, we analyze a useful circuit called the *integrating amplifier*. We begin with the natural response of an *RL* circuit.

8.1 THE NATURAL RESPONSE OF AN *RL* CIRCUIT

The natural response of an *RL* circuit refers to the absence of independent sources in the circuit during the time stored energy is being released. Independent sources are present only during the time energy is being stored. We cover the analogous situation for the natural response of an *RC* circuit in Section 8.2.

The natural response of the *RL* circuit can best be described in terms of the circuit shown in Fig. 8.3. We assume that the independent current source generates a constant current of I_s amperes and that the switch has been in a closed position for a long time. (We define the phrase *a long time* more accurately later in this section.) For now it means that all currents and voltages have reached a constant value. Thus only constant, or dc, cur-

---

[†] We discuss second-order circuits involving both capacitors and inductors in the same circuit in Chapter 9 and multiple time-constant circuits in Chapter 16.

rents can exist in the circuit just prior to the switch being opened. Hence the inductor appears as a short circuit ($L\,di/dt = 0$) prior to the release of the stored energy.

Note that if the voltage across the inductive branch is zero, there can be no current in either R_o or R. Therefore all the source current I_S appears in the inductive branch. The problem is to find the voltage and current at the terminals of the resistor after the switch has been opened. If we let $t = 0$ denote the instant when the switch is opened, the problem becomes one of finding $v(t)$ and $i(t)$ for $t \geq 0$. For $t \geq 0$ the circuit shown in Fig. 8.3 reduces to that shown in Fig. 8.4.

FIGURE 8.4 The circuit shown in Fig. 8.3 for $t \geq 0$.

DERIVING THE EXPRESSION FOR THE CURRENT

To find $i(t)$ we use Kirchhoff's voltage law to obtain an expression involving i, R, and L. Summing the voltages around the closed loop gives

$$L\frac{di}{dt} + Ri = 0, \tag{8.1}$$

where we used the passive sign convention. Equation (8.1) is known as a first-order ordinary differential equation, so called because it contains terms involving the ordinary derivative of the unknown, that is, di/dt. The highest order derivative appearing in the equation is 1; hence the term *first-order*.

We can go one step further in describing this equation. The coefficients in the equation, R and L, are constants; that is, they are not functions of either the dependent variable i or the independent variable t. Thus the equation also can be described as an ordinary differential equation with constant coefficients. (The solution of ordinary differential equations is a well-established practice in mathematics. However, we assume that you have not yet studied the solutions of such equations, and therefore, whenever possible in this book, we solve equations using ordinary calculus.)

To solve Eq. (8.1), we divide through by L, transpose the term involving i to the right-hand side of the equation, and then multiply both sides by a differential time dt. The result is

$$\frac{di}{dt}dt = -\frac{R}{L}i\,dt. \tag{8.2}$$

Next, we recognize the left-hand side of Eq. (8.2) as a differential change in the current i, that is, di. We now divide through by i, getting

$$\frac{di}{i} = -\frac{R}{L}dt. \tag{8.3}$$

We obtain an explicit expression for i as a function of t by integrating both sides of Eq. (8.3). Using x and y as variables of integration yields

$$\int_{i(t_0)}^{i(t)} \frac{dx}{x} = -\frac{R}{L} \int_{t_0}^{t} dy, \tag{8.4}$$

in which $i(t_0)$ is the current corresponding to time t_0 and $i(t)$ is the current corresponding to time t. Here, $t_0 = 0$. Therefore carrying out the indicated integration gives

$$\ln \frac{i(t)}{i(0)} = -\frac{R}{L}t. \tag{8.5}$$

Based on the definition of the natural logarithm,

$$i(t) = i(0)e^{-(R/L)t}. \tag{8.6}$$

We noted earlier that the current in the inductor is I_S just before the switch is opened. Now recall from Chapter 7 that an instantaneous change of current cannot occur in an inductor. Therefore, in the first instant after the switch has been opened, the current in the inductor remains at I_S. If we use 0^- to denote the time just prior to switching and 0^+ the time immediately following switching, then

$$i(0^-) = i(0^+) = I_S = I_0,$$

where, as in Fig. 8.1, I_0 denotes the initial current in the inductor. The initial current in the inductor is oriented in the same direction as the reference direction of i. Hence Eq. (8.6) becomes

$$i(t) = I_0 e^{-(R/L)t}, \qquad t \geq 0, \tag{8.7}$$

which shows that the current starts from an initial value I_0 and decreases exponentially toward zero as t increases. Figure 8.5 shows this response graphically.

The coefficient of t, namely, R/L determines the rate at which the current approaches zero. The reciprocal of this ratio is the *time constant* of the circuit, denoted τ:

$$\tau = \text{time constant} = \frac{L}{R}. \tag{8.8}$$

Using the time-constant concept, we write Eq. (8.7) as

$$i(t) = I_0 e^{-t/\tau}, \qquad t \geq 0. \tag{8.9}$$

The time constant is an important parameter for this type of circuit, so mentioning several of its characteristics is worthwhile. First, thinking of the time elapsed after switching has occurred in terms of integral multiples of τ is convenient. Thus 1 time constant after the inductor has begun to release its stored

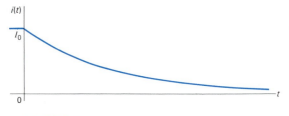

FIGURE 8.5 The current response for the circuit shown in Fig. 8.4.

TABLE 8.1

VALUES OF $e^{-t/\tau}$ FOR t EQUAL TO INTEGRAL MULTIPLES OF τ

t	$e^{-t/\tau}$	t	$e^{-t/\tau}$
τ	3.6788×10^{-1}	6τ	2.4788×10^{-3}
2τ	1.3534×10^{-1}	7τ	9.1188×10^{-4}
3τ	4.9787×10^{-2}	8τ	3.3546×10^{-4}
4τ	1.8316×10^{-2}	9τ	1.2341×10^{-4}
5τ	6.7379×10^{-3}	10τ	4.5400×10^{-5}

energy to the resistance the current has been reduced to e^{-1}, or approximately 0.37 of its initial value. Table 8.1 gives the value of $e^{-t/\tau}$ for integral multiples of τ from 1 to 10. Note that when the elapsed time exceeds 5 time constants, the current is less than 1 percent of its initial value. Thus we sometimes say that 5 time constants after switching has occurred the currents and voltages have, for most practical purposes, reached their final values. Hence for single time-constant circuits and with 1 percent accuracy, the phrase *a long time* implies that 5 or more time constants have elapsed.

A second characteristic of the time constant is that it gives the time required for a current to reach its final value if it continued to change at its initial rate. To illustrate, we evaluate di/dt at 0^+ and assume that the current continues to change at this rate:

$$\frac{di}{dt}(0^+) = -\frac{R}{L}I_0 = -\frac{I_0}{\tau}. \quad (8.10)$$

Now, if i starts as I_0 and decreases at a constant rate of I_0/τ amperes per second, the expression for i becomes

$$i = I_0 - \frac{I_0}{\tau}t. \quad (8.11)$$

Equation (8.11) indicates that i would reach its final value of zero in τ seconds. Figure 8.6 shows this characteristic of the time constant. Sometimes this graphic interpretation of the time constant is useful in estimating the time constant of a circuit from an oscilloscope trace of its natural response.

We derive the voltage across the resistor in the circuit shown in Fig. 8.4 from a direct application of Ohm's law:

$$v = iR = I_0Re^{-(R/L)t} = I_0Re^{-t/\tau}, \qquad t \geq 0^+. \quad (8.12)$$

Note that in contrast to the expression for the current shown in Eq. (8.9) the voltage is defined only for $t > 0$, not at $t = 0$. The reason is that a step change occurs in the voltage at zero. Note that for $t < 0$, the derivative of the current was zero, so

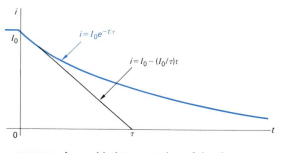

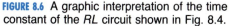

FIGURE 8.6 A graphic interpretation of the time constant of the *RL* circuit shown in Fig. 8.4.

the voltage also was zero. (This result follows from $v = L\,di/dt = 0$.) Thus

$$v(0^-) = 0, \tag{8.13}$$

and

$$v(0^+) = I_0 R, \tag{8.14}$$

where $v(0^+)$ is obtained from Eq. (8.12) with $t = 0^+$.[†] With this step change at an instant in time, the value of the voltage at $t = 0$ is unknown. Thus we use $t \geq 0^+$ in defining the region of validity for these solutions.

We derive the power dissipated in the resistor from any of the following expressions:

$$p = vi, \qquad p = i^2 R, \qquad \text{and} \qquad p = \frac{v^2}{R}. \tag{8.15}$$

Whichever form is used, the resulting expression can be reduced to

$$p = I_0^2 Re^{-2t/\tau}, \qquad t \geq 0^+. \tag{8.16}$$

The energy delivered to the resistor during any interval of time after the switch has been opened is

$$w = \int_0^t p\,dx = \int_0^t I_0^2 Re^{-2x/\tau}\,dx$$

$$= \frac{\tau}{2} I_0^2 R(1 - e^{-2t/\tau})$$

$$= \frac{1}{2} L I_0^2 (1 - e^{-2t/\tau}), \qquad t \geq 0. \tag{8.17}$$

Note from Eq. (8.17) that as t becomes infinite, the energy dissipated in the resistor approaches the initial energy stored in the inductor.

We have mentioned that the current starts at an initial value of I_0 amperes and then decays exponentially to zero. Although ideal circuit components require infinite time for the current to decay to zero, we have already noted that after about 5 time constants, the current is usually a negligible fraction of its initial value. Thus the existence of current in the *RL* circuit shown in Fig. 8.1(a) is a momentary event and therefore is also referred to as the *transient response* of the circuit. The response that exists a long time after the switching has taken place is called the *steady-*

[†] We can define the expressions 0^- and 0^+ more formally. The expression $x(0^-)$ refers to the limit of the variable x as $t \to 0$ from the left, or from negative time. The expression $x(0^+)$ refers to the limit of the variable x as $t \to 0$ from the right, or from the direction of positive time.

state response. In our circuit, the steady-state response is zero. Note that prior to the switch being opened, we assumed that the circuit shown in Fig. 8.3 was operating in a steady-state mode. That is, we assumed that the switch had been in a closed position for a long time so that the current in the inductor had reached its steady-state value of I_0 amperes. In Section 8.3 we discuss the problems of determining how the current in the inductor reached this steady-state value.

Note that the key to analyzing the natural response of the *RL* circuit is to find the initial current in the inductor and the time constant of the circuit. All subsequent calculations follow from knowing $i(t)$. Examples 8.1 and 8.2 illustrate the numerical calculations associated with the natural response of the *RL* circuit.

PSpice The PSpice command .IC can be used to compute a circuit's initial conditions prior to simulating the natural response: Section 7.4

E X A M P L E   8.1

The switch in the circuit shown in Fig. 8.7 has been closed for a long time before it is opened at $t = 0$. Find

a) $i_L(t)$ for $t \geq 0$;

b) $i_o(t)$ for $t \geq 0^+$;

c) $v_o(t)$ for $t \geq 0^+$; and

d) the percentage of the total energy stored in the 2-H inductor that is dissipated in the 10 Ω resistor.

FIGURE 8.7 The circuit for Example 8.1.

S O L U T I O N

a) The switch has been closed for a long time prior to $t = 0$, so we know the voltage across the inductor must be zero at $t = 0^-$. Therefore the initial current in the inductor is 20 A at $t = 0^-$. Hence $i_L(0^+)$ also is 20 A because an instantaneous change in the current cannot occur in an inductor. We replace the resistive circuit connected to the terminals of the inductor with a single resistor of 10 Ω:

$$R_{eq} = 2 + (40 \| 10) = 10 \ \Omega.$$

The time constant of the circuit is L/R_{eq}, or 0.2 s, giving the expression for the inductor current as

$$i_L(t) = 20e^{-5t} \text{ A}, \quad t \geq 0.$$

b) We find the current in the 40-Ω resistor most easily by using current division. That is

$$i_o = -i_L \frac{10}{10 + 40}.$$

Note that this expression is valid for $t \geq 0^+$ because $i_o = 0$ at $t = 0^-$. The inductor behaves as a short circuit prior to the switch being opened, producing an instantaneous change in the current i_o. Then,

$$i_o(t) = -4e^{-5t} \text{ A}, \qquad t \geq 0^+.$$

c) We find the voltage v_o by direct application of Ohm's law:

$$v_o(t) = 40i_o = -160e^{-5t} \text{ V}, \qquad t \geq 0^+.$$

d) The power dissipated in the 10-Ω resistor is

$$p_{10\Omega}(t) = \frac{v_o^2}{10} = 2560e^{-10t} \text{ W}, \qquad t \geq 0^+.$$

The total energy dissipated in the 10-Ω resistor is

$$w_{10\Omega}(t) = \int_0^\infty 2560e^{-10t} \, dt = 256 \text{ J}.$$

The initial energy stored in the 2-H inductor is

$$w(0) = \frac{1}{2}Li^2(0) = \frac{1}{2}(2)(400) = 400 \text{ J}.$$

Therefore the percentage of energy dissipated in the 10-Ω resistor is

$$\text{percent dissipated} = \frac{256}{400}(100) = 64\%.$$

EXAMPLE 8.2

In the circuit shown in Fig. 8.8, the initial currents in inductors L_1 and L_2 have been established by sources not shown. The switch is opened at $t = 0$.

a) Find i_1, i_2, and i_3 for $t \geq 0$.

b) Calculate the initial energy stored in the parallel inductors.

c) Determine how much energy is trapped in the inductors as $t \to \infty$.

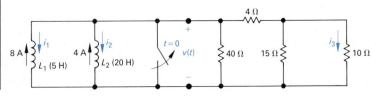

FIGURE 8.8 The circuit for Example 8.2.

d) Show that the total energy delivered to the resistive network equals the difference between the results obtained in (b) and (c).

S O L U T I O N

a) The key to finding currents i_1, i_2, and i_3 lies in knowing the voltage $v(t)$. We can easily find $v(t)$ if we reduce the circuit shown in Fig. 8.8 to the equivalent form shown in Fig. 8.9. The paralleled inductors simplify to an equivalent inductance of 4 H, carrying an initial current of 12 A. The resistive network reduces to a single resistance of 8 Ω. Hence the initial value of $i(t)$ is 12 A and the time constant is $\frac{4}{8}$, or 0.5 s. Therefore

$$i(t) = 12e^{-2t} \text{ A}, \qquad t \ge 0.$$

Now $v(t)$ is simply the product $8i$, so

$$v(t) = 96e^{-2t} \text{ V}, \qquad t \ge 0^+.$$

The circuit shows that $v(t) = 0$ at $t = 0^-$, so the expression for $v(t)$ is valid for $t \ge 0^+$. After obtaining $v(t)$, we can calculate i_1, i_2, and i_3:

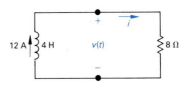

FIGURE 8.9 A simplification of the circuit shown in Fig. 8.8.

$$i_1 = \frac{1}{5} \int_0^t 96e^{-2x} \, dx - 8$$
$$= 1.6 - 9.6e^{-2t} \text{ A}, \qquad t \ge 0;$$

$$i_2 = \frac{1}{20} \int_0^t 96e^{-2x} \, dx - 4$$
$$= -1.6 - 2.4e^{-2t} \text{ A}, \qquad t \ge 0;$$

$$i_3 = \frac{v(t)}{10} \frac{15}{25} = 5.76e^{-2t} \text{ A}, \qquad t \ge 0^+.$$

Note that the expressions for the inductor currents i_1 and i_2 are valid for $t \ge 0$, whereas the expression for the resistor current i_3 is valid for $t \ge 0^+$.

b) The initial energy stored in the inductors is

$$w = \frac{1}{2}(5)(64) + \frac{1}{2}(20)(16) = 320 \text{ J}.$$

c) As $t \to \infty$, $i_1 \to 1.6$ A and $i_2 \to -1.6$ A. Therefore a long time after the switch has been opened, the energy stored in the two inductors is

$$w = \frac{1}{2}(5)(1.6)^2 + \frac{1}{2}(20)(-1.6)^2 = 32 \text{ J}.$$

d) We obtain the total energy delivered to the resistive network by integrating the expression for the instantaneous power from zero to infinity:

$$w = \int_0^\infty p\,dt = \int_0^\infty 1152e^{-4t}\,dt$$

$$= 1152\frac{e^{-4t}}{-4}\Big|_0^\infty = 288 \text{ J.}$$

This result is the difference between the initially stored energy (320 J) and the energy trapped in the parallel inductors (32 J). The equivalent inductor for the parallel inductors (which predicts the terminal behavior of the parallel combination) has an initial energy of 288 J. That is, the energy stored in the equivalent inductor represents the amount of energy that will be delivered to the resistive network at the terminals of the original inductors.

DRILL EXERCISES

8.1 The switch in the circuit shown has been closed for a long time and is opened at $t = 0$.

a) Calculate the initial value of i.

b) Calculate the initial energy stored in the inductor.

c) What is the time constant of the circuit for $t > 0$?

d) What is the numerical expression for $i(t)$ for $t \geq 0$?

e) What percentage of the initial energy stored has been dissipated in the 4-Ω resistor 5 ms after the switch has been opened?

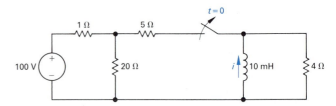

ANSWER: (a) -16 A; (b) 1.28 J; (c) $t = 2.5$ ms; (d) $-16e^{-400t}$ A; (e) 98.17%.

8.2 At $t = 0$, the switch in the circuit shown moves instantaneously from position a to position b.

a) Calculate v_0 for $t \geq 0^+$.

b) What percentage of the initial energy stored in the inductor is eventually dissipated in the 4-Ω resistor?

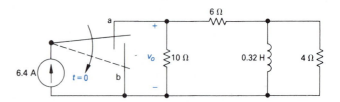

ANSWER: (a) $-8e^{-10t}$ V; (b) 80%.

8.2 THE NATURAL RESPONSE OF AN *RC* CIRCUIT

As mentioned in Section 8.1, the natural response of an *RC* circuit is analogous to that of the *RL* circuit. Consequently, we don't treat the *RC* circuit in the same detail as we did the *RL* circuit. Again, all independent sources are removed when the natural response commences.

We can find the natural response of the *RC* circuit shown in Fig. 8.1(b) from the circuit shown in Fig. 8.10. We begin by assuming that the switch has been in position a for a long time, allowing the loop made up of the dc voltage source V_g, the resistor R_1, and the capacitor C to reach a steady-state condition. In this case the constant voltage source cannot sustain a current in the capacitor; therefore the capacitor must be charged to the source voltage of V_g volts. (Recall from Chapter 7 that a capacitor behaves as an open circuit to a constant voltage.) In Section 8.3 we discuss the problem of determining how the voltage on the capacitor builds to V_g volts. The important point here is that when the switch is moved from position a to position b (at $t = 0$), the voltage on the capacitor is V_g volts. Because there can be no instantaneous change in the voltage at the terminals of a capacitor, the problem reduces to solving the circuit shown in Fig. 8.11.

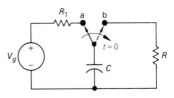

FIGURE 8.10 An *RC* circuit.

PSpice   Switches can be incorporated in PSpice circuit simulations: Chapter 8

FIGURE 8.11 The circuit shown in Fig. 8.10, after switching.

DERIVING THE EXPRESSION FOR THE VOLTAGE

We can easily find the voltage $v(t)$ by thinking in terms of node voltages. Using the lower junction between R and C as the reference node and summing the currents away from the upper junction between R and C gives

$$C\frac{dv}{dt} + \frac{v}{R} = 0. \tag{8.18}$$

Comparing Eq. (8.18) with Eq. (8.1) shows that the same mathematical techniques can be used to obtain the solution for $v(t)$. We leave to you to show that

$$v(t) = v(0)e^{-t/RC}, \qquad t \geq 0. \tag{8.19}$$

As we have already noted, the initial voltage on the capacitor equals the voltage source voltage V_g, or

$$v(0^-) = v(0) = v(0^+)$$

$$= V_g = V_0, \tag{8.20}$$

where V_0 denotes the initial voltage on the capacitor. The time constant for the *RC* circuit equals the product of the resistance

and capacitance, namely,

$$\tau = RC. \tag{8.21}$$

Substituting Eqs. (8.20) and (8.21) into Eq. (8.19) yields

$$v(t) = V_0 e^{-t/\tau}, \qquad t \geq 0, \tag{8.22}$$

which indicates that the natural response of the *RC* circuit is an exponential decay of the initial voltage. The time constant *RC* governs the rate of decay. Figure 8.12 shows the plot of Eq. (8.22) and the graphic interpretation of the time constant.

After determining $v(t)$, we can easily derive the expressions for i, p, and w:

$$i(t) = \frac{v(t)}{R} = \frac{V_0}{R} e^{-t/\tau}, \qquad t \geq 0^+; \tag{8.23}$$

$$p = vi = \frac{V_0^2}{R} e^{-2t/\tau}, \qquad t \geq 0^+; \tag{8.24}$$

$$w = \int_0^t p\,dx = \int_0^t \frac{V_0^2}{R} e^{-2x/\tau}\,dx$$

$$= \frac{1}{2} CV_0^2 (1 - e^{-2t/\tau}), \qquad t \geq 0. \tag{8.25}$$

The key to analyzing the natural response of the *RC* circuit shown in Fig. 8.1(b) is to find the initial voltage across the capacitor and the time constant of the circuit. All subsequent calculations follow from knowing $v(t)$. Examples 8.3 and 8.4 illustrate the numerical calculations associated with the natural response of the *RC* circuit.

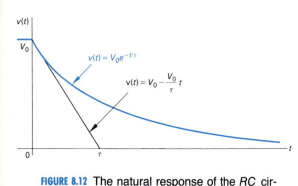

FIGURE 8.12 The natural response of the *RC* circuit.

E X A M P L E 8.3

The switch in the circuit shown in Fig. 8.13 has been in position *x* for a long time. At $t = 0$, the switch moves instantaneously to position *y*. Find

a) $v_c(t)$ for $t \geq 0$;

b) $v_o(t)$ for $t \geq 0^+$;

c) $i_o(t)$ for $t \geq 0^+$;

d) the total energy dissipated in the 60-Ω resistor.

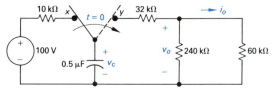

FIGURE 8.13 The circuit for Example 8.3.

S O L U T I O N

a) Because the switch has been in position *x* for a long time, the 0.5-μF capacitor will charge to 100 V and be positive at the upper terminal. We can replace the resistive network connected to the capacitor at $t = 0^+$ with an equivalent resistance of 80 kΩ. Hence the time constant of the circuit is $(0.5 \times 10^{-6})(80 \times 10^3)$ or 40 ms. Then,

$$v_C(t) = 100e^{-25t} \text{ V}, \qquad t \geq 0.$$

b) The easiest way to find $v_o(t)$ is to note that the resistive circuit forms a voltage divider across the terminals of the capacitor. Thus

$$v_o(t) = \frac{48}{80} v_C(t) = 60e^{-25t} \text{ V}, \qquad t \geq 0^+.$$

This expression for $v_o(t)$ is valid for $t \geq 0^+$ because $v_o(0^-)$ is zero. Thus we have an instantaneous change in the voltage across the 240-kΩ resistor.

c) We find the current $i_o(t)$ from Ohm's law:

$$i_o(t) = \frac{v_o(t)}{60 \times 10^3} = e^{-25t} \text{ mA}, \qquad t \geq 0^+.$$

d) The power dissipated in the 60-kΩ resistor is

$$p_{60k\Omega}(t) = i_o^2(t)(60 \times 10^3) = 60e^{-50t} \text{ mW}, \qquad t \geq 0^+.$$

The total energy dissipated is

$$w_{60k\Omega} = \int_0^\infty i_o^2(t)(60 \times 10^3)dt = 1.2 \text{ mJ}.$$

E X A M P L E   8.4

The initial voltages on capacitors C_1 and C_2 in the circuit shown in Fig. 8.14 have been established by sources not shown. The switch is closed at $t = 0$.

a) Find $v_1(t)$, $v_2(t)$, and $v(t)$, for $t \geq 0$, and $i(t)$ for $t \geq 0^+$.

b) Calculate the initial energy stored in the capacitors C_1 and C_2.

c) Determine how much energy is trapped in the capacitors as $t \to \infty$.

d) Show that the total energy delivered to the 250-kΩ resistor is the difference between the results obtained in (b) and (c).

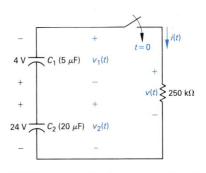

FIGURE 8.14 The circuit for Example 8.4.

S O L U T I O N

a) Once we know $v(t)$, we can obtain the current $i(t)$ from Ohm's law. After determining $i(t)$, we can calculate $v_1(t)$ and

$v_2(t)$, because the voltage across a capacitor is a function of the capacitor current. To find $v(t)$ we replace the series-connected capacitors with an equivalent capacitor. The equivalent capacitor has a capacitance of 4 μF and is charged to a voltage of 20 V. Therefore the circuit shown in Fig. 8.14 reduces to that shown in Fig. 8.15. It reveals that the initial value of $v(t)$ is 20 V and that the time constant of the circuit is $(4)(250) \times 10^{-3}$, or 1 s. Thus the expression for $v(t)$ is

$$v(t) = 20e^{-t} \text{ V}, \qquad t \geq 0.$$

The current $i(t)$ is

$$i(t) = \frac{v(t)}{250,000} = 80e^{-t} \ \mu\text{A}, \qquad t \geq 0^+.$$

Knowing $i(t)$, we calculate the expressions for $v_1(t)$ and $v_2(t)$:

$$v_1(t) = -\frac{10^6}{5} \int_0^t 80 \times 10^{-6} e^{-x} dx - 4$$

$$= (16e^{-t} - 20) \text{ V}, \qquad t \geq 0;$$

$$v_2(t) = -\frac{10^6}{20} \int_0^t 80 \times 10^{-6} e^{-x} dx + 24$$

$$= (4e^{-t} + 20) \text{ V}, \qquad t \geq 0.$$

b) The initial energy stored in C_1 is

$$w_1 = \frac{1}{2} (5 \times 10^{-6})(16) = 40 \ \mu\text{J}.$$

The initial energy stored in C_2 is

$$w_2 = \frac{1}{2} (20 \times 10^{-6})(576) = 5760 \ \mu\text{J}.$$

The total initial energy stored in the two capacitors is

$$w_o = 40 + 5760 = 5800 \ \mu\text{J}.$$

c) As $t \to \infty$,

$$v_1 \to -20 \text{ V} \qquad \text{and} \qquad v_2 \to +20 \text{ V}.$$

Therefore the energy trapped in the two capacitors is

$$w_\infty = \frac{1}{2}(5 + 20) \times 10^{-6}(400) = 5000 \ \mu\text{J}.$$

d) The total energy delivered to the 250-kΩ resistor is

$$w = \int_0^\infty p \, dt = \int_0^\infty \frac{400e^{-2t}}{250,000} \, dt = 800 \ \mu\text{J}.$$

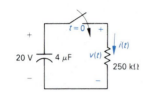

FIGURE 8.15 A simplification of the circuit shown in Fig. 8.14.

Comparing the results obtained in (b) and (c) shows that

$$800 \ \mu J = (5800 - 5000) \ \mu J.$$

The energy stored in the equivalent capacitor in Fig. 8.15 is $\frac{1}{2}(4 \times 10^{-6})(400)$, or $800 \ \mu J$. Because the equivalent capacitor predicts the terminal behavior of the original series-connected capacitor, the energy stored in the equivalent capacitor is the energy that is delivered to the 250-kΩ resistor.

DRILL EXERCISES

8.3 The switch in the circuit shown has been closed for a long time and is opened at $t = 0$. Find

a) the initial value of $v(t)$;

b) the time constant for $t > 0$;

c) the numerical expression for $v(t)$ after the switch has been opened;

d) the initial energy stored in the capacitor;

e) the length of time required to dissipate 75% of the initially stored energy.

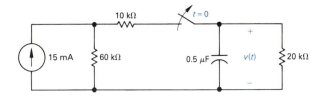

ANSWER: (a) 200 V; (b) 10 ms; (c) $200e^{-100t}$ V; (d) 10 mJ; (e) 6.93 ms.

8.4 The switch in the circuit shown has been closed for a long time before being opened at $t = 0$.

a) Find $v_o(t)$ for $t \geq 0$.

b) What percentage of the initial energy stored in the circuit has been dissipated after the switch has been open for 60 ms?

ANSWER: (a) $8e^{-25t} + 4e^{-10t}$ V; (b) 81.05%.

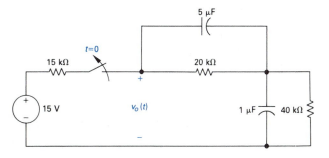

8.3 **THE STEP RESPONSE OF *RL* AND *RC* CIRCUITS**

We are now ready to discuss the problem of finding currents and voltages generated in first-order *RL* or *RC* circuits when either dc voltage or current sources are suddenly applied. The response of

PSpice ⤳   Use PSpice to simulate a circuit's
step response: Chapter 9

a circuit to the sudden application of a constant voltage or current source is referred to as the *step response* of the circuit. In presenting the step response we show how the circuit responds when energy is being stored in the inductor or capacitor. We begin with the step response of an *RL* circuit.

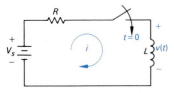

FIGURE 8.16 A circuit used to illustrate the step response of a first-order *RL* circuit.

THE STEP RESPONSE OF AN *RL* CIRCUIT

We use the circuit shown in Fig. 8.16 to describe the step response of an *RL* circuit. It is based on the definition of first-order circuits as summarized in Fig. 8.2. Energy stored in the inductor at the time the switch is closed is given in terms of a nonzero initial current $i(0)$. The task is to find the expression for the current in the circuit and the expression for the voltage across the inductor after the switch has been closed. The procedure is the same as that used in Section 8.1. We use circuit analysis to derive the differential equation that describes the circuit in terms of the variable of interest and then use the techniques of elementary calculus to solve the equation.

After the switch in the circuit shown in Fig. 8.16 has been closed, Kirchhoff's voltage law requires that

$$V_s = Ri + L\frac{di}{dt}, \qquad (8.26)$$

which can be solved for the current by separating the variables i and t and then integrating. The first step in this approach is to solve Eq. (8.26) for the derivative di/dt:

$$\frac{di}{dt} = \frac{-Ri + V_s}{L} = \frac{-R}{L}\left(i - \frac{V_s}{R}\right). \qquad (8.27)$$

Next, we multiply both sides of Eq. (8.27) by a differential time dt. This step reduces the left-hand side of the equation to a differential change in the current. Thus

$$\frac{di}{dt}dt = \frac{-R}{L}\left(i - \frac{V_s}{R}\right)dt,$$

or

$$di = \frac{-R}{L}\left(i - \frac{V_s}{R}\right)dt. \qquad (8.28)$$

We now separate the variables in Eq. (8.28) to get

$$\frac{di}{i - (V_s/R)} = \frac{-R}{L}dt, \qquad (8.29)$$

and then integrate both sides of Eq. (8.29). Using x and y as variables for the integration, we obtain

$$\int_{I_0}^{i(t)} \frac{dx}{x - (V_s/R)} = \frac{-R}{L} \int_0^t dy, \qquad (8.30)$$

where I_0 is the current at $t = 0$ and $i(t)$ is the current at any $t > 0$. Performing the integration called for in Eq. (8.30) generates the expression:

$$\ln \frac{i(t) - (V_s/R)}{I_0 - (V_s/R)} = \frac{-R}{L}t, \qquad (8.31)$$

from which

$$\frac{i(t) - (V_s/R)}{I_0 - (V_s/R)} = e^{-(R/L)t},$$

or

$$i(t) = \frac{V_s}{R} + \left(I_0 - \frac{V_s}{R}\right)e^{-(R/L)t}. \qquad (8.32)$$

When the initial energy in the inductor is zero, I_0 is zero. Thus Eq. (8.32) reduces to

$$i(t) = \frac{V_s}{R} - \frac{V_s}{R}e^{-(R/L)t}. \qquad (8.33)$$

Equation (8.33) indicates that after the switch has been closed, the current increases exponentially from zero to a final value of V_s/R. The time constant of the circuit, L/R, determines the rate of increase. One time constant after the switch has been closed, the current will have reached approximately 63 percent of its final value, or

$$i(\tau) = \frac{V_s}{R} - \frac{V_s}{R}e^{-1} \cong 0.6321\frac{V_s}{R}. \qquad (8.34)$$

If the current were to continue to increase at its initial rate, it would reach its final value at $t = \tau$. That is, because

$$\frac{di}{dt} = \frac{-V_s}{R}\left(\frac{-1}{\tau}\right)e^{-t/\tau} = \frac{V_s}{L}e^{-t/\tau}, \qquad (8.35)$$

the initial rate at which $i(t)$ increases is

$$\frac{di}{dt}(0) = \frac{V_s}{R}. \qquad (8.36)$$

If the current were to continue to increase at this rate, the expression for i would be

$$i = \frac{V_s}{L}t, \qquad (8.37)$$

from which, at $t = \tau$,

$$i = \frac{V_s}{L}\frac{L}{R} = \frac{V_s}{R}. \qquad (8.38)$$

Equations (8.33) and (8.37) are plotted in Fig. 8.17. The values given by Eqs. (8.34) and (8.38) are also shown in Fig. 8.17.

The voltage across an inductor is $L\, di/dt$, so from Eq. (8.32), for $t \ge 0^+$,

$$v = L\left(\frac{-R}{L}\right)\left(I_0 - \frac{V_s}{R}\right)e^{-(R/L)t} = (V_s - I_0 R)e^{-(R/L)t}. \quad (8.39)$$

The voltage across the inductor is zero prior to the switch being closed. Equation (8.39) indicates that the inductor voltage jumps to $V_s - I_0 R$ at the instant the switch is closed and then decays exponentially to zero.

Does the value of v at $t = 0^+$ make sense? Because the initial current is I_0 and the inductor prevents an instantaneous change in current, the current is I_0 during the instant after the switch has been closed. The voltage drop across the resistor is $I_0 R$, and the voltage impressed across the inductor is the source voltage minus this voltage, that is, $V_s - I_0 R$.

When the initial inductor current is zero, Eq. (8.39) simplifies to

$$v = V_s e^{-(R/L)t}. \qquad (8.40)$$

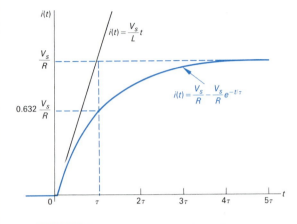

FIGURE 8.17 The step response of the *RL* circuit shown in Fig. 8.16 when $I_0 = 0$.

If the initial current is zero, the voltage across the inductor jumps to V_s. We also expect the inductor voltage to approach zero as t increases because the current in the circuit is approaching the constant value of V_sR. Figure 8.18 shows the plot of Eq. (8.40) and the relationship between the time constant τ and the initial rate at which the inductor voltage is decreasing.

If there is an initial current in the inductor, Eq. (8.32) gives the solution for the current. The algebraic sign of I_0 is positive if the initial current is in the same direction as i; otherwise, I_0 carries a negative sign. Example 8.5 illustrates application of Eq. (8.32) to a specific circuit.

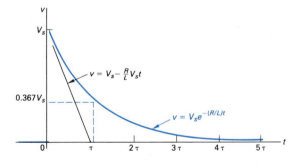

FIGURE 8.18 Inductor voltage versus time.

E X A M P L E   8.5

The switch in the circuit shown in Fig. 8.19 has been in position a for a long time. At $t = 0$, the switch moves from position a to position b. The switch is a *make-before-break* type, so there is no interruption of the inductor current.

a) Find the expression for $i(t)$ for $t \ge 0$.

b) What is the initial voltage across the inductor just after the switch has been moved to position b?

c) Does this initial voltage make sense in terms of circuit behavior?

d) How many milliseconds after the switch has been moved to position b does the inductor voltage equal 24 V?

e) Plot both $i(t)$ and $v(t)$ versus t.

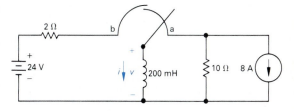

FIGURE 8.19 The circuit for Example 8.5.

S O L U T I O N

a) The switch has been in position a for a long time, so the 200-mH inductor is a short circuit across the 8-A current source. Therefore the inductor carries an initial current of 8 A. This initial current is oriented opposite to the reference direction for i; thus I_0 is -8A. When the switch is in position b, the final value of i will be $\frac{24}{2}$, or 12 A. The time constant of the circuit is $\frac{200}{2}$, or 100 ms. Substituting these values into Eq. (8.32) gives

$$i = 12 + (-8 - 12)e^{-t/0.1}$$

$$= 12 - 20e^{-10t} \text{ A}, \qquad t \geq 0.$$

b) The voltage across the inductor is

$$v = L\frac{di}{dt} = 0.2(200e^{-10t}) = 40e^{-10t} \text{ V}, \qquad t \geq 0^+.$$

The initial inductor voltage is

$$v(0^+) = 40 \text{ V}.$$

c) Yes; in the instant after the switch has been moved to position b, the inductor sustains a current of 8 A counterclockwise around the newly formed closed path. This current causes a 16-V drop across the 2-Ω resistor. This voltage drop adds to the drop across the source, producing a 40-V drop across the inductor.

d) We find the time at which the inductor voltage equals 24 V by solving the expression

$$24 = 40e^{-10t}$$

for t:

$$t = \frac{1}{10} \ln \frac{40}{24} = 51.08 \times 10^{-3} = 51.08 \text{ ms}.$$

e) Figure 8.20 shows the graphs of $i(t)$ and $v(t)$ versus t. Note that the instant of time when the current equals zero corresponds to the instant of time when the inductor voltage equals the source voltage of 24 V.

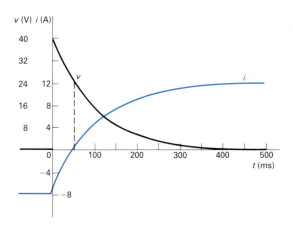

FIGURE 8.20 Current and voltage waveform for Example 8.5.

DRILL EXERCISES

8.5 Assume that the switch in the circuit shown in Figure 8.19 has been in position b for a long time, and at $t = 0$ it moves to position a. Find (a) $i(0^+)$; (b) $v(0^+)$; (c) τ, $t > 0$; (d) $i(t)$, $t \geq 0$; and (e) $v(t)$, $t \geq 0^+$.

ANSWER: (a) 12 A; (b) -200 V; (c) 20 ms; (d) $-8 + 20e^{-50t}$ A, $t \geq 0$; (e) $-200e^{-50t}$ V, $t \geq 0^+$.

We can also describe the behavior of the circuit shown in Fig. 8.16 in terms of the voltage across the inductor, $v(t)$. We begin by noting that the voltage across the resistor is the difference between the source voltage and the inductor voltage and write

$$i(t) = \frac{V_s}{R} - \frac{v(t)}{R}, \qquad (8.41)$$

where V_s is a constant. Differentiating both sides with respect to time yields

$$\frac{di}{dt} = -\frac{dv}{R dt}. \qquad (8.42)$$

Then, if we multiply each side of Eq. (8.42) by the inductance L, we get an expression for the voltage across the inductor on the left-hand side, or

$$v = -\frac{L dv}{R dt}. \qquad (8.43)$$

Putting Eq. (8.43) into standard form yields

$$\frac{dv}{dt} + \frac{R}{L}v = 0. \qquad (8.44)$$

We leave to you verification in Drill Exercise 8.6 that the solution to Eq. (8.44) is identical to that given in Eq. (8.39).

At this point a general observation about the step response of an *RL* circuit, which will prove helpful later, is pertinent. When we derived the differential equation for the inductor current, we obtained Eq. (8.26). We now rewrite Eq. (8.26) as

$$\frac{di}{dt} + \frac{R}{L}i = \frac{V_s}{L}. \qquad (8.45)$$

Observe that Eqs. (8.44) and (8.45) have the same form. Specifically, each equates the sum of the first derivative of the variable and a constant times the variable to a constant value. In Eq. (8.44), the constant on the right-hand side happens to be zero. Hence it took on the same form as the natural response equations in Section 8.1. In both equations, the constant multiplying the dependent variable is the reciprocal of the time constant, that is, $R/L = 1/\tau$. We encounter a similar situation in the derivations for the step response of the *RC* circuit.

DRILL EXERCISES

8.6 a) Derive Eq. (8.44) by first converting the Thévenin equivalent in Fig. 8.16 to a Norton equivalent and then summing the currents away from the upper node using the inductor voltage v as the variable of interest.

b) Use the separation of variables technique to find the solution to Eq. (8.44). Verify that your solution agrees with the solution given in Eq. (8.39).

ANSWER: (a) Derivation; (b) verification.

STEP RESPONSE OF AN *RC* CIRCUIT

We can find the step response of a first-order *RC* circuit by analyzing the circuit shown in Fig. 8.21. We chose the Norton equivalent of the network connected to the equivalent capacitor for mathematical convenience. Summing the currents away from the top node in Fig. 8.21 generates the differential equation:

$$C\frac{dv_C}{dt} + \frac{v_C}{R} = I_s. \qquad (8.46)$$

Division of Eq. (8.46) by C gives

$$\frac{dv_C}{dt} + \frac{v_C}{RC} = \frac{I_s}{C}. \qquad (8.47)$$

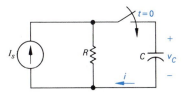

FIGURE 8.21 A circuit used to illustrate the step response of a first-order *RC* circuit.

Comparing Eq. (8.47) with Eq. (8.45) reveals that the form of the solution for v_C is the same as that for the current in the inductive circuit, namely, Eq. (8.32). Therefore, by simply substituting the appropriate variables and coefficients, we can write the solution for v_C directly. The translation requires that I_s replace V_s, C replace L, $1/R$ replace R, and V_0 replace I_0. We get

$$v_C = I_sR + (V_0 - I_sR)e^{-t/RC}; \qquad t \geq 0. \qquad (8.48)$$

A similar derivation for the current in the capacitor yields the differential equation

$$\frac{di}{dt} + \frac{1}{RC}i = 0. \qquad (8.49)$$

Equation (8.49) has the same form as Eq. (8.44), hence the solution for i is obtained by using the same translations that were used to obtain the solution of Eq. (8.47). Thus

$$i = \left(I_s - \frac{V_0}{R}\right) e^{-t/RC}; \qquad t \geq 0^+, \qquad \textbf{(8.50)}$$

where V_0 is the initial value of v_C, the voltage across the capacitor.

We obtained Eqs. (8.48) and (8.50) by using a mathematical analogy with the solution for the step response of the inductive circuit. Let's see whether these solutions for the *RC* circuit make sense in terms of known circuit behavior. From Eq. (8.48), note that the initial voltage across the capacitor is V_0, the final voltage across the capacitor is $I_s R$, and the time constant of the circuit is RC. Also note that the solution for v_C is valid for $t \geq 0$. These observations are consistent with the behavior of a capacitor in parallel with a resistor when driven by a constant current source.

Equation (8.50) predicts that the current in the capacitor at $t = 0^+$ is $I_s - V_0/R$. This prediction makes sense because the capacitor voltage cannot change instantaneously and therefore the initial current in the resistor is V_0/R. The capacitor branch current changes instantaneously from 0 at $t = 0^-$ to $I_s - V_0/R$ at $t = 0^+$. The capacitor current is zero at $t = \infty$, which agrees with the capacitor voltage reaching a constant value as $t \to \infty$. Also note that the final value of i, which is equal to zero, agrees with the final value of $v = I_s R$.

Example 8.6 illustrates how to use Eqs. (8.48) and (8.50) to find the step response of a first-order *RC* circuit.

EXAMPLE 8.6

The switch in the circuit shown in Fig. 8.22 has been in position 1 for a long time. At $t = 0$, the switch moves instantaneously to position 2. Find

a) $v_o(t)$ for $t \geq 0$;

b) $i_o(t)$ for $t \geq 0^+$.

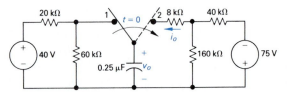

FIGURE 8.22 The circuit for Example 8.6.

SOLUTION

a) The switch has been in position 1 for a long time, so the initial value of v_o is 40(60/80), or 30 V. To take advantage of Eqs. (8.48) and (8.50), we find the Norton equivalent with respect to the terminals of the capacitor for $t \geq 0$. We leave to you verification that the Norton equivalent is a 1.5-mA current source in parallel with a 40-kΩ resistor, as shown in Fig. 8.23. From Fig. 8.23, $I_s R = -60$ V and $RC = 10$ ms.

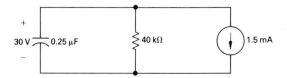

FIGURE 8.23 The equivalent circuit for $t > 0$ for the circuit shown in Fig. 8.22.

We have already noted that $v_o(0) = 30$ V, so the solution for v_o is

$$v_o = -60 + [30 - (-60)]e^{-100t}$$
$$= -60 + 90e^{-100t} \text{ V}, \qquad t \geq 0.$$

b) We write the solution for i_o directly from Eq. (8.50) by noting that $I_s = -1.5$ mA and $V_0/R = (30/40) \times 10^{-3}$ or 0.75 mA:

$$i_o = -2.25e^{-100t} \text{ mA}, \qquad t \geq 0^+.$$

We check the consistency of the solutions for v_o and i_o by noting that

$$i_o = C\frac{dv_o}{dt} = (0.25 \times 10^{-6})(-9000e^{-100t}) = -2.25e^{-100t} \text{ mA}.$$

Because $dv_o(0^-)/dt = 0$, the expression for i_o clearly is valid only for $t \geq 0^+$.

DRILL EXERCISES

8.7 a) Derive Eq. (8.44) by first converting the Norton equivalent circuit shown in Fig. 8.21 to a Thévenin equivalent and then summing the voltages around the closed loop using the capacitor current i as the relevant variable.

b) Use the separation of variables technique to find the solution to Eq. (8.49). Verify that your solution agrees with that of Eq. (8.50).

ANSWER: (a) Derivation; (b) verification.

8.8 a) Find the expression for the voltage across the 160-kΩ resistor in the circuit shown in Fig. 8.22. Let this voltage be denoted v_A and assume that the reference polarity for the voltage is positive at the upper terminal of the 160-kΩ resistor.

b) Specify the interval of time for which the expression obtained in (a) is valid.

ANSWER: (a) $v_A = -60 + 72e^{-100t}$ V; (b) $t \geq 0^+$.

8.4 GENERAL SOLUTION FOR STEP AND NATURAL RESPONSES

The general approach to finding either the natural response or the step response of the first-order RL and RC circuits shown in Fig. 8.24 is based on the differential equations that describe the behavior of the circuits being the same (compare Eqs. 8.44–8.49). To generalize the solution of these four possible circuits, we let $x(t)$ represent the unknown quantity, giving $x(t)$

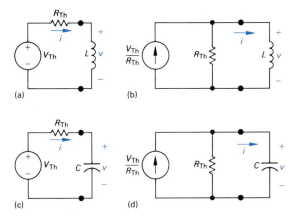

FIGURE 8.24 Four possible first-order circuits: (a) Inductor connected to a Thévenin equivalent; (b) inductor connected to a Norton equivalent; (c) capacitor connected to a Thévenin equivalent; (d) capacitor connected to a Norton equivalent.

four possible values. It can represent either the current or voltage at the terminals of an inductor, or it can represent either the current or voltage at the terminals of a capacitor. From Eqs. (8.44), (8.45), (8.47), and (8.49), we know that the differential equation that describes any one of the four circuits in Fig. 8.24 takes the form

$$\frac{dx}{dt} + \frac{x}{\tau} = K, \qquad (8.51)$$

where the value of the constant K can be zero.

Because the sources in the circuit are constant voltages and/or currents, the final value of x will be constant. That is, the final value must satisfy Eq. (8.51), and, when x reaches its final value, the derivative dx/dt must be zero. Hence

$$x_f = K\tau, \qquad (8.52)$$

where x_f represents the final value of the variable.

We solve Eq. (8.51) by separating the variables, beginning by solving for the first derivative:

$$\frac{dx}{dt} = \frac{-x}{\tau} + K = \frac{-(x - K\tau)}{\tau} = \frac{-(x - x_f)}{\tau}. \qquad (8.53)$$

In writing Eq. (8.53), we used the observation based on Eq. (8.52). We now multiply both sides of Eq. (8.53) by dt to obtain

$$\frac{dx}{x - x_f} = \frac{-1}{\tau} dt. \qquad (8.54)$$

We next integrate Eq. (8.54). In order to obtain as general a solution as possible, we use time t_0 as the lower limit and t as the upper limit. Time t_0 corresponds to the time of switching or other change. Previously we assumed that $t_0 = 0$, but this change allows the switching to take place at any time. Using u

and v as symbols of integration, we get

$$\int_{x(t_0)}^{x(t)} \frac{du}{u - x_f} = -\frac{1}{\tau}\int_{t_0}^{t} dv. \tag{8.55}$$

Carrying out the integration called for in Eq. (8.55) gives

$$x(t) = x_f + [x(t_0) - x_f]e^{-(t-t_0)/\tau}. \tag{8.56}$$

The importance of Eq. (8.56) becomes apparent if we write it out in verbal form:

$$\begin{pmatrix} \text{The unknown} \\ \text{variable as a} \\ \text{function of time} \end{pmatrix} = \begin{pmatrix} \text{the final} \\ \text{value of the} \\ \text{variable} \end{pmatrix}$$

$$+ \left[\begin{pmatrix} \text{the initial} \\ \text{value of the} \\ \text{variable} \end{pmatrix} - \begin{pmatrix} \text{the final} \\ \text{value of the} \\ \text{variable} \end{pmatrix}\right] \times e^{\frac{[t - \text{(time of switching)}]}{\text{(time constant)}}}. \tag{8.57}$$

In many cases, the time of switching, that is t_0, is zero.

Equation (8.57) reduces the task of finding the step response of a single time constant RL or RC circuit to one involving the computation of three quantities: (1) the final value of the variable; (2) the initial value of the variable; and (3) the time constant. Note that we can find the natural response solution in the same way. In the natural response we have to find only two quantities because we know at the outset that the final value will be zero.

When applying Eq. (8.57) you must recognize that the initial value refers to the value of the variable at $t = t_0^+$.[†] For capacitive voltages and inductive currents, this condition isn't important because the values at $t = t_0^-$ and $t = t_0^+$ are equal. For other variables the values at $t = t_0^-$ and $t = t_0^+$ may be different because they can have discontinuities. While acquiring experience with these techniques, you should first find the solution for inductive currents or capacitive voltages by using Eq. (8.57). After determining the inductive current or capacitive voltage, you can find the other variables of interest by using the circuit analysis techniques introduced in Chapters 3 and 4 and the rules presented in Chapter 7. After gaining some experience with Eq. (8.57), you should be able to solve directly for the current or voltage of interest.

Examples 8.7, 8.8, and 8.9 illustrate how to use Eq. (8.57) to find the step response of an RC or RL circuit.

[†] The expressions t_0^- and t_0^+ are analogous to 0^- and 0^+. Thus $x(t_0^-)$ is the limit of $x(t)$ as $t \to t_0$ from the left, and $x(t_0^+)$ is the limit of $x(t)$ as $t \to t_0$ from the right.

E X A M P L E   8.7

The switch in the circuit shown in Fig. 8.25 has been in position a for a long time. At $t = 0$ the switch is moved to position b. What is the

a) initial value of v_C?

b) final value of v_C?

c) time constant of the circuit when the switch is in position b?

d) expression for $v_C(t)$ when $t \geq 0$?

e) expression for $i(t)$ when $t \geq 0^+$?

f) How long after the switch is in position b does the capacitor voltage pass through zero?

g) Plot $v_C(t)$ and $i(t)$ versus t.

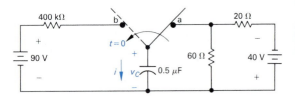

FIGURE 8.25 The circuit for Example 8.7.

S O L U T I O N

a) The switch has been in position a for a long time, so the capacitor looks like an open circuit. Therefore the voltage across the capacitor is the voltage across the 60-Ω resistor. From the voltage divider rule, the voltage across the 60-Ω resistor is $40 \times [60/(60 + 20)]$, or 30 V. As the reference for v_C is positive at the upper terminal of the capacitor, we have $v(0) = -30$ V.

b) After the switch has been in position b for a long time, the capacitor will look like an open circuit in terms of the 90-V source. Thus the final value of the capacitor voltage is $+90$ V.

c) The time constant is

$$\tau = RC = (400 \times 10^3)(0.5) \times 10^{-6} = 0.2 \text{ s.}$$

d) Substituting the appropriate values for v_f, $v(0)$, and t into Eq. (8.57) yields

$$v_C(t) = 90 + (-30 - 90)e^{-5t}$$

$$= 90 - 120e^{-5t} \text{ V,} \qquad t \geq 0.$$

e) Here the value for τ doesn't change. Thus we need to find only the initial and final values for the current in the capacitor. When obtaining the initial value, we must get the value of $i(0^+)$ because the current in the capacitor can have a step jump. The current in the capacitor is equal to the current in the resistor, which from Ohm's law is $[90 - (-30)]$ V $\div$ 400 kΩ, or 300 μA. Note that in applying Ohm's law we recognized that the capacitor voltage cannot change instantaneously. The final value of $i(t) = 0$, so

$$i(t) = \frac{90 - (-30)}{0.4}e^{-5t} = 300e^{-5t}\ \mu A, \qquad t \geq 0^+.$$

Note that we could have obtained this solution by differentiating the solution in (d) and multiplying by the capacitance. You may want to do so for yourself. Note that this alternative approach to finding $i(t)$ also predicts the discontinuity at $t = 0$.

f) To find how long the switch must be in position b before the capacitor voltage becomes zero, we solve the equation derived in (d) for the time when $v_C(t) = 0$:

$$120e^{-5t} = 90 \qquad \text{or} \qquad e^{5t} = \frac{120}{90},$$

so

$$t = \frac{1}{5}\ln\left(\frac{4}{3}\right) = 57.54 \text{ ms.}$$

Note that, when $v_C = 0$, $i = 225$ mA and that the voltage drop across the 400-kΩ resistor is 90 V.

g) Figure 8.26 shows the graphs of $v_C(t)$ and $i(t)$ versus t.

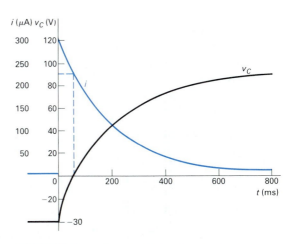

FIGURE 8.26 Current and voltage waveforms for Example 8.7.

EXAMPLE 8.8

The switch in the circuit shown in Fig. 8.27 has been open for a long time. The initial charge on the capacitor is zero. At $t = 0$, the switch is closed. Find the expression for

a) $i(t)$ for $t \geq 0^+$, and

b) $v(t)$ when $t \geq 0^+$.

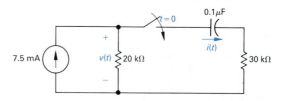

FIGURE 8.27 The circuit for Example 8.8.

SOLUTION

a) Because the initial voltage on the capacitor is zero, at the instant when the switch is closed the current in the 30-kΩ branch will be

$$i(0^+) = \frac{(7.5)(20)}{50} = 3 \text{ mA.}$$

The final value of the capacitor current will be zero because the capacitor eventually will appear as an open circuit in terms of dc current. Thus $i_f = 0$. The time constant of the circuit will equal the product of the Thévenin resistance, seen from the capacitor terminals and the capacitor. Therefore $\tau = (20 + 30)10^3(0.1) \times 10^{-6} = 5$ ms. Substituting these values into Eq. (8.57) generates the expression:

$$i(t) = 0 + (3 - 0)e^{-t/5 \times 10^{-3}} = 3e^{-200t} \text{ mA}, \qquad t \geq 0^+.$$

b) To find $v(t)$, we note from the circuit that it equals the sum of the voltage across the capacitor and the voltage across the 30-kΩ resistor. To find the capacitor voltage (which is a drop in the direction of the current), we note that its initial value is zero and its final value is $(7.5)(20)$, or 150 V. The time constant is the same as before, or 5 ms. Therefore we use Eq. (8.57) to write

$$v_C(t) = 150 + (0 - 150)e^{-200t}$$

$$= (150 - 150e^{-200t}) \text{ V}, \qquad t \geq 0.$$

Hence the expression for the voltage $v(t)$ is

$$v(t) = 150 - 150e^{-200t} + (30)(3)e^{-200t}$$

$$= (150 - 60e^{-200t}) \text{ V}, \qquad t \geq 0^+.$$

As one check on the expression obtained for $v(t)$, note that it predicts the initial value of the voltage across the 20-kΩ resistor as $150 - 60$, or 90 V. The instant the switch is closed the current in the 20-kΩ resistor is $(7.5)(30/50)$, or 4.5 mA. This current produces a 90-V drop across the 20-kΩ resistor, confirming the value predicted by the solution.

EXAMPLE 8.9

The switch in the circuit shown in Fig. 8.28 has been open for a long time. At $t = 0$ the switch is closed. Find the expression for

a) $v(t)$ when $t \geq 0^+$, and

b) $i(t)$ when $t \geq 0$.

SOLUTION

a) The switch has been open for a long time, so the initial current in the inductor is 5 A, oriented from top to bottom. Immediately after the switch has been closed, the current still is

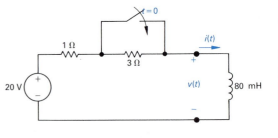

FIGURE 8.28 The circuit for Example 8.9.

5 A, and therefore the initial voltage across the inductor becomes $20 - 5(1)$, or 15 V. The final value of the inductor voltage is 0 V. With the switch closed, the time constant is 80/1, or 80 ms. We use Eq. (8.57) to write the expression for $v(t)$:

$$v(t) = 0 + (15 - 0)e^{-t/80 \times 10^{-3}} = 15e^{-12.5t} \text{ V}, \qquad t \geq 0^+.$$

b) We have already noted that the initial value of the inductor current is 5 A. After the switch has been closed for a long time, the inductor current reaches 20/1, or 20 A. The circuit time constant is 80 ms, so the expression for $i(t)$ is

$$i(t) = 20 + (5 - 20)e^{-12.5t}$$

$$= (20 - 15e^{-12.5t})\text{A}, \qquad t \geq 0.$$

We determine that the solutions for $v(t)$ and $i(t)$ agree by noting that

$$v(t) = L\frac{di}{dt} = 80 \times 10^{-3}[15(12.5)e^{-12.5t}]$$

$$= 15e^{-12.5t} \text{ V}, \qquad t \geq 0^+.$$

DRILL EXERCISES

8.9 Assume that the switch in the circuit shown has been in position b for a long time and that at $t = 0$ it is moved to position a. Find (a) $v_C(0^+)$; (b) $v_C(\infty)$; (c) τ for $t > 0$; (d) $i(0^+)$; (e) v_C, $t \geq 0$; and (f) i, $t \geq 0^+$.

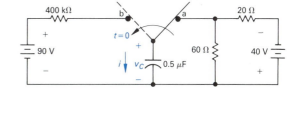

ANSWER: (a) 90 V; (b) -30 V; (c) 7.5 μs; (d) -8 A; (e) $(-30 + 120e^{-(400,000/3)t})$ V, $t \geq 0$; (f) $-8e^{-(400,000/3)t}$ A, $t \geq 0^+$.

8.10 The switch in the circuit shown has been in position a for a long time. At $t = 0$ the switch is moved to position b. Calculate (a) the initial voltage on the capacitor; (b) the final voltage on the capacitor; (c) the time constant (in microseconds) for $t > 0$; and (d) the length of time (in microseconds) required for the capacitor voltage to reach zero after the switch is moved to position b.

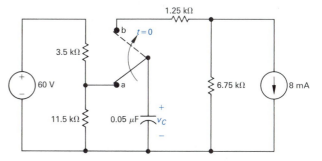

ANSWER: (a) 46 V; (b) -54 V; (c) 400 μs; (d) 246.47 μs.

8.11 After the switch in the circuit shown has been open for a long time, it is closed at $t = 0$. Calculate (a) the initial value of i; (b) the final value of i; (c) the time constant for $t \geq 0$; and (d) the numerical expression for $i(t)$ when $t \geq 0$.

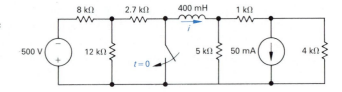

ANSWER: (a) -20 mA; (b) 40 mA; (c) 160 μs; (d) $i = (40 - 60e^{-6250t})$ mA, $t \geq 0$.

8.5 SEQUENTIAL SWITCHING

In this section we apply the general solution to the situation when switching takes place more than once. This situation is called *sequential switching*. It occurs whenever a single switch is moved in sequence to two or more alternative positions or when two or more switches are opened and closed in sequence. In either case, the time reference for all switchings cannot be $t = 0$. We determine the voltages and currents generated by a switching sequence by using the techniques described previously in this chapter. We derive the expressions for $v(t)$ and $i(t)$ for a given position of the switch or switches and then use these solutions to determine the initial conditions for the next position of the switch or switches.

PSpice Sequential switches can be simulated in PSpice circuits: Section 8.2

With sequential switching problems, a premium is placed on obtaining the initial value $x(t_0)$. Recall that anything but inductive currents and capacitive voltages can change instantaneously at the time of switching. Thus solving first for inductive currents and capacitive voltages is even more pertinent in sequential switching problems.

Examples 8.10 and 8.11 illustrate the technique. The former is a natural response problem with two switching times, and the latter is a step response problem.

EXAMPLE 8.10

The two switches in the circuit shown in Fig. 8.29 have been closed for a long time. At $t = 0$, switch 1 is opened. Thirty-five milliseconds later, switch 2 is opened.

a) Find $i_L(t)$ for $0 \leq t \leq 35$ ms.

b) Find $i_L(t)$ for $t \geq 35$ ms.

c) What percentage of the initial energy stored in the 150-mH inductor is dissipated in the 18-Ω resistor?

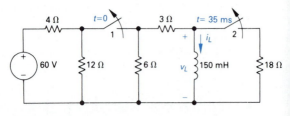

FIGURE 8.29 The circuit for Example 8.10.

d) Repeat (c) for the 3-Ω resistor.

e) Repeat (c) for the 6-Ω resistor.

SOLUTION

a) For $t < 0$ the 150-mH inductor short-circuits the 18-Ω resistor. We determine the initial current in the inductor by solving for $i_L(0^-)$ in the circuit shown in Fig. 8.30. After making several source transformations, we find $i_L(0^-)$ to be 6 A.

Figure 8.31 shows the circuit for $0 \leq t \leq 35$ ms. Note that the equivalent resistance across the terminals of the inductor is the parallel combination of 9 and 18 Ω, or 6 Ω. The time constant of the circuit is $(150/6) \times 10^{-3}$, or 25 ms. Therefore the expression for i_L is

$$i_L = 6e^{-40t} \text{ A}, \qquad 0 \leq t \leq 35 \text{ ms}.$$

b) When $t = 35$ ms, the value of the inductor current is

$$i_L = 6e^{-1.4} = 1.48 \text{ A}.$$

Thus when switch 2 is opened, the circuit reduces to that shown in Fig. 8.32. When switch 2 is opened, the time constant changes to $(150/9) \times 10^{-3}$, or 16.67 ms. The expression for i_L becomes

$$i_L = 1.48e^{-60(t-0.035)} \text{ A}, \qquad t \geq 0.035 \text{ s}.$$

Note that the exponential function is shifted to the right by 35 ms. Also note that, in solving sequential switching problems, drawing the circuit that pertains to each time interval being considered is helpful.

c) The 18-Ω resistor is in the circuit only during the first 35 ms of the switching sequence. During this interval, the voltage across the resistor is

$$v_L = 0.15\frac{d}{dt}(6e^{-40t})$$

$$= -36e^{-40t} \text{ V}, \qquad 0 < t < 0.035 \text{ s}.$$

The power dissipated in the 18-Ω resistor is

$$p = \frac{v_L^2}{18} = 72e^{-80t} \text{ W}, \qquad 0 < t < 0.035 \text{ s}.$$

Hence the energy dissipated is

$$w = \int_0^{0.035} 72e^{-80t} \, dt = \frac{72}{-80}e^{-80t}\Big|_0^{0.035}$$

$$= 0.9(1 - e^{-2.8}) = 845.27 \text{ mJ}.$$

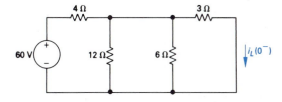

FIGURE 8.30 The circuit shown in Fig. 8.29 for $t < 0$.

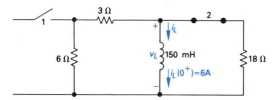

FIGURE 8.31 The circuit shown in Fig. 8.29 for $0 \leq t \leq 35$ ms.

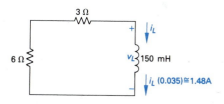

FIGURE 8.32 The circuit shown in Fig. 8.29 for $t \geq 35$ ms.

The initial energy stored in the 150-mH inductor is

$$w_i = \frac{1}{2}(0.15)(36) = 2.7 \text{ J} = 2700 \text{ mJ}.$$

Therefore $(845.27/2700) \times 100$, or 31.31%, of the initial energy stored in the 150-mH inductor is dissipated in the 18-Ω resistor.

d) For $0 < t < 0.035$ s, the voltage across the 3-Ω resistor is

$$v_{3\Omega} = \left(\frac{v_L}{9}\right)(3) = \frac{1}{3}v_L = -12e^{-40t} \text{ V.}$$

Therefore the energy dissipated in the 3-Ω resistor in the first 35 ms is

$$w_{3\Omega} = \int_0^{0.035} \frac{144e^{-80t}}{3} dt$$

$$= 0.6(1 - 3^{-2.8}) = 563.51 \text{ mJ.}$$

For $t > 0.035$ s, the current in the 3-Ω resistor is

$$i_{3\Omega} = i_L = (6e^{-1.4})e^{-60(t - 0.035)} \text{ A.}$$

Hence the energy dissipated in the 3-Ω resistor for $t > 0.035$ s is

$$w_{3\Omega} = \int_{0.035}^{\infty} i_{3\Omega}^2 \times 3\, dt$$

$$= \int_{0.035}^{\infty} 3(36)e^{-2.8}e^{-120(t-0.035)}\, dt$$

$$= 108e^{-2.8} \times \left. \frac{e^{-120(t-0.035)}}{-120} \right|_{0.035}^{\infty}$$

$$= \frac{108}{120}e^{-2.8} = 54.73 \text{ mJ.}$$

The total energy dissipated in the 3-Ω resistor is

$$w_{3\Omega}(\text{total}) = 563.51 + 54.73 = 618.24 \text{ mJ.}$$

The percentage of the initial energy stored is

$$\frac{618.24}{2700} \times 100 = 22.90\%.$$

e) As the 6-Ω resistor is in series with the 3-Ω resistor, the energy dissipated and the percentage of the initial energy stored will be twice that of the 3-Ω resistor:

$$w_{6\Omega}(\text{total}) = 1236.48 \text{ mJ,}$$

and the percentage of the initial energy stored is 45.80%. We check these calculations by observing that

$$1236.48 + 618.24 + 845.27 = 2699.99 \text{ mJ}$$

and

$$31.31 + 22.90 + 45.80 = 100.01\%.$$

The small discrepancies in the summations are the result of round-off errors.

EXAMPLE 8.11

The uncharged capacitor in the circuit shown in Fig. 8.33 is initially switched to terminal a of the three-position switch. At $t = 0$ the switch is moved to position b, where it remains for 15 ms. After the 15-ms delay, the switch is moved to position c, where it remains indefinitely.

a) Derive the numerical expression for the voltage across the capacitor.

b) Plot the capacitor voltage versus time.

c) When will the voltage on the capacitor equal 200 V?

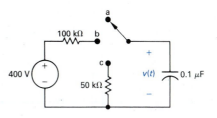

FIGURE 8.33 The circuit for Example 8.11.

SOLUTION

a) At the instant the switch is moved to position b, the initial voltage on the capacitor is zero. If the switch were to remain in position b, the capacitor would eventually charge to 400 V. The time constant of the circuit when the switch is in position b is 10 ms. Therefore we can use Eq. (8.56) with $t_0 = 0$ to write the expression for the capacitor voltage:

$$v = 400 + (0 - 400)e^{-100t}$$

$$= (400 - 400e^{-100t}) \text{ V}, \qquad 0 \le t \le 15 \text{ ms}.$$

Note that, because the switch remains in position b for only 15 ms, this expression is valid only for the time interval from 0 to 15 ms. After the switch has been in position b for 15 ms, the voltage on the capacitor will be

$$v(15 \text{ ms}) = 400 - 400e^{-1.5} = 310.75 \text{ V}.$$

Therefore, when the switch is moved to position c, the initial voltage on the capacitor is 310.75 V. With the switch in position c, the final value of the capacitor voltage is zero and the time constant is 5 ms. Again, we use Eq. (8.56) to write the expression for the capacitor voltage:

$$v = 0 + (310.75 - 0)e^{-200(t-0.015)}$$

$$= 310.75e^{-200(t-0.015)} \text{ V}, \qquad 15 \text{ ms} \le t.$$

In writing the expression for v, we recognized that $t_0 = 15$ ms and that this expression is valid only for $t \ge 15$ ms.

b) Figure 8.34 shows the plot of v versus t.

c) The plot in Fig. 8.34 reveals that the capacitor voltage will equal 200 V at two different times: once in the interval between 0 and 15 ms and once after 15 ms. We find the first time that $v = 200$ V by solving the expression

$$200 = 400 - 400e^{-100t_1},$$

which yields $t_1 = 6.93$ ms. We find the second time that $v = 200$ V by solving the expression

$$200 = 310.75e^{-200(t_2-0.015)}.$$

In this case, $t_2 = 17.20$ ms.

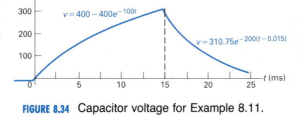

FIGURE 8.34 Capacitor voltage for Example 8.11.

DRILL EXERCISES

8.12 In the circuit shown, switch 1 has been closed and switch 2 has been open for a long time. At $t = 0$, switch 1 is opened. Fifty milliseconds later switch 2 is closed. Find:

a) $v_C(t)$ for $0 \le t \le 0.05$ s;

b) $v_C(t)$ for $t \ge 0.05$ s;

c) the total energy dissipated in the 50-kΩ resistor; and

d) the total energy dissipated in the 200-kΩ resistor.

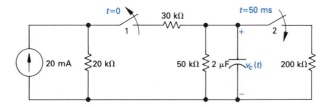

ANSWER: (a) $200e^{-10t}$ V; (b) $121.31e^{-12.5(t-0.05)}$ V; (c) 37.06 mJ; (d) 2.94 mJ.

8.13 Switch a in the circuit shown has been open for a long time and switch b has been closed for a long time. Switch a is closed at $t = 0$ and, after remaining closed for 1 s, is opened again. Switch b is opened simultaneously, and both switches remain open indefinitely. Determine the expression for the inductor current i that is valid when (a) $0 \le t \le 1$ s and (b) $t \ge 1$ s.

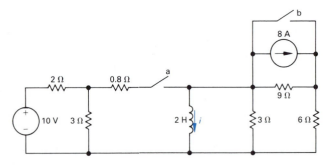

ANSWER: (a) $i(t) = (3 - 3e^{-0.5t})$ A, $0 \le t \le 1$ s; (b) $i(t) = (-4.8 + 5.98e^{-1.25(t-1)})$ A, $t \ge 1$ s.

8.6 UNBOUNDED RESPONSE

A circuit response may grow, rather than decay, exponentially with time. This type of response is possible if the circuit contains dependent sources. In that case, the Thévenin equivalent resistance with respect to the terminals of either an inductor or a capacitor may be negative. This negative resistance generates a positive exponent, and the resulting currents and voltages increase without limit. In an actual circuit, the exponentially increasing response eventually reaches a limiting value when a component breaks down or goes into a saturation state, prohibiting further increases in voltage or current.

When we consider unbounded responses, the concept of a final value is confusing. Hence, rather than using the step response solution given in Eq. (8.56), we derive the differential equation that describes the circuit containing the negative resistance and then solve it using the separation of variables technique. Example 8.12 presents an example of an exponentially growing response in terms of the voltage across a capacitor.

E X A M P L E   8.12

a) When the switch is closed in the circuit shown in Fig. 8.35, the voltage on the capacitor is 10 V. Find the expression for v_o for $t \geq 0$.

b) Assume that the capacitor short-circuits when its terminal voltage reaches 150 V. How many milliseconds are required to short-circuit the capacitor?

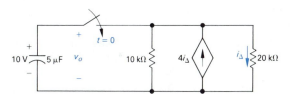

FIGURE 8.35 The circuit for Example 8.12.

S O L U T I O N

a) To find the Thévenin equivalent resistance with respect to the capacitor terminals, we use the test-source method described in Chapter 4. Figure 8.36 shows the resulting circuit, where v_T is the test voltage and i_T is the test current.

From Fig. 8.32 for v_T expressed in volts, we obtain

$$i_T = \frac{v_T}{10} - 4\left(\frac{v_T}{20}\right) + \frac{v_T}{20} \text{ mA}.$$

Solving for the ratio v_T/i_T yields the Thévenin resistance:

$$R_{Th} = \frac{v_T}{i_T} = -20 \text{ k}\Omega.$$

With this Thévenin resistance, we can simplify the circuit shown in Fig. 8.35 to that shown in Fig. 8.37.

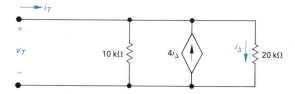

FIGURE 8.36 Test-source method used to find R_{Th}.

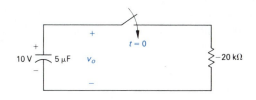

FIGURE 8.37 A simplification of the circuit shown in Fig. 8.35.

For $t \geq 0$, the differential equation describing the circuit shown in Fig. 8.37 is

$$(5 \times 10^{-6})\frac{dv_o}{dt} - \frac{v_o}{20} \times 10^{-3} = 0.$$

Dividing by the coefficient of the first derivative yields

$$\frac{dv_o}{dt} - 10v_o = 0.$$

We now use the separation of variables technique to find $v_o(t)$:

$$v_o(t) = 10e^{10t} \text{ V}, \qquad t \geq 0.$$

b) $v_o = 150$ V when $e^{10t} = 15$. Therefore $10t = \ln 15$, and $t = 270.81$ ms.

The fact that interconnected circuit elements may lead to ever-increasing currents and voltages is a topic of interest to engineers. They must be concerned with the stability of a system when feedback, either deliberate or parasitic, is present.

8.7 THE INTEGRATING AMPLIFIER

Recall from the introduction to Chapter 6 that one reason for our interest in the operational amplifier is its use as an integrating amplifier. We are now ready to analyze the integrating-amplifier circuit, which is shown in Fig. 8.38. The purpose of the integrating-amplifier circuit is to generate an output voltage proportional to the integral of the input voltage. In Fig. 8.38 we added the branch currents i_f and i_s, along with the node voltages v_1 and v_2, to aid the analysis.

In analyzing the circuit we assume that the operational amplifier is ideal. Thus we take advantage of the constraints

$$i_f + i_s = 0 \tag{8.58}$$

and

$$v_1 = v_2 = 0. \tag{8.59}$$

Because $v_1 = 0$,

$$i_s = \frac{v_s}{R_s} \tag{8.60}$$

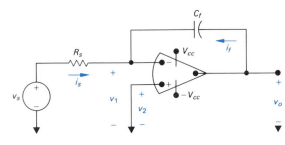

FIGURE 8.38 An integrating amplifier.

and

$$i_f = C_f \frac{dv_o}{dt}. \tag{8.61}$$

Hence, from Eqs. (8.58), (8.60), and (8.61),

$$\frac{dv_o}{dt} = -\frac{1}{R_s C_f} v_s. \tag{8.62}$$

Multiplying both sides of Eq. (8.62) by a differential time dt and then integrating from t_o to t generates the equation

$$v_o(t) = -\frac{1}{R_s C_f} \int_{t_o}^{t} v_s \, dy + v_o(t_o). \tag{8.63}$$

In Eq. (8.63), t_o represents the instant in time when we begin the integration. Thus $v_o(t_o)$ is the value of the output voltage at that time. Also, as $v_1 = v_2 = 0$, $v_o(t_o)$ is identical to the initial voltage on the feedback capacitor C_f.

Equation (8.63) states that the output voltage of an integrating amplifier equals the initial value of the voltage on the capacitor plus an inverted (minus sign), scaled $(1/R_s C_f)$ replica of the integral of the input voltage. If no energy is stored in the capacitor when integration commences, Eq. (8.63) reduces to

$$v_o(t) = -\frac{1}{R_s C_f} \int_{t_o}^{t} v_s \, dy. \tag{8.64}$$

If v_s is a step change in a dc voltage level, the output voltage will vary linearly with time. For example, assume that the input voltage is the rectangular voltage pulse shown in Fig. 8.39. Assume also that the initial value of $v_o(t)$ is zero at the instant v_s steps from 0 to V_m. A direct application of Eq. (8.63) yields

$$v_o = -\frac{1}{R_s C_f} V_m t + 0, \qquad 0 \leq t \leq t_1. \tag{8.65}$$

When t lies between t_1 and $2t_1$,

$$v_o = -\frac{1}{R_s C_f} \int_{t_1}^{t} (-V_m) dy - \frac{1}{R_s C_f} V_m t_1$$

$$= \frac{V_m}{R_s C_f} t - \frac{2V_m}{R_s C_f} t_1, \qquad t_1 \leq t \leq 2t_1. \tag{8.66}$$

Figure 8.40 shows a sketch of $v_o(t)$ versus t. Clearly the output voltage is an inverted, scaled replica of the integral of the input voltage.

The output voltage is proportional to the integral of the input voltage only if the operational amplifier operates within its linear range, that is, if it doesn't saturate. Examples 8.13 and 8.14 further illustrate analysis of the integrating amplifier.

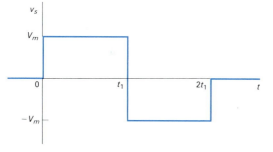

FIGURE 8.39 Input voltage signal.

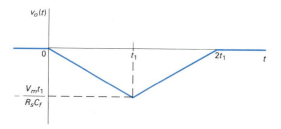

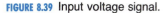

FIGURE 8.40 Output voltage of the integrating amplifier.

E X A M P L E   8.13

At the instant the switch makes contact with terminal a in the circuit shown in Fig. 8.41, the voltage on the 0.1-μF capacitor is 5 V. The switch remains at terminal a for 9 ms and then moves instantaneously to position b. How many milliseconds after making contact with terminal a does the operational amplifier saturate?

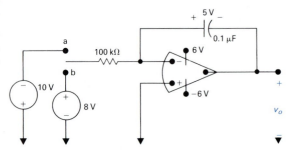

FIGURE 8.41 The circuit for Example 8.13.

S O L U T I O N

The expression for the output voltage during the time the switch is in position a is

$$v_o = -5 - \frac{1}{10^{-2}} \int_0^t (-10) \, dy = (-5 + 1000t) \text{ V.}$$

Nine milliseconds after the switch makes contact with terminal a the output voltage is $-5 + 9$, or 4 V.

The expression for the output voltage after the switch moves to position b is

$$v_o = 4 - \frac{1}{10^{-2}} \int_{9 \times 10^{-3}}^t 8 \, dy$$

$$= 4 - 800(t - 9 \times 10^{-3}) = (11.2 - 800t) \text{ V.}$$

During this time interval, the voltage is decreasing, and the operational amplifer eventually saturates at -6 V. Therefore we set the expression for v_o equal to -6 V to obtain the saturation time t_s:

$$11.2 - 800t_s = -6 \quad \text{or} \quad t_s = 21.5 \text{ ms.}$$

Thus the integrating amplifier saturates 21.5 ms after making contact with terminal a.

E X A M P L E   8.14

Assume that the numerical values for the signal voltage shown in Fig. 8.39 are $V_m = 50$ mV and $t_1 = 1$ s. This signal voltage is applied to the integrating-amplifier circuit shown in Fig. 8.38.

The circuit parameters of the amplifier are $R_s = 100$ kΩ, $C_f = 0.1$ μF, and $V_{CC} = 6$ V. The initial voltage on the capacitor is zero.

a) Calculate $v_o(t)$.

b) Plot $v_o(t)$ versus t.

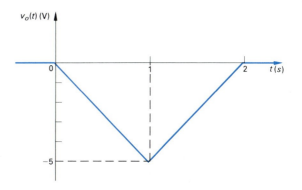

SOLUTION

a) For $0 \leq t \leq 1$ s,

$$v_o = \frac{-1}{(100 \times 10^3)(0.1 \times 10^{-6})} 50 \times 10^{-3}t + 0$$

$$= -5t \text{ V}, \qquad 0 \leq t \leq 1 \text{ s}.$$

For $1 \leq t \leq 2$ s, $v_o = (5t - 10)$ V.

b) Figure 8.42 shows a plot of $v_o(t)$ versus t.

FIGURE 8.42 The output voltage for Example 8.14.

INTEGRATING AMPLIFIER WITH A FEEDBACK RESISTOR

Figure 8.43 shows a variation of the integrating-amplifier circuit illustrated in Fig. 8.38. Here, a resistor R_f is in parallel with the feedback capacitor C_f. The feedback resistor is added to prevent the integrator from saturating because of charge accumulation on the feedback capacitor.

To find the relationship between the output voltage, v_o, and the input voltage, v_s, we sum the currents at the inverting input terminal of the ideal operational amplifier. Because $v_1 = v_2 = 0$, we obtain

$$\frac{0 - v_s}{R_s} + \frac{0 - v_o}{R_f} + C_f \frac{d}{dt}(0 - v_o) = 0. \qquad (8.67)$$

Rearranging the terms in Eq. (8.67) yields

$$\frac{dv_o}{dt} + \frac{1}{R_f C_f} v_o = -\frac{v_s}{R_s C_f}. \qquad (8.68)$$

Comparing Eq. (8.68) with Eq. (8.62) reveals that a simple integral relationship between v_o and v_s no longer exists.

To find out how the feedback resistor affects the relationship between v_o and v_s, we again determine the output voltage when the input voltage has the waveform shown in Fig. 8.38. For

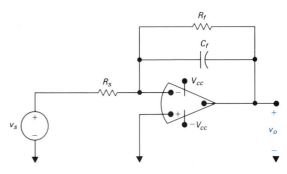

FIGURE 8.43 An integrating amplifier with a feedback resistor.

$0 \le t \le t_1$ the input voltage equals V_m, and Eq. (8.68) becomes

$$\frac{dv_o}{dt} + \frac{1}{R_f C_f} v_o = \frac{-V_m}{R_s C_f}. \tag{8.69}$$

If we assume that the initial charge on the capacitor is zero, that is, that $v_o(0) = 0$, the solution of Eq. (8.69) is

$$v_o = -\frac{R_f}{R_s} V_m(1 - e^{-t/R_f C_f}), \qquad 0 \le t \le t_1. \tag{8.70}$$

When t lies between t_1 and $2t_1$, the input signal is $-V_m$ and Eq. (8.68) is

$$\frac{dv_o}{dt} + \frac{1}{R_f C_f} v_o = \frac{V_m}{R_s C_f}. \tag{8.71}$$

The solution of Eq. (8.71) is

$$v_o = \frac{V_m R_f}{R_s} + \left[v_o(t_1) - \frac{V_m R_f}{R_s} \right] e^{-(t-t_1)/R_f C_f}, \tag{8.72}$$

where

$$v_o(t_1) = -\frac{V_m R_f}{R_s}(1 - e^{-t_1/R_f C_f}). \tag{8.73}$$

Substituting Eq. (8.73) into Eq. (8.72) yields

$$v_o(t) = \frac{V_m R_f}{R_s} - \frac{V_m R_f}{R_s}(2 - e^{-t_1/R_f C_f})e^{-(t-t_1)/R_f C_f}, \qquad t_1 \le t \le 2t_1. \tag{8.74}$$

From Eq. (8.74),

$$v_o(2t_1) = \frac{V_m R_f}{R_s}(1 - e^{-t_1/R_f C_f})^2. \tag{8.75}$$

Note also from Eq. (8.74) that $v_o(t) = 0$ when

$$t_o = R_f C_f \ln (2e^{t_1/R_f C_f} - 1). \tag{8.76}$$

Finally, when t is greater than $2t_1$, $v_s = 0$ and Eq. (8.68) reduces to

$$\frac{dv_o}{dt} + \frac{1}{R_f C_f} v_o = 0. \tag{8.77}$$

From Eq. (8.77),

$$v_o = v_o(2t_1)e^{-(t-2t_1)/R_f C_f}. \tag{8.78}$$

Using the result given in Eq. (8.75), we write Eq. (8.78) as

$$v_o = \frac{V_m R_f}{R_s}(1 - e^{-t_1/R_f C_f})^2 \, e^{-(t-2t_1)/R_f C_f}, \qquad 2t_1 \le t \le \infty. \tag{8.79}$$

DRILL EXERCISES

8.14 Derive Eqs. (8.74), (8.75), and (8.76).   **ANSWER:** Derivations.

Example 8.15 compares numerically the integrating amplifier of Fig. 8.38 with that of Fig. 8.43.

EXAMPLE 8.15

The feedback capacitor in the integrating-amplifier circuit in Example 8.14 is shunted by a 5-MΩ resistor. The input voltage signal is the same as that of Example 8.14.

a) Find the expressions for v_o.

b) Plot v_o.

c) Compare this v_o with that obtained in Example 8.14.

SOLUTION

a) We know that $R_f/R_s = 50$ and $1/R_fC_f = 2.0$, so

$$v_o = -2.5(1 - e^{-2t}) \text{ V}, \qquad 0 \le t \le 1 \text{ s}$$

$$= 2.5 - 2.5(2 - e^{-2})e^{-2(t-1)};$$

$$v_o = 2.5 - 4.66e^{-2(t-1)} \text{ V}, \qquad 1 \text{ s} \le t \le 2 \text{ s}$$

$$= 2.5(1 - e^{-2})^2 \, e^{-2(t-2)};$$

$$v_o = 1.87e^{-2(t-2)} \text{ V}, \qquad 2 \text{ s} \le t \le \infty.$$

b) From the solutions in (a), we have

$$v_o(t_1) = v_o(1) = -2.16 \text{ V};$$

$$v_o(2t_1) = v_o(2) = 1.87 \text{ V};$$

$$t_o = 0.5 \ln (2e^2 - 1) = 1.31 \text{ s}.$$

Figure 8.44 shows a plot of $v_o(t)$ versus t.

c) In order to compare the response of the integrating amplifier without a feedback resistor (Example 8.14) with that of an amplifier having a feedback resistor (Example 8.15), we plot both output voltages, as shown in Fig. 8.44. Each output voltage corresponds to an identical input voltage—namely, that in Fig. 8.39 with $V_m = 50$ mV and $t_1 = 1$ s. Figure 8.44

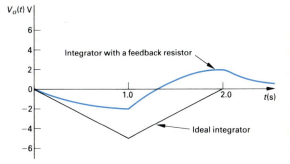

FIGURE 8.44 Plot of $v_o(t)$ versus t for Examples 8.14 and 8.15.

indicates that an exponential response in Example 8.15 replaces the linear response in Example 8.14. The exponential response approaches the linear response as R_f increases.

DRILL EXERCISES

8.15 Rework Example 8.15 with $R_f = 50$ MΩ.

ANSWER: $v_o = -25(1 - e^{-0.2t})$ V, $\quad 0 \leq t \leq 1$ s;

$\qquad\qquad v_o = 25 - 29.53e^{-0.2(t-1)}$ V, $\quad 1$ s $\leq t \leq 2$ s;

$\qquad\qquad v_o = 0.82e^{-0.2(t-1)}$ V, $\quad 2$ s $\leq t \leq \infty$.

Even with nonideal operational amplifiers we can design integrating amplifiers that, within specified limits, perform the integration function very well. Moreoever, we can convert the integrating amplifier to a differentiating amplifier by interchanging the input resistant R_s and the feedback capacitor C_f. Then,

$$v_o = -R_sC_f\frac{dv_s}{dt}. \qquad \textbf{(8.80)}$$

We leave derivation of Eq. (8.80) as an exercise for you. The differentiating amplifier is seldom used because in practice it is a source of unwanted or noisy signals.

Finally, we can design both integrating and differentiating amplifier circuits by using an inductor instead of a capacitor. However, fabricating capacitors for integrated-circuit devices is much easier, so inductors are rarely used in integrating amplifiers.

SUMMARY

The discussion centered on determining the natural and step responses of single time-constant or first-order circuits. We presented the following important concepts:

- A first-order circuit may be reduced to a Thévenin (or Norton) equivalent connected to either a single equivalent inductor or capacitor.

- The time constant of an inductive circuit equals the equivalent inductance divided by the Thévenin resistance as viewed from the terminals of the equivalent inductor.

- The time constant of a capacitive circuit equals the equivalent capacitance times the Thévenin resistance as viewed from the terminals of the equivalent capacitor.

- The natural response corresponds to finding the currents and voltages that exist when stored energy is released to a circuit that contains no independent sources.

- The step response corresponds to finding the currents and voltages that result from abrupt changes in dc sources connected to the circuit. Stored energy may or may not be present at the time the abrupt change takes place.

- The solution for either the natural or step response of both *RL* and *RC* circuits involves finding the initial and final value of the current or voltage of interest and the time constant of the circuit. Equations (8.56) and (8.57) summarize this approach.

- An unbounded response occurs when the Thévenin resistance is negative, which is possible when the first-order circuit contains dependent sources.

- Sequential switching in first-order circuits is analyzed by dividing the analysis into time intervals corresponding to specific switch positions. Initial values for a particular interval are determined from the solution corresponding to the immediately preceding interval.

- The techniques discussed are used to analyze an integrating amplifier consisting of an ideal op amp, a capacitor in the negative feedback branch, and a resistor in series with the signal source.

PROBLEMS

8.1 In the circuit in Fig. P8.1, the voltage and current expressions are

$$v = 160e^{-10t} \text{ V}, t \geq 0^+;$$

$$i = 6.4e^{-10t} \text{ A}, t \geq 0.$$

Find (a) R, (b) τ (ms), (c) L, (d) the initial energy stored in the inductor, and (e) the time (ms) it takes to dissipate 60% of the initial stored energy.

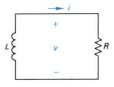

FIGURE P8.1

8.2 In the circuit shown in Fig. P8.2 the switch makes contact with position b just before breaking contact with position a. This is known as a "make-before-break" switch and is designed so that the switch does not interrupt the current in an inductive circuit. The interval of time between "making" and "breaking" is assumed to be negligible. The switch has been in the a position for a long time. At $t = 0$ the switch is thrown from position a to position b.

a) Determine the initial current in the inductor.

b) Determine the time constant of the circuit for $t > 0$.

c) Find i, v_1, and v_2 for $t \geq 0$.

d) What percentage of the initial energy stored in the inductor is dissipated in the 72-Ω resistor 15 ms after the switch is thrown from position a to position b?

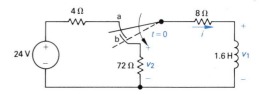

FIGURE P8.2

8.3 The switch in the circuit in Fig. P8.3 has been open for a long time. At $t = 0$ the switch is closed.

a) Determine $i_o(0^+)$ and $i_o(\infty)$.

b) Determine $i_o(t)$ for $t \geq 0^+$.

c) How many microseconds after the switch has been closed will the current in the switch equal 3A?

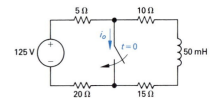

FIGURE P8.3

8.4 The switch in the circuit in Fig. P8.4 has been closed a long time. At $t = 0$ it is opened. Find $v_o(t)$ for $t \geq 0^+$.

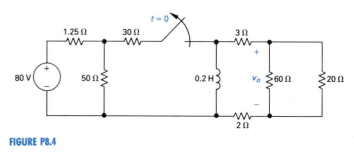

FIGURE P8.4

8.5 Assume that the switch in the circuit in Fig. P8.4 has been open for two time constants. At this instant, what percentage of the total energy stored in the 0.2-H inductor has been dissipated in the 2-Ω resistor?

8.6 The switch in the circuit in Fig. P8.6 has been in position 1 for a long time. At $t = 0$, the switch moves instantaneously to position 2. Find $v_o(t)$ for $t \geq 0^+$.

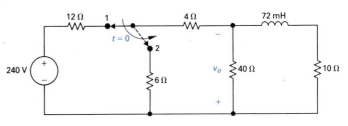

8.7 For the circuit of Fig. P8.6, what percentage of the initial energy stored in the inductor is eventually dissipated in the 40-Ω resistor?

8.8 In the circuit in Fig. P8.8 the switch has been closed for a long time before opening at $t = 0$.

a) Find the value of L so that $v_o(t)$ equals $0.5 \, v_o(0^+)$ when $t = 1$ ms.

b) Find the percentage of the stored energy that has been dissipated in the 10-Ω resistor when $t = 1$ ms.

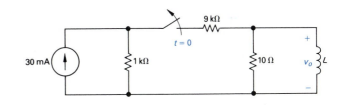

8.9 The switch in the circuit in Fig. P8.9 has been closed for a long time before opening at $t = 0$. Find:

a) $i_1(0^-)$ and $i_2(0^-)$;

b) $i_1(0^+)$ and $i_2(0^+)$;

c) $i_1(t)$ for $t \geq 0$;

d) $i_2(t)$ for $t \geq 0^+$; and

e) Explain why $i_2(0^-) \neq i_2(0^+)$.

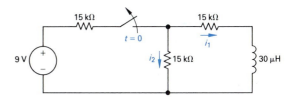

8.10 The switch in the circuit seen in Fig. P8.10 has been in position 1 for a long time. At $t = 0$ the switch moves instantaneously to position 2. Find the value of R so that 0.10 of the initial energy stored in the 10-mH inductor is dissipated in R in 10 μs.

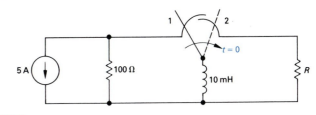

8.11 In the circuit shown in Fig. P8.11 the switch has been in position a for a long time. At $t = 0$, it moves instantaneously from a to b.

a) Find $v_o(t)$ for $t \geq 0^+$.

b) Does it take more than or less than two time constants to dissipate 98% of the energy stored in the circuit at $t = 0^+$? Justify your answer.

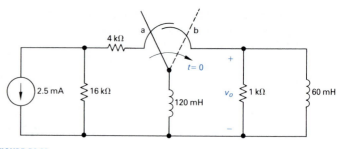

FIGURE P8.11

8.12 The two switches shown in the circuit in Fig. P8.12 operate simultaneously. Prior to $t = 0$ each switch has been in its indicated position for a long time. At $t = 0$ the two switches move instantaneously to their new positions. Find

a) $v_o(t)$, $t \geq 0^+$,

b) $i_o(t)$, $t \geq 0$.

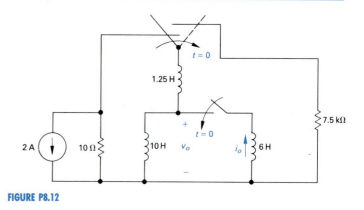

FIGURE P8.12

8.13 For the circuit seen in Fig. P8.12 find

a) the total energy dissipated in the 7.5-kΩ resistor; and

b) the energy trapped in the ideal inductors.

8.14 The switch in the circuit in Fig. P8.14 has been closed for a long time before opening at $t = 0$. Find $v_o(t)$ for $t \geq 0^+$.

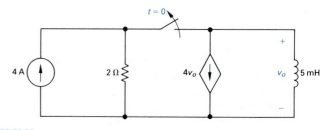

FIGURE P8.14

8.15 The switch in Fig. P8.15 has been closed for a long time before opening at $t = 0$. Find

a) $i_L(t)$, $t \geq 0$;

b) $v_L(t)$, $t \geq 0^+$; and

c) $i_\Delta(t)$, $t \geq 0^+$.

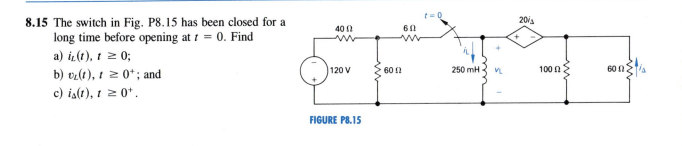

FIGURE P8.15

8.16 What percentage of the total energy dissipated in the two resistors in the circuit in Fig. P8.15 is supplied by the dependent voltage source?

8.17 The 240-V, 2-Ω source in the circuit in Fig. P8.17 is inadvertently short-circuited at its terminals a, b. At the time the fault occurs, the circuit has been in operation for a long time.

a) What is the initial value of the current i_{ab} in the short-circuit connection between terminals a, b?

b) What is the final value of the current i_{ab}?

c) How many microseconds after the short-circuit has occurred is the current in the short equal to 114 A?

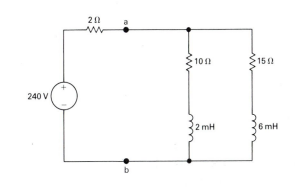

FIGURE P8.17

8.18 In the circuit in Fig. P8.18 the voltage and current expressions are

$$v = 72e^{-500t} \text{ V}, \ t \geq 0;$$

$$i = 9e^{-500t} \text{ mA}, \ t \geq 0^+.$$

Find (a) R, (b) C, (c) τ (ms), (d) the initial energy stored in the capacitor, and (e) how many milliseconds it takes to dissipate 68% of the initial energy stored in the capacitor.

FIGURE P8.18

8.19 The switch in the circuit in Fig. P8.19 has been in position a for a long time. At $t = 0$ the switch is thrown to position b.

a) Find $i_o(t)$ for $t \geq 0^+$.

b) What percentage of the initial energy stored in the capacitor is dissipated in the 3-kΩ resistor 500 μs after the switch has been thrown?

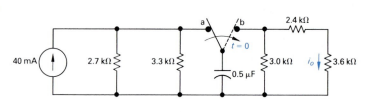

FIGURE P8.19

8.20 The switch in the circuit in Fig. P8.20 is closed at $t = 0$ after being open for a long time.

a) Find $i_1(0^-)$ and $i_2(0^-)$.

b) Find $i_1(0^+)$ and $i_2(0^+)$.

c) Explain why $i_1(0^-) = i_1(0^+)$.

d) Explain why $i_2(0^-) \neq i_2(0^+)$.

e) Find $i_1(t)$ for $t \geq 0$.

f) Find $i_2(t)$ for $t \geq 0^+$.

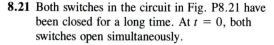

FIGURE P8.20

8.21 Both switches in the circuit in Fig. P8.21 have been closed for a long time. At $t = 0$, both switches open simultaneously.

a) Find $i_o(t)$ for $t \geq 0^+$.

b) Find $v_o(t)$ for $t \geq 0$.

c) Calculate the energy (in microjoules) trapped in the circuit.

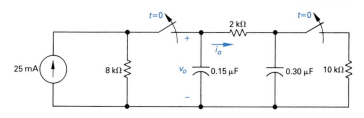

FIGURE P8.21

8.22 In the circuit shown in Fig. P8.22, switches 1 and 2 operate together—that is, they either open or close at the same time. The switches are closed a long time before opening at $t = 0$.

a) How many millijoules of energy have been dissipated in the 12-kΩ resistor 12 ms after the switches open?

b) How long does it take to dissipate 75% of the initially stored energy?

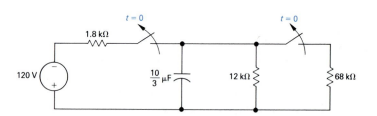

FIGURE P8.22

8.23 The two switches in the circuit seen in Fig. P8.23 are synchronized. The switches have been closed for a long time before opening at $t = 0$.

a) How many microseconds after the switches are open is the energy dissipated in the 10-kΩ resistor 20% of the initial energy stored in the 4.4-H inductor?

b) At the time calculated in part (a), what percentage of the total energy stored in the inductor has been dissipated?

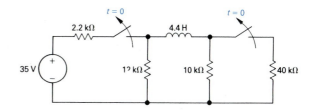

FIGURE P8.23

8.24 The switch in the circuit in Fig. P8.24 has been in position 1 for a long time before moving to position 2 at $t = 0$. Find $i_o(t)$ for $t \geq 0^+$.

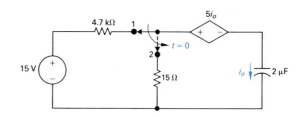

FIGURE P8.24

8.25 The switch in the circuit seen in Fig. P8.25 has been in position x for a long time. At $t = 0$ the switch moves instantaneously to position y.

a) Find α so that the time constant for $t > 0$ is 40 ms.

b) For the α found in part (a), find v_ϕ.

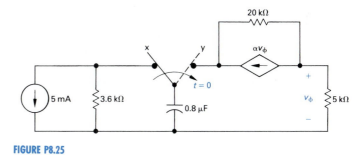

FIGURE P8.25

8.26 a) In Problem 8.25 how many microjoules of energy are generated by the dependent current source during the time the capacitor discharges to 0 V?

b) Show that for $t \geq 0$ the total energy stored and generated in the capacitive circuit equals the total energy dissipated.

8.27 The switch in the circuit in Fig. P8.27 has been in position a for a long time. At $t = 0$ the switch is thrown to position b.

a) Calculate i, v_1, and v_2 for $t \geq 0^+$.

b) Calculate the energy stored in the capacitor at $t = 0$.

c) Calculate the energy trapped in the circuit and the total energy dissipated in the 5-kΩ resistor if the switch remains in position b indefinitely.

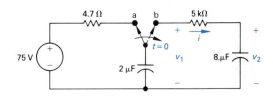

FIGURE P8.27

8.28 At the time the switch is closed in the circuit shown in Fig. P8.28, the capacitors are charged as shown.

a) Find $v_o(t)$ for $t \geq 0^+$.

b) What percentage of the total energy initially stored in the three capacitors is dissipated in the 50-kΩ resistor?

c) Find $v_1(t)$ for $t \geq 0$.

d) Find $v_2(t)$ for $t \geq 0$.

e) Find the energy (microjoules) trapped in the ideal capacitors.

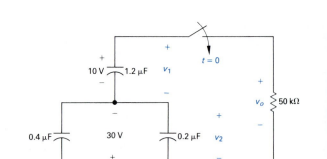

FIGURE P8.28

8.29 At the time the switch is closed in the circuit in Fig. P8.29 the voltage across the paralleled capacitors is 50 V and the voltage on the 0.25-μF capacitor is 40 V.

a) What percentage of the initial energy stored in the three capacitors is dissipated in the 24-kΩ resistor?

b) Repeat part (a) for the 0.4 and 16-kΩ resistors.

c) What percentage of the initial energy is trapped in the capacitors?

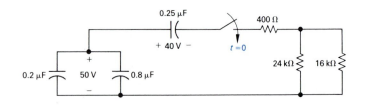

FIGURE P8.29

8.30 After the circuit in Fig. P8.30 has been in operation for a long time a screwdriver was inadvertently connected across the terminals a, b. Assume the resistance of the screwdriver is negligible.

a) Find the current in the screwdriver at $t = 0^+$ and $t = \infty$.

b) Derive the expression for the current in the screwdriver for $t \geq 0^+$.

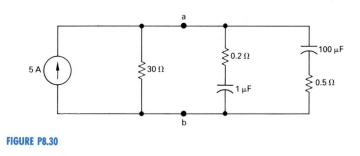

FIGURE P8.30

8.31 The current and voltage at the terminals of the inductor in the circuit in Fig. 8.16 are

$$i(t) = (25 - 25e^{-400t}) \text{ A}, \qquad t \geq 0$$

$$v(t) = 100e^{-400t} \text{ V}, \qquad t \geq 0^+.$$

a) Specify the numerical values of V_s, R, and L.

b) How many milliseconds after the switch has been closed does the energy stored in the inductor reach 25% of its final value?

8.32 The switch in the circuit shown in Fig. P8.32 has been in position a for a long time. At $t = 0$ the switch moves instantaneously to position b.

a) Find the numerical expression for $i_o(t)$ when $t \geq 0$.

b) Find the numerical expression for $v_o(t)$ for $t \geq 0^+$.

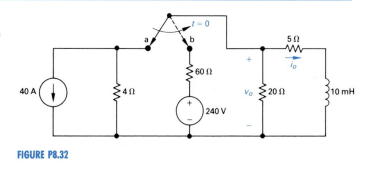

FIGURE P8.32

8.33 The switch in the circuit shown in Fig. P8.33 has been closed for a long time before opening at $t = 0$.

a) Find the numerical expressions for $i_L(t)$ and $v_o(t)$ for $t \geq 0$.

b) Find the numerical values of $v_L(0^+)$ and $v_o(0^+)$.

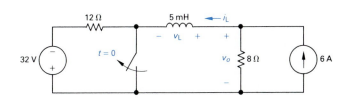

FIGURE P8.33

8.34 The switch in the circuit seen in Fig. P8.34 has been closed for a long time. The switch opens at $t = 0$. Find the numerical expressions for $i_o(t)$ and $v_o(t)$ when $t \geq 0^+$.

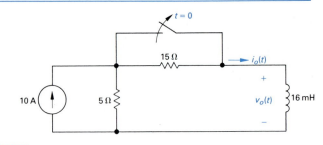

FIGURE P8.34

8.35 The switch in the circuit shown in Fig. P8.35 has been closed for a long time. The switch opens at $t = 0$. For $t \geq 0^+$,

a) find $v_o(t)$ as a function of I_g, R_1, R_2, and L;

b) verify your expression by using it to find $v_o(t)$ in the circuit of Fig. P8.34;

c) explain what happens to $v_o(t)$ as R_2 gets larger and larger;

d) find v_{sw} as a function of I_g, R_1, R_2, and L; and

e) explain what happens to v_{sw} as R_2 gets larger and larger.

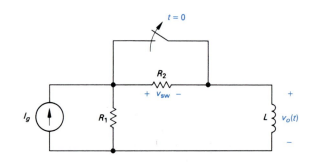

FIGURE P8.35

8.36 The switch in the circuit in Fig. P8.36 has been closed for a long time. A student abruptly opens the switch and reports to her instructor that when the switch opened, an electric arc with noticeable persistence was established across the switch and at the same time the voltmeter placed across the coil was damaged. On the basis of your analysis of the circuit in Problem 8.35, can you explain to the student why this happened?

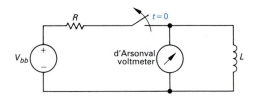

FIGURE P8.36

8.37 The switch in the circuit in Fig. P8.37 has been open a long time before closing at $t = 0$. Find $i_o(t)$ for $t \geq 0$.

FIGURE P8.37

8.38 The "make-before-break" switch in the circuit of Fig. P8.38 has been in position a for a long time. At $t = 0$ the switch moves instantaneously to position b. Find

a) $v_o(t)$, $t \geq 0^+$,

b) $i_1(t)$, $t \geq 0$, and

c) $i_2(t)$, $t \geq 0$.

FIGURE P8.38

8.39 The switch in the circuit in Fig. P8.39 has been open a long time before closing at $t = 0$. Find $v_o(t)$ for $t \geq 0^+$.

FIGURE P8.39

8.40 There is no energy stored in the inductors L_1 and L_2 at the time the switch is opened in the circuit shown in Fig. P8.40.

a) Derive the expressions for the currents $i_1(t)$ and $i_2(t)$ for $t \geq 0$.

b) Use the expressions derived in part (a) to find $i_1(\infty)$ and $i_2(\infty)$.

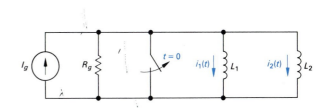

FIGURE P8.40

8.41 The current and voltage at the terminals of the capacitor in the circuit in Fig. 8.21 are

$$i(t) = 25e^{-500t} \text{ mA} \qquad (t \geq 0^+),$$

$$v(t) = (200 - 200e^{-500t}) \text{ V} \qquad (t \geq 0).$$

a) Specify the numerical values of I_s, R, C, and τ.

b) How many milliseconds after the switch has been closed does the energy stored in the capacitor reach 36% of its final value?

8.42 The switch in the circuit shown in Fig. P8.42 has been closed a long time before opening at $t = 0$. For $t \geq 0^+$ find:

a) $v_o(t)$;

b) $i_o(t)$; c) $i_1(t)$;

d) $i_2(t)$; and

e) $i_1(0^+)$.

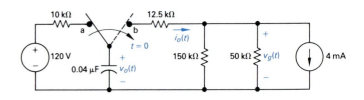

FIGURE P8.42

8.43 The switch in the circuit seen in Fig. P8.43 has been in position a for a long time. At $t = 0$ the switch moves instantaneously to position b. For $t \geq 0^+$ find:

a) $v_o(t)$;

b) $i_o(t)$;

c) $v_g(t)$; and

d) $v_g(0^+)$.

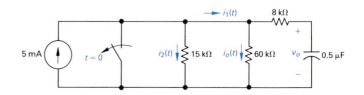

FIGURE P8.43

8.44 The switch in the circuit shown in Fig. P8.44 has been closed a long time before opening at $t = 0$.

a) What is the initial value of $i_o(t)$?

b) What is the final value of $i_o(t)$?

c) What is the time constant of the circuit for $t \geq 0$?

d) What is the numerical expression for $i_o(t)$ when $t \geq 0^+$?

e) What is the numerical expression for $v_o(t)$ when $t \geq 0^+$?

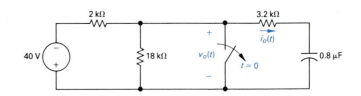

FIGURE P8.44

8.45 The switch in the circuit seen in Fig. P8.45 has been in position a for a long time. At $t = 0$ the switch moves instantaneously to position b. Find $v_o(t)$ and $i_o(t)$ for $t \geq 0^+$.

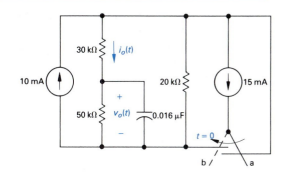

FIGURE P8.45

8.46 The switch in the circuit shown in Fig. P8.46 has been in the OFF position for a long time. At $t = 0$ the switch moves instantaneously to the ON position. Find $v_o(t)$ for $t \geq 0$.

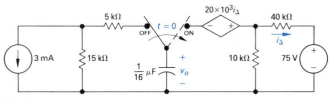

FIGURE P8.46

8.47 Assume that the switch in the circuit of Fig. P8.46 has been in the ON position for a long time before switching instantaneously to the OFF position at $t = 0$. Find $v_o(t)$ for $t \geq 0$.

8.48 At $t = 0$ the voltage source in the circuit seen in Fig. P8.48 drops instantaneously from 100 to 25 V. At the same instant, the current source reverses direction. Find $v_o(t)$ for $t \geq 0$.

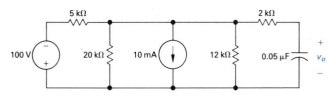

FIGURE P8.48

8.49 There is no energy stored in the capacitors C_1 and C_2 at the time the switch is closed in the circuit seen in Fig. P8.49.

a) Derive the expressions for $v_1(t)$ and $v_2(t)$ for $t \geq 0$.

b) Use the expressions derived in part (a) to find $v_1(\infty)$ and $v_2(\infty)$.

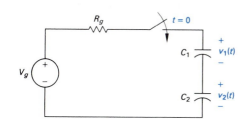

FIGURE P8.49

8.50 The switch in the circuit of Fig. P8.50 has been in position a for a long time. At $t = 0$ it moves instantaneously to position b. For $t \geq 0^+$ find

a) $v_o(t)$,

b) $i_o(t)$,

c) $v_1(t)$,

d) $v_2(t)$,

e) the energy trapped in the capacitors as $t \to \infty$.

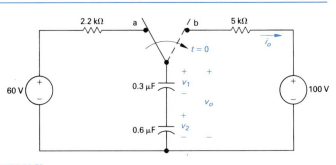

FIGURE P8.50

8.51 The switch in the circuit shown in Fig. P8.51 opens at $t = 0$ after being closed for a long time. How many milliseconds after the switch opens is the energy stored in the capacitor 36% of its final value?

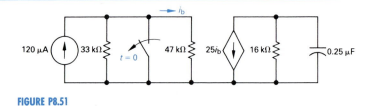

FIGURE P8.51

8.52 The switch in the circuit in Fig. P8.52 has been open a long time before closing at $t = 0$. Find $v_o(t)$ for $t \geq 0^+$.

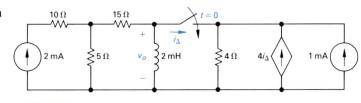

FIGURE P8.52

8.53 The switch in the circuit shown in Fig. P8.53 has been in position a for a long time. At $t = 0$ the switch is moved to position b, where it remains for 800 μs. The switch is then moved to position c, where it remains indefinitely.

a) Find $i(0^+)$.

b) Find $i(300 \ \mu s)$.

c) Find i (1 ms).

d) Find v (800^- μs).

e) Find v (800^+ μs).

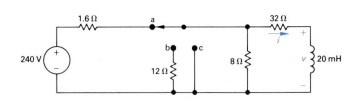

FIGURE P8.53

8.54 In the circuit in Fig. P8.54 switch A has been open and switch B has been closed for a long time. At $t = 0$ switch A closes. One second after switch A closes switch B opens. Find $i_L(t)$ for $t \geq 0$.

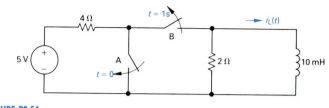

FIGURE P8.54

8.55 The action of the two switches in the circuit seen in Fig. P8.55 is as follows. For $t < 0$, switch 1 is in position a and switch 2 is open. This state has existed for a long time. At $t = 0$, switch 1 moves instantaneously from position a to position b while switch 2 remains open. Ten milliseconds after switch 1 operates switch 2 closes, remains closed for 10 ms, and then opens. Find $v_o(t)$ 25 ms after switch 1 moves to position b.

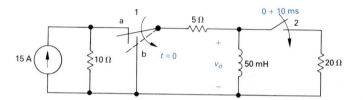

FIGURE P8.55

8.56 For the circuit in Fig. P8.55, how many milliseconds after switch 1 moves to position b is the energy stored in the inductor 4% of its initial value?

8.57 The switch in the circuit in Fig. P8.57 has been in position a for a long time. At $t = 0$ it moves instantaneously to position b where it remains for 5 s before moving instantaneously to position c. Find v_o for $t \geq 0$.

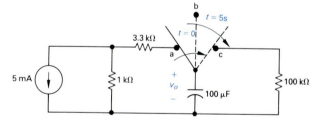

FIGURE P8.57

8.58 There is no energy stored in the capacitor in the circuit in Fig. P8.58 when switch 1 closes at $t = 0$. Three microseconds later switch 2 closes. Find $v_o(t)$ for $t \geq 0$.

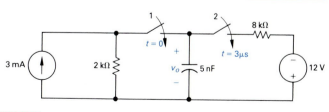

FIGURE P8.58

8.59 The capacitor in the circuit seen in Fig. P8.59 has been charged to 300 V. At $t = 0$, switch 1 closes, causing the capacitor to discharge into the resistive network. Switch 2 closes 200 μs after switch 1 closes. Find the magnitude and direction of the current in the second switch 300 μs after switch 1 closes.

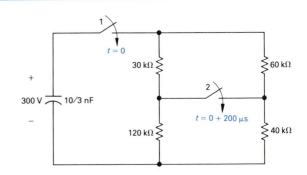

FIGURE P8.59

8.60 In the circuit in Fig. P8.60 switch 1 has been in position a, and switch 2 has been closed for a long time. At $t = 0$, switch 1 moves instantaneously to position b. Eight hundred microseconds later switch 2 opens, remains open for 300 μs and then recloses. Find v_o 1.5 ms after switch 1 makes contact with terminal b.

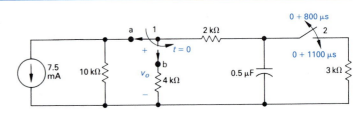

FIGURE P8.60

8.61 For the circuit in Fig. P8.60 what percentage of the initial energy stored in the 0.5-μF capacitor is dissipated in the 3-kΩ resistor?

8.62 The voltage waveform shown in Fig. P8.62(a) is applied to the circuit of Fig. P8.62(b). The initial voltage on the capacitor is zero.

a) Calculate $v_o(t)$.

b) Make a sketch of $v_o(t)$ versus t.

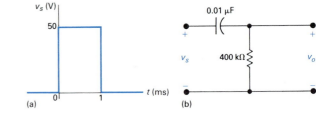

FIGURE P8.62

8.63 The voltage waveform shown in Fig. P8.63(a) is applied to the circuit of Fig. P8.63(b). The initial current in the inductor is zero.

a) Calculate $v_o(t)$.

b) Make a sketch of $v_o(t)$ versus t.

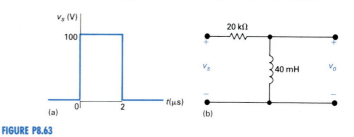

FIGURE P8.63

8.64 The current source in the circuit in Fig. P8.64(a) generates the current pulse shown in Fig. P8.64(b). There is no energy stored at $t = 0$.

a) Derive the numerical expressions for $v_o(t)$ for the time intervals $t < 0, 0 < t < 75 \ \mu s$, and $75 \ \mu s < t < \infty$.

b) Calculate $v_o(75^- \ \mu s)$ and $v_o(75^+ \ \mu s)$.

c) Calculate $i_o(75^- \ \mu s)$ and $i_o(75^+ \ \mu s)$.

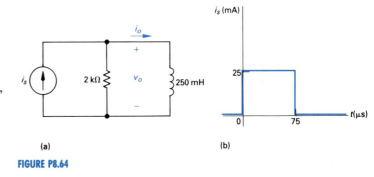

FIGURE P8.64

8.65 The current source in the circuit in Fig. P8.65(a) generates the current pulse shown in Fig. P8.65(b). There is no energy stored at $t = 0$.

a) Derive the expressions for $i_o(t)$ and $v_o(t)$ for the time intervals $t < 0; 0 < t < 0.002 \ s$; and $0.002 \ s < t < \infty$.

b) Calculate $i_o(0^-)$; $i_o(0^+)$; $i_o(0.002^-)$; and $i_o(0.002^+)$.

c) Calculate $v_o(0^-)$; $v_o(0^+)$; $v_o(0.002^-)$; $v_o(0.002^+)$.

d) Sketch $i_o(t)$ versus t for the interval $-1 \ ms < t < 4 \ ms$.

e) Sketch $v_o(t)$ versus t for the interval $-1 \ ms < t < 4 \ ms$.

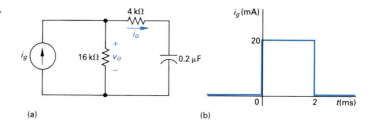

FIGURE P8.65

8.66 The voltage signal source in the circuit in Fig. P8.66(a) is generating the signal shown in Fig. P8.66(b). There is no stored energy at $t = 0$.

a) Derive the expressions for $v_o(t)$ that apply in the intervals $t < 0$; $0 \le t \le 4$ ms; 4 ms $\le t \le 8$ ms; and 8 ms $\le t \le \infty$.

b) Sketch v_o and v_s on the same coordinate axes.

c) Repeat parts (a) and (b) with R reduced to 50 kΩ.

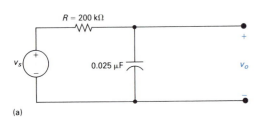

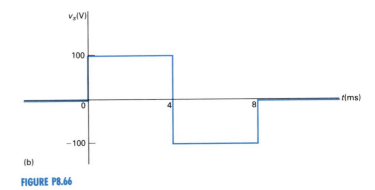

(a)

(b)

FIGURE P8.66

8.67 The circuit shown in Fig. P8.67 is used to close the switch between a and b for a predetermined length of time. The electric relay holds its contact arms down so long as the voltage across the relay coil exceeds 5 V. When the coil voltage equals 5 V, the relay contacts return to their initial position by a mechanical spring action. The switch between a and b is initially closed by momentarily pressing the push button. Assume that the capacitor is fully charged when the push button is first pushed down. The resistance of the relay coil is 25 kΩ, and the inductance of the coil is negligible.

a) How long will the switch between a and b remain closed?

b) Write the numerical expression for i from the time when the relay contacts first close to the time when the capacitor is completely charged.

c) How many milliseconds (after the circuit between a and b is interrupted) does it take the capacitor to reach 85% of its final value?

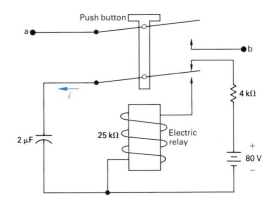

FIGURE P8.67

8.68 In the circuit of Fig. P8.68, the lamp starts to conduct whenever the lamp voltage reaches 15 V. During the time when the lamp conducts, it can be modeled as a 10-kΩ resistor. Once the lamp conducts, it will continue to conduct until the lamp voltage drops to 5 V. When the lamp

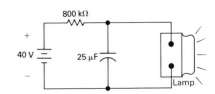

FIGURE P8.68

is not conducting, it appears as an open circuit. Assume that the circuit has been in operation for a long time. Let $t = 0$ at the instant when the lamp stops conducting.

a) Derive the expression for the voltage across the lamp for one full cycle of operation.

b) How many times per minute will the lamp turn on?

c) The 800-kΩ resistor is replaced with a variable resistor R. The resistance is adjusted until the lamp "flashes" 12 times per minute. What is the value of R?

8.69 The capacitor in the circuit shown in Fig. P8.69 is charged to 20 V at the time the switch is closed. If the capacitor ruptures when its terminal voltage equals or exceeds 20 kV, how long does it take to rupture the capacitor?

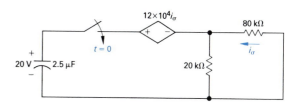

FIGURE P8.69

8.70 The inductor current in the circuit in Fig. P8.70 is 25 mA at the instant the switch is opened. The inductor will malfunction whenever the magnitude of the inductor current equals or exceeds 5 A. How long after the switch is opened does the inductor malfunction?

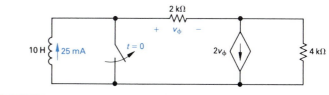

FIGURE P8.70

8.71 The gap in the circuit seen in Fig. P8.71 will arc over whenever the voltage across the gap reaches 36 kV. The initial current in the inductor is zero. The value of β is adjusted so that the Thévenin resistance with respect to the terminals of the inductor is -3 kΩ. How many milliseconds after the switch has been closed will the gap arc over?

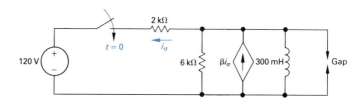

FIGURE P8.71

8.72 The switch in the circuit in Fig. P8.72 has been closed for a long time. The maximum voltage rating of the 1.6 μF capacitor is 14,400 V. How long after the switch is opened does the voltage across the capacitor reach the maximum voltage rating?

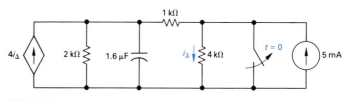

FIGURE P8.72

8.73 The energy stored in the capacitor in the circuit shown in Fig. P8.73 is zero at the instant the switch is closed. The ideal operational amplifier reaches saturation in 15 ms. What is the numerical value of R in kilohms?

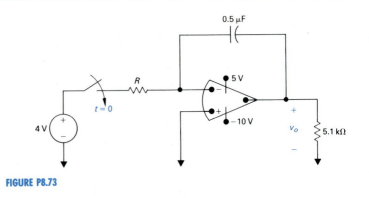

FIGURE P8.73

8.74 At the instant the switch is closed in the circuit of Fig. P8.73, the capacitor is charged to 6 V, positive at the left-hand terminal. If the ideal operational amplifier saturates in 40 ms, what is the value of R?

8.75 There is no energy stored in the capacitor at the time the switch in the circuit of Fig. P8.75 makes contact with terminal a. The switch remains at position a for 32 ms and then moves instantaneously to position b. How many milliseconds after making contact with terminal a does the op amp saturate?

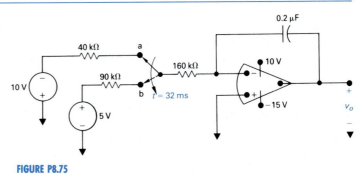

FIGURE P8.75

8.76 At the instant the switch of Fig. P8.76 is closed, the voltage on the capacitor is 56 V. Assume an ideal operational amplifier. How many milliseconds after the switch is closed will the output voltage v_o equal zero?

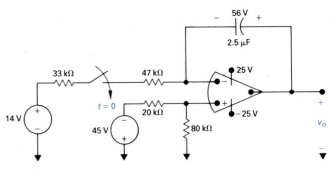

FIGURE P8.76

8.77 a) When the switch closes in the circuit seen in Fig. P8.77, there is no energy stored in the capacitor. How long does it take to saturate the op amp?

b) Repeat part (a) with an initial voltage on the capacitor of 1.0 V, positive at the upper terminal.

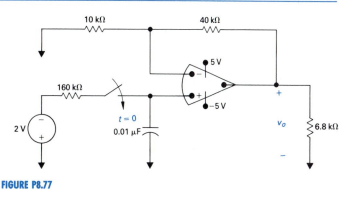

FIGURE P8.77

8.78 There is no energy stored in the capacitors in the circuit shown in Fig. P8.78 at the instant the two switches close.

a) Find v_o as a function of v_a, v_b, R, and C.

b) On the basis of the result obtained in part (a), describe the operation of the circuit.

c) How long will it take to saturate the amplifier if $v_a = 10$ mV; $v_b = 60$ mV; $R = 40$ kΩ; $C = 25$ nF; and $V_{CC} = 12$ V?

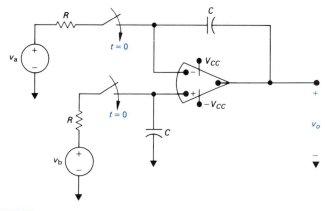

FIGURE P8.78

8.79 At the time the double-pole switch in the circuit shown in Fig. P8.79 is closed, the initial voltages on the capacitors are 12 and 4 V, as shown. Find the numerical expressions for $v_o(t)$, $v_2(t)$, and $v_f(t)$ that are applicable as long as the ideal op amp operates in its linear range.

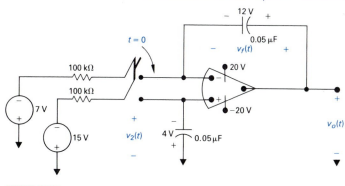

FIGURE P8.79

8.80 There is no charge on the capacitor in the circuit of Fig. P8.80 when the switch makes contact with terminal a. The switch remains at terminal a for 20 ms and then moves instantaneously to terminal b. The switch then remains at terminal b. Derive the equations that describe $v_o(t)$ when the op amp operates in its linear range.

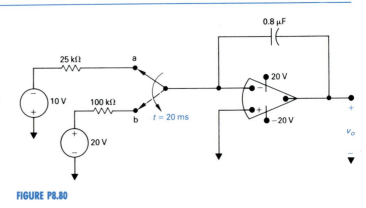

FIGURE P8.80

8.81 The capacitor in the circuit seen in Fig. P8.81 has an initial voltage of 5 V at the instant the switch is closed. How many milliseconds after the switch is closed does the op amp saturate?

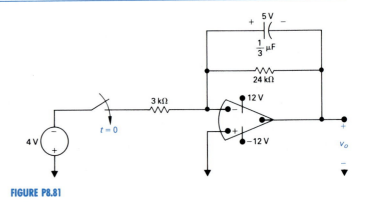

FIGURE P8.81

8.82 The voltage pulse shown in Fig. P8.82(a) is applied to the ideal integrating amplifier shown in Fig. P8.82(b). Derive the numerical expressions for $v_o(t)$ for the time intervals (a) $t < 0$; (b) $0 \le t \le 250$ ms; (c) 250 ms $\le t \le$ 500 ms; and (d) 500 ms $\le t \le \infty$ when $v_o(0) = 0$.

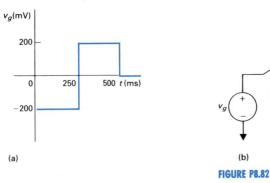

(a)

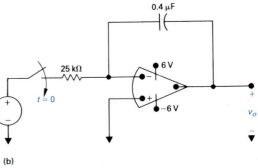

(b)

FIGURE P8.82

8.83 Repeat Problem 8.82 with a 5-MΩ resistor placed across the 0.4-μF feedback capacitor.

8.84 a) At the time the switch makes contact with terminal a in the circuit seen in Fig. P8.84 the charge on the capacitor is zero. The switch remains at terminal a for 2 ms and then moves instantaneously to terminal b. How long after the switch makes contact with terminal a is the output voltage zero?

b) If the switch remains in contact with terminal b, how many milliseconds after making contact with terminal b does the op amp saturate?

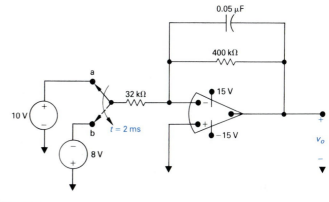

FIGURE P8.84

8.85 The voltage source in the circuit in Fig. P8.85(a) is generating the triangular waveform shown in Fig. P8.85(b). Assume the energy stored in the capacitor is zero at $t = 0$.

a) Derive the numerical expressions for $v_o(t)$ for the following time intervals: $0 \le t \le 1\ \mu s$; $1\ \mu s \le t \le 3\ \mu s$; and $3\ \mu s \le t \le 4\ \mu s$.

b) Sketch the output waveform between 0 and 4 μs.

c) If the triangular input voltage continues to repeat itself for $t > 4\ \mu s$ what would you expect the output voltage to be? Explain.

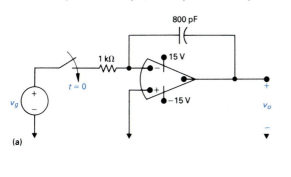

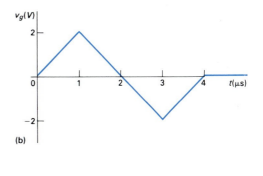

FIGURE P8.85

8.86 The circuit shown in Fig. P8.86 is known as an *astable multivibrator* and finds wide application in pulse circuits. The purpose of this problem is to relate the charging and discharging of the capacitors to the operation of the circuit. The key to analyzing the circuit is to understand the be-

havior of the ideal transistor switches T_1 and T_2. The circuit is designed so that the switches automatically alternate between ON and OFF. When T_1 is OFF, T_2 is ON and vice versa. Thus in the analysis of this circuit we assume a switch is either ON or OFF. We also assume that the ideal

(continued)

transistor switch can change its state instantaneously. In other words, it can snap from OFF to ON and vice versa.

When a transistor switch is ON, (1) the base current i_b is greater than zero, (2) the terminal voltage v_{be} is zero, and (3) the terminal voltage v_{ce} is zero. Thus when a transistor switch is ON, it presents a short circuit between the terminals b, e and c, e.

When a transistor switch is OFF, (1) the terminal voltage v_{be} is negative, (2) the base current is zero, and (3) there is an open circuit between the terminals c, e. Thus when a transistor switch is OFF, it presents an open circuit between the terminal sets b, e and c, e.

Assume that T_2 has been ON and has just snapped OFF, while T_1 has been OFF and has just snapped ON. You may assume that at this instance C_2 is charged to the supply voltage V_{CC} and the charge on C_1 is zero. Also assume $C_1 = C_2$ and $R_1 = R_2 = 10R_L$.

a) Derive the expression for v_{be2} during the interval T_2 is OFF.

b) Derive the expression for v_{ce2} during the interval T_2 is OFF.

c) Find the length of time T_2 is OFF.

d) Find the value of v_{ce2} at the end of the interval that T_2 is OFF.

e) Derive the expression for i_{b1} during the interval T_2 is OFF.

f) Find the value of i_{b1} at the end of the interval that T_2 is OFF.

g) Sketch v_{ce2} versus t during the interval T_2 is OFF.

h) Sketch i_{b1} versus t during the interval T_2 is OFF.

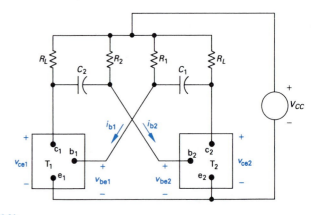

FIGURE P8.86

8.87 The component values in the circuit of Fig. P8.86 are $V_{CC} = 10$ V; $R_L = 1$ kΩ; $C_1 = C_2 = 1$ nF; and $R_1 = R_2 = 14.43$ kΩ.

a) How long is T_2 in the OFF state during one cycle of operation?

b) How long is T_2 in the ON state during one cycle of operation?

c) Repeat part (a) for T_1.

d) Repeat part (b) for T_1.

e) At the first instant after T_1 turns ON, what is the value of i_{b1}?

f) At the instant just before T_1 turns OFF, what is the value of i_{b1}?

g) What is the value of v_{ce2} at the instant just before T_2 turns ON?

8.88 Repeat Problem 8.87 with $C_1 = 1$ nF and $C_2 = 0.8$ nF. All other component values are unchanged.

8.89 An astable multivibrator circuit is to satisfy the following criteria: (1) One transistor switch is to be ON for 48 μs and OFF for 36 μs for each cycle; (2) $R_L = 2$ kΩ; (3) $V_{CC} = 5$ V; (4) $R_1 = R_2$; and (5) $6R_L \leq R_1 \leq 50 R_L$. What are the limiting values for the capacitors C_1 and C_2?

8.90 The circuit shown in Fig. P8.90 is known as a *monostable multivibrator*. The adjective "monostable" is used to describe the fact that the circuit has one stable state. That is, if left alone the electronic switch T_2 will be ON and T_1 will be OFF. (The operation of the ideal transistor switch is described in Problem 8.86.) T_2 can be turned OFF by momentarily closing the switch S. After S returns to its open position, T_2 will return to its ON state.

a) Show that if T_2 is ON, T_1 is OFF and will stay OFF.

b) Explain why T_2 is turned OFF when S is momentarily closed.

c) Show that T_2 will stay OFF for $RC \ln 2$ s.

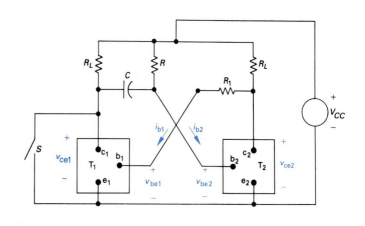

FIGURE P8.90

8.91 The parameter values in the circuit in Fig. P8.90 are $V_{CC} = 6$ V; $R_1 = 5.0$ kΩ; $R_L = 20$ kΩ; $C = 250$ pF; and $R = 23,083$ Ω.

a) Sketch v_{ce2} versus t assuming that after S is momentarily closed it remains open until the circuit has reached its stable state. Assume S is closed at $t = 0$. Sketch v_{ce2} versus t for the interval $-5 \le t \le 10$ μs.

b) Repeat part (a) for i_{b2} versus t.

NATURAL AND STEP RESPONSES OF *RLC* CIRCUITS

CHAPTER 9

In this chapter, discussion of the natural response and step response of circuits containing both inductors and capacitors is limited to two simple structures: the parallel *RLC* circuit and the series *RLC* circuit. Finding the natural response of a parallel *RLC* circuit consists of finding the voltage created across the parallel branches by the release of energy that has been stored in the inductor or capacitor or both. The task is defined in terms of the circuit shown in Fig. 9.1. The initial voltage on the capacitor, V_0, represents the initial energy stored in the capacitor. The initial current through the inductor, I_0, represents the initial energy stored in the inductor. If the individual branch currents are of interest, you can find them after determining the terminal voltage.

We derive the step response of a parallel *RLC* circuit by using the circuit shown in Fig. 9.2. We are interested in the voltage that appears across the parallel branches as a result of the sudden application of a dc, or constant, current source. Energy may or may not be stored in the circuit when the current source is applied to the circuit.

Finding the natural response of a series *RLC* circuit consists of finding the current generated in the series-connected elements by the release of initially stored energy in either the inductor or capacitor or both. The task is defined by the circuit shown in Fig. 9.3. As before, the initial inductor current, I_0, and the initial capacitor voltage, V_0, represent the initially stored energy. If any of the individual element voltages are of interest, you can find them after determining the current.

We describe the step response of a series *RLC* circuit in terms of the circuit shown in Fig. 9.4. We are interested in the current resulting from sudden application of the dc voltage source. En-

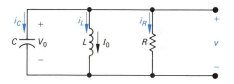

FIGURE 9.1 A circuit used to illustrate the natural response of the parallel *RLC* circuit.

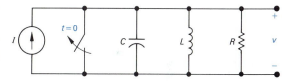

FIGURE 9.2 A circuit used to illustrate the step response of the parallel *RLC* circuit.

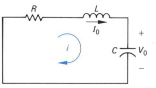

FIGURE 9.3 A circuit used to illustrate the natural response of the series *RLC* circuit.

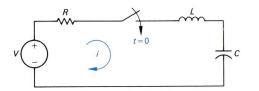

FIGURE 9.4 A circuit used to illustrate the step response of the series *RLC* circuit.

ergy may or may not be stored in the circuit when the switch is closed.

If you have not studied ordinary differential equations, derivation of the natural and step responses of parallel and series *RLC* circuits may be a bit difficult to follow. However, the results are important enough to warrant presentation at this time. We begin with the natural response of a parallel *RLC* circuit and cover this material over two sections: one to discuss the solution of the differential equation that describes the circuit, and one to present the three distinct forms that the solution can take. After introducing the forms for solving the natural response of the parallel *RLC* circuit, we show that the same forms apply to the natural response of the series *RLC* circuit and the step responses of both the series and parallel *RLC* circuits.

9.1 INTRODUCTION TO THE NATURAL RESPONSE OF A PARALLEL *RLC* CIRCUIT

The first step in finding the natural response of the circuit shown in Fig. 9.1 is to derive the differential equation that the voltage v must satisfy. Recall that we choose to find the voltage first, because it is the same for each component. We can find a branch current once we have found the voltage by using the current–voltage relationship for the branch component. We easily obtain the differential equation for the voltage by summing the currents away from the top node, where each current is expressed as a function of the unknown voltage v:

$$\frac{v}{R} + \frac{1}{L}\int_0^t v\,d\tau + I_0 + C\frac{dv}{dt} = 0. \tag{9.1}$$

We eliminate the integral in Eq. (9.1) by differentiating once with respect to t, and, as I_0 is a constant, we get

$$\frac{1}{R}\frac{dv}{dt} + \frac{v}{L} + C\frac{d^2v}{dt^2} = 0. \tag{9.2}$$

We now divide through Eq. (9.2) by the capacitance C and arrange the derivatives in descending order:

$$\frac{d^2v}{dt^2} + \frac{1}{RC}\frac{dv}{dt} + \frac{v}{LC} = 0. \tag{9.3}$$

Comparing Eq. (9.3) with the differential equations derived in Chapter 8 reveals that they differ by the presence of the term involving the second derivative. Equation (9.3) is described as an *ordinary, second-order differential equation with constant coefficients*. Whenever a circuit contains both types of storage elements, the differential equation describing the circuit is of second order or higher.

GENERAL SOLUTION OF THE SECOND-ORDER DIFFERENTIAL EQUATION

We can't solve Eq. (9.3) by separating the variables and integrating as we were able to do with the first-order equations in Chapter 8. The classical approach to solving Eq. (9.3) is to assume that the solution is of exponential form, that is, to assume that the voltage is of the form

$$v = Ae^{st}, \tag{9.4}$$

where A and s are unknown constants.

Before showing how this assumption leads to the solution of Eq. (9.3), we need to show that the assumption contained in Eq. (9.4) is a rational one. The strongest argument we can make in favor of Eq. (9.4) is to note from Eq. (9.3) that the second derivative of the solution, plus a constant times the first derivative plus a constant times the solution itself, must sum to zero for all values of t. This result can occur only if higher order derivatives of the solution have the same form as the solution. The exponential function satisfies this criterion. A second argument can be made in favor of Eq. (9.4) by noting that the solutions of all the first-order equations we derived in Chapter 8 were exponential. It seems reasonable to assume that solution of the second-order equation also involves the exponential function.

To show that the assumption contained in Eq. (9.4) leads to the solution of Eq. (9.3), we proceed as follows. If Eq. (9.4) is a solution of Eq. (9.3), it must satisfy Eq. (9.3) for all values of t. Substituting Eq. (9.4) into Eq. (9.3) generates the expression:

$$As^2e^{st} + \frac{As}{RC}e^{st} + \frac{Ae^{st}}{LC} = 0,$$

or

$$Ae^{st}\left(s^2 + \frac{s}{RC} + \frac{1}{LC}\right) = 0, \tag{9.5}$$

which can be satisfied for all values of t only if A is zero or the parenthetical term is zero, as $e^{st} \neq 0$ for any finite values of st. We cannot use $A = 0$ as a general solution because to do so implies that the voltage is zero for all time—a physical impossibil-

ity if energy is stored in either the inductor or capacitor. Therefore, in order for Eq. (9.4) to be a solution of Eq. (9.3), the
parenthetical term in Eq. (9.5) must be zero, or

$$s^2 + \frac{s}{RC} + \frac{1}{LC} = 0. \tag{9.6}$$

Equation (9.6) is called the *characteristic equation of the differential equation* because the roots of this quadratic equation determine the mathematical character of $v(t)$.

The two roots of Eq. (9.6) are

$$s_1 = -\frac{1}{2RC} + \sqrt{\left(\frac{1}{2RC}\right)^2 - \frac{1}{LC}} \tag{9.7}$$

and

$$s_2 = -\frac{1}{2RC} - \sqrt{\left(\frac{1}{2RC}\right)^2 - \frac{1}{LC}}. \tag{9.8}$$

If either root is substituted into Eq. (9.4), the assumed solution
satisfies the given differential equation, that is, Eq. (9.3). Note
from Eq. (9.5) that this result holds regardless of the value of A.
Therefore

$$v = A_1 e^{s_1 t} \quad \text{or} \quad v = A_2 e^{s_2 t}$$

each satisfy Eq. (9.3). Denoting these two solutions v_1 and v_2,
respectively, we can show that their sum also is a solution.
Specifically, if we let

$$v = v_1 + v_2 = A_1 e^{s_1 t} + A_2 e^{s_2 t}, \tag{9.9}$$

then

$$\frac{dv}{dt} = A_1 s_1 e^{s_1 t} + A_2 s_2 e^{s_2 t} \tag{9.10}$$

and

$$\frac{d^2 v}{dt^2} = A_1 s_1^2 e^{s_1 t} + A_2 s_2^2 e^{s_2 t}. \tag{9.11}$$

Substituting Eqs. (9.9), (9.10), and (9.11) into Eq. (9.3) gives

$$A_1 e^{s_1 t}\left(s_1^2 + \frac{1}{RC} s_1 + \frac{1}{LC}\right) + A_2 e^{s_2 t}\left(s_2^2 + \frac{1}{RC} s_2 + \frac{1}{LC}\right) = 0. \tag{9.12}$$

But each parenthetical term is zero because by definition s_1 and
s_2 are roots of the characteristic equation. Hence the natural response of the parallel *RLC* circuit shown in Fig. 9.1 is of the

form

$$v = A_1 e^{s_1 t} + A_2 e^{s_2 t}. \tag{9.13}$$

Equation (9.13) is a repeat of the assumption made in Eq. (9.9). We have shown v_1 is a solution, v_2 is a solution, and $v_1 + v_2$ is a solution. Therefore, the general solution of Eq. (9.3) has the form given in Eq. (9.13). The roots of the characteristic equation (s_1 and s_2) are determined by the circuit parameters R, L, and C. The initial conditions determine the values of the constants A_1 and A_2.[†]

The behavior of $v(t)$ depends on the roots s_1 and s_2. Therefore the first step in finding the natural response is to determine the roots of the characteristic equation. Thus we return to Eqs. (9.7) and (9.8) and rewrite them using a notation that is widely used in the literature:

$$s_1 = -\alpha + \sqrt{\alpha^2 - \omega_o^2} \tag{9.14}$$

and

$$s_2 = -\alpha - \sqrt{\alpha^2 - \omega_o^2}, \tag{9.15}$$

where

$$a = \frac{1}{2RC} \tag{9.16}$$

and

$$\omega_o = \frac{1}{\sqrt{LC}}. \tag{9.17}$$

The exponent of e must be dimensionless, so both s_1 and s_2 (and hence α and ω_o) must have the dimension of the reciprocal of time, or frequency. To distinguish among the frequencies s_1, s_2, α, and ω_o we use the following terminology: s_1 and s_2 are referred to as complex frequencies, α is called the neper frequency, and ω_o is the resonant radian frequency. The full significance of this terminology unfolds as we move through the remaining chapters of this book. All these frequencies have the dimension of angular frequency per time, typically radians per second (rad/s).

The nature of the roots s_1 and s_2 depend on the values of α and ω_o. There are three possible outcomes. First, if $\omega_o^2 < \alpha^2$, both roots will be real and distinct. For reasons to be discussed later,

---

[†] The form of Eq. (9.13) must be modified if the two roots s_1 and s_2 are equal. We discuss the modification when we turn to the critically damped voltage response in Section 9.2.

the voltage response is said to be *overdamped* (od) when $\omega_o^2 < \alpha^2$. Second, if $\omega_o^2 > \alpha^2$, both s_1 and s_2 will be complex and, in addition, will be conjugates of each other. In this situation, the voltage response is said to be *underdamped* (ud). The third possible outcome is that ω_o^2 equals α^2. In this case, s_1 and s_2 will be real and equal. Here the voltage response is said to be *critically damped* (cd). We discuss each case separately in Section 9.2.

However, before doing so we need to make a general observation that underlies the work to follow. The solution for the natural response of any second-order circuit—regardless of the type of damping—involves finding two unknown coefficients, such as A_1 and A_2 in Eq. (9.13). To find the two unknown coefficients we use circuit analysis to establish the initial value of the current (or voltage) and the initial value of the first derivative of the current (or voltage). To determine the step response for a second-order circuit we use circuit analysis to find the initial value of the variable, the initial value of the first derivative of the variable, and the final value of the variable.

Example 9.1 illustrates how the numerical values of s_1 and s_2 are determined by the values of R, L, and C.

E X A M P L E   9.1

a) Find the roots of the characteristic equation that governs the transient behavior of the voltage shown in Fig. 9.5 if $R = 200 \ \Omega$, $L = 50$ mH, and $C = 0.2 \ \mu$F.

b) Will the response be over-, under-, or critically damped?

c) Repeat (a) and (b) for $R = 312.5 \ \Omega$.

d) What value of R causes the response to be critically damped?

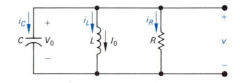

FIGURE 9.5 A circuit used to illustrate the natural response of the parallel *RLC* circuit.

S O L U T I O N

a) For the given values of R, L, and C,

$$\alpha = \frac{1}{2RC} = \frac{10^6}{(400)(0.2)} = 1.25 \times 10^4 \text{ rad/s},$$

$$\omega_o^2 = \frac{1}{LC} = \frac{(10^3)(10^6)}{(50)(0.2)} = 10^8 \text{ rad}^2/\text{s}^2.$$

From Eqs. (9.14) and (9.15),

$$s_1 = -1.25 \times 10^4 + \sqrt{1.5625 \times 10^8 - 10^8}$$

$$= -12,500 + 7500 = -5000 \text{ rad/s}$$

and

$$s_2 = -1.25 \times 10^4 - \sqrt{1.5625 \times 10^8 - 10^8}$$

$$= -12,500 - 7500 = -20000 \text{ rad/s}.$$

b) The voltage response is overdamped because $\omega_o^2 < \alpha^2$.

c) For $R = 312.5 \ \Omega$,

$$\alpha = \frac{10^6}{(625)(0.2)} = 8000 \text{ rad/s},$$

$$\alpha^2 = 64 \times 10^6 = 0.64 \times 10^8 \text{ rad}^2/\text{s}^2.$$

As ω_o^2 remains at $10^8 \text{ rad}^2/\text{s}^2$,[†]

$$s_1 = -8000 + j6000 \text{ rad/s},$$

and

$$s_2 = -8000 - j6000 \text{ rad/s}.$$

In this case, the voltage response is underdamped since $\omega_o^2 > \alpha^2$.

d) For critical damping, $\alpha^2 = \omega_o^2$, so

$$\left(\frac{1}{2RC}\right)^2 = \frac{1}{LC} = 10^8 \qquad \text{or} \qquad \frac{1}{2RC} = 10^4$$

and

$$R = \frac{10^6}{(2 \times 10^4)(0.2)} = 250 \ \Omega.$$

DRILL EXERCISES

9.1 The resistance and inductance of the circuit in Fig. 9.5 are 200 Ω and 10 mH, respectively.

a) Find the value of C that makes the voltage response critically damped.

b) If C is adjusted to give a neper frequency of 10 krad/s, find the value of C and the roots of the characteristic equation.

c) If C is adjusted to give a resonant frequency of 50 krad/s, find the value of C and the roots of the characteristic equation.

ANSWER: (a) $C = 62.5$ nF; (b) $C = 0.25 \ \mu$F, $s_1 = -10,000 + j17,320.51$ rad/s, $s_2 = -10,000 - j17,320.51$ rad/s; (c) $C = 40$ nF, $s_1 = -25,000$ rad/s, $s_2 = -100,000$ rad/s.

[†] In electrical engineering, the imaginary number $\sqrt{-1}$ is represented by the letter j, because the letter i represents current.

9.2 THE FORMS OF THE NATURAL RESPONSE OF A PARALLEL *RLC* CIRCUIT

As indicated in Section 9.1, the behavior of the second-order RLC circuit depends on the values of s_1 and s_2. The values of s_1 and s_2 in turn depend on the circuit parameters R, L, and C. Therefore the first step in finding the response of an RLC circuit is to determine whether the response is over-, under-, or critically damped. After ascertaining the nature of the damping, we find the unknown coefficients, which completes the description of the response. Keep in mind that the evaluation of these unknown coefficients is based on matching the solution to the initial conditions imposed by the circuit. We analyze the response form for each of the three kinds of damping, beginning with the overdamped response.

OVERDAMPED VOLTAGE RESPONSE

When the roots of the characteristic equation are real and distinct, the voltage response of the circuit in Fig. 9.5 is said to be overdamped. The solution for the voltage is of the form

$$v = A_1 e^{s_1 t} + A_2 e^{s_2 t}, \tag{9.18}$$

where s_1 and s_2 are the roots of the characteristic equation. The constants A_1 and A_2 are determined by the initial conditions, specifically from the values of $v(0^+)$, and $dv(0^+)/dt$. They are determined from the initial voltage on the capacitor, V_0, and the initial current in the inductor, I_0.

Next, we show how to use the initial voltage on the capacitor and initial current in the inductor to find A_1 and A_2. First we note from Eq. (9.18) that

$$v(0^+) = A_1 + A_2 \tag{9.19}$$

and

$$\frac{dv(0^+)}{dt} = s_1 A_1 + s_2 A_2. \tag{9.20}$$

With s_1 and s_2 known, the task of finding A_1 and A_2 reduces to finding $v(0^+)$ and $dv(0^+)/dt$. The value of $v(0^+)$ is the initial voltage on the capacitor V_0. We get the initial value of dv/dt by first finding the current in the capacitor branch at $t = 0^+$. Then,

$$\frac{dv(0^+)}{dt} = \frac{i_C(0^+)}{C}. \tag{9.21}$$

We use Kirchhoff's current law to find the initial current in the capacitor branch. We know that the sum of the three branch currents at $t = 0^+$ must be zero. At $t = 0^+$ the current in the resistive branch is the initial voltage V_0 divided by the resistance, and the current in the inductive branch is I_0. Using the reference system depicted in Fig. 9.5, we obtain

$$i_C(0^+) = \frac{-V_0}{R} - I_0. \qquad (9.22)$$

After finding the numerical value of $i_C(0^+)$, we use Eq. (9.21) to find the initial value of dv/dt.

We can summarize the thought process for finding $v(t)$ as follows. First, knowing the values of R, L, and C, we find the roots of the characteristic equation, s_1 and s_2. Next, we find $v(0^+)$ and $dv(0^+)/dt$ by straightforward circuit analysis. We then determine the values of A_1 and A_2 from Eqs. (9.23) and (9.24):

$$v(0^+) = A_1 + A_2 \qquad (9.23)$$

and

$$\frac{dv(0^+)}{dt} = \frac{i_C(0^+)}{C} = s_1A_1 + s_2A_2. \qquad (9.24)$$

Knowing s_1, s_2, A_1, and A_2, we determine the expression for $v(t)$ for $t \geq 0$ from Eq. (9.18).

Examples 9.2 and 9.3 illustrate the conceptual approach involved in finding the overdamped response of a parallel *RLC* circuit.

EXAMPLE 9.2

The initial voltage across the capacitor in the circuit in Fig. 9.5 is 12 V and the initial current in the inductor is 30 mA. The circuit element values are 0.2 μF, 50 mH, and 200 Ω, respectively.

a) Find the initial current in each branch of the circuit.
b) Find the initial value of dv/dt.
c) Find the expression for $v(t)$.
d) Sketch $v(t)$ in the interval $0 \leq t \leq 250$ μs.

SOLUTION

a) The inductor prevents an instantaneous change in its current, so the initial value of the inductor current is 30 mA:

$$i_L(0^-) = i_L(0) = i_L(0^+) = 30 \text{ mA}.$$

The capacitor holds the initial voltage across the parallel elements to 12 V. Thus the initial current in the resistive branch,

$i_R(0^+)$, is 12/200, or 60 mA. Kirchhoff's current law requires the sum of the currents leaving the top node to equal zero in every instant. Hence

$$i_C(0^+) = -i_L(0^+) - i_R(0^+)$$

$$= -90 \text{ mA}.$$

Note that if we assume that the inductor current and capacitor voltage had reached their dc values at the instant energy begins to be released, $i_C(0^-) = 0$. In other words, there is an instantaneous change in the capacitor current at $t = 0$.

b) Because $i_C = C(dv/dt)$,

$$\frac{dv(0^+)}{dt} = \frac{-90 \times 10^{-3}}{0.2 \times 10^{-6}} = -450 \text{ kV/s}.$$

c) The roots of the characteristic equation come from the values of R, L, and C. For the values specified and from Eqs. (9.14) and (9.15),

$$s_1 = -1.25 \times 10^4 + \sqrt{1.5625 \times 10^8 - 10^8}$$

$$= -12,500 + 7500 = -5000 \text{ rad/s}$$

and

$$s_2 = -1.25 \times 10^4 - \sqrt{1.5625 \times 10^8 - 10^8}$$

$$= -12,500 - 7500 = -20,000 \text{ rad/s}.$$

As the roots are real and distinct, we know that response is overdamped and hence has the form of Eq. (9.18). We find the coefficients A_1 and A_2 from Eqs. (9.23) and (9.24). We've already determined s_1, s_2, $v(0^+)$, and $dv(0^+)/dt$, so

$$12 = A_1 + A_2$$

and

$$-450 \times 10^3 = -5000A_1 - 20,000A_2.$$

We solve two equations for A_1 and A_2 to obtain $A_1 = -14$ V and $A_2 = 26$ V. Substituting these values into Eq. (9.18) yields the overdamped voltage response:

$$v(t) = (-14e^{-5000t} + 26e^{-20,000t}) \text{ V}, \qquad t \geq 0.$$

As a check on these calculations, we note that the solution yields $v(0) = 12$ V and $dv(0^+)/dt = -450,000$ V/s.

d) Figure 9.6 shows a plot of $v(t)$ versus t over the interval $0 \leq t \leq 250$ μs.

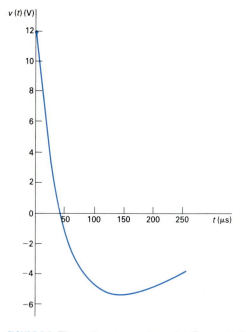

FIGURE 9.6 The voltage response for Example 9.2.

E X A M P L E   9.3

Derive the expressions that describe the three branch currents i_R, i_L, and i_C in the circuit in Example 9.2 during the time the stored energy is being released.

S O L U T I O N

We know the voltage across the three branches from the solution in Example 9.2, namely,

$$v(t) = (-14e^{-5000t} + 26e^{-20,000t})\text{V}, \qquad t \ge 0.$$

The current in the resistive branch then is

$$i_R(t) = \frac{v(t)}{200} = (-70e^{-5000t} + 130e^{-20,000t}) \text{ mA}, \qquad t \ge 0.$$

There are two ways to find the current in the inductive branch. One way is to use the integral relationship that exists between the current and the voltage at the terminals of an inductor:

$$i_L(t) = \frac{1}{L} \int_0^t v_L(x)dx + I_0.$$

A second approach is to find the current in the capacitive branch first and then use the fact that $i_R + i_L + i_C = 0$. Let's use this approach. The current in the capacitive branch is

$$i_C(t) = C\frac{dv}{dt}$$

$$= 0.2 \times 10^{-6}(70,000e^{-5000t} - 520,000e^{-20,000t})$$

$$= (14e^{-5000t} - 104e^{-20,000t}) \text{ mA}, t \ge 0^+.$$

Note that $i_C(0^+) = -90$ mA, which agrees with the result obtained in Example 9.2.

Now we obtain the inductive branch current from the relationship

$$i_L(t) = -i_R(t) - i_C(t)$$

$$= (56e^{-5000t} - 26e^{-20,000t}) \text{ mA}, \qquad t \ge 0.$$

We leave to you, in Drill Exercise 9.2, to show that the integral relationship alluded to leads to the same result. Also note that the expression for i_L agrees with the initial inductor current, as it must.

DRILL EXERCISES

9.2 Use the integral relationship between i_L and v to find the expression for i_L in Example 9.2.

ANSWER: $i_L(t) = (56e^{-5000t} - 26e^{-20,000t})$ mA, $t \geq 0$.

9.3 The element values in the circuit shown are $R = 400$ Ω, $L = 50$ mH, and $C = 50$ nF. The initial current I_0 in the inductor is -4 A, and the initial voltage on the capacitor is 0 V. The output signal is the voltage v. Find (a) $i_R(0^+)$; (b) $i_C(0^+)$; (c) $dv(0^+)/dt$; (d) A_1; (e) A_2; and (f) $v(t)$ when $t \geq 0$.

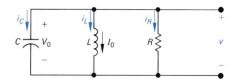

ANSWER: (a) 0; (b) 4 A; (c) 8×10^7 V/s; (d) 8000/3 V; (e) $-8000/3$ V; (f) $(8000/3)(e^{-10,000t} - e^{-40,000t})$ V when $t \geq 0$.

UNDERDAMPED VOLTAGE RESPONSE

When $\omega_o^2 > \alpha^2$, the roots of the characteristic equation are complex and the response is called underdamped. For convenience when discussing the underdamped response, we express the roots s_1 and s_2 as

$$s_1 = -\alpha + \sqrt{-(\omega_0^2 - \alpha^2)}$$
$$= -\alpha + j\sqrt{\omega_0^2 - \alpha^2}$$
$$= -\alpha + j\omega_d \tag{9.25}$$

and

$$s_2 = -\alpha - j\omega_d, \tag{9.26}$$

where

$$\omega_d = \sqrt{\omega_0^2 - \alpha^2} \tag{9.27}$$

and ω_d is called the *damped radian frequency*. We explain later the reason for this terminology.

The underdamped voltage response of a parallel *RLC* circuit is

$$v(t) = B_1 e^{-\alpha t} \cos \omega_d t + B_2 e^{-\alpha t} \sin \omega_d t, \tag{9.28}$$

from Eq. (9.18). In making the transition from Eq. (9.18) to Eq. (9.28), we use the Euler identity:

$$e^{\pm j\theta} = \cos \theta \pm j \sin \theta. \tag{9.29}$$

Thus

$$v(t) = A_1 e^{(-\alpha + j\omega_d)t} + A_2 e^{-(\alpha + j\omega_d)t}$$

$$= A_1 e^{-\alpha t} e^{j\omega_d t} + A_2 e^{-\alpha t} e^{-j\omega_d t}$$

$$= e^{-\alpha t}(A_1 \cos \omega_d t + jA_1 \sin \omega_d t + A_2 \cos \omega_d t$$
$$\quad - jA_2 \sin \omega_d t)$$

$$= e^{-\alpha t}[(A_1 + A_2)\cos \omega_d t + j(A_1 - A_2) \sin \omega_d t].$$

At this point in the transition from Eq. (9.18) to (9.28), we replace the arbitrary constants $A_1 + A_2$ and $j(A_1 - A_2)$ with new arbitrary constants denoted B_1 and B_2, to get

$$v = e^{-\alpha t}(B_1 \cos \omega_d t + B_2 \sin \omega_d t)$$

$$= B_1 e^{-\alpha t} \cos \omega_d t + B_2 e^{-\alpha t} \sin \omega_d t.$$

The constants B_1 and B_2 are real, not complex, because the voltage is a real function. Don't be misled by the fact that $B_2 = j(A_1 - A_2)$. In this underdamped case, A_1 and A_2 are complex conjugates, and thus B_1 and B_2 are real. (See Problems 9.49 and 9.50.) The reason for defining the underdamped response in terms of the coefficients B_1 and B_2 is that it yields a simpler expression for the voltage, v.

We determine the two arbitrary constants B_1 and B_2 by the initial energy stored in the circuit, in the same way that we found A_1 and A_2 for the overdamped response: by evaluating v at $t = 0^+$, and its derivative at $t = 0^+$. As with s_1 and s_2 before, α and ω_d are fixed by the circuit parameters R, L, and C.

For the underdamped response, the two simultaneous equations for the determination of B_1 and B_2 are

$$v(0^+) = V_0 = B_1 \qquad \textbf{(9.30)}$$

and

$$\frac{dv(0^+)}{dt} = \frac{i_C(0^+)}{C} = -\alpha B_1 + \omega_d B_2. \qquad \textbf{(9.31)}$$

Let's look at the general nature of the underdamped response. First, the trigonometric functions indicate that the underdamped response is oscillatory; that is, the voltage alternates between positive and negative values. The rate at which the voltage oscillates is fixed by ω_d. Second, the amplitude of the oscillation decreases exponentially. The rate at which the amplitude falls off is determined by α. Because α determines how quickly the oscillations subside, it also is referred to as the *damping factor*, or *damping coefficient*. That explains why ω_d is called the damped radian frequency. If there is no damping, $\alpha = 0$ and the frequency of oscillation is ω_o. Whenever there is a dissipative element, R, in the circuit, α is not zero and the frequency of oscillation, ω_d is less than ω_o. Thus when α is not zero, the frequency of oscillation is said to be damped.

The oscillatory behavior is possible because of the presence of the two types of energy-storage elements in the circuit: the inductor and capacitor. (A mechanical analogy of this electric circuit is that of a mass suspended on a spring, where oscillation is possible because energy can be stored in both the spring and the moving mass.) We say more about the characteristics of the underdamped response following Example 9.4.

E X A M P L E   9.4

In the circuit shown in Fig. 9.7, $V_0 = 0$, and $I_0 = -12.25$ mA. The circuit parameters are $R = 20$ kΩ, $L = 8$H, and $C = 0.125$ μF.

a) Calculate the roots of the characteristic equation.

b) Calculate v and dv/dt at $t = 0^+$.

c) Calculate the voltage response for $t \geq 0$.

d) Plot $v(t)$ versus t for the time interval $0 \leq t \leq 11$ ms.

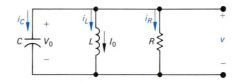

FIGURE 9.7 A circuit used to illustrate the natural response of the parallel *RLC* circuit.

S O L U T I O N

a) Because

$$\alpha = \frac{1}{2RC} = \frac{10^6}{2(20)10^3(0.125)} = 200 \text{ rad/s,}$$

and

$$\omega_0 = \frac{1}{\sqrt{LC}} = \sqrt{\frac{10^6}{(8)(0.125)}} = 10^3 \text{ rad/s,}$$

we have

$$\omega_0^2 > \alpha^2.$$

Therefore, the response is underdamped. Now,

$$\omega_d = \sqrt{\omega_0^2 - \alpha^2} = \sqrt{10^6 - 4 \times 10^4} = 100\sqrt{96}$$

$$= 979.80 \text{ rad/s;}$$

$$s_1 = -\alpha + j\omega_d = -200 + j979.80 \text{ rad/s;}$$

$$s_2 = -\alpha - j\omega_d = -200 - j979.80 \text{ rad/s.}$$

For the underdamped case, we do not ordinarily solve for s_1 and s_2 because we do not use them explicitly. However, this example emphasizes why s_1 and s_2 are known as complex frequencies.

b) Because v is the voltage across the terminals of a capacitor, we have

$$v(0) = v(0^+) = V_0 = 0.$$

As $v(0^+) = 0$, the current in the resistive branch is zero at $t = 0^+$. Hence the current in the capacitor at $t = 0^+$ is the negative of the inductor current:

$$i_C(0^+) = -(-12.25) = 12.25 \text{ mA}.$$

Therefore the initial value of the derivative is

$$\frac{dv(0^+)}{dt} = \frac{(12.25)(10^{-3})}{(0.125)(10^{-6})} = 98,000 \text{ V/s}.$$

c) From Eqs. (9.30) and (9.31), $B_1 = 0$ and

$$B_2 = \frac{98,000}{\omega_d} \approx 100 \text{ V}.$$

Substituting the numerical values of α, ω_d, B_1, and B_2 into the expression for $v(t)$ gives

$$v(t) = 100e^{-200t} \sin 979.80t \text{ V}, \qquad t \geq 0.$$

d) Figure 9.8 shows the plot of $v(t)$ versus t for the first 11 ms after the stored energy is released. It clearly indicates the damped oscillatory nature of the underdamped response. The voltage $v(t)$ approaches its final value, alternating between values that are greater than and less than the final value. Furthermore, these swings about the final value decrease exponentially with time.

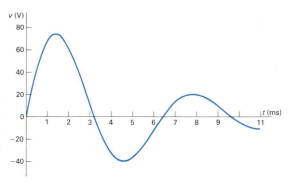

FIGURE 9.8 The voltage response for Example 9.4.

CHARACTERISTICS OF UNDERDAMPED RESPONSE

This type of response has several important characteristics. First, as the dissipative losses in the circuit decrease, the persistence of the oscillations increases and the frequency of the oscillations approaches ω_0. That is, as $R \to \infty$, the dissipation in the circuit in Fig. 9.7 approaches zero because $p = v^2/R$. As $R \to \infty$, $\alpha \to 0$, which tells us that $\omega_d \to \omega_0$. When $\alpha = 0$, the maximum amplitude of the voltage remains constant; thus the oscillation at ω_0 is sustained. In Example 9.3, if R were increased to infinity, the solution for $v(t)$ would become

$$v(t) = 98 \sin 1000t \text{ V}, \qquad t \geq 0.$$

Thus in this case the oscillation is sustained, the maximum amplitude of the voltage is 98 V, and the frequency of oscillation is 1000 rad/s.

The *period of oscillation* is the length of time required for the response to go through one complete cycle—for example, from one positive maximum value to the next positive maximum value. This time interval, denoted T_d, can be calculated directly from ω_d:

$$\omega_d = \frac{2\pi}{T_d} \tag{9.32}$$

or

$$T_d = \frac{2\pi}{\omega_d}. \tag{9.33}$$

The reciprocal of the period of oscillation, known as the *frequency*, is

$$f_d = \frac{1}{T_d}, \tag{9.34}$$

and is typically given with units of cycles per second, or hertz. In Example 9.3, the period of oscillation is

$$T_d = \frac{2\pi}{100\sqrt{96}} = 6.41 \text{ ms},$$

and the frequency of oscillation in hertz is

$$f_d = \frac{10^3}{6.41} = 155.94 \text{ Hz}.$$

We may now describe qualitatively the difference between an underdamped and an overdamped response. In an underdamped system, the response oscillates, or "bounces," about its final value. This oscillation also is referred to as ringing. In an overdamped system, the response approaches its final value without ringing, or in what is sometimes described as a "sluggish" manner.

DRILL EXERCISES

9.4 A 10-mH inductor, a 1-μF capacitor, and a variable resistor are connected in parallel in the circuit shown. The resistor is adjusted so that the roots of the characteristic equation are $-8000 \pm j6000$ rad/s. The initial voltage on the capacitor is 10 V, and the initial current

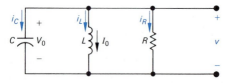

in the inductor is 80 mA. Find (a) R; (b) $dv(0^+)/dt$; (c) B_1 and B_2 in the solution for v; and (d) $i_L(t)$.

ANSWER: (a) 62.5 Ω; (b) $-240{,}000$ V/s; (c) $B_1 = 10$ V, $B_2 = -80/3$ V; (d) $i_L(t) = 10e^{-8000t}[8 \cos 6000t + (82/3) \sin 6000t]$ mA when $t \geq 0$.

CRITICALLY DAMPED VOLTAGE RESPONSE

The second-order circuit in Fig. 9.7 is critically damped when $\omega_0^2 = \alpha^2$, or $\omega_0 = \alpha$. When the circuit is critically damped, the response is on the verge of oscillating. For the critically damped case, the two roots of the characteristic equation are real and equal. That is,

$$s_1 = s_2 = -\alpha = -\frac{1}{2RC}. \qquad (9.35)$$

When the roots of the characteristic equation are real and equal, the solution for the voltage no longer takes the form of Eq. (9.18). Note that Eq. (9.18) breaks down because if $s_1 = s_2 = -\alpha$, it predicts that

$$v = (A_1 + A_2)e^{-\alpha t} = A_0 e^{-\alpha t} \qquad (9.36)$$

where A_0 is an arbitrary constant. Equation (9.36) cannot satisfy two independent initial conditions (V_0, I_0) with only one arbitrary constant A_0. Recall that the circuit parameters R and C fix α.

We can trace this dilemma back to the assumption that the solution takes the form of Eq. (9.18). However, in the case when the roots of the characteristic equation are equal, the solution for the differential equation takes a different form, or

$$v(t) = D_1 t e^{-\alpha t} + D_2 e^{-\alpha t}. \qquad (9.37)$$

Thus in the case of a repeated root, the solution involves a simple exponential term plus the product of a linear term and an exponential term. The justification of Eq. (9.37) is left for an introductory course in differential equations. Finding the solution involves obtaining D_1 and D_2 by following the same pattern set in the overdamped and underdamped cases. We use the initial values of the voltage and the derivative of the voltage with respect to time to write two equations containing D_1 and/or D_2.

From Eq. (9.37), the two simultaneous equations needed to determine D_1 and D_2 are

$$v(0^+) = V_0 = D_2 \qquad (9.38)$$

and

$$\frac{dv(0^+)}{dt} = \frac{i_C(0^+)}{C} = D_1 - \alpha D_2. \qquad (9.39)$$

Example 9.5 illustrates the approach for finding the critically damped response of the parallel *RLC* circuit.

EXAMPLE 9.5

a) For the circuit in Example 9.4, find the value of R that results in a critically damped voltage response.
b) Calculate $v(t)$ for $t \geq 0$.
c) Plot $v(t)$ versus t for $0 \leq t \leq 7$ ms.

SOLUTION

a) From Example 9.3 we know that $\omega_0^2 = 10^6$. Therefore for critical damping

$$\alpha = 10^3 = \frac{1}{2RC}, \quad \text{or} \quad R = \frac{10^6}{(2000)(0.125)} = 4000 \ \Omega.$$

b) From the solution of Example 9.4, we know that $v(0^+) = 0$ and $dv(0^+)/dt = 98{,}000$ V/s. From Eqs. (9.38) and (9.39), $D_2 = 0$ and $D_1 = 98{,}000$ V/s. Substituting these values for α, D_1, and D_2 into Eq. (9.37) gives

$$v(t) = 98{,}000te^{-1000t} \text{ V}, \quad t \geq 0.$$

c) Figure 9.9 shows a plot of $v(t)$ versus t in the interval $0 \leq t \leq 7$ ms.

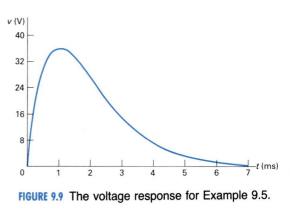

FIGURE 9.9 The voltage response for Example 9.5.

DRILL EXERCISES

9.5 The resistor in the circuit in Drill Exercise 9.4 is adjusted for critical damping. The inductance and capacitance values are 0.4 H and 10 μF, respectively. The initial energy stored in the circuit is 25 mJ and is distributed equally between the inductor and capacitor. Find (a) R; (b) V_0; (c) I_0; (d) D_1 and D_2 in the solution for v; and (e) i_R, $t \geq 0^+$.

ANSWER: (a) 100 Ω; (b) 50 V; (c) 250 mA; (d) $-50{,}000$ V/s, 50 V; (e) $i_R(t) = [-500te^{-500t} + 0.50e^{-500t}]$ A, $t \geq 0^+$.

SUMMARY OF RESULTS

We conclude the discussion of the parallel *RLC* circuit's natural response with a brief summary of the results. The first step in finding the natural response is to calculate the roots of the characteristic equation. After determining the roots, you know immediately whether the response is overdamped, underdamped, or critically damped.

If the roots are real and distinct ($\omega_0^2 < \alpha^2$), the response is overdamped and the voltage is

$$v(t) = A_1 e^{s_1 t} + A_2 e^{s_2 t},$$

where

$$s_1 = -\alpha + \sqrt{\alpha^2 - \omega_0^2}, \qquad s_2 = -\alpha - \sqrt{\alpha^2 - \omega_0^2},$$

$$\alpha = \frac{1}{2RC}, \qquad \text{and} \qquad \omega_0^2 = \frac{1}{LC}.$$

If the roots are complex ($\omega_0^2 > \alpha^2$), the response is underdamped and the voltage is

$$v(t) = B_1 e^{-\alpha t} \cos \omega_d t + B_2 e^{-\alpha t} \sin \omega_d t,$$

where

$$\omega_d = \sqrt{\omega_0^2 - \alpha^2}.$$

If the roots of the characteristic equation are real and equal ($\omega_0^2 = \alpha^2$), the voltage response is

$$v(t) = D_1 t e^{-\alpha t} + D_2 e^{-\alpha t},$$

where α is as in the other solution forms. In any of the three cases, you determine the pairs of constants A_1 and A_2, B_1 and B_2, or D_1 and D_2 by using the initial value of the variable and its first derivative with respect to time.

9.3 THE STEP RESPONSE OF A PARALLEL *RLC* CIRCUIT

Finding the step response of a parallel *RLC* circuit involves finding the voltage that appears across the parallel branches, or the current that appears in the individual branches, as a result of the sudden application of a constant, or dc, current source. There may or may not be energy stored in the circuit when the current source is applied. The task is represented by the circuit shown in Fig. 9.10.

PSpice   Use PSpice and PROBE to simulate and plot the step response of *RLC* circuits: Chapter 9

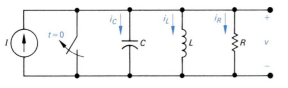

FIGURE 9.10 A circuit used to describe the step response of the parallel *RLC* circuit.

In order to develop a general approach to finding the step response of a second-order circuit—that is, a circuit described by a second-order differential equation—we focus on finding the current in the inductive branch (i_L). This current is of particular interest because it does not approach zero as t increases. Thus after the switch has been open for a long time, the inductor current equals the dc source current I. Because we want to focus on the technique for finding the step response, we assume that the initial energy stored in the circuit is zero. This assumption simplifies the calculations and makes concentrating on how to find the step response easier. (The introduction of initially stored energy enters into calculations involving the determination of the arbitrary constants but doesn't alter the basic thought process involved in finding the step response.)

To find the inductor current i_L, we must solve a second-order differential equation equated to the forcing function I, which we derive as follows. From Kirchhoff's current law, we have

$$i_L + i_R + i_C = I,$$

or

$$i_L + \frac{v}{R} + C\frac{dv}{dt} = I. \tag{9.40}$$

Because

$$v = L\frac{di_L}{dt}, \tag{9.41}$$

we get

$$\frac{dv}{dt} = L\frac{d^2 i_L}{dt^2}. \tag{9.42}$$

Substituting Eqs. (9.41) and (9.42) into Eq. (9.40), gives

$$i_L + \frac{L}{R}\frac{di_L}{dt} + LC\frac{d^2 i_L}{dt^2} = I. \tag{9.43}$$

For convenience, we divide through by LC and rearrange terms:

$$\frac{d^2 i_L}{dt^2} + \frac{1}{RC}\frac{di_L}{dt} + \frac{i_L}{LC} = \frac{I}{LC}. \tag{9.44}$$

Comparing Eq. (9.44) with Eq. (9.3) reveals that the presence of a nonzero term on the right-hand side of the equation alters the task. Before showing how to solve Eq. (9.44) by a direct approach, we obtain the solution indirectly. When we know what the solution of Eq. (9.44) is, explaining the direct approach is easier.

We can solve for i_L indirectly by first finding the voltage v. We find v by the techniques introduced in Section 9.2 because the differential equation that v must satisfy is identical with Eq. (9.3). To see this, we simply return to Eq. (9.40) and express i_L as a function of v; thus

$$\frac{1}{L}\int_0^t v\,d\tau + \frac{v}{R} + C\frac{dv}{dt} = I. \qquad (9.45)$$

Differentiating Eq. (9.42) once with respect to t, reduces the right-hand side to zero because I is a constant. Thus

$$\frac{v}{L} + \frac{1}{R}\frac{dv}{dt} + C\frac{d^2v}{dt^2} = 0,$$

or

$$\frac{d^2v}{dt^2} + \frac{1}{RC}\frac{dv}{dt} + \frac{v}{LC} = 0. \qquad (9.46)$$

As discussed in Section 9.2, the solution for v depends on the roots of the characteristic equation. Thus the three possible solutions are

$$v = A_1 e^{s_1 t} + A_2 e^{s_2 t}, \qquad (9.47)$$

$$v = B_1 e^{-\alpha t}\cos \omega_d t + B_2 e^{-\alpha t}\sin \omega_d t, \qquad (9.48)$$

and

$$v = D_1 t e^{-\alpha t} + D_2 e^{-\alpha t}. \qquad (9.49)$$

A word of caution is called for at this point. Because there is a source in the circuit for $t > 0$, you must take into account the value of the source current at $t = 0^+$ when you evaluate the coefficients in Eqs. (9.47)–(9.49).

To find the three possible solutions for i_L we substitute Eqs. (9.47)–(9.49) into Eq. (9.40). You should be able to verify that, when this has been done, the three solutions for i_L will be of the forms

$$i_L = I + A_1' e^{s_1 t} + A_2' e^{s_2 t}, \qquad (9.50)$$

$$i_L = I + B_1' e^{-\alpha t}\cos \omega_d t + B_2' e^{-\alpha t}\sin \omega_d t, \qquad (9.51)$$

and

$$i_L = I + D_1' t e^{-\alpha t} + D_2' e^{-\alpha t}, \qquad (9.52)$$

where A_1', A_2', B_1', B_2', D_1', and D_2' are arbitrary constants.

In each case, the primed constants can be found in terms of the arbitrary constants associated with the voltage solution. However, this approach is cumbersome. It is much easier to find the primed constants directly in terms of the initial values of the response function. For the circuit being discussed, we would find the primed constants from $i_L(0)$ and $di_L(0)/dt$. We illustrate this approach in Examples 9.6–9.10.

The solution for the second-order differential equation with a constant forcing function equals the forced response plus a response function that is identical in *form* with the natural response. Thus we can always write the solution for the step response in the form

$$i = I_f + \begin{Bmatrix} \text{function of the same form} \\ \text{as the natural response} \end{Bmatrix} \qquad \textbf{(9.53)}$$

or

$$v = V_f + \begin{Bmatrix} \text{function of the same form} \\ \text{as the natural response} \end{Bmatrix}, \qquad \textbf{(9.54)}$$

where I_f and V_f represent the final value of the response function. The final value may be zero, as was, for example, the final value of the voltage v in the circuit in Fig 9.8.

Examples 9.6–9.10 illustrate the technique of finding the step response of a parallel *RLC* circuit.

E X A M P L E   9.6

The initial energy stored in the circuit in Fig. 9.10 is zero. At $t = 0$, a dc current source of 24 mA is applied to the circuit. The circuit elements are a 400-Ω resistor, a 25-mH inductor, and a 25-nF capacitor.

a) What is the initial value of i_L?

b) What is the initial value of di_L/dt?

c) What are the roots of the characteristic equation?

d) What is the numerical expression for $i_L(t)$ when $t \geq 0$?

S O L U T I O N

a) No energy is stored in the circuit prior to application of the dc current source, so the initial current in the inductor is zero. The inductor prohibits an instantaneous change in inductor current; therefore $i_L(0) = 0$ immediately after the switch has been opened.

b) The initial voltage on the capacitor is zero before the switch has been opened; therefore it will be zero immediately after the switch has been opened. Now, as $v = L \, di_L/dt$,

$$\frac{di_L}{dt}(0^+) = 0.$$

c) From the circuit elements, we obtain

$$\omega_0^2 = \frac{1}{LC} = \frac{10^{12}}{(25)(25)} = 16 \times 10^8$$

and

$$\alpha = \frac{1}{2RC} = \frac{10^9}{(2)(400)(25)} = 5 \times 10^4,$$

or

$$\alpha^2 = 25 \times 10^8.$$

Because $\omega_0^2 < \alpha^2$, the roots of the characteristic equation are real and distinct. Thus

$$s_1 = -5 \times 10^4 + 3 \times 10^4 = -20{,}000 \text{ s}^{-1}$$

and

$$s_2 = -5 \times 10^4 - 3 \times 10^4 = -80{,}000 \text{ s}^{-1}.$$

d) Because the roots of the characteristic equation are real and distinct, the inductor current response will be overdamped. Thus $i_L(t)$ takes the form of Eq. (9.50), namely,

$$i_L = I + A_1' e^{s_1 t} + A_2' e^{s_2 t}.$$

Hence, from this solution, the two simultaneous equations for the determination of A_1' and A_2' are

$$i_L(0) = I + A_1' + A_2' = 0$$

and

$$\frac{di_L}{dt}(0) = s_1 A_1' + s_2 A_2' = 0.$$

Solving for A_1' and A_2' gives

$$A_1' = -32 \text{ mA} \qquad \text{and} \qquad A_2' = 8 \text{ mA}.$$

The numerical solution for $i_L(t)$ is

$$i_L(t) = (24 - 32e^{-20{,}000t} + 8e^{-80{,}000t})\text{mA}, \qquad t \geq 0.$$

EXAMPLE 9.7

The resistor in the circuit in Example 9.6 is increased to 625 Ω. Find $i_L(t)$ for $t \geq 0$.

SOLUTION

Because L and C remain fixed, ω_0^2 has the same value as in Example 9.6; that is, $\omega_0^2 = 16 \times 10^8$. Increasing R to 625 Ω decreases α to 3.2×10^4. With $\omega_0^2 > \alpha^2$, the roots of the characteristic equation are complex. Hence

$$s_1 = -3.2 \times 10^4 + j2.4 \times 10^4 \text{ s}^{-1}$$

and

$$s_2 = -3.2 \times 10^4 - j2.4 \times 10^4 \text{ s}^{-1}.$$

The current response is now underdamped and given by Eq. 9.51:

$$i_L(t) = I + B_1' e^{-\alpha t} \cos \omega_d t + B_2' e^{-\alpha t} \sin \omega_d t.$$

Here α is 32,000 s^{-1}, ω_d is 24,000 rad/s, and $I = 24$ mA.

As in Example 9.6, B_1' and B_2' are determined from the initial conditions. Thus the two simultaneous equations for the determination of B_1' and B_2' are

$$i_L(0) = I + B_1' = 0 \quad \text{and} \quad \frac{di_L}{dt}(0) = \omega_d B_2' - \alpha B_1' = 0.$$

Then,

$$B_1' = -24 \text{ mA} \quad \text{and} \quad B_2' = -32 \text{ mA}.$$

The numerical solution for $i_L(t)$ is

$$i_L(t) = (24 - 24e^{-32,000t} \cos 24,000t$$
$$- 32e^{-32,000t} \sin 24,000t) \text{ mA}, \quad t \geq 0.$$

EXAMPLE 9.8

The resistor in the circuit in Example 9.6 is set at 500 Ω. Find $i_L(t)$ for $t \geq 0$.

SOLUTION

We know that ω_0^2 remains at 16×10^8. With R set at 500 Ω, α becomes 4×10^4 s^{-1}, which corresponds to critical damping. Therefore the solution for $i_L(t)$ takes the form of Eq. (9.52):

$$i_L(t) = I + D_1' t e^{-\alpha t} + D_2' e^{-\alpha t}.$$

Again D_1' and D_2' are computed from initial conditions, or

$$i_L(0) = I + D_2' = 0$$

and

$$\frac{di_L}{dt}(0) = D_1' - \alpha D_2' = 0.$$

Thus

$$D_1' = -960,000 \text{ mA/s} \quad \text{and} \quad D_2' = -24 \text{ mA}.$$

The numerical expression for $i_L(t)$ is

$$i_L(t) = (24 - 960,000te^{-40,000t} - 24e^{-40,000t}) \text{ mA}, \quad t \geq 0.$$

EXAMPLE 9.9

a) Plot on a single graph, over a range from 0 to 220 μs, the overdamped, underdamped, and critically damped responses derived in Examples 9.6–9.8.

b) Use the plots of (a) to find the time required for i_L to reach 90% of its final value.

c) On the basis of the results obtained in (b), which response would you describe as being the most "sluggish"?

SOLUTION

a) See Fig. 9.11.

b) The final value of i_L is 24 mA, so we can read the times off the plots corresponding to $i_L = 21.6$ mA. Thus $t_{od} = 130 \ \mu$s, $t_{cd} = 97 \ \mu$s, and $t_{ud} = 74 \ \mu$s.

c) The overdamped response is the most "sluggish," because it takes the longest time for i_L to reach 90% of its final value.

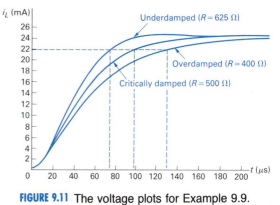

FIGURE 9.11 The voltage plots for Example 9.9.

EXAMPLE 9.10

Energy is stored in the circuit in Example 9.8 at the instant the dc current source is applied. The initial current in the inductor is

29 mA, and the initial voltage across the capacitor is 50 V. Find
(a) $i_L(0)$; (b) $di_L(0)/dt$; (c) $i_L(t)$ for $t \geq 0$.

SOLUTION

a) There cannot be an instantaneous change of current in an inductor, so the initial value of i_L in the first instant after the dc
current source has been applied must be 29 mA.

b) The capacitor holds the initial voltage across the inductor to
50 V. Therefore

$$L\frac{di_L(0^+)}{dt} = 50,$$

or

$$\frac{di_L}{dt}(0^+) = \frac{50}{25} \times 10^3 = 2000 \text{ A/s}.$$

c) From the solution of Example 9.8, we know that the current
response is critically damped. Thus

$$i_L(t) = I + D_1' t e^{-\alpha t} + D_2' e^{-\alpha t},$$

where

$$\alpha = \frac{1}{2RC} = 40,000 \text{ s}^{-1} \qquad \text{and} \qquad I = 24 \text{ mA}.$$

We obtain the constants D_1' and D_2' from the initial conditions, or

$$i_L(0) = I + D_2' = 29 \text{ mA},$$

from which we get

$$D_2' = 29 - 24 = 5 \text{ mA}.$$

The solution for D_1' is

$$\frac{di_L}{dt}(0^+) = D_1' - \alpha D_2' = 2000$$

or

$$D_1' = 2000 + \alpha D_2'$$
$$= 2000 + (40,000)(5 \times 10^{-3})$$
$$= 2200 \text{ A/s} = 2.2 \times 10^6 \text{ mA/s}.$$

Thus the numerical expression for $i_L(t)$ is

$$i_L(t) = (24 + 2.2 \times 10^6 t e^{-40,000t} + 5e^{-40,000t}) \text{ mA}, \quad t \geq 0.$$

DRILL EXERCISES

9.6 In the circuit shown, $R = 250\ \Omega$, $L = 0.32$ H, $C = 2\ \mu$F, $I_0 = 0.5$ A, $V_0 = 80$ V, and $I = -1.5$ A. Find (a) $i_R(0^+)$; (b) $i_C(0^+)$; (c) $di_L(0^+)/dt$; (d) s_1, s_2; and (e) $i_L(t)$ for $t \geq 0$.

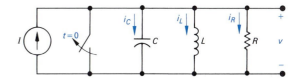

ANSWER: (a) 320 mA; (b) -2.32 A; (c) 250 A/s;
(d) $(-1000 + j750)s^{-1}$, $(-1000 - j750)s^{-1}$;
(e) $i_L(t) = [-1.5 + 2e^{-1000t}\ (\cos 750t + 1.5 \sin 750t)]$ A
for $t \geq 0$.

9.4 THE NATURAL AND STEP RESPONSES OF A SERIES *RLC* CIRCUIT

In Section 9.1 we defined the natural and step responses of a series *RLC* circuit with reference to the circuits shown in Figs. 9.3 and 9.4. The procedures for finding the natural or step responses of a series *RLC* circuit are the same as those used to find the natural or step responses of a parallel *RLC* circuit because both circuits are described by differential equations that have the same form. For example, the differential equation that describes the current in the circuit shown in Fig. 9.12 has the same form as the differential equation that describes the voltage in the circuit shown in Fig. 9.13. We can show this equivalence by summing the voltages around the closed path in the circuit shown in Fig. 9.12. Thus

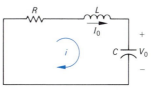

FIGURE 9.12 A circuit used to illustrate the natural response of the series *RLC* circuit.

$$Ri + L\frac{di}{dt} + \frac{1}{C}\int_0^t i\ d\tau + V_0 = 0. \qquad (9.55)$$

We now differentiate Eq. (9.52) once with respect to t to get

$$R\frac{di}{dt} + L\frac{d^2i}{dt^2} + \frac{i}{C} = 0, \qquad (9.56)$$

which we can rearrange as

$$\frac{d^2i}{dt^2} + \frac{R}{L}\frac{di}{dt} + \frac{i}{LC} = 0. \qquad (9.57)$$

Comparing Eq. (9.57) with Eq. (9.3) reveals that they have the same form. Therefore, to find the solution of Eq. (9.57), we follow the same thought process that led us to the solution of Eq. (9.3).

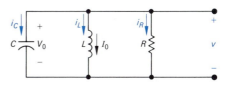

FIGURE 9.13 A circuit used to illustrate the natural response of the parallel *RLC* circuit.

From Eq. (9.57), the characteristic equation for the series *RLC* circuit is

$$s^2 + \frac{R}{L}s + \frac{1}{LC} = 0. \tag{9.58}$$

The roots of the characteristic equation are

$$s_{1,2} = -\frac{R}{2L} \pm \sqrt{\left(\frac{R}{2L}\right)^2 - \frac{1}{LC}} \tag{9.59}$$

or

$$s_{1,2} = -\alpha \pm \sqrt{\alpha^2 - \omega_0^2}. \tag{9.60}$$

The neper frequency (α) for the series *RLC* circuit is

$$\alpha = \frac{R}{2L} \text{ s}^{-1}, \tag{9.61}$$

whereas the expression for the resonant radian frequency is the same as that of the parallel *RLC* circuit:

$$\omega_o = \frac{1}{\sqrt{LC}} \text{ rad/s.} \tag{9.62}$$

The current response will be overdamped, underdamped, or critically damped according to whether $\omega_o^2 < \alpha^2$, $\omega_o^2 > \alpha^2$, or $\omega_o^2 = \alpha^2$, respectively. Thus the three possible solutions for the current are

$$i(t) = A_1 e^{s_1 t} + A_2 e^{s_2 t} \qquad \text{(od);} \tag{9.63}$$
$$i(t) = B_1 e^{-\alpha t} \cos \omega_d t + B_2 e^{-\alpha t} \sin \omega_d t \qquad \text{(ud);} \tag{9.64}$$
$$i(t) = D_1 t e^{-\alpha t} + D_2 e^{-\alpha t} \qquad \text{(cd).} \tag{9.65}$$

When you have obtained the natural current response, you can find the natural voltage response across any circuit element.

To verify that the procedure for finding the step response of a series *RLC* circuit is the same as that for finding the step response of a parallel *RLC* circuit, we show that the differential equation that describes the capacitor voltage in the circuit shown in Fig. 9.14 has the same form as the differential equation that describes the inductor current in the circuit shown in Fig. 9.10. For convenience, we assume that zero energy is stored in the circuit at the instant the switch is closed.

Applying Kirchhoff's voltage law to the circuit shown in Fig. 9.14 gives

$$V = Ri + L\frac{di}{dt} + v_C. \tag{9.66}$$

The current in the circuit (i) is related to the capacitor voltage

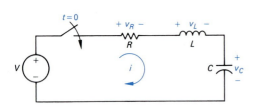

FIGURE 9.14 A circuit used to illustrate the step response of the series *RLC* circuit.

(v_C) by the expression

$$i = C\frac{dv_C}{dt},$$ **(9.67)**

from which

$$\frac{di}{dt} = C\frac{d^2v_C}{dt^2}.$$ **(9.68)**

Substitute Eqs. (9.67) and (9.68) into Eq. (9.66) and write the resulting expression as

$$\frac{d^2v_C}{dt^2} + \frac{R}{L}\frac{dv_C}{dt} + \frac{v_C}{LC} = \frac{V}{LC}.$$ **(9.69)**

Equation (9.69) has the same form as Eq. (9.44); therefore the procedure for finding v_C parallels that for finding i_L. The three possible solutions for v_C are

$$v_C = V_f + A_1'e^{s_1t} + A_2'e^{s_2t}$$ (od);   **(9.70)**

$$v_C = V_f + B_1'e^{-\alpha t}\cos \omega_d t + B_2'e^{-\alpha t}\sin \omega_d t$$ (ud);   **(9.71)**

$$v_C = V_f + D_1'te^{-\alpha t} + D_2'e^{-\alpha t}$$ (cd),   **(9.72)**

where V_f is the final value of v_C. Hence, from the circuit shown in Fig. 9.14, the final value of v_C is the dc source voltage V.

Examples 9.11 and 9.12 illustrate the mechanics of finding the natural and step responses of a series *RLC* circuit.

EXAMPLE 9.11

The 0.1-μF capacitor in the circuit shown in Fig. 9.15 is charged to 100 V. At $t = 0$ the capacitor is discharged through a series combination of a 100-mH inductor and a 560-Ω resistor.

a) Find $i(t)$ for $t \geq 0$.

b) Find $v_C(t)$ for $t \geq 0$.

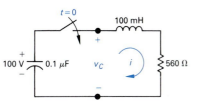

FIGURE 9.15 The circuit for Example 9.11.

SOLUTION

a) The first step to finding $i(t)$ is to calculate the roots of the characteristic equation. For the given element values,

$$\omega_0^2 = \frac{1}{LC} = \frac{(10^3)(10^6)}{(100)(0.1)} = 10^8;$$

$$\alpha = \frac{R}{2L} = \frac{560}{2(100)} \times 10^3 = 2800.$$

Next, we compare α^2 to ω_0^2 and note that $\omega_0^2 > \alpha^2$, because

$$\alpha^2 = 7.84 \times 10^6 = 0.0784 \times 10^8.$$

At this point, we know that the response is underdamped and that the solution for $i(t)$ is of the form

$$i(t) = B_1 e^{-\alpha t} \cos \omega_d t + B_2 e^{-\alpha t} \sin \omega_d t,$$

where $\alpha = 2800$ s^{-1} and $\omega_d = 9600$ rad/s.

The numerical values of B_1 and B_2 come from the initial conditions. The inductor current is zero before the switch has been closed, and hence it is zero immediately after the switch has been closed. Therefore

$$i(0) = 0 = B_1.$$

To find B_2, we evaluate $di(0^+)/dt$. From the circuit, we note that, as $i(0) = 0$ immediately after the switch has been closed, there will be no voltage drop across the resistor. Thus the initial voltage on the capacitor appears across the terminals of the inductor, which leads to the expression

$$L\frac{di(0^+)}{dt} = V_0$$

or

$$\frac{di(0^+)}{dt} = \frac{V_0}{L} = \frac{100}{100} \times 10^3 = 1000 \text{ A/s}.$$

Because $B_1 = 0$,

$$\frac{di}{dt} = 400B_2 e^{-2800t}(24 \cos 9600t - 7 \sin 9600t).$$

Thus

$$\frac{di(0^+)}{dt} = 9600B_2 \quad \text{and} \quad B_2 = \frac{1000}{9600} \cong 0.1042 \text{ A}.$$

The solution for $i(t)$ is

$$i(t) = 0.1042e^{-2800t} \sin 9600t \text{ A}, \qquad t \geq 0.$$

b) To find $v_C(t)$, we can use either of the following relationships:

$$v_C = -\frac{1}{C}\int_0^t i \, d\tau + 100$$

or

$$v_C = iR + L\frac{di}{dt}.$$

Whichever expression is used (the second is recommended)

the result is

$$v_C(t) = (100 \cos 9600t + 29.17 \sin 9600t)e^{-2800t} \text{ V},$$

$$t \geq 0.$$

EXAMPLE 9.12

No energy is stored in the 100-mH inductor or the 0.4-μF capacitor when the switch in the circuit shown in Fig. 9.16 is closed. Find $v_C(t)$ for $t \geq 0$.

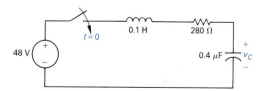

FIGURE 9.16 The circuit for Example 9.12.

SOLUTION

The roots of the characteristic equation are

$$s_1 = -\frac{280}{0.2} + \sqrt{\left(\frac{280}{0.2}\right)^2 - \frac{10^6}{(0.1)(0.4)}}$$

$$= (-1400 + j4800)\text{s}^{-1};$$

$$s_2 = (-1400 - j4800)\text{s}^{-1}.$$

The roots are complex, so the voltage response is underdamped. Thus

$$v_C(t) = 48 + B_1'e^{-1400t} \cos 4800t + B_2'e^{-1400t} \sin 4800t,$$

$$t \geq 0.$$

No energy is stored in the circuit initially, so both $v_C(0)$ and $dv_C(0^+)/dt$ are zero. Then,

$$v_C(0) = 0 = 48 + B_1'$$

and

$$\frac{dv_C(0^+)}{dt} = 0 = 4800B_2' - 1400B_1'.$$

Solving for B_1' and B_2' yields

$$B_1' = -48 \text{ V} \quad \text{and} \quad B_2' = -14 \text{ V}.$$

Therefore the solution for $v_C(t)$ is

$$v_C(t) = (48 - 48e^{-1400t} \cos 4800t - 14e^{-1400t} \sin 4800t) \text{ V},$$

$$t \geq 0.$$

DRILL EXERCISES

9.7 The switch in the circuit shown has been in position a for a long time. At $t = 0$ it moves to position b. Find (a) $i(0^+)$; (b) $v_C(0^+)$; (c) $di(0^+)/dt$; (d) s_1, s_2; and (e) $i(t)$ for $t \geq 0$.

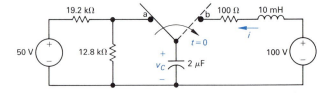

ANSWER: (a) 0; (b) 20 V; (c) 8000 A/s; (d) $(-5000 + j5000)s^{-1}$, $(-5000 - j5000)s^{-1}$; (e) $i(t) = (1.6e^{-5000t} \sin 5000t)$ A for $t \geq 0$.

9.8 Find $v_C(t)$ for $t \geq 0$ for the circuit in Drill Exercise 10.6.

ANSWER: $v_C = [100 - 80e^{-5000t} (\cos 5000t + \sin 5000t)]$ V for $t \geq 0$.

9.5 A CIRCUIT WITH TWO INTEGRATING AMPLIFIERS

PSpice ∿ Use subcircuit models in PSpice when a circuit contains two or more op amps: Section 5.2

A circuit containing two integrating amplifiers connected in cascade[†] also is a second-order circuit; that is, the output voltage of the second integrator is related to the input voltage of the first integrator by a second-order differential equation. We begin analysis of a circuit containing two cascaded amplifiers with the circuit shown in Fig. 9.17.

We assume that the operational amplifiers are ideal. The task is to derive the differential equation that establishes the relationship between v_o and v_g.

We begin the derivation by summing the currents at the inverting input terminal of the first integrator. Because the operational amplifier is ideal,

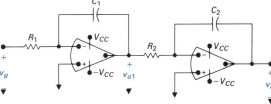

FIGURE 9.17 Two integrating amplifiers connected in cascade.

$$\frac{0 - v_g}{R_1} + C_1 \frac{d}{dt}(0 - v_{o1}) = 0. \tag{9.73}$$

From Eq. (9.73),

$$\frac{dv_{o1}}{dt} = -\frac{1}{R_1 C_1} v_g. \tag{9.74}$$

[†] In a cascade connection, the output signal of the first amplifier (v_{o1} in Fig. 9.17) is the input signal for the second amplifier.

Now we sum the currents away from the inverting input terminal of the second integrating amplifier:

$$\frac{0 - v_{o1}}{R_2} + C_2 \frac{d}{dt}(0 - v_o) = 0, \qquad \text{(9.75)}$$

or

$$\frac{dv_o}{dt} = -\frac{1}{R_2 C_2} v_{o1}. \qquad \text{(9.76)}$$

Differentiating Eq. (9.76) gives

$$\frac{d^2 v_o}{dt^2} = -\frac{1}{R_2 C_2} \frac{dv_{o1}}{dt}. \qquad \text{(9.77)}$$

We find the differential equation that governs the relationship between v_o and v_g by substituting Eq. (9.74) into Eq. (9.77), namely,

$$\frac{d^2 v_o}{dt^2} = \frac{1}{R_1 C_1} \frac{1}{R_2 C_2} v_g. \qquad \text{(9.78)}$$

Example 9.13 illustrates the step response of a circuit containing two cascaded integrating amplifiers.

E X A M P L E   9.13

No energy is stored in the circuit shown in Fig. 9.18 when the input voltage v_g jumps instantaneously from 0 to 25 mV.

a) Derive the expression for $v_o(t)$ for $0 \leq t \leq t_{\text{sat}}$.

b) How long is it before the circuit saturates?

S O L U T I O N

a) Figure 9.18 indicates that the amplifier scaling factors are

$$\frac{1}{R_1 C_1} = \frac{1000}{(250)(0.1)} = 40$$

and

$$\frac{1}{R_2 C_2} = \frac{1000}{(500)(1)} = 2.$$

Now, as $v_g = 25$ mV for $t > 0$, Eq. (9.78) becomes

$$\frac{d^2 v_o}{dt^2} = (40)(2)(25 \times 10^{-3}) = 2.$$

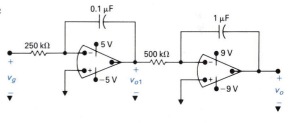

FIGURE 9.18 The circuit for Example 9.13.

To solve for v_o we let

$$g(t) = \frac{dv_o}{dt};$$

then

$$\frac{dg(t)}{dt} = 2 \quad \text{and} \quad dg(t) = 2dt.$$

Hence

$$\int_{g(0)}^{g(t)} dy = 2\int_0^t dx,$$

from which

$$g(t) - g(0) = 2t.$$

However,

$$g(0) = \frac{dv_o(0)}{dt} = 0,$$

because the energy stored in the circuit initially is zero and the operational amplifiers are ideal. (See Drill Exercise 9.9.) Then,

$$\frac{dv_o}{dt} = 2t \quad \text{and} \quad v_o = t^2 + v_o(0).$$

But $v_o(0) = 0$; so the expression for v_o becomes

$$v_o = t^2, \quad 0 \le t \le t_{\text{sat}}.$$

b) The second integrating amplifier saturates when v_o reaches 9 V or $t = 3$ s. Before accepting this length of time for the circuit to saturate, we must check the state of the first stage at $t = 3$ s. From Eq. (9.74),

$$\frac{dv_{o1}}{dt} = -40(25) \times 10^{-3} = -1.$$

Solving for v_{o1} yields

$$v_{o1} = -t.$$

At $t = 3$ s, $v_{o1} = -3$ V, and, as the power supply voltage on the first integrating amplifier is ± 5 V, the circuit reaches saturation when the second amplifier saturates. When one of the operational amplifiers saturates, we no longer can use the linear model to predict the behavior of the circuit.

DRILL EXERCISES

9.9 Show that, if no energy is stored in the circuit shown in Fig. 9.18 at the instant v_g jumps in value, then dv_o/dt equals zero at $t = 0$.

ANSWER: Derivation.

9.10 a) Find the equation for $v_o(t)$ for $0 \leq t \leq t_{sat}$ in the circuit shown in Fig. 9.18 if $v_{o1}(0) = 2$ V and $v_o(0) = 8$ V.

b) How long does the circuit take to reach saturation?

ANSWER: (a) $v_o = t^2 - 4t + 8$ V; (b) $t_{sat} = 4.24$ s.

TWO INTEGRATING AMPLIFIERS WITH FEEDBACK RESISTORS

Figure 9.19 depicts a variation of the circuit shown in Fig. 9.17. Here, a resistor is placed in parallel with each feedback capacitor, that is, C_1 and C_2. The feedback resistors are introduced into the amplifier circuit for practical reasons, as mentioned in Section 8.7.

We begin the derivation of the second-order differential equation that relates v_o to v_g by summing the currents at the inverting input node of the first integrator:

$$\frac{0 - v_g}{R_a} + \frac{0 - v_{o1}}{R_1} + C_1 \frac{d}{dt}(0 - v_{o1}) = 0. \quad \textbf{(9.79)}$$

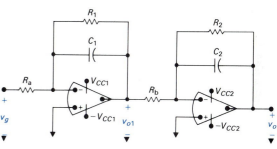

FIGURE 9.19 Cascaded integrating amplifiers with feedback resistors.

We simplify Eq. (9.79) to read

$$\frac{dv_{o1}}{dt} + \frac{1}{R_1 C_1} v_{o1} = \frac{-v_g}{R_a C_1}. \quad \textbf{(9.80)}$$

For convenience, we let $\tau_1 = R_1 C_1$ and write Eq. (9.80) as

$$\frac{dv_{o1}}{dt} + \frac{v_{o1}}{\tau_1} = \frac{v_g}{R_a C_1}. \quad \textbf{(9.81)}$$

The next step is to sum the currents at the inverting input terminal of the second integrator:

$$\frac{0 - v_{o1}}{R_b} + \frac{0 - v_o}{R_2} + C_2 \frac{d}{dt}(0 - v_o) = 0. \quad \textbf{(9.82)}$$

We rewrite Eq. (9.82) as

$$\frac{dv_o}{dt} + \frac{v_o}{\tau_2} = \frac{-v_{o1}}{R_b C_2}, \quad \textbf{(9.83)}$$

where $\tau_2 = R_2 C_2$. Differentiating Eq. (9.83) yields

$$\frac{d^2 v_o}{dt^2} + \frac{1}{\tau_2} \frac{dv_o}{dt} = -\frac{1}{R_b C_2} \frac{dv_{o1}}{dt}. \quad \textbf{(9.84)}$$

From Eq. (9.81),

$$\frac{dv_{o1}}{dt} = \frac{-v_{o1}}{\tau_1} - \frac{v_g}{R_aC_1},$$
(9.85)

and from Eq. (9.83),

$$v_{o1} = -R_bC_2\frac{dv_o}{dt} - \frac{R_bC_2}{\tau_2}v_o.$$
(9.86)

We use Eqs. (9.85) and (9.86) to eliminate dv_{o1}/dt from Eq. (9.84) and obtain the desired relationship:

$$\frac{d^2v_o}{dt^2} + \left(\frac{1}{\tau_1} + \frac{1}{\tau_2}\right)\frac{dv_o}{dt} + \left(\frac{1}{\tau_1\tau_2}\right)v_o = \frac{v_g}{R_aC_1R_bC_2}.$$
(9.87)

From Eq. (9.87), the characteristic equation is

$$s^2 + \left(\frac{1}{\tau_1} + \frac{1}{\tau_2}\right)s + \frac{1}{\tau_1\tau_2} = 0.$$
(9.88)

The roots of the characteristic equation are real, namely,

$$s_1 = \frac{-1}{\tau_1}$$
(9.89)

and

$$s_2 = \frac{-1}{\tau_2}.$$
(9.90)

Example 9.14 illustrates the analysis of the step response of two cascaded integrating amplifiers when the feedback capacitors are shunted with feedback resistors.

E X A M P L E   9.14

The parameters for the circuit shown in Fig. 9.19 are $R_a = 100$ kΩ, $R_1 = 500$ kΩ, $C_1 = 0.1$ μF, $R_b = 25$ kΩ, $R_2 = 100$ kΩ, and $C_2 = 1$ μF. The power supply voltage for each operational amplifier is ± 6 V. The signal voltage (v_g) jumps from 0 to 250 mV at $t = 0$. No energy is stored in the feedback capacitors at the instant the signal is applied.

a) Find the numerical expression of the differential equation for v_o.

b) Find $v_o(t)$ for $t \geq 0$.

c) Find the numerical expression of the differential equation for v_{o1}.

d) Find $v_{o1}(t)$ for $t \geq 0$.

SOLUTION

a) From the numerical values of the circuit parameters, we have $\tau_1 = R_1C_1 = 0.05\,\text{s}; \tau_2 = R_2C_2 = 0.10\,\text{s}; \text{and}\; v_g/R_aC_1R_bC_2 = 1000\;\text{V/s}^2$. Substituting these values into Eq. (9.87) gives

$$\frac{d^2v_o}{dt^2} + 30\frac{dv_o}{dt} + 200v_o = 1000.$$

b) The roots of the characteristic equation are $s_1 = -10$ rad/s and $s_2 = -20$ rad/s. The final value of v_o is $(250 \times 10^{-3})(-500/100)(-100/25)$, or 5 V. The solution for v_o thus takes the form

$$v_o = 5 + A_1'e^{-10t} + A_2'e^{-20t}.$$

With $v_o(0) = 0$ and $dv_o(0)/dt = 0$, the numerical values of A_1' and A_2' are $A_1' = -10$ V and $A_2' = 5$ V. Therefore the solution for v_o is

$$v_o(t) = (5 - 10e^{-10t} + 5e^{-20t})\;\text{V}, \qquad t \geq 0.$$

The solution is based on the assumption that neither operational amplifier saturates. We have already noted that the final value of v_o is 5 V, which is less than 6 V; hence the second op amp does not saturate. The final value of v_{o1} is $(250 \times 10^{-3})(-500/100)$, or -1.25 V. Therefore the first operational amplifier does not saturate and the solution is correct.

c) Substituting the numerical values of the parameters into Eq. (9.81) generates the desired differential equation:

$$\frac{dv_{o1}}{dt} + 20v_{o1} = -25.$$

d) We have already noted the initial and final values of v_{o1}, along with the time constant τ_1. Thus we write the solution in accordance with the technique developed in Section 8.4:

$$v_{o1} = -1.25 + [0 - (-1.25)]e^{-20t}$$

$$= -1.25 + 1.25e^{-20t}\;\text{V}, \qquad t \geq 0.$$

DRILL EXERCISES

9.11 Rework Example 9.14 with feedback resistors R_1 and R_2 removed.

ANSWER: (a) $d^2v_o/dt^2 = 1000$; (b) $v_o = 500t^2$ V, $0 \leq t \leq 0.1095$ s; (c) $dv_{o1}/dt = -25$; (d) $v_{o1} = -25t$ V, $0 \leq t \leq 0.1095$ s.

9.12 Rework Example 9.14 with $v_{o1}(0) = -2$ V and $v_o(0) = 4$ V.

ANSWER: (a) Same as Example 9.14; (b) $v_o = (5 + 2e^{-10t} - 3e^{-20t})$ V, $t \geq 0$; (c) same as Example 9.14; (d) $v_{o1} = -(1.25 + 0.75e^{-20t})$ V, $t \geq 0$.

SUMMARY

To find the natural and step responses of the parallel and series *RLC* circuits you must be able to solve an ordinary, second-order differential equation. An understanding of the following concepts is essential to finding such solutions.

- The nature of the solution depends on the roots of the characteristic equation.

- The characteristic equation for both the parallel and series *RLC* circuits has the form:

$$s^2 + 2\alpha s + \omega_0^2 = 0,$$

  where $\alpha = 1/2RC$ for the parallel circuit, $\alpha = R/2L$ for the series circuit, and $\omega_0^2 = 1/LC$ for both the parallel and series circuits.

- The roots of the characteristic equation are

$$s_{1,2} = -\alpha \pm \sqrt{\alpha^2 - \omega_0^2}.$$

- The three forms of the solution are called

  overdamped when $\alpha^2 > \omega_0^2$,

  underdamped when $\alpha^2 < \omega_0^2$, and

  critically damped when $\alpha^2 = \omega_0^2$.

- The three possible natural responses are

$$x(t) = A_1 e^{s_1 t} + A_2 e^{s_2 t} \quad \text{(od)},$$

$$x(t) = (B_1 \cos \omega_d t + B_2 \sin \omega_d t)e^{-\alpha t} \quad \text{(ud)},$$

  and

$$x(t) = (D_1 t + D_2)e^{-\alpha t} \quad \text{(cd)},$$

  where $x(t)$ represents either a current or voltage response. In the underdamped response, $\omega_d = \sqrt{\omega_0^2 - \alpha^2}$.

- The three possible step responses are

$$x(t) = x_F + A_1' e^{s_1 t} + A_2' e^{s_2 t} \quad \text{(od)},$$

$$x(t) = x_F + (B_1' \cos \omega_d t + B_2' \sin \omega_d t)e^{-\alpha t} \quad \text{(ud)},$$

and

$$x(t) = x_F + (D_1't + D_2')e^{-\alpha t} \qquad \text{(cd)},$$

where x_F is the final value of the desired current or voltage response.

- The unknown coefficients (i.e., the A's, B's, and D's) are obtained by evaluating the initial value of the desired response $[x(0)]$ and the initial value of the first derivative of the desired response $[dx(0)/dt]$.

The terms overdamped, underdamped, and critically damped describe the impact of the dissipative element (R) on the response. The neper frequency or damping factor α reflects the effect of R. The following characteristics define the three types of damping.

- Overdamped: When α is large compared to the resonant frequency ω_0, the voltage or current approaches its final value without oscillation.

- Underdamped: When α is small compared to ω_0, the response oscillates about its final value. The smaller α's value, the longer the oscillation persists. If the dissipative element is removed from the circuit, $\alpha = 0$, and the response becomes a sustained oscillation.

- Critically damped. When $\alpha = \omega_0$ the response is on the verge of oscillating.

When two integrating amplifiers with ideal op amps are connected in cascade the output voltage of the second integrator is related to the input voltage of the first integrator by an ordinary, second-order differential equation. Therefore the techniques developed may be used to analyze the behavior of the cascaded integrators.

PROBLEMS

9.1 The resistance, inductance, and capacitance in a parallel RLC circuit are 1000 Ω, 12.5 H, and 2 μF, respectively.

a) Calculate the roots of the characteristic equation that describe the voltage response of the circuit.

b) Will the response be over-, under-, or critically damped?

c) What value of R will yield a damped frequency of 120 rad/s?

d) What are the roots of the characteristic equation for the value of R found in part (c)?

e) What value of R will result in a critically damped response?

9.2 The initial voltage on the 0.1 μF capacitor in the circuit shown in Fig. 9.1 is 24 V. The initial current in the inductor is zero. The voltage response for $t \geq 0$ is

$$v(t) = -8e^{-250t} + 32e^{-1000t} \text{ V}.$$

a) Determine the numerical values of R, L, α, and ω_0.

b) Calculate $i_R(t)$, $i_L(t)$, and $i_C(t)$ for $t \geq 0^+$.

9.3 The circuit elements in the circuit in Fig. 9.1 are $R = 20$ kΩ, $C = 0.02$ μF, and $L = 50$ H. The initial inductor current is 1.2 mA and the initial capacitor voltage is zero.

a) Calculate the initial current in each branch of the circuit.

b) Find $v(t)$ for $t \geq 0$.

c) Find $i_L(t)$ for $t \geq 0$.

9.4 The natural response for the circuit shown in Fig. 9.1 is known to be

$$v = 3(e^{-100t} + e^{-900t}) \text{ V}, \qquad t \geq 0.$$

If $L = 40/9$ H and $C = 2.5$ μF, find $i_L(0^+)$ in milliamperes.

9.5 The natural voltage response of the circuit of Fig. 9.1 is

$$v = 100e^{-20,000t}(\cos 15,000t - 2 \sin 15,000t) \text{ V},$$
$$t \geq 0,$$

when the capacitor is 0.04 μF. Find (a) R; (b) L; (c) V_0; (d) I_0; and (e) $i_L(t)$.

9.6 The initial value of the voltage v in the circuit in Fig 9.1 is zero and the initial value of the capacitor current $[i_c(0^+)]$ is 15 mA. The expression for the capacitor current is known to be

$$i_c(t) = A_1e^{-160t} + A_2e^{-40t}, \, t \geq 0^+$$

when R is 200 Ω. Find the numerical:

a) value of α, ω_0, L, C, A_1, and A_2

b) expression for $v(t)$, $t \geq 0$

c) expression for $i_R(t) \geq 0$

d) expression for $i_L(t) \geq 0$.

9.7 The voltage response for the circuit in Fig. 9.1 is known to be

$$v(t) = D_1t\,e^{-500t} + D_2e^{-500t}, \, t \geq 0.$$

The initial current in the inductor $[I_0]$ is -10 mA and the initial voltage on the capacitor $[V_0]$ is 8 V. The inductor has an inductance of 4 H.

a) Find the value of R, C, D_1, and D_2.

b) Find $i_c(t)$ for $t \geq 0^+$.

9.8 In the circuit in Fig. 9.1, $R = 12.5$ Ω, $L = 50/101$ H, $C = 0.08$ F, $V_0 = 0$ V, and $I_0 = -4$ A.

a) Find $v(t)$ for $t \geq 0$.

b) Find the first three values of t for which dv/dt is zero. Let these values of t be denoted t_1, t_2, and t_3.

c) Show that $t_3 - t_1 = T_d$.

d) Show that $t_2 - t_1 = T_d/2$.

e) Calculate $v(t_1)$, $v(t_2)$, and $v(t_3)$.

f) Sketch $v(t)$ versus t for $0 \leq t \leq t_2$.

9.9 a) Find $v(t)$ for $t \geq 0$ in the circuit in Problem 9.8 if the 12.5-Ω resistor is removed from the circuit.

b) Calculate the frequency of $v(t)$ in hertz.

c) Calculate the maximum amplitude of $v(t)$ in volts.

9.10 In the circuit shown in Fig. 9.1 a 12.5-H inductor is shunted by 3.2-nF capacitor, the resistor R is adjusted for critical damping, $V_0 = 100$ V and $I_0 = 6.4$ mA.

  a) Calculate the numerical value of R.

  b) Calculate $v(t)$ for $t \geq 0$.

  c) Find $v(t)$ when $i_c(t) = 0$.

  d) What percentage of the initially stored energy remains stored in the circuit at the instant $i_c(t)$ is 0?

9.11 The resistor in the circuit in Example 9.4 is changed to 3200 Ω.

  a) Find the numerical expression for $v(t)$ when $t \geq 0$.

  b) Plot $v(t)$ versus t for the time interval $0 \leq t \leq 7$ ms. Compare this response with that of Example 9.4 ($R = 20$ kΩ) and Example 9.5 ($R = 4$ kΩ). In particular, compare peak values of $v(t)$ and the times when these peak values occur.

9.12 The two switches in the circuit seen in Fig. P9.12 operate synchronously. When switch 1 is in position a, switch 2 is in position d. When switch 1 moves to position b, switch 2 moves to position c, and vice versa. Switch 1 has been in position a for a long time. At $t = 0$ the switches move to their alternate positions. Find $v_o(t)$ for $t \geq 0$.

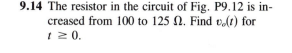

FIGURE P9.12

9.13 The resistor in the circuit of Fig. P9.12 is increased from 100 to 250 Ω. Find $v_o(t)$ for $t > 0$.

9.14 The resistor in the circuit of Fig. P9.12 is increased from 100 to 125 Ω. Find $v_o(t)$ for $t \geq 0$.

9.15 The switch in the circuit of Fig. P9.15 has been in position a for a long time. At $t = 0$ the switch moves instantaneously to position b. Find $v_o(t)$ for $t \geq 0$.

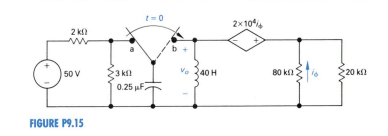

FIGURE P9.15

9.16 For the circuit in Example 9.6, find for $t \geq 0$ (a) $v(t)$; (b) $i_R(t)$; and (c) $i_C(t)$.

9.17 For the circuit in Example 9.7, find for $t \geq 0$ (a) $v(t)$ and (b) $i_C(t)$.

9.18 For the circuit in Example 9.8, find $v(t)$ for $t \geq 0$.

9.19 The switch in the circuit in Fig. P9.19 has been open a long time before closing at $t = 0$. Find $i_L(t)$ for $t \geq 0$.

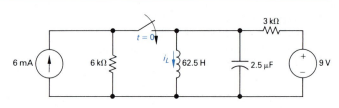

FIGURE P9.19

9.20 The two switches in the circuit in Fig. P9.20 are synchronized. For $t < 0$ switch 1 is in position a and switch 2 is open. At $t = 0$ switch 1 moves to position b and switch 2 closes. Find $v_o(t)$ for $t \geq 0^+$.

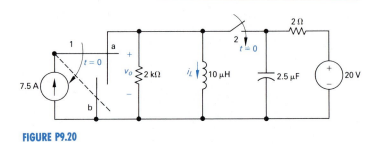

FIGURE P9.20

9.21 For the circuit in Fig. P9.20 find $i_L(t)$ for $t \geq 0$.

9.22 Switch 2 in the circuit shown in Fig. P9.22 has been ON for a long time. At $t = 0$, switch 1 opens and switch 2 moves instantaneously to its OFF position. The initial inductor current is zero. Find $i_L(t)$ for $t \geq 0$.

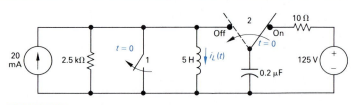

FIGURE P9.22

9.23 Switches 1 and 2 in the circuit in Fig. P9.23 operate synchronously. When switch 1 opens, switch 2 closes and vice versa. Switch 1 has been closed for a long time. At $t = 0$, switch 1 opens.

a) Calculate $i_L(0^+)$.

b) Calculate $di_L(0^+)/dt$.

c) Find $i_L(t)$ for $t \geq 0$.

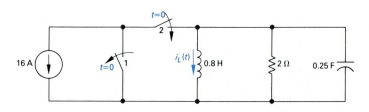

FIGURE P9.23

9.24 Switches 1 and 2 in the circuit shown in Fig. P9.24 operate synchronously. When switch 1 opens, switch 2 closes and vice versa. Switch 1 has been closed for a long time. At $t = 0$, switch 1 opens. Find

a) $v_o(0^+)$;

b) $dv_o(0^+)/dt$;

c) $v_o(t)$ for $t \geq 0$.

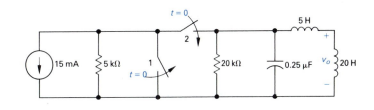

FIGURE P9.24

9.25 Switches 1 and 2 in the circuit in Fig. P9.25 operate synchronously. When switch 1 opens, switch 2 closes and vice versa. Switch 1 has been closed for a long time. At $t = 0$, switch 1 opens. Find

a) $v_o(0^+)$;

b) $dv_o(0^+)/dt$;

c) $v_o(t)$ for $t \geq 0$.

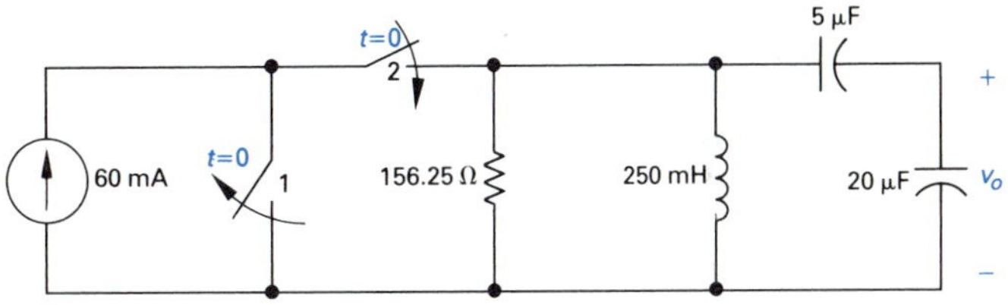

FIGURE P9.25

9.26 In the circuit in Fig. 9.8 the initial energy stored is 11.76 mJ. The initial voltage on the capacitor is 56 V. The dc current source is delivering 7 mA. The circuit elements are $R = 10\ k\Omega$; $L = 20\ H$; and $C = 2.5\ \mu F$.

a) Find the solution for $i_L(t)$ for $t \geq 0$.

b) Find the solution for $v(t)$ for $t \geq 0$.

c) Find the maximum value of $v(t)$.

9.27 The initial energy stored in the 31.25-nF capacitor in the circuit in Fig. P9.27 is 9 μJ. The initial energy stored in the inductor is zero. The roots of the characteristic equation that describes the natural behavior of the current i are -4000 s^{-1} and $-16,000$ s^{-1}.

a) Find the numerical values of R and L.

b) Find the numerical values of $i(0)$ and $di(0)/dt$ immediately after the switch has been closed.

c) Find $i(t)$ for $t \geq 0$.

d) How many microseconds after the switch

closes does the current reach its maximum value?

e) What is the maximum value of i in milliamperes?

f) Find $v_L(t)$ for $t \geq 0$.

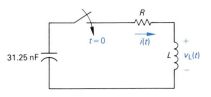

FIGURE P9.27

9.28 The current in the circuit in Fig. 9.3 is known to be

$$i = B_1 e^{-800t} \cos 600t + B_2 e^{-800t} \sin 600t,$$
$$t \geq 0.$$

The capacitor has a value of 500 μF; the initial value of the current is zero; and the initial voltage on the capacitor is 12 V, positive at the upper terminal. Find the values of R, L, B_1, and B_2.

9.29 Find the voltage across the 500-μF capacitor for the circuit described in Problem 9.28. Assume the reference polarity for the capacitor voltage is positive at the upper terminal.

9.30 In the circuit in Fig. P9.30, the resistor is adjusted for critical damping. The initial capacitor voltage is 20 V, and the initial inductor current is 30 mA.

a) Find the numerical value of R.

b) Find the numerical values of i and di/dt immediately after the switch is closed.

c) Find $v_C(t)$ for $t \geq 0$.

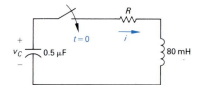

FIGURE P9.30

9.31 The switch in the circuit in Fig. P9.31 has been in position a for a long time. At $t = 0$, the switch moves instantaneously to position b.

a) What is the initial value of v_a?

b) What is the initial value of dv_a/dt?

c) What is the numerical expression for $v_a(t)$ for $t \geq 0$?

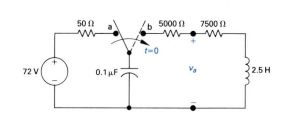

FIGURE P9.31

9.32 The "make-before-break" switch in the circuit shown in Fig. P9.32 has been in position a for a long time. At $t = 0$ the switch is moved instantaneously to position b. Find $i(t)$ for $t \geq 0$.

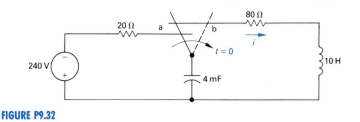

FIGURE P9.32

9.33 The switch in the circuit shown in Fig. P9.33 has been closed for a long time. The switch opens at $t = 0$. Find $v_o(t)$ for $t \geq 0$.

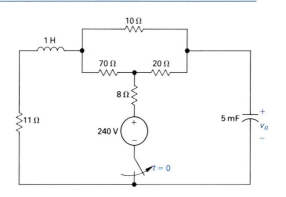

FIGURE P9.33

9.34 The switch in the circuit shown in Fig. P9.34 has been closed for a long time. The switch opens at $t = 0$.

a) Find $i_o(t)$ for $t \geq 0$.

b) Find $v_o(t)$ for $t \geq 0$.

FIGURE P9.34

9.35 The initial energy stored in the circuit in Fig. P9.35 is zero. Find $v_o(t)$ for $t \geq 0$.

FIGURE P9.35

9.36 The two switches in the circuit seen in Fig. P9.36 operate synchronously. When switch 1 is in position a, switch 2 is closed. When switch 1 is in position b, switch 2 is open. Switch 1 has been in position a for a long time. At $t = 0$ it moves instantaneously to position b. Find $v_c(t)$ for $t \geq 0$.

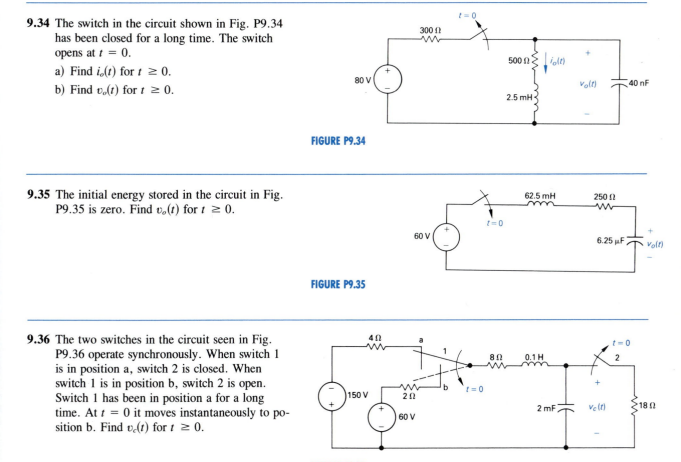

FIGURE P9.36

9.37 The circuit shown in Fig. P9.37 has been in operation for a long time. At $t = 0$ the voltage suddenly jumps to 400 V. Find $v_c(t)$ for $t \geq 0$.

9.38 The switch in the circuit of Fig. P9.38 has been in position a for a long time. At $t = 0$ the switch moves instantaneously to position b.

a) Find $v_o(0^+)$.

b) Find $dv_o(0^+)/dt$.

c) Find $v_o(t)$ for $t \geq 0$.

9.39 The switch in the circuit in Fig. P9.39 has been open for a long time before closing at $t = 0$. Find $v_o(t)$ for $t \geq 0$.

9.40 The switch in the circuit in Fig. P9.40 has been closed for a long time before opening at $t = 0$.

a) Find $i_o(t)$ for $t \geq 0$.

b) Find $v_o(t)$ for $t \geq 0$.

9.41 Assume that the capacitor voltage in the circuit of Fig. 9.10 is underdamped. Also assume that no energy is stored in the circuit elements when the switch is closed.

a) Show that $dv_C/dt = (\omega_0^2/\omega_d)Ve^{-\alpha t} \sin \omega_d t$.

b) Show that $dv_C/dt = 0$ when $t = n\pi/\omega_d$, where $n = 0, 1, 2, \ldots$.

c) Let $t_n = n\pi/\omega_d$ and show that $v_C(t_n) = V - V(-1)^n e^{-\alpha n\pi/\omega_d}$.

d) Show that

$$\alpha = \frac{1}{T_d} \ln \frac{v_C(t_1) - V}{v_C(t_3) - V},$$

where $T_d = t_3 - t_1$.

9.42 The voltage across a 0.1-μF capacitor in the circuit of Fig. 9.10 is described as follows. After the switch has been closed for several seconds, the voltage is constant at 100 V. The first time the voltage exceeds 100 V, it reaches a peak of 163.84 V. This occurs $(\pi/7)$ ms after the switch has been closed. The second time the voltage exceeds 100 V, it reaches a peak of 126.02 V. This second peak occurs $(3\pi/7)$ ms after the switch has been closed. At the time when the switch is closed, there is no energy stored in either the capacitor or the inductor. Find the numerical values of R and L. (*Hint:* Work Problem 9.41 first.)

9.43 The switch in the circuit shown in Fig. P9.43 has been closed for a long time before it is opened at $t = 0$. Assume that the circuit parameters are such that the response is underdamped.

a) Derive the expression for $v_o(t)$ as a function of v_g, α, ω_d, C, and R for $t \geq 0$.

b) Derive the expression for the value of t when the magnitude of v_o is maximum.

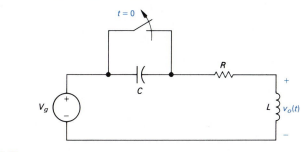

FIGURE P9.43

9.44 The circuit parameters in the circuit of Fig. P9.43 are $R = 600\ \Omega$, $L = 20$ mH, $C = 0.08\ \mu$F, and $v_g = -480$ V.

a) Express $v_o(t)$ numerically for $t \geq 0$.

b) How many microseconds after the switch opens is the inductor voltage maximum?

c) What is the maximum value of the inductor voltage?

d) Repeat parts (a), (b), and (c), with R reduced to 60 Ω.

9.45 The voltage signal of Fig. P9.45(a) is applied to the cascaded integrating amplifiers shown in Fig. P9.45(b). There is no energy stored in the capacitors at the instant the signal is applied.

a) Derive the numerical expressions for $v_o(t)$ and $v_{o1}(t)$ for the time intervals $0 \leq t \leq 0.5$ s and 0.5 s $\leq t \leq t_{\text{sat}}$.

b) Compute the value of t_{sat}.

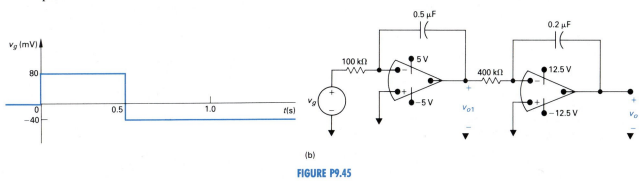

(a)   (b)

FIGURE P9.45

9.46 The circuit in Fig. P9.45(b) is modified by adding a 1-MΩ resistor in parallel with the 0.5-μF capacitor and a 5-MΩ resistor in parallel with the 0.2-μF capacitor. As in Problem 9.45, there is no energy stored in the capacitors at the time the signal is applied. Derive the numerical expressions for $v_o(t)$ and $v_{o1}(t)$ for the time invervals $0 \le t \le 0.5$ s and 0.5 s $\le t \le \infty$.

9.47 a) Derive the differential equation that relates the output voltage to the input voltage for the circuit shown in Fig. P9.47.

b) Compare the result with Eq. (9.86) when $R_1 C_1 = R_2 C_2 = RC$ in Fig. 9.13.

c) What is the advantage of the circuit shown in Fig. P9.47?

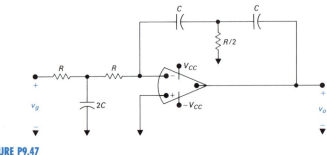

FIGURE P9.47

9.48 We now wish to illustrate how several op-amp circuits can be interconnected to solve a differential equation.

a) Derive the differential equation for the spring-mass system shown in Fig. P9.48(a). Assume that the force exerted by the spring is directly proportional to the spring displacement, that the mass is constant, and that the frictional force is directly proportional to the velocity of the moving mass.

b) Rewrite the differential equation derived in part (a) so that the highest-order derivative is expressed as a function of all the other terms in the equation. Now assume that a voltage equal to d^2x/dt^2 is available and by successive integrations generate dx/dt and x. We can synthesize the coefficients in the equations by scaling amplifiers and combine the terms required to generate d^2x/dt^2 by a summing amplifier. With these ideas in mind, analyze the interconnection shown in Fig. P9.48(b). In particular, describe the purpose of each shaded area in the circuit and describe the signal at the points labeled B, C, D, E, and F, assuming the signal at A represents d^2x/dt^2. Also discuss the parameters R; R_1, C_1; R_2, C_2; R_3, R_4; R_5, R_6; and R_7, R_8 in terms of the coefficients in the differential equation.

9.49 Assume the underdamped voltage response of the circuit in Fig. 9.1 is written as

$$v(t) = (A_1 + A_2)e^{-\alpha t} \cos \omega_d t$$

$$+ j(A_1 + A_2)e^{-\alpha t} \sin \omega_d t.$$

The initial value of the inductor current is I_0 and the initial value of the capacitor voltage is V_0. Show that A_2 is the conjugate of A_1. (*Hint:* Use the same thought process as outlined in the text to find A_1 and A_2.)

9.50 Show that the results obtained from Problem 9.49, that is the expressions for A_1 and A_2 are consistent with Eqs. (9.27) and (9.28) in the text.

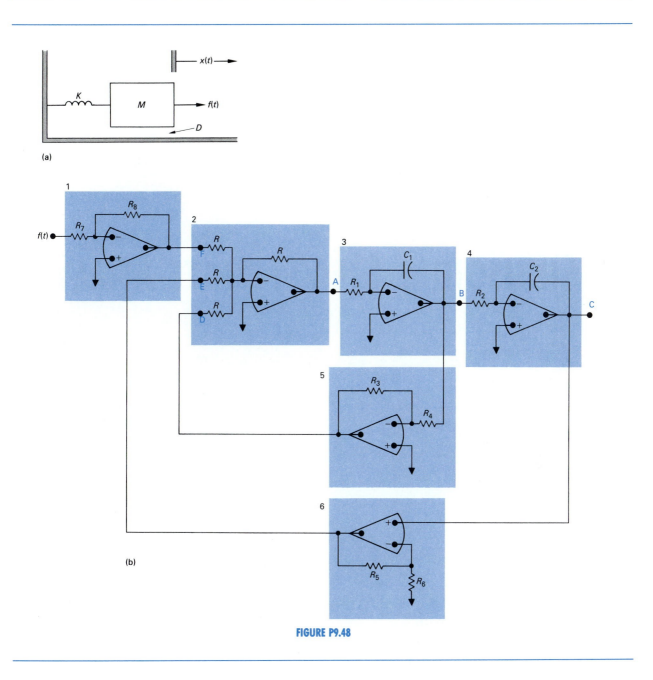

(a)

(b)

FIGURE P9.48

SINUSOIDAL STEADY-STATE ANALYSIS

CHAPTER 10

The steady-state behavior of circuits energized by sinusoidal sources is an important area of study for several reasons. First, the generation, transmission, distribution, and consumption of electric energy occur under essentially sinusoidal steady-state conditions. Second, an understanding of sinusoidal behavior makes possible the prediction of circuit behavior to nonsinusoidal sources. Third, steady-state sinusoidal behavior often simplifies the design of electrical systems. Thus the designer spells out specifications in terms of a desired steady-state sinusoidal response and designs the circuit or system to meet those characteristics. If the device satisfies the specifications, the designer knows that the circuit will respond satisfactorily to nonsinusoidal inputs.

The importance of sinusoidal steady-state behavior cannot be overemphasized. Most of the topics in the subsequent chapters of this book are based on a thorough understanding of the techniques needed to analyze circuits driven by sinusoidal sources. We begin with a review of the important characteristics of the sinusoidal function.

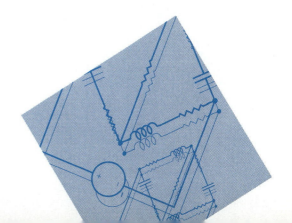

10.1 THE SINUSOIDAL SOURCE

PSpice　　　PSpice can simulate a generalized damped sinusoidal voltage or current source: Section 11.1

A *sinusoidal voltage source* (independent or dependent) produces a voltage that varies sinusoidally with time. A *sinusoidal current source* (independent or dependent) produces a current that varies sinusoidally with time. In reviewing the sinusoidal function we use a voltage source, but the observations we make also apply to a current source.

We can express a sinusoidally varying function with either the sine function or the cosine function. There is no clear-cut choice for the use of either function. Although either works equally well in sinusoidal steady-state analysis, we cannot use both functional forms simultaneously. We chose to use the cosine function throughout the discussion.

Hence we write a sinusoidally varying voltage as

$$v = V_m \cos(\omega t + \phi). \qquad \textbf{(10.1)}$$

To aid discussion of the parameters in Eq. (10.1), we show the voltage versus time plot in Fig. 10.1. Note that the sinusoidal function is repetitive. Such a function is called *periodic*. One of the parameters of interest, therefore, is the length of time required for the sinusoidal function to pass through all its possible values. This time is referred to as the *period* of the function and is denoted T. The reciprocal of T gives the number of cycles per second, or frequency, of the sine function and is denoted f, or

$$f = \frac{1}{T}. \qquad \textbf{(10.2)}$$

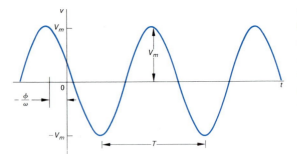

FIGURE 10.1　A sinusoidal voltage.

A cycle per second is referred to as a hertz, abbreviated Hz. (The term *cycles per second* rarely is used in contemporary technical literature.) The coefficient of t in Eq. (10.1) contains the numerical value of T or f. Omega (ω) represents the angular frequency of the sinusoidal function, or

$$\omega = 2\pi f = 2\pi / T \qquad \text{(radians/second)}. \qquad \textbf{(10.3)}$$

Equation 10.3 is based on the fact that the sine (or cosine) function passes through a complete set of values each time its argument passes through 2π rad (360°). From Eq. (10.3), note that, whenever t is an integral multiple of T, the argument ωt increases by an integral multiple of 2π rad.

The coefficient V_m gives the maximum amplitude of the sinusoidal voltage. Because ± 1 bounds the cosine function $\pm V_m$ bounds the amplitude. Figure 10.1 clearly shows these characteristics.

The phase angle of the sinusoidal voltage of Eq. (10.1) is the angle ϕ. It determines the value of the sinusoidal function at $t = 0$; therefore it fixes the point on the periodic wave at which we start measuring time. Changing the phase angle ϕ shifts the sinusoidal function along the time axis, but has no effect on either the amplitude (V_m) or the angular frequency (ω). Note, for example, that reducing ϕ to zero shifts the sinusoidal function shown in Fig. 10.1 ϕ/ω time units to the right. Note also that if ϕ is positive, the sinusoidal function shifts to the left, whereas if ϕ is negative, the function shifts to the right. (See Problem 10.2.)

A comment with regard to the phase angle is in order: ωt and ϕ must carry the same units because they are added together in the argument of the sinusoidal function. With ωt expressed in radians, you would expect ϕ to be also. However, ϕ normally is given in degrees and ωt is converted from radians to degrees before the two quantities are added. We continue this bias toward degrees by expressing the phase angles in degrees.

A fourth important characteristic of the sinusoidal voltage (or current) is its rms value. The rms value of a periodic function is defined as the square *root* of the *mean* value of the *squared* function. Hence, if $v = V_m \cos(\omega t + \phi)$, the rms value of v is

$$V_{\text{rms}} = \sqrt{\frac{1}{T} \int_{t_0}^{t_0 + T} V_m^2 \cos^2(\omega t + \phi)dt)}. \qquad \textbf{(10.4)}$$

Note from Eq. (10.4) that we obtain the mean value of the squared voltage by integrating v^2 over one period, that is from t_0 to $t_0 + T$, and then dividing by the range of integration, T. Note further that the starting point for the integration t_0 is arbitrary.

The quantity under the radical sign in Eq. (10.4) reduces to $V_m^2/2$. (See Problem 10.7.) Hence the rms value of v is

$$V_{\text{rms}} = \frac{V_m}{\sqrt{2}}. \qquad \textbf{(10.5)}$$

The rms value of the sinusoidal voltage depends only on the peak amplitude of v, namely, V_m. The rms value is *not* a function of either the frequency or the phase angle. We stress the importance of the rms value as it relates to power calculations in Chapter 11 (see Section 11.2).

Thus the *maximum amplitude* (or rms value), *frequency,* and *phase angle* completely specify the sinusoidal signal. Examples 10.1, 10.2, and 10.3 illustrate these basic properties of the sinusoidal function. In Example 10.4 we illustrate the calculation of the rms value of a periodic function and in so doing clarify the meaning of *root mean square.*

E X A M P L E   10.1

A sinusoidal current has a maximum amplitude of 20 A. The current passes through one complete cycle in 1.0 ms. The magnitude of the current at zero time is 10 A.

a) What is the frequency of the current in hertz?
b) What is the frequency in radians per second?
c) Write the expression for $i(t)$ using the cosine function. Express ϕ in degrees.
d) What is the rms value of the current?

S O L U T I O N

a) From the statement of the problem, $T = 1$ ms; hence $f = 1/T = 1000$ Hz.

b) $\omega = 2\pi f = 2000\pi$ rad/s.

c) We have $i(t) = I_m \cos(\omega t + \phi) = 20 \cos(2000\pi t + \phi)$, but $i(0) = 10$ A. Therefore $10 = 20 \cos \phi$ and $\phi = 60°$. Thus the expression for $i(t)$ becomes

$$i(t) = 20 \cos(2000\pi t + 60°).$$

d) From the derivation of Eq. (10.5), the rms value of a sinusoidal current is $I_m/\sqrt{2}$. Therefore the rms value is $20\sqrt{2}$, or 14.14 A.

E X A M P L E   10.2

A sinusoidal voltage is given by the expression $v = 300 \cos(120\pi t + 30°)$.

a) What is the period of the voltage in milliseconds?
b) What is the frequency in hertz?
c) What is the magnitude of v at $t = 2.778$ ms?
d) What is the rms value of v?

S O L U T I O N

a) From the expression for v, $\omega = 120\pi$ rad/s. As $\omega = 2\pi/T$, $T = 2\pi/\omega = \frac{1}{60}$ s, or 16.667 ms.

b) The frequency is $1/T$, or 60 Hz.

c) At $t = 2.778$ ms, ωt is nearly 1.047 rad, or 60°. Therefore
$v(2.778$ ms$) = 300 \cos (60° + 30°) = 0$ V.

d) $V_{rms} = 300/\sqrt{2} = 212.13$ V.

E X A M P L E 10.3

We can translate the sine function to the cosine function by subtracting 90° ($\pi/2$ rad) from the argument of the sine function.

a) Verify this translation by showing that

$$\sin (\omega t + \theta) = \cos (\omega t + \theta - 90°).$$

b) Use the result in (a) to express $\sin (\omega t + 30°)$ as a cosine function.

S O L U T I O N

a) Verification involves direct application of the trigonometric identity

$$\cos (\alpha - \beta) = \cos \alpha \cos \beta + \sin \alpha \sin \beta.$$

We let $\alpha = \omega t + \theta$ and $\beta = 90°$. As $\cos 90° = 0$ and $\sin 90° = 1$, we have

$$\cos (\alpha - \beta) = \sin \alpha = \sin (\omega t + \theta) = \cos (\omega t + \theta - 90°).$$

b) From (a) we have

$$\sin(\omega t + 30°) = \cos (\omega t + 30° - 90°) = \cos (\omega t - 60°).$$

E X A M P L E 10.4

Calculate the rms value of the periodic triangular current shown in Fig. 10.2. Express the answer in terms of the peak current I_p.

S O L U T I O N

From Eq. (10.4), the rms value of i is

$$I_{eff} = \sqrt{\frac{1}{T} \int_{t_0}^{t_0+T} i^2 \, dt}.$$

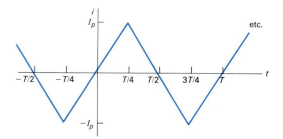

FIGURE 10.2 Periodic triangular current.

Interpreting the integral under the radical sign as the area under the squared function over an interval of one period is helpful for finding the rms value. The squared function with the area between 0 and T shaded is shown in Fig. 10.3, which also indicates that for this particular function, the area under the squared current over an interval of one period is equal to four times the area under the squared current over the interval 0 to $T/4$ seconds; that is,

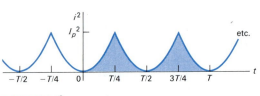

FIGURE 10.3 i^2 versus t.

$$\int_{t_0}^{t_0+T} i^2 \, dt = 4 \int_0^{T/4} i^2 \, dt.$$

The analytical expression for i in the interval 0 to $T/4$ is

$$i = \frac{4I_p}{T} t, \qquad 0 < t < T/4.$$

The area under the squared function over one period is

$$\int_{t_0}^{t_0+T} i^2 \, dt = 4 \int_0^{T/4} \frac{16I_p^2}{T^2} t^2 \, dt = \frac{I_p^2 T}{3}.$$

The mean, or average, value of the function is simply this area over one period divided by the period. Thus

$$i_{\text{mean}} = \frac{1}{T} \frac{I_p^2 T}{3} = \frac{1}{3} I_p^2.$$

The effective, or rms, value of the current is the square root of this mean value. Hence

$$I_{\text{eff}} = \frac{I_p}{\sqrt{3}}.$$

With these attributes of the sinusoidal function in mind, we are now ready to discuss the problem of finding the steady-state sinusoidal response of a circuit.

DRILL EXERCISES

10.1 A sinusoidal voltage is given by the expression

$$v = 40 \cos (2513.27t + 36.87°).$$

Find (a) f in hertz; (b) T in milliseconds; (c) V_m; (d) $v(0)$; (e) ϕ in degrees and radians; (f) the smallest positive value of t at which $v = 0$; and

(g) the smallest positive value of t at which $dv/dt = 0$.

ANSWER: (a) 400 Hz; (b) 2.5 ms; (c) 40 V; (d) 32 V; (e) 36.87°, or 0.6435 rad; (f) 368.96 μs; (g) 993.96 μs.

10.2 Find the rms value of the half-wave rectified sinusoidal voltage shown.

ANSWER: $V_{\text{eff}} = V_m/2$.

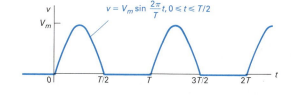

10.2 THE SINUSOIDAL RESPONSE

Before focusing on the steady-state response to sinusoidal sources, let's consider the problem in broader terms. Such an overview will help you keep the steady-state solution in perspective. The circuit shown in Fig 10.4 describes the general nature of the problem. There, v_s is a sinusoidal voltage, or

$$v_s = V_m \cos{(\omega t + \phi)}. \qquad \textbf{(10.6)}$$

For convenience, we assume the initial current in the circuit to be zero and measure time from the moment the switch is closed. The task is to derive the expression for $i(t)$ when $t \geq 0$. It is similar to finding the step response of the RL circuit in Chapter 8. The only difference is that the voltage source is now a time-varying sinusoidal voltage rather than a constant, or dc, voltage. Direct application of Kirchhoff's voltage law to the circuit shown in Fig. 10.4 leads to the ordinary differential equation

$$L\frac{di}{dt} + Ri = V_m \cos{(\omega t + \phi)}, \qquad \textbf{(10.7)}$$

the formal solution of which is discussed in an introductory course in differential equations. We ask those of you who have not yet studied differential equations to accept that the solution for i is

$$i = \frac{-V_m}{\sqrt{R^2 + \omega^2 L^2}} \cos{(\phi - \theta)} e^{-(R/L)t}$$

$$+ \frac{V_m}{\sqrt{R^2 + \omega^2 L^2}} \cos{(\omega t + \phi - \theta)}, \qquad \textbf{(10.8)}$$

where θ is defined as the angle whose tangent is $\omega L/R$. Thus we can easily determine θ for a circuit driven by a sinusoidal source of known frequency.

We check the validity of Eq. (10.8) by determining that it satisfies Eq. (10.7) for all values of $t \geq 0$. We first show that

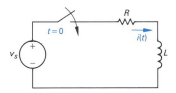

FIGURE 10.4 An RL circuit excited by a sinusoidal voltage source.

PSpice ⟿   Combine transient analysis (Section
7.2) with a sinusoidal source (Section 11.1) in
PSpice to obtain the total response

Eq. (10.8) equals zero for $t = 0$, thus confirming that the initial current is zero, as stipulated by the initial condition. Next, we substitute Eq. (10.8) into the left-hand side of Eq. (10.7) and show that it is equivalent to the right-hand side, thus proving that the solution is valid for all $t > 0$. (See Problem 10.5.)

The first term on the right-hand side of Eq. (10.8) is referred to as the *transient component* of the current because it becomes infinitesimal as time elapses. The second term on the right-hand side of Eq. (10.8) is known as the *steady-state component* of the solution. It exists as long as the switch remains closed and the source continues to supply the sinusoidal voltage. In this chapter we develop a technique for calculating the steady-state response directly, thus avoiding the problem of solving the differential equation. However, in solving for the steady-state solution directly, we forfeit obtaining either the transient component or the total response, which is the sum of the transient and steady-state components.

We now focus on the steady-state portion of Eq. (10.8) because it exhibits the following characteristics of the sinusoidal steady-state, which, in general, are true.

1. The steady-state solution is a sinusoidal function.

2. The frequency of the response signal is identical to the frequency of the source signal. This condition is always true in a linear circuit when the circuit parameters, R, L, and C are constant. (If frequencies in the response signals are not present in the source signals, there is a nonlinear element in the circuit.)

3. The maximum amplitude of the steady-state response, in general, differs from the maximum amplitude of the source. For the circuit being discussed, the maximum amplitude of the response signal is $V_m / \sqrt{R^2 + \omega^2 L^2}$, and the maximum amplitude of the signal source is V_m.

4. The phase angle of the response signal, in general, differs from the phase angle of the source. For the circuit being discussed, the phase angle of the current is $\phi - \theta$ and that of the voltage source is ϕ.

These characteristics are important because they help you understand the motivation for the phasor method, which we begin developing in Section 10.3. In particular, note that once the decision has been made to find only the steady-state response, the task is reduced to finding the maximum amplitude and phase angle of the response signal. The waveform and frequency of the response are already known.

DRILL EXERCISES

10.3 The voltage applied to the circuit shown in Fig. 10.4 at $t = 0$ is $100 \cos (400t + 60°)$. The circuit resistance is 40 Ω, and the initial current in the 75-mH inductor is zero.

a) Find $i(t)$ for $t \geq 0$.

b) Write the expressions for the transient and steady-state components of $i(t)$.

c) Find the numerical value of i 1.875 ms after the switch has been closed.

d) What are the maximum amplitude, frequency (in radians per second), and phase angle of the steady-state current?

e) By how many degrees are the voltage and the steady-state current out of phase?

ANSWER: (a) $-1.84e^{-533.33t} + 2 \cos (400t + 23.13°)$ A; (b) $-1.84e^{-533.33t}$ A, $2 \cos (400t + 23.13)$ A; (c) 133.61 mA; (d) 2 A, 400 rad/s, 23.13°; (e) 36.87°.

10.3 THE PHASOR

The *phasor* is a complex number[†] that carries the amplitude and phase angle information of a sinusoidal function. The phasor concept is rooted in Euler's identity, which relates the exponential function to the trigonometric function

$$e^{\pm j\theta} = \cos \theta \pm j \sin \theta. \qquad (10.9)$$

Equation (10.9) is important here because it gives us another way of expressing the cosine and sine functions.[‡] We can think of the cosine function as the real part of the exponential function and the sine function as the imaginary part of the exponential function; that is,

$$\cos \theta = \mathscr{R}e\{e^{j\theta}\} \qquad (10.10)$$

and

$$\sin \theta = \mathscr{I}m\{e^{j\theta}\}, \qquad (10.11)$$

where $\mathscr{R}e$ represents *the real part of* and $\mathscr{I}m$ represents *the imaginary part of*.

Because we have already chosen to use the cosine function in analyzing the sinusoidal steady state (Section 10.1), we can apply Eq. (10.10) directly. In particular, we write the sinusoidal

---

[†] If you feel a bit uneasy about complex numbers, pause here and peruse Appendix B.

[‡] The letter i is used for current in electrical engineering literature, so the letter j has been adopted to signify the square root of -1; that is, $j = \sqrt{-1}$.

voltage function given by Eq. (10.1) in the form suggested by Eq. (10.10):

$$v = V_m \cos (\omega t + \phi)$$
$$= V_m \mathcal{R}e \{e^{j(\omega t + \phi)}\}$$
$$= V_m \mathcal{R}e \{e^{j\omega t} e^{j\phi}\}. \qquad (10.12)$$

We can move the coefficient V_m inside the argument of the real part of the function without altering the result. We can also reverse the order of the two exponential functions inside the argument and write Eq. (10.12) as

$$v = \mathcal{R}e \{V_m e^{j\phi} e^{j\omega t}\}. \qquad (10.13)$$

In Eq. (10.13) note that the coefficient of the exponential $e^{j\omega t}$ is a complex number that carries the amplitude and phase angle of the given sinusoidal function. This complex number is by definition the *phasor representation*, or *phasor transform*, of the given sinusoidal function. Thus

$$\mathbf{V} = V_m e^{j\phi} = \mathcal{P}\{V_m \cos (\omega t + \phi)\}, \qquad (10.14)$$

where the notation $\mathcal{P}\{V_m \cos (\omega t + \phi)\}$ is read "the phasor transform of $V_m \cos (\omega t + \phi)$." Thus the phasor transform transfers the sinusoidal function from the time domain to the complex-number domain. As in Eq. (10.14), we represent a phasor quantity throughout the book by a boldface letter.

Equation (10.14) is the polar form of a phasor, but we also can express a phasor in rectangular form. Thus we rewrite Eq. (10.14) as

$$\mathbf{V} = V_m \cos \phi + jV_m \sin \phi. \qquad (10.15)$$

Both polar and rectangular forms are useful in circuit applications of the phasor concept.

One additional comment regarding Eq. (10.14) is in order. The frequent occurrence of the exponential function $e^{j\phi}$ has led to a shorthand abbreviation that lends itself to text material. This abbreviation is the angle notation

$$1\underline{/\phi} \equiv 1e^{j\phi}.$$

We use this notation extensively in the material that follows.

INVERSE PHASOR TRANSFORM

So far we have emphasized moving from the sinusoidal function to its phasor transform. However, we may also reverse the process. That is, for a phasor we may write the expression for the sinusoidal function. Thus for $\mathbf{V} = 100\underline{/-26°}$, the expression for

v is $100 \cos (\omega t - 26°)$ because we have decided to use the cosine function for all sinusoids. Observe that we cannot deduce the value of ω from the phasor. *The phasor carries only amplitude and phase information.* The step of going from the phasor transform to the time-domain expression is referred to as *finding the inverse phasor transform* and is formalized by the equation

$$\mathcal{P}^{-1}\{V_m e^{j\phi}\} = \mathcal{R}e\{V_m e^{j\phi} e^{j\omega t}\}, \qquad (10.16)$$

where the notation $\mathcal{P}^{-1}\{V_m e^{j\phi}\}$ is read as "the inverse phasor transform of $V_m e^{j\phi}$." Equation (10.16) indicates that to find the inverse phasor transform we multiply the phasor by $e^{j\omega t}$ and then extract the real part of the product.

The phasor transform is useful in circuit analysis because it reduces the task of finding the maximum amplitude and phase angle of the steady-state sinusoidal response to the algebra of complex numbers. The following observations verify this conclusion.

1. The transient component vanishes as time elapses, so the steady-state component of the solution must also satisfy the differential equation. (See Problem 10.5b.)

2. In a linear circuit driven by sinusoidal sources, the steady-state response also is sinusoidal.

3. Using the notation introduced in Eq. (10.10), we can postulate that the steady-state solution is of the form $\mathcal{R}e\{Ae^{j\beta} e^{j\omega t}\}$, where A is the maximum amplitude of the response and β is the phase angle of the response.

4. When we substitute the postulated steady-state solution into the differential equation, the exponential term $e^{j\omega t}$ cancels out, leaving the solution for A and β in the domain of complex numbers.

We illustrate these observations with the circuit shown in Fig. 10.4. We know that the steady-state solution for the current i is of the form

$$i_{\text{ss}}(t) = \mathcal{R}e\{I_m e^{j\beta} e^{j\omega t}\}, \qquad (10.17)$$

where the subscript "ss" emphasizes that we are dealing with the steady-state solution. When we substitute Eq. (10.17) into Eq. (10.7), we generate the expression

$$\mathcal{R}e\{j\omega L I_m e^{j\beta} e^{j\omega t}\} + \mathcal{R}e\{R I_m e^{j\beta} e^{j\omega t}\} = \mathcal{R}e\{V_m e^{j\phi} e^{j\omega t}\}. \quad (10.18)$$

In deriving Eq. (10.18) we recognized that both differentiation and multiplication by a constant can be taken inside the real part of an operation. We also rewrote the right-hand side of Eq.

(10.7), using the notation of Eq. (10.10). The sum of the real parts is the same as the real part of the sum. Therefore we may reduce the left-hand side of Eq. (10.18) to a single term:

$$\mathcal{R}e\{(j\omega L + R)I_m e^{j\beta} e^{j\omega t}\} = \mathcal{R}e\{V_m e^{j\phi} e^{j\omega t}\}. \qquad \textbf{(10.19)}$$

From Eq. (10.19),

$$(j\omega L + R)I_m e^{j\beta} = V_m e^{j\phi},$$

or

$$I_m e^{j\beta} = \frac{V_m e^{j\phi}}{R + j\omega L}. \qquad \textbf{(10.20)}$$

Note that $e^{j\omega t}$ has been eliminated from determination of the amplitude (I_m) and phase angle (β) of the response. Thus for this circuit the task of finding I_m and β involves the algebraic manipulation of the complex quantities $V_m e^{j\phi}$ and $R + j\omega L$. Note that we encountered both polar and rectangular forms.

The phasor transform also is useful in circuit analysis because it applies directly to the sum of sinusoidal functions. Circuit analysis involves summing currents and voltages, so the importance of this observation is obvious. We can formalize this property as follows. If

$$v = v_1 + v_2 + \cdots + v_n, \qquad \textbf{(10.21)}$$

where all the voltages on the right-hand side are sinusoidal voltages of the same frequency, then

$$\mathbf{V} = \mathbf{V}_1 + \mathbf{V}_2 + \cdots + \mathbf{V}_n. \qquad \textbf{(10.22)}$$

Thus the phasor representation of the sum is the sum of the phasors of the individual terms. We discuss development of Eq. (10.22) in Section 10.5.

Before applying the phasor transform to circuit analysis, we illustrate its usefulness in solving a problem with which you are already familiar: adding sinusoidal functions via trigonometric identities. Example 10.5 shows how the phasor transform greatly simplifies this type of problem.

E X A M P L E   10.5

If $y_1 = 20 \cos(\omega t - 30°)$ and $y_2 = 40 \cos(\omega t + 60°)$, express $y = y_1 + y_2$ as a single sinusoidal function.

a) Solve by using trigonometric identities.

b) Solve by using the phasor concept.

S O L U T I O N

a) First we expand both y_1 and y_2, using the cosine of the sum of two angles, to get

$$y_1 = 20 \cos \omega t \cos 30° + 20 \sin \omega t \sin 30°,$$

$$y_2 = 40 \cos \omega t \cos 60° - 40 \sin \omega t \sin 60°.$$

Adding y_1 and y_2, we obtain

$$y = (20 \cos 30 + 40 \cos 60) \cos \omega t$$
$$+ (20 \sin 30 - 40 \sin 60) \sin \omega t$$
$$= 37.32 \cos \omega t - 24.64 \sin \omega t.$$

To combine these two terms we treat the coefficients of the cosine and sine as sides of a right triangle (Fig. 10.5) and then multiply and divide the right-hand side by the hypotenuse. Our expression for y becomes

$$y = 44.72\left(\frac{37.42}{44.72} \cos \omega t - \frac{24.64}{44.72} \sin \omega t\right)$$

$$= 44.72(\cos 33.43° \cos \omega t - \sin 33.43° \sin \omega t).$$

Again, we invoke the identity involving the cosine of the sum of two angles and write

$$y = 44.72 \cos (\omega t + 33.43°).$$

b) We can solve the problem by using phasors as follows. As

$$y = y_1 + y_2,$$

then, from Eq. (10.22),

$$\mathbf{Y} = \mathbf{Y_1} + \mathbf{Y_2}$$
$$= 20\underline{/-30°} + 40\underline{/60°}$$
$$= (17.32 - j10) + (20 + j34.64)$$
$$= 37.32 + j24.64$$
$$= 44.72\underline{/33.43°}.$$

Once we know the phasor $\mathbf{Y}$, we can write the corresponding trigonometric function for y by taking the inverse phasor transform:

$$y = \mathcal{P}^{-1}\{44.72e^{j33.43}\} = \mathcal{R}e\,\{44.72e^{j33.43}\,e^{j\omega t}\}$$

$$= 44.72 \cos (\omega t + 33.43°).$$

FIGURE 10.5 A right triangle used in the solution for y.

The superiority of the phasor approach to adding sinusoidal functions should be apparent. Note that it requires the ability to move back and forth between the polar and rectangular forms of complex numbers.

DRILL EXERCISES

10.4 Find the phasor transform of each trigonometric function:

a) $v = 170 \cos(377t - 40°)$ V;

b) $i = 10 \sin(1000t + 20°)$ A;

c) $i = [5 \cos(\omega t + 36.87°)$
$\quad\quad + 10 \cos(\omega t - 53.13°)]$ A;

d) $v = [300 \cos(20{,}000\pi t + 45°)$
$\quad\quad - 100 \sin(20{,}000\pi t + 30°)]$ mV.

ANSWER: (a) $\mathbf{V} = 170\underline{/-40°}$ V; (b) $\mathbf{I} = 10\underline{/-70°}$ A; (c) $\mathbf{I} = 11.18\underline{/-26.57°}$ A; (d) $\mathbf{V} = 339.90\underline{/61.51°}$ mV.

10.5 Find the time-domain expression corresponding to each phasor:

a) $\mathbf{V} = 86.3\underline{/+26°}$ V;

b) $\mathbf{I} = (10\underline{/30°} + 25\underline{/60°})$ mA;

c) $\mathbf{V} = (60 + j30 + 100\underline{/-28°})$ V.

ANSWER: (a) $v = 86.3 \cos(\omega t + 26°)$ V; (b) $i = 34.03 \cos(\omega t + 51.55°)$ mA; (c) $v = 149.26 \cos(\omega t - 6.52°)$ V.

10.4 THE PASSIVE CIRCUIT ELEMENTS IN THE PHASOR DOMAIN

The systematic application of the phasor transform in circuit analysis requires two steps. First, we must establish the relationship between the phasor current and the phasor voltage at the terminals of the passive circuit elements. Second, we must develop the phasor-domain version of Kirchhoff's laws, which we discuss in Section 10.5. In this section we establish the relationship between the phasor current and the phasor voltage at the terminals of the resistor, inductor, and capacitor. We begin with the resistor and use the passive sign convention in all the derivations.

V–I RELATIONSHIPS FOR A RESISTOR

From Ohm's law, if the current in a resistor varies sinusoidally with time, that is, if $i = I_m \cos(\omega t + \theta_i)$, the voltage at the terminals of the resistor, as shown in Fig. 10.6 is

$$v = R[I_m \cos(\omega t + \theta_i)]$$
$$= RI_m[\cos(\omega t + \theta_i)], \qquad (10.23)$$

FIGURE 10.6 A resistive element carrying a sinusoidal current.

where I_m is the maximum amplitude of the current in amperes and θ_i is the phase angle of the current.

The phasor transform of this voltage is

$$\mathbf{V} = RI_m e^{j\theta_i} = RI_m\underline{/\theta_i}. \qquad (10.24)$$

But $I_m\underline{/\theta_i}$ is the phasor representation of the sinusoidal current, so we can write Eq. (10.24) as

$$\mathbf{V} = R\mathbf{I}, \qquad (10.25)$$

which states that the phasor voltage at the terminals of a resistor is simply the resistance times the phasor current. Figure 10.7 shows the circuit diagram for a resistor in the phasor domain.

FIGURE 10.7 The phasor-domain equivalent circuit of a resistor.

Equations (10.23) and (10.25) both contain another important piece of information—namely, that at the terminals of a resistor there is no phase shift between the current and voltage. Figure 10.8 depicts this phase relationship. The signals are said to be in time phase because they both reach corresponding values on their respective curves at the same time (for example, they are at their positive maxima at the same instant).

V–I RELATIONSHIP FOR AN INDUCTOR

We derive the relationship between the phasor current and phasor voltage at the terminals of the inductor by assuming a sinusoidal current and using $L \, di/dt$ to establish the corresponding voltage. Thus, for $i = I_m \cos(\omega t + \theta_i)$, the expression for the voltage is

$$v = L\frac{di}{dt} = -\omega L I_m \sin(\omega t + \theta_i). \qquad (10.26)$$

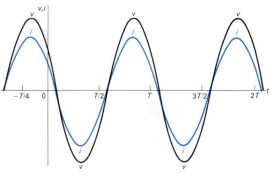

FIGURE 10.8 A plot showing that the voltage and current at the terminals of a resistor are in phase ($\theta_i = 60°$).

We now rewrite Eq. (10.26) using the cosine function:

$$v = -\omega L I_m \cos(\omega t + \theta_i - 90°). \qquad (10.27)$$

The phasor representation of the voltage given by Eq. (10.27) is

$$\mathbf{V} = -\omega L I_m e^{j(\theta_i - 90°)}$$
$$= -\omega L I_m e^{j\theta_i} e^{-j90°}$$
$$= j\omega L I_m e^{j\theta_i}$$
$$= j\omega L \mathbf{I}. \qquad (10.28)$$

FIGURE 10.9 The phasor-domain equivalent circuit for an inductor.

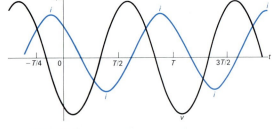

FIGURE 10.10 A plot showing the phase relationship between the current and voltage at the terminals of an inductor ($\theta_i = 60°$).

Note that in deriving Eq. (10.28) we used the identity

$$e^{-j90°} = \cos 90° - j \sin 90° = -j.$$

Equation (10.28) states that the phasor voltage at the terminals of an inductor equals $j\omega L$ times the phasor current. Figure 10.9 shows the phasor-domain equivalent circuit for the inductor.

Equation (10.28) contains the phasor relationship between the current and voltage at the terminals of an inductor. This relationship becomes obvious when we rewrite Eq. (10.28) as

$$\mathbf{V} = (\omega L \underline{/90°})I_m\underline{/\theta_i}$$
$$= \omega L I_m\underline{/\theta_i + 90°}, \qquad \textbf{(10.29)}$$

which indicates that the voltage and current are out of phase by exactly 90°. In particular, the voltage leads the current by 90° or, equivalently, the current lags behind the voltage by 90°. Figure 10.10 illustrates this concept of *voltage leading current* or *current lagging voltage*. Note that as time increases from zero the voltage comes to a particular value on the sine function ahead of the current. For example, it reaches its negative peak exactly 90° before the current reaches its negative peak. The same observation can be made with respect to the *zero-going-positive crossing* or the positive peak. Thus we say that the voltage leads the current by 90° or, equivalently, the current lags behind the voltage by 90°.

We can also express the phase shift in seconds. A phase shift of 90° corresponds to one-fourth of a period; hence the voltage leads the current by $T/4$, or $1/4f$ second.

V–I RELATIONSHIP FOR A CAPACITOR

We obtain the relationship between the phasor current and phasor voltage at the terminals of a capacitor from the derivation of Eq. (10.28). That is, if we note that for a capacitor

$$i = C\frac{dv}{dt}$$

and assume that

$$v = V_m \cos (\omega t + \theta_v),$$

then

$$\mathbf{I} = j\omega C \mathbf{V}. \qquad \textbf{(10.30)}$$

Now if we solve Eq. (10.30) for the voltage as a function of the

current we get

$$V = \left(\frac{1}{j\omega C}\right)I. \qquad (10.31)$$

Equation (10.31) demonstrates that the equivalent circuit for the capacitor in the phasor domain is as shown in Fig. 10.11.

The voltage across the terminals of a capacitor *lags* behind the capacitor current by exactly 90°. We can easily show this relationship by rewriting Eq. (10.31) as

$$V = \left(\frac{1}{\omega C}\right)\underline{/-90°}I_m\underline{/\theta_i}$$

$$= \frac{I_m}{\omega C}\underline{/\theta_i - 90°}. \qquad (10.32)$$

FIGURE 10.11 The phasor-domain equivalent circuit for a capacitor.

The alternative way to express the phase relationship contained in Eq. (10.32) is to say that the current *leads* the voltage by 90°. Figure 10.12 shows the phase relationship between the current and voltage at the terminals of a capacitor.

We conclude this discussion of passive circuit elements in the phasor domain with an important observation. When we compare Eqs. (10.25), (10.28), and (10.31), we note that they are all of the form

$$V = ZI, \qquad (10.33)$$

where Z represents the *impedance* of the circuit element. Thus the impedance of a resistor is R, the impedance of an inductor is $j\omega L$, and the impedance of a capacitor is $1/j\omega C$. In all cases, impedance is measured in ohms. The concept of impedance is crucial in sinusoidal steady-state analysis, and we have much more to say about its usefulness in subsequent sections.

We also introduce a couple of additional terms associated with the impedance of the passive circuit elements at this time. The term *inductive reactance* describes the product ωL (Ω). The reactance of an inductor is ωL (Ω), whereas the impedance of an inductor is $j\omega L$ (Ω). The term *capacitive reactance* describes the quantity $-1/\omega C$. The reactance of a capacitor is $-1/\omega C$ (Ω), whereas the impedance of a capacitor is $1/j\omega C$ or $j(-1/\omega C)(\Omega)$.

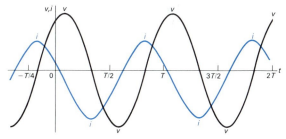

FIGURE 10.12 A plot showing the phase relationship between the current and voltage at the terminals of a capacitor ($\theta_i = 60°$).

One final comment about impedance. Note that, although impedance is a complex number, it is *not* a phasor. Although phasors are complex numbers, not all complex numbers are phasors.

And finally a reminder. If the reference direction for the current in a passive circuit element is in the direction of the voltage rise across the element, you must insert a minus sign into the equation that relates the voltage to the current.

DRILL EXERCISES

10.6 The current in the 75-mH inductor is 4 cos $(40,000t - 38°)$ mA. Calculate (a) the inductive reactance; (b) the impedance of the inductor; (c) the phasor voltage **V**; and (d) the steady-state expression for $v(t)$.

ANSWER: (a) 3000 Ω; (b) $j3000$ Ω; (c) $12\underline{/52°}$ V; (d) 12 cos $(40,000t + 52°)$ V.

10.7 The voltage across the terminals of the 0.2-μF capacitor is 40 cos $(10^5t - 50°)$ V. Calculate (a) the capacitive reactance; (b) the impedance of the capacitor; (c) the phasor current **I**; and (d) the steady-state expression for $i(t)$.

ANSWER: (a) -50 Ω; (b) $-j50$ Ω; (c) $0.8\underline{/40°}$ A; (d) 0.8 cos $(10^5t + 40°)$ A.

10.5 KIRCHHOFF'S LAWS IN THE PHASOR DOMAIN

We pointed out in Section 10.3—with reference to Eqs. (10.21) and (10.22)—that the phasor transform is useful in circuit analysis because it applies to the sum of sinusoidal functions. We illustrated this usefulness in Example 10.5. We now formalize this observation by developing Kirchhoff's laws in the phasor domain.

KIRCHHOFF'S VOLTAGE LAW IN THE PHASOR DOMAIN

We begin by assuming that v_1-v_n represent voltages around a closed path in a circuit. We also assume that the circuit is operating in a sinusoidal steady state. Thus Kirchhoff's voltage law requires that

$$v_1 + v_2 + \cdots + v_n = 0, \qquad \textbf{(10.34)}$$

which in the sinusoidal steady state becomes

$$V_{m_1} \cos (\omega t + \theta_1) + V_{m_2} \cos (\omega t + \theta_2)$$
$$+ \cdots + V_{m_n} \cos (\omega t + \theta_n) = 0. \qquad \textbf{(10.35)}$$

We now use Euler's identity to write Eq. (10.35) as

$$\mathscr{Re}\{V_{m_1}e^{j\theta_1}e^{j\omega t}\} + \mathscr{Re}\{V_{m_2}e^{j\theta_2}e^{j\omega t}\}$$
$$+ \cdots + \mathscr{Re}\{V_{m_n}e^{j\theta_n}e^{j\omega t}\} = 0. \qquad \textbf{(10.36)}$$

From the algebra of complex numbers we know that the sum of the real parts is the same as the real part of the sum. Therefore we rewrite Eq. (10.36) as

$$\mathscr{Re}\{V_{m_1}e^{j\theta_1}e^{j\omega t} + V_{m_2}e^{j\theta_2}e^{j\omega t} + \cdots + V_{m_n}e^{j\theta_n}e^{j\omega t}\} = 0. \quad \textbf{(10.37)}$$

Factoring the term $e^{j\omega t}$ from each term yields

$$\mathscr{Re}\{(V_{m_1}e^{j\theta_1} + V_{m_2}e^{j\theta_2} + \cdots + V_{m_n}e^{j\theta_n})e^{j\omega t}\} = 0,$$

or

$$\mathscr{Re}\{(\mathbf{V}_1 + \mathbf{V}_2 + \cdots + \mathbf{V}_n)e^{j\omega t}\} = 0. \qquad \textbf{(10.38)}$$

But $e^{j\omega t} \neq 0$, so

$$\mathbf{V}_1 + \mathbf{V}_2 + \cdots + \mathbf{V}_n = 0, \qquad \textbf{(10.39)}$$

which is the statement of Kirchhoff's voltage law as it applies to phasor voltages. That is, Eq. (10.34) applies to a set of sinusoidal voltages in the time domain, and Eq. (10.39) is the equivalent statement in the phasor domain.

KIRCHHOFF'S CURRENT LAW IN THE PHASOR DOMAIN

A similar derivation applies to a set of sinusoidal currents. Thus if

$$i_1 + i_2 + \cdots + i_n = 0, \qquad \textbf{(10.40)}$$

then

$$\mathbf{I}_1 + \mathbf{I}_2 + \cdots + \mathbf{I}_n = 0, \qquad \textbf{(10.41)}$$

where $\mathbf{I}_1, \mathbf{I}_2, \ldots, \mathbf{I}_n$ are the phasor representations of the individual currents $i_1, i_2, \ldots, i_n$.

Equations (10.33), (10.39), and (10.41) form the basis for circuit analysis in the phasor domain. Note that Eq. (10.33) has the same algebraic form as Ohm's law and that Eqs. (10.39) and (10.41) state Kirchhoff's laws for phasor quantities. Therefore you may use all the techniques developed for analyzing resistive circuits to find phasor currents and voltages. That is, you need learn no new analytic techniques in order to analyze circuits in the phasor domain. The basic tools of

1. series–parallel simplifications,

2. Δ-to-Y transformations,

3. source transformations,

4. Thévenin–Norton equivalent circuits,

5. superposition,

6. node-voltage analysis, and

7. mesh-current analysis

all can be used to analyze circuits in the phasor domain. Learning phasor circuit analysis consists of two fundamental parts: (1) You must be able to construct the phasor-domain model of a circuit; and (2) you must be able to manipulate complex numbers and/or quantities algebraically to arrive at a solution. We illustrate these attributes of phasor analysis in the discussion that follows.

PSpice Use PSpice to perform circuit analysis in the phasor domain: Section 11.2

DRILL EXERCISES

10.8 Four branches terminate at a common node. The reference direction of each branch current i_1, i_2, i_3, and i_4 is toward the node. If $i_1 = 100 \cos (\omega t + 25°)$ A, $i_2 = 100 \cos (\omega t + 145°)$ A, and $i_3 = 100 \cos (\omega t - 95°)$ A, find i_4.

ANSWER: $i_4 = 0$.

10.6 SERIES, PARALLEL, AND DELTA-TO-WYE SIMPLIFICATIONS

The rules for combining impedances in series or parallel and for making delta-to-wye transformations are the same as those for resistors. The only difference is that combining impedances involves the algebraic manipulation of complex numbers.

COMBINING IMPEDANCES IN SERIES AND PARALLEL

Impedances in series can be combined into a single impedance by simply adding the individual impedances. The circuit shown in Fig. 10.13 defines the problem in general terms. The impedances Z_1, Z_2, . . . , Z_n are connected in series between terminals a, b. When impedances are in series, they carry the same phasor current **I**. From Eq. (10.33), the voltage drop across each

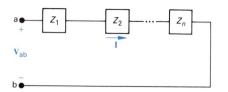

FIGURE 10.13 Impedances in series.

impedance is $Z_1\mathbf{I}, Z_2\mathbf{I}, \ldots, Z_n\mathbf{I}$ and from Kirchhoff's voltage law

$$\mathbf{V}_{ab} = Z_1\mathbf{I} + Z_2\mathbf{I} + \cdots + Z_n\mathbf{I}$$

$$= (Z_1 + Z_2 + \cdots + Z_n)\mathbf{I}. \qquad (10.42)$$

The equivalent impedance between terminals a, b is

$$Z_{ab} = \frac{\mathbf{V}_{ab}}{\mathbf{I}} = Z_1 + Z_2 + \cdots + Z_n. \qquad (10.43)$$

Example 10.6 illustrates a numerical application of Eq. (10.43).

EXAMPLE 10.6

A 90-Ω resistor, a 32-mH inductor, and a 5-μF capacitor are connected in series across the terminals of a sinusoidal voltage source, as shown in Fig. 10.14. The steady-state expression for the source voltage v_s is 750 cos $(5000t + 30°)$.

a) Construct the phasor-domain equivalent circuit.

b) Calculate the steady-state current i by the phasor method.

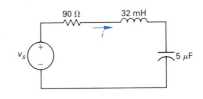

FIGURE 10.14 The circuit for Example 10.6.

SOLUTION

a) From the expression for v_s we have $\omega = 5000$ rad/s. Therefore the impedance of the 32-mH inductor is

$$Z_L = j\omega L = j(5000)(32 \times 10^{-3}) = j160 \ \Omega,$$

and the impedance of the capacitor is

$$Z_C = j\frac{-1}{\omega C} = -j\frac{10^6}{(5000)(5)} = -j40 \ \Omega.$$

The phasor transform of v_s is

$$\mathbf{V}_s = 750\underline{/30°} \text{ V}.$$

Figure 10.15 illustrates the phasor-domain equivalent circuit of the circuit shown in Fig. 10.14.

b) We compute the phasor current simply by dividing the voltage of the voltage source by the equivalent impedance between the terminals a, b. From Eq. (10.43),

$$Z_{ab} = 90 + j160 - j40$$

$$= 90 + j120 = 150\underline{/53.13°} \ \Omega.$$

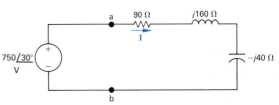

FIGURE 10.15 The phasor-domain equivalent circuit of the circuit shown in Fig. 10.14.

Thus

$$I = \frac{750\underline{/30°}}{150\underline{/53.13°}} = 5\underline{/-23.13°} \text{ A.}$$

We may now write the steady-state expression for i directly:

$$i = 5 \cos (5000t - 23.13°) \text{ A.}$$

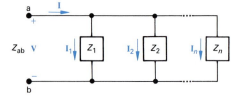

FIGURE 10.16 Impedances in parallel.

Impedances connected in parallel may be reduced to a single equivalent impedance by the reciprocal relationship

$$\frac{1}{Z_{ab}} = \frac{1}{Z_1} + \frac{1}{Z_2} + \cdots + \frac{1}{Z_n}. \tag{10.44}$$

Figure 10.16 depicts the parallel connection of impedances. Note that when impedances are in parallel they have the same voltage across their terminals. We derive Eq. (10.44) directly from Fig. 10.16 by simply combining Kirchhoff's current law with the phasor-domain version of Ohm's law, that is, Eq. (10.33). From Fig. 10.16,

$$\mathbf{I} = \mathbf{I}_1 + \mathbf{I}_2 + \cdots + \mathbf{I}_n$$

or

$$\frac{\mathbf{V}}{Z_{ab}} = \frac{\mathbf{V}}{Z_1} + \frac{\mathbf{V}}{Z_2} + \cdots + \frac{\mathbf{V}}{Z_n}. \tag{10.45}$$

Canceling the common voltage term out of Eq. (10.45) reveals Eq. (10.44).

We can also express Eq. (10.44) in terms of *admittance,* defined as the reciprocal of impedance and denoted Y. Thus

$$Y = \frac{1}{Z} = G + jB \qquad \text{(siemens).} \tag{10.46}$$

Like conductance, admittance is measured in siemens and commonly expressed in mhos.

Using Eq. (10.46) in Eq. (10.44), we get

$$Y_{ab} = Y_1 + Y_2 + \cdots + Y_n. \tag{10.47}$$

Also, from Eq. (10.44) for the special case of just two impedances in parallel,

$$Z_{ab} = \frac{Z_1 Z_2}{Z_1 + Z_2}. \tag{10.48}$$

The admittance of each of the ideal passive circuit elements also is worth noting. In particular, the admittance of a resistor is

$1/R$ $\mho$. We stated in our discussion of resistive circuits that the reciprocal of resistance is called *conductance*. The admittance of an inductor is $1/j\omega L$, or $j(-1/\omega L)$ $\mho$. The quantity $-1/\omega L$ is referred to as the *inductive susceptance* of the inductor. The admittance of a capacitor is $j\omega C$ $\mho$. The quantity ωC is referred to as the *capacitive susceptance* of the capacitor.

Example 10.7 illustrates the application of Eqs. (10.46) and (10.47) to a specific circuit.

EXAMPLE 10.7

The sinusoidal current source in the circuit shown in Fig. 10.17 produces the current $i_s = 8 \cos 200{,}000t$ A.

a) Construct the phasor-domain equivalent circuit.

b) Find the steady-state expressions for v, i_1, i_2, and i_3.

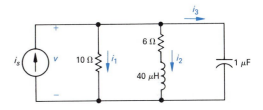

FIGURE 10.17 The circuit for Example 10.7.

SOLUTION

a) The phasor transform of the current source is $8\underline{/0°}$; the resistors transfer directly to the phasor domain as 10 and 6 Ω; the 40-μH inductor has an impedance of $j8$ Ω at the given frequency of 200,000 rad/s; and at this frequency the 1-μF capacitor has an impedance of $-j5$ Ω. Figure 10.18 shows the phasor-domain equivalent circuit and symbols representing the phasor transforms of the unknowns.

b) The circuit shown in Fig. 10.18 indicates that we can easily obtain the voltage across the current source once we know the equivalent impedance of the three parallel branches. Moreover, once we know **V**, we can calculate the three phasor currents $\mathbf{I}_1$, $\mathbf{I}_2$, and $\mathbf{I}_3$ by using Eq. (10.33). To find the equivalent impedance of the three branches we first find the equivalent admittance simply by adding the admittances of each branch. The admittance of the first branch is

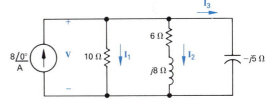

FIGURE 10.18 The phasor-domain equivalent circuit.

$$Y_1 = \frac{1}{10} = 0.1\,\mho;$$

the admittance of the second branch is

$$Y_2 = \frac{1}{6 + j8} = \frac{6 - j8}{100} = 0.06 - j0.08\,\mho;$$

and the admittance of the third branch is

$$Y_3 = \frac{1}{-j5} = j0.2\,\mho.$$

The admittance of the three branches is

$$Y = Y_1 + Y_2 + Y_3 = 0.16 + j0.12 = 0.2\underline{/36.87°} \; \text{℧}.$$

The impedance at the current source is

$$Z = \frac{1}{Y} = 5\underline{/-36.87°} \; \Omega.$$

The voltage **V** is

$$\mathbf{V} = Z\mathbf{I} = 40\underline{/-36.87°} \; V.$$

Hence

$$\mathbf{I}_1 = \frac{40\underline{/-36.87°}}{10} = 4\underline{/-36.87°} = 3.2 - j2.4 \; A;$$

$$\mathbf{I}_2 = \frac{40\underline{/-36.78°}}{6 + j8} = 4\underline{/-90°} = -j4 \; A,$$

and

$$\mathbf{I}_3 = \frac{40\underline{/-36.87°}}{5\underline{/-90°}} = 8\underline{/53.13°} = 4.8 + j6.4 \; A.$$

We check the computations at this point by verifying that

$$\mathbf{I}_1 + \mathbf{I}_2 + \mathbf{I}_3 = \mathbf{I}.$$

Specifically,

$$3.2 - j2.4 - j4 + 4.8 + j6.4 = 8 + j0.$$

The corresponding steady-state time domain expressions are

$$v = 40 \cos (200{,}000t - 36.87°) \; V;$$

$$i_1 = 4 \cos (200{,}000t - 36.87°) \; A;$$

$$i_2 = 4 \cos (200{,}000t - 90°) \; A;$$

$$i_3 = 8 \cos (200{,}000t + 53.13°) \; A.$$

DRILL EXERCISES

10.9 A 100-Ω resistor is connected in parallel with a 50-mH inductor. This parallel combination is connected in series with a 10-Ω resistor and a 10-μF capacitor.

a) Calculate the impedance of this interconnection if the frequency is 1 krad/s.

b) Repeat (a), for a frequency of 4 krad/s.

c) At what finite frequency does the impedance of the interconnection become purely resistive?

d) What is the impedance at the frequency found in (c)?

ANSWER: (a) $30 - j60 \; \Omega$; (b) $90 + j15 \; \Omega$; (c) 2 krad/s; (d) $60 \; \Omega$.

10.10 The interconnection described in Drill Exercise 10.9 is connected across the terminals of a voltage source that is generating $v = 300 \cos 2000t$ V. What is the maximum amplitude of the current in the 50-mH inductor?

ANSWER: 3.54 A.

10.11 Three branches having impedances of $3 + j4$ Ω, $16 - j12$ Ω and $-j4$ Ω, respectively, are connected in parallel. What are the equivalent (a) admittance, (b) conductance, and (c) susceptance of the parallel connection in millisiemens? (d) If the parallel branches are excited from a sinusoidal cur- rent source where $i = 8 \cos \omega t$ A, what is the maximum amplitude of the current in the purely capac- itive branch?

ANSWER: (a) $200\underline{/36.87°}$ mS; (b) 160 mS; (c) 120 mS; (d) 10 A.

10.12 Find the steady-state expression for v_o in the circuit shown if $i_g = 0.8 \cos 4000t$ A.

ANSWER: $v_o = 56 \cos 4000t$ V.

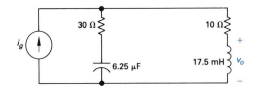

DELTA-TO-WYE TRANSFORMATIONS

The Δ-to-Y transformation that we discussed in Section 3.10 with regard to resistive circuits also applies to impedances. Fig- ure 10.19 defines the Δ-connected impedances along with the Y- equivalent circuit. The Y impedances as functions of the Δ impedances are

$$Z_1 = \frac{Z_b Z_c}{Z_a + Z_b + Z_c}; \qquad (10.49)$$

$$Z_2 = \frac{Z_c Z_a}{Z_a + Z_b + Z_c}; \qquad (10.50)$$

$$Z_3 = \frac{Z_a Z_b}{Z_a + Z_b + Z_c}. \qquad (10.51)$$

The Δ-to-Y transformation also may be reversed. That is, we can start with the Y structure and replace it with an equivalent Δ structure. The Δ impedances as functions of the Y impedances are

$$Z_a = \frac{Z_1 Z_2 + Z_2 Z_3 + Z_3 Z_1}{Z_1}; \qquad (10.52)$$

$$Z_b = \frac{Z_1 Z_2 + Z_2 Z_3 + Z_3 Z_1}{Z_2}; \qquad (10.53)$$

$$Z_c = \frac{Z_1 Z_2 + Z_2 Z_3 + Z_3 Z_1}{Z_3}. \qquad (10.54)$$

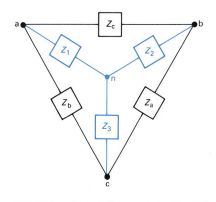

FIGURE 10.19 The delta-wye transformation.

The derivation of Eqs. (10.49)–(10.51) or Eqs. (10.52)–(10.54) follows the same thought process used to derive the corresponding equations for pure resistive circuits. In fact, comparing Eqs. (3.37)–(3.39) with Eqs. (10.49)–(10.51) and Eqs. (3.40)–(3.42) with Eqs. (10.52)–(10.54) reveals that the symbol Z has replaced the symbol R. You may want to review Problem 3.54 concerning the derivation of the Δ-to-Y transformation.

Example 10.8 illustrates the usefulness of the Δ-to-Y transformation in phasor circuit analysis.

E X A M P L E   10.8

Use a Δ-to-Y impedance transformation to find I_0, I_1, I_2, I_3, I_4, I_5, V_1, and V_2 in the circuit in Fig. 10.20.

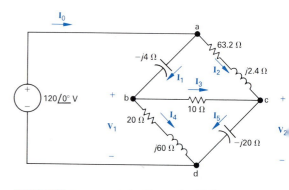

FIGURE 10.20 The circuit for Example 10.8.

S O L U T I O N

First note that the circuit is not amenable to series or parallel simplification as it now stands. A Δ-to-Y impedance transformation allows us to solve for all the branch currents without resorting to either the node-voltage method or the mesh-current method. If we replace either the upper delta (abc) or the lower delta (bcd) with its Y equivalent, we can further simplify the resulting circuit by series–parallel combinations. In deciding which delta to replace, the sum of the impedances around the delta is worth checking because this quantity forms the denominator for the equivalent Y impedances. The sum around the lower delta is $30 + j40$, so we choose to eliminate it from the circuit. The Y impedance connecting to terminal b is

$$Z_1 = \frac{(20 + j60)(10)}{30 + j40} = 12 + j4 \ \Omega;$$

the Y impedance connecting to terminal c is

$$Z_2 = \frac{10(-j20)}{30 + j40} = -3.2 - j2.4 \ \Omega;$$

and the Y impedance connecting to terminal d is

$$Z_3 = \frac{(20 + j60)(-j20)}{30 + j40} = 8 - j24 \ \Omega.$$

Inserting the Y-equivalent impedances into the circuit, we get the circuit shown in Fig. 10.21, which we can now simplify by series–parallel reductions. The impedance of the abn branch is

$$Z_{abn} = 12 + j4 - j4 = 12 \ \Omega,$$

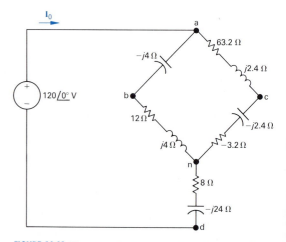

FIGURE 10.21 The circuit shown in Fig. 10.20, with the lower delta replaced by its equivalent wye.

and the impedance of the acn branch is

$$Z_{acn} = 63.2 + j2.4 - j2.4 - 3.2 = 60 \ \Omega.$$

Note that the abn branch is in parallel with the acn branch. Therefore we may replace these two branches with a single branch having an impedance of

$$Z_{an} = \frac{(60)(12)}{72} = 10 \ \Omega.$$

Combining this 10-Ω resistor with the impedance between n and d reduces the circuit shown in Fig. 10.21 to that shown in Fig. 10.22. From the circuit shown in Fig. 10.22,

$$\mathbf{I}_0 = \frac{120\underline{/0°}}{18 - j24} = 4\underline{/53.13°} = 2.4 + j3.2 \text{ A}.$$

Once we know $\mathbf{I}_0$, we can work back through the equivalent circuits to find the branch currents in the original circuit. We begin by noting that $\mathbf{I}_0$ is the current in the branch nd of Fig. 10.21. Therefore

$$\mathbf{V}_{nd} = (8 - j24)\mathbf{I}_0 = 96 - j32 \text{ V}.$$

We may now calculate the voltage $\mathbf{V}_{an}$ because

$$\mathbf{V} = \mathbf{V}_{an} + \mathbf{V}_{nd}$$

and both $\mathbf{V}$ and $\mathbf{V}_{nd}$ are known. Thus

$$\mathbf{V}_{an} = 120 - 96 + j32 = 24 + j32 \text{ V}.$$

We now compute the branch currents $\mathbf{I}_{abn}$ and $\mathbf{I}_{acn}$:

$$\mathbf{I}_{abn} = \frac{24 + j32}{12} = 2 + j\frac{8}{3} \text{ A};$$

$$\mathbf{I}_{acn} = \frac{24 + j32}{60} = \frac{4}{10} + j\frac{8}{15} \text{ A}.$$

In terms of the branch currents defined in Fig. 10.20,

$$\mathbf{I}_1 = \mathbf{I}_{abn} = 2 + j\frac{8}{3} \text{ A};$$

$$\mathbf{I}_2 = \mathbf{I}_{acn} = \frac{4}{10} + j\frac{8}{15} \text{ A}.$$

We check the calculations of $\mathbf{I}_1$ and $\mathbf{I}_2$ by noting that

$$\mathbf{I}_1 + \mathbf{I}_2 = 2.4 + j3.2$$

$$= \mathbf{I}_0.$$

To find the branch currents $\mathbf{I}_3$, $\mathbf{I}_4$, and $\mathbf{I}_5$ we must first calculate

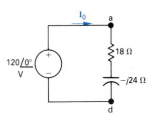

FIGURE 10.22 A simplified version of the circuit shown in Fig. 10.21.

the voltages $\mathbf{V}_1$ and $\mathbf{V}_2$. Referring to Fig. 10.20, we note that

$$\mathbf{V}_1 = 120\underline{/0°} - (-j4)\mathbf{I}_1 = \frac{328}{3} + j8 \text{ V}$$

and

$$\mathbf{V}_2 = 120\underline{/0°} - (63.2 + j2.4)\mathbf{I}_2 = 96 - j\frac{104}{3} \text{ V}.$$

We now calculate the branch currents $\mathbf{I}_3$, $\mathbf{I}_4$, and $\mathbf{I}_5$:

$$\mathbf{I}_3 = \frac{\mathbf{V}_1 - \mathbf{V}_2}{10} = \frac{4}{3} + j\frac{12.8}{3} \text{A};$$

$$\mathbf{I}_4 = \frac{\mathbf{V}_1}{20 + j60} = \frac{2}{3} - j1.6 \text{ A};$$

$$\mathbf{I}_5 = \frac{\mathbf{V}_2}{-j20} = \frac{26}{15} + j4.8 \text{ A}.$$

We check the calculations by noting that

$$\mathbf{I}_4 + \mathbf{I}_5 = \frac{2}{3} + \frac{26}{15} - j1.6 + j4.8 = 2.4 + j3.2 = \mathbf{I}_0;$$

$$\mathbf{I}_3 + \mathbf{I}_4 = \frac{4}{3} + \frac{2}{3} + j\frac{12.8}{3} - j1.6 = 2 + j\frac{8}{3} = \mathbf{I}_1;$$

$$\mathbf{I}_3 + \mathbf{I}_2 = \frac{4}{3} + \frac{4}{10} + j\frac{12.8}{3} + j\frac{8}{15} = \frac{26}{15} + j4.8 = \mathbf{I}_5.$$

DRILL EXERCISES

10.13 Use a Δ-to-Y transformation to find the current **I** in the circuit shown.

ANSWER: I = $4\underline{/+28.07°}$ A.

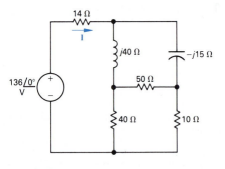

10.7 SOURCE TRANSFORMATIONS AND THÉVENIN–NORTON EQUIVALENT CIRCUITS

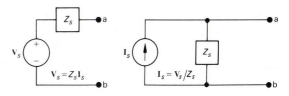

FIGURE 10.23 A source transformation in the phasor domain.

The source transformations introduced in Section 4.9 and the Thévenin–Norton equivalent circuits discussed in Section 4.10 are analytical techniques that also can be applied to phasor-domain circuits. We prove the validity of these techniques by following the same thought process used in Sections 4.9 and 4.10, except that we substitute impedance (Z) for resistance (R). Figure 10.23 shows the source-transformation equivalent circuit but with the nomenclature of the phasor domain.

Figure 10.24 illustrates the phasor-domain version of the Thévenin equivalent circuit. Figure 10.25 shows the Norton equivalent circuit. The techniques for finding the Thévenin equivalent voltage and impedance are identical to those used for resistive circuits, except that the phasor-domain equivalent circuit involves the manipulation of complex quantities. The same holds for finding the Norton equivalent current and impedance.

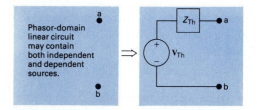

FIGURE 10.24 The phasor-domain version of the Thévenin equivalent.

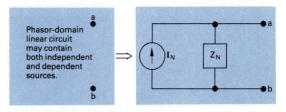

FIGURE 10.25 The phasor-domain version of the Norton equivalent circuit.

Example 10.9 illustrates the application of the source-transformation equivalent circuit to phasor-domain analysis. Example 10.10 illustrates the details of finding a Thévenin equivalent circuit in the phasor domain.

EXAMPLE 10.9

Use the concept of source transformation to find the phasor voltage $\mathbf{V}_0$ in the circuit shown in Fig. 10.26.

SOLUTION

We can replace the series combination of the voltage source ($40\underline{/0°}$) and the impedance of $1 + j3\ \Omega$ with the parallel combination of a current source and the $1 + j3$-Ω impedance. The source current is

$$\mathbf{I} = \frac{40}{1 + j3} = \frac{40}{10}(1 - j3) = 4 - j12 \text{ A.}$$

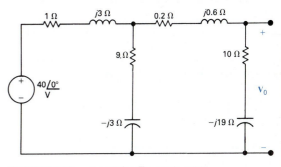

FIGURE 10.26 The circuit for Example 10.9.

Thus we can modify the circuit shown in Fig. 10.26 to that shown in Fig. 10.27. Note that the polarity reference of the 40-V source determines the reference direction for **I**.

Next, we combine the two parallel branches into a single impedance,

$$Z = \frac{(1 + j3)(9 - j3)}{10} = 1.8 + j2.4 \ \Omega,$$

which· is in parallel with the current source of $4 - j12$ A. Another source transformation converts this parallel combination to a series combination consisting of a voltage source in series with the impedance of $1.8 + j2.4 \ \Omega$. The voltage of the voltage source is

$$\mathbf{V} = (4 - j12)(1.8 + j2.4) = 36 - j12 \text{ V}.$$

Using this source transformation, we redraw the circuit shown in Fig. 10.27 as that shown in Fig. 10.28. Note the polarity of the voltage source. We added the current $\mathbf{I}_0$ to the circuit to expedite the solution for $\mathbf{V}_0$.

Note that we have reduced the circuit to a simple series circuit. We calculate the current $\mathbf{I}_0$ by dividing the voltage of the source by the total series impedance:

$$\mathbf{I}_0 = \frac{36 - j12}{12 - j16} = \frac{12(3 - j1)}{4(3 - j4)}$$

$$= \frac{39 + j27}{25} = 1.56 + j1.08 \text{ A}.$$

We now obtain the value of $\mathbf{V}_0$ by multiplying $\mathbf{I}_0$ by the impedance $10 - j19$:

$$\mathbf{V}_0 = (1.56 + j1.08)(10 - j19) = 36.12 - j18.84 \text{ V}.$$

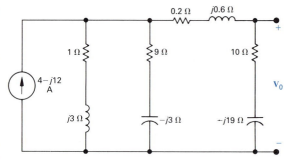

FIGURE 10.27 The first step in reducing the circuit shown in Fig. 10.26.

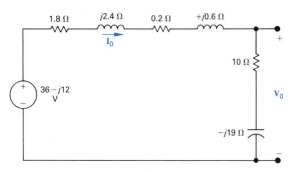

FIGURE 10.28 The second step in simplifying the circuit shown in Fig. 10.26.

E X A M P L E 10.10

Find the Thévenin equivalent circuit with respect to terminals a, b for the circuit shown in Fig. 10.29.

S O L U T I O N

We first determine the Thévenin equivalent voltage. This voltage is the open-circuit voltage appearing at terminals a, b. We choose the reference for the Thévenin voltage as positive at terminal a. We can make two source transformations relative to the 120-V, 12-Ω, and 60-Ω circuit elements in order to simplify this portion of the circuit. At the same time, these transformations must preserve the identity of the controlling voltage $\mathbf{V}_x$ because of the dependent voltage source. We determine the two source transformations by first replacing the series combination of the 120-V source and 12-Ω resistor with a 10-A current source in parallel with 12 Ω. Next, we replace the parallel combination of the 12- and 60-Ω resistors with a single 10-Ω resistor. Finally we replace the 10-A source in parallel with 10 Ω with a 100-V source in series with 10 Ω. Figure 10.30 shows the resulting circuit. We added the current $\mathbf{I}$ to the circuit shown in Fig. 10.30 to aid further discussion. Note that once we know the current $\mathbf{I}$ we can compute the Thévenin voltage. We find $\mathbf{I}$ by summing the voltages around the closed path in the circuit shown in Fig. 10.30. Hence

$$100 = 10\mathbf{I} - j40\mathbf{I} + 120\mathbf{I} + 10\mathbf{V}_x = (130 - j40)\mathbf{I} + 10\mathbf{V}_x.$$

We relate the controlling voltage $\mathbf{V}_x$ to the current $\mathbf{I}$ by noting from Fig. 10.30 that

$$\mathbf{V}_x = 100 - 10\mathbf{I}.$$

Then,

$$\mathbf{I} = \frac{-900}{30 - j40} = 18\underline{/-126.87°}\ \text{A}.$$

We now calculate $\mathbf{V}_x$:

$$\mathbf{V}_x = 100 - 180\underline{/-126.87°} = 208 + j144\ \text{V}.$$

Finally we note from Fig. 10.30 that

$$\mathbf{V}_{\text{Th}} = 10\mathbf{V}_x + 120\mathbf{I}$$

$$= 2080 + j1440 + 120(18)\underline{/-126.87°}$$

$$= 784 - j288 = 835.22\underline{/-20.17°}\ \text{V}.$$

To obtain the Thévenin impedance, we may use any of the techniques previously used to find the Thévenin resistance. We illustrate the test-source method in this example. Recall that in using the test-source method we deactivate all independent sources from the circuit and then apply either a test voltage source or a test current source to the terminals of interest. The ratio of the voltage to the current at the source is the Thévenin impedance. Figure 10.31 shows the result of applying this technique to the circuit shown in Fig. 10.29. Note that we chose a test voltage source $\mathbf{V}_T$. Also note that we deactivated the inde-

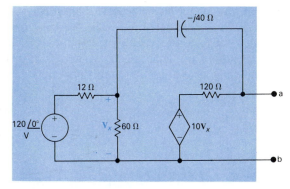

FIGURE 10.29 The circuit for Example 10.10.

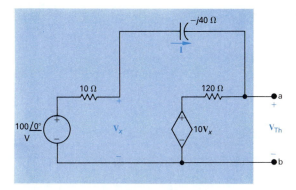

FIGURE 10.30 A simplified version of the circuit shown in Fig. 10.29.

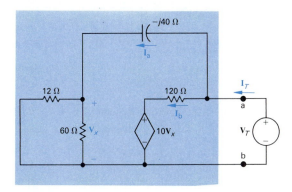

FIGURE 10.31 A circuit for calculating the Thévenin equivalent impedance.

pendent voltage source with an appropriate short circuit and pre-
served the identity of $\mathbf{V}_x$. The branch currents $\mathbf{I}_a$ and $\mathbf{I}_b$ have
been added to the circuit to simplify the calculation of $\mathbf{I}_T$. By
straightforward applications of Kirchhoff's circuit laws, you
should be able to verify the following relationships:

$$\mathbf{I}_a = \frac{\mathbf{V}_T}{10 - j40}; \qquad \mathbf{V}_x = 10\mathbf{I}_a;$$

$$\mathbf{I}_b = \frac{\mathbf{V}_T - 10\mathbf{V}_x}{120}$$

$$= \frac{-\mathbf{V}_T(9 + j4)}{120(1 - j4)};$$

$$\mathbf{I}_T = \mathbf{I}_a + \mathbf{I}_b$$

$$= \frac{\mathbf{V}_T}{10 - j40}\left(1 - \frac{9 + j4}{12}\right)$$

$$= \frac{\mathbf{V}_T(3 - j4)}{12(10 - j40)};$$

$$Z_{\text{Th}} = \frac{\mathbf{V}_T}{\mathbf{I}_T} = 91.2 - j38.4 \ \Omega.$$

Figure 10.32 depicts the Thévenin equivalent circuit.

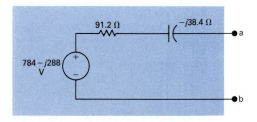

FIGURE 10.32 The Thévenin equivalent for the circuit shown in Fig. 10.29.

DRILL EXERCISES

10.14 Find the Thévenin equivalent with respect to terminals a, b in the circuit shown.

ANSWER: $\mathbf{V}_{\text{Th}} = \mathbf{V}_{ab} = 20/{-90°}$; $Z_{\text{Th}} = 2.5 - j2.5 \ \Omega$.

10.15 Find the Thévenin equivalent with respect to terminals a, b in the circuit shown.

ANSWER: $\mathbf{V}_{\text{Th}} = 10/{-36.87°}$ V; $Z_{\text{Th}} = 1.6 -$
$j1.2 \ \Omega$.

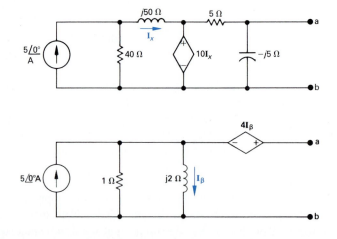

10.8 THE NODE-VOLTAGE METHOD

In Sections 4.2–4.4 we introduced the basic concepts of the node-voltage method of circuit analysis. The same concepts apply when we use the node-voltage method to analyze phasor-domain circuits. Here, we simply illustrate numerically in Example 10.11 the solution of a phasor-domain circuit by the node-voltage technique. Drill Exercise 10.16 and Problems 10.35–10.39 give you an opportunity to use the node-voltage method to solve for steady-state sinusoidal responses.

E X A M P L E   10.11

Use the node-voltage method to find the branch currents I_a, I_b, and I_c in the circuit shown in Fig. 10.33.

S O L U T I O N

We can describe the circuit shown in Fig. 10.33 in terms of two node voltages because the circuit contains three essential nodes. Four branches terminate at the essential node, which stretches across the bottom of Fig. 10.33, so we use it as the reference node. The remaining two essential nodes are labled 1 and 2, and the appropriate node voltages are designated V_1 and V_2. Figure 10.34 reflects the choice of reference node and the terminal labels. Summing the currents away from node 1 yields

$$-10.6 + \frac{V_1}{10} + \frac{V_1 - V_2}{1 + j2} = 0.$$

Multiplying by $1 + j2$ and collecting the coefficients of V_1 and V_2 generates the expression

$$V_1(1.1 + j0.2) - V_2 = 10.6 + j21.2.$$

Summing the currents away from node 2 gives

$$\frac{V_2 - V_1}{1 + j2} + \frac{V_2}{-j5} + \frac{V_2 - 20I_x}{5} = 0.$$

The controlling current I_x is

$$I_x = \frac{V_1 - V_2}{1 + j2}.$$

Substituting this expression for I_x into the node-2 equation, multiplying by $1 + j2$, and collecting coefficients of V_1 and V_2 pro-

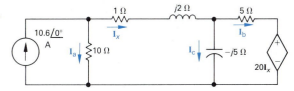

FIGURE 10.33 The circuit for Example 10.11.

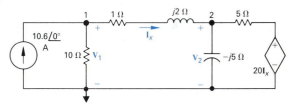

FIGURE 10.34 The circuit shown in Fig. 10.33, with the node voltages defined.

duces the equation

$$-5\mathbf{V}_1 + (4.8 + j0.6)\mathbf{V}_2 = 0.$$

The solutions for $\mathbf{V}_1$ and $\mathbf{V}_2$ are

$$\mathbf{V}_1 = 68.40 - j16.80 \text{ V},$$

and

$$\mathbf{V}_2 = 68 - j26 \text{ V}.$$

Hence the branch currents are

$$\mathbf{I}_a = \frac{\mathbf{V}_1}{10} = 6.84 - j1.68 \text{ A};$$

$$\mathbf{I}_x = \frac{\mathbf{V}_1 - \mathbf{V}_2}{1 + j2} = 3.76 + j1.68 \text{ A};$$

$$\mathbf{I}_b = \frac{\mathbf{V}_2 - 20\mathbf{I}_x}{5} = -1.44 - j11.92 \text{ A};$$

$$\mathbf{I}_c = \frac{\mathbf{V}_2}{-j5} = 5.2 + j13.6 \text{ A}.$$

To check our work we note that

$$\mathbf{I}_a + \mathbf{I}_x = 6.84 - j1.68 + 3.76 + j1.68$$

$$= 10.6 \text{ A}$$

and

$$\mathbf{I}_x = \mathbf{I}_b + \mathbf{I}_c = -1.44 - j11.92 + 5.2 + j13.6$$

$$= 3.76 + j1.68 \text{ A}.$$

DRILL EXERCISES

10.16 Use the node-voltage method to find the steady-state expression for $v(t)$ in the circuit shown. The sinusoidal sources are $i_s = 10 \cos \omega t$ A and $v_s = 100 \sin \omega t$ V, where $\omega = 50$ krad/s.

ANSWER: $v(t) = 31.62 \cos(50{,}000t - 71.57°)$ V.

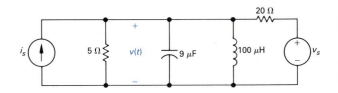

10.9 THE MESH-CURRENT METHOD

We can also use the mesh-current method to analyze phasor-domain circuits. The procedures used in phasor-domain applications of the method are the same as those used in analyzing resistive circuits. In Sections 4.5–4.7 we introduced the basic techniques of the mesh-current method. We demonstrate the extension of the mesh-current method to phasor-domain circuits in Example 10.12.

EXAMPLE 10.12

Use the mesh-current method of circuit analysis to find the voltages V_1, V_2, and V_3 in the circuit shown in Fig. 10.35.

SOLUTION

The circuit has two windows and a dependent voltage source, so we must write two mesh-current equations and a constraint equation. The reference direction for the mesh currents I_1 and I_2 is clockwise, as shown in Fig. 10.36. Once we know I_1 and I_2 we can easily find the unknown voltages. Summing the voltages around mesh 1 gives

$$150 = (1 + j2)I_1 + (12 - j16)(I_1 - I_2)$$

or

$$150 = (13 - j14)I_1 - (12 - j16)I_2.$$

Summing the voltages around mesh 2 generates the equation

$$0 = (12 - j16)(I_2 - I_1) + (1 + j3)I_2 + 39I_x.$$

Figure 10.36 reveals that the controlling current I_x is the difference between I_1 and I_2; that is, the constraint is

$$I_x = I_1 - I_2.$$

Substituting this constraint into the mesh-2 equation and simplifying the resulting expression gives

$$0 = (27 + j16)I_1 - (26 + j13)I_2.$$

Solving for I_1 and I_2 yields

$$I_1 = -26 - j52 \text{ A};$$

$$I_2 = -24 - j58 \text{ A};$$

$$I_x = -2 + j6 \text{ A}.$$

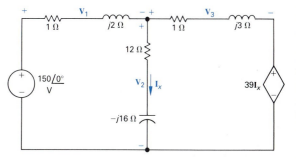

FIGURE 10.35 The circuit for Example 10.12.

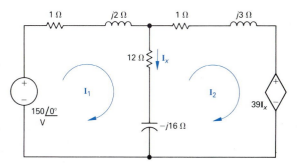

FIGURE 10.36 Mesh currents used to solve the circuit shown in Fig. 10.35.

The three voltages are

$$\mathbf{V}_1 = (1 + j2)\mathbf{I}_1 = 78 - j104 \text{ V};$$

$$\mathbf{V}_2 = (12 - j16)\mathbf{I}_x = 72 + j104 \text{ V};$$

$$\mathbf{V}_3 = (1 + j3)\mathbf{I}_2 = 150 - j130 \text{ V}:$$

$$39\mathbf{I}_x = -78 + j234 \text{ V}.$$

We check these calculations by summing the voltages around closed paths:

$$-150 + \mathbf{V}_1 + \mathbf{V}_2 = -150 + 78 - j104 + 72$$
$$+ j104 = 0;$$

$$-\mathbf{V}_2 + \mathbf{V}_3 + 39\mathbf{I}_x = -72 - j104 + 150 - j130$$
$$- 78 + j234 = 0;$$

$$-150 + \mathbf{V}_1 + \mathbf{V}_3 + 39\mathbf{I}_x = -150 + 78 + j104 + 150$$
$$- j130 - 78 + j234 = 0.$$

DRILL EXERCISES

10.17 Use the mesh-current method to find the phasor current **I** in the circuit shown.

ANSWER: I = 29 + j2 = 29.07$\underline{/3.95°}$ A.

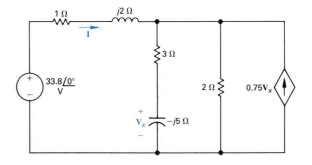

10.10 PHASOR DIAGRAMS

When we are using the phasor method to analyze the steady-state sinusoidal operation of a circuit, a diagram of the phasor currents and voltages may give further insight into the behavior of the circuit. A phasor diagram shows the magnitude and phase angle of each phasor quantity in the complex-number plane. Phase angles are measured counterclockwise from the positive real axis, and magnitudes are measured from the origin of the axes. For example, Fig. 10.37 shows the phasor quantities $10\underline{/30°}$, $12\underline{/150°}$, $5\underline{/-45°}$, and $8\underline{/-170°}$.

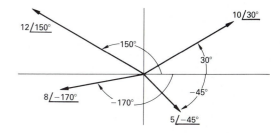

FIGURE 10.37 A graphic representation of phasors.

Constructing phasor diagrams of circuit quantities generally involves both currents and voltages. As a result, two different magnitude scales are necessary, one for currents and one for voltages. The ability to visualize a phasor quantity on the complex-number plane can be useful when you are checking pocket calculator calculations. The typical pocket calculator doesn't offer a printout of the data entered. Visualizing the quadrant in which the number lies is helpful when you are entering complex numbers. Then when the calculated angle is displayed, you can quickly check whether you keyed in the appropriate values. For example, suppose that you are to compute the polar form of $-7 - j3$. Without making any calculations you should anticipate a magnitude greater than 7 and an angle in the third quadrant that is more negative than $-135°$ or less positive than $225°$, as illustrated in Fig. 10.38.

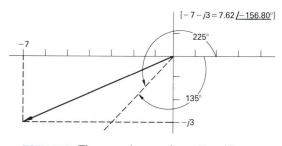

FIGURE 10.38 The complex number $-7 - j3$.

Construction and Use of Phasor Diagrams

To illustrate the usefulness of phasor diagrams in circuit analysis, we construct a phasor diagram for the circuit shown in Fig. 10.39. In this circuit, the parallel combination of R_2 and L_2 represents a load on the output end of a distribution line, which is modeled by the series combination of R_1 and L_1. The sinusoidal voltage source is modeled by v_s.

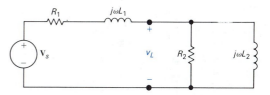

FIGURE 10.39 A circuit illustrating construction of a phasor diagram.

Our analytic task can be described as follows. We want to know the effect that the addition of a capacitor across the terminals of the load has on the amplitude of v_s if we adjust v_s so that the amplitude of v_L remains constant. Utility companies use this technique to control the voltage drop on their lines.

We begin by assuming zero capacitance. After constructing the phasor diagram for the zero-capacitance case, we can add the capacitor to the circuit and study its effect on the amplitude of v_s, holding the amplitude of v_L constant. Figure 10.40 shows the phasor-domain equivalent circuit of the circuit shown in Fig. 10.39. We added the phasor branch currents $\mathbf{I}$, $\mathbf{I}_a$, and $\mathbf{I}_b$ to Fig. 10.40 to aid discussion of the phasor diagram.

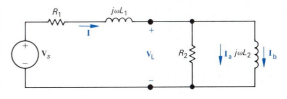

FIGURE 10.40 A phasor-domain equivalent circuit.

Figure 10.41 shows the stepwise evolution of the phasor diagram. Relating the phasor diagram to the circuit shown in Fig. 10.40 reveals the following points:

1. Because we are holding the amplitude of the load voltage constant, we chose $\mathbf{V}_L$ as our reference.

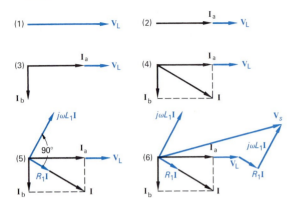

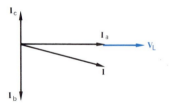

FIGURE 10.41 The step-by-step evolution of the phasor diagram for the circuit shown in Fig. 10.40.

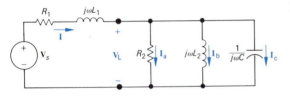

FIGURE 10.42 The addition of a capacitor to the circuit shown in Fig. 10.40.

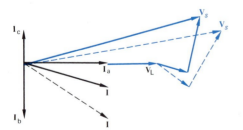

FIGURE 10.43 The effect of the capacitor current I_c on the line current I.

FIGURE 10.44 A phasor diagram showing the effect of adding a load-shunting capacitor to the circuit shown in Fig. 10.40 if V_L is held constant.

2. We know that $\mathbf{I}_a$ is in phase with $\mathbf{V}_L$ and that its magnitude is $|\mathbf{V}_L|/R_2$. (On the phasor diagram, the magnitude scale for the current phasors is independent of the magnitude scale for the voltage phasors.)

3. We know that $\mathbf{I}_b$ lags behind $\mathbf{V}_L$ by $90°$ and that its magnitude is $|\mathbf{V}_L|/\omega L_2$.

4. The line current $\mathbf{I}$ is equal to the sum of $\mathbf{I}_a$ and $\mathbf{I}_b$.

5. The voltage drop across R_1 is in phase with the line current, and the voltage drop across $j\omega L_1$ leads the line current by $90°$.

6. The source voltage is the sum of the load voltage and the drop along the line, that is, $\mathbf{V}_s = \mathbf{V}_L + (R_1 + j\omega L_1)\mathbf{I}$.

Note that the completed phasor diagram shown in step 6 of Fig. 10.41 clearly shows the amplitude and phase angle relationships among all the currents and voltages in the circuit shown in Fig. 10.40.

We now add the capacitor branch shown in Fig. 10.42. We are holding $\mathbf{V}_L$ constant, so we construct the phasor diagram for the circuit in Fig. 10.42 following the same steps as those in Fig. 10.41 except that, at step 4, we add the capacitor current $\mathbf{I}_c$ to the diagram. In so doing, we cause $\mathbf{I}_c$ to lead $\mathbf{V}_L$ by $90°$, with its magnitude being $|\mathbf{V}_L|\omega C$. Figure 10.43 shows the effect of $\mathbf{I}_c$ on the line current; that is, both the magnitude and phase angle of the line current $\mathbf{I}$ change with changes in the magnitude of $\mathbf{I}_c$. As the magnitude and phase angle of $\mathbf{I}$ change, so do the magnitude and phase angle of the voltage drop along the line. As the drop along the line changes, the magnitude and phase angle of $\mathbf{V}_s$ change. The phasor diagram shown in Fig. 10.44 depicts these observations. The dotted phasors represent the pertinent currents and voltages before addition of the capacitor. Thus comparing the dotted phasors of $\mathbf{I}$, $R_1\mathbf{I}$, $j\omega L_1\mathbf{I}$, and $\mathbf{V}_s$ with their solid counterparts clearly shows the effect of adding C to the circuit. In particular, note that by adding the capacitor across the load we reduce the amplitude of the source voltage and still maintain the amplitude of the load voltage. Practically, this result means that, as the load increases (that is, as $\mathbf{I}_a$ and $\mathbf{I}_b$ increase), we can add capacitors to the system (that is, increase $\mathbf{I}_c$) so that under heavy load conditions we can maintain $\mathbf{V}_L$ without increasing the amplitude of the source voltage.

We use phasor diagrams in subsequent chapters whenever such diagrams give additional insight into the steady-state sinusoidal operation of the circuit under investigation. Problem 10.60 shows how a phasor diagram can help explain the operation of a phase-shifting circuit.

DRILL EXERCISES

10.18 The parameters in the circuit shown in Fig. 10.40 are $R_1 = 0.1\ \Omega$, $\omega L_1 = 0.8\ \Omega$, $R_2 = 24\ \Omega$, $\omega L_2 = 32\ \Omega$, and $\mathbf{V}_L = 240 + j0$ V.

a) Calculate the phasor voltage $\mathbf{V}_s$.

b) Connect a capacitor in parallel with the inductor, hold $\mathbf{V}_L$ constant, and adjust the capacitor until the magnitude of $\mathbf{I}$ is a minimum. What is the capacitive reactance? What is the value of $\mathbf{V}_s$?

c) Find the value of the capacitive reactance that keeps the magnitude of $\mathbf{I}$ as small as possible and that at the same time makes

$$|\mathbf{V}_s| = |\mathbf{V}_L| = 240 \text{ V}.$$

ANSWER: (a) $247.11\underline{/1.68°}$ V; (b) $-32\ \Omega$, $241.13\underline{/1.90°}$ V; (c) $-26.90\ \Omega$.

SUMMARY

Finding the sinusoidal steady-state response of linear, lumped-parameter circuits is of great importance in electrical engineering. The best way to find the steady-state voltages and currents that exist in a circuit driven by sinusoidal sources is to transfer the analysis from the time domain to the phasor domain. We discussed the following important features of phasor-domain analysis:

• When making the transformation from the time domain to the phasor domain, you cannot be in both domains at the same time. Going back and forth between the domains is allowed, but straddling them is not allowed when you analyze a circuit.

• The frequency of any response signal is identical to the frequency of the excitation signal.

• All sinusoidal sources are expressed in terms of the cosine function and then transferred to the phasor domain via the phasor transform. That is

$$\mathcal{P}\{A \cos (\omega t + \theta)\} = Ae^{j\theta},$$

where A is the amplitude of the source and θ is the phase angle of the source.

• Each resistor appears in the phasor domain without modification.

• Each inductor appears in the phasor domain as an impedance $j\omega L$, where ωL is the inductive reactance.

- Each capacitor appears in the phasor domain as an impedance $j(-1/\omega C)$ where $(-1/\omega C)$ is the capacitive reactance.

- Kirchhoff's laws hold for phasor currents and voltages.

- The relationship between the phasor current and voltage at the terminals of an impedance Z is

$$\mathbf{V} = Z\mathbf{I},$$

where the reference direction for $\mathbf{I}$ is from the positive to the negative polarity reference for $\mathbf{V}$ (passive sign convention).

- The reciprocal of the impedance (Z) is the admittance (Y):

$$Y = \frac{1}{Z}.$$

- All the techniques used in dc circuit analysis (series–parallel and delta–wye simplifications, node voltages, mesh currents, source transformations, Thévenin/Norton equivalents, and superposition) may be used to analyze a phasor-domain circuit.

- Phasor-domain currents and voltages may be transferred to the time domain by means of the inverse phasor transform:

$$\mathscr{P}^{-1}\{Ae^{j\theta}\} = \mathscr{R}e\{Ae^{j\theta}e^{j\omega t}\}.$$

PROBLEMS

10.1 Consider the sinusoidal voltage

$$v = 100 \cos(400\pi t + 60°) \text{ V}.$$

a) What is the maximum amplitude of the voltage?

b) What is the frequency in hertz?

c) What is the frequency in radians per second?

d) What is the phase angle in radians?

e) What is the phase angle in degrees?

f) What is the period in milliseconds?

g) What is the first time after $t = 0$ that $v = 100$ V?

h) The sinusoidal function is shifted 5/12 ms to the left along the time axis. What is the expression for $v(t)$?

i) What is the minimum number of milliseconds that the function must be shifted to the right if the expression for $v(t)$ is $100 \cos 400\pi t$ V?

j) What is the minimum number of milliseconds that the function must be shifted to the left if the expression for $v(t)$ is $100 \sin 400\pi t$ V?

10.2 In a single graph, sketch $v = 60 \cos (\omega t + \phi)$ versus ωt for $\phi = -60°, -30°, 0°, +30°,$ and $60°$.

a) State whether the voltage function is shifting to the right or left as ϕ becomes more positive.

b) What is the direction of shift if ϕ changes from 0 to $-30°$?

10.3 At $t = -250$ μs a sinusoidal voltage is known to be zero and going positive. The voltage is next zero at $t = 750$ μs. It is also known that the voltage is 100 V at $t = 0$.

a) What is the frequency of v in hertz?

b) What is the expression for v?

10.4 A sinusoidal current is zero at $t = 15$ μs and increasing at a rate of $2 \times 10^5 \pi$ A/s. The maximum amplitude of the current is 10 A.

a) What is the frequency of i in radians per second?

b) What is the expression for i?

10.5 a) Verify that Eq. (10.8) is the solution of Eq. (10.7). This can be done by substituting Eq. (10.8) into the left-hand side of Eq. (10.7) and then noting that it equals the right-hand side for all values to $t > 0$. At $t = 0$, Eq. (10.8) should reduce to the initial value of the current.

b) Since the transient component vanishes as time elapses and since our solution must satisfy the differential equation for all values of t, the steady-state component, by itself, must also satisfy the differential equation. Verify this observation by showing that the steady-state component of Eq. (10.8) satisfies Eq. (10.7).

10.6 Use the concept of the phasor to combine the following sinusoidal functions into a single trigonometric expression:

a) $y = 100 \cos (500t + 30°)$
$+ 50 \cos (500t - 45°)$;

b) $y = 100 \sin (377t + 40°)$
$- 50 \cos (377t + 200°)$;

c) $y = 40 \cos (\omega t + 60°)$
$+ 80 \sin (\omega t + 135°)$
$- 100 \cos (\omega t + 270°)$.

10.7 Show that

$$\int_{t_o}^{t_o+T} V_m^2 \cos^2 (\omega t + \phi)dt = \frac{V_m^2 T}{2}.$$

10.8 The rms value of the sinusoidal voltage supplied to the convenience outlet of a US home is 120 V. What is the maximum value of the voltage at the outlet?

10.9 A 1000-Hz sinusoidal voltage with a maximum amplitude of 200 V at $t = 0$ is applied across the terminals of an inductor. The maximum amplitude of the steady-state current in the inductor is 25 A.

a) What is the frequency of the inductor current?

b) What is the phase angle of the voltage?

c) What is the phase angle of the current?

d) What is the inductive reactance of the inductor?

e) What is the inductance of the inductor in millihenrys?

f) What is the impedance of the inductor?

10.10 A 50-kHz sinusoidal voltage has zero phase angle and a maximum amplitude of 10 mV. When this voltage is applied across the terminals of a capacitor, the resulting steady-state current has a maximum amplitude of 628.32 μA.

a) What is the frequency of the current in radians per second?

b) What is the phase angle of the current?

c) What is the capacitive reactance of the capacitor?

d) What is the capacitance of the capacitor in microfarads?

e) What is the impedance of the capacitor?

10.11 A 10-Ω resistor, a 8-mH inductor, and a 2.5-μF capacitor are connected in series. The series-connected elements are energized by a sinusoidal voltage source whose voltage is 240 cos $(5000t - 40°)$ V.

a) Draw the phasor-domain equivalent circuit.

b) Reference the current in the direction of the voltage rise across the source and find the phasor current.

c) Find the steady-state expression for $i(t)$.

10.12 A 20-Ω resistor and a 1-μF capacitor are connected in parallel. This parallel combination is also in parallel with the series combination of a 1-Ω resistor and a 40-μH inductor. These three parallel branches are driven by a sinusoidal current source whose current is 20 cos $(50,000t - 20°)$ A.

a) Draw the phasor-domain equivalent circuit.

b) Reference the voltage across the current source as a rise in the direction of the source current and find the phasor voltage.

c) Find the steady-state expression for $v(t)$.

10.13 Find the steady-state expression for $i_o(t)$ in the circuit in Fig. P10.13 if $v_s = 100 \sin 50t$ mV.

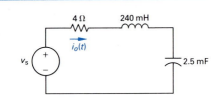

FIGURE P10.13

10.14 The circuit in Fig. P10.14 is operating in the sinusoidal steady state. Find the steady-state expression for $v_o(t)$ if $v_g = 10 \cos 100t$ V.

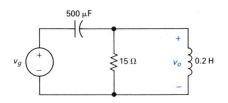

FIGURE P10.14

10.15 Find the impedance Z_{ab} in the circuit seen in Fig. P10.15. Express Z_{ab} in both polar and rectangular form.

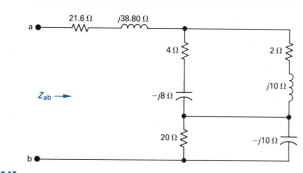

FIGURE P10.15

10.16 Find the admittance Y_{ab} in the circuit seen in Fig. P10.16. Express Y_{ab} in both polar and rectangular form. Give the value of Y_{ab} in millimhos.

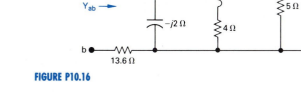

FIGURE P10.16

10.17 Find Z_{ab} in the circuit shown in Fig. P10.17 when the circuit is operating at a frequency of 20 krad/s.

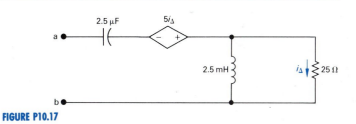

FIGURE P10.17

10.18 a) For the circuit shown in Fig. P10.18, find the frequency (in radians per second) at which the impedance Z_{ab} is purely resistive.

b) Find the value of Z_{ab} at the frequency of part (a).

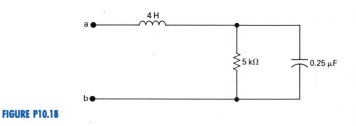

FIGURE P10.18

10.19 a) The source voltage in the circuit in Fig. P10.19 is $v_g = 200 \cos 500t$ V. Find the values of L such that i_g is in phase with v_g when the circuit is operating in the steady state.

b) For the values of L found in part (a) find the steady-state expressions for i_g.

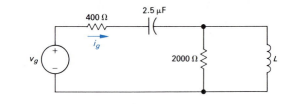

FIGURE P10.19

10.20 The circuit shown in Fig. P10.20 is operating in the sinusoidal steady state. The capacitor is adjusted until the current i_g is in phase with the sinusoidal voltage v_g.

a) Specify the values of capacitance in microfarads if $v_g = 250 \cos 1000t$ V.

b) Give the steady-state expressions for i_g when C has the values found in part (a).

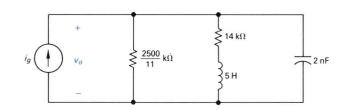

FIGURE P10.20

10.21 The frequency of the sinusoidal current source in the circuit in Fig. P10.21 is adjusted until v_o is in phase with i_g.

a) What is the value of ω in radians per second?

b) If $i_g = 0.25 \cos \omega t$ mA [where ω is the frequency found in part (a)] what is the steady-state expression for v_o?

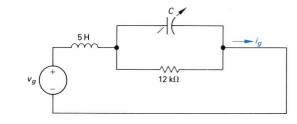

FIGURE P10.21

10.22 The frequency of the sinusoidal voltage source in the circuit in Fig. P10.22 is adjusted until the current i_o is in phase with v_g.

a) Find the frequency in hertz.

b) Find the steady-state expression for i_o [at the frequency found in part (a)] if $v_g = 80 \cos \omega t$ V.

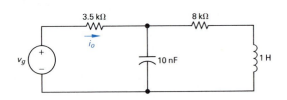

FIGURE P10.22

10.23 The circuit shown in Fig. P10.23 is operating in the sinusoidal steady state. Find the value of ω if

$$i_o = 0.10 \sin(\omega t + 173.13°) \text{ A}$$

and

$$v_g = 500 \cos(\omega t + 30°) \text{ V}.$$

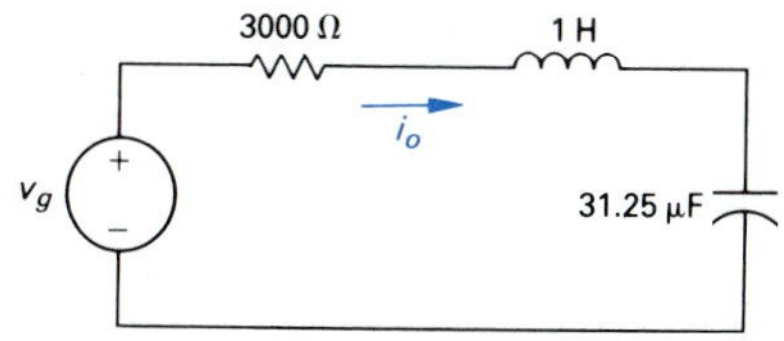

FIGURE P10.23

10.24 Find the steady-state expression for $v_o(t)$ in the circuit shown in Fig. P10.24 if $i_g = 10 \cos 8t$ A.

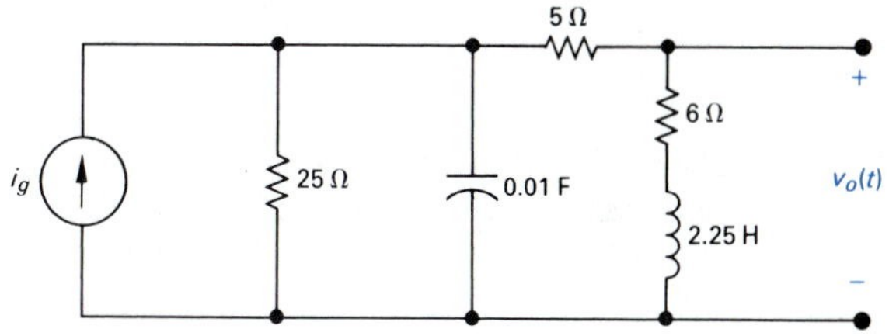

FIGURE P10.24

10.25 The current source in the circuit shown in Fig. P10.25 is generating a sinusoidal waveform such that $i_g = 20 \cos(40{,}000t - 73.74°)$ A. Find the steady-state expression for $v_o(t)$.

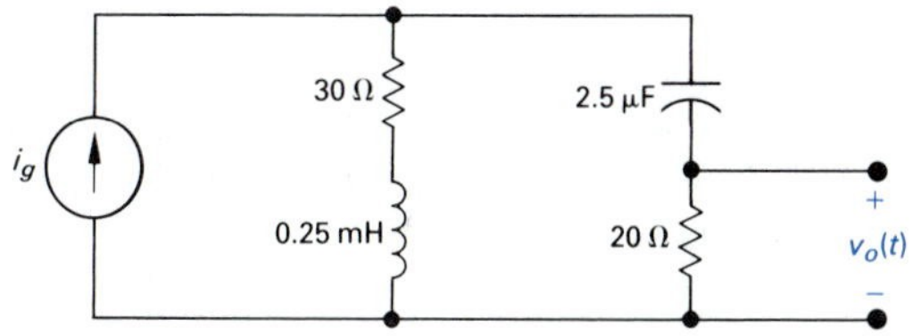

FIGURE P10.25

10.26 The expressions for the steady-state voltage and current at the terminals of the circuit seen in Fig. P10.26 are

$$v_g = 240 \sin(2000\pi t + 210°) \text{ V}$$

and

$$i_g = 12 \cos(2000\pi t + 84°) \text{ A}.$$

a) What is the impedance seen by the source?

b) By how many microseconds is the current out of phase with the voltage?

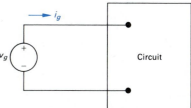

FIGURE P10.26

10.27 a) Show that at a given frequency ω, the circuits in Fig. P10.27(a) and (b) will have the same impedance between the terminals a, b if

$$R_1 = \frac{R_2}{1 + \omega^2 R_2^2 C_2^2} \quad \text{and} \quad C_1 = \frac{1 + \omega^2 R_2^2 C_2^2}{\omega^2 R_2^2 C_2}.$$

b) Find the values of resistance and capacitance that when connected in series will have the same impedance at 40 krad/s as that of a 1-kΩ resistor connected in parallel with a 0.05-μF capacitor.

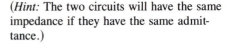

FIGURE P10.27

10.28 a) Show that at a given frequency ω, the circuits in Fig. P10.27(a) and (b) will have the same impedance between the terminals a, b if

$$R_2 = \frac{1 + \omega^2 R_1^2 C_1^2}{\omega^2 R_1 C_1^2} \quad \text{and} \quad C_2 = \frac{C_1}{1 + \omega^2 R_1^2 C_1^2}.$$

(*Hint:* The two circuits will have the same impedance if they have the same admittance.)

b) Find the values of resistance and capacitance that when connected in parallel will give the same impedance at 50 krad/s as that of a 1-kΩ resistor connected in series with a capacitance of 40 nF.

10.29 a) Show that at a given frequency ω, the circuits in Fig. P10.29(a) and (b) will have the same impedance between the terminals a, b if

$$R_1 = \frac{\omega^2 L_2^2 R_2}{R_2^2 + \omega^2 L_2^2} \quad \text{and} \quad L_1 = \frac{R_2^2 L_2}{R_2^2 + \omega^2 L_2^2}.$$

b) Find the values of resistance and inductance that when connected in series will have the same impedance at 8 krad/s as that of a 10-kΩ resistor connected in parallel with a 2.5-H inductor.

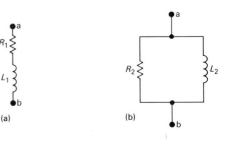

FIGURE P10.29

10.30 a) Show that at a given frequency ω, the circuits in Fig. P10.29(a) and (b) will have the same impedance between the terminals a, b if

$$R_2 = \frac{R_1^2 + \omega^2 L_1^2}{R_1} \quad \text{and} \quad L_2 = \frac{R_1^2 + \omega^2 L_1^2}{\omega^2 L_1}.$$

(*Hint:* The two circuits will have the same impedance if they have the same admittance.)

b) Find the values of resistance and inductance that when connected in parallel will have the same impedance at 4000 rad/s as a 5-kΩ resistor connected in series with a 0.25-H inductor.

10.31 The phasor current $\mathbf{I}_a$ in the circuit shown in Fig. P10.31 is $5\underline{/0°}$ A.

a) Find $\mathbf{I}_b$, $\mathbf{I}_c$, and $\mathbf{V}_g$.

b) If $\omega = 5$ krad/s, write the expressions for $i_b(t)$, $i_c(t)$, and $v_g(t)$.

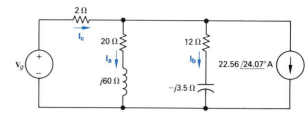

FIGURE P10.31

10.32 Find $\mathbf{I}_b$ and Z in the circuit shown in Fig. P10.32 if $\mathbf{V}_g = 25\underline{/0°}$ V and $\mathbf{I}_a = 5\underline{/90°}$ A.

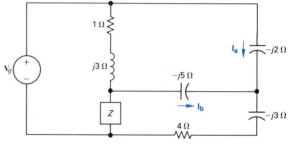

FIGURE P10.32

10.33 Find the steady-state expression for $v_o(t)$ in the circuit seen in Fig. P10.33 by using the technique of source transformations. The sinusoidal voltage sources are

$$v_1 = 400 \cos (5000t + 36.87°) \text{ V}$$

and

$$v_2 = 128 \sin 5000t \text{ V}.$$

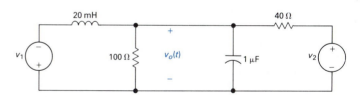

FIGURE P10.33

10.34 The circuit in Fig. P10.34 is operating in the sinusoidal steady state. Find $v_o(t)$ if $i_s(t) = 3 \cos 200t$ mA.

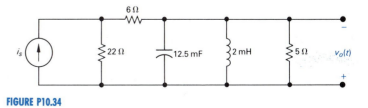

FIGURE P10.34

10.35 Use the node-voltage method to find the steady-state expression for $v_o(t)$ in the circuit seen in Fig. P10.35 if $L = 0.5$ mH, $C = 5$ μF, $v_{g1} = 200 \cos 40{,}000t$ V, and $v_{g2} = 100 \cos (40{,}000t + 36.87°)$ V.

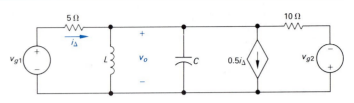

FIGURE P10.35

10.36 Use the node-voltage method to find the steady-state expression for v_o in the circuit seen in Fig. P10.36 if $v_g = 160 \cos 50{,}000t$ mV.

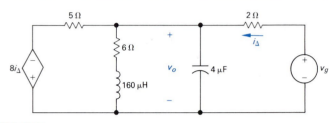

FIGURE P10.36

10.37 Use the node-voltage method to find the phasor voltage $\mathbf{V}_o$ in the circuit shown in Fig. P10.37. Express the voltage in both polar and rectangular form.

FIGURE P10.37

10.38 Use the node-voltage method to find $\mathbf{V}_o$ and $\mathbf{I}_o$ in the circuit seen in Fig. P10.38.

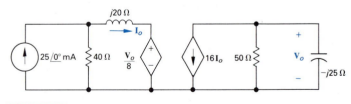

FIGURE P10.38

10.39 In the circuit shown in Fig. P10.39 $v_g = 10 \sin 10{,}000t$ V and $i_g = 4 \cos 10{,}000t$ A. Use the node voltage method to determine the steady-state expressions for v_1 and v_2.

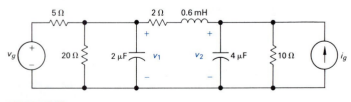

FIGURE P10.39

10.40 Use the mesh-current method to find the phasor branch current I_1 in the circuit shown in Fig. P10.40.

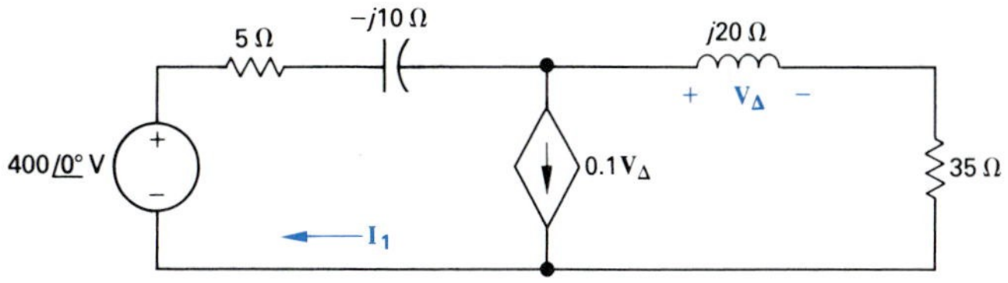

FIGURE P10.40

10.41 Use the mesh-current method to find the phasor current I_g in the circuit shown in Fig. P10.41.

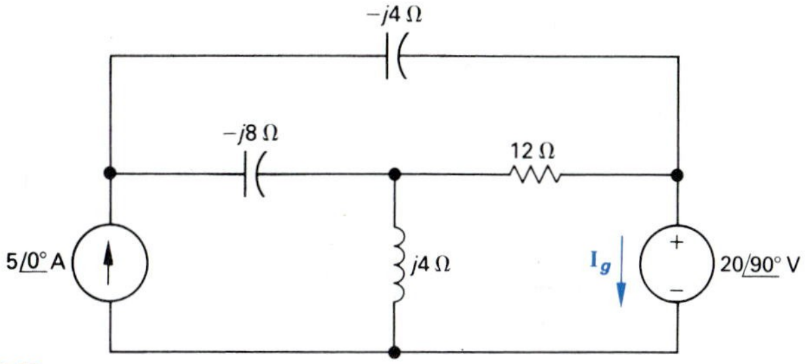

FIGURE P10.41

10.42 Use the mesh-current method to find the phasor branch currents I_a, I_b, and I_c in the circuit shown in Fig. P10.42.

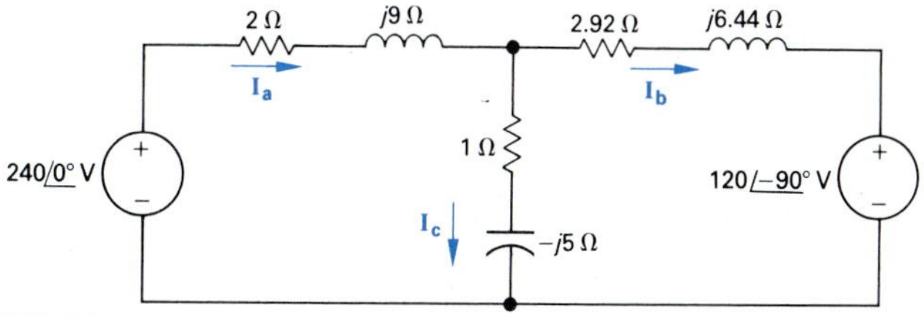

FIGURE P10.42

10.43 Use the mesh-current method to find the branch currents I_a, I_b, I_c, and I_d in the circuit shown in Fig. P10.43.

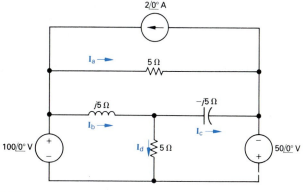

FIGURE P10.43

10.44 Use the mesh-current method to find the steady-state expression for v_o in the circuit seen in Fig. P10.44 if v_g equals $800 \cos 8000t$ V.

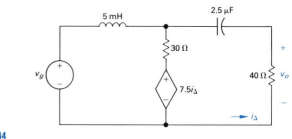

FIGURE P10.44

10.45 In order to introduce you to a circuit configuration that is widely used in residential wiring, we have shown a representation circuit in Fig. P10.45. In this simplified model, the resistor R_3 is used to model a 250-V appliance (such as an electric range), and the resistors R_1 and R_2 to model 125-V appliances (such as a lamp, toaster, and iron). The branches carrying $\mathbf{I}_1$ and $\mathbf{I}_2$ are modeling what electricians refer to as the "hot" conductors in the circuit, and the branch carrying $\mathbf{I}_n$ is modeling the neutral conductor. Our purpose in analyzing the circuit is to show the importance of the neutral conductor in the satisfactory operation of the circuit. You are to choose the method for analyzing the circuit.

a) Show that $\mathbf{I}_n$ is zero if $R_1 = R_2$.

b) Show that $\mathbf{V}_1 = \mathbf{V}_2$ if $R_1 = R_2$.

c) Open the neutral branch and calculate $\mathbf{V}_1$

and $\mathbf{V}_2$ if $R_1 = 40 \ \Omega$, $R_2 = 400 \ \Omega$, and $R_3 = 8 \ \Omega$.

d) Close the neutral branch and repeat part (c).

e) On the basis of your calculations, explain why the neutral conductor is never fused in such a manner that it could open while the "hot" conductors are energized.

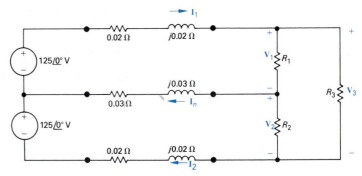

FIGURE P10.45

10.46 Find the steady-state expressions for the branch currents i_a, i_b, and i_c in the circuit seen in Fig. P10.46 if $v_a = 240 \sin 10^5 t$ V and $v_b = 120 \cos 10^5 t$ V.

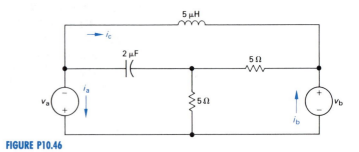

FIGURE P10.46

10.47 a) For the circuit shown in Fig. P10.47, find
the steady-state expression for v_o if
$i_g = 5 \cos (8 \times 10^5 t)$ A.

b) By how many microseconds does v_o lag i_g?

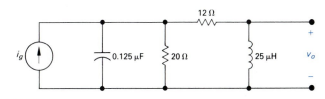

FIGURE P10.47

10.48 Find the value of Z in the circuit seen in Fig.
P10.48 if $\mathbf{V}_g = 100 - j50$ V, $\mathbf{I}_g = 30 + j20$ A, and $\mathbf{V}_b = 140 + j30$ V.

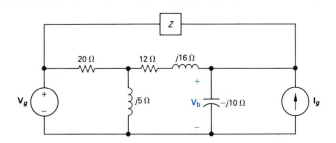

FIGURE P10.48

10.49 Find the Thévenin equivalent circuit with respect to the terminals a, b for the circuit
shown in Fig. P10.49.

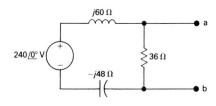

FIGURE P10.49

10.50 Find the Norton equivalent circuit with respect
to the terminals a, b for the circuit shown in
Fig. P10.50.

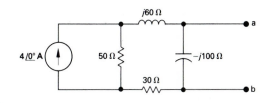

FIGURE P10.50

10.51 Find the Thévenin equivalent circuit with respect to the terminals a, b for the circuit shown in Fig. P10.51.

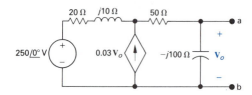

FIGURE P10.51

10.52 Find the Thévenin equivalent circuit with respect to the terminals a, b of the circuit shown in Fig. P10.52.

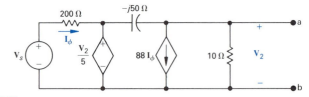

FIGURE P10.52

10.53 Find the Norton equivalent circuit with respect to the terminals a, b for the circuit shown in Fig. P10.53 when $V_s = 5\underline{/0°}$ V.

FIGURE P10.53

10.54 The circuit in Fig. P10.54 is operating in the sinusoidal steady state at a frequency of $(500/\pi)$ Hz.

a) Find the Thévenin impedance seen looking into the terminals a, b.

b) Repeat part (a) for the terminals c, d.

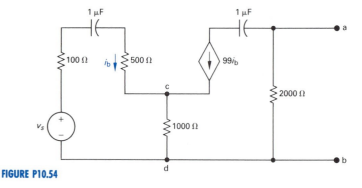

FIGURE P10.54

10.55 Find the Thévenin impedance seen looking into the terminals a, b of the circuit in Fig. P10.55 if the frequency of operation is $(200/\pi)$ Hz.

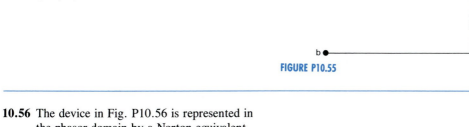

FIGURE P10.55

10.56 The device in Fig. P10.56 is represented in the phasor domain by a Norton equivalent. When an inductor having an impedance of $j100\ \Omega$ is connected across the device the value of $\mathbf{V}_0$ is $100\underline{/120°}$ mV.
When a capacitor having an impedance of $-j100\ \Omega$ is connected across the device the value of $\mathbf{I}_0$ is $-3\underline{/210°}$ mA.
Find the Norton current $\mathbf{I}_n$ and the Norton impedance Z_n.

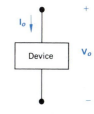

FIGURE P10.56

10.57 The circuit shown in Fig. P10.57 is operating at a frequency of 10 rad/s. Assume σ is real and lies between -10 and $+10$, i.e., $-10 \le \sigma \le 10$.

a) Find the value of σ so that the Thévenin impedance looking into the terminals a, b is purely resistive.

b) What is the value of the Thévenin impedance for the σ found in part (a)?

c) Can σ be adjusted so that the Thévenin

impedance equals $500 - j500\ \Omega$? If so, what is the value of σ?

d) For what values of σ will the Thévenin impedance be inductive?

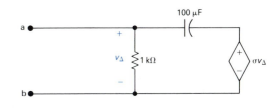

FIGURE P10.57

10.58 The ideal sinusoidal current sources in the circuit in Fig. P10.58 are generating the currents

$$i_{g1} = 6 \cos 25t \text{ A}$$

and

$$i_{g2} = 5 \cos (50t + 30°) \text{ A}.$$

Find the steady-state expression for $v_o(t)$.

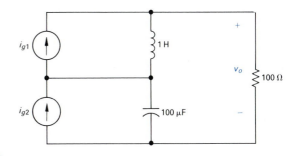

FIGURE P10.58

10.59 a) For the circuit shown in Fig. P10.59, compute $\mathbf{V}_s$ and $\mathbf{V}_l$.

b) Construct a phasor diagram showing the relationship between $\mathbf{V}_s$, $\mathbf{V}_l$, and the load voltage of $240\underline{/0°}$ V.

c) Repeat parts (a) and (b), given that the load voltage remains at $240\underline{/0°}$ V when a capacitive reactance of -5 Ω is connected across the load terminals.

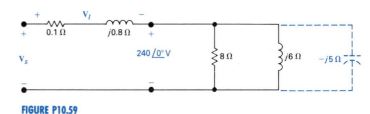

FIGURE P10.59

10.60 Show by using a phasor diagram what happens to the magnitude and phase angle of the voltage v_o in the circuit in Fig. P10.60 as R_x is varied from zero to infinity. The amplitude and phase angle of the source voltage are held constant as R_x varies.

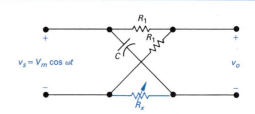

FIGURE P10.60

10.61 The operational amplifier in the circuit in Fig. P10.61 is ideal.

a) Find the steady-state expression for $v_o(t)$.

b) How large can the amplitude of v_g be before the amplifier saturates?

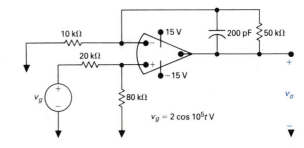

FIGURE P10.61

10.62 The operational amplifier in the circuit seen in Fig. P10.62 is ideal. Find the steady-state expression for $v_o(t)$ when $v_g = 2 \cos 10^6 t$ V.

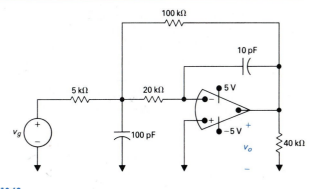

FIGURE P10.62

10.63 The operational amplifier in the circuit shown in Fig. P10.63 is ideal. The voltage of the ideal sinusoidal source is $v_g = 6 \cos 10^5 t$ V.

a) How small can C_o be before the steady-state output voltage no longer has a pure sinusoidal waveform?

b) For the value of C_o found in part (a), write the steady-state expression for v_o.

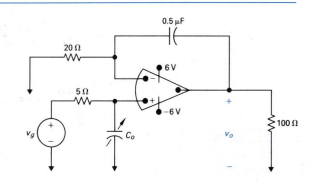

FIGURE P10.63

10.64 The sinusoidal voltage source in the circuit shown in Fig. P10.64 is generating the voltage $v_g = 4 \cos 200t$ V. If the op amp is ideal, what is the steady-state expression for $v_o(t)$?

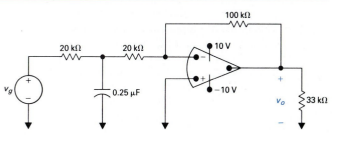

FIGURE P10.64

10.65 The 0.25-μf capacitor in the circuit seen in Fig. P10.64 is replaced with a variable capacitor. The capacitor is adjusted until the output voltage leads the input voltage by 135°.

a) Find the value of C in microfarads.

b) Write the steady-state expression for $v_o(t)$ when C has the value found in part (a).

10.66 a) Find the input impedance Z_{ab} for the circuit in Fig. P10.66. Express Z_{ab} as a function of Z and K where $K = (R_2/R_1)$.

b) If Z is a pure capacitive element what is the capacitance seen looking into the terminals a, b?

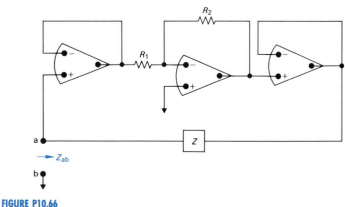

FIGURE P10.66

10.67 You may have the opportunity as an engineering graduate to serve as an expert witness in lawsuits involving either personal injury or property damage. As an illustrative example of the type of problem on which you may be asked to give an opinion, consider the following event.

At the end of a day of fieldwork, a farmer returns to his farmstead, checks his hog-confinement building, and finds to his dismay that the hogs are dead. The problem is traced to a blown fuse that caused a 240-V fan motor to stop. The loss of ventilation led to the suffocation of the livestock. The interrupted fuse is located in the main switch that connects the farmstead to the electrical service.

Before the insurance company settles the claim, they want to know if the electric circuit supplying the farmstead functioned properly. The lawyers for the insurance company are puzzled because the farmer's wife, who was in the house on the day of the accident convalescing from minor surgery, was able to watch TV during the afternoon. Furthermore, when she went to the kitchen to start preparing the

evening meal, the electric clock indicated the correct time.

The lawyers have hired you to explain (1) why the electric clock in the kitchen and the television set in the living room continued to operate after the fuse in the main switch blew and (2) why the second fuse in the main switch didn't blow after the fan motor stalled.

After ascertaining the loads on the three-wire distribution circuit prior to the interruption of fuse A, you are able to construct the circuit model shown in Fig. P10.67. The

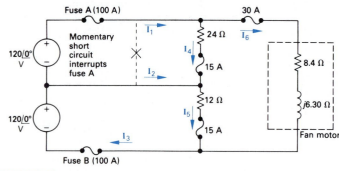

FIGURE P10.67

impedances of the line conductors and the neutral conductor are assumed negligible.

a) Calculate the branch currents I_1, I_2, I_3, I_4, I_5, and I_6 prior to the interruption of fuse A.

b) Calculate the branch currents after the interruption of fuse A. Assume the stalled fan motor behaves as a short circuit.

c) Explain why the clock and television set were not affected by the momentary short circuit that interrupted fuse A.

d) Assume the fan motor is equipped with a thermal cutout that is designed to interrupt the motor circuit if the motor current becomes excessive. Would you expect the thermal cutout to operate? Explain.

e) Explain why fuse B is not interrupted when the fan motor stalls.

SINUSOIDAL STEADY-STATE POWER CALCULATIONS

CHAPTER 11

In Chapter 10 we showed how to determine the steady-state voltages and currents in circuits driven by sinusoidal sources. We now concentrate on power calculations associated with steady-state sinusoidal operation. Our primary interest is to determine the average power that is either delivered to or extracted from a pair of terminals by a sinusoidal voltage and current. The power associated with a pair of terminals in a circuit operating in the sinusoidal steady state is an important calculation. The voltage and current ratings of electrical equipment such as generators, motors, lamps, toasters, and ovens determine how much power the appliance can handle. Because the usefulness of the appliance is determined by its ability to convert energy either to or from the electrical form, the calculation of power is of prime importance.

Figure 11.1 shows the problem graphically. Here, v and i are steady-state sinusoidal signals. Note use of the passive sign convention; therefore the power at any instant of time is

$$p = vi. \tag{11.1}$$

The power is measured in watts when the voltage is in volts and the current is in amperes. First, we write expressions for v and i:

$$v = V_m \cos (\omega t + \theta_v) \tag{11.2}$$

and

$$i = I_m \cos (\omega t + \theta_i), \tag{11.3}$$

where θ_v is the voltage phase angle and θ_i is the current phase angle.

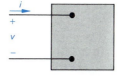

FIGURE 11.1 The basic power calculation is to find the average power associated with the voltage and current at a pair of terminals.

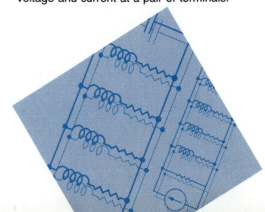

We are operating in the sinusoidal steady state, so we may choose any convenient reference for zero time. Engineers concerned with operating systems involving the transfer of large blocks of power have found that a zero time corresponding to the instant when the current is passing through a positive maximum is convenient. This reference system requires a shift of both the voltage and current by θ_i. Thus Eqs. (11.2) and (11.3) become

$$v = V_m \cos(\omega t + \theta_v - \theta_i); \tag{11.4}$$

$$i = I_m \cos \omega t. \tag{11.5}$$

When we substitute Eqs. (11.4) and (11.5) into Eq. (11.1), the expression for the instantaneous power becomes

$$p = V_m I_m \cos(\omega t + \theta_v - \theta_i) \cos \omega t, \tag{11.6}$$

which is the starting point for developing the expression for average power.

11.1 REAL AND REACTIVE POWER

The average power associated with sinusoidal signals is the average of the instantaneous power over one period, or in equation form,

$$P = \frac{1}{T} \int_{t_0}^{t_0+T} p \, dt, \tag{11.7}$$

where T is the period of the sinusoidal function. The limits on Eq. (11.7) imply that we can initiate the integration process at any convenient time t_0 but that we must terminate the integration exactly one period later. (We could integrate over nT periods, where n is an integer, provided we multiply the integral by $1/nT$.)

We could find the average power by substituting Eq. (11.6) directly into Eq. (11.7) and then performing the integration. More informative, however, is to first expand Eq. (11.6) by the trigonometric identity

$$\cos \alpha \cos \beta = \tfrac{1}{2} \cos(\alpha - \beta) + \tfrac{1}{2} \cos(\alpha + \beta).$$

Letting $\alpha = \omega t + \theta_v - \theta_i$ and $\beta = \omega t$, we write Eq. (11.6) as

$$p = \frac{V_m I_m}{2} \cos(\theta_v - \theta_i) + \frac{V_m I_m}{2} \cos(2\omega t + \theta_v - \theta_i). \tag{11.8}$$

We now use the trigonometric identity

$$\cos(\alpha + \beta) = \cos \alpha \cos \beta - \sin \alpha \sin \beta$$

to expand the second term on the right-hand side of Eq. (11.8), to get

$$p = \frac{V_m I_m}{2} \cos (\theta_v - \theta_i) + \frac{V_m I_m}{2} \cos (\theta_v - \theta_i) \cos 2\omega t$$

$$- \frac{V_m I_m}{2} \sin (\theta_v - \theta_i) \sin 2\omega t. \qquad (11.9)$$

A careful study of Eq. (11.9) reveals the following characteristics of the instantaneous power.

1. The average value of p is given by the first term on the right-hand side of the equation because the integral of either cos $2\omega t$ or sin $2\omega t$ over one period is zero. Thus the average power is

$$P = \frac{V_m I_m}{2} \cos (\theta_v - \theta_i). \qquad (11.10)$$

2. The frequency of the instantaneous power is twice the frequency of the voltage or current, which follows directly from the second two terms on the right-hand side of the equation. Figure 11.2 depicts a representative relationship between v, i, and p, based on the assumptions that $\theta_v = 60°$ and $\theta_i = 0°$. Note that the instantaneous power goes through two complete cycles for every cycle of either the voltage or the current. Also note that the instantaneous power may be negative for a portion of each cycle, even if the network between the terminals is passive. In a completely passive network, negative power implies that energy stored in inductors or capacitors is now being extracted. The fact that the instantaneous power varies with time in the sinusoidal steady-state operation of a circuit explains why some motor-driven appliances (such as a refrigerator) experience vibration, and require resilient mountings to prevent excessive vibration of the appliance itself.

3. If the circuit between the terminals is purely resistive, the voltage and current are in phase, which means that $\theta_v = \theta_i$. Equation (11.9) then reduces to

$$p = P + P \cos 2\omega t. \qquad (11.11)$$

In writing Eq. (11.11), we took advantage of the result given in Eq. (11.10). The instantaneous power expressed in Eq. (11.11) is referred to as the *instantaneous real power*. Note that this designation implies that average power (P) is also referred to as *real power*, a term used to describe power transformed from electrical to nonelectrical form. In the case of the purely resistive network, the electric energy is transformed into thermal energy. Also note from Eq. (11.11) that the instantaneous real power can never be negative. That is, power cannot be extracted from a purely resistive network.

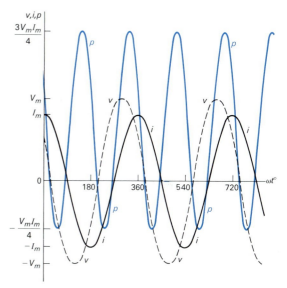

FIGURE 11.2 Instantaneous power, voltage, and current versus ωt for steady-state sinusoidal operation.

4. If the circuit between the terminals is purely inductive, the voltage and current are out of phase by precisely 90°. In particular, the current lags the voltage by 90° (that is, $\theta_i = \theta_v - 90°$); therefore $\theta_v - \theta_i = +90°$. The expression for the instantaneous power then reduces to

$$p = \frac{-V_m I_m}{2} \sin 2\omega t. \tag{11.12}$$

In the case of a purely inductive circuit the average power is zero. Therefore in a purely inductive circuit there is no transformation of energy from electrical to nonelectrical form. The instantaneous power at the terminals in a purely inductive circuit oscillates between the circuit and the source driving the circuit. When p is positive, energy is being stored in the magnetic fields associated with the inductive elements. When p is negative, energy is being extracted from the magnetic fields of the inductive elements.

5. If the circuit between the terminals is purely capacitive, the voltage and current also are precisely 90° out of phase. In this case, the current leads the voltage by 90° (that is, $\theta_i = \theta_v + 90°$); thus $\theta_v - \theta_i = -90°$. The expression for the instantaneous power then becomes

$$p = \frac{V_m I_m}{2} \sin 2\omega t. \tag{11.13}$$

Again, the average power is zero, so there is no transformation of energy from electrical to nonelectrical form. In the purely capacitive circuit, the power oscillates between the source driving the circuit and the electric field associated with the capacitive elements.

The power associated with purely inductive or capacitive circuits is referred to as *reactive power*. Inductors and capacitors are referred to as *reactive elements* in steady-state sinusoidal analysis because their impedances are characterized as inductive reactance and capacitive reactance, respectively. In terms of the general expression for instantaneous power, namely, Eq. (11.9), the coefficient of sin $2\omega t$ is referred to as the reactive power and is denoted Q. Hence

$$Q = \frac{V_m I_m}{2} \sin (\theta_v - \theta_i). \tag{11.14}$$

The concept of reactive power is very useful. If we use the notation for average and reactive power, we can express Eq. (11.9) as

$$p = P + P \cos 2\omega t - Q \sin 2\omega t. \tag{11.15}$$

Three additional comments regarding Eqs. (11.10) and (11.14) are in order. First, P and Q carry the same dimension. However, in order to distinguish between real and reactive power, we use the term *var* for reactive power (var is an acronym for the phrase *volt-amp reactive*).

Second, the decision to use the current as the reference leads to Q being positive for inductors (that is, $\theta_v - \theta_i = 90°$) and negative for capacitors (that is, $\theta_v - \theta_i = -90°$). Power engineers recognize this difference in the algebraic sign of Q for inductors and capacitors by saying that inductors demand, or absorb, magnetizing vars and capacitors furnish, or deliver, magnetizing vars. We say more about this convention later.

Third, the angle $\theta_v - \theta_i$ is referred to as the *power factor angle*. The cosine of this angle is called the *power factor*, abbreviated pf, and the sine of this angle is called the *reactive factor*, abbreviated rf. Thus

$$pf = \cos (\theta_v - \theta_i) \qquad \textbf{(11.16)}$$

and

$$rf = \sin (\theta_v - \theta_i). \qquad \textbf{(11.17)}$$

Once we know the magnitude of the power factor, we know the magnitude of the reactive factor. However, there is an ambiguity regarding the algebraic sign of the reactive factor because $\cos (\theta_v - \theta_i) = \cos (\theta_i - \theta_v)$ and $\sin (\theta_v - \theta_i) = -\sin (\theta_i - \theta_v)$. To completely describe the reactive factor through the power factor, we use the descriptive phrases *lagging power factor* and *leading power factor*. Lagging power factor implies that current lags voltage—hence an inductive load. Leading power factor implies that current leads voltage—hence a capacitive load.

Both the power factor and the reactive factor are convenient quantities to use in describing electrical loads.

Example 11.1 illustrates the interpretation of P and Q on the basis of a numerical calculation.

E X A M P L E   11.1

a) Calculate the average power and the reactive power at the terminals of the network shown in Fig. 11.3 if

$$v = 100 \cos (\omega t + 15°) \text{ V}$$

and

$$i = 4 \sin (\omega t - 15°) \text{ A}.$$

b) State whether the network inside the box is absorbing or delivering average power.

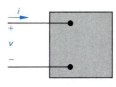

FIGURE 11.3 Pair of terminals used for calculating power.

c) State whether the network inside the box is absorbing or supplying magnetizing vars.

S O L U T I O N

a) As i is expressed in terms of the sine function, the first step in the calculation for P and Q is to rewrite i as a cosine function:

$$i = 4 \cos (\omega t - 105°) \text{ A.}$$

We now calculate P and Q directly from Eqs. (11.10) and (11.14). Thus

$$P = \tfrac{1}{2}(100)(4) \cos [15 - (-105)] = -100 \text{ W}$$

and

$$Q = \tfrac{1}{2}100(4) \sin [15 - (-105)] = 173.21 \text{ VAR.}$$

b) Note from Fig. 11.3 use of the passive sign convention. Therefore the negative value of -100 W means that the network inside the box is delivering average power to the terminals.

c) The passive sign convention means that, as Q is positive, the network inside the box is absorbing magnetizing vars at its terminals.

D R I L L   E X E R C I S E S

11.1 For each of the following sets of voltage and current, calculate the real and reactive power in the line between networks A and B. In each case, state whether the power flow is from A to B or vice versa. Also state whether magnetizing vars are being transferred from A to B or vice versa.

a) $v = 250 \cos (\omega t + 45°)$ V;
   $i = 12 \cos (\omega t - 15°)$ A.

b) $v = 250 \cos (\omega t + 45°)$ V;
   $i = 12 \cos (\omega t + 165°)$ A.

c) $v = 250 \cos (\omega t + 45°)$ V;
   $i = 12 \cos (\omega t + 105°)$ A.

d) $v = 250 \cos \omega t$ V;
   $i = 12 \cos (\omega t - 120°)$ A.

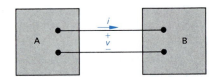

ANSWER: (a) $P = 750$ W (A to B), $Q = 1299.04$ VAR (A to B); (b) $P = -750$ W (B to A), $Q = -1299.04$ VAR (B to A); (c) $P = 750$ W (A to B), $Q = -1299.04$ VAR (B to A); (d) $P = -750$ W (B to A), $Q = 1299.04$ VAR (A to B).

11.2 THE RMS VALUE AND POWER CALCULATIONS

In introducing the rms value of a sinusoidal voltage (or current) in Section 10.1 we mentioned that it would play an important role in power calculations. We can now discuss this role.

Assume that a sinusoidal voltage is applied to the terminals of a resistor, as shown in Fig. 11.4 and that we want to determine the average power delivered to the resistor. From Eq. (11.7),

$$P = \frac{1}{T} \int_{t_0}^{t_0+T} \frac{V_m^2 \cos^2(\omega t + \phi_v)}{R} \, dt$$

$$= \frac{1}{R}\left[\frac{1}{T} \int_{t_0}^{t_0+T} V_m^2 \cos^2(\omega t + \phi_v)dt\right]. \tag{11.18}$$

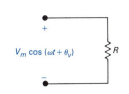

FIGURE 11.4 A sinusoidal voltage applied to the terminals of a resistor.

Comparing Eq. (11.18) with Eq. (10.4) reveals that the average power delivered to R is simply the rms value of the voltage squared divided by R, or

$$P = \frac{V_{\text{rms}}^2}{R} \tag{11.19}$$

If the resistor is carrying a sinusoidal current, say, $I_m \cos(\omega t + \phi_i)$, the average power delivered to the resistor is

$$P = I_{\text{rms}}^2 R. \tag{11.20}$$

The rms value is also referred to as the *effective* value of the sinusoidal voltage (or current). The concept of effective value comes from a desire to compare the ability of a sinusoidal varying voltage (or current) to deliver energy to a resistor with the ability of a constant (dc) voltage (or current) to deliver energy to a resistor. Interestingly, a constant (dc) voltage equal to the rms value of the sinusoidal voltage delivers the same amount of energy every T seconds that the sinusoidal voltage does, assuming, of course, equal resistance. (See Problem 11.8.) For example, a dc voltage of 100 V delivers the same energy in T seconds that a sinusoidal voltage having an rms value of 100 V does if the two resistances are the same. Hence the terms *rms* and *effective* are used interchangeably.

The average power given by Eq. (11.10) and the reactive power given by Eq. (11.14) can be written in terms of effective values:

$$P = \frac{V_m I_m}{2} \cos (\theta_v - \theta_i)$$

$$= \frac{V_m}{\sqrt{2}} \frac{I_m}{\sqrt{2}} \cos (\theta_v - \theta_i)$$

$$= V_{\text{eff}} I_{\text{eff}} \cos (\theta_v - \theta_i) \tag{11.21}$$

and, by similar manipulation,

$$Q = V_{\text{eff}} I_{\text{eff}} \sin (\theta_v - \theta_i). \tag{11.22}$$

The effective value of the sinusoidal signal in power calculations is so widely used that voltage and current ratings of circuits and equipment involved in power utilization are given in terms of rms values. For example, the voltage rating of residential electric wiring is often 240 V/120 V service. These voltage levels are the rms values of the sinusoidal voltages supplied by the utility company. Appliances such as electric lamps, irons, and toasters all carry rms ratings on their nameplates. For example, a 120-V, 100-W lamp has a resistance of $120^2/100$, or 144 Ω, and draws an rms current of $120/144$, or 0.833 A. The peak value of the lamp current is $0.833\sqrt{2}$, or 1.18 A.

The phasor transform of a sinusoidal function also may be expressed in terms of the rms value. The magnitude of the rms phasor is equal to the rms value of the sinusoidal function. If a phasor is based on the rms value, we indicate this by either an explicit statement, a parenthetical rms adjacent to the phasor quantity, or the subscript "eff" as in Eq. (11.21).

DRILL EXERCISES

11.2 The periodic triangular current in Example 10.4 has a peak value of 3 A. Find the average power that this current delivers to a 20-Ω resistor.

ANSWER: 60 W.

11.3 COMPLEX POWER

Before proceeding to the various methods of calculating real and reactive power in circuits operating in the sinusoidal steady state, we need to introduce and define complex power. *Complex power* is the complex sum of average real power and reactive power, or

$$S = P + jQ. \tag{11.23}$$

Dimensionally, complex power is the same as real or reactive power. However, in order to distinguish complex power from either real or reactive power, we use the term *volt amps*. Thus we use volt amps for complex power, watts for average real power, and vars for reactive power. The magnitude of complex power is referred to as *apparent power.* Apparent power also is measured in volt amps.

When working with Eq. (11.23), think of P, Q, and $|S|$ as the sides of a right triangle, as shown in Fig. 11.5. As we show in Section 11.4, the angle θ in the power triangle is the power factor angle $\theta_v - \theta_i$.

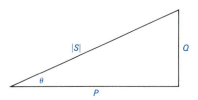

FIGURE 11.5 The power triangle.

11.4 POWER CALCULATIONS

We are now ready to develop equations that can be used to calculate real, reactive, and complex power. We begin by combining Eqs. (11.10), (11.14), and (11.23) to get

$$S = \frac{V_m I_m}{2} \cos (\theta_v - \theta_i) + j\frac{V_m I_m}{2} \sin (\theta_v - \theta_i)$$

$$= \frac{V_m I_m}{2} [\cos (\theta_v - \theta_i) + j \sin (\theta_v - \theta_i)]$$

$$= \frac{V_m I_m}{2} e^{j(\theta_v - \theta_i)} = \tfrac{1}{2} V_m I_m \underline{/\theta_v - \theta_i}. \qquad (11.24)$$

If we use the effective values of the sinusoidal voltage and current, Eq. (11.24) becomes

$$S = V_{\text{eff}} I_{\text{eff}} \underline{/\theta_v - \theta_i}. \qquad (11.25)$$

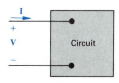 **PSpice** Use PSpice to calculate a circuit's sinusoidal steady-state response (Section 11.2) and then use PROBE to study the power relationships (Chapter 12)

Equations (11.24) and (11.25) are important relationships in power calculations because they show that if the phasor current and voltage are known at a pair of terminals, the complex power associated with that pair of terminals is either one half the product of the voltage and the *conjugate* of the current or the product of the rms phasor voltage and the *conjugate* of the rms phasor current. Thus for the phasor voltage and current at a pair of terminals, as shown in Fig. 11.6, the complex power is

$$S = \tfrac{1}{2} \mathbf{V} \mathbf{I}^* = P + jQ \qquad (11.26)$$

or

$$S = \mathbf{V}_{\text{eff}} \mathbf{I}_{\text{eff}}^* = P + jQ. \qquad (11.27)$$

FIGURE 11.6 Phasor voltage and current associated with a pair of terminals.

Both Eqs. (11.26) and (11.27) are based on the passive sign convention. If the current reference is in the direction of the voltage rise across the terminals, we insert a minus sign on the right-hand side of each equation.

To illustrate the use of Eq. (11.26) in a power calculation, let's utilize the same circuit that we used in Example 11.1. Expressed in terms of the phasor representation of the terminal voltage and current,

$$\mathbf{V} = 100\underline{/15^\circ} \quad \text{and} \quad \mathbf{I} = 4\underline{/-105^\circ} \text{ A}.$$

Therefore

$$S = \tfrac{1}{2}(100\underline{/15^\circ})(4\underline{/+105^\circ}) = 200\underline{/120^\circ}$$

$$= -100 + j173.21 \text{ VA}.$$

Once we calculate the complex power we can read off both the real and reactive power. Thus

$$P = -100 \text{ W} \quad \text{and} \quad Q = 173.21 \text{ VAR}.$$

The interpretations of the algebraic signs on P and Q are identical to those given in the solution of Example 11.1.

ALTERNATIVE FORMS FOR COMPLEX POWER

Equations (11.26) and (11.27) have several useful variations. Here, we use the rms, or effective value, form of the equations with the suggestion that you "think" rms when making power calculations.

The first variation of Eq. (11.27) is to replace the voltage with the product of the current times the impedance. That is, we can always represent the circuit inside the box of Fig. 11.6 by an equivalent impedance, as shown in Fig. 11.7. Then,

$$\mathbf{V}_{\text{eff}} = Z\mathbf{I}_{\text{eff}}. \tag{11.28}$$

Substituting Eq. (11.28) into Eq. (11.27) yields

$$S = Z\mathbf{I}_{\text{eff}}\mathbf{I}_{\text{eff}}^*$$

$$= |\mathbf{I}_{\text{eff}}|^2 Z$$

$$= |\mathbf{I}_{\text{eff}}|^2(R + jX)$$

$$= |\mathbf{I}_{\text{eff}}|^2 R + j|\mathbf{I}_{\text{eff}}|^2 X = P + jQ, \tag{11.29}$$

from which,

$$P = |\mathbf{I}_{\text{eff}}|^2 R = \tfrac{1}{2}I_m^2 R \tag{11.30}$$

and

$$Q = |\mathbf{I}_{\text{eff}}|^2 X = \tfrac{1}{2}I_m^2 X. \tag{11.31}$$

In Eq. (11.31), X is the reactance of either the equivalent inductance or equivalent capacitance of the circuit; it is positive for inductive circuits and negative for capacitive circuits.

A second useful variation of Eq. (11.27) comes from replacing the current with the voltage divided by the impedance:

$$S = \mathbf{V}_{\text{eff}}\left(\frac{\mathbf{V}_{\text{eff}}}{Z}\right)^* = \frac{|\mathbf{V}_{\text{eff}}|^2}{Z^*} = P + jQ. \tag{11.32}$$

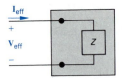

FIGURE 11.7 The general circuit of Fig. 11.6 replaced with an equivalent impedance.

Note that if Z is a pure resistance element,

$$P = \frac{|\mathbf{V}_{\text{eff}}|^2}{R}, \qquad (11.33)$$

and if Z is a pure reactive element,

$$Q = \frac{|\mathbf{V}_{\text{eff}}|^2}{X}. \qquad (11.34)$$

In Eq. (11.34), X is positive for an inductor and negative for a capacitor.

11.5 ILLUSTRATIVE EXAMPLES

We illustrate numerical application of the equations developed in Section 11.4 in Examples 11.2–11.6. These examples demonstrate basic power and reactive power calculations in circuits operating in the sinusoidal steady state.

E X A M P L E   11.2

a) A sinusoidal voltage having a maximum amplitude of 625 V is applied to the terminals of a 50-Ω resistor. Find the average power delivered to the resistor.

b) Repeat (a) by first finding the current in the resistor.

S O L U T I O N

a) The rms, or effective, value of the sinusoidal voltage is $625/\sqrt{2}$, or approximately 441.94 V. From Eq. (11.19), the average power delivered to the 50-Ω resistor is

$$P = \frac{(441.94)^2}{50} = 3906.25 \text{ W}.$$

b) The maximum amplitude of the current in the resistor is $625/50$, or 12.5 A. The rms value of the current is $12.5/\sqrt{2}$, or approximately 8.84 A. Hence the average power delivered to the resistor is

$$P = (8.84)^2 50 = 3906.25 \text{ W}.$$

EXAMPLE 11.3

In the circuit shown in Fig. 11.8, a load having an impedance of $39 + j26\ \Omega$ is fed from a voltage source through a line having an impedance of $1 + j4\ \Omega$. The effective, or rms, value of the source voltage is 250 V.

a) Calculate the load current $\mathbf{I}_L$ and voltage $\mathbf{V}_L$.

b) Calculate the average and reactive power delivered to the load.

c) Calculate the average and reactive power delivered to the line.

d) Calculate the average and reactive power supplied by the source.

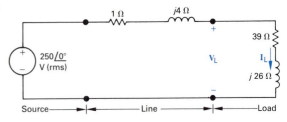

FIGURE 11.8 The circuit for Example 11.3.

SOLUTION

a) The line and load impedances are in series across the voltage source, so the load current equals the voltage divided by the total impedance, or

$$\mathbf{I}_L = \frac{250/0°}{40 + j30} = 4 - j3 = 5/-36.87°\ A \qquad \text{(rms).}$$

Because the voltage is given in terms of its rms value, the current also is rms. The load voltage is the product of the load current and load impedance:

$$\mathbf{V}_L = (39 + j26)\mathbf{I}_L = 234 - j13$$

$$= 234.36/-3.18°\ V \qquad \text{(rms).}$$

b) The average and reactive power delivered to the load can be computed using Eq. (11.27). Therefore

$$S = \mathbf{V}_L \mathbf{I}_L^* = (234 - j13)(4 + j3)$$

$$= 975 + j650\ VA.$$

Thus the load is absorbing an average power of 975 W and a reactive power of 650 VAR.

c) The average and reactive power delivered to the line are most easily calculated from Eqs. (11.30) and (11.31) because the line current is known. Thus

$$P = (5)^2(1) = 25\ W \qquad \text{and} \qquad Q = (5^2)(4) = 100\ VAR.$$

Note that the reactive power associated with the line is positive because the line reactance is inductive.

d) One way to calculate the average and reactive power delivered by the source is to add the complex power delivered to the line to the complex power delivered to the load, or

$$S_s = 25 + j100 + 975 + j650$$
$$= 1000 + j750 \text{ VA.}$$

The complex power at the source can also be calculated from Eq. (11.27):

$$S_s = -250\,\mathbf{I}_L^*.$$

The minus sign is inserted in Eq. (11.27) whenever the current reference is in the direction of a voltage rise. Thus

$$S_s = -250(4 + j3) = -(1000 + j750) \text{ VA.}$$

The minus sign implies that the source is delivering average power and magnetizing reactive power. Note that this result agrees with the previous calculation of S_s, as it must, because the source must furnish all the average and reactive power absorbed by the line and load.

E X A M P L E   11.4

An electrical load operates at 240 V rms. The load absorbs an average power of 8 kW at a lagging power factor of 0.8.

a) Calculate the complex power of the load.

b) Calculate the impedance of the load.

S O L U T I O N

a) The power factor is described as lagging, so we know that the load is inductive and that the algebraic sign of the reactive power is positive. From the power triangle shown in Fig. 11.9,

$$P = |S| \cos \theta \qquad \text{and} \qquad Q = |S| \sin \theta.$$

Now, as $\cos \theta = 0.8$, $\sin \theta = 0.6$. Therefore

$$|S| = \frac{P}{\cos \theta} = \frac{8 \text{ kW}}{0.8} = 10 \text{ kVA},$$

$$Q = 10 \sin \theta = 6 \text{ kVAR},$$

and

$$S_L = 8 + j6 \text{ kVA.}$$

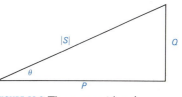

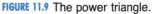

FIGURE 11.9 The power triangle.

b) From Eq. (11.32),

$$Z^* = \frac{|\mathbf{V}|^2}{P + jQ} = \frac{(240)^2}{8000 + j6000}$$

$$= 4.608 - j3.456 \ \Omega.$$

Hence

$$Z = 4.608 + j3.456 \ \Omega = 5.76\underline{/36.87°} \ \Omega.$$

E X A M P L E   11.5

The three loads in the circuit shown in Fig. 11.10 can be described as follows. Load 1 absorbs an average power of 8 kW at a lagging power factor of 0.8. Load 2 absorbs 20 kVA at a leading power factor of 0.6. Load 3 is an impedance of $2.5 + j5.0 \ \Omega$. Derive the steady-state expression for $v_s(t)$ if the frequency of the source is 60 Hz.

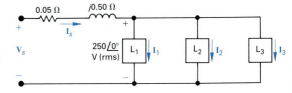

FIGURE 11.10 The circuit for Example 11.5.

S O L U T I O N

We placed the currents $\mathbf{I}_1$, $\mathbf{I}_2$, $\mathbf{I}_3$, and $\mathbf{I}_s$ on the circuit diagram in Fig. 11.10 to aid discussion of the solution. We begin by noting that $\mathbf{I}_s$ is the sum of the three load currents and that $\mathbf{V}_s$ is the load voltage plus the drop across the line impedance of $0.05 + j0.5 \ \Omega$. Hence the first step in the solution is to find the three load currents. We find $\mathbf{I}_1$ and $\mathbf{I}_2$ directly from Eq. (11.27):

$$250\mathbf{I}_1^* = 8000 + j6000$$

or

$$\mathbf{I}_1^* = 32 + j24 \ \text{A} \quad \text{(rms)}.$$

Hence

$$\mathbf{I}_1 = 32 - j24 \ \text{A} \quad \text{(rms)},$$

$$250 \, \mathbf{I}_2^* = 12{,}000 - j16{,}000,$$

or

$$\mathbf{I}_2^* = 48 - j64 \ \text{A} \quad \text{(rms)}.$$

Thus

$$\mathbf{I}_2 = 48 + j64 \ \text{A} \quad \text{(rms)}.$$

Load 3 is described in terms of its impedance, so

$$\mathbf{I}_3 = \frac{250}{2.5 + j5} = 20 - j40 \text{ A} \qquad \text{(rms)}.$$

Kirchhoff's current law gives

$$\mathbf{I}_s = \mathbf{I}_1 + \mathbf{I}_2 + \mathbf{I}_3$$
$$= 100 + j0 \text{ A} \qquad \text{(rms)}.$$

Kirchhoff's voltage law yields

$$\mathbf{V}_s = 250 + (0.05 + j0.50)100$$
$$= 255 + j50$$
$$= 259.86\underline{/11.09°} \text{ V} \qquad \text{(rms)}.$$

Then,

$$v_s(t) = \sqrt{2}(259.86) \cos (120\pi t + 11.09°)$$
$$= 367.49 \cos (377t + 11.09°) \text{ V}.$$

E X A M P L E   11.6

a) Calculate the total average and reactive power delivered to each impedance in the circuit shown in Fig 11.11.

b) Calculate the average and reactive powers associated with each source in the circuit.

c) Verify that the average power delivered equals the average power absorbed and that the magnetizing reactive power delivered equals the magnetizing reactive power absorbed.

S O L U T I O N

a) The complex power delivered to the $(1 + j2)$-Ω impedance is

$$S_1 = \tfrac{1}{2} \mathbf{V}_1 \mathbf{I}_1^* = P_1 + jQ_1$$
$$= \tfrac{1}{2}(78 - j104)(-26 + j52)$$
$$= \tfrac{1}{2}(3380 + j6760)$$
$$= 1690 + j3380 \text{ VA}.$$

Thus this impedance is absorbing an average power of 1690 W and 3380 VAR. The complex power delivered to the

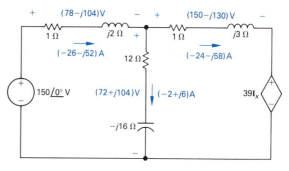

FIGURE 11.11 The circuit, with solution, for Example 11.6.

$(12 - j16)$-Ω impedance is

$$S_2 = \tfrac{1}{2}\mathbf{V}_2\mathbf{I}_x^* = P_2 + jQ_2$$
$$= \tfrac{1}{2}(72 + j104)(-2 - j6)$$
$$= 240 - j320 \text{ VA.}$$

Therefore the impedance in the vertical branch is absorbing 240 W and delivering 320 VAR. The complex power delivered to the $(1 + j3)$-Ω impedance is

$$S_3 = \tfrac{1}{2}\mathbf{V}_3\mathbf{I}_2^* = P_3 + jQ_3$$
$$= \tfrac{1}{2}(150 - j130)(-24 + j58)$$
$$= 1970 + j5910 \text{ VA.}$$

This impedance is absorbing 1970 W and 5910 VAR.

b) The complex power associated with the independent voltage source is

$$S_s = -\tfrac{1}{2}\mathbf{V}_s\mathbf{I}_1^* = P_s + jQ_s$$
$$= -\tfrac{1}{2}(150)(-26 + j52)$$
$$= 1950 - j3900 \text{ VA.}$$

Note that the independent voltage source is absorbing an average power of 1950 W and is delivering 3900 VAR. The complex power associated with the current-controlled voltage source is

$$S_x = \tfrac{1}{2}(39\mathbf{I}_x)(\mathbf{I}_2^*) = P_x + jQ_x$$
$$= \tfrac{1}{2}(-78 + j234)(-24 + j58)$$
$$= -5850 - j5070 \text{ VA.}$$

The dependent source is delivering both average power and magnetizing reactive power.

c) The total power absorbed by the passive impedances and the independent voltage source is

$$P_{\text{absorbed}} = P_1 + P_2 + P_3 + P_s = 5850 \text{ W.}$$

The dependent voltage source is the only circuit element delivering average power. Thus

$$P_{\text{delivered}} = 5850 \text{ W.}$$

Magnetizing reactive power is being absorbed by the two horizontal branches. Thus

$$Q_{\text{absorbed}} = Q_1 + Q_3 = 9290 \text{ VAR.}$$

Magnetizing reactive power is being delivered by the independent voltage source, the capacitor in the vertical impedance branch, and the dependent voltage source. Therefore

$$Q_{\text{delivered}} = 9290 \text{ VAR}.$$

COMMENTS ON APPARENT POWER

The apparent power, or volt-amp, requirement of a device designed to convert electric energy to nonelectrical form is more important than the average power requirement. The device must be insulated to withstand the voltage and must be large enough to carry the current even if the power factor is small. Because the average power represents the useful output of the energy-converting device, operating such devices close to the unity power factor is desirable. Many useful appliances (such as refrigerators, fans, air conditioners, fluorescent lighting fixtures, and washing machines) and most industrial loads operate at a lagging power factor. These loads sometimes are corrected either by adding a capacitor to the device itself or by connecting capacitors across the line that is feeding the load. The capacitors are connected at the load end of the line. The use of capacitors across the line is representative of power-factor correction for large industrial loads. Drill Exercise 11.3 and Problems 11.37 and 11.38 give you a chance to make some calculations that show why connecting a capacitor across the terminals of a lagging-power-factor load improves operation of the circuit.

DRILL EXERCISES

11.3 The load impedance in the circuit shown is shunted by a capacitor having a capacitive reactance of $-52 \ \Omega$. Calculate:

a) the rms phasors $\mathbf{V}_L$ and $\mathbf{I}_L$;

b) the average power and magnetizing reactive power absorbed by the $(39 + j26)$-Ω load impedance;

c) the average power and magnetizing reactive power absorbed by the $(1 + j4)$-Ω line impedance;

d) the average power and magnetizing reactive power delivered by the source;

e) the magnetizing reactive power delivered by the shunting capacitor.

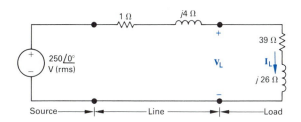

ANSWER: (a) $252.20\underline{/-4.54°}$ V (rms), $5.38\underline{/-38.23°}$ A (rms); (b) 1129.09 W, 752.73 VAR; (c) 23.52 W, 94.09 VAR; (d) 1152.62 W, -376.36 VAR; (e) 1223.18 VAR.

11.4 The rms voltage at the terminals of a load is 440 V. The load is absorbing an average power of 20 kW and a magnetizing reactive power of 10 kVAR. Derive two equivalent impedance models of the load.

11.5 Find the phasor voltage $\mathbf{V}_s$ (rms) in the circuit shown if loads L_1 and L_2 are 13 kVA at 0.8-pf lag and 10 kVA at 0.96-pf lead, respectively. Express $\mathbf{V}_s$ in polar form.

ANSWER: $263.06\underline{/8.75°}$ V.

ANSWER: 7.744 Ω in series with 3.872 Ω of inductive reactance; 9.68 Ω in parallel with 19.36 Ω of inductive reactance.

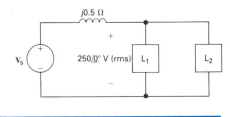

11.6 APPLIANCE RATINGS

The average power rating and estimated annual kilowatt-hour consumption of some common appliances are presented in Table 11.1. The energy consumption in kilowatt-hours is obtained by estimating the number of hours annually that an appliance is in use. For example, a coffee maker has an estimated annual consumption of 140 kWh and an average power consumption during operation of 1.2 kW. Therefore a coffee maker is assumed to be in operation 140/1.2, or 116.67, hours per year, or approximately 19 minutes per day.

TABLE 11.1

ANNUAL ENERGY REQUIREMENTS OF ELECTRIC HOUSEHOLD APPLIANCES

APPLIANCE	AVERAGE WATTAGE	EST. kWh CONSUMED ANNUALLY*
Food preparation		
Blender	300	1
Broiler	1,140	85
Carving knife	92	8
Coffee maker	1,200	140
Deep fryer	1,448	83
Dishwasher	1,201	165
Egg cooker	516	14
Frying pan	1,196	100
Hot plate	1,200	90
Mixer	127	2
Oven, microwave (only)	1,450	190
Range		
with oven	12,200	596
with self-cleaning oven	12,200	626

ANNUAL ENERGY REQUIREMENTS OF ELECTRIC HOUSEHOLD APPLIANCES

APPLIANCE	AVERAGE WATTAGE	EST. kWh CONSUMED ANNUALLY*
Roaster	1,333	60
Sandwich grill	1,161	33
Toaster	1,146	39
Trash compactor	400	50
Waffle iron	1,200	20
Waste dispenser	445	7
Food preservation		
Freezer		
manual defrost		
16 cu. ft.	—	1,050
automatic defrost		
16.5 cu. ft.	—	1,820
Refrigerator/freezer		
manual defrost,		
12.5 cu. ft.	—	1,500
automatic defrost,		
17.5 cu. ft.	—	1,591
Laundry		
Clothes dryer	4,856	993
Iron (hand)	1,100	60
Washing machine		
automatic	512	103
nonautomatic	286	76
Water heater	2,475	4,219
quick-recovery	4,474	4,811
Comfort conditioning		
Air cleaner	50	216
Air conditioner (room)	860	860[†]
Bed covering	177	147
Dehumidifier	257	377
Fan (attic)	370	291
Fan (circulating)	88	43
Fan (rollaway)	171	138
Fan (window)	200	170
Furnace fan	500	650
Heater (portable)	1,322	176
Heating pad	65	10
Humidifier	177	163
Health and beauty		
Germicidal lamp	20	141
Hair dryer	600	25
Heat lamp (infrared)	250	13
Shaver	15	0.5
Sunlamp	279	16
Toothbrush	1.1	1
Vibrator	40	2

(continued)

TABLE 11.1 (continued)

ANNUAL ENERGY REQUIREMENTS OF ELECTRIC HOUSEHOLD APPLIANCES

APPLIANCE	AVERAGE WATTAGE	EST. kWh CONSUMED ANNUALLY*
Home entertainment		
Radio	71	86
Radio/record player	109	109
Television		
black and white		
tube-type	100	220
solid-state	45	100
color		
tube-type	240	528
solid-state	145	320
Housewares		
Clock	2	17
Floor polisher	305	15
Sewing machine	75	11
Vacuum cleaner	630	46

* The figures for the estimated annual kilowatt-hour consumption of the electric appliances listed in this handy reference are based on normal usage. When using these figures for projections, such factors as the size of the specific appliance, the geographical area of use, and individual usage should be taken into consideration. Please note that the wattages are not additive since all units are normally not in operation at the same time.

† Based on 1000 hours of operation per year. This figure will vary widely depending on area and specific size of unit. See EEI-Pub #76-2 "Air Conditioning Usage Study" for an estimate for your location.

Source: Edison Electric Institute, 1111 19th Street N.W., Washington, D.C.

EXAMPLE 11.7

The branch circuit supplying the outlets in a typical home kitchen is wired with #12 conductor and is protected by either a 20-A fuse or a 20-A circuit breaker. Assume that the following 120-V appliances are in operation at the same time: coffee maker, egg cooker, frying pan, and toaster. Will the circuit be interrupted by the protective device?

SOLUTION

From Table 11.1, the total average power demanded by the four appliances is

$$P = 1200 + 516 + 1196 + 1146 = 4058 \text{ W}.$$

The total current in the protective device is

$$I_{\text{eff}} = \frac{4058}{120} \cong 33.82 \text{ A}.$$

Yes, the protective device will interrupt the circuit.

DRILL EXERCISES

11.6 a) A university student is drying her hair under a sunlamp while watching a soap opera on a color tube-type television set. At the same time, her roommate is vacuuming the rug in their air-conditioned bedroom. If all these appliances are supplied from a 120-V branch circuit that is protected by a 15-A circuit breaker, will the breaker interrupt the soap opera?

11.7 a) A personal computer that has a built-in monitor and keyboard requires 40 W at 115 V (rms). Calculate the rms value of the current carried by its power cord.

b) A flexible disk drive for the personal computer in part (a) is rated at 90 W at 115 V (rms).

b) Will the student be able to watch television if she turns off the sunlamp and her roommate turns off the vacuum cleaner?

ANSWER: (a) Yes, the breaker current is approximately 22 A. (b) Yes, if she can locate the distribution panel and reclose the breaker. The current will be approximately 14 A.

If the disk drive is plugged into the same wall outlet as the computer, what is the rms value of the current drawn from the outlet?

ANSWER: (a) 0.35 A; (b) 1.13 A.

11.7 MAXIMUM POWER TRANSFER

When information is transmitted by means of electric signals, the ability to deliver as much power as possible to the load is important sometimes. Transmission efficiency may be of secondary importance. For a network operating in the sinusoidal steady state, maximum average power is delivered to a load when the load impedance is the conjugate of the Thévenin impedance of the network as viewed from the terminals of the load. We describe the problem of maximum power transfer in terms of Fig. 11.12. Here, a linear network is operating in the sinusoidal steady state, and we must determine the load impedance Z_L that results in delivery of maximum average power to terminals a and

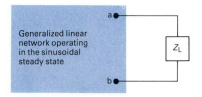

FIGURE 11.12 A circuit describing maximum power transfer.

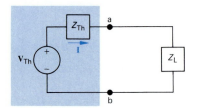

FIGURE 11.13 The circuit shown in Fig. 11.12, with the network replaced by its Thévenin equivalent.

b. Any linear network may be viewed from the terminals of the load in terms of a Thévenin equivalent circuit. Thus the task reduces to finding the value of Z_L that results in maximum average power delivered to Z_L in the circuit shown in Fig. 11.13.

For maximum average power transfer to the load impedance, Z_L must equal the conjugate of the Thévenin impedance; that is,

$$Z_L = Z_{Th}^*. \tag{11.35}$$

We derive Eq. (11.35) by straightforward application of elementary calculus. We begin by expressing Z_{Th} and Z_L in rectangular form:

$$Z_{Th} = R_{Th} + jX_{Th} \tag{11.36}$$

and

$$Z_L = R_L + jX_L. \tag{11.37}$$

In both Eqs. (11.36) and (11.37), the reactance term carries its own algebraic sign—positive for inductance and negative for capacitance. Because we are making an average-power calculation, we assume that the amplitude of the Thévenin voltage is expressed in terms of its rms value. We also use the Thévenin voltage as the reference phasor. Then, from Fig. 11.13, the rms value of the load current $\mathbf{I}$ is

$$\mathbf{I} = \frac{\mathbf{V}_{Th}\underline{/0°}}{(R_{Th} + R_L) + j(X_{Th} + X_L)}. \tag{11.38}$$

The average power delivered to the load is

$$P = |\mathbf{I}|^2 R_L. \tag{11.39}$$

Substituting Eq. (11.38) into Eq. (11.39) yields

$$P = \frac{|\mathbf{V}_{Th}|^2 R_L}{(R_{Th} + R_L)^2 + (X_{Th} + X_L)^2}. \tag{11.40}$$

When working with Eq. (11.40) always remember that $\mathbf{V}_{Th}$, R_{Th}, and X_{Th} are fixed quantities, whereas R_L and X_L are independent variables. Therefore, to maximize P, we must find the values of R_L and X_L, where $\partial P/\partial R_L$ and $\partial P/\partial X_L$ both are zero. From Eq. (11.40),

$$\frac{\partial P}{\partial X_L} = \frac{-|\mathbf{V}_{Th}|^2 2R_L(X_L + X_{Th})}{[(R_L + R_{Th})^2 + (X_L + X_{Th})^2]^2}; \tag{11.41}$$

$$\frac{\partial P}{\partial R_L} = \frac{|\mathbf{V}_{Th}|^2[(R_L + R_{Th})^2 + (X_L + X_{Th})^2 - 2R_L(R_L + R_{Th})]}{[(R_L + R_{Th})^2 + (X_L + X_{Th})^2]^2}. \tag{11.42}$$

From Eq. (11.41), $\partial P / \partial X_L$ is zero when

$$X_L = -X_{Th}. \tag{11.43}$$

From Eq. (11.42), $\partial P / \partial R_L$ is zero when

$$R_L = \sqrt{R_{Th}^2 + (X_L + X_{Th})^2}. \tag{11.44}$$

Note that, when we combine Eq. (11.43) with Eq. (11.44), both derivatives are zero when $Z_L = Z_{Th}^*$.

MAXIMUM VALUE OF AVERAGE POWER ABSORBED

The maximum average power that can be delivered to Z_L when it is set equal to the conjugate of Z_{Th} is calculated directly from the circuit in Fig. 11.13. When $Z_L = Z_{Th}^*$, the rms load current is $\mathbf{V}_{Th}/2R_L$ and the maximum average power delivered to the load is

$$P_{max} = \frac{|\mathbf{V}_{Th}|^2 R_L}{4R_L^2} = \frac{1}{4} \cdot \frac{|\mathbf{V}_{Th}|^2}{R_L}. \tag{11.45}$$

If the Thévenin voltage is expressed in terms of its maximum amplitude rather than its rms amplitude, Eq. (11.45) becomes

$$P_{max} = \frac{1}{8} \frac{V_m^2}{R_L}. \tag{11.46}$$

MAXIMUM POWER TRANSFER WHEN Z IS RESTRICTED

Maximum average power can be delivered to Z_L *only* if Z_L can be set equal to the conjugate of Z_{Th}. There are situations in which this is not possible. First R_L and X_L may be restricted to a limited range of values. In this situation, the optimum condition for R_L and X_L is to adjust X_L as near $-X_{Th}$ as possible and then adjust R_L as close to $\sqrt{R_{Th}^2 + (X_L + X_{Th})^2}$ as possible (see Example 11.8). A second type of restriction occurs when the magnitude of Z_L can be varied but its phase angle cannot. Under this restriction, the greatest amount of power is transferred to the load when the magnitude of Z_L is set equal to the magnitude of Z_{Th}, that is, when

$$|Z_L| = |Z_{Th}|. \tag{11.47}$$

The proof of Eq. (11.47) is left to you as Problem 11.33.

For purely resistive networks the condition for maximum power transfer is simply that the load resistance equal the Thévenin resistance. Note that we first derived this result in the introduction to maximum power transfer in Chapter 4.

Examples 11.8, 11.9, and 11.10 illustrate the problem of obtaining maximum power transfer in the basic situations discussed above.

E X A M P L E   11.8

a) For the circuit shown in Fig. 11.14, determine the impedance Z_L that results in maximum average-power transfer to Z_L.

b) What is the maximum average power transferred to the load impedance determined in (a)?

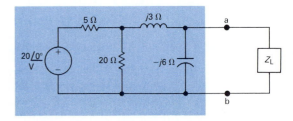

FIGURE 11.14 The circuit for Example 11.8.

S O L U T I O N

a) We begin by determining the Thévenin equivalent with respect to the load terminals a, b. After two source transformations involving the 20-V source, the 5-Ω resistor, and the 20-Ω resistor, we simplify the circuit shown in Fig. 11.14 to that shown in Fig. 11.15. Then,

$$\mathbf{V}_{Th} = \frac{16\underline{/0°}}{4 + j3 - j6}(-j6)$$

$$= 19.2\underline{/-53.13°} = 11.52 - j15.36 \text{ V}$$

and

$$Z_{Th} = \frac{(-j6)(4 + j3)}{4 + j3 - j6} = 5.76 - j1.68 \ \Omega.$$

For maximum average-power transfer, the load impedance must be the conjugate of Z_{Th}, so

$$Z_L = 5.76 + j1.68 \ \Omega.$$

b) We calculate the maximum average power delivered to Z_L from the circuit shown in Fig. 11.16, where we replaced the original network with its Thévenin equivalent. From the circuit shown in Fig. 11.16, the rms magnitude of the load current $\mathbf{I}$ is

$$I_{eff} = \frac{19.2/\sqrt{2}}{2(5.76)} = 1.1785 \text{ A}.$$

The average power delivered to the load is

$$P = I_{eff}^2(5.76) = 8 \text{ W}.$$

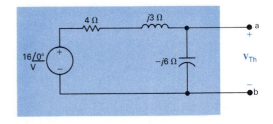

FIGURE 11.15 A simplification of Fig. 11.14 by source transformations.

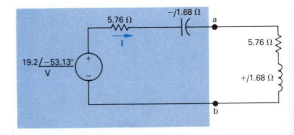

FIGURE 11.16 The circuit shown in Fig. 11.14, with the original network replaced by its Thévenin equivalent.

EXAMPLE 11.9

a) For the circuit shown in Fig. 11.17, what value of Z_L results in maximum average-power transfer to Z_L? What is the maximum power in milliwatts?

b) Assume that the load resistance can be varied between 0 and 4000 Ω and that the capacitive reactance of the load can be varied between 0 and -2000 Ω. What settings of R_L and X_L transfer the most average power to the load? What is the maximum power that can be transferred to the load under these restrictions?

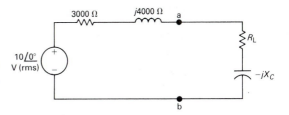

FIGURE 11.17 The circuit for Examples 11.9 and 11.10.

SOLUTION

a) If there are no restrictions on R_L and X_L, the load impedance is set equal to the conjugate of the output or Thévenin impedance. Therefore we set

$$R_L = 3000 \ \Omega \quad \text{and} \quad X_C = -4000 \ \Omega,$$

or

$$Z_L = 3000 - j4000 \ \Omega.$$

As the source voltage is given in terms of its rms value, the average power delivered to Z_L is

$$P = \frac{1}{4} \frac{10^2}{3000} = \frac{25}{3} \text{ mW} = 8.33 \text{ mW}.$$

b) As R_L and X_L are restricted, we first set X_C as close to -4000 Ω as possible, or -2000 Ω. Next, we set R_L as close to $\sqrt{R_{Th}^2 + (X_L + X_{Th})^2}$ as possible. Thus

$$R_L = \sqrt{3000^2 + (-2000 + 4000)^2} = 3605.55 \ \Omega.$$

Now, because R_L can be varied from 0 to 4000 Ω, we can set R_L to 3605.55 Ω. Therefore the load impedance is adjusted to a value of

$$Z_L = 3605.55 - j2000 \ \Omega.$$

With Z_L set at this value, the value of the load current is

$$\mathbf{I}_{eff} = \frac{10/0°}{6605.55 + j2000} = 1.4489/-16.85° \text{ mA}.$$

The average power delivered to the load is

$$P = (1.4489 \times 10^{-3})^2(3605.55) = 7.567 \text{ mW}.$$

This quantity is the maximum power that we can deliver to a load, given the restrictions on R_L and X_L. Note that this power is less than can be delivered if there are no restrictions on R_L and X_L; recall that in (a) we found that we can deliver 8.33 mW.

E X A M P L E 11.10

A load impedance having a constant phase angle of $-36.87°$ is connected across the load terminals a, b in the circuit shown in Fig. 11.17. The magnitude of Z_L is varied until the average power delivered is the most possible under the given restriction.

a) Specify Z_L in rectangular form.
b) Calculate the average power delivered to Z_L.

S O L U T I O N

a) From Eq. (11.47), we know that the magnitude of Z_L must equal the magnitude Z_{Th}. Therefore

$$|Z_L| = |Z_{Th}| = |3000 + j4000| = 5000 \ \Omega.$$

Now, as we know that the phase angle of Z_L is $-36.87°$, we have

$$Z_L = 5000\underline{/-36.87°} = 4000 - j3000 \ \Omega.$$

b) With Z_L set equal to $4000 - j3000 \ \Omega$, the load current is

$$\mathbf{I}_{eff} = \frac{10}{7000 + j1000} = 1.4142\underline{/-8.13°} \text{ mA},$$

and the average power delivered to the load is

$$P = 1.4142^2(4) = 8 \text{ mW}.$$

This quantity is the maximum power that can be delivered by this circuit to a load impedance whose angle is constant at $-36.87°$. Again, this quantity is less than the maximum power that can be delivered if there are no restrictions on Z_L.

DRILL EXERCISES

11.8 The source current in the circuit shown is
5 cos 8000*t* A.

a) What impedance should be connected across terminals a, b for maximum average-power transfer?

b) What is the average power transferred to the impedance in (a)?

c) Assume that the load is restricted to pure resistance. What size resistor connected across a, b will result in the maximum average power transferred?

d) What is the average power transferred to the resistor in (c)?

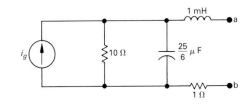

ANSWER: (a) $10 - j5 \ \Omega$; (b) 28.125 W; (c) 11.18 Ω; (d) 26.56 W.

SUMMARY

Most of the generation, transmission distribution, and utilization of electric power is done with circuits operating in the sinusoidal steady state. Thus power calculations involving sinusoidal currents and voltages are an important aspect of circuit analysis. We discussed the following concepts used in power calculations:

- Instantaneous power is the product of the instantaneous terminal voltage and current, or $p = \pm vi$. The positive sign is used when the reference direction for the current is from the positive to the negative reference polarity of the voltage. The frequency of the instantaneous power is twice the frequency of the voltage (current).

- Average, or real, power is the average value of the instantaneous power over one period. It is the power converted from electrical to nonelectrical form and vice versa. This conversion is the reason that average power also is referred to as real power. Average power, with the passive sign convention, is expressed as

$$P = \frac{1}{2} V_m I_m \cos (\theta_v - \theta_i)$$

$$= V_{\text{eff}} I_{\text{eff}} \cos (\theta_v - \theta_i).$$

- Reactive power is the electrical power that oscillates between the magnetic field of an inductor and the electrical field of a capacitor. Reactive power is never converted to nonelectrical power. Reactive power, with the passive sign convention, is expressed as

$$Q = \frac{1}{2} V_m I_m \sin (\theta_v - \theta_i)$$

$$= V_{eff} I_{eff} \sin (\theta_v - \theta_i).$$

- The power factor is the cosine of the phase angle between the voltage and the current:

$$pf = \cos (\theta_v - \theta_i).$$

The terms *lagging* and *leading* added to the description of the power factor indicate whether the current is lagging or leading the voltage.

- The reactive factor is the sine of the phase angle between the voltage and the current:

$$rf = \sin (\theta_v - \theta_i).$$

- Complex power is the complex sum of the real and reactive power, or

$$S = P + jQ$$

$$= \frac{1}{2} \mathbf{V} \mathbf{I}^* = \mathbf{V}_{eff} \mathbf{I}_{eff}^*$$

$$= I_{eff}^2 Z = \frac{V_{eff}^2}{Z^*}.$$

- Apparent power is the magnitude of the complex power:

$$|S| = \sqrt{P^2 + Q^2}.$$

- The watt is used as the unit for both instantaneous and real power.

- The VAR (volt amp reactive) is used as the unit for reactive power.

- The VA (volt amp) is used as the unit for complex and apparent power.

- Maximum power transfer occurs in circuits operating in the sinusoidal steady state when the load impedance is the conjugate of the Thévenin impedance as viewed from the terminals of the load impedance. (See Section 11.7 for a discussion of maximum power transfer when there are restrictions on the load impedance.)

PROBLEMS

11.1 The following sets of values for v and i pertain to the circuit seen in Fig. 11.1. For each set of values, calculate P and Q and state whether the circuit inside the box is absorbing or delivering (1) average power and (2) magnetizing vars.

a) $v = 100 \cos(\omega t + 50°)$ V
   $i = 10 \cos(\omega t + 15°)$ A

b) $v = 40 \cos(\omega t - 15°)$ V
   $i = 20 \cos(\omega t + 60°)$ A

c) $v = 400 \cos(\omega t + 30°)$ V
   $i = 10 \sin(\omega t + 240°)$ A

d) $v = 200 \sin(\omega t + 250°)$ V
   $i = 5 \cos(\omega t + 40°)$ A

11.2 Show that the maximum value of the instantaneous power given by Eq. (11.15) is $P + \sqrt{P^2 + Q^2}$ and the minimum value is $P - \sqrt{P^2 + Q^2}$.

11.3 A load consisting of a 14.7-kΩ resistor in parallel with a 5.04-H inductor is connected across the terminals of a sinusoidal voltage source v_g, where $v_g = 840 \cos 10^4 t$ V.

a) What is the peak value of the instantaneous power delivered by the source?

b) What is the peak value of the instantaneous power absorbed by the source?

c) What is the average power delivered to the load?

d) What is the reactive power?

e) Does the load absorb or generate magnetizing vars?

f) What is the power factor of the load?

g) What is the reactive factor of the load?

11.4 The three loads in the circuit in Fig. P11.4 can be described as follows: Load 1 is a 10-Ω resistor in series with a 10-mH inductor; load 2 is a 10-μF capacitor in series with a 100-Ω resistor; and load 3 is a 125-Ω resistor in series with the parallel combination of a 10-H inductor and a 10-μF capacitor. Give the power factor and reactive factor of each load if the frequency of the voltage source is 60 Hz.

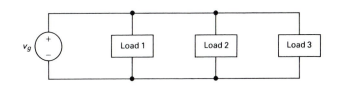

FIGURE P11.4

11.5 Find the rms value of the periodic current shown in Fig. P11.5.

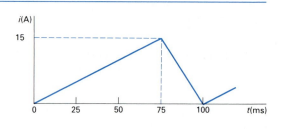

FIGURE P11.5

11.6 The periodic current shown in Fig. P11.5 dissipates an average power of 3000 W in a resistor. What is the value of the resistor?

11.7 a) Find the rms value of the periodic voltage shown in Fig. P11.7.

b) If this voltage is applied to the terminals of a 2.25-Ω resistor, what is the average power dissipated in the resistor?

FIGURE P11.7

11.8 A dc voltage equal to V_{dc} volts is applied to a resistor of R ohms. A sinusoidal voltage equal to v_s volts is also applied to a resistor of R ohms. Show that the dc voltage will deliver the same amount of energy in T seconds (where T is the period of the sinusoidal voltage) as the sinusoidal voltage provided V_{dc} equals the rms value of v_s. (*Hint:* Equate the two expressions for the energy delivered to the resistor.)

11.9 The voltage V_g in the phasor-domain circuit shown in Fig. P11.9 is $170\underline{/0°}$ V.

a) Find the average and reactive powers at the terminals of the voltage source.

b) Is the voltage source absorbing or delivering average power?

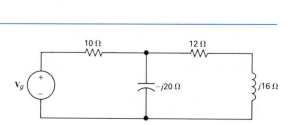

FIGURE P11.9

c) Is the voltage source absorbing or delivering magnetizing vars?

d) Find the average and reactive powers associated with each impedance branch in the circuit.

e) Check the balance between delivered and absorbed average power.

f) Check the balance between delivered and absorbed magnetizing vars.

11.10 Find the average power, the reactive power, and the complex power absorbed by the load in the circuit in Fig. P11.10 if i_g equals $30 \cos 100t$ mA.

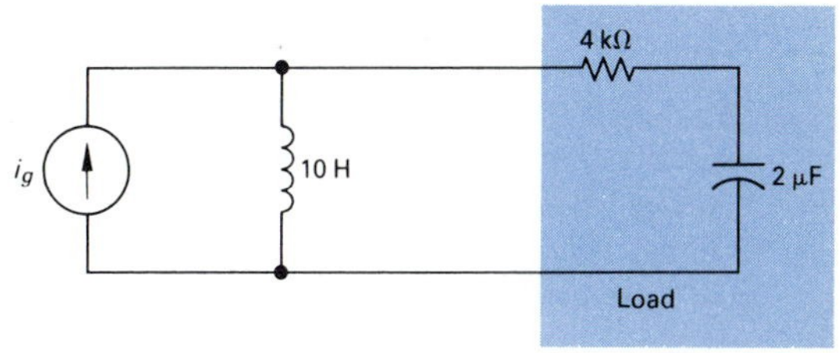

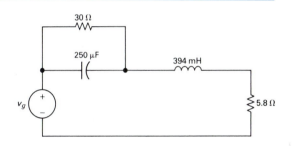

FIGURE P11.10

11.11 Find the average power, the reactive power, and the complex power supplied by the voltage source in the circuit in Fig. P11.11 if $v_g = 10 \cos 100t$ V.

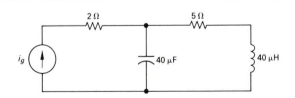

FIGURE P11.11

11.12 Find the average power delivered by the ideal current source in the circuit in Fig. P11.12 if $i_g = 10 \cos 25,000t$ mA.

FIGURE P11.12

11.13 a) Calculate the real and reactive power associated with each circuit element in the circuit in Fig. P10.41.

b) Verify that the average power generated equals the average power absorbed.

c) Verify that the magnetizing vars generated equal the magnetizing vars absorbed.

11.14 Repeat Problem 11.13 for the circuit shown in Fig. P10.44.

11.15 Find the average power dissipated in the 40-Ω resistor in the circuit seen in Fig. P11.15 if $i_g = 4 \cos 10^5 t$ A.

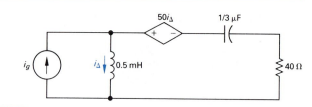

FIGURE P11.15

11.16 The capacitive load impedance in Fig. P11.16 has a magnitude of 100 Ω and absorbs an average power of 6144 W. The sinusoidal voltage source supplies an average power of 6400 W. Find the inductive reactance of the line.

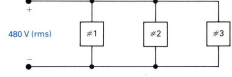

FIGURE P11.16

11.17 Three loads are connected in parallel across a 480-V (rms) line as shown in Fig. P11.17. Load 1 absorbs 25 kW and 25 kVAR. Load 2 absorbs 15 kVA at 0.8 pf lead. Load 3 absorbs 11 kW at unity power factor.

a) Find the impedance that is equivalent to the three parallel loads.

b) Find the power factor of the equivalent load as seen from the line's input terminals.

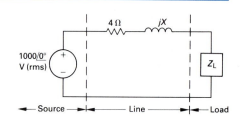

FIGURE P11.17

11.18 Two 250-V (rms) loads are connected in parallel. The two loads draw a total average power of 19,200 W at a power factor of 0.80 leading.

One of the loads draws 10 kVA at a power factor of 0.96 lagging. What is the reactive factor of the other load?

11.19 The two loads shown in Fig. P11.19 can be described as follows: Load 1 absorbs an average power of 218.24 kW and 56.32 kVAR magnetizing reactive power; load 2 has an impedance of $40 - j30$ Ω. The voltage at the terminals of the loads is $4400\sqrt{2} \cos 120\pi t$ V.

a) Find the rms value of the source voltage.

b) By how many microseconds is the load voltage out of phase with the source voltage?

c) Does the load voltage lead or lag the source voltage?

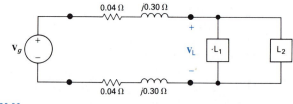

FIGURE P11.19

11.20 The three parallel loads in the circuit shown in Fig. P11.20 can be described as follows: Load 1 is absorbing an average power of 60 kW and 40 kVAR of magnetizing vars; load 2 is absorbing an average power of 20 kW and is generating 10 kVAR of magnetizing reactive power; load 3 consists of a 144-Ω resistor in parallel with an inductive reactance of 96 Ω.

Find the rms magnitude and the phase angle of $\mathbf{V}_g$ if $\mathbf{V}_o = 2400\underline{/0°}$ V (rms).

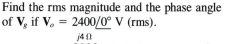
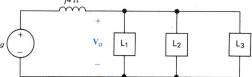

FIGURE P11.20

11.21 The three loads in the circuit seen in Fig. P11.21 are described as follows: Load 1 is absorbing 1.2 kW and 240 VAR; load 2 is 1 kVA at a 0.96-pf lead; load 3 is a 6.25-Ω resistor in parallel with an inductor that has a reactance of 25 Ω. Calculate the average power and the magnetizing reactive power delivered by each source if $\mathbf{V}_{g1} = \mathbf{V}_{g2} = 125\underline{/0°}$ V (rms).

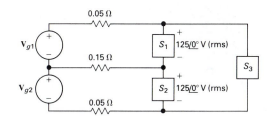

FIGURE P11.21

11.22 The three loads in the circuit shown in Fig. P11.22 are $S_1 = 4 + j1$ kVA, $S_2 = 5 + j2$ kVA, and $S_3 = 10 + j0$ kVA.

a) Calculate the complex power associated with each voltage source $\mathbf{V}_{g1}$ and $\mathbf{V}_{g2}$.

b) Verify that the total real and reactive power delivered by the sources equals the total real and reactive power absorbed by the network.

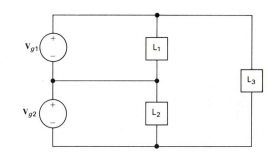

FIGURE P11.22

11.23 The three loads in Problem 11.17 are fed from a line having a series impedance $0.03 + j0.24$ Ω as shown in Fig. P11.23.

a) Calculate the rms value of the voltage ($\mathbf{V}_s$) at the sending end of the line.

b) Calculate the average and reactive powers associated with the line impedance.

c) Calculate the average and reactive powers at the sending end of the line.

d) Calculate the efficiency (η) of the line if the efficiency is defined as

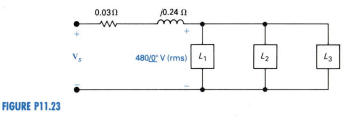

FIGURE P11.23

11.24 A group of small appliances on a 60-Hz system requires 20 kVA at 0.85 pf lagging when operated at 125 V (rms). The impedance of the feeder supplying the appliances is 0.01+ $j0.08$ Ω. The voltage at the load end of the feeder is 125 V.

   a) What is the rms magnitude of the voltage at the source end of the feeder?

   b) What is the average power loss in the feeder?

   c) What size capacitor (in microfarads) at the load end of the feeder is needed to improve the load power factor to unity?

   d) After the capacitor is installed, what is the rms magnitude of the voltage at the source end of the feeder if the load voltage is maintained at 125 V?

   e) What is the average power loss in the feeder for part (d)?

11.25 The circuit shown in Fig. P11.21 represents a residential distribution circuit in which the impedances of the service conductors are negligible and $\mathbf{V}_{g1} = \mathbf{V}_{g2} = 125\underline{/0°}$ V (rms). The three loads in the circuit are L_1: a dishwasher, a mixer, and a microwave oven; L_2: a portable heater and a sun lamp; and L_3: a clothes dryer, water heater, and a range with an oven. Assume that all these appliances are in operation at the same time. The service conductors are protected with 100-A circuit breakers. Will the service to this residence be interrupted? Explain.

11.26 a) Find V (rms) and θ for the circuit in Fig. P11.26 if the load absorbs 2500 VA at a lagging power factor of 0.8.

   b) Construct a phasor diagram of each solution obtained in part (a).

FIGURE P11.26

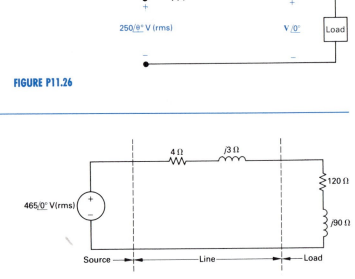

11.27 a) Find the average power dissipated in the line in the circuit in Fig. P11.27.

   b) Find the capacitive reactance that when connected in parallel with the load will make the load look purely resistive.

   c) What is the equivalent impedance of the load in part (b)?

   d) Find the average power dissipated in the line when the capacitive reactance is connected across the load.

FIGURE P11.27

11.28 The steady-state voltage drop between the load and the sending end of the line seen in Fig. P11.28 is excessive. A capacitor is placed in parallel with the 192-kVA load and is adjusted until the steady-state voltage at the sending end of the line has the same magnitude as the

voltage at the load end, that is, 4800 V (rms).
The 192-kVA load is operating at a power factor of 0.8 lag. Calculate the size of the capacitor in microfarads if the circuit is operating at
60 Hz. In selecting the capacitor, keep in
mind the need to keep the power loss in the
line at a reasonable level.

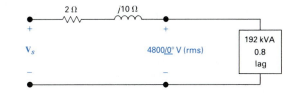

FIGURE P11.28

11.29 a) Determine the load impedance for the circuit shown in Fig. P11.29 that will result in
maximum average power being transferred
to the load if $\omega = 10k$ rad/s.

b) Determine the maximum average power if
$v_g = 120 \cos 10^4 t$ V.

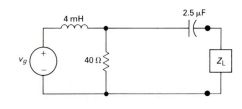

FIGURE P11.29

11.30 The peak amplitude of the sinusoidal voltage
source in the circuit shown in Fig. P11.30 is
$100\sqrt{2}$ V and its period is 250π μs. The
load resistor can be varied from 0 to 200 Ω,
and the load capacitor can be varied from 1 to
4 μF.

a) Calculate the average power delivered to
the load when $R_L = 100$ Ω and $C_L = 4$ μF.

b) Determine the settings of R_L and C_L that
will result in the most average power being
transferred to R_L.

c) What is the most average power in part
(b)? Is it greater than the power in part (a)?

d) If there are no constraints on R_L and C_L,

what is the maximum average power that
can be delivered to a load?

e) What are the values of R_L and C_L for the
condition of part (d)?

f) Is the average power calculated in part (d)
larger than that calculated in part (c)?

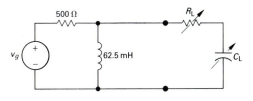

FIGURE P11.30

11.31 a) Assume that R_L in Fig. P11.30 can be
varied between 0 and 500 Ω. Repeat parts
(b) and (c) of Problem 11.30.

b) Is the new average power calculated in part

(a) greater than that found in Problem
11.30?

c) Is the new average power calculated in part
(a) less than that found in 11.30(d)?

11.32 The phasor voltage $\mathbf{V}_{ab}$ in the circuit shown in
Fig. P11.32 is $480/\underline{0°}$ V (rms) when no external load is connected to the terminals a, b.

When a load having an impedance of $200 - j150$ Ω is connected across a, b, the value of
$\mathbf{V}_{ab}$ is $440 - j80$ V (rms).   (*continued*)

a) Find the impedance that should be connected across a, b for maximum average power transfer.

b) Find the maximum average power transferred to the load of part (a).

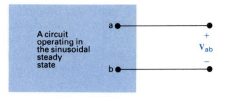

FIGURE P11.32

11.33 Prove that if only the magnitude of the load impedance can be varied, the most average power is transferred to the load when $|Z_L| = |Z_{Th}|$. (*Hint:* In deriving the expression for the average load power, write the load impedance (Z_L) in the form $Z_L = |Z_L| \cos \theta + j|Z_L| \sin \theta$ and note that only $|Z_L|$ is variable.)

11.34 The variable resistor in the circuit shown in Fig. P11.34 is adjusted until the average power it absorbs is maximum.

a) Find R.

b) Find the maximum average power.

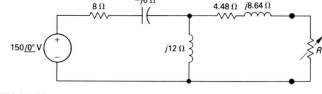

FIGURE P11.34

11.35 The load impedance Z_L for the circuit shown in Fig. P11.35 is adjusted until maximum average power is delivered to Z_L.

a) Find the maximum average power delivered to Z_L.

b) What percentage of the total power developed in the circuit is delivered to Z_L?

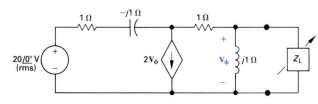

FIGURE P11.35

11.36 The variable resistor R_σ in the circuit shown in Fig. P11.36 is adjusted until maximum average power is delivered to R_σ.

a) What is the value of R_σ in ohms?

b) Calculate the average power delivered to R_σ.

c) If R_σ is replaced with a variable impedance Z_σ, what is the maximum average power that can be delivered to Z_σ?

d) In part (c), what percentage of the circuit's developed power is delivered to the load Z_σ?

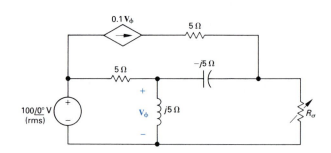

FIGURE P11.36

11.37 A factory has an electrical load of 1200 kW at a lagging power factor of 0.8. An additional variable power factor load is to be added to the factory. The new load will add 240 kW to the real power load of the factory. The power factor of the added load is to be adjusted so that the overall power factor of the factory is 0.96 lagging.

a) Specify the reactive power associated with the added load.

b) Does the added load absorb or deliver magnetizing vars?

c) What is the power factor of the additional load?

d) Assume that the rms voltage at the input to the factory is 2500 V. What is the rms magnitude of the current into the factory before the variable power factor load is added?

e) What is the rms magnitude of the current into the factory after the variable power factor load has been added?

11.38 The sending end voltage in the circuit seen in Fig. P11.38 is adjusted so that the rms value of the load voltage is always 13,200 V. The variable capacitor is adjusted until the average power dissipated in the line resistance is minimum.

a) If the frequency of the sinusoidal source is 60 Hz, what is the value of the capacitance in microfarads?

b) If the capacitor is removed from the circuit, what percentage increase in the magnitude of $\mathbf{V}_s$ is necessary to maintain 13,200 V at the load?

c) If the capacitor is removed from the circuit, what is the percentage increase in line loss?

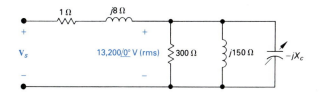

FIGURE P11.38

11.39 The operational amplifier in the circuit shown in Fig. P11.39 is ideal. Calculate the average power delivered to the 1000-Ω resistor when $v_g = 1 \cos 1000t$ V.

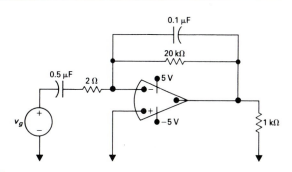

FIGURE P11.39

BALANCED THREE-PHASE CIRCUITS

CHAPTER 12

The generation, transmission, distribution, and utilization of large blocks of electric power are accomplished by means of three-phase circuits. The comprehensive analysis of three-phase systems is a field of study in its own right, which we cannot hope to cover in a single chapter. Fortunately, an understanding of only the steady-state sinusoidal behavior of *balanced* three-phase circuits is sufficient for engineers who do not specialize in power systems. We define what we mean by a balanced circuit later in the discussion. For the moment, we note that there are two reasons for restricting our introduction to balanced operation. First, for economic reasons, three-phase systems are designed to operate in the balanced state. That is, under normal operating conditions the three-phase circuit is so close to being balanced that we are justified in finding the solution that assumes perfect balance. Second, problems involving some types of unbalanced operating conditions can be solved by a technique known as the *method of symmetrical components,* which relies heavily on a thorough understanding of balanced operation. Although we do not discuss the method of symmetrical components, an understanding of balanced operation is a starting point for an advanced technique used to analyze certain types of unbalanced conditions.

The basic structure of a three-phase system consists of voltage sources connected to loads by means of transformers[†] and transmission lines. We can reduce the problem to the analysis of a circuit consisting of a voltage source connected to a load via a line. The omission of the transformer as an element in the system simplifies the discussion without jeopardizing a basic understanding of the calculations involved. Figure 12.1 shows the basic circuit.

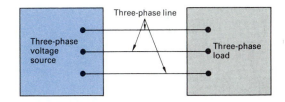

FIGURE 12.1 A basic three-phase circuit.

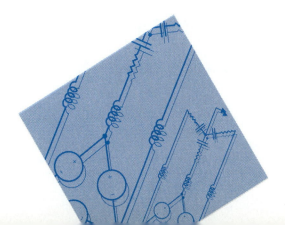

[†]We introduce transformers in Chapter 13.

To begin analyzing a circuit of this type requires an understanding of the characteristics of a balanced three-phase set of sinusoidal voltages.

12.1 BALANCED THREE-PHASE VOLTAGES

A set of balanced three-phase voltages consists of three sinusoidal voltages that have identical amplitudes and frequency but are out of phase with each other by exactly 120°. In discussing three-phase circuits, standard practice is to refer to the three phases as a, b, and c. Furthermore, the a-phase is almost always used as the reference phase. The three voltages that comprise the three-phase set are referred to as the *a-phase voltage*, the *b-phase voltage*, and the *c-phase voltage*.

Because the phase voltages are out of phase by 120°, two possible phase relationships can exist between the a-phase voltage and the b- and c-phase voltages. One possibility is for the b-phase voltage to lag the a-phase voltage by 120°, in which case the c-phase voltage must lead the a-phase voltage by 120°. This phase relationship is known as the *abc, or positive, phase sequence*. The only other possibility is for the b-phase voltage to lead the a-phase voltage by 120°, in which case the c-phase voltage must lag the a-phase voltage by 120°. This phase relationship is known as the *acb, or negative, phase sequence*. In phasor notation, the two possible sets of balanced phase voltages are

$$\mathbf{V}_a = V_m \underline{/0°};$$
$$\mathbf{V}_b = V_m \underline{/-120°}; \qquad \textbf{(12.1)}$$
$$\mathbf{V}_c = V_m \underline{/+120°};$$

and

$$\mathbf{V}_a = V_m \underline{/0°};$$
$$\mathbf{V}_b = V_m \underline{/+120°}; \qquad \textbf{(12.2)}$$
$$\mathbf{V}_c = V_m \underline{/-120°}.$$

The phase sequence of the voltages given by Eqs. (12.1) is the abc, or positive, sequence. The phase sequence of the voltages given by Eqs. (12.2) is the acb, or negative, sequence. Figure 12.2 shows the phasor diagram representations of the voltage sets in Eqs. (12.1) and (12.2). The phase sequence is the clockwise order of the subscripts around the diagram from $\mathbf{V}_a$. The possibility that a three-phase circuit can have one of two phase

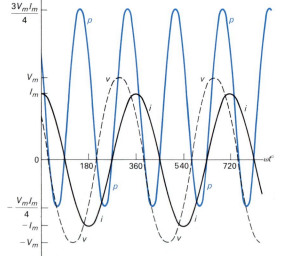

FIGURE 12.2 Phasor diagrams of a balanced set of three-phase voltages: (a) abc (positive) sequence; (b) acb (negative) sequence.

sequences must be taken into account whenever two separate circuits are operated in parallel. The two circuits can operate in parallel only if they have the same phase sequence.

Another important characteristic of a set of balanced three-phase voltages is that the sum of the voltages is zero. Thus, from either Eqs. (12.1) or Eqs. (12.2),

$$\mathbf{V}_a + \mathbf{V}_b + \mathbf{V}_c = 0. \qquad (12.3)$$

Because the sum of the phasor voltages is zero, the sum of the instantaneous voltages also is zero; that is,

$$v_a + v_b + v_c = 0. \qquad (12.4)$$

If we know the *phase sequence and one voltage in the set, we know the entire set*. Thus for a balanced three-phase system, we can focus on determining the voltage (or current) in one phase, because once we know one phase quantity we automatically know the corresponding quantity in the other two phases.

DRILL EXERCISES

12.1 What is the phase sequence of each of the following sets of voltages?

a) $v_a = 208 \cos (\omega t + 76°)$ V,
$v_b = 208 \cos (\omega t + 316°)$ V,
$v_c = 208 \cos (\omega t - 164°)$ V.

b) $v_a = 4160 \cos (\omega t - 49°)$ V,
$v_b = 4160 \cos (\omega t - 289°)$ V,
$v_c = 4160 \cos (\omega t + 191°)$ V.

ANSWER: (a) abc; (b) acb.

12.2 THREE-PHASE VOLTAGE SOURCES

Three-phase voltage sources consist of generators that have three separate windings distributed around the periphery of the stator. Each winding comprises one phase of the generator. The rotor of the generator is an electromagnet driven at synchronous speed by a prime mover, such as a steam or gas turbine. Rotation of the electromagnet past the three windings induces a sinusoidal voltage in each winding. The phase windings are designed so that the sinusoidal voltages induced in them are equal in amplitude and out of phase with each other by 120°. The phase windings are stationary with respect to the rotating electromagnet, so the frequency of the voltage induced in each winding is the same.

PSpice   Use three PSpice ac sources with different phases to simulate three-phase circuit behavior: Section 11.2

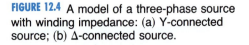

FIGURE 12.3 The two basic connections of an ideal three-phase source: (a) Y-connected source; (b) Δ-connected source.

Normally, the impedance of each phase winding on a three-phase generator is small compared to the other impedances in the circuit. Therefore, to an approximation, we can model each phase winding in an electric circuit by an ideal sinusoidal voltage source. There are two ways of interconnecting the separate phase windings to form a three-phase source. The windings can be connected in either a wye (Y) or a delta (Δ) configuration. Figure 12.3 shows the wye and delta connections, with ideal voltage sources used to model the phase windings of the three-phase generator. The common terminal in the Y-connected source, labeled n in Fig. 12.3(a), is called the *neutral terminal* of the source. The neutral terminal may or may not be available for external connections.

If the impedance of each phase winding is not negligible, we model the three-phase source by placing the winding impedance in series with an ideal sinusoidal voltage source. All windings on the machine are of the same construction, so we assume the winding impedances to be identical. The winding impedance of three-phase generators is inductive. Figure 12.4 shows the model of a three-phase source, including winding impedance, in which R_w is the winding resistance and X_w is the inductive reactance of the winding.

Because a three-phase voltage source can be either Y-connected or Δ-connected—and the three-phase loads can also be either Y-connected or Δ-connected—the basic circuit in Fig. 12.1 can take four different configurations. The four possible arrangements are (1) a Y-connected source and a Y-connected load; (2) a Y-connected source and Δ-connected load; (3)

FIGURE 12.4 A model of a three-phase source with winding impedance: (a) Y-connected source; (b) Δ-connected source.

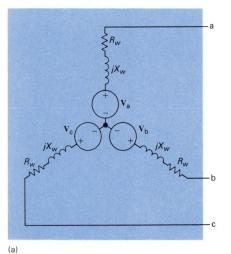

(a)

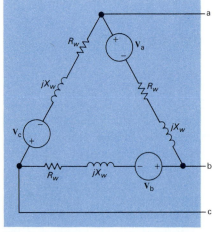

(b)

a Δ-connected source and a Y-connected load; and (4) a Δ-connected source and a Δ-connected load.

We begin our analysis of three-phase circuits with arrangement (1). After analyzing the Y–Y circuit, we show for balanced circuits how to reduce the remaining three arrangements to a Y–Y equivalent circuit. In other words, analysis of the Y–Y circuit is the key to solving all balanced three-phase arrangements.

12.3 ANALYSIS OF THE WYE–WYE CIRCUIT

We begin analysis of the Y–Y circuit by assuming that the circuit is *not* balanced! We do so to show what we mean by a balanced three-phase circuit and its consequences for circuit analysis. Figure 12.5 illustrates the general Y–Y circuit, in which we included a fourth conductor that connects the source neutral to the load neutral. A fourth conductor is possible only in the Y–Y arrangement. (More about this later.) For convenience in drawing the diagram, we transformed the Y-connections into "tipped-over tees." In Fig. 12.5, Z_{ga}, Z_{gb}, and Z_{gc} represent the internal impedance associated with each phase winding of the voltage source; Z_{1a}, Z_{1b}, and Z_{1c} represent the impedance of each phase conductor of the line connecting the source to the load; Z_o is the impedance of the neutral conductor that connects the source neutral to the load neutral; and Z_A, Z_B, and Z_C represent the impedance of each phase of the load.

We can describe the circuit shown in Fig. 12.5 with a single node-voltage equation. Using the source neutral as the reference node and letting V_N denote the node voltage between the nodes

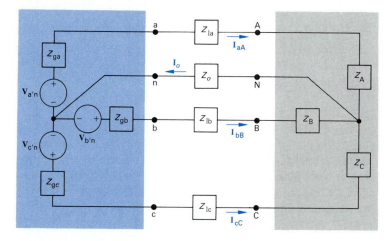

FIGURE 12.5 A three-phase Y–Y system.

N and n, we find that the node-voltage equation is

$$\frac{\mathbf{V_N}}{Z_o} + \frac{\mathbf{V_N} - \mathbf{V_{a'n}}}{Z_A + Z_{1a} + Z_{ga}} + \frac{\mathbf{V_N} - \mathbf{V_{b'n}}}{Z_B + Z_{1b} + Z_{gb}}$$

$$+ \frac{\mathbf{V_N} - \mathbf{V_{c'n}}}{Z_C + Z_{1c} + Z_{gc}} = 0. \qquad (12.5)$$

Before we pursue Eq. (12.5), note that the circuit analysis techniques discussed in the earlier chapters are directly applicable to three-phase circuits. Thus introducing new analytical techniques in order to analyze three-phase circuits is unnecessary. However, a balanced three-phase circuit allows taking some significant analytic shortcuts to predict the behavior of the system.

The circuit shown in Fig. 12.5 is a balanced three-phase circuit if it satisfies *all* the following criteria:

1. $\mathbf{V_{a'n}}$, $\mathbf{V_{b'n}}$, and $\mathbf{V_{c'n}}$ form a set of balanced three-phase voltages;

2. $Z_{ga} = Z_{gb} = Z_{gc}$;

3. $Z_{1a} = Z_{1b} = Z_{1c}$; and

4. $Z_A = Z_B = Z_C$.

There is no restriction on the impedance of the neutral conductor (Z_o); its value has no effect on whether the system is balanced.

If the system is balanced, Eq. (12.5) tells us that $\mathbf{V_N}$ must be zero. To demonstrate, we let

$$Z_\phi = Z_A + Z_{1a} + Z_{ga} \qquad (12.6)$$

and then rewrite Eq. (12.5) as

$$\mathbf{V_N}\left(\frac{1}{Z_o} + \frac{3}{Z_\phi}\right) = \frac{\mathbf{V_{a'n}} + \mathbf{V_{b'n}} + \mathbf{V_{c'n}}}{Z_\phi}. \qquad (12.7)$$

The right-hand side of Eq. (12.7) is zero because by hypothesis the numerator is a set of balanced three-phase voltages and Z_ϕ is *not* zero. The only value of $\mathbf{V_N}$ that satisfies Eq. (12.7) is zero. Therefore for a balanced three-phase circuit,

$$\mathbf{V_N} = 0. \qquad (12.8)$$

Equation (12.8) is extremely important. If $\mathbf{V_N}$ is zero, there is no difference in potential between the source neutral, n, and the load neutral, N; consequently, the current in the neutral conductor is zero. Hence we may either remove the neutral conductor from a balanced Y–Y configuration ($I_0 = 0$) or replace it with a perfect short circuit between the nodes n and N ($\mathbf{V_N} = 0$). We find both equivalents convenient to use when modeling balanced three-phase circuits.

We now turn to the effect that balanced conditions have on the three line currents. With reference to Fig. 12.5, when the system is balanced the three line currents are

$$\mathbf{I}_{aA} = \frac{\mathbf{V}_{a'n} - \mathbf{V}_N}{Z_A + Z_{1a} + Z_{ga}} = \frac{\mathbf{V}_{a'n}}{Z_\phi}; \qquad (12.9)$$

$$\mathbf{I}_{bB} = \frac{\mathbf{V}_{b'n} - \mathbf{V}_N}{Z_B + Z_{1b} + Z_{gb}} = \frac{\mathbf{V}_{b'n}}{Z_\phi}; \qquad (12.10)$$

$$\mathbf{I}_{cC} = \frac{\mathbf{V}_{c'n} - \mathbf{V}_N}{Z_c + Z_{1c} + Z_{gc}} = \frac{\mathbf{V}_{c'n}}{Z_\phi}. \qquad (12.11)$$

That is, in a balanced system the three line currents form a balanced set of three-phase currents. Thus the current in each line is equal in amplitude and frequency and 120° out of phase with the other two line currents. Thus, if we calculate the current $\mathbf{I}_{aA}$, we can write the line currents $\mathbf{I}_{bB}$ and $\mathbf{I}_{cC}$ without further computations. This statement implies that the phase sequence is known.

We can use Eq. (12.9) to construct a single-phase equivalent circuit of the balanced three-phase Y–Y circuit. From Eq. (12.9), the current in the a-phase conductor line is simply the voltage generated in the a-phase winding of the generator divided by the total impedance in the a-phase of the circuit. Thus Eq. (12.9) describes the simple circuit shown in Fig. 12.6, in which the neutral conductor has been replaced by a perfect short circuit. A word of caution here. The current in the neutral conductor in Fig. 12.6 is *not* the current in the neutral conductor of a balanced three-phase circuit. The current in the neutral conductor is

$$\mathbf{I}_o = \mathbf{I}_{aA} + \mathbf{I}_{bB} + \mathbf{I}_{cC}, \qquad (12.12)$$

whereas the current in the neutral conductor in Fig. 12.6 is $\mathbf{I}_{aA}$. Thus the circuit shown in Fig. 12.6 gives the correct value of the line current but only the a-phase component of the neutral current. Whenever the single-phase equivalent circuit shown in Fig. 12.6 is applicable, the line currents form a balanced three-phase set, and the right-hand side of Eq. (12.12) sums to zero.

Once we know the line currents in the circuit in Fig. 12.5, calculating any voltages of interest is relatively simple. Of particular interest is the relationship between the line-to-line voltages and the line-to-neutral voltages. We establish this relationship at the load terminals, but our observations also apply at the source terminals. The line-to-line voltages at the load terminals

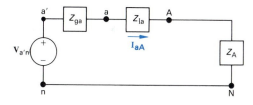

FIGURE 12.6 A single-phase equivalent circuit.

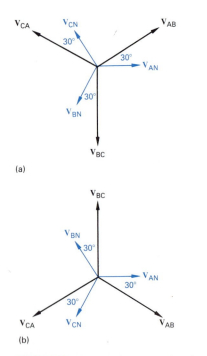

in terms of the line-to-neutral load voltages are

$$\mathbf{V}_{AB} = \mathbf{V}_{AN} - \mathbf{V}_{BN}, \tag{12.13}$$

$$\mathbf{V}_{BC} = \mathbf{V}_{BN} - \mathbf{V}_{CN}, \tag{12.14}$$

and

$$\mathbf{V}_{CA} = \mathbf{V}_{CN} - \mathbf{V}_{AN}. \tag{12.15}$$

The double subscript notation in voltage equations indicates a voltage drop from the first to the second subscript. Figure 12.7 shows the relationships expressed by Eqs. (12.13)–(12.15). Because we are interested in the balanced state, we omitted the neutral conductor from Fig. 12.7.

To show the relationship between the line-to-line voltages and the line-to-neutral voltages, we assume a positive, or abc, sequence. We arbitrarily choose the line-to-neutral voltage of the a-phase as the reference. Thus

$$\mathbf{V}_{AN} = V_{\phi}\underline{/0°}, \tag{12.16}$$

$$\mathbf{V}_{BN} = V_{\phi}\underline{/-120°}, \tag{12.17}$$

and

$$\mathbf{V}_{CN} = V_{\phi}\underline{/+120°}, \tag{12.18}$$

where V_{ϕ} represents the magnitude of the line-to-neutral voltage. Substituting Eqs. (12.16)–(12.18) into Eqs. (12.13)–(12.15), respectively, yields

$$\mathbf{V}_{AB} = V_{\phi} - V_{\phi}\underline{/-120°} = \sqrt{3}\,V_{\phi}\underline{/30°}, \tag{12.19}$$

$$\mathbf{V}_{BC} = V_{\phi}\underline{/-120°} - V_{\phi}\underline{/120°} = \sqrt{3}\,V_{\phi}\underline{/-90°}, \tag{12.20}$$

and

$$\mathbf{V}_{CA} = V_{\phi}\underline{/120°} - V_{\phi}\underline{/0°} = \sqrt{3}\,V_{\phi}\underline{/150°}. \tag{12.21}$$

Equations (12.19)–(12.21) reveal that (1) the magnitude of the line-to-line voltage is $\sqrt{3}$ times the magnitude of the line-to-neutral voltage, (2) the line-to-line voltages form a balanced three-phase set of voltages, and (3) the set of line-to-line voltages leads the set of line-to-neutral voltages by 30°. We leave to you demonstration that for a negative, or acb, sequence the only change is that the set of line-to-line voltages lags the set of line-to-neutral voltages by 30°. The phasor diagrams shown in Fig. 12.8 summarize these observations. Thus, in a balanced system, if you know the line-to-neutral voltage at some point in the circuit, you also know the line-to-line voltages at the same point, and vice versa.

DRILL EXERCISES

12.2 The voltage from B to N in a balanced three-phase circuit is $120\underline{/60°}$ V. If the phase sequence is positive, what is the value of $\mathbf{V}_{BC}$?

ANSWER: $207.85\underline{/+90°}$ V.

12.3 The c-phase voltage of a balanced, three-phase, Y-connected system is $660\underline{/160°}$ V. If the phase sequence is negative, what is the value of $\mathbf{V}_{AB}$?

ANSWER: $1143.15\underline{/-110°}$ V.

Before illustrating balanced three-phase calculations with a numerical example, we must comment further on terminology. In the Y–Y system, the line-to-neutral voltage also is called the *phase voltage*. For brevity, the line-to-line voltage also is called the *line voltage*. The *phase current* is the current in each phase of the load or, at the source end of the circuit, the current in each phase of the generator. The *line current* is the current in each phase of the line. For the Y–Y arrangement, the phase current and line current are identical. Because three-phase systems are designed to handle large blocks of electric power, all voltage and current specifications and calculations are given as rms values. Thus when a three-phase transmission line is rated at 345 kV, the nominal value of the rms line-to-line voltage is 345,000 V. In this chapter *we express all voltages and currents as rms values*. Finally, the Greek letter phi (ϕ) is widely used in the literature to denote a per phase quantity. Thus $\mathbf{V}_\phi$, $\mathbf{I}_\phi$, Z_ϕ, P_ϕ, and Q_ϕ are interpreted as voltage/phase, current/phase, impedance/phase, power/phase, and reactive power/phase, respectively.

Example 12.1 shows how to use the observations made so far to solve a balanced three-phase Y–Y circuit.

E X A M P L E   12.1

A three-phase, positive-sequence, Y-connected generator has an impedance of $0.2 + j0.5$ Ω/ϕ. The internal phase voltage of the generator is 120 V. The generator feeds a balanced, three-phase, Y-connected load having an impedance of $39 + j28$ Ω/ϕ. The impedance of the line connecting the generator to the load is $0.8 + j1.5$ Ω/ϕ. The a-phase internal voltage of the generator is specified as the reference phasor.

a) Construct a single-phase equivalent circuit of the three-phase system.

b) Calculate the three line currents $\mathbf{I}_{aA}$, $\mathbf{I}_{bB}$, and $\mathbf{I}_{cC}$.

c) Calculate the three line-to-neutral voltages at the load $\mathbf{V}_{AN}$, $\mathbf{V}_{BN}$, $\mathbf{V}_{CN}$.

d) Calculate the line voltages $\mathbf{V}_{AB}$, $\mathbf{V}_{BC}$, and $\mathbf{V}_{CA}$ at the terminals of the load.

e) Calculate the line-to-neutral voltages at the terminals of the generator $\mathbf{V}_{an}$, $\mathbf{V}_{bn}$, $\mathbf{V}_{cn}$.

f) Calculate the line voltages $\mathbf{V}_{ab}$, $\mathbf{V}_{bc}$, and $\mathbf{V}_{ca}$ at the terminals of the generator.

g) Repeat (a)–(f) for a negative phase sequence.

S O L U T I O N

a) Figure 12.9 shows the single-phase equivalent circuit.

b) The a-phase line current is

$$\mathbf{I}_{aA} = \frac{120/0°}{(0.2 + 0.8 + 39) + j(0.5 + 1.5 + 28)}$$

$$= \frac{120/0°}{40 + j30} = 2.4/-36.87° \,\text{A}.$$

For a positive phase sequence,

$$\mathbf{I}_{bB} = 2.4/-156.87° \,\text{A};$$

$$\mathbf{I}_{cC} = 2.4/83.13° \,\text{A}.$$

c) The line-to-neutral voltage at the A terminal of the load is

$$\mathbf{V}_{AN} = (39 + j28)(2.4/-36.87°)$$

$$= 115.22/-1.19° \,\text{V}.$$

For a positive phase sequence,

$$\mathbf{V}_{BN} = 115.22/-121.19° \,\text{V};$$

$$\mathbf{V}_{CN} = 115.22/+118.81° \,\text{V}.$$

d) For a positive phase sequence, the line-to-line voltages lead the line-to-neutral voltages by 30°; thus

$$\mathbf{V}_{AB} = (\sqrt{3}/30°)\,\mathbf{V}_{AN}$$

$$= 199.58/28.81° \,\text{V};$$

$$\mathbf{V}_{BC} = 199.58/-91.19° \,\text{V};$$

$$\mathbf{V}_{CA} = 199.58/148.81° \,\text{V}.$$

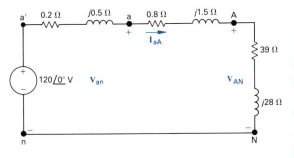

FIGURE 12.9 The single-phase equivalent circuit for Example 12.1.

e) The line-to-neutral voltage at the a terminal of the source is

$$\mathbf{V}_{an} = 120 - (0.2 + j0.5)(2.4\underline{/-36.87°})$$

$$= 120 - 1.29\underline{/31.33°}$$

$$= 118.90 - j0.67$$

$$= 118.90\underline{/-0.32°}\,V.$$

For a positive phase sequence,

$$\mathbf{V}_{bn} = 118.90\underline{/-120.32°};$$

$$\mathbf{V}_{cn} = 118.90\underline{/119.68°}\,V.$$

f) The line-to-line voltages at the source terminals are

$$\mathbf{V}_{ab} = (\sqrt{3}\underline{/30°})\,\mathbf{V}_{an}$$

$$= 205.94\underline{/29.68°}\,V;$$

$$\mathbf{V}_{bc} = 205.94\underline{/-90.32°}\,V;$$

$$\mathbf{V}_{ca} = 205.94\underline{/149.68°}\,V.$$

g) Changing the phase sequence has no effect on the single-phase equivalent circuit. The three line currents are

$$\mathbf{I}_{aA} = 2.4\underline{/-36.87°}\,A;$$

$$\mathbf{I}_{bB} = 2.4\underline{/83.13°}\,A;$$

$$\mathbf{I}_{cC} = 2.4\underline{/-156.87°}\,A.$$

The line-to-neutral voltages at the load are

$$\mathbf{V}_{AN} = 115.22\underline{/-1.19°}\,V;$$

$$\mathbf{V}_{BN} = 115.22\underline{/+118.81°}\,V;$$

$$\mathbf{V}_{CN} = 115.22\underline{/-121.19°}\,V.$$

For a negative phase sequence, the line-to-line voltages lag the line-to-neutral voltages by 30°:

$$\mathbf{V}_{AB} = (\sqrt{3}\underline{/-30°}\,\mathbf{V}_{AN}$$

$$= 199.58\underline{/-31.19°}\,V;$$

$$\mathbf{V}_{BC} = 199.58\underline{/88.81°}\,V;$$

$$\mathbf{V}_{CA} = 199.58\underline{/-151.19°}\,V.$$

The line-to-neutral voltages at the terminals of the generator are

$$\mathbf{V}_{an} = 118.90\underline{/-0.32°}\,V;$$

$$\mathbf{V}_{bn} = 118.90\underline{/119.68°}\,V;$$

$$\mathbf{V}_{cn} = 118.90\underline{/-120.32°}\,V.$$

The line-to-line voltages at the terminals of the generator are

$$\begin{aligned} \mathbf{V}_{ab} &= (\sqrt{3}\underline{/-30°})\mathbf{V}_{an} \\ &= 205.94\underline{/-30.32°}\,\text{V}; \\ \mathbf{V}_{bc} &= 205.94\underline{/89.68°}\,\text{V}; \\ \mathbf{V}_{ca} &= 205.94\underline{/-150.32°}\,\text{V}. \end{aligned}$$

In Example 12.1 note that calculation of the a-phase quantity allows tabulation of the corresponding b- and c-phase values simply by shifting the a-phase value by 120°. For a positive phase sequence, the b-phase lags the a-phase by 120°, whereas the c-phase leads the a-phase by 120°. For a negative phase sequence, the b-phase leads the a-phase by 120° and the c-phase lags the a-phase by 120°. Thus calculating line-to-line voltages is easy once you know the line-to-neutral voltages.

DRILL EXERCISES

12.4 The line-to-neutral voltage at the terminals of a balanced, three-phase, Y-connected load is 2400 V. The load has an impedance of $16 + j12\ \Omega/\phi$ and is fed from a line having an impedance of $0.10 + j0.80\ \Omega/\phi$. The Y-connected source at the sending end of the line has a phase sequence of acb and an internal impedance of $0.02 + j0.16\ \Omega/\phi$. Use the a-phase line-to-neutral voltage at the load as the reference and calculate (a) the line currents $\mathbf{I}_{aA}$, $\mathbf{I}_{bB}$, $\mathbf{I}_{cC}$; (b) the line-to-line voltages at the source $\mathbf{V}_{ab}$, $\mathbf{V}_{bc}$, $\mathbf{V}_{ca}$; and (c) the internal phase-to-neutral voltages at the source $\mathbf{V}_{a'n}$, $\mathbf{V}_{b'n}$, $\mathbf{V}_{c'n}$.

ANSWER: (a) $\mathbf{I}_{aA} = 120\underline{/-36.87°}\,\text{A}$, $\mathbf{I}_{bB} = 120\underline{/83.13°}\,\text{A}$, $\mathbf{I}_{cC} = 120\underline{/-156.87°}\,\text{A}$; (b) $\mathbf{V}_{ab} = 4275.02\underline{/-28.38°}\,\text{V}$, $\mathbf{V}_{bc} = 4275.02\underline{/91.62°}\,\text{V}$, $\mathbf{V}_{ca} = 4275.02\underline{/-148.38°}\,\text{V}$; (c) $\mathbf{V}_{a'n} = 2482.05\underline{/1.93°}\,\text{V}$, $\mathbf{V}_{b'n} = 2482.05\underline{/121.93°}\,\text{V}$, $\mathbf{V}_{c'n} = 2482.05\underline{/-118.07°}\,\text{V}$.

12.4 ANALYSIS OF THE WYE-DELTA CIRCUIT

If the load in a three-phase circuit is connected in a delta, it can be transformed into a wye by using the delta-to-wye transformation discussed in Section 10.6. When the load is balanced, the impedance of each leg of the wye is one third the impedance of each leg of the delta, or

$$Z_Y = \frac{Z_\Delta}{3}, \tag{12.22}$$

which follows directly from Eq. (10.49)–(10.51). After the Δ-load has been replaced by its Y-equivalent, the Y-source, Δ-load, three-phase circuit can be modeled by the single-phase equivalent circuit shown in Fig. 12.10.

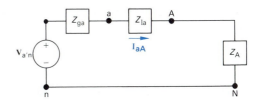

FIGURE 12.10 A single-phase equivalent circuit.

After we use the single-phase equivalent circuit to calculate the line currents, we obtain the current in each leg of the original Δ-load simply by dividing the line currents by $\sqrt{3}$ and shifting them through 30°. This relationship between the line currents and phase currents in the delta can be derived by using the circuit shown in Fig. 12.11.

When a load, or source, is connected in a delta, the current in each leg of the delta is the phase current, and the voltage across each leg is the phase voltage. Figure 12.11 shows that, in the Δ-configuration, the phase voltage is identical to the line voltage.

To demonstrate the relationship between the phase currents and line currents, we assume a positive phase sequence and let I_ϕ represent the magnitude of the phase current. Then

$$\mathbf{I}_{AB} = I_\phi\underline{/0°}\,, \tag{12.23}$$

$$\mathbf{I}_{BC} = I_\phi\underline{/-120°}\,, \tag{12.24}$$

and

$$\mathbf{I}_{CA} = I_\phi\underline{/+120°}\,. \tag{12.25}$$

In writing these equations, we arbitrarily selected $\mathbf{I}_{AB}$ as the reference phasor.

We can write the line currents in terms of the phase currents by direct application of Kirchhoff's current law:

$$\begin{aligned}\mathbf{I}_{aA} &= \mathbf{I}_{AB} - \mathbf{I}_{CA} = I_\phi\underline{/0°} - I_\phi\underline{/120°} \\ &= \sqrt{3}I_\phi\underline{/-30°}\,; \tag{12.26}\end{aligned}$$

$$\begin{aligned}\mathbf{I}_{bB} &= \mathbf{I}_{BC} - \mathbf{I}_{AB} = I_\phi\underline{/-120°} - I_\phi\underline{/0°} \\ &= \sqrt{3}I_\phi\underline{/-150°}\,; \tag{12.27}\end{aligned}$$

$$\begin{aligned}\mathbf{I}_{cC} &= \mathbf{I}_{CA} - \mathbf{I}_{BC} = I_\phi\underline{/120°} - I_\phi\underline{/-120°} \\ &= \sqrt{3}I_\phi\underline{/90°}\,. \tag{12.28}\end{aligned}$$

Comparing Eqs. (12.26)–(12.28) with Eqs. (12.23)–(13.25) reveals that the magnitude of the line currents is $\sqrt{3}$ times the magnitude of the phase currents and that the set of line currents *lags* the set of phase currents by 30°.

We leave to you to verify that, for a negative phase sequence, the line currents are $\sqrt{3}$ times larger than the phase currents and *lead* the phase currents by 30°.

Figure 12.12 summarizes the relationship between the line currents and the phase currents of a Δ-connected load.

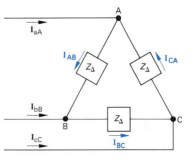

FIGURE 12.11 A circuit used to establish the relationship between line currents and phase currents in a balanced Δ-load.

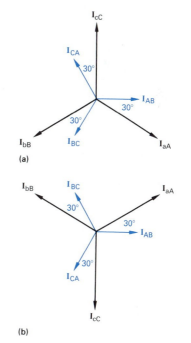

FIGURE 12.12 Phasor diagrams showing the relationship between line currents and phase currents in a Δ-connected load: (a) positive sequence; (b) negative sequence.

DRILL EXERCISES

12.5 The current $\mathbf{I}_{CA}$ in a balanced, three-phase, Δ-connected load is $15/38°$A. If the phase sequence is positive, what is the value of $\mathbf{I}_{cC}$?

ANSWER: $25.98/8°$A.

12.6 A balanced, three-phase, Δ-connected load is fed from a balanced three-phase circuit. The reference for the b-phase line current is toward the load. The value of the current is $26/-50°$A. If the phase sequence is negative, what is the value of $\mathbf{I}_{AB}$?

ANSWER: $15.01/160°$A.

Example 12.2 illustrates the calculations involved in analyzing a balanced three-phase circuit having a Y-connected source and a Δ-connected load.

E X A M P L E   12.2

The Y-connected source in Example 12.1 feeds a Δ-connected load through a distribution line having an impedance of $0.3 + j0.9 \ \Omega/\phi$. The load impedance is $118.5 + j85.8 \ \Omega/\phi$. Use the a-phase internal voltage of the generator as the reference.

a) Construct a single-phase equivalent circuit of the three-phase ·system.

b) Calculate the line currents $\mathbf{I}_{aA}$, $\mathbf{I}_{bB}$, and $\mathbf{I}_{cC}$.

c) Calculate the phase voltages at the load terminals.

d) Calculate the phase currents of the load.

e) Calculate the line voltages at the source terminals.

S O L U T I O N

a) Figure 12.13 shows the single-phase equivalent circuit. The load impedance of the Y-equivalent is

$$\left(\frac{1}{3}\right)(118.5 + j85.8) \qquad \text{or} \qquad 39.5 + j28.6 \ \Omega/\phi.$$

b) The a-phase line current is

$$\mathbf{I}_{aA} = \frac{120/0°}{(0.2 + 0.3 + 39.5) + j(0.5 + 0.9 + 28.6)}$$

$$= \frac{120/0°}{40 + j30} = 2.4/-36.87°A.$$

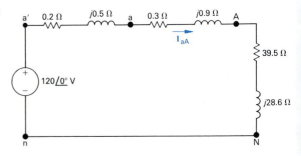

FIGURE 12.13 The single-phase equivalent circuit for Example 12.2.

Hence

$$\mathbf{I}_{bB} = 2.4\underline{/-156.87°}\,\mathrm{A};$$

$$\mathbf{I}_{cC} = 2.4\underline{/83.13°}\,\mathrm{A}.$$

c) As the load is Δ-connected, the phase voltages are the same as the line voltages. To calculate the line voltages, we first calculate $\mathbf{V}_{AN}$:

$$\mathbf{V}_{AN} = (39.5 + j28.6)(2.4\underline{/-36.87°})$$

$$= 117.04\underline{/-0.96°}\,\mathrm{V}.$$

Because the phase sequence is positive, the line voltage $\mathbf{V}_{AB}$ is

$$\mathbf{V}_{AB} = \sqrt{3}\underline{/30°}\,\mathbf{V}_{AN}$$

$$= 202.72\underline{/29.04°}\,\mathrm{V}.$$

Therefore

$$\mathbf{V}_{BC} = 202.72\underline{/-90.96°}\,\mathrm{V};$$

$$\mathbf{V}_{CA} = 202.72\underline{/149.04°}\,\mathrm{V}.$$

d) The phase currents of the load may be calculated directly from the line currents:

$$\mathbf{I}_{AB} = \frac{1}{\sqrt{3}}\underline{/30°}\,\mathbf{I}_{aA}$$

$$= 1.39\underline{/-6.87°}\,\mathrm{A}.$$

Once we know $\mathbf{I}_{AB}$, we also know the other load phase currents:

$$\mathbf{I}_{BC} = 1.39\underline{/-126.87°}\,\mathrm{A};$$

$$\mathbf{I}_{CA} = 1.39\underline{/113.13°}\,\mathrm{A}.$$

Note that we can check the calculation of $\mathbf{I}_{AB}$ by using the previously calculated $\mathbf{V}_{AB}$ and the impedance of the Δ-connected load. That is,

$$\mathbf{I}_{AB} = \frac{\mathbf{V}_{AB}}{Z_\phi} = \frac{202.72\underline{/29.04°}}{118.5 + j85.8}$$

$$= 1.39\underline{/-6.87°}\,\mathrm{A}.$$

(Alternative methods of calculation help eliminate errors, and we highly recommend their use in all work involving analysis and design.)

e) To calculate the line voltage at the terminals of the source, we first calculate $\mathbf{V}_{an}$. Figure 12.13 shows that $\mathbf{V}_{an}$ is the voltage drop across the line impedance plus the load impedance, so

$$\mathbf{V}_{an} = (39.8 + j29.5)2.4\underline{/-36.87°}$$

$$= 118.90\underline{/-0.32°}\,\text{V}.$$

The line voltage $\mathbf{V}_{ab}$ is

$$\mathbf{V}_{ab} = \sqrt{3}\underline{/30°}\,\mathbf{V}_{an} \quad \text{or} \quad \mathbf{V}_{ab} = 205.94\underline{/29.68°}\,\text{V}.$$

Therefore

$$\mathbf{V}_{bc} = 205.94\underline{/-90.32°}\,\text{V};$$

$$\mathbf{V}_{ca} = 205.94\underline{/+149.68°}\,\text{V}.$$

DRILL EXERCISES

12.7 The line-to-line voltage $\mathbf{V}_{AB}$ at the terminals of a balanced, three-phase, Δ-connected load is $4160\underline{/0°}\,\text{V}$. The line current $\mathbf{I}_{aA}$ is $69.28\underline{/-10°}\,\text{A}$.

  a) Calculate the per phase impedance of the load if the phase sequence is positive.

  b) Repeat (a) for a negative phase sequence.

ANSWER: (a) $104\underline{/-20°}\,\Omega$; (b) $104\underline{/+40°}\,\Omega$.

12.8 The line voltage at the terminals of a balanced, Δ-connected load is 208 V. Each phase of the load consists of a 5.2-Ω resistor in parallel with a 6.933-Ω inductor. What is the magnitude of the current in the line feeding the load?

ANSWER: 86.60 A.

12.5 ANALYSIS OF THE DELTA–WYE CIRCUIT

In the Δ–Y three-phase circuit, the source is Δ-connected and the load is Y-connected. We obtain the single-phase equivalent circuit by replacing the balanced Δ-connected source with a Y-equivalent. We obtain the Y-equivalent of the source by dividing the internal phase voltages of the Δ-source by $\sqrt{3}$ and shifting this set of three-phase voltages by $-30°$ if the phase sequence is positive and by $+30°$ if the phase sequence is negative. The internal impedance of the Y-equivalent is one third the internal impedance of the Δ-source. Figure 12.14 illustrates the Y-equivalent circuit of a positive-sequence Δ-connected source.

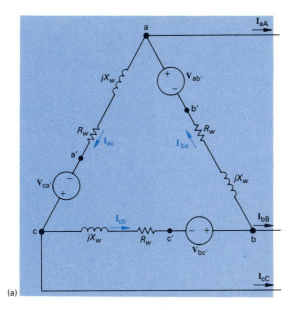

(a)

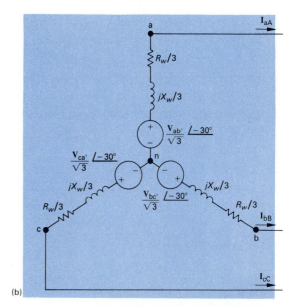

(b)

FIGURE 12.14 The Y-equivalent of a balanced, three-phase, Δ-connected source (positive sequence): (a) Δ-source; (b) Y-equivalent.

For a positive phase sequence, the set of Δ-source phase currents (I_{ba}, I_{cb}, and I_{ac} in Fig. 12.14) lead the set of line currents I_{aA}, I_{bB}, and I_{cC} by 30°. The magnitude of the phase currents is $1/\sqrt{3}$ times the magnitude of the line currents. For a negative phase sequence, the phase currents in the source lag the line currents by 30°.

To show that the Y-source in Fig. 12.14(b) is equivalent to the Δ-source of Fig. 12.14(a), we need to show only that the two circuits produce the same terminal conditions for any balanced *external* connections applied to the terminals a, b, and c. The two test conditions that are easiest to prove are open circuit and short circuit. For open-circuit conditions, the three line currents are zero, and the two circuits are equivalent if they deliver the same voltages between the terminals a, b, and c. For an external short circuit connecting the terminals a, b, and c, the line voltages are zero and the two circuits are equivalent if they deliver the same line currents. We leave verification that these two circuits are equivalent to you. (See Problem 12.14.)

Example 12.3 illustrates the numerical analysis of a Δ–Y three-phase circuit.

EXAMPLE 12.3

A balanced, negative-sequence, Δ-connected source has an internal impedance of $0.018 + j0.162 \ \Omega/\phi$. At no load, the terminal voltage of the source is 600 V. The source is connected to a Y-connected load, having an impedance of $7.92 - j6.35$ Ω/ϕ, through a distribution line having an impedance of $0.074 + j0.296\ \Omega/\phi$.

a) Construct a single-phase equivalent circuit of the system using $\mathbf{V}_{ab'}$ as the reference.

b) Calculate the magnitude of the line voltage at the terminals of the load.

c) Calculate the three line currents $\mathbf{I}_{aA}$, $\mathbf{I}_{bB}$, and $\mathbf{I}_{cC}$.

d) Calculate the phase currents $\mathbf{I}_{ba}$, $\mathbf{I}_{cb}$, and $\mathbf{I}_{ac}$ of the source.

e) Calculate the magnitude of the line voltage at the terminals of the source.

SOLUTION

a) At no load, the terminal voltage equals the internal voltage source. Therefore the internal voltage of the Δ-source is 600 V. Using $\mathbf{V}_{ab'}$ as a reference, we determine that the internal a-phase voltage of the Y-equivalent source is

$$\mathbf{V}_{a'n} = \frac{\mathbf{V}_{ab'}}{\sqrt{3}}\underline{/30°} = \frac{600}{\sqrt{3}}\underline{/30°}$$

$$\cong 346.41\underline{/30°}\text{V}.$$

The internal impedance of the equivalent Y-generator is $(1/3)(0.018 + j0.162)$, or $0.006 + j0.054\ \Omega/\phi$. Therefore the single-phase equivalent circuit is as shown in Fig. 12.15.

b) From the circuit shown in Fig. 12.15,

$$\mathbf{I}_{aA} = \frac{346.41\underline{/30°}}{8.00 - j6.00} = 34.64\underline{/66.87°}\ \text{A}$$

and

$$\mathbf{V}_{AN} = (7.92 - j6.35)(34.64\underline{/66.87°})$$

$$= 351.65\underline{/28.15°}\text{V}.$$

At the load the line voltage is

$$|\mathbf{V}_{AB}| = \sqrt{3}\,|\mathbf{V}_{AN}| = 609.08\ \text{V}.$$

c) Using the results of (b), we obtain the three line currents:

$$\mathbf{I}_{aA} = 34.64\underline{/66.87°}\ \text{A};$$

$$\mathbf{I}_{bB} = 34.64\underline{/186.87°}\ \text{A};$$

$$\mathbf{I}_{cC} = 34.64\underline{/-53.13°}\ \text{A}.$$

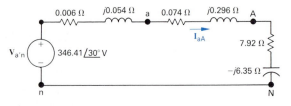

FIGURE 12.15 The single-phase equivalent circuit for Example 12.3.

d) We calculate the phase currents of the generator directly from the line currents. The phase sequence is negative, so

$$\mathbf{I}_{ba} = \frac{1}{\sqrt{3}} \underline{/-30°}\, \mathbf{I}_{aA}$$

$$= 20\underline{/36.87°}\, \text{A};$$

$$\mathbf{I}_{cb} = 20\underline{/156.87°}\, \text{A};$$

$$\mathbf{I}_{ac} = 20\underline{/-83.13°}\, \text{A}.$$

e) From the circuit shown in Fig. 12.15,

$$\mathbf{V}_{an} = (7.994 - j6.054)\mathbf{I}_{aA}$$

$$= (7.994 - j6.054)34.64\underline{/66.87°}$$

$$= 347.37\underline{/29.73°}\, \text{V}.$$

The line voltage at the source is

$$|\mathbf{V}_{ab}| = \sqrt{3}\,|\mathbf{V}_{an}| = 601.66 \text{ V}.$$

DRILL EXERCISES

12.9 A balanced, positive-sequence , Δ-connected source has an internal impedance of $0.09 + j0.81 \ \Omega/\phi$. The source is feeding a balanced load through a balanced line. The b-phase line current $\mathbf{I}_{bB}$ is $6\underline{/-120°}$ A, and the line voltage $\mathbf{V}_{ab}$ is $480\underline{/60°}$ V. Calculate the internal source voltage $\mathbf{V}_{ab'}$.

ANSWER: $481.68\underline{/60.27°}$ V.

12.6 ANALYSIS OF THE DELTA–DELTA CIRCUIT

In the Δ–Δ circuit, both source and load are Δ-connected. Replacing both source and load with their Y-equivalents produces the single-phase equivalent circuit of a balanced Δ–Δ system. As before, we use the Y-equivalent circuit to solve for line currents and line-to-neutral voltages. Once we know the line currents, we find the phase currents in both load and source by using the techniques described in Sections 12.4 and 12.5. We convert the line-to-neutral voltages to line-to-line voltages as described in

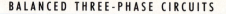

Section 12.3. We illustrated all these techniques in Examples 12.1, 12.2, and 12.3. You can gain additional experience with these types of calculations by solving Problems 12.12–12.17.

12.7 POWER CALCULATIONS IN BALANCED THREE-PHASE CIRCUITS

So far, we have limited analysis of balanced three-phase circuits to determining currents and voltages in a circuit. We now discuss three-phase power calculations. We begin by considering the average power delivered to a balanced, Y-connected load.

AVERAGE POWER IN A BALANCED Y-LOAD

Figure 12.16 shows a Y-connected load, along with its pertinent currents and voltages. We calculate the average power associated with any one phase by using the techniques introduced in Chapter 11. With Eq. (11.21) as a starting point, we express the average power associated with the a-phase of the load as

$$P_A = |\mathbf{V}_{AN}||\mathbf{I}_{aA}| \cos (\theta_{vA} - \theta_{iA}), \quad \textbf{(12.29)}$$

where θ_{vA} and θ_{iA} denote the phase angles of $\mathbf{V}_{AN}$ and $\mathbf{I}_{aA}$, respectively. Using the notation introduced in Eq. (12.29), we find the power associated with the b- and c-phases:

$$P_B = |\mathbf{V}_{BN}||\mathbf{I}_{bB}| \cos (\theta_{vB} - \theta_{iB}); \quad \textbf{(12.30)}$$

$$P_C = |\mathbf{V}_{CN}||\mathbf{I}_{cC}| \cos (\theta_{vC} - \theta_{iC}). \quad \textbf{(12.31)}$$

In Eqs. (12.29)–(12.31), all phasor currents and voltages are written in terms of the rms value of the sinusoidal function that they represent.

In a balanced three-phase system, the magnitude of each line-to-neutral voltage is the same, as is the magnitude of each phase current. The argument of the cosine functions also is the same for all three phases. We emphasize these observations by introducing the following notation to aid further discussion of power calculations in balanced three-phase circuits:

$$V_\phi = |\mathbf{V}_{AN}| = |\mathbf{V}_{BN}| = |\mathbf{V}_{CN}|, \quad \textbf{(12.32)}$$

$$I_\phi = |\mathbf{I}_{aA}| = |\mathbf{I}_{bB}| = |\mathbf{I}_{cC}|, \quad \textbf{(12.33)}$$

and

$$\theta_\phi = \theta_{vA} - \theta_{iA} = \theta_{vB} - \theta_{iB} = \theta_{vC} - \theta_{iC}. \quad \textbf{(12.34)}$$

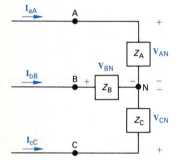

FIGURE 12.16 A balanced Y-load used to introduce average power calculations in three-phase circuits.

Moreover, for a balanced system, the power delivered to each phase of the load is the same, so

$$P_A = P_B = P_C = P_\phi = V_\phi I_\phi \cos \theta_\phi, \qquad \textbf{(12.35)}$$

where P_ϕ represents average power per phase.

The total average power delivered to the balanced Y-connected load is simply three times the power per phase, or

$$P_T = 3P_\phi = 3V_\phi I_\phi \cos \theta_\phi. \qquad \textbf{(12.36)}$$

Expressing the total power in terms of the *rms magnitude* of the line voltage and the *rms magnitude* of the line current also is desirable. If we let V_L represent the rms magnitude of the line voltage and I_L represent the rms magnitude of the line current, we can modify Eq. (12.36) as follows:

$$P_T = 3\left(\frac{V_L}{\sqrt{3}}\right) I_L \cos \theta_\phi$$

$$= \sqrt{3} V_L I_L \cos \theta_\phi. \qquad \textbf{(12.37)}$$

In deriving Eq. (12.37), we recognized that, for a balanced Y-connected load, the magnitude of the phase voltage is the magnitude of the line voltage divided by $\sqrt{3}$ and that the magnitude of the line current is equal to the magnitude of the phase current. When using Eq. (12.37) to calculate the total power delivered to the load, remember that θ_ϕ is the phase angle between the *phase voltage* and the *phase current*.

COMPLEX POWER IN A BALANCED Y-LOAD

We can also calculate the reactive power and complex power associated with any one phase of a Y-connected load by using the techniques introduced in Chapter 11. For the balanced load the expressions for the reactive power are

PSpice   PROBE can plot the power relationships in three-phase circuits after PSpice simulation is completed: Chapter 12

$$Q_\phi = V_\phi I_\phi \sin \theta_\phi; \qquad \textbf{(12.38)}$$

$$Q_T = 3Q_\phi = \sqrt{3} V_L I_L \sin \theta_\phi. \qquad \textbf{(12.39)}$$

Equation (11.27) is the basis for expressing the complex power associated with any phase. For a balanced load,

$$S = \mathbf{V}_{AN} \mathbf{I}_{aA}^* = \mathbf{V}_{BN} \mathbf{I}_{bB}^* = \mathbf{V}_{CN} \mathbf{I}_{cC}^* = \mathbf{V}_\phi \mathbf{I}_\phi^*, \qquad \textbf{(12.40)}$$

where $\mathbf{V}_\phi$ and $\mathbf{I}_\phi$ represent a phase voltage and current taken from the same phase. Thus, in general,

$$S_\phi = P_\phi + jQ_\phi = \mathbf{V}_\phi \mathbf{I}_\phi^*; \qquad \textbf{(12.41)}$$

$$S_T = 3S_\phi = \sqrt{3} V_L I_L \underline{/\theta_\phi}. \qquad \textbf{(12.42)}$$

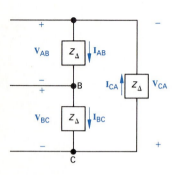

FIGURE 12.17 A Δ-connected load used to discuss power calculations.

POWER CALCULATIONS IN A BALANCED Δ-LOAD

If the load is Δ-connected, the calculation of power—reactive power or complex power—is basically the same as that for the Y-connected load. Figure 12.17 shows a Δ-connected load, along with its pertinent currents and voltages. The power associated with each phase is

$$P_A = |\mathbf{V}_{AB}||\mathbf{I}_{AB}| \cos(\theta_{vAB} - \theta_{iAB}); \tag{12.43}$$

$$P_B = |\mathbf{V}_{BC}||\mathbf{I}_{BC}| \cos(\theta_{vBC} - \theta_{iBC}); \tag{12.44}$$

$$P_C = |\mathbf{V}_{CA}||\mathbf{I}_{CA}| \cos(\theta_{vCA} - \theta_{iCA}). \tag{12.45}$$

For a balanced load,

$$|\mathbf{V}_{AB}| = |\mathbf{V}_{BC}| = |\mathbf{V}_{CA}| = V_\phi, \tag{12.46}$$

$$|\mathbf{I}_{AB}| = |\mathbf{I}_{BC}| = |\mathbf{I}_{CA}| = I_\phi, \tag{12.47}$$

$$\theta_{vAB} - \theta_{iAB} = \theta_{vBC} - \theta_{iBC} = \theta_{vCA} - \theta_{iCA} = \theta_\phi, \tag{12.48}$$

and

$$P_A = P_B = P_C = P_\phi = V_\phi I_\phi \cos\theta_\phi. \tag{12.49}$$

Note that Eq. (12.49) is the same as Eq. (12.35). That is, in a balanced load the average power per phase is equal to the product of the rms magnitude of the phase voltage, the rms magnitude of the phase current, and the cosine of the angle between the phase voltage and phase current.

The total power delivered to a balanced Δ-connected load is

$$P_T = 3P_\phi = 3V_\phi I_\phi \cos\theta_\phi$$

$$= 3V_L\left(\frac{I_L}{\sqrt{3}}\right) \cos\theta_\phi$$

$$= \sqrt{3}\,V_L I_L \cos\theta_\phi. \tag{12.50}$$

Note that Eq. (12.50) is the same as Eq. (12.37).

The expressions for reactive power and complex power also have the same form as those developed for the Y-load:

$$Q_\phi = V_\phi I_\phi \sin\theta_\phi; \tag{12.51}$$

$$Q_T = 3Q_\phi = 3V_\phi I_\phi \sin\theta_\phi; \tag{12.52}$$

$$S_\phi = P_\phi + jQ_\phi = \mathbf{V}_\phi \mathbf{I}_\phi^*; \tag{12.53}$$

$$S_T = 3S_\phi = \sqrt{3}V_L I_L \underline{/\theta_\phi}. \tag{12.54}$$

Examples 12.4–12.6 illustrate power calculations in balanced three-phase circuits.

E X A M P L E   12.4

---

a) Calculate the average power per phase delivered to the Y-connected load of Example 12.1.

b) Calculate the total average delivered to the load.

c) Calculate the total average power lost in the line.

d) Calculate the total average power lost in the generator.

e) Calculate the total number of magnetizing vars absorbed by the load.

f) Calculate the total complex power delivered by the source.

S O L U T I O N

---

a) From Example 12.1, $V_\phi = 115.22$ V, $I_\phi = 2.4$ A, and $\theta_\phi = -1.19 - (-36.87) = 35.68°$. Therefore

$$P_\phi = (115.22)(2.4) \cos 35.68°$$

$$= 224.64 \text{ W.}$$

The power per phase also may be calculated from $I_\phi^2 R_\phi$, or

$$P_\phi = (2.4)^2(39) = 224.64 \text{ W.}$$

b) The total average power delivered to the load is $P_T = 3P_\phi = 673.92$ W. We calculated the line voltage in Example 12.1, so we also may use Eq. (12.37):

$$P_T = \sqrt{3}(199.58)(2.4) \cos 35.68°$$

$$= 673.92 \text{ W.}$$

c) The total power lost in the line is

$$P_{\text{line}} = 3(2.4)^2(0.8) = 13.824 \text{ W.}$$

d) The total internal power loss in the generator is

$$P_{\text{gen}} = 3(2.4)^2(0.2) = 3.456 \text{ W.}$$

e) The total number of magnetizing vars absorbed by the load is

$$Q_T = \sqrt{3}(199.58)(2.4) \sin 35.68°$$

$$= 483.84 \text{ VAR.}$$

f) The total complex power associated with the source is

$$S_T = 3S_\phi = -3(120)(2.4)\underline{/36.87°}$$

$$= -691.20 - j518.40 \text{ VA.}$$

The minus sign indicates that the internal power and magnetizing reactive power are being delivered to the circuit. We check this result by calculating the total power and reactive power absorbed by the circuit:

$$P = 673.92 + 13.824 + 3.456$$

$$= 691.20 \text{ W} \quad (\text{check});$$

$$Q = 483.84 + 3(2.4)^2(1.5) + 3(2.4)^2(0.5)$$

$$= 483.84 + 25.92 + 8.64$$

$$= 518.40 \text{ VAR} \quad (\text{check}).$$

EXAMPLE 12.5

a) Calculate the total complex power delivered to the Δ-connected load of Example 12.2.

b) What percentage of the average power at the sending end of the line is delivered to the load?

SOLUTION

a) Using the a-phase values from the solution of Example 12.2, we obtain

$$\mathbf{V}_\phi = \mathbf{V}_{AB} = 202.72\underline{/29.04°} \text{ V};$$

$$\mathbf{I}_\phi = \mathbf{I}_{AB} = 1.39\underline{/-6.87°} \text{ A}.$$

Using Eqs. (12.53) and (12.54) we have

$$S_T = 3(202.72\underline{/29.04°})(1.39\underline{/+6.87°})$$

$$= 682.56 + j494.208 \text{ VA}.$$

b) The total power at the sending end of the distribution line equals the total power delivered to the load plus the total power lost in the line; therefore

$$P_{\text{input}} = 682.56 + 3(2.4)^2(0.3)$$

$$= 687.744 \text{ W}.$$

The percentage of the average power at the input of the line reaching the load is 682.56/687.744, or 99.25%.

EXAMPLE   12.6

A balanced three-phase load requires 480 kW at a lagging power factor of 0.8. The load is fed from a line having an impedance of $0.005 + j0.025 \ \Omega/\phi$. The line voltage at the terminals of the load is 600 V.

a) Construct a single-phase equivalent circuit of the system.

b) Calculate the magnitude of the line current.

c) Calculate the magnitude of the line voltage at the sending end of the line.

d) Calculate the power factor at the sending end of the line.

SOLUTION

a) Figure 12.18 shows the single-phase equivalent circuit. We arbitrarily selected the line-to-neutral voltage at the load as the reference.

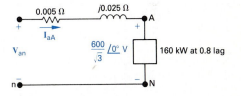

FIGURE 12.18 The single-phase equivalent circuit for Example 12.6.

b) The line current $\mathbf{I}_{aA}$ is

$$\left(\frac{600}{\sqrt{3}}\right) \mathbf{I}_{aA}^* = (160 + j120)10^3$$

or

$$\mathbf{I}_{aA}^* = 577.35 \underline{/36.87°} \ A.$$

Therefore $\mathbf{I}_{aA} = 577.35\underline{/-36.87°} \, A$. The magnitude of the line current is the magnitude of $\mathbf{I}_{aA}$:

$$I_L = 577.35 \ A.$$

We obtain an alternative solution for I_L from the expression

$$P_T = \sqrt{3}V_L I_L \cos \theta_p$$
$$= \sqrt{3}(600)I_L(0.8) = 480,000 \ W;$$

$$I_L = \frac{480,000}{\sqrt{3}(600)(0.8)} = \frac{1000}{\sqrt{3}} = 577.35 \ A.$$

c) To calculate the magnitude of the line voltage at the sending end of the line, we first calculate $\mathbf{V}_{an}$. From Fig. 12.18,

$$\mathbf{V}_{an} = \mathbf{V}_{an} + Z_L \mathbf{I}_{aA}$$

$$= \frac{600}{\sqrt{3}} + (0.005 + j0.025)(577.35\underline{/-36.87°})$$

$$= 357.51\underline{/1.57°}V.$$

Thus

$$V_L = \sqrt{3}|\mathbf{V}_{an}| = 619.23 \text{ V.}$$

d) The power factor at the sending end of the line is the cosine of the phase angle between $\mathbf{V}_{an}$ and $\mathbf{I}_{aA}$:

$$\text{pf} = \cos [1.57° - (-36.87°)]$$

$$= \cos 38.44° = 0.783 \text{ lagging.}$$

An alternative method for calculating the power factor is to first calculate the complex power at the sending end of the line:

$$S_\phi = (160 + j120)10^3 + (577.35)^2(0.005 + j0.025)$$

$$= 161.67 + j128.33 \text{ kVA}$$

$$= 206.41\underline{/38.44°} \text{ kVA.}$$

The power factor is

$$\text{pf} = \cos 38.44° = 0.783 \text{ lagging.}$$

Finally, if we calculate the total complex power at the sending end of the line, after first calculating the magnitude of the line current, we may use this value to calculate V_L. That is,

$$\sqrt{3} \, V_L I_L = 3(206.41) \times 10^3;$$

$$V_L = \frac{3(206.41) \times 10^3}{\sqrt{3}(577.35)} = 619.23 \text{ V.}$$

INSTANTANEOUS POWER IN THREE-PHASE CIRCUITS

Although we are primarily interested in average, reactive, and complex power calculations, the computation of the total *instantaneous* power also is important. The total instantaneous power in a balanced three-phase circuit has an interesting property: it is invariant with time!

To demonstrate this property we let the instantaneous line-to-neutral voltage v_{AN} be the reference and, as before, θ_ϕ be the phase angle $\theta_{vA} - \theta_{iA}$. Then, for a positive phase sequence, the instantaneous power in each phase is

$$p_A = v_{AN}i_{aA} = V_\phi I_\phi \cos \omega t \cos (\omega t - \theta_\phi),$$

$$p_B = v_{BN}i_{bB} = V_\phi I_\phi \cos (\omega t - 120°) \cos (\omega t - \theta_\phi - 120°),$$

and

$$p_C = v_{CN}i_{cC} = V_\phi I_\phi \cos (\omega t + 120°) \cos (\omega t - \theta_\phi + 120°),$$

where V_ϕ and I_ϕ represent the rms values of the line-to-neutral voltage and line current, respectively. The instantaneous total power is the sum of the instantaneous phase powers, which reduces to $1.5V_\phi I_\phi \cos \theta_\phi$; that is,

$$p_T = p_A + p_B + p_C = 1.5V_\phi I_\phi \cos \theta_\phi.$$

We leave this reduction to you. (See Problem 12.29.)

An important property of three-phase circuits is that the total instantaneous power is constant. Thus the torque developed at the shaft of a three-phase motor is constant, which in turn means less vibration in machinery powered by three-phase motors.

DRILL EXERCISES

12.10 The three-phase average power rating of the central processing unit (CPU) on a mainframe digital computer is 22,659 W. The three-phase line supplying the computer has a line voltage rating of 208 V (rms). The line current is 73.8 A (rms).

a) Calculate the total magnetizing reactive power absorbed by the CPU.

b) Calculate the power factor.

ANSWER: (a) 13,909.50 var; (b) 0.852 lagging.

12.11 The complex power associated with each phase of a balanced load is $384 + j288$ kVA. The line voltage at the terminals of the load is 4160 V.

a) What is the magnitude of the line current feeding the load?

b) The load is delta-connected, and the impedance of each phase consists of a resistance in parallel with a reactance. Calculate R and X.

c) The load is wye-connected, and the impedance of each phase consists of a resistance in series with a reactance. Calculate R and X.

ANSWER: (a) 199.85 A; (b) $R = 45.07$ Ω, $X = 60.09$ Ω; (c) $R = 9.61$ Ω, $X = 7.21$ Ω.

12.12 A balanced bank of delta-connected capacitors is connected in parallel with the load described in Drill Exercise 12.11. The line voltage at the terminals of the load remains at 4160 V. The circuit is operating at a frequency of 60 Hz. The capacitors are adjusted so that the magnitude of the line current feeding the parallel combination of the load and capacitor bank is at its minimum.

a) What is the size of each capacitor in microfarads?

b) Repeat (a) for wye-connected capacitors.

c) What is the magnitude of the line current?

ANSWER: (a) 44.14 μF; (b) 132.42 μF; (c) 159.88 A.

SUMMARY

We discussed the analysis of balanced three-phase circuits operating in the sinusoidal steady state. Three-phase circuit analysis comprises the following important characteristics.

- *Analytic short cuts:* The key to all analytic short cuts is to transform a balanced three-phase circuit to a Y–Y structure and then replace the Y–Y structure with a single phase equivalent circuit.

- *Single-phase equivalent circuit:* The single-phase equivalent circuit is used to calculate the line current and the line-to-neutral voltage in one phase of the Y–Y structure. The a-phase is normally chosen as the reference phase.

- *Translating single-phase calculations:* The line current and any line-to-neutral voltage calculated from the a-phase equivalent circuit may be used to find any current or voltage in the balanced three-phase circuit, based on the following facts.

1. In a balanced system, b- and c-phase currents and voltages are identical to the corresponding a-phase current and voltage except for 120° shift in phase. In a positive sequence circuit, the b-phase quantity lags the a-phase quantity by 120°, and the c-phase quantity leads the a-phase quantity by 120°. For a negative sequence circuit, phases b and c are interchanged with respect to phase a.

2. The set of line voltages is out of phase with the set of line-to-neutral voltages by ±30°. The plus or minus sign corresponds to positive and negative sequence, respectively.

3. The magnitude of a line voltage is $\sqrt{3}$ times the magnitude of a line-to-neutral voltage.

4. The set of line currents is out of phase with the set of phase currents in Δ-connected sources and loads by ∓30°. The minus or plus sign corresponds to positive and negative sequence, respectively.

5. The magnitude of a line current is $\sqrt{3}$ times the magnitude of a phase current in the Δ-connected source or load.

Power calculations in three-phase circuits involve the following methods.

- *Per phase power:* The techniques for calculating per phase average power, reactive power, and complex power are identical to those introduced in Chapter 11.

- *Total power:* The total real, reactive, and complex power can be determined either by multiplying the corresponding per phase quantity by 3 or by using the expressions based on line current and line voltage as given by Eqs. (12.37), (12.39), and (12.42).

- *Instantaneous power:* The total instantaneous power in a balanced three-phase circuit is constant and equals 1.5 times the average power per phase.

PROBLEMS

All phasor voltages are stated in terms of the rms value.

12.1 For each set of voltages given below, state whether or not the voltages form a balanced three-phase set. If the set is a balanced set, state whether the phase sequence is positive or negative. If the set is not balanced, explain why.

a) $v_a = 339 \cos 377t$ V
$v_b = 339 \cos (377t - 120°)$ V
$v_c = 339 \cos (377t + 120°)$ V

b) $v_a = 679 \cos 377t$ V
$v_b = 679 \cos (377t + 120°)$ V
$v_c = 679 \cos (377t + 240°)$ V

c) $v_a = 933 \sin 377t$ V
$v_b = 933 \cos (377t + 150°)$ V
$v_c = 933 \sin (377t + 120°)$ V

d) $v_a = 170 \sin (\omega t + 45°)$ V
$v_b = -170 \cos (\omega t - 105°)$ V
$v_c = 170 \cos (\omega t - 165°)$ V

e) $v_a = 339 \cos \omega t$ V
$v_b = 339 \cos (\omega t - 120°)$ V
$v_c = 311 \cos (\omega t + 120°)$ V

f) $v_a = 1697 \sin (\omega t + 70°)$ V
$v_b = 1697 \cos (\omega t + 100°)$ V
$v_c = 1697 \cos (\omega t - 120°)$ V

12.2 Verify that Eq. (12.3) is true for either Eq. (12.1) or Eq. (12.2).

12.3 The time-domain expressions for three line-to-neutral voltages at the terminals of a Y-connected load are

$$v_{AN} = 169.71 \cos (\omega t + 60°) \text{ V},$$

$$v_{BN} = 169.71 \cos (\omega t + 180°) \text{ V},$$

$$v_{CN} = 169.71 \cos (\omega t - 60°) \text{ V}.$$

What are the time-domain expressions for the three line-to-line voltages v_{AB}, v_{BC}, and v_{CA}?

12.4 Refer to the circuit in Fig. 12.4(b). Assume that there are no external connections to the terminals a, b, c. Assume further that the three windings are from a balanced three-phase generator. How much current will circulate in the Δ-connected generator?

12.5 a) Is the circuit in Fig. P12.5 a balanced or unbalanced three-phase system? Explain.

b) Find $\mathbf{I}_o$.

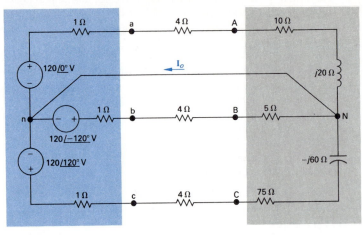

FIGURE P12.5

12.6 a) Find $\mathbf{I}_o$ in the circuit in Fig. P12.6.

b) Find $\mathbf{V}_{BN}$.

c) Find $\mathbf{V}_{BC}$.

d) Is the circuit a balanced or unbalanced three-phase system?

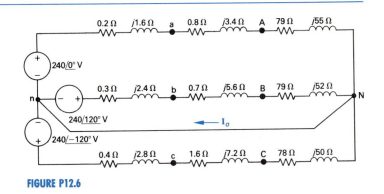

FIGURE P12.6

12.7 Find the rms value of $\mathbf{I}_o$ in the unbalanced three-phase circuit seen in Fig. P12.7.

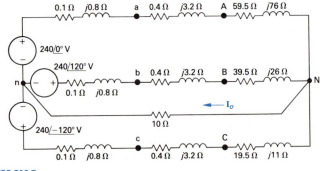

FIGURE P12.7

12.8 The magnitude of the phase voltage of an ideal balanced, three-phase, Y-connected source is 240 V. The source is connected to a balanced Y-connected load by a distribution line that has an impedance of $0.2 + j1.6 \ \Omega/\phi$. The load impedance is $23.8 + j5.4 \ \Omega/\phi$. The phase sequence of the source is abc. Use the a-phase voltage of the source as the reference. Specify the magnitude and phase angle of the following quantities: (a) the three line currents, (b) the three line voltages at the source, (c) the three phase voltages at the load, and (d) the three line voltages at the load.

12.9 The magnitude of the line voltage at the terminals of a balanced Y-connected load is 762 V. The load impedance is $48 + j14 \ \Omega/\phi$. The load is fed from a line that has an impedance of $0.1 + j0.8 \ \Omega/\phi$.

a) What is the magnitude of the line current?

b) What is the magnitude of the line voltage at the source?

12.10 A balanced Δ-connected load has an impedance of $432 - j126 \ \Omega/\phi$. The load is fed through a line having an impedance of $1 + j8 \ \Omega/\phi$. The phase voltage at the terminals of the load is 13,200 V. The phase sequence is positive. Use $\mathbf{V}_{AB}$ as the reference.

a) Calculate the three phase currents of the load.

b) Calculate the three line currents.

c) Calculate the three line voltages at the sending end of the line.

12.11 A balanced Δ-connected load having an impedance of $216 + j63 \ \Omega/\phi$ is connected in parallel with a balanced Y-connected load having an impedance of $50\underline{/0°} \ \Omega/\phi$. The paralleled loads are fed from a line having an impedance of $0.5 + j4.0 \ \Omega/\phi$. The magnitude of the line-to-neutral voltage of the Y-load is 750 V.

a) Calculate the magnitude of the current in the line feeding the loads.

b) Calculate the magnitude of the phase current in the Δ-connected load.

c) Calculate the magnitude of the phase current in the Y-connected load.

d) Calculate the magnitude of the line voltage at the sending end of the line.

12.12 For the circuit shown in Fig. P12.12, find

a) the phase currents $\mathbf{I}_{AB}$, $\mathbf{I}_{BC}$, and $\mathbf{I}_{CA}$;

b) the line currents $\mathbf{I}_{aA}$, $\mathbf{I}_{bB}$, and $\mathbf{I}_{cC}$;

c) the phase currents $\mathbf{I}_{ba}$, $\mathbf{I}_{ac}$, and $\mathbf{I}_{cb}$; when $z_1 = 2.4 - j0.7 \ \Omega$, $z_2 = 8 + j6 \ \Omega$, and $z_3 = 20 + j0 \ \Omega$.

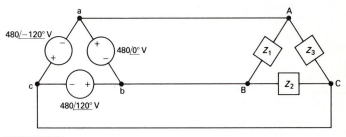

FIGURE P12.12

12.13 The impedance Z in the balanced three-phase circuit in Fig. P12.13 is $200 + j150 \ \Omega$. Find:

a) $\mathbf{I}_{AB}$, $\mathbf{I}_{BC}$, and $\mathbf{I}_{CA}$;

b) $\mathbf{I}_{aA}$, $\mathbf{I}_{bB}$, and $\mathbf{I}_{cC}$; and

c) $\mathbf{I}_{ba}$, $\mathbf{I}_{cb}$, and $\mathbf{I}_{ac}$.

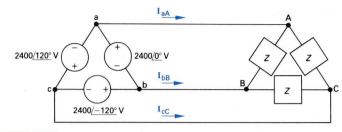

FIGURE P12.13

12.14 A balanced, three-phase, Δ-connected source is shown in Fig. P12.14.

a) Find the Y-connected equivalent circuit.

b) Show that the Y-connected equivalent circuit delivers the same open-circuit voltage as the original Δ-connected source.

c) Apply an external short circuit to the terminals A, B, and C. Use the Δ-connected source to find the three line currents I_{aA}, I_{bB}, and I_{cC}.

d) Repeat part (c) but use the Y-equivalent source to find the three line currents.

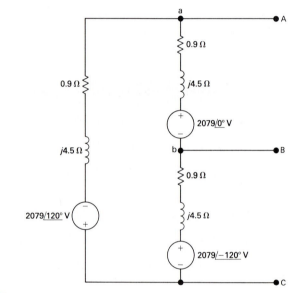

FIGURE P12.14

12.15 The Δ-connected source of Problem 12.14 is connected to a Y-connected load by means of a balanced, three-phase distribution line. The load impedance is $318 - j255 \ \Omega/\phi$ and the line impedance is $1.7 + j13.5 \ \Omega/\phi$.

a) Construct a single-phase equivalent circuit of the system.

b) Determine the magnitude of the line voltage at the terminals of the load.

c) Determine the magnitude of the phase current in the Δ-source.

d) Determine the magnitude of the line voltage at the terminals of the source.

12.16 In Example 12.3, the solution for $\mathbf{V}_{an}$ is $347.37\underline{/29.73°}$ V, where $\mathbf{V}_{an}$ is the line-to-neutral voltage of the equivalent Y-source.

a) What are the magnitude and the phase angle of the line voltage $\mathbf{V}_{ab}$?

b) Calculate the line voltage using the appropriate phase current, internal voltage, and internal impedance of the Δ-source.

c) Do these two alternative calculations for $\mathbf{V}_{ab}$ check?

12.17 A three-phase, Δ-connected generator has an internal impedance of $0.6 + j6.0\ \Omega/\phi$. When the load is removed from the generator, the magnitude of the terminal voltage is 13,800 V. The generator feeds a Δ-connected load through a transmission line with an impedance of $0.8 + j5.0\ \Omega/\phi$. The per-phase impedance of the load is $141 + j21\ \Omega$.

a) Construct a single-phase equivalent circuit.

b) Calculate the magnitude of the line current.

c) Calculate the magnitude of the line voltage at the terminals of the load.

d) Calculate the magnitude of the line voltage at the terminals of the source.

e) Calculate the magnitude of the phase current in the load.

f) Calculate the magnitude of the phase current in the source.

12.18 a) Find the rms magnitude and the phase angle of $\mathbf{I}_{BC}$ in the circuit shown in Fig. P12.18.

b) What percentage of the average power delivered by the three-phase source is dissipated in the three-phase load?

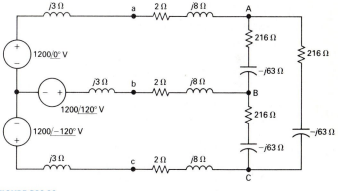

FIGURE P12.18

12.19 A three-phase line has an impedance of $0.15 + j1.2\ \Omega/\phi$. The line feeds two balanced three-phase loads that are connected in parallel. The first load is absorbing a total of 240 kW and delivering 180 kVAR magnetizing vars. The second load is Δ-connected and has an impedance of $288 + j84\ \Omega/\phi$. The line-to-neutral voltage at the load end of the line is 4800 V. What is the magnitude of the line voltage at the source end of the line?

12.20 Calculate the complex power in each phase of the unbalanced load in Problem 12.12.

12.21 Three balanced three-phase loads are connected in parallel. Load 1 is Y-connected with an impedance of $150 + j50\ \Omega/\phi$; load 2 is Δ-connected with an impedance of $900 - j1200\ \Omega/\phi$; and load 3 is 150 kVA at 0.8 pf lagging. The loads are fed from a distribution line with an impedance of $1 + j2\ \Omega/\phi$. The magnitude of the line-to-neutral voltage at the load end of the line is 2.5 kV.

a) Calculate the total complex power at the sending end of the line.

b) What percent of the average power at the sending end of the line is delivered to the loads?

12.22 The line-to-neutral voltage at the terminals of the balanced three-phase load in the circuit shown in Fig. P12.22 is 2500 V. At this voltage the load is absorbing 750 kVA at 0.96 pf lag.

a) Use $\mathbf{V}_{AN}$ as the reference and express $\mathbf{I}_{na}$ in polar form.

b) Calculate the complex power associated with the ideal three-phase source.

c) Check that the total average power delivered equals the total average power absorbed.

d) Check that the total magnetizing reactive power delivered equals the total magnetizing reactive power absorbed.

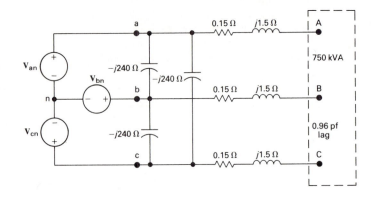

FIGURE P12.22

12.23 A balanced three-phase source is supplying 90 kVA at 0.8 pf lag to two balanced, Y-connected, parallel loads. The distribution line connecting the source to the load has negligible impedance. Load 1 is purely resistive and absorbs 60 kW.

a) Determine the impedance per phase of load 2 if the line voltage is 415.69 V and the impedance components are in series.

b) Repeat part (a), with the impedance components in parallel.

12.24 The three pieces of computer equipment described below are installed as part of a computation center. Each piece of equipment is a balanced three-phase load rated at 208 V. Calculate (a) the magnitude of the line current supplying these three devices and (b) the power factor of the combined load.

Disk: 4.864 kW at 0.79 pf lag,

Drum: 17.636 kVA at 0.96 pf lag,

CPU: line current 73.8 A, 13.853 kVAR.

12.25 A balanced three-phase distribution line has an impedance of $1 + j8 \ \Omega/\phi$. This line is used to supply three balanced three-phase loads that are connected in parallel. The three loads are

$L_1 = 120$ kVA at 0.96 pf lead,

$L_2 = 180$ kVA at 0.80 pf lag,

$L_3 = 100.8$ kW and 15.6 kVAR (magnetizing).

The magnitude of the line voltage at the terminals of the loads is $2400\sqrt{3}$ V.

a) What is the magnitude of the line voltage at the sending end of the line?

b) What is the percent efficiency of the distribution line with respect to average power?

12.26 The output of the balanced, positive-sequence, three-phase source in Fig. P12.26 is 60 kVA at a lagging power factor of 0.96. The line voltage at the source is 680 V.

a) Find the line voltage at the load.

b) Find the total complex power at the terminals of the load.

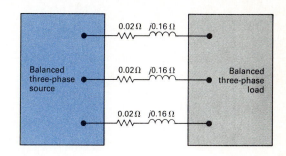

FIGURE P12.26

12.27 The total power delivered to a balanced three-phase load when operating at a line voltage of $4800\sqrt{3}$ V is 900 kW at a lagging power factor of 0.60. The impedance of the distribution line supplying the load is $0.6 + j4.8$ Ω/ϕ. Under these operating conditions, the drop in the magnitude of the line voltage between the sending end and the load end of the line is excessive. To compensate for the excessive voltage drop, a bank of Δ-connected capacitors is placed in parallel with the load. The capacitor bank is designed to furnish 1200 kVAR of magnetizing reactive power when operated at a line voltage of $4800\sqrt{3}$V.

a) What is the magnitude of the voltage at the sending end of the line when the load is operating at a line voltage of $4800\sqrt{3}$ V and the capacitor bank is disconnected?

b) Repeat part (a), with the capacitor bank connected.

c) What is the average power efficiency of the line in part (a)?

d) What is the average power efficiency in part (b)?

e) If the system is operating at a frequency of 60 Hz, what is the size of each capacitor in microfarads?

12.28 A balanced three-phase load absorbs 1200 kVA at a lagging power factor of 0.8 when the line voltage at the terminals of the load is 12,600 V. Find four equivalent circuits that can be used to model this load.

12.29 Show that the total instantaneous power in a balanced three-phase circuit is constant and equal to $1.5V_\phi I_\phi \cos \theta_\phi$, where V_ϕ and I_ϕ represent the maximum amplitudes of the phase voltage and phase current, respectively.

MUTUAL INDUCTANCE

CHAPTER 13

When introducing the inductor in Chapter 7, we pointed out that it is a circuit element based on phenomena associated with magnetic fields. In particular, we noted that the source of the magnetic field is current and that a time-varying current produces a time-varying field. The time-varying field, in turn, induces a voltage in any conductor linked by the field. We expressed the relationship between the time-varying current and the resulting voltage induced by the time-varying field in terms of the inductance parameter L, or

$$v = L\frac{di}{dt}.$$

We now consider the situation in which a time-varying current in one circuit produces a time-varying magnetic field that links a second circuit, inducing a voltage in the second circuit. The voltage induced in the second circuit can be related to the time-varying current in the first circuit by means of an inductance parameter known as *mutual inductance*. Mutual inductance (also measured in henrys) is denoted M. Because the inductance parameter L relates a voltage induced in a circuit to a time-varying current in the same circuit, it is also called *self-inductance*.

Two circuits linked by a magnetic field are said to be *magnetically coupled*. Magnetic coupling is an important physical phenomenon that is used effectively in both communication and power circuits. The transformer is a device designed entirely around the concept of magnetic coupling. In Section 13.5 we discuss the transformer as a circuit element.

In order to describe completely the concept of mutual inductance, we must expand the concept of a magnetic field and flux

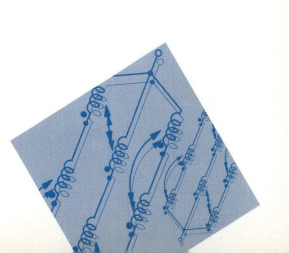

linkage beyond the verbal description given earlier. In Section 13.1, we review the concept of self-inductance, using a more quantitative approach than we used in Chapter 7. Following this review, we discuss the concept of mutual inductance in Section 13.2.

13.1 A REVIEW OF SELF-INDUCTANCE

The concept of inductance can be traced to Michael Faraday, who did pioneering work in this area early in the 1800s. Faraday postulated that the magnetic field consists of lines of force surrounding the current-carrying conductor. Visualize these lines of force as energy-storing elastic bands that close upon themselves. As the current increases and decreases, the elastic bands spread and collapse about the conductor. The voltage induced in the conductor is proportional to the number of lines that collapse into, or cut, the conductor. This image of induced voltage is expressed by what is called *Faraday's law*; that is,

$$v = \frac{d\lambda}{dt}, \qquad (13.1)$$

where λ is referred to as the flux linkage and is measured in weber-turns.

You can see how Eq. (13.1) leads to the concept of self-inductance by looking at the single N-turn coil shown in Fig. 13.1. The lines threading the N turns and labeled ϕ show graphically the magnetic lines of force. The strength of the magnetic field depends on the strength of the current, and the spatial orientation of the field depends on the direction of the current. The right-hand rule relates the orientation of the field to the direction of the current: When the fingers of the right hand are wrapped around the coil so that the fingers point in the direction of the current, the thumb points in the direction of the magnetic field *inside* the coil. The flux linkage is the product of the magnetic field (ϕ), measured in webers (Wb), and the number of turns linked by the field (N):

$$\lambda = N\phi. \qquad (13.2)$$

A more detailed picture of flux linkage introduces partial linkages, that is, portions of the magnetic field linking a fraction of the total turns. However, these details are not necessary here. The important point is that the flux ϕ in Eq. (13.2) is an equiva-

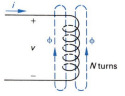

FIGURE 13.1 Representation of a magnetic field linking an N-turn coil.

lent flux that yields the same flux linkage that you would obtain by including the partial linkages in the determination of the total flux linkage λ.

The magnitude of the flux, ϕ, depends on the magnitude of the coil current, the number of turns on the coil, and the magnetic properties of the space occupied by the flux. The relationship between ϕ and i is

$$\phi = \mathscr{P}Ni, \qquad (13.3)$$

where $\mathscr{P}$ is the permeance of the space occupied by the field. The *permeance* is a function of the physical dimensions and the permeability of the space occupied by the flux. When the space is nonmagnetic, the permeance is constant, giving a linear relationship between ϕ and i. When the space is made up of magnetic materials (such as iron, nickel, and cobalt), the permeance varies with the flux, giving a nonlinear relationship between ϕ and i. Here, we assume that the core material is nonmagnetic. We do not delve further into the factors that determine the permeance of the space occupied by the magnetic flux. We simply accept permeance as a positive constant of proportionality that relates the strength of the magnetic field to the current that creates it. Note from Eq. (13.3) that the flux also is proportional to the number of turns on the coil.

Substituting Eqs. (13.2) and (13.3) into Eq. (13.1) yields

$$v = \frac{d\lambda}{dt} = \frac{d(N\phi)}{dt}$$

$$= N\frac{d\phi}{dt} = N\frac{d}{dt}(\mathscr{P}Ni)$$

$$= N^2\mathscr{P}\frac{di}{dt} = L\frac{di}{dt}, \qquad (13.4)$$

which shows that self-inductance is proportional to the square of the number of turns on the coil. We make use of this observation later.

The polarity of the induced voltage in the circuit in Fig. 13.1 reflects the reaction of the field against the current creating the field. For example, when i is increasing, di/dt is positive and v is positive. Thus energy is required to establish the magnetic field. The product vi gives the rate at which energy is stored in the field. When the field collapses, di/dt is negative, and again the polarity of the induced voltage is in opposition to the change. As the field collapses about the coil, energy is returned to the circuit.

With this further insight into the concept of self-inductance in mind, we now turn to the concept of mutual inductance.

13.2 THE CONCEPT OF MUTUAL INDUCTANCE

Mutual inductance is the circuit parameter that relates the voltage induced in one circuit to a time-varying current in another circuit. This situation arises whenever a common magnetic field links two or more circuits. We restrict our discussion to magnetic coupling of only two circuits.

Figure 13.2 shows two magnetically coupled coils. The number of turns on each coil are N_1 and N_2, respectively. Coil 1 is energized by a time-varying current source that establishes the current i_1 in the N_1 turns. Coil 2 is not energized and is open. The coils are wound on a nonmagnetic core. The flux produced by the current i_1 can be divided into two components, labeled ϕ_{11} and ϕ_{21} in Fig. 13.2. The flux component ϕ_{11} is the flux produced by i_1, which links only the N_1 turns. The component ϕ_{21} is the flux produced by i_1, which links the N_2 turns as well as the N_1 turns. The first digit in the subscript to the flux gives the coil number, and the second digit refers to the coil current. Thus ϕ_{11} is a flux linking coil 1 produced by a current in coil 1, whereas ϕ_{21} is a flux linking coil 2 produced by a current in coil 1.

The total flux linking coil 1 is ϕ_1, the sum of ϕ_{11} and ϕ_{21}:

$$\phi_1 = \phi_{11} + \phi_{21}. \tag{13.5}$$

The flux ϕ_1 and its components ϕ_{11} and ϕ_{21} are related to the coil current i_1 as follows:

$$\phi_1 = \mathcal{P}_1 N_1 i_1, \tag{13.6}$$

$$\phi_{11} = \mathcal{P}_{11} N_1 i_1, \tag{13.7}$$

and

$$\phi_{21} = \mathcal{P}_{21} N_1 i_1, \tag{13.8}$$

where $\mathcal{P}_1$ is the permeance of the space occupied by the flux ϕ_1, $\mathcal{P}_{11}$ is the permeance of the space occupied by the flux ϕ_{11}, and $\mathcal{P}_{21}$ is the permeance of the space occupied by the flux ϕ_{21}. Substituting Eqs. (13.6), (13.7), and (13.8) into Eq. (13.5) yields the relationship between the permeance of the space occupied by the total flux ϕ_1 and the permeances of the spaces occupied by its components ϕ_{11} and ϕ_{21}:

$$\mathcal{P}_1 = \mathcal{P}_{11} + \mathcal{P}_{21}. \tag{13.9}$$

We use Faraday's law to derive expressions for v_1 and v_2,

$$v_1 = \frac{d\lambda_1}{dt} = \frac{d(N_1\phi_1)}{dt} = N_1 \frac{d}{dt}(\phi_{11} + \phi_{21})$$

$$= N_1^2(\mathcal{P}_{11} + \mathcal{P}_{21})\frac{di_1}{dt} = N_1^2 \mathcal{P}_1 \frac{di_1}{dt} = L_1 \frac{di_1}{dt} \tag{13.10}$$

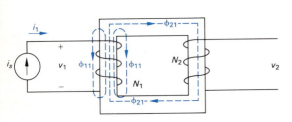

FIGURE 13.2 Two magnetically coupled coils.

and

$$v_2 = \frac{d\lambda_2}{dt} = \frac{d}{dt}(N_2\phi_{21}) = N_2\frac{d}{dt}(\mathcal{P}_{21}N_1 i_1)$$

$$= N_2 N_1 \mathcal{P}_{21}\frac{di_1}{dt}. \qquad (13.11)$$

The coefficient of di_1/dt in Eq. (13.10) is the self-inductance of coil 1. The coefficient of di_1/dt in Eq. (13.11) is the mutual inductance between coils 1 and 2. Thus

$$M_{21} = N_2 N_1 \mathcal{P}_{21}. \qquad (13.12)$$

The subscript on M specifies an inductance that relates the voltage induced in coil 2 to the current in coil 1.

The coefficient of mutual inductance gives

$$v_2 = M_{21}\frac{di_1}{dt}. \qquad (13.13)$$

Note that no polarity references are assigned to v_2 in Fig. 13.2. We discuss determining the polarity of mutually induced voltages in Section 13.4.

For the coupled coils in Fig. 13.2, exciting coil 2 from a time-varying current source (i_2) and leaving coil 1 open produces the circuit arrangement shown in Fig. 13.3. In this case, no polarity references are assigned to v_1.

The total flux linking coil 2 is

$$\phi_2 = \phi_{22} + \phi_{12}. \qquad (13.14)$$

The flux ϕ_2 and its components ϕ_{22} and ϕ_{12} are related to the coil current i_2 as follows:

$$\phi_2 = \mathcal{P}_2 N_2 i_2, \qquad (13.15)$$

$$\phi_{22} = \mathcal{P}_{22} N_2 i_2, \qquad (13.16)$$

and

$$\phi_{12} = \mathcal{P}_{12} N_2 i_2. \qquad (13.17)$$

The voltages v_2 and v_1 are

$$v_2 = \frac{d\lambda_2}{dt} = \mathcal{P}_2 N_2^2\frac{di_2}{dt} = L_2\frac{di_2}{dt} \qquad (13.18)$$

and

$$v_1 = \frac{d\lambda_{12}}{dt} = \frac{d}{dt}(N_1\phi_{12}) = N_1 N_2 \mathcal{P}_{12}\frac{di_2}{dt}. \qquad (13.19)$$

The coefficient of mutual inductance that relates the voltage induced in coil 1 to the time-varying current in coil 2 is the

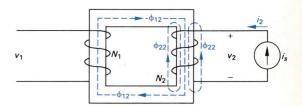

FIGURE 13.3 The magnetically coupled coils of Fig. 13.2, with coil 2 excited and coil 1 open.

coefficient of di_2/dt in Eq. (13.19):

$$M_{12} = N_1 N_2 \mathscr{P}_{12}. \tag{13.20}$$

For nonmagnetic materials, the permeances $\mathscr{P}_{12}$ and $\mathscr{P}_{21}$ are equal and therefore

$$M_{12} = M_{21} = M. \tag{13.21}$$

Hence for linear circuits with just two magnetically coupled coils, attaching subscripts to the coefficient of mutual inductance is not necessary.

MUTUAL INDUCTANCE IN TERMS OF SELF-INDUCTANCES

The coefficient of mutual inductance is a function of the self-inductances. We derive this relationship as follows. From Eqs. (13.10) and (13.18), respectively,

$$L_1 = N_1^2 \mathscr{P}_1 \tag{13.22}$$

and

$$L_2 = N_2^2 \mathscr{P}_2. \tag{13.23}$$

From Eqs. (13.22) and (13.23),

$$L_1 L_2 = N_1^2 N_2^2 \mathscr{P}_1 \mathscr{P}_2. \tag{13.24}$$

We now use Eq. (13.9) and the corresponding expression for $\mathscr{P}_2$ to write

$$L_1 L_2 = N_1^2 N_2^2 (\mathscr{P}_{11} + \mathscr{P}_{21})(\mathscr{P}_{22} + \mathscr{P}_{12}). \tag{13.25}$$

But for a linear system, $\mathscr{P}_{21} = \mathscr{P}_{12}$, so Eq. (13.25) becomes

$$L_1 L_2 = (N_1 N_2 \mathscr{P}_{12})^2 \left(1 + \frac{\mathscr{P}_{11}}{\mathscr{P}_{12}}\right)\left(1 + \frac{\mathscr{P}_{22}}{\mathscr{P}_{12}}\right)$$

$$= M^2 \left(1 + \frac{\mathscr{P}_{11}}{\mathscr{P}_{12}}\right)\left(1 + \frac{\mathscr{P}_{22}}{\mathscr{P}_{12}}\right). \tag{13.26}$$

Replacing the two terms involving permeances by a single constant expresses Eq. (13.26) in a more meaningful form:

$$\frac{1}{k^2} = \left(1 + \frac{\mathscr{P}_{11}}{\mathscr{P}_{12}}\right)\left(1 + \frac{\mathscr{P}_{22}}{\mathscr{P}_{12}}\right). \tag{13.27}$$

Substituting Eq. (13.27) into Eq. (13.26) yields

$$M^2 = k^2 L_1 L_2$$

or

$$M = k\sqrt{L_1 L_2}, \tag{13.28}$$

where the constant k is called the coefficient of coupling. According to Eq. (13.27), $1/k^2$ must be greater than 1, which means that k must be less than 1. In fact, the coefficient of coupling must lie between 0 and 1, or

$$0 \le k \le 1. \qquad (13.29)$$

The coefficient of coupling is zero when the two coils have no common flux, that is, when $\phi_{12} = \phi_{21} = 0$. This condition implies that $\mathcal{P}_{12} = 0$, and Eq. (13.27) indicates that $1/k^2 = \infty$, or $k = 0$. If there is no flux linkage between the coils, obviously M is zero.

The coefficient of coupling is equal to 1 when ϕ_{11} and ϕ_{22} are 0. This condition implies that all the flux that links coil 1 also links coil 2. In terms of Eq. (13.27), $\mathcal{P}_{11} = \mathcal{P}_{22} = 0$, which obviously represents an ideal state. The reason is that winding two coils so that they share precisely the same flux is physically impossible.

We mention in passing that magnetic materials (such as alloys of iron, cobalt, and nickel) are important. They create a space with high permeance and are used to establish coefficients of coupling that approach unity. (We say more about this later.)

In this section, we showed how to extend the concept of inductance to relate the voltage induced in one circuit to a time-varying current in another circuit. We also showed how inductance relates to the number of turns on the magnetically coupled coils and a constant, known as permeance, which characterizes the magnetic properties of the space occupied by the flux. For linearly coupled coils, Eqs. (13.21), (13.28), and (13.29) contain the important results for circuit analysis. Next we discuss the importance of determining the polarity of the mutually induced voltages.

DRILL EXERCISES

13.1 Two magnetically coupled coils are wound on a non-magnetic core. The self-inductance of coil 1 is 6 H, the mutual inductance is 9.6 H, the coefficient of coupling is 0.8, and the physical structure of the coils is such that $\mathcal{P}_{11} = \mathcal{P}_{22}$.

a) Find L_2 and the turns ratio N_1/N_2.

b) If $N_1 = 800$, what is the value of $\mathcal{P}_1$ and $\mathcal{P}_2$?

ANSWER: (a) 24 H, 0.5; (b) 9.375×10^{-6} Wb/A.

13.2 The self-inductances of two magnetically coupled coils are $L_1 = 50$ mH and $L_2 = 72$ mH. The coupling medium is nonmagnetic. If coil 1 has 500 turns and coil 2 has 600 turns, find $\mathcal{P}_{11}$ and $\mathcal{P}_{21}$ (in microwebers per ampere) when the coefficient of coupling is 0.85.

ANSWER: 0.03 μWb/A, 0.17 μWb/A.

13.3 POLARITY OF MUTUALLY INDUCED VOLTAGES (THE DOT CONVENTION)

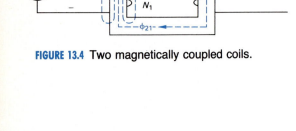

FIGURE 13.4 Two magnetically coupled coils.

The polarity of a mutually induced voltage also reflects a reaction against the time-varying flux that creates the voltage. For example, in the circuit in Fig. 13.4, when i_1 is increasing in the reference direction, the polarity of v_2 is positive at the upper terminal of coil 2. The reason is that the polarity of v_2 would establish a current out of the upper terminal of the N_2 coil to produce a flux in opposition to ϕ_{21}. Thus when v_1 is positive at the upper terminal of coil 1, v_2 is positive at the upper terminal of coil 2, and vice versa. (Note that exciting coil 2 as in Fig. 13.5 yields the same conclusion concerning the relative polarities of v_1 and v_2.)

Having deduced the polarity of v_2 in relation to the current i_1, we now express v_2 as a function of i_1 with the proper algebraic sign. If the reference for v_2 is positive at the top terminal of coil 2,

$$v_2 = M\frac{di_1}{dt}. \qquad (13.30)$$

But if the reference for v_2 is positive at the lower terminal of coil 2,

$$v_2 = -M\frac{di_1}{dt}. \qquad (13.31)$$

Similarly, for the circuit in Fig. 13.5 where coil 2 is energized by i_2 and coil 1 is open,

$$v_1 = \pm M\frac{di_2}{dt}, \qquad (13.32)$$

where the positive sign is a positive reference for v_1 at the top terminal of coil 1, and the minus sign otherwise.

In general, showing the details of mutually coupled windings is too cumbersome, so we do not use the "reaction" method for determining polarity signs in writing circuit equations. Instead, we keep track of the polarities of the mutually induced voltages by a method known as the *dot convention*, in which a dot is placed on one terminal of each winding. These dots carry the sign information regarding the mutually induced voltages. For example, for the coils in Fig. 13.4 either the two upper terminals or the two lower terminals receive a dot marking. Figure 13.6 shows the two possible markings. Note that the dot markings allow drawing the coils schematically rather than showing how the coils wrap around a core structure.

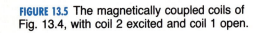

FIGURE 13.5 The magnetically coupled coils of Fig. 13.4, with coil 2 excited and coil 1 open.

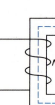

FIGURE 13.6 Dot markings for the coils of Fig. 13.4.

Two types of problems are associated with polarity dots. One is to determine a proper set of dot markings when the physical arrangement of the coupled coils is given. The other is to determine how the dots are used in writing the circuit equations that describe magnetically coupled coils. We begin the discussion of the dot convention by describing a systematic method for determining the dot markings.

PROCEDURE FOR DETERMINING DOT MARKINGS

Assume that you have a set of two magnetically coupled coils and that you know the physical arrangement of the two coils and the mode of each winding as, for example, those shown in Fig. 13.7. Then follow six steps to determine a set of dot markings.

1. Arbitrarily select one terminal—say, the D terminal—of one coil and give it a dot.

2. Assign a current *into* the arbitrarily selected dotted terminal and label it i_D.

3. Use the right-hand rule to determine the direction of the magnetic field established by i_D *inside* the coupled coils and label this field ϕ_D.

4. Arbitrarily pick one terminal of the second coil—say, the terminal A—and assign a current *into* this terminal and show the current as i_A.

5. Use the right-hand rule to determine the direction of the flux established by i_A *inside* the coupled coils and label this flux ϕ_A.

6. Compare the directions of the two fluxes ϕ_D and ϕ_A. If the fluxes are additive, place a dot on the terminal of the second coil where the test current (i_A) enters. (In Fig. 13.7, the fluxes ϕ_D and ϕ_A are additive and therefore a dot goes on terminal A.) If the fluxes are subtractive, place a dot on the terminal of the second coil where the test current leaves.

The relative polarities of magnetically coupled coils also can be determined experimentally. This capability is important, because in some situations determining how the coils are wound on the core is impossible. One way to ascertain experimentally the polarities of coupled coils is to connect a dc voltage source, a resistor, a switch, and a dc voltmeter to the pair of coils, as shown in Fig. 13.8. The shaded box covering the coils implies that physical inspection of the coils in order to determine the polarity markings is not possible. The resistor R limits the magnitude of the current supplied by the dc voltage source. The procedure for

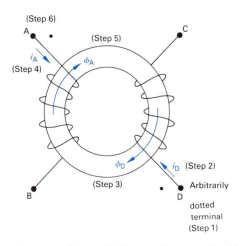

FIGURE 13.7 A set of coils showing a method for determining a set of dot markings.

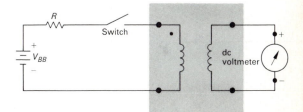

FIGURE 13.8 An experimental setup for determining polarity marks.

determining the relative polarity marks is as follows. The coil terminal connected to the positive terminal of the dc source via the switch and limiting resistor receives a polarity mark, as shown in Fig. 13.8. When the switch is *closed,* the voltmeter deflection is observed. If the momentary deflection is *upscale,* the coil terminal connected to the positive terminal of the voltmeter receives the polarity mark. If the momentary deflection is *downscale,* the coil terminal connected to the negative terminal of the voltmeter receives the polarity mark.

DRILL EXERCISES

13.3 Assume that the magnetic flux is confined to the core material in the structure shown. Which terminal of coil 2 should be given a dot marking?

ANSWER: b.

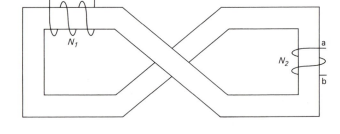

USE OF DOT MARKINGS IN CIRCUIT ANALYSIS

We now turn to the second type of problem: learning how to use the dot markings when writing circuit equations. The easiest way to analyze circuits containing mutual inductance is with mesh currents. Another possibility is the node-voltage method, but it is somewhat clumsy to use. The following analysis shows the mesh-current approach, with Fig. 13.9 illustrating the procedure.

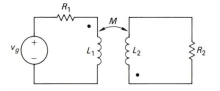

FIGURE 13.9 A circuit illustrating how dot markings are used in writing circuit equations.

The circuit in Fig. 13.9 schematically represents two magnetically coupled coils with dot markings as shown. The self-inductance of each coil is labeled L_1 and L_2 and the mutual inductance is labeled M. The double-headed arrow adjacent to M indicates the pair of coils with this value of mutual inductance. This notation is needed particularly in circuits containing more than one pair of magnetically coupled coils. The problem is to write the circuit equations that describe the circuit in terms of the coil currents.

At this point in the analysis of the circuit, choose the reference direction for each coil current. Figure 13.10 shows arbitrarily selected reference currents.

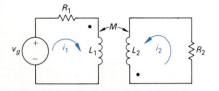

FIGURE 13.10 Coil currents i_1 and i_2 used to describe the circuit shown in Fig. 13.9.

After choosing the reference directions for i_1 and i_2, sum the voltages around each closed path. Because of the mutual inductance M, there will be two voltages across each coil, namely, a self-induced voltage and a mutually induced voltage. The self-induced voltage is a voltage drop in the direction of the current producing the voltage. Then determine the polarity of the mutually induced voltage as follows. When the reference direction for a current *enters* the dotted terminal of a coil, the reference polarity of the voltage that it induces in the *other* coil is positive at its dotted terminal. Alternatively, when the reference direction for a current leaves the dotted terminal of a coil, the reference polarity of the voltage that it induces in the other coil is negative at its dotted terminal.

This rule indicates that the reference polarity for the voltage induced in coil 1 by the current i_2 is negative at the dotted terminal of 1. This voltage ($M\, di_2/dt$) is a voltage rise with respect to i_1. The voltage induced in coil 2 by the current i_1 is $M\, di_1/dt$, and its reference polarity is positive at the dotted terminal of coil 2. This voltage is a voltage rise in the direction of i_2. Figure 13.11 shows the self- and mutually induced voltages across coils 1 and 2, along with their polarity marks.

Now sum the voltages around each closed loop. In Eqs. (13.33) and (13.34), voltage rises in the reference direction of a current are negative:

$$-v_g + i_1 R_1 + L_1 \frac{di_1}{dt} - M \frac{di_2}{dt} = 0, \qquad (13.33)$$

$$i_2 R_2 + L_2 \frac{di_2}{dt} - M \frac{di_1}{dt} = 0. \qquad (13.34)$$

Example 13.1 shows how to use the dot markings to formulate a set of circuit equations.

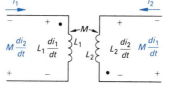

FIGURE 13.11 The self- and mutually induced voltages appearing across the coils shown in Fig. 13.10.

EXAMPLE 13.1

a) Write a set of mesh-current equations that describe the circuit in Fig. 13.12 in terms of the currents i_1, i_2, and i_3.

b) What is the coefficient of coupling of the magnetically coupled coils?

SOLUTION

a) (In the following set of mesh-current equations, voltage drops appear as positive quantities on the right-hand side of each

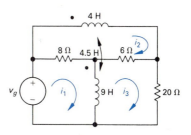

FIGURE 13.12 The circuit for Example 13.1.

equation.) Summing the voltages around the first mesh yields

$$v_g = 8(i_1 - i_2) + 9\frac{d}{dt}(i_1 - i_3) + 4.5\frac{di_2}{dt}.$$

The second mesh equation is

$$0 = 4\frac{di_2}{dt} + 4.5\frac{d}{dt}(i_1 - i_3) + 6(i_2 - i_3) + 8(i_2 - i_1).$$

The third mesh equation is

$$0 = 9\frac{d}{dt}(i_3 - i_1) - 4.5\frac{di_2}{dt} + 6(i_3 - i_2) + 20i_3.$$

Note that the voltage induced across the 4-H coil by the current $(i_1 - i_3)$; that is, $4.5\, d(i_1 - i_3)/dt$ is a voltage drop in the reference direction of i_2. The voltage induced in the 9-H coil by the current i_2, that is, $4.5\, di_2/dt$, is a voltage drop in the reference direction of i_1 but a voltage rise in the reference direction of i_3.

b) $$k = \frac{M}{\sqrt{L_1 L_2}} = \frac{4.5}{\sqrt{(9)(4)}} = \frac{4.5}{6} = 0.75.$$

So far, we have written equations describing circuits that contain mutual inductance in differential form. We do not discuss the solution of these equations at this time. We illustrate the transient response of circuits containing mutual inductance after we introduce the Laplace transform method of solving differential equations (see Chapter 16). For now we limit our analysis of circuits containing mutual inductance to the sinusoidal steady-state response. Before pursuing the sinusoidal analysis, however, we present the basic energy relationships that pertain to magnetically coupled coils.

DRILL EXERCISES

13.4 Write a set of mesh-current equations for the circuit in Example 13.1, with the dot on the 9-H inductor at the lower terminal and the reference direction of i_3 reversed.

ANSWER:

$$v_g = 8(i_1 - i_2) + 9\frac{d}{dt}(i_1 + i_3) - 4.5\frac{di_2}{dt},$$

$$0 = 4\frac{di_2}{dt} - 4.5\frac{d}{dt}(i_1 + i_3) + 6(i_2 + i_3) + 8(i_2 - i_1),$$

$$0 = 20i_3 + 6(i_3 + i_2) + 9\frac{d}{dt}(i_3 + i_1) - 4.5\frac{di_2}{dt}.$$

13.4 ENERGY CALCULATIONS

There are three important reasons for our discussion of the total energy stored in magnetically coupled coils. First, we want to show how the energy stored in a pair of coils coupled by a linear magnetic medium relates to the coil currents, the self-inductances, and the mutual inductance. Second, we want to show that for linear magnetic coupling the computation of stored energy confirms $M_{12} = M_{21}$. Third, we can use the energy calculation to show that the mutual inductance cannot exceed the square root of the product of the self-inductances, thus supporting our earlier contention that $M = k\sqrt{L_1 L_2}$, where $0 \le k \le 1$.

We use the circuit shown in Fig. 13.13 to derive the expression for the total energy stored in the magnetic fields associated with a pair of linearly coupled coils. We begin by assuming that the currents i_1 and i_2 are zero and that this zero-current state corresponds to zero energy stored in the coils. Now we let i_1 increase from zero to some arbitrary value I_1 and compute the energy stored when $i_1 = I_1$. As $i_2 = 0$, the total power input into the pair of coils is $v_1 i_1$, and the energy stored is

$$\int_0^{W_1} dw = L_1 \int_0^{I_1} i_1 \, di_1;$$

$$W_1 = \tfrac{1}{2} L_1 I_1^2. \tag{13.35}$$

Now we hold i_1 constant at I_1 and increase i_2 from zero to some arbitrary value I_2. During this time interval, the voltage induced in coil 2 by i_1 is zero because I_1 is constant. The voltage induced in coil 1 by i_2 is $M_{12} \, di_2/dt$. Therefore the power input to the pair of coils is

$$p = I_1 M_{12} \frac{di_2}{dt} + i_2 v_2.$$

The total energy stored in the pair of coils when $i_2 = I_2$ is

$$\int_{W_1}^{W} dw = \int_0^{I_2} I_1 M_{12} \, di_2 + \int_0^{I_2} L_2 i_2 \, di_2$$

or

$$W = W_1 + I_1 I_2 M_{12} + \tfrac{1}{2} L_2 I_2^2$$
$$= \tfrac{1}{2} L_1 I_1^2 + \tfrac{1}{2} L_2 I_2^2 + I_1 I_2 M_{12}. \tag{13.36}$$

If we reverse the procedure, that is, if we first increase i_2 from zero to I_2 and then increase i_1 from zero to I_1, the total energy stored is

$$W = \tfrac{1}{2} L_1 I_1^2 + \tfrac{1}{2} L_2 I_2^2 + I_1 I_2 M_{21}. \tag{13.37}$$

Equations (13.36) and (13.37) express the total energy stored in

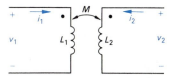

FIGURE 13.13 The circuit used to derive the basic energy relationships.

a pair of linearly coupled coils as a function of the coil currents, the self-inductances, and the mutual inductance. Note that the only difference in these equations is the coefficient of the current product $I_1 I_2$. We use Eq. (13.36) if I_1 is established first and Eq. (13.37) if I_2 is established first.

When the coupling medium is linear—that is, nonmagnetic—the total energy stored is the same regardless of the order used to establish I_1 and I_2. The reason is that in a linear magnetic medium the resultant magnetic flux depends only on the final values of i_1 and i_2—not on how the currents reached their final values. If the resultant flux is the same, the stored energy is the same. Therefore for linear coupling, $M_{12} = M_{21}$. Also, because I_1 and I_2 are arbitrary values of i_1 and i_2, respectively, we represent the coil currents by their instantaneous values i_1 and i_2. Thus at any instant of time, the total energy stored in the coupled coils is

$$w(t) = \tfrac{1}{2}L_1 i_1^2 + \tfrac{1}{2}L_2 i_2^2 + M i_1 i_2. \tag{13.38}$$

We derived Eq. (13.38) by assuming that both coil currents entered polarity-marked terminals. We leave it to you to verify that, if one current enters a polarity-marked terminal while the other leaves a polarity-marked terminal, the algebraic sign of the term $M i_1 i_2$ reverses. Thus, in general,

$$w(t) = \tfrac{1}{2}L_1 i_1^2 + \tfrac{1}{2}L_2 i_2^2 \pm M i_1 i_2. \tag{13.39}$$

We use Eq. (13.39) to show that M cannot exceed $\sqrt{L_1 L_2}$. The magnetically coupled coils are passive elements, so the total energy stored can never be negative. If $w(t)$ can never be negative, Eq. (13.39) indicates that the quantity

$$\tfrac{1}{2}L_1 i_1^2 + \tfrac{1}{2}L_2 i_2^2 - M i_1 i_2$$

must be greater than or equal to zero when i_1 and i_2 are either both positive or both negative. The limiting value of M corresponds to setting the quantity equal to zero:

$$\tfrac{1}{2}L_1 i_1^2 + \tfrac{1}{2}L_2 i_2^2 - M i_1 i_2 = 0. \tag{13.40}$$

To find the limiting value of M we add and subtract the term $i_1 i_2 \sqrt{L_1 L_2}$ to the left-hand side of Eq. (13.40). Doing so generates a term that is a perfect square:

$$\left(\sqrt{\frac{L_1}{2}} i_1 - \sqrt{\frac{L_2}{2}} i_2 \right)^2 + i_1 i_2 (\sqrt{L_1 L_2} - M) = 0. \tag{13.41}$$

The squared term in Eq. (13.41) can never be negative, but it can be zero. Therefore $w(t) \geq 0$ only if

$$\sqrt{L_1 L_2} \geq M, \tag{13.42}$$

which is another way of saying that

$$M = k\sqrt{L_1 L_2} \qquad (0 \le k \le 1).$$

We derived Eq. (13.42) by assuming that i_1 and i_2 are either both positive or both negative. However, we get the same result if i_1 and i_2 are of opposite sign, because in this case we obtain the limiting value of M by selecting the plus sign in Eq. (13.39).

DRILL EXERCISES

13.5 The self-inductances of the coils in Fig. 13.13 are $L_1 = 5$ mH and $L_2 = 33.8$ mH. If the coefficient of coupling is 0.96, calculate the energy stored in the system in millijoules when (a) $i_1 = 10$ A, $i_2 = 5$ A; (b) $i_1 = -10$ A, $i_2 = -5$ A; (c) $i_1 = -10$ A, $i_2 = 5$ A; (d) $i_1 = 10$ A, $i_2 = -5$ A.

ANSWER: (a) 1296.50 mJ; (b) 1296.50 mJ; (c) 48.50 mJ; (d) 48.50 mJ.

13.6 The coefficient of coupling in Drill Exercise 13.5 is increased to 1.0.
a) If i_1 equals 10 A, what value of i_2 results in zero stored energy?
b) Is there any physically realizable value of i_2 that can make the stored energy negative?

ANSWER: (a) -3.846 A; (b) no.

13.5 THE LINEAR TRANSFORMER

As we mentioned earlier, the transformer is a device that uses magnetic coupling as the basis for its operation. We now analyze the sinusoidal steady-state response of the linear transformer. The term *linear* calls attention to the fact that the windings are wound on a nonmagnetic core. The presence of a nonmagnetic core means that there is a linear relationship between the magnetic flux and the winding currents.

Figure 13.14 shows the basic circuit using the transformer as a coupling device, where the transformer connects the load to the source. The problem is to determine how the transformer affects the relationship between the load and the source. We can make one observation immediately: The transformer prevents any dc components of current (or voltage) in the source from reaching the load. Thus one reason for using transformers is to isolate portions of a circuit from dc currents and voltages. We show how the transformer affects steady-state sinusoidal currents and

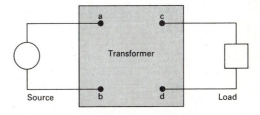

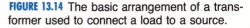

FIGURE 13.14 The basic arrangement of a transformer used to connect a load to a source.

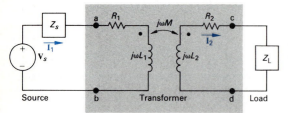

FIGURE 13.15 The phasor-domain circuit model for the system of Fig. 13.14.

PSpice ~~~→ Use PSpice to simulate circuits with mutually coupled inductor coils: Chapter 13

voltages with a phasor-domain circuit model of the source, transformer, and load.

ANALYSIS OF A LINEAR TRANSFORMER CIRCUIT

Figure 13.15 shows the phasor-domain circuit model of the system in Fig. 13.14. In discussing this circuit, we refer to the transformer winding connected to the source as the *primary* winding and the winding connected to the load as the *secondary* winding. Based on this terminology, the transformer circuit parameters are

R_1 = the resistance of the primary winding,

R_2 = the resistance of the secondary winding,

L_1 = the self-inductance of the primary winding,

L_2 = the self-inductance of the secondary winding, and

M = the mutual inductance.

The internal voltage of the sinusoidal source is $\mathbf{V}_s$, and the internal impedance of the source is Z_s. The impedance Z_L represents the load connected to the secondary winding of the transformer. The phasor currents $\mathbf{I}_1$ and $\mathbf{I}_2$ represent the primary and secondary currents of the transformer, respectively.

The analysis of the circuit in Fig. 13.15 consists of finding $\mathbf{I}_1$ and $\mathbf{I}_2$ as functions of the circuit parameters $\mathbf{V}_s$, Z_s, R_1, L_1, R_2, L_2, M, Z_L, and ω. We also are interested in finding the impedance seen looking into the terminals a, b. To find $\mathbf{I}_1$ and $\mathbf{I}_2$ we first write the two mesh-current equations that describe the circuit:

$$\mathbf{V}_s = (Z_s + R_1 + j\omega L_1)\mathbf{I}_1 - j\omega M\,\mathbf{I}_2, \qquad \textbf{(13.43)}$$

$$0 = -j\omega M\,\mathbf{I}_1 + (R_2 + j\omega L_2 + Z_L)\mathbf{I}_2. \qquad \textbf{(13.44)}$$

To facilitate the algebraic manipulation of Eqs. (13.43) and (13.44), we let

$$Z_{11} = Z_s + R_1 + j\omega L_1 \qquad \textbf{(13.45)}$$

and

$$Z_{22} = R_2 + j\omega L_2 + Z_L, \qquad \textbf{(13.46)}$$

where Z_{11} is the total self-impedance of the mesh containing the primary winding of the transformer and Z_{22} is the total self-impedance of the mesh containing the secondary winding of the transformer. Based on the notation introduced in Eqs. (13.45) and (13.46), the solutions for $\mathbf{I}_1$ and $\mathbf{I}_2$ from Eqs. (13.43) and (13.44) are

$$\mathbf{I}_1 = \frac{Z_{22}}{Z_{11}Z_{22} + \omega^2 M^2}\mathbf{V}_s \qquad \textbf{(13.47)}$$

and

$$\mathbf{I}_2 = \frac{j\omega M}{Z_{11} Z_{22} + \omega^2 M^2} \mathbf{V}_s = \frac{j\omega M}{Z_{22}} \mathbf{I}_1. \qquad (13.48)$$

To the internal source voltage $\mathbf{V}_s$ the impedance appears as $\mathbf{V}_s/\mathbf{I}_1$, or

$$\frac{\mathbf{V}_s}{\mathbf{I}_1} = Z_{\text{int}} = \frac{Z_{11} Z_{22} + \omega^2 M^2}{Z_{22}} = Z_{11} + \frac{\omega^2 M^2}{Z_{22}}. \qquad (13.49)$$

The impedance at the terminals of the source is $Z_{\text{int}} - Z_s$, so

$$Z_{ab} = Z_{11} + \frac{\omega^2 M^2}{Z_{22}} - Z_s$$

$$= R_1 + j\omega L_1 + \frac{\omega^2 M^2}{(R_2 + j\omega L_2 + Z_L)}. \qquad (13.50)$$

The impedance given by Eq. (13.50) is of particular interest because it shows how the transformer modifies how the load appears at the source. In other words, without the transformer the load at the source appears as Z_L, but with the transformer the load at the source appears as a modified version of Z_L.

REFLECTED IMPEDANCE

The impedance Z_{ab} is independent of the magnetic polarity of the transformer. The reason is that the mutual inductance appears in Eq. (13.50) as a squared quantity. Moreover, the last term in Eq. (13.50) represents the impedance reflected into the primary side of the transformer. That is, if the two windings are decoupled, M becomes zero and Z_{ab} reduces to the self-impedance of the primary winding.

To consider reflected impedance in more detail, we first express the load impedance in rectangular form:

$$Z_L = R_L + jX_L, \qquad (13.51)$$

where the load reactance X_L carries its own algebraic sign. In other words, X_L is a positive number if the load is inductive and a negative number if the load is capacitive. We now use Eq. (13.51) to write the reflected impedance (Z_r), the last term on the right in Eq. (13.50), in rectangular form:

$$Z_r = \frac{\omega^2 M^2}{R_2 + R_L + j(\omega L_2 + X_L)}$$

$$= \frac{\omega^2 M^2 [(R_2 + R_L) - j(\omega L_2 + X_L)]}{(R_2 + R_L)^2 + (\omega L_2 + X_L)^2}$$

$$= \frac{\omega^2 M^2}{|Z_{22}|^2} [(R_2 + R_L) - j(\omega L_2 + X_L)]. \qquad (13.52)$$

The derivation of Eq. (13.52) recognizes that, when Z_L is written in rectangular form, the self-impedance of the mesh containing the secondary winding is

$$Z_{22} = R_2 + R_L + j(\omega L_2 + X_L). \qquad (13.53)$$

Now observe from Eq. (13.52) that the self-impedance of the secondary circuit is reflected into the primary circuit by a scaling factor of $(\omega M / |Z_{22}|)^2$ and that the sign of the reactive component $(\omega L_2 + X_L)$ is reversed. Thus the linear transformer reflects the *conjugate* of the self-impedance of the secondary circuit (Z_{22}^*) into the primary winding by a scalar multiplier. The multiplier is equal to the square of the ratio of the mutual reactance to the magnitude of the secondary-circuit self-impedance $[(\omega M / |Z_{22}|)^2]$. Before describing Eq. (13.52) further, we illustrate numerically in Example 13.2 the results discussed so far.

EXAMPLE 13.2

The parameters of a certain linear transformer are $R_1 = 200 \ \Omega$, $R_2 = 100 \ \Omega$, $L_1 = 9$ H, $L_2 = 4$ H, and $k = 0.5$. The transformer couples an impedance consisting of an 800-Ω resistor in series with a 1-μF capacitor to a sinusoidal voltage source. The 300-V (rms) source has an internal impedance of $500 + j100 \ \Omega$ and a frequency of 400 rad/s.

a) Construct a phasor-domain equivalent circuit of the system.

b) Calculate the self-impedance of the primary circuit.

c) Calculate the self-impedance of the secondary circuit.

d) Calculate the impedance reflected into the primary winding.

e) Calculate the scaling factor for the reflected impedance.

f) Calculate the impedance seen looking into the primary terminals of the transformer.

g) Calculate the rms value of the primary and secondary current.

h) Calculate the rms value of the voltage at the terminals of the load and source.

i) Calculate the average power delivered to the 800-Ω resistor.

j) What percentage of the average power delivered to the transformer is delivered to the load?

SOLUTION

a) Figure 13.16 shows the phasor-domain equivalent circuit. Note that the internal voltage of the source serves as the ref-

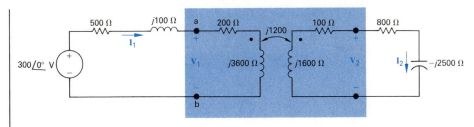

FIGURE 13.16 The phasor-domain equivalent circuit for Example 13.2.

erence phasor and V_1 and V_2 represent the terminal voltages of the transformer. In constructing the circuit in Fig. 13.16, we made the following calculations:

$$j\omega L_1 = j(400)(9) = j3600 \ \Omega;$$
$$j\omega L_2 = j(400)(4) = j1600 \ \Omega;$$
$$M = 0.5\sqrt{(9)(4)} = 3 \ \text{H};$$
$$j\omega M = j(400)(3) = j1200 \ \Omega;$$
$$\frac{1}{j\omega C} = \frac{10^6}{j400} = -j2500 \ \Omega.$$

b) The self-impedance of the primary circuit is

$$Z_{11} = 500 + j100 + 200 + j3600 \ \Omega = 700 + j3700 \ \Omega.$$

c) The self-impedance of the secondary circuit is

$$Z_{22} = 100 + j1600 + 800 - j2500 = 900 - j900 \ \Omega.$$

d) The impedance reflected into the primary winding is

$$Z_r = \left(\frac{1200}{|900 - j900|}\right)^2 (900 + j900)$$
$$= \frac{8}{9}(900 + j900) = 800 + j800 \ \Omega.$$

e) The scaling factor by which Z_{22}^* is reflected is $8/9$.

f) The impedance seen looking into the primary terminals of the transformer is the impedance of the primary winding plus the reflected impedance; thus

$$Z_{ab} = 200 + j3600 + 800 + j800 = 1000 + j4400 \ \Omega.$$

g) Knowing the input impedance to the transformer, we can easily calculate I_1:

$$I_1 = \frac{V_s}{Z_s + Z_{ab}} = \frac{300/0°}{1500 + j4500}$$
$$= 20 - j60 = 63.25/-71.57° \ \text{mA rms}.$$

We calculate $\mathbf{I}_2$ from Eq. (14.48):

$$\mathbf{I}_2 = \frac{j(1200)}{900 - j900}\mathbf{I}_1 = 59.63\underline{/63.43°} \text{ mA rms.}$$

h) The voltages at the terminals of the transformer are

$$\mathbf{V}_2 = (800 - j2500)\,\mathbf{I}_2 = 156.52\underline{/-8.82°} \text{ V rms}$$

and

$$\mathbf{V}_1 = Z_{ab}\mathbf{I}_1$$
$$= (1000 + j4400)\mathbf{I}_1$$
$$= 285.38\underline{/5.63°} \text{ V rms.}$$

i) The average power delivered to the load is

$$P = (800)|\mathbf{I}_2|^2$$
$$= 2.84 \text{ W.}$$

j) The average power delivered to the transformer is

$$P_{ab} = (1000)|\mathbf{I}_1|^2 = 4.00 \text{ W;}$$

therefore

$$\eta = \frac{2.84}{4.00} \times 100 = 71.11\%.$$

INTRODUCTION TO THE IDEAL TRANSFORMER

We now are interested in what happens to the input impedance Z_{ab} as L_1 and L_2 each become infinitely large and, at the same time, the coefficient of coupling approaches its limiting value of unity. We are interested in this behavior because transformers wound on ferromagnetic cores can approach this condition. Even though magnetic-core transformers are nonlinear, we can obtain some useful information about such devices by constructing an ideal model that ignores the nonlinearities.

To show how Z_{ab} changes when $k = 1$ and L_1 and L_2 approach infinity, we first introduce the notation

$$Z_{22} = R_2 + R_L + j(\omega L_2 + X_L) = R_{22} + jX_{22}$$

and then rearrange Eq. (13.50):

$$Z_{ab} = R_1 + \frac{\omega^2 M^2 R_{22}}{R_{22}^2 + X_{22}^2} + j\left(\omega L_1 - \frac{\omega^2 M^2 X_{22}}{R_{22}^2 + X_{22}^2}\right)$$
$$= R_{ab} + jX_{ab}. \tag{13.54}$$

At this point, we must be careful with the coefficient of j in Eq. (13.54) because, as L_1 and L_2 approach infinity, this coefficient is the difference between two large quantities. Thus before letting L_1 and L_2 increase, we write the coefficient as

$$X_{ab} = \omega L_1 - \frac{(\omega L_1)(\omega L_2)X_{22}}{R_{22}^2 + X_{22}^2} = \omega L_1\left(1 - \frac{\omega L_2 X_{22}}{R_{22}^2 + X_{22}^2}\right), \quad \textbf{(13.55)}$$

where we recognized that, when $k = 1$, $M^2 = L_1 L_2$. Putting the term multiplying ωL_1 over a common denominator gives

$$X_{ab} = \omega L_1\left(\frac{R_{22}^2 + \omega L_2 X_L + X_L^2}{R_{22}^2 + X_{22}^2}\right). \quad \textbf{(13.56)}$$

Factoring ωL_2 out of the numerator and denominator of Eq. (13.56) yields

$$X_{ab} = \frac{L_1}{L_2}\frac{X_L + (R_{22}^2 + X_L^2)/\omega L_2}{(R_{22}/\omega L_2)^2 + [1 + (X_L/\omega L_2)]^2}. \quad \textbf{(13.57)}$$

As $k \to 1.0$, the ratio L_1/L_2 approaches the constant value of $(N_1/N_2)^2$, which follows from Eqs. (13.22) and (13.23). The reason is that, as the coupling becomes extremely tight, the two permeances $\mathcal{P}_1$ and $\mathcal{P}_2$ become equal. Equation (13.57) then reduces to

$$X_{ab} = \left(\frac{N_1}{N_2}\right)^2 X_L, \quad \textbf{(13.58)}$$

as $L_1 \to \infty$, $L_2 \to \infty$, and $k \to 1.0$.

The same reasoning leads to simplification of the reflected resistance in Eq. (13.54):

$$\frac{\omega^2 M^2 R_{22}}{R_{22}^2 + X_{22}^2} = \frac{L_1}{L_2}R_{22} = \left(\frac{N_1}{N_2}\right)^2 R_{22}. \quad \textbf{(13.59)}$$

Applying the results given by Eqs. (13.58) and (13.59) to Eq. (13.54) yields

$$Z_{ab} = R_1 + \left(\frac{N_1}{N_2}\right)^2 R_2 + \left(\frac{N_1}{N_2}\right)^2 (R_L + jX_L). \quad \textbf{(13.60)}$$

Equation (13.60) tells us that, when the coefficient of coupling approaches unity and the self-inductances of the coupled coils approach infinity, the transformer transfers the secondary winding resistance and the load impedance to the primary side by a scaling factor equal to the turns ratio (N_1/N_2) squared. By lumping the winding resistance R_2 with the load, we can use this special type of transformer to raise or lower a load's impedance level. In Section 13.6 we expand on this desirable feature by defining the ideal transformer.

DRILL EXERCISES

13.7 A linear transformer couples a load consisting of a 360-Ω resistor in series with a 0.25-H inductor to a sinusoidal voltage source, as shown. The voltage source has an internal impedance of $184 + j0$ Ω and a maximum voltage of 245.20 V, and is operating at 800 rad/s. The transformer parameters are $R_1 = 100$ Ω, $L_1 = 0.5$ H, $R_2 = 40$ Ω, $L_2 = 0.125$ H, and $k = 0.4$. Calculate (a) the reflected impedance; (b) the primary current; (c) the secondary current; and (d) the average power delivered to the primary terminals of the transformer.

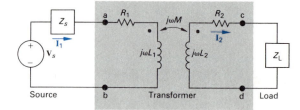

ANSWER: (a) $10.24 - j7.68$ Ω;
(b) $0.5 \cos(800t - 53.13°)$ A; (c) $0.08 \cos 800t$;
(d) 13.78 W.

13.6 THE IDEAL TRANSFORMER

The ideal transformer consists of two magnetically coupled coils, having N_1 and N_2 turns, respectively, that exhibit three properties:

1. the coefficient of coupling is unity ($k = 1$),

2. the self-inductance of each coil is infinite ($L_1 = L_2 = \infty$), and

3. the coil losses are negligible.

Hence we may describe the terminal behavior of the ideal transformer in terms of two characteristics. First, the magnitude of the volts per turn is the same for each coil, or

$$\left| \frac{v_1}{N_1} \right| = \left| \frac{v_2}{N_2} \right|. \tag{13.61}$$

Second, the magnitude of the ampere turns is the same for each coil, or

$$|i_1 N_1| = |i_2 N_2|. \tag{13.62}$$

We are forced to use magnitude signs in Eqs. (13.61) and (13.62) because we have not yet established reference polarities for the currents and voltages; we discuss the removal of the magnitude signs shortly. Figure 13.17 shows two lossless ($R_1 = R_2 = 0$) magnetically coupled coils. We use Fig. 13.17 to validate Eqs. (13.61) and (13.62). In Fig. 13.17(a) coil 2 is open, and in Fig. 13.17(b) coil 2 is shorted. Although we carry out the following analysis in terms of sinusoidal steady-state operation, the results also apply to instantaneous values of v and i.

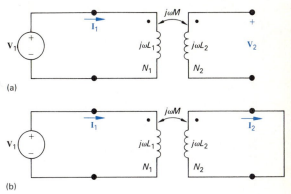

(a)

(b)

FIGURE 13.17 The circuits used to verify the volts-per-turn and ampere-turn relationships for the ideal transformer.

DETERMINING THE VOLTAGE AND CURRENT RATIOS

Note in Fig. 13.17(a) that the voltage at the terminals of the open-circuit coil is entirely the result of the current in coil 1; therefore

$$\mathbf{V}_2 = j\omega M\,\mathbf{I}_1. \tag{13.63}$$

The current in coil 1 is

$$\mathbf{I}_1 = \frac{\mathbf{V}_1}{j\omega L_1}. \tag{13.64}$$

From Eqs. (13.63) and (13.64),

$$\mathbf{V}_2 = \frac{M}{L_1}\mathbf{V}_1. \tag{13.65}$$

For unity coupling the mutual inductance equals $\sqrt{L_1 L_2}$, so Eq. (13.65) becomes

$$\mathbf{V}_2 = \sqrt{\frac{L_2}{L_1}}\,\mathbf{V}_1. \tag{13.66}$$

For unity coupling, the flux linking winding 1 is the same as the flux linking winding 2, so we need only one permeance in order to describe the self-inductance of each winding. Thus Eq. (13.66) becomes

$$\mathbf{V}_2 = \sqrt{\frac{N_2^2 \mathscr{P}}{N_1^2 \mathscr{P}}}\,\mathbf{V}_1 = \frac{N_2}{N_1}\mathbf{V}_1 \tag{13.67}$$

or

$$\frac{\mathbf{V}_1}{N_1} = \frac{\mathbf{V}_2}{N_2}. \tag{13.68}$$

Summing the voltages around the shorted coil of Fig. 13.17(b) yields

$$0 = -j\omega M\,\mathbf{I}_1 + j\omega L_2\,\mathbf{I}_2, \tag{13.69}$$

from which, for $k = 1$,

$$\frac{\mathbf{I}_1}{\mathbf{I}_2} = \frac{L_2}{M} = \frac{L_2}{\sqrt{L_1 L_2}} = \sqrt{\frac{L_2}{L_1}} = \frac{N_2}{N_1}. \tag{13.70}$$

FIGURE 13.18 The graphic symbol for an ideal transformer.

Equation (13.70) is equivalent to

$$\mathbf{I}_1 N_1 = \mathbf{I}_2 N_2. \qquad (13.71)$$

Figure 13.18 shows the graphic symbol for the ideal transformer. The vertical bars imply that coils wound on ferromagnetic cores can approximate an ideal transformer. These cores frequently consist of laminated sheets of magnetic material, and the vertical bars symbolize the laminated construction.

An ideal transformer can approximate reasonably well coils wrapped on a ferromagnetic core for several reasons. The ferromagnetic material creates a space with high permeance. Thus most of the magnetic flux is trapped inside the core material, establishing tight magnetic coupling between coils that share the same core. High permeance also means high self-inductance, as $L = N^2 \mathcal{P}$. Finally, ferromagnetically coupled coils efficiently transfer power from one coil to the other. Efficiencies in excess of 95 percent are common, so neglecting losses is not a crippling approximation for many applications.

DETERMINING THE POLARITY OF THE VOLTAGE AND CURRENT RATIOS

We now turn to the removal of the magnitude signs from Eqs. (13.61) and (13.62). Note that magnitude signs did not show up in the derivations of Eqs. (13.68) and (13.71). We did not need them there because we had established reference polarities for voltages and reference directions for currents. In addition, we knew the magnetic polarity dots of the two coupled coils. The rules for assigning the proper algebraic sign to Eqs. (13.61) and (13.62) are as follows:

1. If the coil voltages v_1 and v_2 are *both* positive or negative at the dot-marked terminal, use a plus sign in Eq. (13.61). Otherwise, use a negative sign.

2. If the coil currents i_1 and i_2 are *both* directed into or out of the dot-marked terminal, use a minus sign in Eq. (13.62). Otherwise, use a plus sign.

The four circuits shown in Figure 13.19 illustrate these rules.

FIGURE 13.19 Circuits that show the proper algebraic signs for relating the terminal voltages and terminal currents of the ideal transformer.

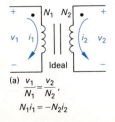

(a) $\dfrac{v_1}{N_1} = \dfrac{v_2}{N_2}$,

$N_1 i_1 = -N_2 i_2$

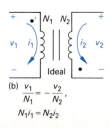

(b) $\dfrac{v_1}{N_1} = -\dfrac{v_2}{N_2}$,

$N_1 i_1 = N_2 i_2$

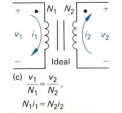

(c) $\dfrac{v_1}{N_1} = \dfrac{v_2}{N_2}$,

$N_1 i_1 = N_2 i_2$

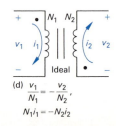

(d) $\dfrac{v_1}{N_1} = -\dfrac{v_2}{N_2}$,

$N_1 i_1 = -N_2 i_2$

The ratio of turns on the two windings is an important parameter of the ideal transformer. The turns ratio is defined as either N_1/N_2 or N_2/N_1; both ratios appear in various writings. In this text, we use a to denote the ratio N_2/N_1, or

$$a = \frac{N_2}{N_1}. \qquad (13.72)$$

Figure 13.20 shows three (of several) ways to represent the turns ratio of an ideal transformer on the circuit symbol. Figure 13.20(a) shows the number of turns in each coil explicitly. Figure 13.20(b) shows that the ratio N_2/N_1 is 5 to 1, and Fig. 13.20(c) shows that N_2/N_1 is 1 to $\frac{1}{5}$.

Example 13.3 illustrates the analysis of a circuit containing an ideal transformer.

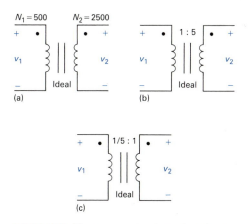

FIGURE 13.20 Three ways to show that the turns ratio of an ideal transformer is 5.

E X A M P L E   13.3

a) Find the average power delivered by the sinusoidal current source in the circuit shown in Fig. 13.21.

b) Find the average power delivered to the 20-Ω resistor.

S O L U T I O N

a) The ideal transformer encourages use of the mesh-current method of analysis. Therefore we begin by transforming the current source to an equivalent voltage source. Figure 13.22 shows the transformation, along with the phasor mesh currents and transformer terminal voltages used in the solution. Summing the voltages around meshes 1 and 2 generates the equations:

$$300 = 60\mathbf{I}_1 + \mathbf{V}_1 + 20(\mathbf{I}_1 - \mathbf{I}_2);$$

$$0 = 20(\mathbf{I}_2 - \mathbf{I}_1) + \mathbf{V}_2 + 40\mathbf{I}_2.$$

In addition to these two mesh-current equations, we need the constraint equations imposed by the ideal transformer. From the circuit diagram, $N_2/N_1 = 1/4$. In terms of the currents and voltages defined at the terminals of the ideal transformer,

$$\mathbf{V}_2 = \tfrac{1}{4}\mathbf{V}_1 \quad \text{and} \quad \mathbf{I}_2 = -4\mathbf{I}_1.$$

We now have four equations and four unknowns. The solutions for $\mathbf{V}_1$, $\mathbf{V}_2$, $\mathbf{I}_1$, and $\mathbf{I}_2$ are

$$\mathbf{V}_1 = 260 \text{ V rms}; \qquad \mathbf{I}_1 = 0.25 \text{ A rms};$$

$$\mathbf{V}_2 = 65 \text{ V rms}; \qquad \mathbf{I}_2 = -1.0 \text{ A rms}.$$

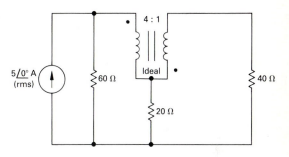

FIGURE 13.21 The circuit for Example 13.3.

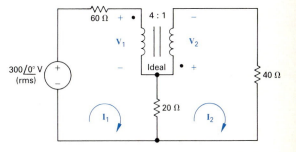

FIGURE 13.22 The circuit for the solution of Example 13.3.

The voltage across the 5-A current source in the circuit in Fig. 13.21 is

$$\mathbf{V}_{5A} = \mathbf{V}_1 + 20(\mathbf{I}_1 - \mathbf{I}_2)$$
$$= 260 + 20[0.25 - (-1)]$$
$$= 285 \text{ V rms.}$$

Note that $\mathbf{V}_{5A}$ is a voltage rise in the direction of the source current. The average power associated with the current source is

$$P = -(285)(5) = -1425 \text{ W.}$$

The minus sign means that the source is delivering power to the circuit.

b) To find the average power delivered to the 20-Ω resistor, we first calculate the current in the resistor. From the circuit shown in Fig. 13.22, the current oriented down through the 20-Ω resistor is

$$\mathbf{I}_{20} = \mathbf{I}_1 - \mathbf{I}_2 = 0.25 - (-1)$$
$$= 1.25 \text{ A rms.}$$

Therefore the average power dissipated in the resistor is

$$P_{20\Omega} = (1.25)^2(20) = 31.25 \text{ W.}$$

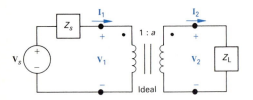

FIGURE 13.23 Using an ideal transformer to couple a load to a source.

USE OF AN IDEAL TRANSFORMER FOR IMPEDANCE MATCHING

At the end of Section 13.5, we implied that the ideal transformer can be used to raise or lower the impedance level of a load. The circuit shown in Fig. 13.23 confirms this fact. The impedance seen by the voltage source $\mathbf{V}_s$ is $\mathbf{V}_1/\mathbf{I}_1$. The voltage and current at the terminals of the load impedance ($\mathbf{V}_2$ and $\mathbf{I}_2$) are related to $\mathbf{V}_1$ and $\mathbf{I}_1$ by the transformer turns ratio; thus

$$\mathbf{V}_1 = \frac{\mathbf{V}_2}{a} \tag{13.73}$$

and

$$\mathbf{I}_1 = a\mathbf{I}_2. \tag{13.74}$$

Therefore the impedance seen by the source is

$$Z_{\text{IN}} = \frac{\mathbf{V}_1}{\mathbf{I}_1} = \frac{1}{a^2} \frac{\mathbf{V}_2}{\mathbf{I}_2}. \tag{13.75}$$

But the ratio $\mathbf{V}_2/\mathbf{I}_2$ is the load impedance Z_L, so Eq. (13.75) becomes

$$Z_{IN} = \frac{1}{a^2} Z_L. \qquad (13.76)$$

Note that the ideal transformer changes the magnitude of Z_L but does not affect its phase angle. Whether Z_{IN} is greater or less than Z_L depends on the turns ratio a.

The ideal transformer—or its practical counterpart, the ferromagnetic core transformer—can be used to match the magnitude of Z_L to the magnitude of Z_s, thus improving the amount of average power transferred from the source to the load.

PSpice Modify the PSpice representation for mutual inductance to get an ideal transformer: Chapter 14

DRILL EXERCISES

13.8 Make the following changes in the ideal transformer in the circuit shown in Fig. 13.21: (1) Place the dot on the right-hand side of the transformer at the upper terminal and (2) change the turns ratio from 4 : 1 to 3 : 1. Calculate the average power delivered to the 20-Ω resistor.

ANSWER: 28.8 W.

13.9 Find the average power delivered to the 4-kΩ resistor in the circuit shown.

ANSWER: 160 W.

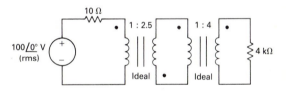

13.10 The ideal transformer connected to the 4-kΩ load in Drill Exercise 13.9 is replaced with an ideal transformer that has a turns ratio of 1 : a.

a) What value of a results in maximum average power being delivered to the 4-kΩ resistor?

b) What is the maximum average power?

ANSWER: (a) 8; (b) 250 W.

13.7 EQUIVALENT CIRCUITS FOR MAGNETICALLY COUPLED COILS

At times modeling magnetically coupled coils with an equivalent circuit that does not involve magnetic coupling is convenient. Consider the two magnetically coupled coils shown in Fig. 13.24. The resistances R_1 and R_2 represent the winding resistance of each coil. The goal is to replace the magnetically coupled coils inside the shaded area with a set of inductors that are not magnetically coupled. Before deriving the equivalent cir-

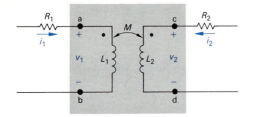

FIGURE 13.24 The circuit used to develop an equivalent circuit for magnetically coupled coils.

cuits, we must point out an important restriction: The voltage between terminals b and d must be zero. In other words, if terminals b and d can be shorted together without disturbing the voltages and currents in the original circuit in which the magnetically coupled coils are embedded, the equivalent circuits derived in the material that follows can be used to model the coils (see Problem 13.39).

We begin development of the circuit models by writing the two equations that relate the terminal voltages v_1 and v_2 to the terminal currents i_1 and i_2. For the given references and polarity dots,

$$v_1 = L_1 \frac{di_1}{dt} + M \frac{di_2}{dt} \tag{13.77}$$

and

$$v_2 = M \frac{di_1}{dt} + L_2 \frac{di_2}{dt}. \tag{13.78}$$

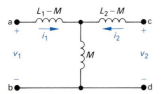

FIGURE 13.25 The T-equivalent circuit for the magnetically coupled coils of Fig. 13.24.

T-EQUIVALENT CIRCUIT

In order to arrive at an equivalent circuit for these two magnetically coupled coils, we seek an arrangement of inductors that can be described by a set of equations equivalent to Eqs. (13.77) and (13.78). The key to finding the arrangement is to regard Eqs. (13.77) and (13.78) as mesh-current equations with i_1 and i_2 as the mesh variables. Then we need one mesh with a total inductance of L_1 henrys and a second mesh with a total inductance of L_2 henrys. Furthermore, the two meshes must have a common inductance of M henrys. The T-arrangement of coils shown in Fig. 13.25 satisfies these requirements. We leave to you verification that the equations relating v_1 and v_2 to i_1 and i_2 in the circuit in Fig. 13.25 reduce to Eqs. (13.77) and (13.78). Note the absence of magnetic coupling between the inductors in the circuit shown in Fig. 13.25 and the zero voltage between b and d.

When we use the T-equivalent circuit shown in Fig. 13.25 to model the magnetically coupled coils shown in Fig. 13.24, the equivalent circuit that includes the winding resistances R_1 and R_2 appears as shown in Fig. 13.26.

π-EQUIVALENT CIRCUIT

We can derive a π-equivalent circuit for the magnetically coupled coils shown in Fig. 13.24. This derivation is based on solving Eqs. (13.77) and (13.78) for the derivatives di_1/dt and di_2/dt and then regarding the resulting expressions as a pair of node-voltage equations. Using Cramer's method for solving simulta-

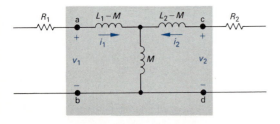

FIGURE 13.26 An equivalent circuit that includes the winding resistances R_1 and R_2.

neous equations, we obtain expressions for di_1/dt and di_2/dt:

$$\frac{di_1}{dt} = \frac{\begin{vmatrix} v_1 & M \\ v_2 & L_2 \end{vmatrix}}{\begin{vmatrix} L_1 & M \\ M & L_2 \end{vmatrix}} = \frac{L_2}{L_1 L_2 - M^2} v_1 - \frac{M}{L_1 L_2 - M^2} v_2; \qquad \textbf{(13.79)}$$

$$\frac{di_2}{dt} = \frac{\begin{vmatrix} L_1 & v_1 \\ M & v_2 \end{vmatrix}}{L_1 L_2 - M^2} = \frac{-M}{L_1 L_2 - M^2} v_1 + \frac{L_1}{L_1 L_2 - M^2} v_2. \qquad \textbf{(13.80)}$$

Now we solve for i_1 and i_2 by multiplying both sides of Eqs. (13.79) and (13.80) by dt and then integrating:

$$i_1 = i_1(0) + \frac{L_2}{L_1 L_2 - M^2} \int_0^t v_1 \, d\tau - \frac{M}{L_1 L_2 - M^2} \int_0^t v_2 \, d\tau \qquad \textbf{(13.81)}$$

and

$$i_2 = i_2(0) - \frac{M}{L_1 L_2 - M^2} \int_0^t v_1 \, d\tau + \frac{L_1}{L_1 L_2 - M^2} \int_0^t v_2 \, d\tau. \qquad \textbf{(13.82)}$$

If we regard v_1 and v_2 as node voltages, Eqs. (13.81) and (13.82) describe a circuit of the form shown in Fig. 13.27.

All that remains to be done in deriving the π-equivalent circuit is to find L_A, L_B, and L_C as functions of L_1, L_2, and M. We easily do so by writing the equations for i_1 and i_2 in the circuit shown in Fig. 13.27 and then comparing them with Eqs. (13.81) and (13.82). Thus

$$i_1 = i_1(0) + \frac{1}{L_A} \int_0^t v_1 \, d\tau + \frac{1}{L_B} \int_0^t (v_1 - v_2) \, d\tau$$

$$= i_1(0) + \left(\frac{1}{L_A} + \frac{1}{L_B}\right) \int_0^t v_1 \, d\tau - \frac{1}{L_B} \int_0^t v_2 \, d\tau, \qquad \textbf{(13.83)}$$

$$i_2 = i_2(0) + \frac{1}{L_C} \int_0^t v_2 \, d\tau + \frac{1}{L_B} \int_0^t (v_2 - v_1) \, d\tau$$

$$= i_2(0) - \frac{1}{L_B} \int_0^t v_1 \, d\tau + \left(\frac{1}{L_B} + \frac{1}{L_C}\right) \int_0^t v_2 \, d\tau. \qquad \textbf{(13.84)}$$

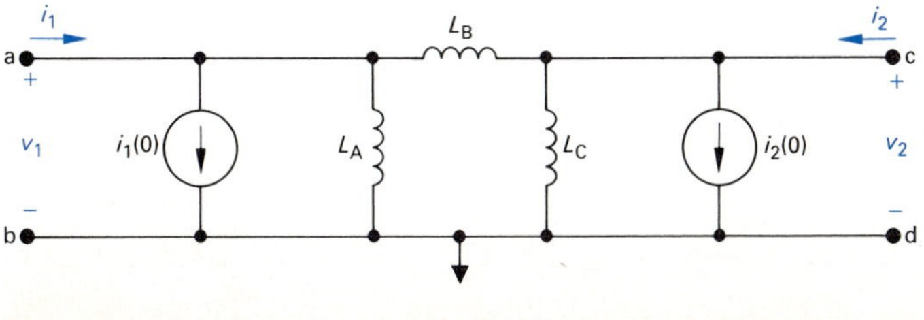

FIGURE 13.27 The circuit used to derive the π-equivalent circuit for magnetically coupled coils.

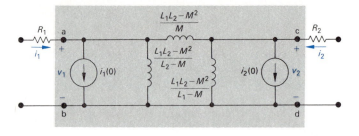

FIGURE 13.28 The π-equivalent circuit for the magnetically coupled coils of Fig. 13.24.

Then

$$\frac{1}{L_B} = \frac{M}{L_1 L_2 - M^2},$$

$$(13.85)$$

$$\frac{1}{L_A} = \frac{L_2 - M}{L_1 L_2 - M^2},$$

$$(13.86)$$

$$\frac{1}{L_C} = \frac{L_1 - M}{L_1 L_2 - M^2}.$$

$$(13.87)$$

When we incorporate Eqs. (13.85)–(13.87) into the circuit shown in Fig. 13.27, the π-equivalent circuit for the magnetically coupled coils shown in Fig. 13.24 is as shown in Fig. 13.28.

Note that the initial values of i_1 and i_2 are explicit in the π-equivalent circuit but are implicit in the T-equivalent circuit. In this chapter we are focusing on the sinusoidal steady-state behavior of circuits containing mutual inductance, so we can assume that the initial values of i_1 and i_2 are zero and thus eliminate the current sources in the π-equivalent circuit. Thus for sinusoidal steady-state analysis, the circuit shown in Fig. 13.28 simplifies to that shown in Fig. 13.29.

The mutual inductance carries its own algebraic sign in the T- and π-equivalent circuits. That is, if the magnetic polarity of the coupled coils is reversed from that given in Fig. 13.24, the algebraic sign of M reverses. A reversal in magnetic polarity requires moving one polarity dot without changing the reference polarities of the terminal currents and voltages. Example 13.4 illustrates the application of the T-equivalent circuit.

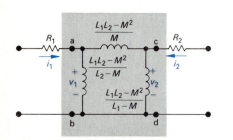

FIGURE 13.29 The π-equivalent circuit used for sinusoidal steady-state analysis.

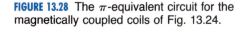

E X A M P L E   13.4

a) Use the T-equivalent circuit for the magnetically coupled coils described in Example 13.2 to find the phasor currents $\mathbf{I}_1$ and $\mathbf{I}_2$. The phasor currents $\mathbf{I}_1$ and $\mathbf{I}_2$ are defined in the circuit shown in Fig. 13.30.

b) Repeat (a), but with the polarity dot on the secondary winding moved to the lower terminal.

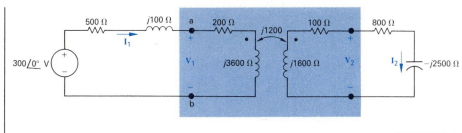

FIGURE 13.30 The phasor-domain equivalent circuit for Example 13.4.

S O L U T I O N

a) For the polarity dots shown in Fig. 13.30, M carries a value of $+3$ H in the T-equivalent circuit. Therefore the three inductances in the equivalent circuit are

$$L_1 - M = 9 - 3 = 6 \text{ H};$$
$$L_2 - M = 4 - 3 = 1 \text{ H};$$
$$M = 3 \text{ H}.$$

Figure 13.31 shows the T-equivalent circuit. Figure 13.32 shows the phasor-domain equivalent circuit at a frequency of 400 rad/s. Figure 13.33 shows the phasor-domain circuit for the original system. Here the magnetically coupled coils are modeled by the circuit shown in Fig. 13.32. To find the phasor currents $\mathbf{I}_1$ and $\mathbf{I}_2$, we first find the node voltage across the 1200-Ω inductive reactance. If we use the lower node as the reference, the single node-voltage equation is

$$\frac{\mathbf{V} - 300}{700 + j2500} + \frac{\mathbf{V}}{j1200} + \frac{\mathbf{V}}{900 - j2100} = 0.$$

Solving for $\mathbf{V}$ yields

$$\mathbf{V} = 136 - j8 = 136.24\underline{/-3.37°} \text{ V rms}.$$

Then

$$\mathbf{I}_1 = \frac{300 - (136 - j8)}{700 + j2500} = 63.25\underline{/-71.57°} \text{ mA rms}$$

and

$$\mathbf{I}_2 = \frac{136 - j8}{900 - j2100} = 59.63\underline{/63.43°} \text{ mA rms}.$$

Checking against those obtained in Example 13.2 shows that they are identical.

b) When the polarity dot is moved to the lower terminal of the secondary coil, M carries a value of -3 H in the T-equivalent circuit. Before carrying out the solution with the new T-equivalent circuit, we note from Example 13.2 that reversing

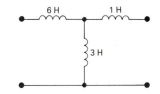

FIGURE 13.31 The T-equivalent circuit for the magnetically coupled coils in Example 13.2.

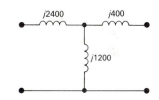

FIGURE 13.32 The phasor-domain model of the T-equivalent circuit at 400 rad/s.

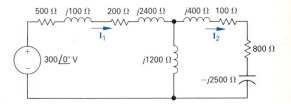

FIGURE 13.33 The circuit of Fig. 13.30 with the magnetically coupled coils replaced by their T-equivalent circuit.

the algebraic sign of M has no effect on the solution for $\mathbf{I}_1$ and shifts $\mathbf{I}_2$ by $180°$. Therefore we anticipate that

$$\mathbf{I}_1 = 63.25\underline{/-71.57°} \text{ mA rms};$$

$$\mathbf{I}_2 = 59.63\underline{/-116.57°} \text{ mA rms}.$$

We now proceed to find these solutions by using the new T-equivalent circuit. With $M = -3$ H, the three inductances in the equivalent circuit are

$$L_1 - M = 9 - (-3) = 12 \text{ H};$$

$$L_2 - M = 4 - (-3) = 7 \text{ H};$$

$$M = -3 \text{ H}.$$

At an operating frequency of 400 rad/s, the phasor-domain equivalent circuit requires two inductors and a capacitor, as shown in Fig. 13.34. The resulting phasor-domain circuit for the original system appears in Fig. 13.35. As before, we first find the node voltage across the center branch, which in this case is a capacitive reactance of $-j1200 \ \Omega$. If we use the lower node as reference, the node-voltage equation is

$$\frac{\mathbf{V} - 300}{700 + j4900} + \frac{\mathbf{V}}{-j1200} + \frac{\mathbf{V}}{900 + j300} = 0.$$

Solving for $\mathbf{V}$ gives

$$\mathbf{V} = -8 - j56 = 56.57\underline{/-98.13°} \text{ V rms}.$$

Then

$$\mathbf{I}_1 = \frac{300 - (-8 - j56)}{700 + j4900} = 63.25\underline{/-71.57°} \text{ mA rms}$$

and

$$\mathbf{I}_2 = \frac{-8 - j56}{900 + j300} = 59.63\underline{/-116.57°} \text{ mA rms}.$$

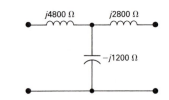

FIGURE 13.34 The phasor-domain equivalent circuit for $M = -3$ H and $\omega = 400$ rad/s.

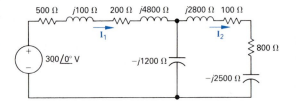

FIGURE 13.35 The phasor-domain equivalent circuit for Example 13.4(b).

DRILL EXERCISES

13.11 a) Show that, if the reference direction for i_2 is reversed in both circuits shown at the top of the next page, the T-equivalent circuit shown on the right is still valid.

    b) Show that if the dot on the L_2 coil in the circuit shown on the left is moved to the lower terminal,

the three inductors in the circuit shown on the right are $L_1 + M$, $L_2 + M$, and $-M$.

    c) Will the circuit derived in (b) be valid if the reference direction for i_2 is reversed in both circuits?

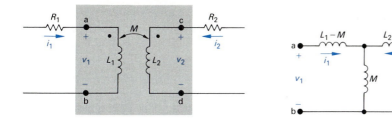

ANSWER: (a) Show; (b) show; (c) yes.

13.12 Use the T-equivalent circuit for the linear transformer in Drill Exercise 13.7 to find $\mathbf{I}_1$ and $\mathbf{I}_2$. The phasor currents $\mathbf{I}_1$ and $\mathbf{I}_2$ are defined as shown.

ANSWER: $\mathbf{I}_1 = 500\underline{/-53.13°}$ mA; $\mathbf{I}_2 = 80\underline{/0°}$ mA.

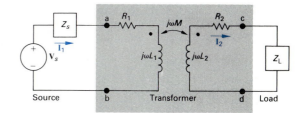

13.8 THE NEED FOR IDEAL TRANSFORMERS IN THE EQUIVALENT CIRCUITS

The inductors in the T- and π-equivalent circuits of magnetically coupled coils can have negative values. For example, if $L_1 = 3$ mH, $L_2 = 12$ mH, and $M = 5$ mH, the T-equivalent circuit requires an inductor of -2 mH and the π-equivalent circuit requires an inductor of -4.5 mH. These negative inductance values are not troublesome when you are using the equivalent circuits in computations. However, if you are to build the equivalent circuits with circuit components, the negative inductors can be bothersome. The reason is that whenever the frequency of the sinusoidal source changes, you must change the capacitor used to simulate the negative reactance. For example, at a frequency of 50 krad/s, a -2-mH inductor has an impedance of $-j100$ Ω. This impedance can be modeled with a capacitor having a capacitance of 0.2 μF. If the frequency changes to 25 krad/s, the -2-mH inductor impedance changes to $-j50$ Ω. At 25 krad/s this requires a capacitor with a capacitance of 0.8 μF. Obviously, in a situation where the frequency is varied continuously, the use of a capacitor to simulate negative inductance is practically worthless.

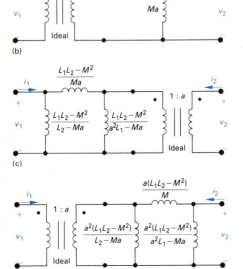

(a)

(b)

(c)

(d)

FIGURE 13.36 The four ways of using an ideal transformer in the T- and π-equivalent circuit for magnetically coupled coils.

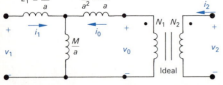

FIGURE 13.37 The circuit of Fig. 13.36(a) with i_o and v_o defined.

You can circumvent the problem of dealing with negative inductances by introducing an ideal transformer into the equivalent circuit. This doesn't completely solve the modeling problem, because ideal transformers can only be approximated. However, in some situations the approximation to an ideal transformer is good enough to warrant a discussion of using the ideal transformer in the T- and π-equivalent circuits of magnetically coupled coils.

An ideal transformer can be used in two different ways in either the T-equivalent or the π-equivalent circuit for magnetically coupled coils. Figure 13.36 shows the two arrangements for each type of equivalent circuit.

Verifying any of the equivalent circuits in Fig. 13.36 requires showing only that for any circuit the equations relating v_1 and v_2 to di_1/dt and di_2/dt are identical with Eqs. (13.77) and (13.78). We verify the validity of the circuit shown in Fig. 13.36(a) and leave verification of the circuits in Fig. 13.36(b), (c), and (d) as exercises. To aid the discussion, we redrew the circuit shown in Fig. 13.36(a) as Fig. 13.37, adding the variables i_0 and v_0. From this circuit,

$$v_1 = \left(L_1 - \frac{M}{a}\right)\frac{di_1}{dt} + \frac{M}{a}\frac{d}{dt}(i_1 + i_0); \qquad \textbf{(13.88)}$$

$$v_0 = \left(\frac{L_2}{a_2} - \frac{M}{a}\right)\frac{di_0}{dt} + \frac{M}{a}\frac{d}{dt}(i_0 + i_1). \qquad \textbf{(13.89)}$$

The ideal transformer imposes constraints on v_0 and i_0:

$$v_0 = \frac{v_2}{a}; \qquad \textbf{(13.90)}$$

$$i_0 = ai_2. \qquad \textbf{(13.91)}$$

Substituting Eqs. (13.90) and (13.91) into Eqs. (13.88) and (13.89) gives

$$v_1 = L_1\frac{di_1}{dt} + \frac{M}{a}\frac{d}{dt}(ai_2), \qquad \textbf{(13.92)}$$

$$\frac{v_2}{a} = \frac{L_2}{a^2}\frac{d}{dt}(ai_2) + \frac{M}{a}\frac{di_1}{dt}. \qquad \textbf{(13.93)}$$

From Eqs. (13.92) and (13.93),

$$v_1 = L_1\frac{di_1}{dt} + M\frac{di_2}{dt}, \qquad \textbf{(13.94)}$$

$$v_2 = M\frac{di_1}{dt} + L_2\frac{di_2}{dt}. \qquad \textbf{(13.95)}$$

Equations (13.94) and (13.95) are identical to Eqs. (13.77) and

(13.78); thus insofar as terminal behavior is concerned, the circuit shown in Fig. 13.37 is equivalent to the magnetically coupled coils shown inside the box in Fig. 13.24.

In showing that the circuit in Fig. 13.37 is equivalent to the magnetically coupled coils in Fig. 13.24, we placed *no restrictions* on the turns ratio a. Therefore, an infinite number of equivalent circuits are possible. Furthermore, we can always find a turns ratio to make all the inductances positive. Three values of a are of particular interest:

$$a = \frac{M}{L_1}, \qquad (13.96)$$

$$a = \frac{L_2}{M}, \qquad (13.97)$$

and

$$a = \sqrt{\frac{L_2}{L_1}}. \qquad (13.98)$$

The value of a given by Eq. (13.96) eliminates the inductances $L_1 - M/a$ and $a^2 L_1 - aM$ from the T-equivalent circuits and the inductances $(L_1 L_2 - M^2)/(a^2 L_1 - aM)$ and $a^2(L_1 L_2 - M^2)/(a^2 L_1 - aM)$ from the π-equivalent circuits.

The value of a given by Eq. (13.97) eliminates the inductances $(L_2/a^2) - (M/a)$ and $L_2 - aM$ from the T-equivalent circuits and the inductances $(L_1 L_2 - M^2)/(L_2 - aM)$ and $a^2(L_1 L_2 - M^2)/(L_2 - aM)$ from the π-equivalent circuits.

Also note that when $a = M/L_1$, the circuits in Fig. 13.36(a) and (c) become identical, and when $a = L_2/M$, the circuits in Fig. 13.36(b) and (d) become identical. Figures 13.38 and 13.39, respectively, summarize these observations. In deriving the expressions for the inductances there, we used the relationship $M = k\sqrt{L_1 L_2}$. Expressing the inductances as functions of the self-inductances L_1 and L_2 and the coefficient of coupling k allows the values of a given by Eqs. (13.96) and (13.97) not only to reduce the number of inductances needed in the equivalent circuit, but also to guarantee that all the inductances will be positive. We leave to you investigation of the consequences of choosing the value of a given by Eq. (13.98).

The values of a given by Eqs. (13.96)–(13.98) can be determined experimentally. The ratio M/L_1 is obtained by driving the coil designated as having N_1 turns by a sinusoidal voltage source. The source frequency is set high enough that $\omega L_1 \gg R_1$, and the N_2 coil is left open. Figure 13.40 shows this arrangement.

With the N_2 coil open,

$$\mathbf{V}_2 = j\omega M \mathbf{I}_1. \qquad (13.99)$$

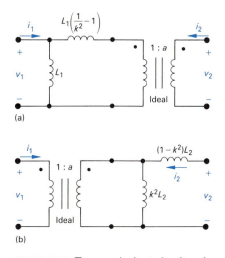

(a)

(b)

FIGURE 13.38 Two equivalent circuits when $a = M/L_1$.

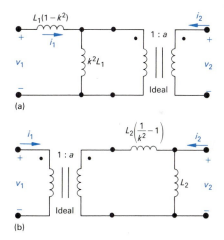

(a)

(b)

FIGURE 13.39 Two equivalent circuits when $a = L_2/M$.

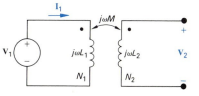

FIGURE 13.40 Experimental determination of the ratio M/L_1.

Now, as $j\omega L_1 \gg R_1$, the current $\mathbf{I}_1$ is

$$\mathbf{I}_1 = \frac{\mathbf{V}_1}{j\omega L_1}. \tag{13.100}$$

Substituting Eq. (13.100) into Eq. (13.99) yields

$$\left(\frac{\mathbf{V}_2}{\mathbf{V}_1}\right)_{\mathbf{I}_2=0} = \frac{M}{L_1}, \tag{13.101}$$

in which the ratio M/L_1 is the terminal voltage ratio corresponding to coil 2 being open; that is, $\mathbf{I}_2 = 0$.

We obtain the ratio L_2/M by reversing the procedure. That is, coil 2 is energized and coil 1 is left open. Then

$$\frac{L_2}{M} = \left(\frac{\mathbf{V}_2}{\mathbf{V}_1}\right)_{\mathbf{I}_1=0}. \tag{13.102}$$

Finally, we observe that the value of a given by Eq. (13.98) is the geometric mean of these two voltage ratios; thus

$$\sqrt{\left(\frac{\mathbf{V}_2}{\mathbf{V}_1}\right)_{\mathbf{I}_2=0}\left(\frac{\mathbf{V}_2}{\mathbf{V}_1}\right)_{\mathbf{I}_1=0}} = \sqrt{\frac{M}{L_1}\frac{L_2}{M}} = \sqrt{\frac{L_2}{L_1}}. \tag{13.103}$$

For coils wound on nonmagnetic cores the voltage ratio is *not* the same as the turns ratio, as it very nearly is for coils wound on ferromagnetic cores. Because the self-inductances vary as the square of the number of turns, Eq. (13.103) reveals that the turns ratio is approximately equal to the geometric mean of the two voltage ratios, or

$$\sqrt{\frac{L_2}{L_1}} = \frac{N_2}{N_1} = \sqrt{\left(\frac{\mathbf{V}_2}{\mathbf{V}_1}\right)_{\mathbf{I}_2=0}\left(\frac{\mathbf{V}_2}{\mathbf{V}_1}\right)_{\mathbf{I}_1=0}}. \tag{13.104}$$

DRILL EXERCISES

13.13 The circuit shown is the equivalent circuit of a loss-less linear transformer. Compute (a) L_1, (b) L_2, (c) M, and (d) k.

ANSWER: (a) 8 H; (b) 2 H; (c) 1.6 H; (d) 0.4.

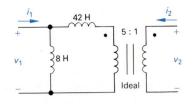

SUMMARY

Analysis of circuits containing coils linked by a common time-varying magnetic field is facilitated by introducing the concept of mutual inductance. We covered the following important points.

- *Mutual inductance:* This circuit parameter relates the voltage induced in one winding to a time-varying current in another winding.

- *Linear magnetic medium:* Coils are wound on nonmagnetic cores, making the magnetic flux directly proportional to the current. In such a medium, $M_{12} = M_{21} = M$.

- *Coefficient of coupling:* Denoted k, this measure of the degree of magnetic coupling, by definition, must lie between 0 and 1.

- $M = k\sqrt{L_1 L_2}$: This equation expresses the relationship between the self inductance of each winding and the mutual inductance between the windings.

- *Dot convention:* This method allows keeping track of the polarities of mutually induced voltages.

- *Stored energy:* The energy stored in magnetically coupled coils in a linear medium is related to the coil currents and inductances by the relationship

$$\frac{1}{2}L_1 i_1^2 + \frac{1}{2}L_2 i_2^2 \pm M i_1 i_2.$$

- *Two-winding linear transformer:* This coupling device is made up of two coils wound on a nonmagnetic core comprising a

  primary winding, which is the winding connected to the source side of the transformer;

  secondary winding, which is the winding connected to the load side of the transformer; and

  reflected impedance, which is the reflection of the conjugate of the self-impedance of the secondary circuit into the primary circuit by the scaling factor $[\omega M / |Z_{zz}|]^2$.

- *Two-winding ideal transformer:* This lossless coupling device is characterized by perfect coupling ($k = 1$) and infinite inductances ($L_1 = L_2 = M = \infty$). The circuit behavior is governed by the turns ratio $a = N_2/N_1$. In particular, the volts per turn is the same for each winding, or

$$\frac{v_1}{N_1} = \pm \frac{v_2}{N_2},$$

and the ampere turns are the same for each winding, or

$$N_1 i_1 = \pm N_2 i_2.$$

- *Reflected impedance:*

  The impedance reflected from the N_2 side to the N_1 side, or vice versa;

  Looking into coil 1, the impedance connected across coil 2 multiplied by $1/a^2$; or

  Looking into coil 2, the impedance connected across coil 1 multiplied by a^2.

- *Equivalent circuits:* Magnetically coupled coils can always be replaced by three uncoupled coils arranged in either a T or π configuration. Ideal transformers are used in T- or π-equivalent circuits to eliminate the need for negative inductors.

PROBLEMS

13.1 Two magnetically coupled coils have self-inductances of 60 and 9.6 mH, respectively. The mutual inductance between the coils is 22.8 mH.

a) What is the coefficient of coupling?

b) For these two coils, what is the largest value that M can have?

c) Assume that the physical structure of these coupled coils is such that $\mathcal{P}_1 = \mathcal{P}_2$. What is the turns ratio N_1/N_2 if N_1 is the number of turns on the 60-mH coil?

13.2 The self-inductances of two magnetically coupled coils are 72 and 40.5 mH, respectively. The 72-mH coil has 250 turns and the coefficient of coupling between the coils is (2/3). The coupling medium is nonmagnetic. When coil 1 is excited, with coil 2 open, the flux linking coil 1 only is 0.2 as large as the flux linking coil 2.

a) How many turns does coil 2 have?

b) What is the value of $\mathcal{P}_2$ in nanowebers per ampere?

c) What is the value of $\mathcal{P}_{11}$ in nanowebers per ampere?

d) What is the ratio (ϕ_{22}/ϕ_{12})?

13.3 The physical construction of four pairs of magnetically coupled coils is shown in Fig. P13.3. Assume that the magnetic flux is confined to the core material in each structure. Show two possible locations for the dot markings on each pair of coils.

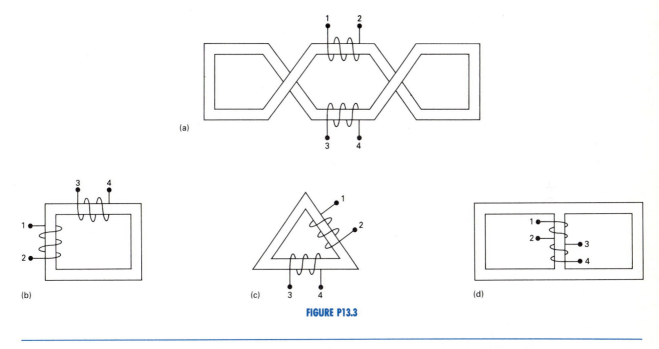

FIGURE P13.3

13.4 The polarity markings on two coils are to be determined experimentally. The experimental setup is shown in Fig. P13.4. Assume that the terminal connected to the negative terminal of the battery has been given a polarity mark as shown. When the switch is *closed*, the dc voltmeter "kicks" up-scale. Where should the polarity mark be placed on the coil connected to the voltmeter?

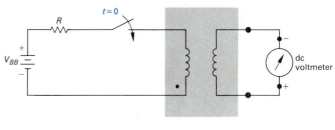

FIGURE P13.4

13.5 a) Starting with Eq. (13.27), show that the coefficient of coupling can also be expressed as

$$k = \sqrt{\left(\frac{\phi_{21}}{\phi_1}\right)\left(\frac{\phi_{12}}{\phi_2}\right)}.$$

b) On the basis of the fractions ϕ_{21}/ϕ_1 and ϕ_{12}/ϕ_2, explain why k is less than 1.0.

13.6 a) Show that the two coupled coils in Fig. P13.6 can be replaced by a single coil having an inductance of $L_{ab} = L_1 + L_2 + 2M$. (*Hint:* Express v_{ab} as a function of i_{ab}.)

b) Show that if the connections to the terminals of the coil labeled L_2 are reversed, $L_{ab} = L_1 + L_2 - 2M$.

FIGURE P13.6

13.7 a) Show that the two magnetically coupled coils in Fig. P13.7 can be replaced by a single coil having an inductance of

$$L_{ab} = \frac{L_1 L_2 - M^2}{L_1 + L_2 - 2M}.$$

(*Hint:* Let i_1 and i_2 be clockwise mesh currents in the left and right "windows" of Fig. P13.7, respectively. Sum the voltages around the two meshes. In mesh 1 let v_{ab} be the unspecified applied voltage. Solve for di_1/dt as a function of v_{ab}.)

b) Show that if the magnetic polarity of coil 2 is reversed, then

$$L_{ab} = \frac{L_1 L_2 - M^2}{L_1 + L_2 + 2M}.$$

FIGURE P13.7

13.8 A series combination of a 300-Ω resistor and a 100-mH inductor is connected to a sinusoidal voltage source by a linear transformer. The source is operating at a frequency of 1 krad/s. At this frequency the internal impedance of the source is $100 + j13.74 \ \Omega$. The rms voltage at the terminals of the source is 50 V when it is not loaded. The parameters of the linear transformer are: $R_1 = 41.68 \ \Omega$, $L_1 = 180$ mH, $R_2 = 500 \ \Omega$, $L_2 = 500$ mH, and $M = 270$ mH.

a) What is the value of the impedance reflected into the primary?

b) What is the value of the impedance seen from the terminals of the practical source?

c) What is the rms magnitude of the voltage across the load impedance?

d) What percentage of the average power developed by the practical source is delivered to the load impedance?

13.9 The value of k in the circuit in Fig. P13.9 is adjusted so that Z_{ab} is purely resistive when $\omega = 5$ krad/s. Find Z_{ab}.

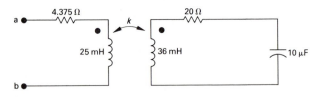

FIGURE P13.9

13.10 The sinusoidal voltage source in the circuit seen in Fig. P13.10 is operating at a frequency of 40 krad/s. The coefficient of coupling is adjusted until the peak amplitude of i_1 is maximum.

a) What is the value of k?

b) What is the peak amplitude of i_1 if $v_g = 500 \cos (4 \times 10^4 t)$ V?

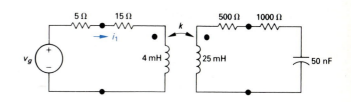

FIGURE P13.10

13.11 Find the average power delivered to the 16-Ω resistor in the circuit shown in Fig. P13.11 if $v_g = 100 \cos 2000t$ V.

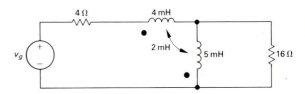

FIGURE P13.11

13.12 Find the average power delivered to the 15-Ω resistor in the circuit shown in Fig. P13.12 if $v_g = 300 \cos 1000t$ V.

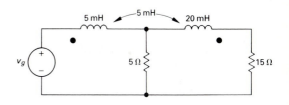

FIGURE P13.12

13.13 The impedance Z_L in the circuit in Fig. P13.13 is adjusted for maximum average power transfer to Z_L. The internal impedance of the sinusoidal voltage source is $4 + j7 \, \Omega$.

a) What is the maximum average power delivered to Z_L?

b) What percentage of the average power delivered to the linear transformer is delivered to Z_L?

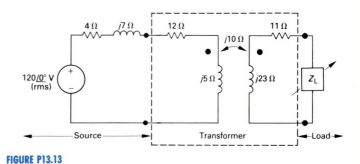

FIGURE P13.13

13.14 The variable resistor R_o in the circuit in Fig. P13.14 is adjusted until it dissipates the maximum average power possible.

a) Find the maximum average power possible if $i_g = \sqrt{21.20} \cos 5000t$ A.

b) What percentage of the average power developed by the ideal current source is delivered to R_o?

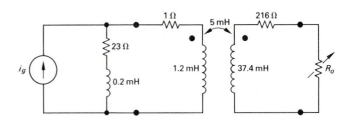

FIGURE P13.14

13.15 The values of the parameters in the circuit shown in Fig. P13.15 are $L_1 = 8$ mH; $L_2 = 2$ mH; $k = 0.75$; $R_g = 1\ \Omega$; and $R_L = 7\ \Omega$. If $v_g = 54\sqrt{2} \cos 1000t$ V, find

a) the rms magnitude of v_o;

b) the average power delivered to R_L; and

c) the percentage of the average power generated by the ideal voltage source that is delivered to R_L.

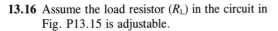

FIGURE P13.15

13.16 Assume the load resistor (R_L) in the circuit in Fig. P13.15 is adjustable.

a) What value of R_L will result in the maximum average power being transferred to R_L?

b) What is the value of the maximum power transferred?

13.17 Assume the coefficient of coupling in the circuit in Fig. P13.15 is adjustable. Find the value of k that reduces the magnitude of v_o to zero.

13.18 For the phasor-domain circuit in Fig. P13.18 calculate:

a) the rms magnitude of $\mathbf{V}_o$;

b) the average power dissipated in the 9-Ω resistor; and

c) the percentage of the average power generated by the ideal voltage source that is delivered to the 9-Ω load resistor.

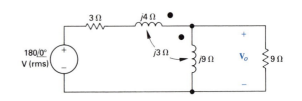

FIGURE P13.18

13.19 The 9-Ω resistor in the circuit in Fig. P13.18 is replaced with a variable impedance Z_o. Assume Z_o is adjusted for maximum average power transfer to Z_o.

a) What is the maximum average power that can be delivered to Z_o?

b) What is the average power developed by the ideal voltage source when maximum average power is delivered to Z_o?

13.20 Find Z_{ab} in the circuit in Fig. P13.20.

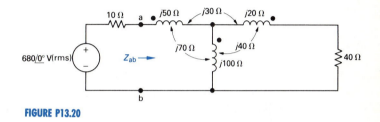

FIGURE P13.20

13.21 Find the impedance seen by the ideal voltage source in the circuit in Fig. P13.21 when Z_o is adjusted for maximum average power transfer to Z_o.

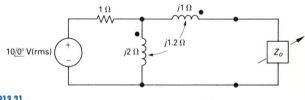

FIGURE P13.21

13.22 a) Find the six branch currents $\mathbf{I}_a$ through $\mathbf{I}_f$ in the circuit in Fig. P13.22.

b) Find the complex power in each branch of the circuit.

c) Check your calculations by verifying that the average power developed equals the average power dissipated.

d) Check your calculations by verifying that the magnetizing vars generated equals the magnetizing vars absorbed.

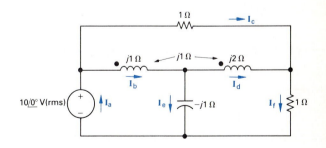

FIGURE P13.22

13.23 a) Show that the impedance seen looking into the terminals a, b in the circuit in Fig. P13.23 is given by the expression

$$Z_{ab} = \left(1 + \frac{N_1}{N_2}\right)^2 Z_L.$$

b) Show that if the polarity terminal of either one of the coils is reversed that

$$Z_{ab} = \left(1 - \frac{N_1}{N_2}\right)^2 Z_L.$$

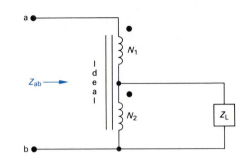

FIGURE P13.23

13.24 a) Show that the impedance seen looking into the terminals a, b in the circuit in Fig. P13.24 is given by the expression

$$Z_{ab} = \frac{Z_L}{\left(1 + \frac{N_1}{N_2}\right)^2}.$$

b) Show that if the polarity terminal of either one of the coils is reversed that

$$Z_{ab} = \frac{Z_L}{\left(1 - \frac{N_1}{N_2}\right)^2}.$$

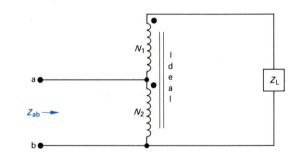

FIGURE P13.24

13.25 Find the average power dissipated in the 1-Ω resistor in the circuit in Fig. P13.25.

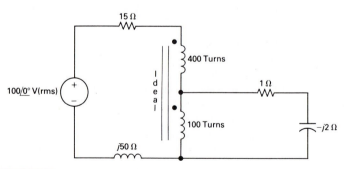

FIGURE P13.25

13.26 a) If N_2 equals 1000 turns, how many turns should be placed on the N_1 winding of the ideal transformer in the circuit seen in Fig. P13.26 so that maximum average power is delivered to the 6800-Ω load?

b) Find the average power delivered to the 6800-Ω resistor.

c) What percentage of the average power delivered by the ideal voltage source is dissipated in the linear transformer?

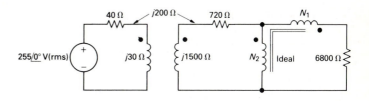

FIGURE P13.26

13.27 a) Find the turns ratio N_1/N_2 for the ideal transformer in Fig. P13.27 so that maximum average power is transferred to the 400-Ω load.

b) Find the average power delivered to the 400-Ω load.

c) Find the rms magnitude of the voltage at the input of the ideal transformer, i.e., $\mathbf{V}_i$.

d) What percentage of the power developed by the ideal current source is delivered to the 400-Ω resistor?

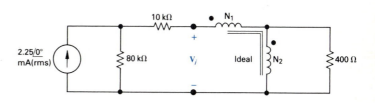

FIGURE P13.27

13.28 The load impedance Z_L in the circuit in Fig. P13.28 is adjusted until maximum average power is transferred to Z_L.

a) Specify the value of Z_L if $N_1 = 9000$ turns and $N_2 = 1500$ turns.

b) Specify the values of $\mathbf{I}_L$ and $\mathbf{V}_L$ when Z_L is absorbing maximum average power.

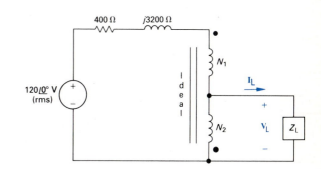

FIGURE P13.28

13.29 The variable load resistor R_L in the circuit shown in Fig. P13.29 is adjusted for maximum average power transfer to R_L.

  a) Find the maximum average power.

  b) What percentage of the average power developed by the ideal voltage source is delivered to R_L when R_L is absorbing maximum average power?

  c) Test your solution by showing the power developed by the ideal voltage source equals the power dissipated in the circuit.

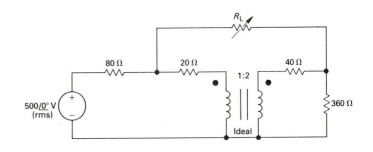

FIGURE P13.29

13.30 The sinusoidal current source in the circuit in Fig. P13.30 is operating at a frequency of 20 krad/s. The variable capacitor in the circuit is adjusted until the average power delivered to the 100-Ω resistor is as large as is possible.

  a) Find the value of C in μF.

  b) When C has the value found in part (a) what is the average power delivered to the 100-Ω resistor?

  c) Replace the 100-Ω resistor with a variable resistor R_o. Specify the value of R_o so that maximum average power is delivered to R_o.

  d) What is the maximum average power that can be delivered to R_o?

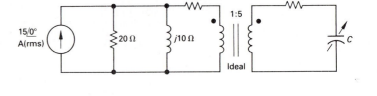

FIGURE P13.30

13.31 Find the impedance Z_{ab} in the circuit in Fig. P13.31 if $Z_L = 160\underline{/30°}\ \Omega$.

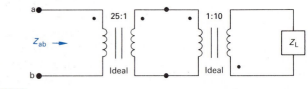

FIGURE P13.31

13.32 The sinusoidal voltage source in the circuit in Fig. P13.32 is developing an rms voltage of 2000 V. The 4-Ω load in the circuit is absorbing four times as much average power as the 25-Ω load. The two loads are matched to the sinusoidal source that has an internal impedance of $500\underline{/0°}$ Ω.

   a) Specify the numerical values of a_1 and a_2.

   b) Calculate the power delivered to the 25-Ω load.

   c) Calculate the rms value of the voltage across the 4-Ω resistor.

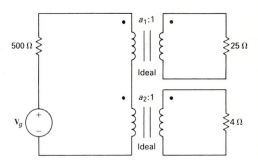

FIGURE P13.32

13.33 a) Find the T-equivalent circuit for the magnetically coupled coils shown in Fig. P13.33.

   b) Find the π-equivalent circuit for the same set of coils. Assume that $i_1 0 = i_2(0) = 0$.

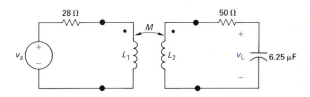

FIGURE P13.33

13.34 Show that the phasor-domain π-equivalent circuit for the magnetically coupled coils in Fig. 13.22 can be derived from the T-equivalent circuit by simply making a Y-to-Δ transformation.

13.35 A sinusoidal voltage source with an internal resistance of 28 Ω is coupled to a load by means of a lossless linear transformer, as shown in Fig. P13.35. The transformer inductances are $L_1 = 20$ mH and $L_2 = 10$ mH. The transformer coefficient of coupling is $0.5\sqrt{2}$.

   a) Specify the phasor-domain T-equivalent circuit for the lossless transformer when the source frequency is 4 krad/s.

   b) Use the T-equivalent circuit of part (a) to find the steady-state expression for v_L when $v_g = 240 \cos 4000t$ V.

   c) Repeat part (b), given that the polarity dot on the secondary side of the transformer is shifted to the lower terminal.

FIGURE P13.35

13.36 The equivalent circuit of a lossless linear transformer is shown in Fig. P13.36.

a) Find L_1, L_2, and M.

b) Find the coefficient of coupling k.

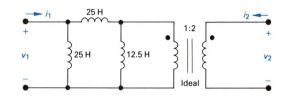

FIGURE P13.36

13.37 The following measurements were made on a lossless linear transformer. With the high-voltage side open, the inductance measured looking into the low-voltage side is 5 mH. With a 5-V rms sinusoidal voltage applied to the low-voltage winding, the open-circuit voltage measured on the high-voltage winding is 16 V rms. With a 20-V rms sinusoidal voltage applied to the high-voltage side, the open-circuit voltage measured on the low-voltage side is 4 V.

a) Specify the numerical values of L_1, L_2, M, and k for the transformer.

b) Calculate the turns ratio of the transformer.

c) Construct two possible equivalent circuits for the linear transformer if a is chosen to equal M/L_1.

d) Repeat part (c), given that a is chosen equal to L_2/M.

13.38 a) Use one (your choice) of the equivalent circuits derived in Problem 13.37 to calculate the rms voltage at the terminals of the low-voltage winding when the transformer is used in the circuit in Fig. P13.38. The 2165-V (rms) sinusoidal voltage source is operating at a frequency of 5000 rad/s.

b) Verify your calculation in part (a) by finding the same voltage without using an equivalent circuit for the linear transformer.

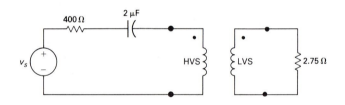

FIGURE P13.38

13.39 The purpose of this problem is to illustrate a circuit structure in which the T- or π-equivalent circuits derived in Section 13.7 cannot be used because the voltage V_{bd} in Fig. 13.22 is not zero. With this in mind, calculate the voltage V_{bd} in the circuit in Fig. P13.39.

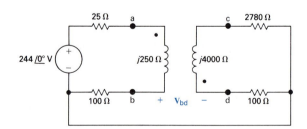

FIGURE P13.39

13.40 The circuit shown in Fig. P13.40 is a balanced, three-phase system. The load is connected to the source through ideal transformers. As can be seen from the circuit diagram, the primary windings of the three transformers are connected in wye and the secondary windings are connected in delta. The primary and secondary windings of each transformer are identified by a circled number. For example, the primary winding of transformer 1 is connected between terminals A and N and the secondary winding of transformer 1 is connected between terminals A' and B'. Each

transformer reduces the load side voltage by a factor of 2, that is, $a = 1/2$.

a) Calculate $\mathbf{V}_{AB}$.

b) Calculate $\mathbf{V}_{A'B'}$.

c) Calculate $\mathbf{I}_{A'B'}$.

d) Calculate the total average power delivered to the balanced, Δ-connected, 15-Ω load.

e) Calculate the total average power delivered by the balanced 3-φ source.

f) What percentage of the average power delivered by the source reaches the load?

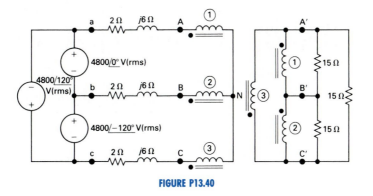

FIGURE P13.40

SERIES AND PARALLEL RESONANCE

CHAPTER 14

So far in our analysis of circuits excited by sinusoidal sources, we have concentrated on finding the steady-state currents and voltages for constant source frequency. We now consider another important aspect of sinusoidal circuit analysis; determining what happens to the amplitude and phase of the steady-state currents and voltages when source frequency varies. In doing so we are determining the circuit's *frequency response*.

We are interested in the frequency response of a circuit for two primary reasons. First, circuits that exhibit discriminatory characteristics insofar as frequency is concerned transmit signals at some frequencies noticeably better than at other frequencies. Designers use this characteristic to filter out, or eliminate, the signals in an unwanted frequency range. The ability to design frequency selective circuits makes radio, telephone, and television communication possible.

Second, knowing the frequency response we can predict the response of the circuit to any other input. We say more about this topic in Chapter 17, but several comments on its significance are in order here. Because frequency response is related to other responses, the designer can design circuits or devices in terms of frequency specifications, knowing that the constraints placed on the frequency response also exert control over the response to nonsinusoidal inputs. Another important point is that the researcher can measure the frequency response in the laboratory and from these data formulate a model of the circuit or device. This laboratory derived model allows prediction of the device's response to other types of inputs. Thus the study of frequency response is an important part of circuit analysis and design.

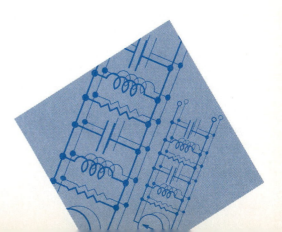

We introduce the general topic of frequency response in this chapter by describing the frequency selectivity characteristics of two specific circuit structures. The first is the *parallel-resonant circuit* and the second as the *series-resonant circuit*.

14.1 INTRODUCTION TO PARALLEL RESONANCE

We begin the study of circuit resonance with the parallel structure shown in Fig. 14.1. We are interested in how the steady-state amplitude and phase of the output voltage v_o vary as the frequency of the sinusoidal current source varies.

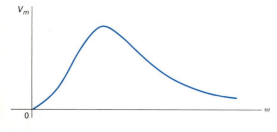

FIGURE 14.1 The parallel-resonant circuit.

QUALITATIVE REASONING

Before proceeding with a quantitative analysis of the circuit, we need to give you a feel for the behavior of the voltage by analyzing the circuit qualitatively. At low frequencies, the inductive reactance of the inductor is small and so the inductor appears as a small impedance across the output. Thus the output voltage is small at low frequencies. At high frequencies, the capacitive reactance of the capacitor is small and therefore appears as a small impedance across the output. Consequently, the output voltage is small at high frequencies. At intermediate frequencies the impedance of the three parallel branches has a nonzero value. The output voltage v_o has a nonzero value whenever the impedance has a nonzero value. Hence a sketch of the amplitude of v_o versus the frequency ω has the general shape shown in Fig. 14.2.

Because the amplitude of v_o behaves as shown in Fig. 14.2, we face several important questions. At what frequency is V_m maximum? What is the maximum value of V_m? How sharp is the peak in the neighborhood of the maximum value? In order to answer these questions, and others, we must turn to a quantitative description of V_m as a function of ω.

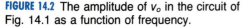

FIGURE 14.2 The amplitude of v_o in the circuit of Fig. 14.1 as a function of frequency.

QUANTITATIVE REASONING

To find the analytical expression for V_m as a function of ω, we resort to a phasor-domain description of v_o. Figure 14.3 shows the phasor-domain equivalent of the circuit shown in Fig. 14.1. Because the three branches of the parallel-resonant circuit are in parallel, we can more easily relate $\mathbf{V}_o$ to $\mathbf{I}_s$ through an admit-

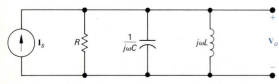

FIGURE 14.3 The phasor-domain equivalent circuit for the circuit of Fig. 14.1.

tance function, specifically,

$$\mathbf{V}_o = \frac{\mathbf{I}_s}{Y} = \frac{\mathbf{I}_s}{\dfrac{1}{R} + \dfrac{1}{j\omega L} + j\omega C} = \frac{\mathbf{I}_s}{\dfrac{1}{R} + j\left(\omega C - \dfrac{1}{\omega L}\right)}. \quad (14.1)$$

Using the current source as the reference, we express Eq. (14.1) as

$$\mathbf{V}_o = \frac{I_m \underline{/0^\circ}}{\sqrt{\dfrac{1}{R^2} + \left(\omega C - \dfrac{1}{\omega L}\right)^2} \underline{/\phi}}, \quad (14.2)$$

from which the amplitude of v_o is

$$V_m = \frac{I_m}{\sqrt{1/R^2 + [\omega C - (1/\omega L)]^2}} \quad (14.3)$$

and the phase angle between i_s and v_o is $-\phi$, where

$$\tan \phi = \left(\omega C - \frac{1}{\omega L}\right) R. \quad (14.4)$$

For a given source and circuit I_m, R, L, and C are fixed, and Eqs. (14.3) and (14.4) indicate how the amplitude and phase of v_o change as ω changes.

The *resonant frequency* for the circuit in Fig. 14.1 is defined as the frequency that makes the impedance across the current source purely resistive. This frequency makes the corresponding admittance purely conductive, because by definition $Y = 1/Z$.

From Eq. (14.1) the resonant frequency (ω_0) is

$$\omega_0 C = \frac{1}{\omega_0 L} \quad \text{or} \quad \omega_0 = \frac{1}{\sqrt{LC}}. \quad (14.5)$$

From Eq. (14.3) the amplitude of v_o—that is, V_m—is maximum when the magnitude of the admittance is minimum. This condition occurs at the resonant frequency ω_0. Equation (14.3) also shows that when ω equals ω_0, the maximum value of V_m is $I_m R$. Adding these quantitative results to Fig. 14.2 produces the curve shown in Fig. 14.4.

If i_s is an input signal and v_0 is an output signal, Fig. 14.4 indicates that the parallel circuit structure shown in Fig. 14.1 transmits a signal whose frequency is close to ω_0 much better than a signal whose frequency is much lower or higher than ω_0. This observation raises the question: How close to ω_0 must the signal frequency be in order to get an acceptable output voltage? In order to answer this question quantitatively, we must introduce the concept of circuit bandwidth and its relationship to resonant frequency.

PSpice Use the PSpice .STEP command and PROBE to simulate and plot the effects of varying component values on circuit performance: Chapter 9

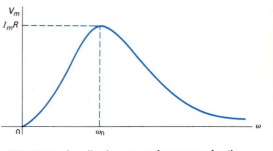

FIGURE 14.4 Amplitude versus frequency for the circuit of Fig. 14.1.

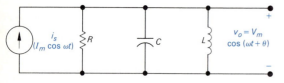

FIGURE 14.5 The parallel-resonant circuit.

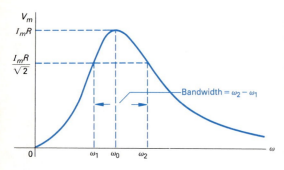

FIGURE 14.6 Bandwidth of the parallel-resonant circuit.

14.2 BANDWIDTH AND QUALITY FACTOR

The *bandwidth* for the parallel-resonant circuit of Fig. 14.5 is defined as the range of frequencies in which the amplitude of the output voltage is equal to or greater than the maximum value divided by $\sqrt{2}$. Thus an acceptable output voltage for this circuit is one whose amplitude is at least $1/\sqrt{2}$ times the maximum amplitude that can be transmitted by the circuit. The bandwidth is defined on the amplitude versus frequency plot shown in Fig 14.6. We chose the factor $1/\sqrt{2}$ because when the amplitude of v_o is reduced by this factor, the average power delivered to the resistor is one-half its maximum value. Therefore the circuit is described as transmitting a signal if the average power delivered to R is at least one-half the maximum possible value.

We find the limiting frequencies ω_1 and ω_2, which define the bandwidth, by noting from Eq. (14.3) that they correspond to the values of ω that make

$$\sqrt{\frac{1}{R^2} + \left(\omega C - \frac{1}{\omega L}\right)^2} = \frac{\sqrt{2}}{R};$$

$$\left(\omega C - \frac{1}{\omega L}\right)^2 = \frac{1}{R^2}. \qquad \textbf{(14.6)}$$

Because Eq. (14.6) is a fourth-order polynomial in ω, four values of ω yield an output voltage with an amplitude equal to $I_m R/\sqrt{2}$. Two of the four values are negative. These negative values simply mirror the positive values and have no physical significance. The two positive values of interest are

$$\omega_1 = -\frac{1}{2RC} + \sqrt{\left(\frac{1}{2RC}\right)^2 + \frac{1}{LC}} \qquad \textbf{(14.7)}$$

and

$$\omega_2 = \frac{1}{2RC} + \sqrt{\left(\frac{1}{2RC}\right)^2 + \frac{1}{LC}}. \qquad \textbf{(14.8)}$$

Hence the bandwidth of the circuit is

$$\beta = \omega_2 - \omega_1 = \frac{1}{RC}. \qquad \textbf{(14.9)}$$

With Eqs. (14.5) and (14.9) in mind, we use the inductor and capacitor to fix the resonant frequency and then the value of R to adjust the bandwidth. Large values of R correspond to a narrow bandwidth, which in turn implies that the circuit is selective as to which frequencies it transmits. For a circuit intended to be frequency selective, the sharpness of the selectivity is a measure of the circuit's quality. The quality of the frequency response is described quantitatively in terms of the ratio of the resonant frequency to the bandwidth This ratio is called the Q of the circuit and

$$Q = \frac{\omega_0}{\beta}. \qquad (14.10)$$

For the parallel-resonant structure in Fig. 14.5, we can express the quality factor in several ways:

$$Q = \omega_0 RC = \frac{R}{\omega_0 L} = R\sqrt{\frac{C}{L}}. \qquad (14.11)$$

The definition of Q given by Eq. (14.10) lends itself to laboratory measurements because measuring both resonant frequency, ω_0, and bandwidth, β, is possible. For applications outside the realm of circuits, we may also define Q in terms of an energy ratio:

$$Q = 2\pi \frac{\text{maximum energy stored}}{\text{total energy lost per period}}. \qquad (14.12)$$

For the circuit shown in Fig. 14.5, Eq. (14.12) leads to the same result. To find the Q of the parallel circuit in Fig. 14.5 from Eq. (14.12), we first compute the total energy stored in the inductor and capacitor at the resonant frequency ω_0. Because the output voltage at resonance is

$$v_o = I_m R \cos \omega_0 t, \qquad (14.13)$$

the current in the inductor is

$$i_L = \frac{I_m R}{\omega_0 L} \cos (\omega_0 t - 90°) = \frac{I_m R}{\omega_0 L} \sin \omega_0 t. \qquad (14.14)$$

Thus, from Eqs. (14.13) and (14.14),

$$w_C(t) = \tfrac{1}{2} C I_m^2 R^2 \cos^2 \omega_0 t \qquad (14.15)$$

and

$$w_L(t) = \tfrac{1}{2} L \frac{I_m^2 R^2}{\omega_0^2 L^2} \sin^2 \omega_0 t. \qquad (14.16)$$

At resonance, $C = 1/\omega_0^2 L$, so we write Eq. (14.16) as

$$w_L(t) = \tfrac{1}{2} C I_m^2 R^2 \sin^2 \omega_0 t. \qquad (14.17)$$

Adding Eqs. (14.15) and (14.17) gives the total energy stored in the circuit:

$$w_T = w_C(t) + w_L(t) = \tfrac{1}{2} C I_m^2 R^2, \qquad (14.18)$$

which states that the total energy stored in the circuit at resonance is a constant and that this constant value must also be the maximum value of the energy stored.

The total energy lost per period is the product of the average power dissipated in R and the period. Thus at resonance,

$$\text{total energy lost per period} = \frac{I_m^2}{2}R\left(\frac{1}{f_0}\right). \qquad (14.19)$$

Substituting Eqs. (14.18) and (14.19) into Eq. (14.12), yields

$$Q = 2\pi\frac{CI_m^2R^2/2}{I_m^2R/2f_0} = 2\pi f_0 RC = \frac{\omega_0}{\beta}. \qquad (14.20)$$

We use the definition of Q expressed by Eq. (14.10). The alternative definition given in Eq. (14.12) is not a useful definition for our work. (See Problem 14.35.)

Example 14.1 illustrates the computations involved in quantitatively describing the frequency response of a parallel resonant circuit.

E X A M P L E   14.1

The amplitude of the sinusoidal current source in the circuit in Fig. 14.7 is 50 mA. The circuit parameters are $R = 2$ kΩ, $L = 40$ mH, and $C = 0.25$ μF.

a) Find ω_0, Q, ω_1, ω_2, and the amplitude of the output voltage at ω_0, ω_1, and ω_2.

b) What value of R produces a bandwidth of 500 rad/s?

c) What is the value of Q in (b)?

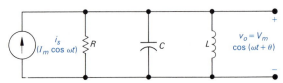

FIGURE 14.7 The parallel-resonant circuit.

S O L U T I O N

a) $$\omega_0 = \sqrt{\frac{1}{LC}} = \sqrt{\frac{10^9}{40(0.25)}} = \sqrt{10^8} = 10^4 \text{ rad/s};$$

$$Q = \omega_0 RC = (10^4)(2000)(0.25 \times 10^{-6}) = 5;$$

$$\beta = \frac{\omega_0}{Q} = \frac{10^4}{5} = 2000 \text{ rad/s};$$

$$\omega_1 = -\frac{1}{2RC} + \sqrt{\left(\frac{1}{2RC}\right)^2 + \frac{1}{LC}}$$

$$= -1000 + \sqrt{10^6 + 10^8}$$

$$= -1000 + 1000\sqrt{101}$$

$$= -1000 + 10{,}049.88 = 9049.88 \text{ rad/s};$$

$$\omega_2 = \frac{1}{2RC} + \sqrt{\left(\frac{1}{2RC}\right)^2 + \frac{1}{LC}}$$

$$= 1000 + 10{,}049.88 = 11{,}049.88 \text{ rad/s};$$

$$V_m(\omega_0) = I_m R = 50 \times 10^{-3}(2000) = 100 \text{ V};$$

$$V_m(\omega_1) = V_m(\omega_2) = V_m(\omega_0)/\sqrt{2} = 70.7 \text{ V}.$$

b) $R = \dfrac{1}{\beta C} = \dfrac{10^6}{(500)(0.25)} = 8000 \ \Omega.$

c) $Q = \dfrac{\omega_0}{\beta} = \dfrac{10^4}{500} = \dfrac{100}{5} = 20.$

In Example 14.1 the resonant frequency is not in the center of the passband. That is, the frequency range from ω_1 to ω_0 is not the same as the range from ω_0 to ω_2. Specifically, $\omega_0 - \omega_1 = 10{,}000 - 9049.88 = 950.12$ rad/s, but $\omega_2 - \omega_0 = 11{,}049.88 - 10{,}000 = 1049.88$ rad/s. As Q gets larger, ω_0 approaches the center of the passband. You can verify this result by calculating ω_1 and ω_2 in Example 14.1 when $R = 8000 \ \Omega$; we verify it analytically in Section 14.4. The exact relationship among ω_1, ω_2, and ω_0 can be obtained from Eqs. (14.7) and (14.8). Multiplying these two equations together gives

$$\omega_1 \omega_2 = \omega_0^2.$$

Therefore the *resonant frequency is the geometric mean of half-power frequencies* ω_1 *and* ω_2; that is,

$$\omega_0 = \sqrt{\omega_1 \omega_2}. \tag{14.21}$$

The frequencies ω_1 and ω_2 are called the half-power frequencies, because at either of these frequencies the power dissipated in R is exactly one half the power dissipated in R at the resonant frequency.

DRILL EXERCISES

14.1 The resonant frequency of the circuit in Fig. 14.7 is 10 Mrad/s. The bandwidth of the circuit is 100 krad/s. If R is 100 kΩ, calculate (a) Q; (b) L; (c) C; (d) ω_1; and (e) ω_2.

ANSWER: (a) 100; (b) 100 μH; (c) 100 pF; (d) 9.95 Mrad/s; (e) 10.05 Mrad/s.

14.2 The following components are available for a parallel-resonant circuit: $L_1 = 7.5$ mH, $L_2 = 15$ mH, $C_1 = 6$ μF, $C_2 = 3$ μF, and $R = 2$ kΩ. Design a circuit that has the highest possible resonant frequency. Specify (a) ω_0; (b) β; (c) Q; (d) ω_1; and (e) ω_2.

ANSWER: (a) 10^4 rad/s; (b) 250 rad/s; (c) 40; (d) 9875rad/s; (e) 10,125 rad/s.

14.3 Use the components in Drill Exercise 14.2 to design a parallel-resonant circuit with the lowest possible resonant frequency. Specify (a) ω_0; (b) β; (c) Q; (d) ω_1; and (e) ω_2.

ANSWER: (a) 2222.22 rad/s; (b) 55.56 rad/s; (c) 40; (d) 2194.62 rad/s; (e) 2250.17 rad/s.

14.3 FURTHER ANALYSIS OF PARALLEL RESONANCE

The purpose of the parallel-resonant structure is to be frequency selective, so describing the behavior of the circuit in terms of ω_0, β, and Q is convenient. With this in mind, we return to Eqs. (14.9) and (14.10) and note that

$$\frac{1}{2RC} = \frac{\beta}{2} = \frac{\omega_0}{2Q}.$$
(14.22)

By substituting Eq. (14.22) into Eqs. (14.7) and (14.8), we express the half-power frequencies as functions of the resonant frequency and the Q of the circuit:

$$\omega_1 = -\frac{\omega_0}{2Q} + \sqrt{\left(\frac{\omega_0}{2Q}\right)^2 + \omega_0^2}$$

$$= \omega_0\left[-\frac{1}{2Q} + \sqrt{\left(\frac{1}{2Q}\right)^2 + 1}\right];$$
(14.23)

$$\omega_2 = \omega_0\left[\frac{1}{2Q} + \sqrt{\left(\frac{1}{2Q}\right)^2 + 1}\right].$$
(14.24)

Equations (14.23) and (14.24) show that, when Q is 10 or greater,

$$\omega_1 \cong -\frac{1}{2}\frac{\omega_0}{Q} + \omega_0 = \omega_0 - \frac{\beta}{2};$$
(14.25)

$$\omega_2 \cong \frac{1}{2}\frac{\omega_0}{Q} + \omega_0 = \omega_0 + \frac{\beta}{2}.$$
(14.26)

Thus for high-Q circuits the resonant frequency ω_0 lies close to the center of the passband. Equations (14.25) and (14.26) yield the half-power frequencies with less than 1 percent error whenever $Q \geq 5$. For example, when $Q = 5$, the exact ex-

PSpice〜〜〜 Combine the PSpice .STEP command with PSpice frequency analysis to examine the effects of varying component values on the circuit's frequency response: Section 15.1

pressions (Eqs. 14.23 and 14.24) predict that $\omega_1/\omega_0 = 0.905$ and that $\omega_2/\omega_0 = 1.105$, whereas the approximate expressions (Eqs. 14.25 and 14.26) predict that $\omega_1/\omega_0 = 0.900$ and $\omega_2/\omega_0 = 1.100$.

Expressing the admittance in terms of the resonant frequency ω_0 and the bandwidth β highlights the behavior of the output voltage near the resonant frequency. From Eq. (14.1),

$$Y = \frac{1}{R} + j\left(\omega C - \frac{1}{\omega L}\right). \qquad (14.27)$$

Performing a series of algebraic manipulations,

$$
\begin{aligned}
Y &= \frac{1}{R}\left[1 + j\left(\omega RC - \frac{R}{\omega L}\right)\right] \\
&= \frac{1}{R}\left[1 + j\left(\frac{\omega \omega_0 RC}{\omega_0} - \frac{\omega_0 R}{\omega \omega_0 L}\right)\right] \\
&= \frac{1}{R}\left[1 + jQ\left(\frac{\omega}{\omega_0} - \frac{\omega_0}{\omega}\right)\right] \\
&= \frac{1}{R}\left[1 + j\frac{\omega_0}{\beta}\left(\frac{\omega^2 - \omega_0^2}{\omega \omega_0}\right)\right], \qquad (14.28)
\end{aligned}
$$

leads to

$$Y = \frac{1}{R}\left[1 + j\frac{(\omega + \omega_0)(\omega - \omega_0)}{\beta \omega}\right]. \qquad (14.29)$$

Equation (14.29) is an exact expression for the admittance for all values of ω because its derivation involves only legitimate algebraic manipulations of the original expression, Eq. (14.27). It also shows more clearly how Y varies in the neighborhood of ω_0. As $\omega \to \omega_0$, the ratio $(\omega + \omega_0)/\omega \to 2$ and the expression for Y can be approximated as

$$Y \cong \frac{1}{R}\left(1 + j\frac{\omega - \omega_0}{\beta/2}\right). \qquad (14.30)$$

The error from Eq. (14.30) is less than 5 percent in both magnitude and phase, provided that ω lies within the passband; that is, $\omega_1 \le \omega \le \omega_2$.

Example 14.2 uses the circuit in Example 14.1 to demonstrate the usefulness of Eqs. (14.29) and (14.30).

EXAMPLE 14.2

For the circuit in Example 14.1

a) find $\mathbf{V}_o$ at 10,500 rad/s and 9500 rad/s using the exact formula for Y;

b) repeat (a) using the approximate expression for Y; and

c) find the percent error in the magnitude of $\mathbf{V}_o$ that results from using the approximate expression for Y.

S O L U T I O N

a) At 10,500 rad/s,

$$Y = \frac{1}{2000}\left[1 + j\frac{(20,500)(500)}{2000(10,500)}\right]$$

$$= \frac{1.1128\underline{/26.02°}}{2000} = 0.5564\underline{/26.02°}\ \text{m}\Omega;$$

$$\mathbf{V}_o = \frac{50 \times 10^{-3}(2000)}{1.1128\underline{/26.02°}} = 89.87\underline{/-26.02°}\ \text{V.}$$

At 9500 rad/s,

$$Y = \frac{1}{2000}\left[1 + j\frac{19,500(-500)}{2000(9500)}\right]$$

$$= \frac{1.1240\underline{/-27.16°}}{2000} = 0.5620\underline{/-27.16°}\ \text{m}\Omega;$$

$$\mathbf{V}_o = \frac{50 \times 10^{-3}(2000)}{1.1240\underline{/-27.16°}} = 88.97\underline{/27.16°}\ \text{V.}$$

b) Using the approximate formula for Y, we get the following results. At $\omega = 10,500$ rad/s,

$$Y \cong \frac{1}{2000}\left(1 + j\frac{10,500 - 10,000}{2000/2}\right)$$

$$= \frac{1 + j0.5}{2000} = \frac{1.1180\underline{/26.57°}}{2000}$$

$$= 0.5590\underline{/26.57°}\ \text{m}\Omega;$$

$$\mathbf{V}_o = \frac{50 \times 10^{-3}(2000)}{1.1180\underline{/26.57°}} = 89.44\underline{/-26.57°}\ \text{V.}$$

At $\omega = 9500$ rad/s,

$$Y = \frac{1}{2000}\left(1 + j\frac{9500 - 10,000}{1000}\right)$$

$$= \frac{1 - j0.5}{2000} = 0.5590\underline{/-26.57°}\ \text{m}\Omega;$$

$$\mathbf{V}_o = 89.44\underline{/26.57°}\ \text{V.}$$

c) The percent error in the magnitude of V_o at 10,500 rad/s is

$$\% \text{ error} = \frac{89.44 - 89.87}{89.87} \times 100 = -0.48\%.$$

The percent error at 9500 rad/s is

$$\% \text{ error} = \frac{89.44 - 88.97}{88.97} \times 100 = 0.53\%.$$

Our primary interest in studying the parallel-resonant circuit is to establish the relationship between the output voltage and the source current. Note, however, that at resonance the magnitude of the current in either the inductive or capacitive branch is Q times the magnitude of the source current. We leave verification of this statement to you as Problem 14.6. You must take into account these large internal currents when selecting components.

The variation with frequency of the phase angle between the output signal and the input signal also is of interest. For the parallel circuit in Fig. 14.8 the phase angle of v_o with respect to i_s is the negative to the phase angle of the admittance Y:

$$\theta = -\phi = -\tan^{-1}\frac{(\omega + \omega_0)(\omega - \omega_0)}{\beta\omega}. \qquad (14.31)$$

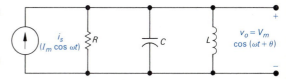

FIGURE 14.8 The parallel-resonant circuit.

We may also express the phase angle θ in terms of Q because $\beta = \omega_0/Q$:

$$\theta = -\tan^{-1}\frac{(\omega + \omega_0)(\omega - \omega_0)Q}{\omega_0\omega}, \qquad (14.32)$$

which indicates that, when $\omega < \omega_0$, the phase angle is positive and, as $\omega \to 0$, $\theta \to +90°$. For $\omega > \omega_0$, θ is negative and, as $\omega \to \infty$, $\theta \to -90°$. We know that $\theta = 0$ at $\omega = \omega_0$, $+45°$ at $\omega = \omega_1$, and $-45°$ at $\omega = \omega_2$. Figure 14.9 shows these characteristics of θ with respect to ω.

We can show the effect of Q on the phase-angle characteristic most easily by investigating the slope of the θ versus ω curve at ω_0. Thus by letting

$$\theta = -\tan^{-1}\frac{(\omega^2 - \omega_0^2)Q}{\omega\omega_0} = -\tan^{-1}u,$$

we find that

$$\frac{d\theta}{d\omega} = \frac{d\theta}{du}\frac{du}{d\omega} = -\frac{1}{1 + u^2}\frac{du}{d\omega}. \qquad (14.33)$$

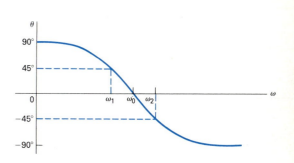

FIGURE 14.9 The phase angle θ versus ω for the parallel-resonant structure of Fig. 14.8.

But

$$\frac{du}{d\omega} = \frac{\omega\omega_0(2\omega)Q - (\omega^2 - \omega_0^2)Q\omega_0}{\omega^2\omega_0^2}$$

$$= \frac{\omega_0 Q(\omega^2 + \omega_0^2)}{\omega^2\omega_0^2}. \tag{14.34}$$

Substituting Eq. (14.34) into Eq. (14.33) gives

$$\frac{d\theta}{d\omega} = -\frac{\omega_0 Q(\omega^2 + \omega_0^2)}{\omega^2\omega_0^2 + (\omega^2 - \omega_0^2)^2 Q^2}. \tag{14.35}$$

At $\omega = \omega_0$, Eq. (14.34) reduces to

$$\frac{d\theta}{d\omega} = -\frac{2Q}{\omega_0} = -2RC. \tag{14.36}$$

Thus as Q is increased by increasing R, the slope at ω_0 increases. Note that if R is made infinitely large, the phase angle snaps from $+90°$ to $-90°$ at the resonant frequency. Figure 14.10 shows the effect of Q on the phase shift.

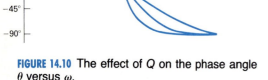

FIGURE 14.10 The effect of Q on the phase angle θ versus ω.

DRILL EXERCISES

14.4 The components in the circuit shown in Fig 14.8 are $R = 40$ kΩ, $L = 0.5$ mH, and $C = 500$ pF.

a) At what frequency is the amplitude of v_o maximum?

b) At what frequencies is the amplitude of v_o 92.5% of its maximum value?

ANSWER: (a) 318.31 kHz; (b) 316.68 kHz, 319.95 kHz.

14.4 THE FREQUENCY RESPONSE VERSUS THE NATURAL RESPONSE OF THE PARALLEL *RLC* CIRCUIT

In the introduction to this chapter, we mentioned that we are interested in the frequency response of a circuit because it can be used to predict the time-domain behavior of the circuit. Now that we have discussed the frequency response of the parallel *RLC* circuit, we show how attributes of the frequency-response behavior relate to attributes of the natural time-domain response. Recall that we investigated the natural response of the parallel *RLC* circuit in Chapter 9. For the underdamped case expressed by Eq. (9.28), repeated here for convenience,

$$v(t) = B_1 e^{-\alpha t} \cos \omega_d t + B_2 e^{-\alpha t} \sin \omega_d t, \qquad \textbf{(14.37)}$$

$$\alpha = \frac{1}{2RC}, \qquad \textbf{(14.38)}$$

and

$$\omega_d = \sqrt{\omega_0^2 - \alpha^2}. \qquad \textbf{(14.39)}$$

Initial conditions determine the coefficients B_1 and B_2, which are not germane here.

We relate the damping coefficient α and the damped frequency of oscillation ω_d to the Q of the circuit by

$$\alpha = \frac{\beta}{2} = \frac{\omega_0}{2Q} \qquad \textbf{(14.40)}$$

and

$$\omega_d = \omega_0 \sqrt{1 - \frac{1}{4Q^2}}. \qquad \textbf{(14.41)}$$

The transition from the overdamped to the underdamped response occurs when $\omega_0^2 = \alpha^2$, or when $Q = \frac{1}{2}$. Therefore the natural response is underdamped whenever the Q of the circuit is greater than $\frac{1}{2}$ or, alternatively, whenever the bandwidth is less than $2\omega_0$. Equation (14.40) indicates that in high-Q circuits, the natural response oscillations persist longer than in low-Q circuits. Equation (14.41) indicates that the frequency of oscillation in the underdamped response approaches the resonant frequency ω_0 as Q increases.

We summarize these observations as follows: If the frequency response of the parallel RLC structure shows a sharp resonant peak, the natural response is underdamped, the frequency of oscillation approximates the resonant frequency, and the oscillations persist over a relatively long interval of time.

PSpice   Use PROBE to plot a circuit's natural response and frequency response to compare them graphically: Sections 7.4 and 15.2

DRILL EXERCISES

14.5 The natural response of the circuit shown is

$$v_o = 100e^{-100t} \sin 100\sqrt{99}\, t \text{ V},$$

when $C = 1\ \mu\text{F}$. Calculate (a) ω_0; (b) β; (c) Q; (d) R; and (e) L.

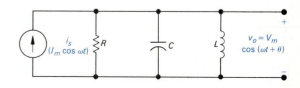

ANSWER: (a) 1000 rad/s; (b) 200 rad/s; (c) 5; (d) 5 kΩ; (e) 1H.

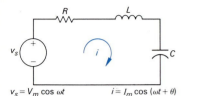

$v_s = V_m \cos \omega t$ $i = I_m \cos(\omega t + \theta)$

FIGURE 14.11 The series-resonant circuit.

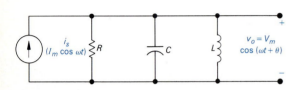

FIGURE 14.12 The parallel-resonant circuit.

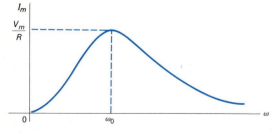

FIGURE 14.13 Current amplitude versus frequency for the circuit of Fig. 14.11.

14.5 SERIES RESONANCE

Figure 14.11 shows the basic series-resonant circuit. Here we are interested in how the amplitude and phase angle of the current vary with the frequency of the sinusoidal voltage source. As the frequency of the source changes, the maximum amplitude of source voltage (V_m) remains constant.

Analysis of the series-resonant circuit follows closely the analysis of the parallel-resonant structure shown in Fig. 14.12. First, note that the amplitude of the current approaches zero at both small and large values of ω. The capacitor blocks the passage of current at low frequencies, and the inductor blocks the passage of current at high frequencies. The current amplitude peaks at V_m/R, when the inductive reactance cancels the capacitive reactance. The frequency at which the reactive impedances cancel is the resonant frequency of the circuit. Thus $\omega_0 L = 1/\omega_0 C$, or

$$\omega_0 = \frac{1}{\sqrt{LC}}. \qquad (14.42)$$

The expression for the resonant frequency of the circuit shown in Fig. 14.11 is the same as that for the circuit shown in Fig. 14.12, namely, Eq. (14.5).

The current amplitude, I_m, versus the frequency, ω, plot takes the form shown in Fig. 14.13. We obtain a more detailed analysis of the curve by expressing the amplitude of i as a function of ω. Again, we use the phasor-domain method of analysis. Therefore

$$\mathbf{I} = I_m\underline{/\theta} = \frac{\mathbf{V}_s}{Z} = \frac{V_m\underline{/0°}}{|Z|\underline{/\phi}}. \qquad (14.43)$$

The series impedance of the circuit shown in Fig. 14.11 is

$$Z = R + j\left(\omega L - \frac{1}{\omega C}\right).$$

The magnitude and angle of the impedance are, respectively,

$$|Z| = \sqrt{R^2 + \left(\omega L - \frac{1}{\omega C}\right)^2}; \qquad (14.44)$$

$$\phi = \tan^{-1}\frac{\omega L - (1/\omega C)}{R}. \qquad (14.45)$$

Hence, from Eqs. (14.43)–(14.45),

$$I_m = \frac{V_m}{\sqrt{R^2 + [\omega L - (1/\omega C)]^2}}, \qquad (14.46)$$

and

$$\theta = -\phi = -\tan^{-1}\frac{\omega L - (1/\omega C)}{R}. \qquad (14.47)$$

The bandwidth of the series circuit is defined as the range of frequencies in which the amplitude of the current is equal to or greater than $1/\sqrt{2}$ times the maximum amplitude V_m/R. Thus the frequencies at the edge of the passband are the frequencies where the magnitude of Z equals $\sqrt{2}\,R$. Setting

$$\sqrt{R^2 + \left(\omega L - \frac{1}{\omega C}\right)^2} = \sqrt{2}\,R$$

and solving for the two positive values yields

$$\omega_1 = -\frac{R}{2L} + \sqrt{\left(\frac{R}{2L}\right)^2 + \frac{1}{LC}}; \qquad (14.48)$$

$$\omega_2 = \frac{R}{2L} + \sqrt{\left(\frac{R}{2L}\right)^2 + \frac{1}{LC}}. \qquad (14.49)$$

Then from Eqs. (14.48) and (14.49), the resonant frequency is the geometric mean of the half-power frequencies ($\omega_0 = \sqrt{\omega_1\omega_2}$), and the bandwidth of the series circuit is

$$\beta = \omega_2 - \omega_1 = \frac{R}{L}. \qquad (14.50)$$

Figure 14.14 shows these characteristics of the amplitude response of the series circuit shown in Fig. 14.11.

By definition Q is the ratio of the resonant frequency to the bandwidth, so the Q of the series circuit is

$$Q = \frac{\omega_0}{\beta} = \frac{\omega_0 L}{R} = \frac{1}{\omega_0 CR} = \frac{1}{R}\sqrt{\frac{L}{C}}. \qquad (14.51)$$

We express the half-power frequencies and the impedance of the series circuit in terms of the resonant frequency, ω_0, and the Q of the circuit:

$$\omega_1 = -\frac{R}{2L} + \sqrt{\left(\frac{R}{2L}\right)^2 + \frac{1}{LC}}$$

$$= -\frac{R\omega_0}{2\omega_0 L} + \sqrt{\left(\frac{R}{2\omega_0 L}\right)^2 \omega_0^2 + \omega_0^2}$$

$$= \omega_0\left[-\frac{1}{2Q} + \sqrt{1 + \left(\frac{1}{2Q}\right)^2}\right] \qquad (14.52)$$

and

$$\omega_2 = \omega_0\left[\frac{1}{2Q} + \sqrt{1 + \left(\frac{1}{2Q}\right)^2}\right]. \qquad (14.53)$$

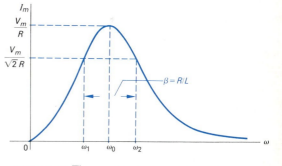

FIGURE 14.14 The amplitude response of the series-resonant circuit.

The expression for the impedance is

$$Z = R + j\left(\omega L - \frac{1}{\omega C}\right)$$

$$= R\left[1 + j\left(\frac{\omega L}{R} - \frac{1}{\omega RC}\right)\right]$$

$$= R\left[1 + j\left(\frac{\omega_0 L}{R}\frac{\omega}{\omega_0} - \frac{1}{\omega_0 RC}\frac{\omega_0}{\omega}\right)\right]$$

$$= R\left[1 + jQ\left(\frac{\omega}{\omega_0} - \frac{\omega_0}{\omega}\right)\right]$$

$$= R\left[1 + jQ\frac{(\omega - \omega_0)(\omega + \omega_0)}{\omega\omega_0}\right]$$

$$= R\left[1 + j\frac{(\omega - \omega_0)(\omega + \omega_0)}{\beta\omega}\right]. \qquad (14.54)$$

For high-Q circuits, Eqs. (14.52) and (14.53) reduce to

$$\omega_1 \approx \omega_0 - \frac{\omega_0}{2Q} = \omega_0 - \frac{\beta}{2}; \qquad (14.55)$$

$$\omega_2 \approx \omega_0 + \frac{\omega_0}{2Q} = \omega_0 + \frac{\beta}{2}. \qquad (14.56)$$

Thus for high-Q circuits the resonant frequency approaches the center of the passband.

For values of ω near resonance, we simplify Eq. (14.54) to

$$Z \approx R\left(1 + j\frac{\omega - \omega_0}{\beta/2}\right). \qquad (14.57)$$

Equation (14.47) gives the phase angle between the source voltage and the current, which we express in terms of ω_0 and β by using Eq. (14.54):

$$\theta = -\tan^{-1}\frac{(\omega - \omega_0)(\omega + \omega_0)}{\beta\omega}. \qquad (14.58)$$

Comparing Eqs. (14.58) and (14.31) clearly shows that the behavior of the phase angle between the output current i and the input voltage v_s in the series circuit is identical to that of the phase angle between the output voltage v_o and the input current i_s in the parallel circuit. Thus the phase angle versus frequency characteristic curves shown in Fig. 14.10 also apply to the series circuit.

The amplitude of the voltage across either the inductor or the capacitor at the resonant frequency ω_0 is Q times the amplitude of the source voltage. We verify this relation by noting that

$$|V_L| = I_m \omega_0 L = \frac{V_m}{R} \omega_0 L = QV_m \qquad \textbf{(14.59)}$$

and

$$|V_c| = \frac{I_m}{\omega_0 C} = \frac{V_m}{\omega_0 CR} = QV_m. \qquad \textbf{(14.60)}$$

In high-Q circuits, you must take into account these large voltages when selecting components.

Problems 14.12–14.15 are designed to give you some practice in analyzing the series-resonant circuit.

DRILL EXERCISES

14.6 The series-resonant circuit shown is designed to have a resonant frequency of 5 Mrad/s and a lower half-power frequency of 4.5 Mrad/s. If the capacitor in the circuit equals 0.0l μF, calculate (a) Q; (b) β; (c) L; and (d) R.

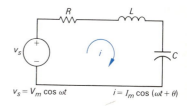

$v_s = V_m \cos \omega t \qquad\qquad i = I_m \cos(\omega t + \theta)$

ANSWER: (a) 4.74; (b) 1.0556 Mrad/s; (c) 4 μH; (d) 4.22 Ω.

14.7 The components in the preceding circuit are $R = 10\ \Omega$, $L = 0.2$ mH, and $C = 5$ nF. Calculate (a) ω_0; (b) Q; (c) β; and (d) the frequencies when the amplitude of the current is 80% of its maximum value.

ANSWER: (a) 1 Mrad/s; (b) 20; (c) 50 krad/s; (d) 1.019 Mrad/s, 0.981 Mrad/s.

14.6 MORE ON PARALLEL RESONANCE

So far we have limited our discussion of parallel resonance to a simple structure in which the maximum amplitude of the response signal occurred at the resonant frequency ω_0. We now consider a more realistic structure, in which the maximum amplitude of the response does not coincide with unity power-factor resonance. The parallel-resonant structure shown in Fig. 14.15 is a more practical model because it accounts for the losses in the inductor—represented by the resistor R.

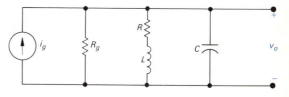

FIGURE 14.15 A parallel-resonant circuit containing a low-loss inductor.

ANALYSIS OF A PRACTICAL PARALLEL RESONANT CIRCUIT

We determine the unity power-factor resonant frequency ω_0 by finding the frequency that makes the admittance of the circuit conductive. The expression for the admittance is

$$Y = \frac{1}{R_g} + \frac{1}{R + j\omega L} + j\omega C. \tag{14.61}$$

To find the unity power-factor frequency, we first rewrite Eq. (14.61) in rectangular form:

$$Y = \frac{1}{R_g} + \frac{R}{R^2 + \omega^2 L^2} - \frac{j\omega L}{R^2 + \omega^2 L^2} + j\omega C, \tag{14.62}$$

which indicates that the admittance is purely conductive when the j terms cancel. That occurs at

$$\omega_0 = \sqrt{\frac{1}{LC} - \left(\frac{R}{L}\right)^2}. \tag{14.63}$$

Note that as the resistance of the inductor approaches zero, the unity power-factor resonant frequency approaches $1/\sqrt{LC}$.

We find the value of the admittance at ω_0 by substituting Eq. (14.63) into Eq. (14.62), noting that at ω_0,

$$R^2 + \omega_0^2 L^2 = \frac{L}{C} \tag{14.64}$$

and, at the same time, the j terms vanish. Thus

$$Y(\omega_0) = \frac{1}{R_g} + \frac{RC}{L} = \frac{L + R_g RC}{LR_g}, \tag{14.65}$$

from which

$$Z(\omega_0) = \frac{1}{Y(\omega_0)} = \frac{R_g L}{R_g RC + L} \tag{14.66}$$

follows. The amplitude of the output voltage at unity power-factor resonance is

$$|\mathbf{V}_o| = \frac{|\mathbf{I}_g| R_g L}{R_g RC + L}. \tag{14.67}$$

Analysis of the amplitude of the output voltage as a function of frequency reveals that the amplitude is not maximum at ω_0. The analysis requires that we express the amplitude of $\mathbf{V}_o$ as a function of ω, differentiate this expression with respect to ω, and then find the value of ω that makes the derivative zero. The algebra involved becomes unwieldy, so we simply state the result and then illustrate the calculations involved. The frequency corresponding to the maximum amplitude of $\mathbf{V}_o$ is

$$\omega_m = \sqrt{\sqrt{\left(\frac{1}{LC}\right)^2\left(1 + \frac{2R}{R_g}\right) + \left(\frac{R}{L}\right)^2\left(\frac{2}{LC}\right)} - \left(\frac{R}{L}\right)^2}. \qquad \textbf{(14.68)}$$

Comparing Eqs. (14.68) and (14.63) reveals that $\omega_m \rightarrow \omega_0$ as $R \rightarrow 0$. Examples 14.3 and 14.4 illustrate the calculation of ω_m and ω_0 when the coil resistance is not negligible.

EXAMPLE 14.3

The circuit shown in Fig. 14.15 is driven by a sinusoidal current source. The maximum amplitude of the current is 4 mA; the internal resistance of the source is 10 kΩ; the 25-mH inductor has a resistance of 800 Ω; and the losses associated with the 10-nF capacitor are negligible.

a) Calculate the resonant frequency of the circuit.

b) Calculate the amplitude of the output voltage at resonance.

c) Calculate the frequency at which the amplitude of the output voltage is maximum.

d) Calculate the maximum amplitude of v_o.

SOLUTION

a) $\omega_0 = \sqrt{\dfrac{1}{LC} - \left(\dfrac{R}{L}\right)^2} = \sqrt{40 \times 10^8 - \left(\dfrac{800}{0.025}\right)^2}$

$\qquad = 10^4\sqrt{40 - 10.24} = 54{,}552.73$ rad/s.

b) $V_m = |\mathbf{V}_o| = \dfrac{I_m R_g L}{R_g RC + L}$

$\qquad = \dfrac{(4 \times 10^{-3})(10^4)(25 \times 10^{-3})}{(10^4)(800)(0.01) \times 10^{-6} + 25 \times 10^{-3}}$

$\qquad = \dfrac{(250)(4)10^{-3}}{0.105} = 9.52$ V.

c) Before substituting into Eq. (15.68), we make the following preliminary calculations:

$\left(\dfrac{1}{LC}\right)^2 = 16 \times 10^{18}, \qquad 1 + \dfrac{2R}{R_g} = 1.16;$

$\left(\dfrac{R}{L}\right)^2 = 10.24 \times 10^8, \qquad \dfrac{2}{LC} = 80 \times 10^8;$

and

$$\omega_m = \sqrt{\sqrt{16 \times 10^{18}(1.16) + (10.24)(80)10^{16}} - 10.24 \times 10^8}$$

$$= \sqrt{51.70 \times 10^8 - 10.24 \times 10^8}$$

$$= 64{,}406.78 \text{ rad/s.}$$

d) To find the amplitude of v_o at this frequency, we first calculate the admittance:

$$Y(\omega_m) = \frac{1}{10^4} + \frac{1}{800 + j\omega_m 25 \times 10^{-3}} + j\omega_m 10 \times 10^{-9}$$

$$= 100 + 247.48 - j498.10 + j644.07 \ \mu\Omega$$

$$= 376.89\underline{/22.79°} \ \mu\Omega.$$

We now calculate V_m:

$$V_m = |\mathbf{V}_o| = \frac{I_m}{|Y(\omega_m)|}$$

$$= \frac{4 \times 10^{-3}}{376.89} \times 10^6 = 10.61 \text{ V.}$$

E X A M P L E   14.4

Repeat the calculations of Example 14.3 for a 25-mH inductor resistance of 80 Ω.

S O L U T I O N

a) $\omega_0 = \sqrt{40 \times 10^8 - 10.24 \times 10^6}$

$\qquad = 1000\sqrt{4000 - 10.24}$

$\qquad = 63{,}164.55 \text{ rad/s.}$

b) $V_m = \dfrac{(4 \times 10^{-3})(10^4)(25 \times 10^{-3})}{(10^4)(80)(10) \times 10^{-9} + 25 \times 10^{-3}}$

$\qquad = \dfrac{1}{8 \times 10^{-3} + 25 \times 10^{-3}} = \dfrac{1000}{33} = 30.30 \text{ V.}$

c) $\omega_m = \sqrt{\sqrt{16 \times 10^{18}(1.016 + (80 \times 10^8)(10.24 \times 10^6)} - 10.24 \times 10^6}$

$\qquad = \sqrt{40.42 \times 10^8 - 0.1024 \times 10^8}$

$\qquad = 63{,}496.29 \text{ rad/s.}$

d) $Y(\omega_m) = \dfrac{1}{10^4} + \dfrac{1}{80 + j\omega_m 25 \times 10^{-3}} + j\omega_m \times 10^{-8}$

$\qquad = (100 + 31.67 - j628.36 + j634.96)\ \mu\Omega$

$\qquad = 131.83\underline{/2.87^\circ}\ \mu\Omega;$

$V_m = \dfrac{4 \times 10^{-3}}{131.83 \times 10^{-6}} = 30.34$ V.

Comparing the results obtained in Example 14.4 with those obtained in Example 14.3 shows that, as the coil resistance decreases, the unity power-factor frequency, ω_0, approaches the frequency at which the amplitude of v_o is maximum, or ω_m. Note that the amplitude of v_o at ω_0 is nearly the same as the amplitude of v_o at ω_m. As the coil resistance approaches zero, the maximum amplitude approaches its largest possible value. The largest possible amplitude of v_o is $I_m R_g$. For the circuit in Examples 14.3 and 14.4, this magnitude is $(4 \times 10^{-3})(10 \times 10^3)$, or 40 V.

Q OF A COIL

Manufacturers of coils intended for frequency-selective circuits specify the Q of the coil at specific frequencies. The Q of the coil in Example 14.3 at unity power-factor resonance is

$$Q_{\text{coil}} = \frac{\omega_0 L}{R} \cong \frac{(54.55)(25)}{800} = 1.70,$$

whereas the Q of the coil in Example 14.4 at ω_0 is

$$Q_{\text{coil}} \cong \frac{(63.16)(25)}{80} = 19.74.$$

Thus the higher Q coil results in a larger peak output voltage. For high-Q coils the bandwidth of the circuit in Fig. 14.16 can be approximated by

$$\beta \cong \frac{R}{L} + \frac{1}{R_g C}. \qquad \textbf{(14.69)}$$

We leave the derivation of Eq. (14.69) to you as Problem 14.25.
We can use Eq. (14.69) to estimate the bandwidth of the circuit in Example 14.4 because the Q of the coil is greater than 5. For the circuit in Example 14.4,

$$\beta = \frac{80}{25} \times 10^3 + \frac{10^9}{10^4(10)} = 3200 + 10^4 = 13,200 \text{ rad/s.}$$

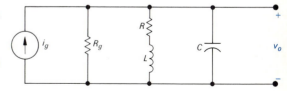

FIGURE 14.16 A parallel-resonant circuit containing a low-loss inductor.

We now calculate the Q for the circuit from the defining equation:

$$Q = \frac{\omega_0}{\beta} = \frac{63,164.55}{13,200} = 4.79.$$

In this instance, even though the Q of the coil is relatively high, the relatively low internal resistance of the source R_g significantly reduces the Q of the circuit.

SIMPLIFIED ANALYSIS WHEN Q COIL IS HIGH

There is a second method for estimating the behavior of the circuit in Fig. 14.16 when the $Q \geq 5$. It makes use of the equivalent circuit developed in Problem 10.30 and shown in Fig. 14.17. The equations that relate R_2 and L_2 to R_1 and L_1 given in Problem 10.30 are

$$R_2 = \frac{R_1^2 + \omega^2 L_1^2}{R_1}; \tag{14.70}$$

$$L_2 = \frac{R_1^2 + \omega^2 L_1^2}{\omega^2 L_1}. \tag{14.71}$$

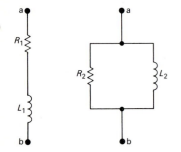

FIGURE 14.17 Series–parallel *RL* equivalent circuits.

Before showing how to use the parallel equivalent circuit in Fig. 14.17 to analyze the frequency behavior of the circuit in Fig. 14.16, we need to make some preliminary comments. First, the two circuits are exactly equivalent at a single frequency. If ω changes, R_2 and L_2 must be changed in order for the parallel circuit to represent properly the original series circuit at the new frequency. However, by restricting ω to values close to ω_0, we may assume that R_2 and L_2 are essentially constant. In calculating the parameters R_2 and L_2 at the resonant frequency ω_0, we express them in terms of the Q of the series circuit for convenience. Thus if we let

$$Q_s = \frac{\omega_0 L_1}{R_1}, \tag{14.72}$$

then

$$R_2 = R_1(1 + Q_s^2) \tag{14.73}$$

and

$$L_2 = L_1\left(1 + \frac{1}{Q_s^2}\right). \tag{14.74}$$

Example 14.5 illustrates how the parallel equivalent circuit in Fig. 14.17, along with Eqs. (14.73) and (14.74), may be used to estimate the bandwidth of the circuit in Example 14.4.

E X A M P L E   14.5

a) Find the parallel equivalent circuit for the inductor in the circuit in Example 14.4.

b) Estimate the bandwidth of the circuit by using the circuit derived in (a).

c) Estimate the value of the half-power frequencies ω_1 and ω_2.

d) Calculate $|V_o(\omega_1)|$ and $|V_o(\omega_2)|$. Compare these magnitudes with $|V_o(\omega_0)|/\sqrt{2}$.

S O L U T I O N

a) From Example 14.4, $R = 80 \ \Omega$, $L = 25$ mH, and $\omega_0 = 63{,}164.55$ rad/s. The Q of series-connected coil calculated in the discussion following Example 14.4 is 19.74. Therefore the resistance in the parallel equivalent circuit is

$$R_2 = 80[1 + (19.74)^2] = 31{,}253.41 \ \Omega,$$

and the inductance is

$$L_2 = 25[1 + (1/19.74)^2] = 25.06 \cong 25 \text{ mH}.$$

b) The bandwidth of the circuit is

$$\beta = \frac{1}{R_{eq} C} = \frac{10^8}{R_{eq}},$$

where R_{eq} is the parallel combination of the source resistance R_g and R_2. Thus

$$R_{eq} = \frac{(31{,}253.41)(10{,}000)}{41{,}253.41} = 7575.96 \ \Omega.$$

Therefore

$$\beta = \frac{10^8}{7575.96} = 13{,}200 \text{ rad/s}.$$

Note that this value for β is the same as the one obtained earlier from Eq. (14.69). (See the discussion following Eq. 14.69 and Problem 14.25.)

c) Because of the high-Q coil in the circuit, we approximate the half-power frequencies by assuming that ω_0 lies in the center of the passband. Thus

$$\omega_1 = \omega_0 - \frac{\beta}{2} = 63{,}164.55 - 6600 = 56{,}564.55 \text{ rad/s};$$

$$\omega_2 = \omega_0 + \frac{\beta}{2} = 63{,}164.55 + 6600 = 69{,}764.55 \text{ rad/s}.$$

d) We calculate the magnitude of the output voltage from the relationship

$$|V_o(\omega)| = \frac{I_m}{|Y(\omega)|}.$$

Thus

$$Y(\omega_1) = \frac{1}{10^4} + \frac{1}{80 + j1414.11} + j5.656 \times 10^{-4}$$

$$= 1.97\underline{/-44.88°} \times 10^{-4} \ \Omega;$$

$$|V_o(\omega_1)| = \frac{4 \times 10^{-3} \times 10^4}{1.97} = 20.26 \ \text{V};$$

$$Y(\omega_2) = \frac{1}{10^4} + \frac{1}{80 + j1744.11} + j697.65 \times 10^{-6}$$

$$= 1.78\underline{/44.83°} \times 10^{-4} \ \Omega;$$

$$|V_o(\omega_2)| = \frac{4 \times 10^{-3} \times 10^4}{1.78} = 22.47 \ \text{V}.$$

From Example 14.4, $|V_o(\omega_0)| = 30.30$. Therefore

$$\frac{|V_o(\omega_0)|}{\sqrt{2}} = \frac{30.30}{\sqrt{2}} = 21.43 \ \text{V}.$$

The magnitudes of $|V_o(\omega_1)|$ and $|V_o(\omega_2)|$ are close to $|V_o(\omega_0)|/\sqrt{2}$, which means that our estimate of the half-power frequencies is accurate.

SERIES EQUIVALENT CIRCUIT IN TERMS OF THE PARALLEL EQUIVALENT CIRCUIT OF A COIL

Equations (14.73) and (14.74) give the parameters of the parallel equivalent circuit in terms of the series circuit parameters. We can reverse the process, that is, express R_1 and L_1 as functions of R_2 and L_2. The basic relationships stated in Problem 10.29 are

$$R_1 = \frac{\omega^2 L_2^2 R_2}{R_2^2 + \omega^2 L_2^2}; \tag{14.75}$$

$$L_1 = \frac{R_2^2 L_2}{R_2^2 + \omega^2 L_2^2}. \tag{14.76}$$

Rewriting Eqs. (14.75) and (14.76) in terms of the Q of the original parallel circuit is more convenient. Equation (14.11) contains the Q of the parallel combination of R_2 and L_2, so

$$Q_p = \frac{R_2}{\omega_0 L_2}. \tag{14.77}$$

Substituting Eq. (14.77) into Eqs. (14.75) and (14.76) yields

$$R_1 = \frac{R_2}{1 + Q_p^2}; \tag{14.78}$$

$$L_1 = \frac{Q_p^2 L_2}{1 + Q_p^2}. \tag{14.79}$$

Equations (14.75) and (14.76) show that if we calculate the ratio $\omega_0 L_1/R_1$ we get

$$\frac{\omega_0 L_1}{R_1} = \frac{\omega_0 R_2^2 L_2}{R_2^2 + \omega_0^2 L_2^2} \frac{R_2^2 + \omega_0^2 L_2^2}{\omega_0^2 L_2^2 R_2}$$

$$= \frac{R_2}{\omega_0 L_2},$$

or

$$Q_s = Q_p. \tag{14.80}$$

In other words, we may calculate the Q of the equivalent circuits from either the series or parallel equivalent.

SERIES–PARALLEL EQUIVALENTS FOR AN *RC* CIRCUIT

The series–parallel *RC* circuit of Problems 10.27 and 10.28 also may be expressed in terms of the Q of the circuits. Figure 14.18 shows the circuit; the appropriate relationships are

$$Q_p = \omega_0 R_2 C_2; \tag{14.81}$$

$$Q_s = \frac{1}{\omega_0 R_1 C_1}; \tag{14.82}$$

$$R_1 = \frac{R_2}{1 + Q_p^2}; \tag{14.83}$$

$$C_1 = \frac{1 + Q_p^2}{Q_p^2} C_2; \tag{14.84}$$

$$R_2 = (1 + Q_s^2)R_1; \tag{14.85}$$

$$C_2 = \frac{Q_s^2 C_1}{1 + Q_s^2}. \tag{14.86}$$

As in the case of the *RL* circuits, $Q_s = Q_p$; hence we may calculate Q from either equivalent circuit.

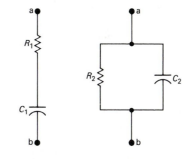

FIGURE 14.18 Series–parallel *RC* equivalent circuits.

DRILL EXERCISES

14.8 The internal resistance of the current source in the circuit shown is infinite, that is, $R_g = \infty$. The 16-mH coil has a resistance of 100 Ω. The lossless capacitor is 100 nF. The peak amplitude of the current source is 5 mA at all frequencies. Calculate (a) ω_m; (b) ω_0; (c) $V_m(\omega_0)$; and (d) $V_m(\omega_m)$.

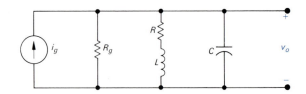

ANSWER: (a) 24.976.99 rad/s; (b) 24.206.15; (c) 8 V; (d) 8.25 V.

14.9 A 40-mH coil has a resistance of 3.2 kΩ. The coil is shunted by a 6.25-pF lossless capacitor. This parallel-resonant circuit is driven by a sinusoidal current source with an internal impedance of $8 + j0$ MΩ. Calculate (a) ω_0; (b) f_0; (c) β; (d) the Q of the circuit; and (e) the Q of the coil.

ANSWER: (a) 199.84×10^4 rad/s; (b) 318.06 kHz; (c) 100 krad/s; (d) 19.98; (e) 24.98.

14.7 MORE ON SERIES RESONANCE

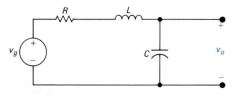

FIGURE 14.19 A series-resonant circuit that uses the capacitor voltage as the output signal.

In the series-resonant circuit shown in Fig. 14.19 the output voltage is the voltage across the capacitor. At resonance the amplitude of the voltage across the capacitor is Q times the amplitude of the source voltage (Eq. 14.60). The maximum voltage across the capacitor does not occur at the resonant frequency ω_0. To verify this condition, we express the amplitude of v_o as a function of ω and then proceed to find the value of ω that causes the amplitude to be a maximum. Using phasor circuit analysis, we obtain

$$\mathbf{V}_o = \frac{1/j\omega C}{R + j[\omega L - (1/\omega C)]} \mathbf{V}_g$$

$$= \frac{\mathbf{V}_g}{1 - \omega^2 LC + j\omega RC}. \tag{14.87}$$

The magnitude of the output voltage is

$$|\mathbf{V}_o| = \frac{V_m}{\sqrt{(1 - \omega^2 LC)^2 + \omega^2 R^2 C^2}}, \tag{14.88}$$

where V_m is the amplitude of the sinusoidal input voltage v_g.

To find the frequency at which the magnitude of v_o is maximum, we differentiate Eq. (14.88) with respect to ω and then find the value of ω that makes the derivative zero. Thus

$$\frac{d|\mathbf{V}_o|}{d\omega} = -\frac{V_m(1/2)[2(1 - \omega^2 LC)(-2\omega LC) + 2\omega R^2 C^2]}{[(1 - \omega^2 LC)^2 + \omega^2 R^2 C^2]^{3/2}},$$

(14.89)

and the value of ω greater than zero that makes the derivative equal to zero is

$$\omega_m = \sqrt{\frac{1}{LC} - \frac{1}{2}\left(\frac{R}{L}\right)^2}.$$

(14.90)

We substitute Eq. (14.90) into Eq. (14.88) to find the maximum value of $|\mathbf{V}_o|$:

$$|\mathbf{V}_o(\omega_m)| = \frac{V_m}{\sqrt{\dfrac{R^2 C}{L}\left(1 - \dfrac{R^2 C}{4L}\right)}}.$$

(14.91)

We also may express Eq. (14.91) in terms of the circuit's Q. From Eq. (14.51),

$$\frac{R^2 C}{L} = \frac{1}{Q^2},$$

(14.92)

from which

$$|\mathbf{V}_o(\omega_m)| = \frac{QV_m}{\sqrt{1 - (1/2Q)^2}}.$$

(14.93)

Equations (14.90) and (14.93) tell us that for high-Q circuits the frequency at which the maximum occurs, ω_m, approaches the resonant frequency, ω_0, and the maximum value of $|\mathbf{V}_o|$ approaches QV_m. The relationship between the magnitude of the capacitor voltage in a series RLC circuit and the frequencies ω_0 and ω_m is illustrated numerically in Example 14.6.

E X A M P L E   14.6

The sinusoidal voltage source in the circuit shown in Fig. 14.19 delivers a voltage of $10 \cos \omega t$ V to a 50-Ω resistor, a 5-mH inductor, and a 0.5-μF capacitor.

a) Calculate ω_0 and Q.

b) What is the peak amplitude of v_o at ω_0?

c) At what frequency is the peak amplitude of v_o maximum?

d) What is the peak amplitude of v_o at the frequency in (c)?

e) Sketch $|\mathbf{V}_o|$ versus ω.

f) Repeat (a)–(e) for $R = 10 \ \Omega$.

SOLUTION

a) $\omega_0 = 1/\sqrt{LC} = 20{,}000$ rad/s; $Q = \omega_0 L/R = 2$.

b) $|V_o| = QV_m = 2(10) = 20$ V.

c) $\omega_m = \sqrt{4 \times 10^8 - 0.5 \times 10^8} = 18{,}708.29$ rad/s.

d) $|\mathbf{V}_o| = 2(10)/\sqrt{1 - 1/16} = 20.66$ V.

e) Figure 14.20(a) shows a sketch of $|\mathbf{V}_o|$ versus ω and the relationship between the magnitude of $\mathbf{V}_o$ at ω_m and ω_0.

f) $\omega_0 = 20{,}000$ rad/s; $Q = \omega_0 L/R = 10$; $|\mathbf{V}_o(\omega_0)| = (10)(10) = 100$ V; $\omega_m = \sqrt{4 \times 10^8 - 10^8/50} = 19{,}949.94$; $|\mathbf{V}_o(\omega_m)| = 100\sqrt{1 - 1/400} = 100.13$ V. Figure 14.20(b) shows the sketch.

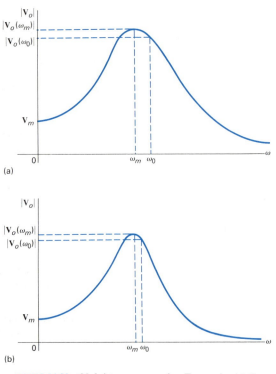

(a)

(b)

FIGURE 14.20 $|\mathbf{V}_o(\omega)|$ versus ω for Example 14.6: (a) $R = 50 \ \Omega$; (b) $R = 10 \ \Omega$.

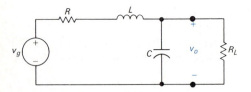

FIGURE 14.21 The series-resonant circuit of Fig. 14.19 with a resistive load.

EFFECT OF LOADING THE CAPACITOR ON THE RESPONSE

The effect of "loading" the capacitor in the circuit shown in Fig. 14.19 is of practical importance. That is, the circuit, or device, driven from the capacitor shunts the capacitor with its input impedance. This impedance typically is resistive and represents a load across the output terminals of the series-resonant circuit. In Fig. 14.21 R_L indicates the loading effect.

The RC equivalent circuit shown in Fig. 14.18 aids the analysis of the circuit shown in Fig. 14.21 at, or near, the resonant frequency. We replace the parallel combination of R_L and C with a series combination of R_1 and C_1 by means of Eqs. (14.83) and (14.84). So long as the frequency does not deviate greatly from ω_0, the equivalent resistance, R_1, and capacitance, C_1, can be assumed constant. Note from Eq. (14.84) that, if the Q of the parallel combination is greater than 5, $C_1 \cong C$ and R_L does not greatly disturb the resonant frequency ω_0.

We determine the effect of R_L on the magnitude of the output voltage (assuming that $\omega_0 R_L C > 5$) as follows. If we use the series-equivalent circuit for R_L and C, the phasor-domain equivalent circuit is as shown in Fig. 14.22. As $C_1 \cong C$, the phasor current at resonance is nearly $\mathbf{V}_g/(R + R_1)$, and the output voltage is

$$\mathbf{V}_o(\omega_0) \cong \frac{\mathbf{V}_g}{R + R_1}\left(R_1 - j\frac{1}{\omega_0 C}\right). \qquad (14.94)$$

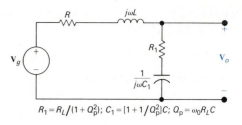

$R_1 = R_L/(1 + Q_p^2); \; C_1 = [1 + 1/Q_p^2]C; \; Q_p = \omega_0 R_L C$

FIGURE 14.22 A phasor-domain equivalent circuit for the circuit of Fig. 14.21.

For $Q_p > 5$, we replace R_1 with R_L/Q_p^2.

The coefficient of j in Eq. (14.94) is equivalent to $R/\omega_0 RC$ or QR, where Q is the quality factor of the series-resonant circuit *before* loading. Therefore we write Eq. (14.94) as

$$\mathbf{V}_o(\omega_0) = \frac{\mathbf{V}_g Q_p^2}{RQ_p^2 + R_L}\left(\frac{R_L}{Q_p^2} - jQR\right) = \frac{\mathbf{V}_g}{RQ_p^2 + R_L}(R_L - jQRQ_p^2).$$

$$(14.95)$$

The magnitude of the output voltage at resonance is

$$\begin{aligned}
|\mathbf{V}_o(\omega_0)| &= \frac{|\mathbf{V}_g|}{RQ_p^2 + R_L}\sqrt{R_L^2 + Q^2 R^2 Q_p^4} \\
&= \frac{|\mathbf{V}_g|QRQ_p^2\sqrt{1 + (R_L^2/Q^2 R^2 Q_p^4)}}{RQ_p^2[1 + (R_L/RQ_p^2)]} \\
&= Q|\mathbf{V}_g|\frac{\sqrt{1 + (R_L/Q_p^2 QR)^2}}{1 + (R_L/RQ_p^2)}. \qquad (14.96)
\end{aligned}$$

The magnitude of $\mathbf{V}_o$ at resonance before loading is $Q|\mathbf{V}_g|$, so we use Eq. (14.96) to assess the effect of the load resistor on the magnitude of the output voltage at ω_0. Example 14.7 illustrates the calculations used to analyze the circuit shown in Fig. 14.22.

E X A M P L E   14.7

Assume that the circuit in Example 14.6 (with $R = 10 \, \Omega$) is loaded with a resistance of 10 kΩ.

a) Calculate the Q of the parallel RC portion of the circuit.

b) Calculate the equivalent circuit parameters R_1 and C_1.

c) Calculate the peak amplitude of v_o at ω_0.

d) Repeat (a), (b), and (c) for $R_L = 100$ kΩ.

S O L U T I O N

a) We assume that Q_p is greater than 5 so that R_L has little effect on ω_0. (If we get a contradiction to this assumption, we have to calculate ω_0 from the definition of unity power-factor resonance—see Problem 14.28—and the analytical techniques based on a highly frequency selective circuit have to be abandoned.) We have

$$Q_p = \omega_0 R_L C = (2 \times 10^4)(10^4)(0.5 \times 10^{-6}) = 100.$$

b) The equivalent circuit parameters are

$$R_1 = \frac{R_L}{1 + Q_p^2} \cong \frac{10^4}{10^4} = 1\ \Omega;$$

$$C_1 = \left(1 + \frac{1}{Q_p^2}\right)C \cong 0.5\ \mu F.$$

c) In anticipation of using Eq. (14.96) to calculate $|V_o(\omega_o)|$, we make the following preliminary calculations:

$$Q = 10 \qquad \text{(see Example 14.6f)};$$

$$\left(\frac{R_L}{Q_p^2 QR}\right)^2 = \left[\frac{10^4}{(10^4)(10)(10)}\right]^2 = (10^{-2})^2 = 10^{-4};$$

$$\frac{R_L}{RQ_p^2} = \frac{10^4}{(10)(10^4)} = 10^{-1};$$

$$|V_o(\omega_0)| = \frac{(10)(10)\sqrt{1 + 10^{-4}}}{1 + 0.1}$$

$$= \frac{100\sqrt{1.0001}}{1.1} = 90.91\ V.$$

· Although R_L has an almost negligible effect on the resonant frequency, it does cause about a 9 percent drop in the peak amplitude of v_o.

d) When $R_L = 100\ k\Omega$, we get $Q_p = 1000$, $R_1 = 0.1\ \Omega$, $C_1 = 0.5\ \mu F$, and $|V_o(\omega_0)| = (10)(10)/1.01 = 99.01\ V.$

The circuit shown in Fig. 14.21 also is amenable to exact analysis. If interested, you may pursue it in Problem 14.29.

DRILL EXERCISES

14.10 The circuit elements shown are $R = 80\ \Omega$, $L = 50\ mH$, and $C = 0.2\ \mu F$.

a) If the circuit is excited from a sinusoidal voltage source $20 \cos \omega_0 t$ V, what is the rms value of $v_o(t)$?

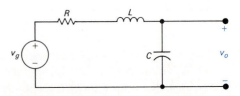

b) The rms value of v_o is measured with a digital voltmeter that has a resistance of 500 kΩ. What is the rms reading of the voltmeter?

ANSWER: (a) 88.39 V; (b) 87.84 V.

14.8 SCALING

In the design and analysis of frequency-selective circuits, working with element values such as 1 Ω, 1 H, and 1 F is convenient. Although these values are unrealistic in terms of practical components, they greatly simplify computations. After making the computations (using the unrealistic values of R, L, and C), the designer can transform the element values into realistic values by the process of *scaling*.

There are two types of scaling: *magnitude* and *frequency*. We scale a circuit in magnitude by multiplying the impedance at a given frequency by the scaling factor k_m. Thus we multiply all resistors and inductors by k_m and all capacitors by $1/k_m$. If we let the subscripts 1 and 2 represent the initial and final values of the parameters, respectively, we have

$$R_2 = k_m R_1, \qquad L_2 = k_m L_1 \qquad \text{and} \qquad C_2 = C_1/k_m, \quad \textbf{(14.97)}$$

where k_m is by definition a positive real number. It can be either less than or greater than 1.

In frequency scaling, we change the circuit parameters so that at the new frequency the impedance of each element is the same as it was at the original frequency. Because resistance values are assumed to be independent of frequency, resistors are unaffected by frequency scaling. If we let k_f denote the frequency scaling factor, both inductors and capacitors are multiplied by $1/k_f$. Thus for frequency scaling,

$$R_2 = R_1, \qquad L_2 = L_1/k_f, \qquad \text{and} \qquad C_2 = C_1/k_f. \quad \textbf{(14.98)}$$

The frequency scaling factor k_f also is a positive real number that can be less than or greater than unity.

A circuit can be scaled simultaneously in both magnitude and frequency. The "new" values (2) in terms of the "old" (1) values are

$$R_2 = k_m R_1, \qquad L_2 = \frac{k_m}{k_f} L_1, \qquad \text{and} \qquad C_2 = \frac{1}{k_m k_f} C_1. \quad \textbf{(14.99)}$$

Example 14.8 illustrates the scaling process.

EXAMPLE 14.8

Scale the circuit in Example 14.1 to shift the resonant frequency to 20 krad/s and the impedance at resonance to 10 kΩ.

SOLUTION

From Example 14.1 we know that the original resonant frequency is 10 krad/s and that the impedance at this resonant frequency is 2 kΩ. Then

$$k_m = \frac{10 \times 10^3}{2 \times 10^3} = 5 \quad \text{and} \quad k_f = \frac{20 \times 10^3}{10 \times 10^3} = 2.$$

The new values of R, L, and C are

$$R = 5(2) = 10 \text{ k}\Omega,$$

$$L = \frac{5}{2}(40) = 100 \text{ mH},$$

and

$$C = \frac{0.25}{(5)(2)} = 0.025 \ \mu\text{F}.$$

USE OF SCALING IN DESIGN

To use the concept of scaling in the design of a frequency-selective circuit, select the resonant frequency as 1 rad/s and the magnitude of the impedance at resonance as 1 Ω. Then calculate the values of L and C on the basis of the desired Q. After computing the values of L and C on the basis of 1 Ω, 1 rad/s, and Q, scale the circuit to give the desired resonant frequency and impedance level. The circuit with a resonant frequency of 1 rad/s and an impedance of 1 Ω at resonance is referred to as a *universal* resonant circuit.

For the universal parallel-resonant circuit,

$$Q = \frac{R}{\omega_0 L} = \frac{1}{L} \tag{14.100}$$

or

$$Q = \omega_0 RC = C. \tag{14.101}$$

Therefore after Q has been specified, L and C can be computed. Observe that, as ω_0 is unity, $Q = 1/\beta$ and thus either Q or β can be used to define L or C.

For the universal series-resonant circuit,

$$Q = \frac{\omega_0 L}{R} = L \qquad (14.102)$$

or

$$Q = \frac{1}{\omega_0 CR} = \frac{1}{C}. \qquad (14.103)$$

Example 14.9 illustrates how scaling is used in conjunction with universal-parallel resonant circuit. Example 14.10 discusses an application of scaling in filter design.

EXAMPLE 14.9

a) Compute the values of L and C for a universal parallel-resonant circuit having a Q of 20.

b) Scale the circuit in (a) so that the resonant frequency is 1.04 MHz and the impedance at resonance is 25 kΩ.

SOLUTION

a) From Eqs. (14.100) and (14.101), $L = 1/20 = 0.05$ H and $C = 20$ F.

b) The magnitude scaling factor is 25×10^3, and the frequency scaling factor is $2.08\pi \times 10^6$. Therefore the circuit values are

$$R = 25 \text{ k}\Omega;$$

$$L = \frac{25 \times 10^3}{2.08\pi \times 10^6}(0.05) = 191.29 \ \mu\text{H};$$

$$C = \frac{20}{(25 \times 10^3)(2.08\pi \times 10^6)} = 122.43 \text{ pF}.$$

EXAMPLE 14.10

Scaling is particularly useful in filter design. There are many handbooks that tabulate R, L and C values for thousands of filters. Low-pass filters are nearly always *normalized* to pass frequencies from 0 to 1 rad/s and require a 1-Ω load resistance. The filter designer scales the circuit to the passband and load resistance required by an application or to a customer's specification.

For example, Fig. 14.23 shows a handbook, low-pass filter that works with a 1-Ω source and a 1-Ω load and passes frequencies between 0 and 1 rad/s. That is the half-power frequency of the filter is 1 rad/s.

Assume a telephone circuit requires a filter to operate between a 600-Ω source and a 600-Ω load. Furthermore, the filter is to pass frequencies between 0 and 3 kH. Use scaling to specify the values of L and C to be used in the filter.

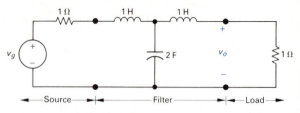

FIGURE 14.23 A low-pass (0 to 1 rad/s) filter design to operate between a 1-Ω source and a 1-Ω load.

S O L U T I O N

By hypothesis

$$k_m = \frac{600}{1} = 600 \quad \text{and} \quad k_f = \frac{2\pi(3000)}{1} = 6000\pi.$$

It follows that

$$L = \frac{600}{6000\pi} \cdot 1 = 31.83 \text{ mH}$$

and

$$C = \frac{2}{(600)(6000\pi)} = 0.1768 \ \mu F.$$

The scaled filter circuit is shown in Fig. 14.24.

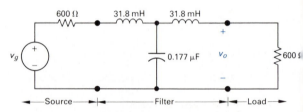

FIGURE 14.24 The filter of Fig. 14.23 scaled to operate between a 600-Ω source and a 600-Ω load and to pass frequencies between 0 and 3 kH.

D R I L L   E X E R C I S E S

14.11 Scale the circuit in Drill Exercise 14.1 to shift the resonant frequency to 8 Mrad/s and the impedance at resonance to 400 kΩ. Specify the values of (a) k_m; (b) k_f; (c) R_2; (d) L_2; (e) C_2; (f) Q; and (g) β.

ANSWER: (a) 4; (b) 0.8; (c) 400 kΩ; (d) 500 μH; (e) 31.25 pF; (f) 100; (g) 80 krad/s.

14.13 The circuit shown is to be scaled so that $\omega_0 = 7.5$ Mrad/s and $C = 10$ pF. Specify (a) k_f; (b) k_m; (c) R; (d) L; (e) Q; (f) β.

ANSWER: (a) 25 × 10⁶; (b) 4000; (c) 4000 Ω; (d) 1.6 mH; (e) 3; (f) 2.5 Mrad/s.

14.12 A universal parallel-resonant circuit has a bandwidth of 0.0625 rad/s.

a) Compute the values of L and C.

b) Scale the circuit in part (a) so that the resonant frequency is 1.35 MHz and the impedance at resonance is 10 kΩ.

ANSWER: (a) 62.5 mH, 16 F; (b) 10 kΩ, 73.68 μH, 188.63 pF.

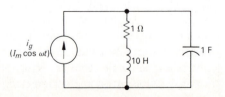

14.14 Assume the source voltage in the circuit in Fig. 14.23 is given by the expression $v_g = V_m \cos \omega t$

    a) What is the amplitude of the output voltage v_o when $\omega = 0$?

    b) Repeat (a) for $\omega = 1$ rad/s.

    c) Why is the frequency in (b) referred to as the half-power frequency?

    d) Sketch the amplitude of the output voltage for $0 \le \omega \le 2$ rad/s.

ANSWER: (a) $0.5\ V_m$; (b) $0.354\ V_m$; (c) because the power delivered to the 1-Ω load at 1 rad/s is one half the maximum power that is delivered to the 1-Ω load when $\omega = 0$ rad/s; (d) is the Figure.

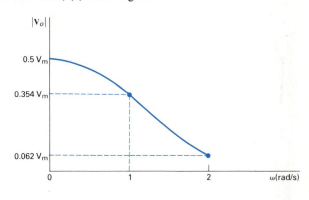

SUMMARY

Frequency-selective circuits are important in communication, computer, control, and signal-processing systems. We introduced the concept of frequency discrimination by studying the frequency response of the parallel and series *RLC* circuits. We discussed the following important concepts.

- Frequency response is the variation in the amplitude and phase angle of an *output signal* when only the frequency of the *input signal* is varied.

- Resonance is the condition when the network appears purely resistive to the input signal source.

- Resonant frequency is the frequency at which resonance occurs.

- Half-power frequency is a frequency where the amplitude of the output signal is $1/\sqrt{2}$ times its maximum amplitude.

- Bandwidth is the range of frequencies where the amplitude of the output signal is equal to or greater than $1/\sqrt{2}$ times its maximum amplitude.

- Quality factor is the ratio of the resonant frequency to the bandwidth.

- Scaling is the process of calculating new values for the circuit parameters R, L, and C based on specifying the magnitude of an impedance at a given frequency and/or maintaining the magnitude of an impedance at a new frequency.

TABLE 14.1

PARALLEL OR SERIES *RLC* CIRCUIT FREQUENCY RESPONSE RELATIONSHIPS

QUANTITY	SYMBOL	RELATIONSHIP
Resonant frequency (parallel or series)	ω_o	$1/\sqrt{LC}$ $\quad$ or $\quad$ $\sqrt{\omega_1\omega_2}$
Bandwidth	β	$1/RC$ $\quad$ (parallel) $\\$ R/L $\quad$ (series)
Quality factor (parallel or series)	Q	ω_o/β $\\$ $\omega_o RC$ $\quad$ (parallel) $\\$ $\omega_o L/R$ $\quad$ (series)
Corner frequency (parallel or series)	ω_1	$\omega_o\left[-\dfrac{1}{2Q} + \sqrt{1 + \left(\dfrac{1}{2Q}\right)^2}\right]$
	ω_2	$\omega_o\left[\dfrac{1}{2Q} + \sqrt{1 + \left(\dfrac{1}{2Q}\right)^2}\right]$
Admittance (parallel)	$Y(\omega)$	$\dfrac{1}{R}\left[1 + j\dfrac{(\omega + \omega_o)(\omega - \omega_o)}{\beta\omega}\right]$
Impedance (series)	$Z(\omega)$	$R\left[1 + j\dfrac{(\omega + \omega_o)(\omega - \omega_o)}{\beta\omega}\right]$

Table 14.1 summarizes the important relationships when the parallel or series *RLC* circuit contains a lossless inductor and capacitor.

The following summary pertains to the parallel *RLC* circuit when the inductor has losses.

- Resonant frequency (ω_o): See Eq. (14.63).

- Amplitude of the output voltage at ω_o: See Eq. (14.67).

- Frequency at which the amplitude of the output voltage is maximum (ω_m): See Eq. (14.68).

- Bandwidth (high Q coil): See Eq. (14.69).

- Equivalent circuits: For a small range of frequencies, a series *RL* circuit can be replaced by a parallel *RL* circuit. The same holds for an *RC* circuit.

PROBLEMS

14.1 Design a parallel-resonant structure like that shown in Fig. 14.1 to meet the following specifications: $L = 0.5\ \mu H$, $\omega_0 = 10^{10}$ rad/s, and $Q = 16$.

a) Specify the numerical values of R and C.
b) Calculate the numerical values of ω_1 and ω_2.

14.2 The parallel-resonant circuit in Fig. 14.1 has a resonant frequency of 600 krad/s. The lower half-power frequency is 570 krad/s. Find (a) the upper half-power frequency; (b) the bandwidth in kilohertz; and (c) the Q of the circuit.

14.3 A sinusoidal current source delivers a current having an amplitude of 75 μA. The internal impedance of the source is $400 + j0$ kΩ. This current source is used with a 25-mH inductor and a 100-kΩ resistor to form a parallel-resonant circuit. The resonant frequency is 200 krad/s.

a) Specify the numerical value of the circuit capacitor.

b) What is the Q of the circuit without the source?

c) What is the Q of the circuit with the source?

d) What is the bandwidth of the circuit with and without the source?

e) What is the maximum output voltage that the source can deliver?

f) What is the maximum value of the capacitor current at resonance?

14.4 The frequency of the sinusoidal voltage source in the circuit in Fig. P14.4 is adjusted until the amplitude of the sinusoidal output voltage is maximum. The maximum amplitude of the source voltage is 600 V.

a) What is the frequency of v_s in hertz?

b) What is the amplitude of v_o at the frequency given in part (a)?

c) What is the bandwidth of the circuit?

d) What is the Q of the circuit?

e) At what frequencies will the amplitude of v_o be $1/\sqrt{2}$ times its maximum value?

f) If the 20-kΩ resistor represents the internal resistance of the source, how much does this source resistance lower the Q of the circuit?

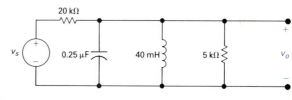

FIGURE 14.4

14.5 a) For the circuit in Example 14.1(b), calculate the magnitude of V_o at 10,500 rad/s and at 9500 rad/s using the exact formula for Y.

b) Repeat part (a) using the approximate formula for Y.

c) Find the percent error in the magnitude of V_o that results from using the approximate expression for Y.

14.6 Show that when $\omega = \omega_0$ in the circuit in Fig. 14.1 the magnitude of the current in both the inductive and capacitive branches is QI_m.

14.7 Select the circuit values for the parallel-resonant structure in Fig. 14.1 so that $\omega_0 = 1.25$ Mrad/s, $\beta = 156.25$ krad/s, $V_m(\omega_0) = 250$ V, and $|I_C(\omega_0)| = 100$ mA. Give the values of R, L, C, and I_m.

14.8 Show that Eq. (14.29) predicts $\theta = +45°$ when $\omega = \omega_1$ and $-45°$ when $\omega = \omega_2$. (*Hint:* Use the fact that $\omega_0 = \sqrt{\omega_1\omega_2}$.)

14.9 A parallel-resonant structure as shown in Fig. 14.1 is to satisfy the following specifications: $f_0 = 1$ MHz, $Z(f_0) = 250$ kΩ, $|Z(f_a)| = 200$ kΩ, and $f_a = 1060$ kHz.

a) Specify R, L, and C.

b) What is the Q of the circuit?

c) What is the bandwidth of the circuit?

d) If this circuit is driven by a sinusoidal current source having an internal impedance of $1 + j0$ MΩ, what is the Q of the energized circuit?

e) What is the bandwidth of the energized circuit?

14.10 A parallel RLC circuit is designed to have an impedance magnitude of 20 kΩ at the resonant frequency and 6.8 kΩ at a frequency of 10 Mrad/s below resonance. If the resonant frequency is 50 Mrad/s, calculate the value of L in microhenries.

14.11 A parallel RLC circuit has energy stored in the inductor but not in the 10-nF capacitor. When the stored energy is released, the voltage appearing across the circuit is

$$v = 12.5e^{-5000t} \sin 80{,}000t \text{ V}, \qquad t \geq 0.$$

a) Compute the bandwidth, Q, and resonant frequency of the circuit.

b) Compute the half-power frequencies ω_1 and ω_2.

c) If the parallel RLC circuit is driven from a sinusoidal current source having an internal resistance of 70 kΩ, what are the Q, bandwidth, and resonant frequency?

14.12 The values of the circuit parameters for the circuit in Fig. 14.8 are $R = 4$ Ω, $L = 16$ μH, and $C = 2.6$ nF.

a) Calculate Q, β, ω_0, ω_1, and ω_2.

b) Assume that the circuit is driven from an ideal sinusoidal voltage source having a maximum amplitude of 10 V. Calculate the amplitude of the voltage across the capacitor at resonance.

c) Repeat parts (a) and (b), given that the sinusoidal voltage source has an internal resistance of 1 Ω.

14.13 For the circuit in Problem 14.12, calculate the magnitude and phase angle of the impedance seen by the voltage source at ω_0 and at frequencies 100 krad/s larger and smaller than ω_0.

14.14 A coil having a resistance of 20 Ω and an inductance of 10 μH is used in the circuit in Fig. P14.14. The resonant frequency of the circuit is $(10/\pi)$ MHz and the bandwidth is $(1.25/\pi)$ MHz. The internal resistance of the sinusoidal voltage source is 1 Ω.

a) Specify the numerical values of R and C.

b) What is the Q of the coil at the resonant frequency ω_0? $(Q_{\text{coil}} = \omega_0 L/R_{\text{coil}})$

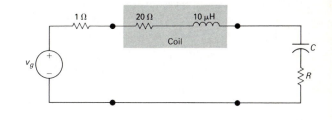

FIGURE P14.14

14.15 The sinusoidal voltage source in the circuit in Fig. P14.15 has a maximum amplitude of 50 V. The internal impedance of the source is negligible.

a) At what frequency is the amplitude of v_o maximum?

b) What is the maximum amplitude of v_o?

c) Over what range of frequencies will the amplitude of v_o be equal to or greater than 0.80 of its maximum value?

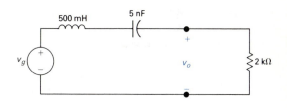

FIGURE P14.15

14.16 In the circuit in Fig. 14.11, $R_g = 50$ kΩ, $R = 350$ Ω, $L = 680$ μH, and $C = 56$ pF. The peak amplitude of the sinusoidal current source is 1 mA.

a) Calculate the resonant frequency ω_0.

b) Calculate the amplitude of the output voltage at resonance.

c) Calculate the frequency at which the amplitude of v_o is maximum.

d) Calculate the maximum amplitude of v_o.

e) Estimate the bandwidth of the circuit.

f) What is the Q of the coil at ω_0?

g) What is the Q of the circuit at ω_0?

14.17 A 39-μH inductor has a Q of 50 at 2.52 MHz. This inductor is to be used in the parallel-resonant circuit in Fig. 14.11. The resonant frequency of the circuit is to be 2.52 MHz.

a) Specify the value of the capacitor C. (*Hint:* Use the parallel equivalent circuit for the inductor.)

b) Specify the value of R_g so that the band-width of the circuit is 380 krad/s.

c) What is the Q of the circuit?

14.18 The frequency of the sinusoidal current source in the circuit shown in Fig. P14.18 is adjusted for unity power-factor resonance.

a) Find the quality factor of the circuit.

b) Find the amplitude of v_o when $i_g = 400 \cos \omega_0 t \ \mu$A.

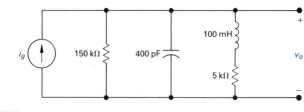

FIGURE P14.18

14.19 The current source in the circuit shown in Fig. P14.19 is generating a sinusoidal current whose amplitude is 50 mA at all frequencies. Calculate

a) the unity power-factor resonant frequency;

b) the peak amplitude of v_o at unity power-factor resonance;

c) the Q of the circuit;

d) the bandwidth of the circuit;

e) the frequency at which the peak amplitude of v_o is maximum; and

f) the maximum value of the peak amplitude.

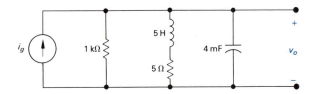

FIGURE P14.19

14.20 The resonant frequency of the circuit shown in Fig. P14.20 is 40 Mrad/s. The Q of the 1-mH coil at this frequency is 25. The circuit is driven by a sinusoidal current source that has an internal impedance of $1 + j0$ MΩ.

a) Calculate C in picofarads.

b) Calculate the Q of the circuit.

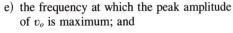

FIGURE P14.20

14.21 The frequency of the sinusoidal voltage source in the circuit in Fig. P14.21 is adjusted to unity power-factor resonance.

a) What is the Q of the coil at the resonant frequency?

b) What is the peak amplitude of v_o at the resonant frequency?

c) What is the Q of the circuit?

d) What is the bandwidth of the circuit?

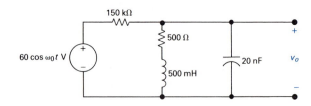

FIGURE P14.21

14.22 Design a parallel-resonant RLC circuit that satisfies the following criteria: $\omega_o = 0.8$ Mrad/s; $Z(\omega_0) = 625$ kΩ; $|Z(\omega_a)| = 500$ kΩ; and $\omega_a = 0.808$ Mrad/s. Specify (a) Q; (b) β in kiloradians per second; (c) R in kilohms; (d) L in millihenrys; and (e) C in picofarads.

14.23 A 33-mH coil with a Q of 60 at 79 kHz is used in the circuit in Fig. P14.23 to design a parallel-resonant circuit. The resonant frequency is specified to be 79 kHz.

a) Specify the value of C in picofarads.

b) Specify the value of R_g so that the Q of the circuit is equal to 50.

c) Specify the bandwidth of the circuit in kiloradians per second.

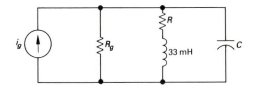

FIGURE P14.23

14.24 A cathode-ray oscilloscope (CRO) is used to measure the voltage across a coil having an inductance of 1 mH and a resistance of 1 kΩ. The laboratory setup is illustrated in Fig. P14.24. The input circuit of the CRO consists of a 1.2-MΩ resistor in parallel with a 20-pF capacitor. A sinusoidal voltage source operating at a frequency of 7 Mrad/s and having an internal resistance of 12 kΩ is used to excite the coil. The peak amplitude of the sinusoidal voltage source is 50 V when the source is operating open circuit.

a) Use the internal voltage of the sinusoidal voltage source as the reference signal. What is the expression for v_g?

b) Construct a circuit model of the setup used to measure the coil voltage.

c) What is the percent error in measuring the amplitude of the voltage across the coil?

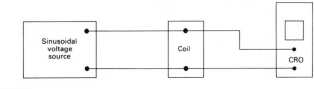

FIGURE P14.24

14.25 Derive Eq. (14.69). [*Hint:* For a parallel-resonant circuit, $\beta = 1/RC$. For the circuit in Fig. 14.11, R takes the value of R_g in parallel with the equivalent resistor $R_1(1 + Q_s^2)$.]

14.26 The internal impedance of the sinusoidal voltage source in Fig. P14.26 is $10 + j0$ Ω. The frequency of the source is set at the unity power-factor resonant frequency of the circuit. The amplitude of the source voltage is set to yield an rms value of 20 mV.

a) What is the rms value of v_o at resonance?

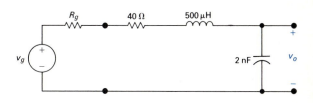

FIGURE P14.26

(*continued*)

b) If v_o is measured by an rms voltmeter having a resistance of 50 kΩ, what will the voltmeter read?

c) Repeat part (b), given that the meter resistance is 250 kΩ.

14.27 The circuit shown in Fig. P14.27 is operating at unity power-factor resonance. The peak amplitude of the voltage generated by the source is 7.071 V. The voltage across the capacitor is measured with an rms voltmeter that has an internal impedance of $160 + j0$ kΩ. What is the percent error in the rms reading?

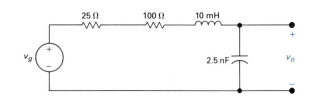

FIGURE P14.27

14.28 a) Show that the impedance seen by the voltage source in the circuit in Fig. 14.16 is

$$Z_{in} = R + j\omega L = \frac{R_L}{1 + j\omega R_L C}.$$

b) Show that the impedance found in part (a) is purely resistive when

$$\omega_0 = \sqrt{\frac{1}{LC} - \frac{1}{R_L^2 C^2}}.$$

c) Show that

$$Z_{in}(\omega_0) = R + \frac{L}{R_L C} + j0 \ \Omega.$$

d) Show that

$$|\mathbf{V}_o(\omega_0)| = \frac{|\mathbf{V}_g| R_L \sqrt{LC}}{L + R_L RC}.$$

(*Hint:* Use the fact that $\omega_0^2 R_L^2 C^2 + 1 = R_L^2 C/L$.)

e) Use the relationship given in part (d) to calculate $|\mathbf{V}_o(\omega_0)|$ for the circuit in Example 14.7(c) and compare this value to that obtained in Example 14.7(c).

14.29 a) For the circuit in Fig. 14.16 show that the magnitude of $\mathbf{V}_o$ can be expressed as

$$|\mathbf{V}_o(\omega)| = \frac{V_m R_L}{\sqrt{(R + R_L - \omega^2 LCR_L)^2 + \omega^2(L + RR_L C)^2}},$$

where V_m is the peak amplitude of the voltage source.

b) Show that $|\mathbf{V}_o(\omega)|$ is maximum when

$$\omega = \omega_m = \sqrt{\frac{R + R_L}{R_L LC} - \frac{(L + RR_L C)^2}{2(LCR_L)^2}}$$

$$= \sqrt{\frac{1}{LC}\left\{1 - \frac{1}{2LC}\left[R^2 C^2 + \left(\frac{L}{R_L}\right)^2\right]\right\}}.$$

c) Show that the expression for $|\mathbf{V}_o(\omega_m)|$ is

$$|\mathbf{V}_o(\omega_m)| = \frac{V_m R_L \sqrt{LC}}{(L + RR_L C)\sqrt{1 - \frac{1}{LC}\left(\frac{L}{2R_L} - \frac{RC}{2}\right)^2}}.$$

[*Hint:* Use the fact that $R + R_L - \omega_m^2 LCR_L = (L + RR_L C)^2/2LCR_L$.]

d) Use the results of parts (b) and (c) to calculate ω_m and $|\mathbf{V}_o(\omega_m)|$ for the circuit used in Example 14.7.

14.30 In the circuit in Fig. P14.30, the peak amplitude and frequency of the sinusoidal voltage source are constant. The capacitor is adjusted until the peak amplitude of v_o has its maximum value.

a) What is the expression for C in terms of R, L, and ω?

b) What is the value of $|\mathbf{V}_o|_{max}$?

c) What value of C will put the circuit into series resonance?

d) How do the answers for parts (a) and (c) compare for a high-Q circuit?

e) How does $|\mathbf{V}_o|_{max}$ compare to $|\mathbf{V}_o(\omega_0)|$ for a high-Q circuit?

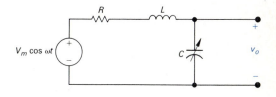

FIGURE P14.30

14.31 The frequency of the sinusoidal voltage source in the circuit in Fig. P14.31 is adjusted until the circuit is operating at unity power factor.

a) Calculate ω_0.

b) Calculate v_o at the unity power-factor frequency.

c) Calculate v_o at $\omega = 0$ rad/s.

d) Calculate v_o at $\omega = \infty$ rad/s.

e) Do you think the amplitude of v_o is maximum at $\omega = \omega_0$?

f) Calculate v_o when $\omega = 7.27$ krad/s.

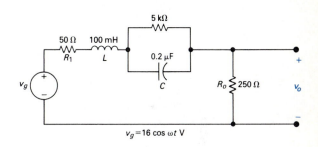

FIGURE P14.31

14.32 In the circuit shown in Fig. P14.32, the voltage appearing at the input terminals is

$$v_i = 50 \cos 20,000\pi t + 50 \cos 40,000\pi t \text{ V}.$$

The circuit capacitors C_1 and C_2 are to be chosen so that the circuit effectively transmits the 10-kHz signal and at the same time effectively blocks the 20-kHz signal.

a) Specify the numerical values of C_1 and C_2.

b) Specify the output voltage v_o.

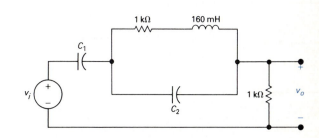

FIGURE P14.32

14.33 In the circuit in Fig. P14.33, v_g is a sinusoidal signal of fixed peak amplitude and frequency. The capacitance is varied until the peak amplitude of v_o is maximum.

a) Derive the expression for C.

b) If $R_g = 600 \ \Omega$, $R = 100 \ \Omega$, $L = 5$ mH, and $\omega = 100$ krad/s, what value of C will make the peak amplitude of v_o maximum?

c) What is the power factor of the circuit when C is adjusted so that v_o has its maximum peak amplitude?

d) For the numerical values of part (b), what is the ratio of the peak amplitude of v_o to the peak amplitude of v_g?

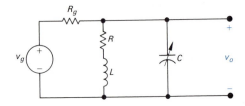

FIGURE P14.33

14.34 a) Show that if $R_L = R_C = \sqrt{L/C}$, the circuit shown in Fig. P14.34 will be in resonance for all values of ω.

b) What is the impedance seen by the current source if $R_L = R_C = \sqrt{L/C}$?

c) If $R_L = R_C = \sqrt{L/C}$ and $L = 50$ mH, $C = 2$ nF, $R_g = 1.25$ kΩ, and $i_g = 25 \cos \omega t$ mA, what is the steady-state expression for v_o?

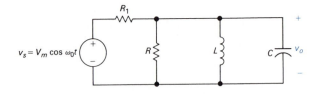

FIGURE P14.34

14.35 The definition of Q given in Eq. (14.12) is not generally useful in defining the frequency selectivity of a circuit. It is true that Eq. (14.12) gave the same result as Eq. (14.10) for the RLC circuit in Fig. 14.1. However, in general, the two definitions do not lead to the same result. The circuit in Fig. P14.35 illustrates this fact.

a) Calculate the Q of the circuit using Eq. (14.10).

b) Calculate the Q of the circuit using Eq. (14.12). {*Hint:* Note that at resonance the voltage across R is $[V_m R/(R_1 + R)] \cos \omega_0 t$

and the voltage across R_1 is $[V_m R_1/(R + R_1)] \cos \omega_0 t$.}

c) Let $R_1 \rightarrow 0$ and comment on the ability of each Q to predict the frequency selectivity of the circuit.

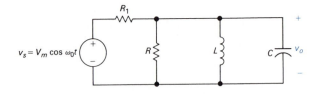

FIGURE P14.35

14.36 A universal parallel-resonant circuit has a Q of 25.

a) Find L and C.

b) Scale the circuit in part (a) so that $\omega_0 = 10$ Mrad/s and the impedance at resonance is 5 kΩ. Specify the values of R, L, and C.

14.37 a) Calculate ω_0, Q, and β for the circuit shown in Fig. P14.37.

b) Specify the values of k_m, k_f, R, L, and C so that the resonant frequency is changed to 560 krad/s and the impedance at resonance is 250 kΩ.

c) Calculate Q and β for the circuit in part (b).

FIGURE P14.37

14.38 A circuit is scaled in both magnitude and frequency.

a) Show that the Q of the scaled circuit is the same as the Q of the original circuit.

b) Show that the bandwidth of the scaled circuit is k_f times the bandwidth of the original circuit.

14.39 Scale the circuit in Fig. P14.39 so that the unity power-factor resonant frequency is 60 krad/s and the impedance seen by the ideal sinusoidal current source at resonance is 200 kΩ. Specify the scaled values of R_g, R, L, and C.

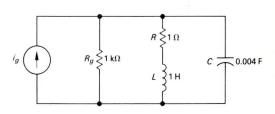

FIGURE P14.39

14.40 The impedance of a series RLC circuit is $2 + j30$ Ω at a frequency of 10 krad/s. The circuit is scaled in magnitude and frequency by the same factor, and when the impedance is measured at 4 krad/s, it is found to be $10 - j609$ Ω.

a) Calculate the original values of R, L, and C.

b) Calculate the resonant frequency, Q, and β of the scaled circuit.

14.41 Assume that the ideal op amp in the circuit in Fig. P14.41 is operating in its linear range. The sinusoidal voltage source v_g is generating the signal $100 \cos \omega t$ mV. Let the expression for v_o be written as $V_m \cos (\omega t + \theta)$.

a) At what frequency will v_o be 180° out of phase with v_g?

b) What is the value of V_m at the frequency found in part (a)?

c) At what frequencies will V_m be reduced by a factor of $1/\sqrt{2}$ from its value in part (b)?

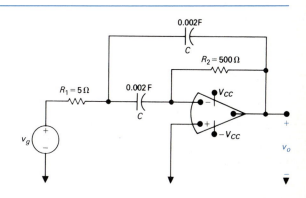

FIGURE P14.41

(continued)

d) What is the bandwidth of the circuit in radians per second?

e) What is the Q of the circuit?

f) Scale the circuit so that $R_2 = 400\ \text{k}\Omega$ and

the frequency found in part (a) is shifted to 5 krad/s. What are the values of R_1 and C?

14.42 The operational amplifier in the circuit shown in Fig. P14.42 is ideal. The circuit is energized from a variable-frequency sinusoidal voltage source. The peak amplitude of the voltage source (V_m) is held constant.

a) What happens to the magnitude of the voltage ratio V_o/V_g as $\omega \to 0$?

b) What happens to this ratio as $\omega \to \infty$?

c) At what frequency is this ratio maximum?

d) What is the maximum value of this ratio?

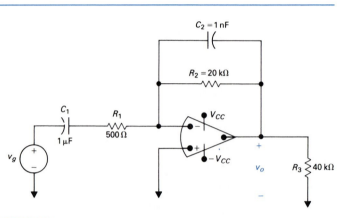

FIGURE P14.42

14.43 a) The output signal in the circuit in Fig. P14.43(a) is the voltage across the 4-Ω resistor. Calculate ω_0, Q, β, ω_1, and ω_2.

b) Repeat part (a) for the circuit shown in Fig. P14.43(b), where v_b is the output signal.

[*Hint:* Find V_b/V_g for the circuit in Fig. P14.43(b) and compare it with V_o/V_g for the circuit in Fig. P14.43(a).] Note that there is no inductor in the circuit in Fig. P14.43(b).

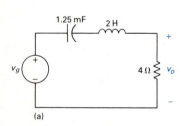

(a)

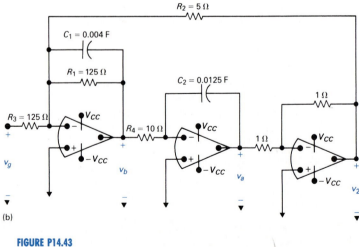

(b)

FIGURE P14.43

14.44 The operational amplifier in the circuit shown in Fig. P14.44 is ideal. If $v_g = 50 \cos \omega t$ mV, at what frequency will the operational amplifier saturate?

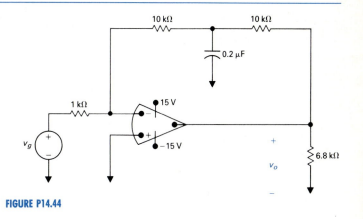

FIGURE P14.44

14.45 a) Show that if $R_1 = 1/2K$ and $R_2 = 2K$ in the circuit in Fig. P14.45 that

$$\frac{V_o}{V_g} = \frac{-2K^2}{1 + jK\left(\omega - \dfrac{1}{\omega}\right)}.$$

b) Sketch $|V_o/V_g|$ versus ω.

c) Based on the analysis of the parallel-resonant circuit in Fig. 14.1 how would you define the resonant frequency; the bandwidth; and the quality factor of the circuit in Fig. P14.45?

d) Using your definition what is the resonant frequency in rad/s?

e) Find the bandwidth of the circuit in terms of K.

f) What is the quality factor of the circuit in terms of K?

g) Design a circuit using the structure given in Fig. P14.45 to meet the following specifications: $f_0 = 1$ kHz; $C = 10$ nF; and $Q = 10$.

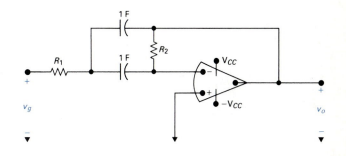

FIGURE P14.45

INTRODUCTION TO THE LAPLACE TRANSFORM

CHAPTER 15

We now introduce a powerful analytic technique that is widely used to study the behavior of linear, lumped-parameter circuits. The method is based on the Laplace transform, which we define in Section 15.1. Before doing so, we need to explain why another analytic technique is needed. In presenting the transient behavior of circuits (Chapters 8 and 9), we restricted the discussion to simple circuit structures that could be described by either a single first-order or a single second-order differential equation. We now extend the discussion to sets of simultaneous ordinary differential equations. By doing so we may consider the transient behavior of circuits whose description requires more than a single node-voltage differential equation or a single mesh-current differential equation. In Chapters 8 and 9 we also restricted sources to simple step changes. That is, the sources were allowed to change abruptly only from one dc level to another. The Laplace transform approach allows determination of the transient response when the signal sources that are switched in or out of a circuit vary with time in ways more involved than simple level jumps.

A second reason for introducing the use of the Laplace transform in circuit analysis is that it is a systematic way of relating the time-domain behavior of a circuit to its frequency-domain behavior. The ability to work simultaneously in both the time and frequency domains is greatly enhanced.

As you begin to study the Laplace transform, keep in mind that you have already used the concept of a mathematical transformation to simplify the solution of a problem. For example, the mathematical operations of multiplication and division are transformed into the simpler operations of addition and subtraction by means of the logarithm transform. Thus we use the logarithm to transform

$$A = BC \qquad \textbf{(15.1)}$$

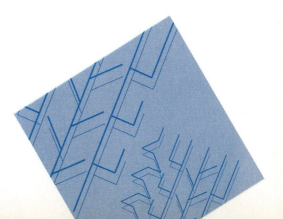

into an addition problem. Specifically,

$$\log A = \log BC = \log B + \log C. \qquad (15.2)$$

To find A, we must be able to carry out the inverse logarithm or antilogarithm operation:

$$A = \text{antilog} \, (\log B + \log C). \qquad (15.3)$$

The phasor method introduced in Chapter 10 is a second example of using a mathematical transformation to simplify the solution of a problem. There we reduced the problem of finding the steady-state sinusoidal response of a circuit to the algebraic manipulation of complex numbers. The phasor transform converted the sinusoidal signal to a complex number, which carried the information about the amplitude and phase angle of the signal. After determining the phasor value of a signal, we transformed it back to its time-domain expression.

Both of these illustrations point out the essential features of a mathematical transformation. The transformation is designed to create a new domain to make the mathematical manipulations easier to carry out. After finding the unknown in the new domain, we inverse-transform it back to the original domain. In circuit analysis we use the Laplace transform to transform a set of integrodifferential equations from the time domain to a set of algebraic equations in the frequency domain. We therefore reduce the solution for an unknown quantity to the manipulation of algebraic equations. After obtaining the frequency-domain expression for the unknown, we inverse-transform it back to the time domain.

In this chapter, we introduce the Laplace transform, discuss its pertinent characteristics, and present a systematic method for finding inverse transforms. In Chapters 16 and 17, we show how to use the Laplace transform in circuit analysis. We begin with the definition of the transform.

15.1 DEFINITION OF THE LAPLACE TRANSFORM

The Laplace transform of a function is given by the expression

$$\mathcal{L}\{f(t)\} = \int_0^\infty f(t)e^{-st} \, dt, \qquad (15.4)$$

where the symbol $\mathcal{L}\{f(t)\}$ is read "the Laplace transform of $f(t)$."

The Laplace transform of $f(t)$ also is denoted $F(s)$; that is,

$$F(s) = \mathcal{L}\{f(t)\}. \qquad (15.5)$$

This notation emphasizes that when the integral in Eq. (15.4) has been evaluated, the resulting expression is a function of s. In our applications, t represents the time domain, and, as the exponent of e in the integral of Eq. (15.4) must be dimensionless, s must have the dimension of reciprocal time, or frequency. In circuit applications, the Laplace transform transforms the problem from the time domain to the frequency domain.

Before we illustrate some of the important properties of the Laplace transform, some general comments are in order. First, note that the integral in Eq. (15.4) is improper because the upper limit is infinite. Thus we are confronted immediately with the question of whether the integral converges. That is, does a given $f(t)$ have a Laplace transform? Obviously, the functions of primary interest in engineering analysis have Laplace transforms; otherwise we would not be interested in the transform. In linear circuit analysis we excite circuits with sources that have Laplace transforms. Excitation functions such as t^t or e^{t^2}, which do not have Laplace transforms, are of no interest here.

Second, because the lower limit on the integral is zero, the Laplace transform ignores $f(t)$ for negative values of t. Or to put it another way, $F(s)$ is determined only by the behavior of $f(t)$ for positive values of t. To emphasize that the lower limit is zero, Eq. (15.4) is frequently referred to as the *one-sided,* or *unilateral,* Laplace transform. In the two-sided, or bilateral, Laplace transform, the lower limit is $-\infty$. We do not use the bilateral form here; hence $F(s)$ is understood to be the one-sided transform. Another point to be made regarding the lower limit concerns the situation when $f(t)$ has a discontinuity at the origin. If $f(t)$ is continuous at the origin, as, for example, in Fig. 15.1(a), $f(0)$ is not ambiguous. However, if $f(t)$ has a finite discontinuity at the origin, as, for example, in Fig. 15.1(b), the question arises as to whether the Laplace transform integral should include or exclude the discontinuity. That is, should we make the lower limit 0^- and include the discontinuity, or should we exclude the discontinuity by making the lower limit 0^+? (We use the notation 0^- and 0^+ to denote values of t just to the left and right of the origin, respectively.) Actually, we may choose either so long as we are consistent. For reasons explained later, we choose 0^- as the lower limit.

Because we are using 0^- as the lower limit on the Laplace transform integral, we note immediately that the integration from 0^- to 0^+ is zero *except when there is an impulse function at the origin.* (We introduce the concept of an impulse function in Section 15.3 and discuss its significance in circuit analysis in Chapter 16.) The important point now is that the two functions shown in Fig. 15.1 have the same unilateral Laplace transform because there is no impulse function at the origin.

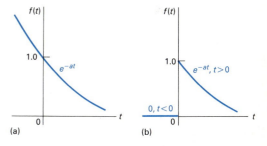

FIGURE 15.1 A continuous and discontinuous function at the origin; (a) $f(t)$ is continuous at the origin and (b) $f(t)$ is discontinuous at the origin.

The one-sided Laplace transform ignores $f(t)$ for $t < 0^-$. What happens prior to 0^- is accounted for by the initial conditions. Thus we use the Laplace transform to predict the response to a disturbance that occurs after the initial conditions have been established.

We now are ready to demonstrate the usefulness of Eq. (15.6). We begin by noting that Laplace transforms are divided into two types: *functional transforms* and *operational transforms.* The first involves finding the Laplace transform of a specific function, such as sin ωt, t, e^{-at}, and so on. The second is concerned with transforms involving mathematical operations of $f(t)$, such as finding the transform of the derivative of $f(t)$. We discuss these two types of transforms in Sections 15.3 and 15.5, respectively. Before considering functional and operational transforms, we need to introduce the step and impulse functions.

15.2 THE STEP FUNCTION

As noted in the introduction to the unilateral Laplace transform, we encounter functions that have a discontinuity, or jump, at the origin. We also know from the earlier discussion of transient behavior (Chapters 8 and 9) that switching operations create abrupt changes in currents and voltages. We accommodate these discontinuities mathematically by introducing the step and impulse functions.

Figure 15.2 illustrates the step function. It is zero for $t < 0$ and has a constant value·of K for $t > 0$. The mathematical symbol for the step function is $Ku(t)$. Thus

$$Ku(t) = 0, \qquad t < 0;$$
$$Ku(t) = K, \qquad t > 0. \tag{15.6}$$

If K is 1, the function defined by Eq. (15.6) is the unit step.

The step function is not defined at $t = 0$. In situations where we need to define the transition between 0^- and 0^+, we assume that it is linear and that

$$Ku(0) = 0.5K. \tag{15.7}$$

FIGURE 15.2 The step function.

As before, 0^- and 0^+ represent symmetric points arbitrarily close to the left and right of the origin. Figure 15.3 illustrates the linear transition from 0^- and 0^+.

A step that occurs at $t = a$ is expressed as $Ku(t - a)$. Thus

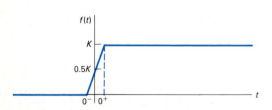

FIGURE 15.3 The linear approximation to the step function.

$$Ku(t - a) = 0, \qquad t < a;$$
$$Ku(t - a) = K, \qquad t > a. \tag{15.8}$$

If $a > 0$, the step occurs to the right of the origin, and if $a < 0$, the step occurs to the left of the origin. Figure 15.4 illustrates Eq. (15.8). Note that the step function is 0 when the argument $t - a$ is negative and is K when the argument is positive.

A step function that is equal to K for $t < a$ and 0 for $t > a$ is written as $Ku(a - t)$. Thus

$$Ku(a - t) = K, \qquad t < a;$$
$$Ku(a - t) = 0, \qquad t > a. \tag{15.9}$$

The discontinuity is to the left of the origin when $a < 0$ and to the right of the origin when $a > 0$. Equation (15.9) is shown in Fig. 15.5.

One application of the step function is to use it to write the mathematical expression for a function that is of finite duration, as illustrated in Example 15.1. We introduce other applications in relation to the Laplace transform method in the work that follows.

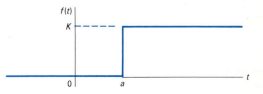

FIGURE 15.4 A step function occurring at $t = a$ when $a > 0$.

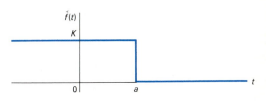

FIGURE 15.5 The step function $Ku(a - t)$ for $a > 0$.

E X A M P L E   15.1

Use step functions to write an expression for the function illustrated in Fig. 15.6.

S O L U T I O N

The function shown in Fig. 15.6 is made up of linear segments with break points at 0, 1, 3, and 4 s. To construct this function we must add and subtract linear functions of the proper slope. We use the step function to initiate these linear segments at the proper time. Figure 15.7 shows the breakdown of the function in Fig. 15.6 into its linear segments. We use the step function to initiate a straight line with the following slopes: $+2$ at $t = 0$; -4 at $t = 1$; $+4$ at $t = 3$; and -2 at $t = 4$. Note that when a straight line with a slope of -4 is added to a straight line with a slope of $+2$, the resulting line has a slope of -2. The step function is used to start this addition at $t = 1$. At 3 s we must add back a straight line with a slope of $+4$ to give a straight line with a slope of $+2$. Finally, we make the function identically zero for all $t \geq 4$ s by adding in a straight line with a slope of -2 at $t = 4$ s. The expression for $f(t)$ is

$$f(t) = 2tu(t) - 4(t - 1)u(t - 1) + 4(t - 3)u(t - 3)$$
$$- 2(t - 4)u(t - 4).$$

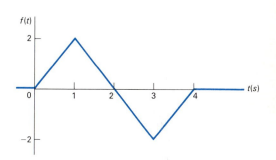

FIGURE 15.6 The function for Example 15.1.

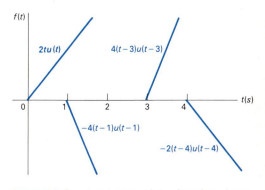

FIGURE 15.7 Decomposition of the function shown in Fig. 15.6.

DRILL EXERCISES

15.1 Use step functions to write the expression for each function shown.

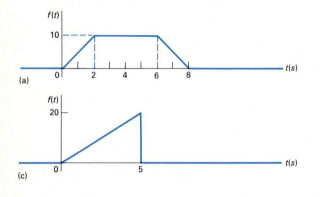

(a)

(c)

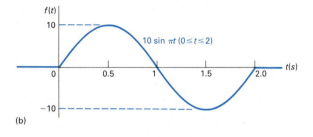

(b)

ANSWER:
(a) $f(t) = 5[tu(t) - (t - 2)u(t - 2) - (t - 6)u(t - 6) + (t - 8)u(t - 8)]$;
(b) $f(t) = 10 \sin \pi t[u(t) - u(t - 2)]$;
(c) $f(t) = 4tu(t) - 4(t - 5)u(t - 5) - 20u(t - 5)$.

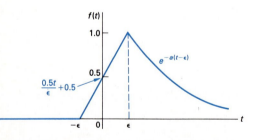

FIGURE 15.8 A magnified view of the discontinuity in Fig. 15.1(b), assuming a linear transition between $-\epsilon$ and $+\epsilon$.

15.3 THE IMPULSE FUNCTION

When we have a finite discontinuity in a function, such as that illustrated in Fig. 15.1(b), the derivative of the function is not defined at the point of the discontinuity. The concept of an impulse function enables us to define the derivative at a discontinuity. That is, we can define the Laplace transform of the derivative of a function that has a discontinuity. We can also define the transform of the higher-order derivatives of the function. As we demonstrate later, the impuse function is a useful concept in circuit analysis.

In order to define the derivative of a function at a discontinuity, we first assume that the function varies linearly across the discontinuity, as shown in Fig. 15.8, where we observe that as $\epsilon \to 0$ an abrupt discontinuity occurs at the origin. When we differentiate the function shown in Fig. 15.8, the derivative between $-\epsilon$ and $+\epsilon$ is constant at a value of $1/2\epsilon$. For $t < -\epsilon$ the derivative is zero and for $t > \epsilon$ the derivative is $-ae^{-a(t-\epsilon)}$. Figure 15.9 shows these observations graphically.

As ϵ approaches zero, the value of $f'(t)$ between $\pm\epsilon$ approaches infinity. At the same time that the value of $f'(t)$ is becoming infinite, the duration of this large value is approaching zero. Furthermore, the area under $f'(t)$ between $\pm\epsilon$ remains constant as $\epsilon \to 0$. In this case, the area is unity. As ϵ approaches zero, we say that the function between $\pm\epsilon$ approaches a *unit impulse function*, denoted $\delta(t)$. Thus the derivative of $f(t)$ at the origin approaches a unit impulse function as ϵ approaches zero, or

$$f'(0) \to \delta(t) \qquad \text{as} \qquad \epsilon \to 0.$$

The impulse function is referred to as a *unit* impulse in this case because the *area* under the impulse-generating curve is unity. The strength of the impulse function is determined by this area. An impulse of strength K is denoted $K\,\delta(t)$, where K is the area under the impulse-generating function. The impulse function also is known as the *Dirac delta function*.

We summarize the creation of an impulse function as follows. An impulse function is created by defining a function in terms of a variable parameter and then allowing this parameter to approach zero. The variable-parameter function generates an impulse if it exhibits the following three characteristics as the parameter approaches zero:

1. the amplitude approaches infinity,

2. the duration of the function approaches zero, and

3. the area under the variable-parameter function is constant as the parameter changes.

Another example of a function that generates an impulse function is the exponential function:

$$f(t) = \frac{K}{2\epsilon}\, e^{-|t|/\epsilon}. \tag{15.10}$$

As ϵ approaches zero, the function becomes infinite at the origin and at the same time decays to zero in an infinitesimal length of time. Figure 15.10 illustrates the character of $f(t)$ as $\epsilon \to 0$. To show that an impulse function is created as $\epsilon \to 0$, we must also show that the area under the function is independent of ϵ. Thus

$$\text{Area} = \int_{-\infty}^{0} \frac{K}{2\epsilon} e^{t/\epsilon}\, dt + \int_{0}^{\infty} \frac{K}{2\epsilon} e^{-t/\epsilon}\, dt$$

$$= \frac{K}{2\epsilon} \cdot \frac{e^{t/\epsilon}}{1/\epsilon}\bigg|_{-\infty}^{0} + \frac{K}{2\epsilon} \cdot \frac{e^{-t/\epsilon}}{-1/\epsilon}\bigg|_{0}^{\infty}$$

$$= \frac{K}{2} + \frac{K}{2} = K, \tag{15.11}$$

which tells us that the area under the curve is constant and equal to K units. Therefore as $\epsilon \to 0$, $f(t) \to K\,\delta(t)$.

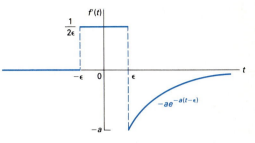

FIGURE 15.9 The derivative of the function shown in Fig. 15.8.

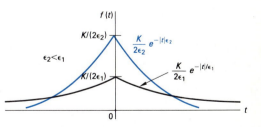

FIGURE 15.10 An impulse-generating function.

Mathematically, the concept of the impulse function is

$$\int_{-\infty}^{\infty} K\,\delta(t)\,dt = K; \tag{15.12}$$

$$\delta(t) = 0, \qquad t \neq 0. \tag{15.13}$$

Equation (15.12) states that the area under the impulse function is constant. This area represents the strength of the impulse. Equation (15.13) states that the impulse is zero everywhere except at $t = 0$. An impulse that occurs at $t = a$ is denoted $K\,\delta(t - a)$.

The graphic symbol for the impulse function is simply an arrow. The strength of the impulse is given parenthetically next to the head of the arrow. Figure 15.11 shows the impulses $K\,\delta(t)$ and $K\,\delta(t - a)$.

An important property of the impulse function is the *sifting property*, which is expressed as

$$\int_{-\infty}^{\infty} f(t)\,\delta(t - a)\,dt = f(a), \tag{15.14}$$

where the function $f(t)$ is assumed to be continuous at $t = a$, that is, at the location of the impulse. Equation (15.14) shows that the impulse function sifts out everything except the value of $f(t)$ at $t = a$. The validity of Eq. (15.14) follows from noting that $\delta(t - a)$ is zero everywhere except at $t = a$, and hence the integral can be written

$$I = \int_{-\infty}^{\infty} f(t)\,\delta(t - a)\,dt = \int_{a-\epsilon}^{a+\epsilon} f(t)\,\delta(t - a)\,dt. \tag{15.15}$$

But as $f(t)$ is continuous at a, it takes on the value $f(a)$ as $t \to a$, so

$$I = \int_{a-\epsilon}^{a+\epsilon} f(a)\,\delta(t - a)\,dt = f(a) \int_{a-\epsilon}^{a+\epsilon} \delta(t - a)\,dt = f(a). \tag{15.16}$$

We use the sifting property of the impulse function to find its Laplace transform:

$$\mathcal{L}\{\delta(t)\} = \int_{0^-}^{\infty} \delta(t)e^{-st}\,dt = \int_{0^-}^{\infty} \delta(t)\,dt = 1, \tag{15.17}$$

which is an important Laplace transform pair that we make good use of in applying the transform to circuit analysis.

We also can define the derivatives of the impulse function and the Laplace transform of these derivatives. We discuss the first derivative, along with its transform, and then state the result for the higher-order derivatives.

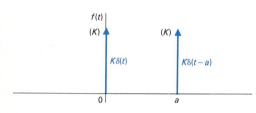

FIGURE 15.11 Graphic representation of the impulse $K\,\delta(t)$ and $K\,\delta(t - a)$.

The function illustrated in Fig. 15.12(a) generates an impulse function as $\epsilon \rightarrow 0$. Figure 15.12(b) shows the derivative of this impulse-generating function, which is defined as the derivative of the impulse $[\delta'(t)]$ as $\epsilon \rightarrow 0$. The derivative of the impulse function sometimes is referred to as a *moment function,* or *unit doublet.*

To find the Laplace transform of $\delta'(t)$ we simply apply the defining integral to the function shown in Fig. 15.12(b), and after integrating, let $\epsilon \rightarrow 0$. Then

$$\mathcal{L}\{\delta'(t)\} = \lim_{\epsilon \to 0} \left[\int_{-\epsilon}^{0^-} \frac{1}{\epsilon^2} e^{-st}\, dt + \int_{0^+}^{\epsilon} \left(-\frac{1}{\epsilon^2}\right) e^{-st}\, dt \right]$$

$$= \lim_{\epsilon \to 0} \frac{e^{s\epsilon} + e^{-s\epsilon} - 2}{s\epsilon^2}$$

$$= \lim_{\epsilon \to 0} \frac{se^{s\epsilon} - se^{-s\epsilon}}{2\epsilon s}$$

$$= \lim_{\epsilon \to 0} \frac{s^2 e^{s\epsilon} + s^2 e^{-s\epsilon}}{2s}$$

$$= s. \tag{15.18}$$

In deriving Eq. (15.18), we had to use l'Hôpital's rule twice to evaluate the indeterminate form $0/0$.

Higher-order derivatives may be generated in a manner similar to that used to generate the first derivative (see Problem 15.5), and the defining integral may then be used to find its Laplace transform. For the nth derivative of the impulse function, we find that its Laplace transform simply is s^n; that is,

$$\mathcal{L}\{\delta^{(n)}(t)\} = s^n. \tag{15.19}$$

Finally, an impulse function can be thought of as a derivative of a step function; that is,

$$\delta(t) = \frac{du(t)}{dt}. \tag{15.20}$$

Figure 15.13 presents the graphic interpretation of Eq. (15.20). The function shown in Fig. 15.13(a) approaches a unit step function as $\epsilon \rightarrow 0$. The function shown in Fig. 15.13(b)—the derivative of the function illustrated in Fig. 15.13(a)—approaches a unit impulse as $\epsilon \rightarrow 0$.

The impulse function is an extremely useful concept in circuit analysis, and we say more about it in the following chapters. We introduced the concept here so that we can include discontinuities at the origin in our definition of the Laplace transform.

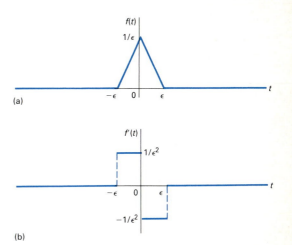

FIGURE 15.12 The first derivative of the impulse function: (a) impulse-generating function used to define the first derivative of the impulse; (b) first derivative of the impulse-generating function that approaches $\delta'(t)$ as $\epsilon \rightarrow 0$.

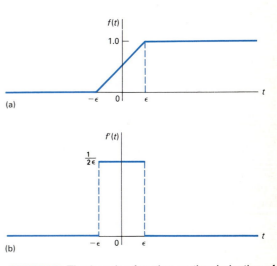

FIGURE 15.13 The impulse function as the derivative of the step function: (a) $f(t) \rightarrow u(t)$ as $\epsilon \rightarrow 0$; (b) $f'(t) \rightarrow \delta(t)$ as $\epsilon \rightarrow 0$.

DRILL EXERCISES

15.2 a) Find the area under the function shown in Fig. 15.12(a).

b) What is the duration of the function when $\epsilon = 0$?

c) What is the magnitude of $f(t)$ when $\epsilon = 0$?

ANSWER: (a) 1; (b) 0; (c) ∞.

15.4 Find $f(t)$ if

$$f(t) = \frac{1}{2\pi} \int_{-\infty}^{\infty} F(\omega)e^{jt\omega} \, d\omega$$

and

$$F(\omega) = \frac{3 + j\omega}{4 + j\omega} \pi \, \delta(\omega).$$

ANSWER: 3/8.

15.3 Evaluate the following integrals:

a) $I = \int_{-2}^{4} (t^3 + 4)[\delta(t) + 4\delta(t - 2)] \, dt$;

b) $I = \int_{-3}^{4} t^2[\delta(t) + \delta(t + 2.5) + \delta(t - 5)] \, dt$.

ANSWER: (a) 52; (b) 6.25.

15.4 FUNCTIONAL TRANSFORMS

A functional transform is simply the Laplace transform of a specified function of t. Because we are limiting our introduction to the unilateral, or one-sided, Laplace transform, we define all functions to be zero for $t < 0^-$. That is, $f(t)$ is uniquely defined by stating how $f(t)$ varies for $t > 0^-$.

We derived one functional transform pair in Section 15.3, where we showed that the Laplace transform of the unit impulse function equals 1, that is, Eq. (15.17). A second illustration is the unit step function of Fig. 15.13(a), where

$$\mathcal{L}\{u(t)\} = \int_{0^-}^{\infty} f(t)e^{-st} \, dt = \int_{0^+}^{\infty} 1e^{-st} \, dt$$

$$= \frac{e^{-st}}{-s}\Bigg|_{0^+}^{\infty} = \frac{1}{s}. \tag{15.21}$$

Equation (15.21) shows that the Laplace transform of the unit step function is $1/s$.

The Laplace transform of the decaying exponential function

shown in Fig. 15.14 is

$$\mathcal{L}\{e^{-at}\} = \int_{0^+}^{\infty} e^{-at}e^{-st}\, dt = \int_{0^+}^{\infty} e^{-(a+s)t}\, dt = \frac{1}{s+a}. \quad \textbf{(15.22)}$$

In deriving Eqs. (15.21) and (15.22) we used the fact that integration across the discontinuity at the origin is zero.

A third illustration of finding a functional transform is the sinusoidal function shown in Fig. 15.15. The expression for $f(t)$ for $t > 0^-$ is $\sin \omega t$; hence the Laplace transform is

$$\mathcal{L}\{\sin \omega t\} = \int_{0}^{\infty} (\sin \omega t)e^{-st}\, dt$$

$$= \int_{0^-}^{\infty} \left(\frac{e^{j\omega t} - e^{-j\omega t}}{2j} \right) e^{-st}\, dt$$

$$= \int_{0^-}^{\infty} \frac{e^{-(s-j\omega)t} - e^{-(s+j\omega)t}}{2j}\, dt$$

$$= \frac{1}{2j} \left(\frac{1}{s - j\omega} - \frac{1}{s + j\omega} \right)$$

$$= \frac{\omega}{s^2 + \omega^2}. \quad \textbf{(15.23)}$$

Table 15.1 gives an abbreviated list of Laplace transform pairs. It includes the functions of most interest in an introductory course on circuit applications.

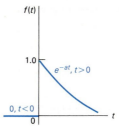

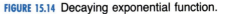

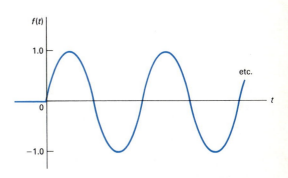

FIGURE 15.14 Decaying exponential function.

FIGURE 15.15 A sinusoidal function for $t > 0$.

TABLE 15.1

AN ABBREVIATED LIST OF LAPLACE TRANSFORM PAIRS

$f(t)\,(t > 0^-)$	TYPE	$F(s)$
$\delta(t)$	(impulse)	1
$u(t)$	(step)	$\dfrac{1}{s}$
t	(ramp)	$\dfrac{1}{s^2}$
e^{-at}	(exponential)	$\dfrac{1}{s+a}$
$\sin \omega t$	(sine)	$\dfrac{\omega}{s^2 + \omega^2}$
$\cos \omega t$	(cosine)	$\dfrac{s}{s^2 + \omega^2}$
te^{-at}	(damped ramp)	$\dfrac{1}{(s+a)^2}$
$e^{-at} \sin \omega t$	(damped sine)	$\dfrac{\omega}{(s+a)^2 + \omega^2}$
$e^{-at} \cos \omega t$	(damped cosine)	$\dfrac{s+a}{(s+a)^2 + \omega^2}$

DRILL EXERCISES

15.5 Use the defining integral to find the Laplace transform of (a) t and (b) $\cosh \beta t$.

ANSWER: (a) $1/s^2$; (b) $s/(s^2 - \beta^2)$.

15.5 OPERATIONAL TRANSFORMS

Operational transforms indicate how mathematical operations performed on either $f(t)$ or $F(s)$ are translated into the opposite domain. The operations of primary interest are (1) multiplication by a constant; (2) addition (subtraction); (3) differentiation; (4) integration; (5) translation in the time domain; (6) translation in the frequency domain; and (7) scale changing.

MULTIPLICATION BY A CONSTANT

From the defining integral, if

$$\mathcal{L}\{f(t)\} = F(s),$$

then

$$\mathcal{L}\{Kf(t)\} = KF(s). \qquad (15.24)$$

Thus multiplication of $f(t)$ by a constant corresponds to multiplying $F(s)$ by the same constant.

ADDITION (SUBTRACTION)

Addition (subtraction) in the time domain translates into addition (subtraction) in the frequency domain. Thus if

$$\mathcal{L}\{f_1(t)\} = F_1(s),$$

$$\mathcal{L}\{f_2(t)\} = F_2(s),$$

and

$$\mathcal{L}\{f_3(t)\} = F_3(s),$$

then

$$\mathcal{L}\{f_1(t) + f_2(t) - f_3(t)\} = F_1(s) + F_2(s) - F_3(s), \qquad (15.25)$$

which is derived by simply substituting the algebraic sum of time-domain functions into the defining integral.

DIFFERENTIATION

Differentiation in the time domain corresponds to multiplying $F(s)$ by s and then subtracting the initial value of $f(t)$, that is, $f(0^-)$, from this product:

$$\mathcal{L}\left\{\frac{df(t)}{dt}\right\} = sF(s) - f(0^-), \qquad \textbf{(15.26)}$$

which is obtained directly from the definition of the Laplace transform, or

$$\mathcal{L}\left\{\frac{df(t)}{dt}\right\} = \int_{0^-}^{\infty}\left[\frac{df(t)}{dt}\right]e^{-st}\,dt. \qquad \textbf{(15.27)}$$

We evaluate the integral in Eq. (15.27) by integrating by parts. Letting $u = e^{-st}$ and $dv = [df(t)/dt]dt$ yields

$$\mathcal{L}\left\{\frac{df(t)}{dt}\right\} = e^{-st}f(t)\,\Big|_{0^-}^{\infty} - \int_{0^-}^{\infty}f(t)(-se^{-st}\,dt). \qquad \textbf{(15.28)}$$

Because we are assuming that $f(t)$ is Laplace transformable, the evaluation of $e^{-st}f(t)$ at $t = \infty$ is zero. Therefore the right-hand side of Eq. (15.28) reduces to

$$-f(0^-) + s\int_{0^-}^{\infty}f(t)e^{-st}\,dt \qquad \text{or} \qquad sF(s) - f(0^-).$$

This observation completes the derivation of Eq. (15.26). It is an important result because it states that differentiation in the time domain reduces to an algebraic operation in the s domain.

We determine the Laplace transform of higher-order derivatives by using Eq. (15.26) as the starting point. For example, to find the Laplace transform of the second derivative of $f(t)$, we first let

$$g(t) = \frac{df(t)}{dt}. \qquad \textbf{(15.29)}$$

Now we use Eq. (15.26) to write

$$G(s) = sF(s) - f(0^-). \qquad \textbf{(15.30)}$$

But as

$$\frac{dg(t)}{dt} = \frac{d^2f(t)}{dt^2},$$

we write

$$\mathcal{L}\left\{\frac{dg(t)}{dt}\right\} = \mathcal{L}\left\{\frac{d^2f(t)}{dt^2}\right\} = sG(s) - g(0^-). \qquad \textbf{(15.31)}$$

Substituting Eq. (15.30) into the right-hand side of Eq. (15.31) gives

$$\mathcal{L}\left\{\frac{d^2 f(t)}{dt^2}\right\} = s^2 F(s) - sf(0^-) - \frac{df(0^-)}{dt}. \qquad \textbf{(15.32)}$$

In writing Eq. (15.32), we also used Eq. (15.29) to express the value of $g(0^-)$.

We find the Laplace transform of the nth derivative by successively applying the preceding thought process, which leads to the general result

$$\mathcal{L}\left\{\frac{d^n f(t)}{dt^n}\right\} = s^n F(s) - s^{n-1} f(0^-) - s^{n-2}\frac{df(0^-)}{dt}$$

$$- s^{n-3}\frac{d^2 f(0^-)}{dt^2} - \cdots - \frac{d^{n-1} f(0^-)}{dt^{n-1}}. \qquad \textbf{(15.33)}$$

INTEGRATION

Integration in the time domain corresponds to dividing by s in the s domain. As before, we establish the relationship by the defining integral:

$$\mathcal{L}\left\{\int_{0^-}^{t} f(x)\, dx\right\} = \int_{0^-}^{\infty}\left[\int_{0^-}^{t} f(x)\, dx\right] e^{-st}\, dt. \qquad \textbf{(15.34)}$$

We evaluate the integral on the right-hand side of Eq. (15.34) by integrating by parts, first letting

$$u = \int_{0^-}^{t} f(x)\, dx \qquad \text{and} \qquad dv = e^{-st}\, dt.$$

Then

$$du = f(t)\, dt \qquad \text{and} \qquad v = -\frac{e^{-st}}{s}.$$

The integration-by-parts formula yields

$$\mathcal{L}\left\{\int_{0^-}^{t} f(x)\, dx\right\} = uv\,\Big|_{0^-}^{\infty} - \int_{0^-}^{\infty} v\, du$$

$$= -\frac{e^{-st}}{s}\int_{0^-}^{t} f(x)\, dx\,\Big|_{0^-}^{\infty} + \int_{0^-}^{\infty}\frac{e^{-st}}{s} f(t)\, dt. \qquad \textbf{(15.35)}$$

The first term on the right-hand side of Eq. (15.35) is zero at both the upper and lower limits. The evaluation at the lower limit obviously is zero, whereas the evaluation at the upper limit is zero because we are assuming that $f(t)$ has a Laplace transform. The second term on the right-hand side of Eq. (15.35) is

$F(s)/s$; therefore

$$\mathcal{L}\left\{\int_{0^-}^{t} f(x)\,dx\right\} = \frac{F(s)}{s}, \qquad \textbf{(15.36)}$$

which reveals that the operation of integration in the time domain is transformed to the algebraic operation of multiplying by $1/s$ in the s domain.

Equations (15.33) and (15.36) form the basis of the earlier statement that the Laplace transform translates a set of integrodifferential equations into a set of algebraic equations.

TRANSLATION IN THE TIME DOMAIN

Translation in the time domain corresponds to multiplication by an exponential in the frequency domain. Thus

$$\mathcal{L}\{f(t-a)u(t-a)\} = e^{-as}F(s), \qquad a > 0. \qquad \textbf{(15.37)}$$

For example, knowing that

$$\mathcal{L}\{t\} = \frac{1}{s^2},$$

Eq. (15.37) permits writing the Laplace transform of $(t-a)u(t-a)$ directly:

$$\mathcal{L}\{(t-a)u(t-a)\} = \frac{e^{-as}}{s^2}.$$

The proof of Eq. (15.37) follows from the defining integral:

$$\mathcal{L}\{f(t-a)u(t-a)\} = \int_{0^-}^{\infty} u(t-a)f(t-a)e^{-st}\,dt$$

$$= \int_{a}^{\infty} f(t-a)e^{-st}\,dt. \qquad \textbf{(15.38)}$$

In writing Eq. (15.38), we took advantage of $u(t-a) = 1$ for $t > a$. Now we change the variable of integration. Specifically, we let $x = t - a$. Then $x = 0$ when $t = a$, $x = \infty$ when $t = \infty$, and $dx = dt$. Thus we write the integral in Eq. (15.38) as

$$\mathcal{L}\{f(t-a)u(t-a)\} = \int_{0}^{\infty} f(x)e^{-s(x+a)}\,dx$$

$$= e^{-sa}\int_{0}^{\infty} f(x)e^{-sx}\,dx$$

$$= e^{-as}F(s),$$

which is what we set out to prove.

TRANSLATION IN THE FREQUENCY DOMAIN

Translation in the frequency domain corresponds to multiplication by an exponential in the time domain:

$$\mathcal{L}\{e^{-at}f(t)\} = F(s + a), \tag{15.39}$$

which follows from the defining integral. Derivation of Eq. (15.39) is left to Problem 15.12.

We may use the relationship in Eq. (15.39) to derive new transform pairs. Thus knowing that

$$\mathcal{L}\{\cos \omega t\} = \frac{s}{s^2 + \omega^2},$$

we use Eq. (15.39) to deduce that

$$\mathcal{L}\{e^{-at} \cos \omega t\} = \frac{s + a}{(s + a)^2 + \omega^2}.$$

SCALE CHANGING

The scale-change property gives the relationship between $f(t)$ and $F(s)$ when the time variable is multiplied by a positive constant:

$$\mathcal{L}\{f(at)\} = \frac{1}{a}F\left(\frac{s}{a}\right), \qquad a > 0, \tag{15.40}$$

the derivation of which is left to Problem 15.16.

The scale-change property is particularly useful in experimental work, especially where time-scale changes are made to facilitate building a model of a system.

We use Eq. (15.40) to formulate new transform pairs. Thus knowing that

$$\mathcal{L}\{\cos t\} = \frac{s}{s^2 + 1},$$

we deduce from Eq. (15.40) that

$$\mathcal{L}\{\cos \omega t\} = \frac{1}{\omega} \frac{s/\omega}{(s/\omega)^2 + 1}$$

$$= \frac{s}{s^2 + \omega^2}.$$

Table 15.2 gives an abbreviated list of operational transforms. Some entries were not discussed in this section, but you will become more familiar with them by working Problems 15.17 and 15.18.

TABLE 15.2

AN ABBREVIATED LIST OF OPERATIONAL TRANSFORMS

$f(t)$	$F(s)$
$Kf(t)$	$KF(s)$
$f_1(t) + f_2(t) - f_3(t) + \cdots$	$F_1(s) + F_2(s) - F_3(s) + \cdots$
$\dfrac{df(t)}{dt}$	$sF(s) - f(0^-)$
$\dfrac{d^2f(t)}{dt^2}$	$s^2F(s) - sf(0^-) - \dfrac{df(0^-)}{dt}$
$\dfrac{d^nf(t)}{dt^n}$	$s^nF(s) - s^{n-1}f(0^-) - s^{n-2}\dfrac{df(0^-)}{dt}$
	$\quad -s^{n-3}\dfrac{d^2f(0^-)}{dt} - \cdots - \dfrac{d^{n-1}f(0^-)}{dt}$
$\displaystyle\int_0^t f(x)\,dx$	$\dfrac{F(s)}{s}$
$f(t - a)u(t - a),\ a > 0$	$e^{-as}F(s)$
$e^{-at}f(t)$	$F(s + a)$
$f(at),\ a > 0$	$\dfrac{1}{a}F\!\left(\dfrac{s}{a}\right)$
$tf(t)$	$-\dfrac{dF(s)}{ds}$
$t^nf(t)$	$(-1)^n\dfrac{d^nF(s)}{ds^n}$
$\dfrac{f(t)}{t}$	$\displaystyle\int_s^\infty F(u)\,du$

DRILL EXERCISES

15.6 Use the appropriate operational transform from Table 15.2 to find the Laplace transform of each function:

(a) t^2e^{-at};   (b) $\dfrac{d}{dt}(e^{-at}\cosh \beta t)$;

(c) $t\cos \omega t$.

ANSWER: (a) $\dfrac{2}{(s + a)^3}$; (b) $\dfrac{s(s + a)}{(s + a)^2 - \beta^2}$;

(c) $\dfrac{s^2 - \omega^2}{(s^2 + \omega^2)^2}$.

15.6 AN ILLUSTRATIVE EXAMPLE

We now illustrate how to use the Laplace transform to solve the ordinary integrodifferential equations that describe the behavior of lumped-parameter circuits. Consider the circuit shown in Fig.

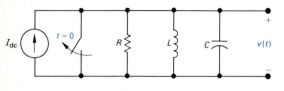

FIGURE 15.16 A parallel *RLC* circuit.

15.16. We assume that no initial energy is stored in the circuit at the instant when the switch, which is shorting the dc current source, is opened. The problem is to find the time-domain expression for $v(t)$ when $t \geq 0$.

We begin by writing the integrodifferential equation that $v(t)$ must satisfy. We need only a single node-voltage equation to describe the circuit. Summing the currents away from the top node in the circuit generates the equation:

$$\frac{v(t)}{R} + \frac{1}{L} \int_0^t v(x)\,dx + C\frac{dv(t)}{dt} = I_{dc}u(t). \qquad \textbf{(15.41)}$$

Note that in writing Eq. (15.41) we indicated the opening of the switch in the step jump of the source current from zero to I_{dc}.

After deriving the integrodifferential equations (in this example, just one), we transform the equations to the s domain. In transforming Eq. (15.41), we use three operational transforms and one functional transform to obtain

$$\frac{V(s)}{R} + \frac{1}{L}\frac{V(s)}{s} + C[sV(s) - v(0^-)] = I_{dc}\left(\frac{1}{s}\right), \qquad \textbf{(15.42)}$$

an algebraic equation where $V(s)$ is the unknown variable. We are assuming that the circuit parameters R, L, and C, as well as the source current I_{dc}, are known. [The initial voltage on the capacitor $v(0^-)$ is zero because the initial energy stored in the circuit is zero.] Thus we reduced the problem to solving an algebraic equation. In general, of course, the problem is reduced to solving a set of algebraic equations.

Next we solve the algebraic equations for the unknowns. That is, we solve Eq. (15.42) for $V(s)$:

$$V(s)\left(\frac{1}{R} + \frac{1}{sL} + sC\right) = \frac{I_{dc}}{s}$$

$$= \frac{I_{dc}/C}{s^2 + (1/RC)s + (1/LC)}. \qquad \textbf{(15.43)}$$

To find $v(t)$ we must inverse-transform the expression for $V(s)$. We denote this inverse operation

$$v(t) = \mathcal{L}^{-1}\{V(s)\}. \qquad \textbf{(15.44)}$$

We now determine the inverse transform of the s-domain unknowns. Finding the inverse transform of s-domain expressions such as that in Eq. (15.43) is the subject of Section 15.7. However, before proceeding to find inverse transforms, we need to mention an additional step in the procedure.

This last step is not unique to the Laplace transform method. It is one that conscientious and prudent engineers always incorporate into their analyses—namely, to test any derived solution to be sure that it makes sense in terms of the known behavior of the system being analyzed. We demonstrate this aspect of analysis in the following material.

Simplifying the notation now is advantageous. We do so by dropping the parenthetical t in time-domain expressions and the parenthetical s in frequency-domain expressions. We imply the time domain by using lowercase letters for all time-domain variables and represent the corresponding s-domain variables by uppercase letters. Thus

$$\mathscr{L}\{v\} = V \quad \text{or} \quad v = \mathscr{L}^{-1}\{V\},$$

$$\mathscr{L}\{i\} = I \quad \text{or} \quad i = \mathscr{L}^{-1}\{I\},$$

$$\mathscr{L}\{f\} = F \quad \text{or} \quad f = \mathscr{L}^{-1}\{F\},$$

and so on.

DRILL EXERCISES

15.7 In the circuit shown in Fig. 15.16, the dc current source is replaced with a sinusoidal source that delivers a current of $1.2 \cos t$ A. The circuit components are $R = 1\ \Omega$, $C = 0.625$ F, and $L = 1.6$ H. Find the numerical expression for V.

ANSWER: $V = \dfrac{1.92s^2}{(s^2 + 1.6s + 1)(s^2 + 1)}$.

15.7 INVERSE TRANSFORMS

The expression for $V(s)$ in Eq. (15.43) is a *rational* function of s—that is, one that can be expressed in the form of a ratio of two polynomials in s such that no nonintegral powers of s appear in the polynomials. This observation about Eq. (15.43) is important because for linear, lumped-parameter, time-invariant circuits, the s-domain expressions for the unknown voltages and currents always are rational functions of s. (You may verify this observation by working Problems 15.19, 15.21, and 15.26.) Therefore, if we can inverse-transform rational functions of s, we can solve for the time-domain expressions for the voltages and currents. The purpose of this section is to present a straight-forward and systematic technique for finding the inverse transform of a rational function.

In general, we face the problem of finding the inverse transform of a quotient of two polynomials in s, that is, a function that has the form:

$$F(s) = \frac{N(s)}{D(s)} = \frac{a_n s^n + a_{n-1} s^{n-1} + \cdots + a_1 s + a_0}{b_m s^m + b_{m-1} s^{m-1} + \cdots + b_1 s + b_0}. \quad \textbf{(15.45)}$$

The terms a and b are real constants, and the terms m and n are positive integers. The ratio $N(s)/D(s)$ is called a *proper* rational function if $m > n$ and an *improper* rational function if $m \leq n$. Only a proper rational function can be expanded as a sum of partial fractions. This restriction poses no problem, as we show at the end of this section.

PARTIAL FRACTION EXPANSION: PROPER RATIONAL FUNCTIONS

A proper rational function is expanded into a sum of partial fractions by writing a term or a series of terms for each root of $D(s)$. Thus $D(s)$ must be in factored form before we can make a partial-fraction expansion. For each distinct root of $D(s)$, a single term appears in the sum of partial fractions. For each multiple root of $D(s)$ of multiplicity r, the expansion contains r terms. For example, in the rational function

$$\frac{s + 6}{s(s + 3)(s + 1)^2},$$

the denominator has four roots. Two of these roots are distinct—namely, at $s = 0$ and $s = -3$; and a multiple root of multiplicity 2 occurs at $s = -1$. Thus the partial fraction expansion of this function takes the form:

$$\frac{s + 6}{s(s + 3)(s + 1)^2} \equiv \frac{K_1}{s} + \frac{K_2}{s + 3} + \frac{K_3}{(s + 1)^2} + \frac{K_4}{s + 1}. \quad \textbf{(15.46)}$$

Note that the key to the partial fraction technique for finding inverse transforms lies in recognizing the $f(t)$ corresponding to each term in the sum of partial fractions. From Table 15.1 (p. 611), you should be able to verify that

$$\mathcal{L}^{-1}\left\{\frac{s + 6}{s(s + 3)(s + 1)^2}\right\} = (K_1 + K_2 e^{-3t} + K_3 t e^{-t} + K_4 e^{-t}) u(t).$$

$$\textbf{(15.47)}$$

All that remains to be done is to establish a technique for determining the coefficients $(K_1, K_2, K_3, \ldots)$ generated by making a partial fraction expansion. We break this problem into three steps. First, we consider the case when the roots of $D(s)$ are real and distinct. Second, we consider the case when some of the roots of $D(s)$ are complex and distinct. Finally, we consider the case when $D(s)$ has repeated roots. Before we do so, however, some general comments are in order.

We used the identity sign $\equiv$ in Eq. (15.46) to emphasize that the operation of expanding a rational function into a sum of partial fractions establishes an identical equation. Thus both sides of the equation must be the same for all values of the variable s. The identity relationship must hold when both sides are subjected to the same mathematical operation. Therefore differentiating both sides or multiplying both sides by the same quantity preserves the identity. These characteristics of an identical equation are pertinent to determining the coefficients. Also, be sure to verify that the rational function is proper. This check is important because nothing in the procedure for finding the various K's will alert you to nonsense results if the rational function is improper. We present a procedure for checking the K's, but you can avoid wasted effort by forming the habit of asking yourself, "Is $F(s)$ a proper rational function?"

PARTIAL FRACTION EXPANSION: REAL AND DISTINCT ROOTS OF $D(s)$

We first consider determining the coefficients in a partial fraction expansion when all the roots of $D(s)$ are real and distinct. To find a K associated with a term that arises because of a distinct root of $D(s)$, we multiply both sides of the identity by a factor equal to the denominator beneath the desired K. Then when we evaluate both sides of the identity at the root corresponding to the multiplying factor, the right-hand side always is the desired K and the left-hand side always is its numerical value. For example,

$$F(s) = \frac{96(s + 5)(s + 12)}{s(s + 8)(s + 6)} \equiv \frac{K_1}{s} + \frac{K_2}{s + 8} + \frac{K_3}{s + 6}. \quad \textbf{(15.48)}$$

To find the value of K_1, we multiply both sides by s and then evaluate both sides at $s = 0$:

$$\frac{96(s + 5)(s + 12)}{(s + 8)(s + 6)}\bigg|_{s=0} \equiv K_1 + \frac{K_2 s}{s + 8}\bigg|_{s=0} + \frac{K_3 s}{s + 6}\bigg|_{s=0}$$

or

$$\frac{96(5)(12)}{8(6)} \equiv K_1 = 120. \quad \textbf{(15.49)}$$

To find the value of K_2, we multiply both sides by $s + 8$ and then evaluate both sides at $s = -8$:

$$\left.\frac{96(s + 5)(s + 12)}{s(s + 6)}\right|_{s=-8} \equiv \left.\frac{K_1(s + 8)}{s}\right|_{s=-8}$$
$$+ K_2 + \left.\frac{K_3(s + 8)}{(s + 6}\right|_{s=-8}$$

or

$$\frac{96(-3)(4)}{(-8)(-2)} = K_2 = -72. \tag{15.50}$$

Then K_3 is

$$\left.\frac{96(s + 5)(s + 12)}{s(s + 8)}\right|_{s=-6} = K_3 = 48. \tag{15.51}$$

From Eq. (15.48) and the K values obtained,

$$\frac{96(s + 5)(s + 12)}{s(s + 8)(s + 6)} \equiv \frac{120}{s} + \frac{48}{s + 6} - \frac{72}{s + 8}. \tag{15.52}$$

At this point, testing the result as a protection against making computational errors is a good idea. As we already mentioned, a partial fraction expansion creates an identity; thus both sides of Eq. (15.52) must be the same for all s values. The choice of test values is completely open; hence we choose values that are easy to verify. For example, in Eq. (15.52) testing at either -5 or -12 is attractive because in both cases the left-hand side reduces to zero. Choosing -5 yields

$$\frac{120}{-5} + \frac{48}{1} - \frac{72}{3} = -24 + 48 - 24 = 0,$$

whereas testing -12 gives

$$\frac{120}{-12} + \frac{48}{-6} - \frac{72}{-4} = -10 - 8 + 18 = 0.$$

Now confident that the numerical values of the various K's are correct, we proceed to find the inverse transform:

$$\mathcal{L}^{-1}\left\{\frac{96(s + 5)(s + 12)}{s(s + 8)(s + 6)}\right\} = (120 + 48e^{-6t} - 72e^{-8t})u(t).$$

$$\tag{15.53}$$

DRILL EXERCISES

15.8 Find $f(t)$ if

$$F(s) = \frac{16(s^2 + 3s + 9)}{(s + 2)(s + 4)(s + 6)}.$$

ANSWER: $f(t) = (14e^{-2t} - 52e^{-4t} + 54e^{-6t})u(t).$

15.9 Find $f(t)$ if

$$F(s) = \frac{2s + 12}{(s + 1)(s^2 + 5s + 6)}.$$

ANSWER: $f(t) = (5e^{-t} - 8e^{-2t} + 3e^{-3t})u(t)$.

PARTIAL FRACTION EXPANSION: DISTINCT AND COMPLEX ROOTS OF $D(s)$

The procedure for finding the coefficients associated with distinct complex roots is the same as that for finding those associated with distinct real roots. The only difference is that the algebra involves the manipulation of complex numbers. We illustrate by expanding the rational function,

$$F(s) = \frac{100(s + 3)}{(s + 6)(s^2 + 6s + 25)}. \qquad (15.54)$$

We begin by noting that $F(s)$ is a proper rational function. Next we must find the roots of the quadratic term $s^2 + 6s + 25$:

$$s^2 + 6s + 25 = (s + 3 - j4)(s + 3 + j4). \qquad (15.55)$$

With the denominator in factored form, we proceed as before:

$$\frac{100(s + 3)}{(s + 6)(s^2 + 6s + 25)} \equiv \frac{K_1}{s + 6} + \frac{K_2}{s + 3 - j4} + \frac{K_3}{s + 3 + j4}. \qquad (15.56)$$

To find K_1, K_2, and K_3 we use the same thought process as previously:

$$K_1 = \frac{100(s + 3)}{s^2 + 6s + 25}\bigg|_{s=-6} = \frac{100(-3)}{25} = -12; \qquad (15.57)$$

$$K_2 = \frac{100(s + 3)}{(s + 6)(s + 3 + j4)}\bigg|_{s=-3+j4} = \frac{100(j4)}{(3 + j4)(j8)}$$

$$= 6 - j8 = 10e^{-j53.13°}; \qquad (15.58)$$

$$K_3 = \frac{100(s + 3)}{(s + 6)(s + 3 - j4)}\bigg|_{s=-3-j4} = \frac{100(-j4)}{(3 - j4)(-j8)}$$

$$= 6 + j8 = 10e^{j53.13°}. \qquad (15.59)$$

Then

$$\frac{100(s + 3)}{(s + 6)(s^2 + 6s + 25)} = \frac{-12}{s + 6} + \frac{10\underline{/-53.13°}}{s + 3 - j4} + \frac{10\underline{/53.13°}}{s + 3 + j4}. \qquad (15.60)$$

Again, we need to make some observations. First, in physically realizable circuits, complex roots always appear in conjugate pairs. Second, the coefficients associated with these conjugate pairs are themselves conjugates. Note, for example, that K_3 (Eq. 15.69) is the conjugate of K_2 (Eq. 15.68). Thus for complex conjugate roots, you actually need to calculate only half the coefficients.

Before inverse-transforming Eq. (15.60), we check the partial fraction expansion numerically. Testing at -3 is attractive because the left-hand side reduces to zero at this value:

$$F(s) = \frac{-12}{3} + \frac{10\underline{/-53.13°}}{-j4} + \frac{10\underline{/53.13°}}{j4}$$

$$= -4 + 2.5\underline{/36.87°} + 2.5\underline{/-36.87°}$$

$$= -4 + 2.0 + j1.5 + 2.0 - j1.5 = 0.$$

We now proceed to inverse-transform Eq. (15.60):

$$\mathscr{L}^{-1}\left\{\frac{100(s+3)}{(s+6)(s^2+6s+25)}\right\} = (-12e^{-6t} + 10e^{-j53.13°}\, e^{-(3-j4)t}$$

$$+ 10e^{j53.13°}\, e^{-(3+j4)t})u(t). \qquad \textbf{(15.61)}$$

In general, to have the function in the time domain contain imaginary components is undesirable. We can easily take care of this problem, because the terms involving imaginary components always come in conjugate pairs. Hence we eliminate the imaginary components simply by adding the pairs:

$$10e^{-j53.13°}\, e^{-(3-j4)t} + 10e^{j53.13°}\, e^{-(3+j4)t}$$

$$= 10e^{-3t}\left(e^{j(4t-53.13°)} + e^{-j(4t-53.13°)}\right)$$

$$= 20e^{-3t}\cos(4t - 53.13°), \qquad \textbf{(15.62)}$$

which enables us to simplify Eq. (15.60):

$$\mathscr{L}^{-1}\left\{\frac{100(s+3)}{(s+6)(s^2+6s+25)}\right\}$$

$$= [-12e^{-6t} + 20e^{-3t}\cos(4t - 53.13°)]u(t). \qquad \textbf{(15.63)}$$

Because distinct complex roots appear rather frequently in lumped-parameter, linear circuit analysis, we need to summarize these results with a new transform pair. Whenever $D(s)$ contains distinct complex roots, that is, factors of the form $(s + \alpha - j\beta)(s + \alpha + j\beta)$, a pair of terms of the form

$$\frac{K}{s + \alpha - j\beta} + \frac{K^*}{s + \alpha + j\beta} \qquad \textbf{(15.64)}$$

appears in the partial fraction expansion, where the partial fraction coefficient is, in general, a complex number. In polar form,

$$K = |K|e^{j\theta} = |K|\underline{/\theta}, \qquad \textbf{(15.65)}$$

where $|K|$ denotes the magnitude of the complex coefficient. Then

$$K^* = |K|e^{-j\theta} = |K|\underline{/-\theta}. \qquad (15.66)$$

The complex conjugate pair in Eq. (15.64) always inverse-transforms as

$$\mathcal{L}^{-1}\left\{\frac{K}{s + \alpha - j\beta} + \frac{K^*}{s + \alpha + j\beta}\right\} = 2|K|e^{-\alpha t}\cos(\beta t + \theta).$$

$$(15.67)$$

In applying Eq. (15.67) note that K is defined as the coefficient associated with the denominator term $s + \alpha - j\beta$.

DRILL EXERCISES

15.10 Find $f(t)$ if
$$F(s) = 10(s^2 + 119)/[(s + 5)(s^2 + 10s + 169)].$$

ANSWER: $f(t) = (10e^{-5t} - 8.33e^{-5t}\sin 12t)u(t).$

15.11 Find $v(t)$ in Drill Exercise 15.7.

ANSWER:
$v(t) = [2e^{-0.8t}\cos(0.6t + 233.13°) + 1.2\cos t]u(t).$

PARTIAL FRACTION EXPANSION: REAL AND REPEATED OR COMPLEX AND REPEATED ROOTS OF D(s)

WHEN ROOTS ARE REAL AND REPEATED To find the coefficients associated with the terms generated by a multiple root of multiplicity r, we multiply both sides of the identity by the multiple root raised to its rth power. We find the K appearing over the factor raised to the rth power by evaluating both sides of the identity at the multiple root. To find the remaining $(r - 1)$ coefficients associated with the multiple root, we differentiate both sides of the identity $(r - 1)$ times. At the end of each differentiation, we evaluate both sides of the identity at the multiple root. The right-hand side of the identity always is the desired K, and the left-hand side always is its numerical value. For example,

$$\frac{180(s + 30)}{s(s + 5)(s + 3)^2} = \frac{K_1}{s} + \frac{K_2}{s + 5} + \frac{K_3}{(s + 3)^2} + \frac{K_4}{s + 3}. \qquad (15.68)$$

We find K_1 and K_2 as previously described; that is,

$$K_1 = \left.\frac{180(s + 30)}{(s + 5)(s + 3)^2}\right|_{s=0} = \frac{180(30)}{(5)(9)} = 120 \qquad (15.69)$$

and

$$K_2 = \left. \frac{180(s + 30)}{s(s + 3)^2} \right|_{s=-5} = \frac{180(25)}{(-5)(4)} = -225. \quad \text{(15.70)}$$

To find K_3, we multiply both sides by $(s + 3)^2$ and then evaluate both sides at -3:

$$\left. \frac{180(s + 30)}{s(s + 5)} \right|_{s=-3} = \left. \frac{K_1(s + 3)^2}{s} \right|_{s=-3} + \left. \frac{K_2(s + 3)^2}{s + 5} \right|_{s=-3}$$

$$+ K_3 + K_4(s + 3) \Big|_{s=-3} ; \quad \text{(15.71)}$$

$$\frac{180(27)}{(-3)(2)} = K_1 \times 0 + K_2 \times 0 + K_3 + K_4 \times 0$$

$$= K_3 = -810. \quad \text{(15.72)}$$

To find K_4 we first must multiply both sides of Eq. (15.68) by $(s + 3)^2$. Next we differentiate both sides once with respect to s and then evaluate at $s = -3$:

$$\frac{d}{ds} \left[\frac{180(s + 30)}{s(s + 5)} \right]_{s=-3} = \frac{d}{ds} \left[\frac{K_1(s + 3)^2}{s} \right]_{s=-3}$$

$$+ \frac{d}{ds} \left[\frac{K_2(s + 3)^2}{s + 5} \right]_{s=-3}$$

$$+ \frac{d}{ds} (K_3)_{s=-3}$$

$$+ \frac{d}{ds} [K_4(s + 3)] \Big|_{s=-3} ,$$

$$\text{(15.73)}$$

$$180 \left[\frac{s(s + 5) - (s + 30)(2s + 5)}{s^2(s + 5)^2} \right]_{s=-3} = K_4, \quad \text{(15.74)}$$

or

$$180 \left[\frac{(-3)(2) - (27)(-1)}{(9)(4)} \right] = K_4 = 105. \quad \text{(15.75)}$$

Then

$$\frac{180(s + 30)}{s(s + 5)(s + 3)^2} = \frac{120}{s} - \frac{225}{s + 5} - \frac{810}{(s + 3)^2} + \frac{105}{s + 3} \quad \text{(15.76)}$$

and therefore

$$\mathscr{L}^{-1} \left\{ \frac{180(s + 30)}{s(s + 5)(s + 3)^2} \right\}$$

$$= (120 - 225e^{-5t} - 810te^{-3t} + 105e^{-3t})u(t). \quad \text{(15.77)}$$

DRILL EXERCISES

15.12 Find $f(t)$ if $F(s) = (4s^2 + 7s + 1)/[s(s + 1)^2]$.   **ANSWER:** $f(t) = (1 + 2te^{-t} + 3e^{-t})u(t)$.

WHEN ROOTS ARE COMPLEX AND REPEATED   We handle repeated complex roots in the same way that we did repeated real roots. The only difference is that the algebra involves complex numbers. Recall that complex roots always appear in conjugate pairs and that the coefficients associated with a conjugate pair also are conjugates, *so that only half the K's need be evaluated.* For example,

$$F(s) = \frac{768}{(s^2 + 6s + 25)^2}. \tag{15.78}$$

After factoring the denominator polynomial, we write

$$F(s) = \frac{768}{(s + 3 - j4)^2(s + 3 + j4)^2}$$

$$= \frac{K_1}{(s + 3 - j4)^2} + \frac{K_2}{s + 3 - j4}$$

$$+ \frac{K_1^*}{(s + 3 + j4)^2} + \frac{K_2^*}{s + 3 + j4}. \tag{15.79}$$

Now we need to evaluate only K_1 and K_2, because K_1^* and K_2^* are conjugate values. The value of K_1 is

$$K_1 = \frac{768}{(s + 3 + j4)^2}\bigg|_{s=-3+j4}$$

$$= \frac{768}{(j8)^2} = -12. \tag{15.80}$$

The value of K_2 is

$$K_2 = \frac{d}{ds}\left[\frac{768}{(s + 3 + j4)^2}\bigg|_{s=-3+j4}\right]$$

$$= -\frac{2(768)}{(s + 3 + j4)^3}\bigg|_{s=-3+j4}$$

$$= -\frac{2(768)}{(j8)^3}$$

$$= -j3 = 3\underline{/-90°}. \tag{15.81}$$

From Eqs. (15.80) and (15.81),

$$K_1^* = -12 \tag{15.82}$$

and

$$K_2^* = j3 = 3\underline{/90°}. \tag{15.83}$$

We now group the partial fraction expansion by conjugate terms to obtain

$$F(s) = \left[\frac{-12}{(s + 3 - j4)^2} + \frac{-12}{(s + 3 + j4)^2}\right]$$

$$+ \left(\frac{3\underline{/-90°}}{s + 3 - j4} + \frac{3\underline{/90°}}{s + 3 + j4}\right). \tag{15.84}$$

We now write the inverse transform of $F(s)$:

$$f(t) = [-24te^{-3t}\cos 4t + 6e^{-3t}\cos (4t - 90°)]u(t). \tag{15.85}$$

Note that if $F(s)$ has a real root a of multiplicity r in its denominator, the term in a partial fraction expansion is of the form:

$$\frac{K}{(s + a)^r}.$$

The inverse transform of this term is

$$\mathcal{L}^{-1}\left\{\frac{K}{(s + a)^r}\right\} = \frac{Kt^{r-1}e^{-at}}{(r - 1)!} u(t). \tag{15.86}$$

If $F(s)$ has a complex root of $\alpha + j\beta$ of multiplicity r in its denominator, the term in partial fraction expansion is the conjugate pair

$$\frac{K}{(s + \alpha - j\beta)^r} + \frac{K^*}{(s + \alpha + j\beta)^r}.$$

The inverse transform of this pair is

$$\mathcal{L}^{-1}\left\{\frac{K}{(s + \alpha - j\beta)^r} + \frac{K^*}{(s + \alpha + j\beta)^r}\right\}$$

$$= \left[\frac{2|K|t^{r-1}}{(r - 1)!} e^{-\alpha t}\cos (\beta t + \theta)\right]u(t). \tag{15.87}$$

Equations (15.86) and (15.87) are the key to being able to inverse-transform any partial fraction expansion by inspection. One further note regarding these two equations: In most circuit analysis problems, r seldom is greater than 2. Therefore, for most problems the inverse transform of a rational function can be handled with four transform pairs. Table 15.3 lists these transform pairs.

TABLE 15.3

FOUR USEFUL TRANSFORM PAIRS

PAIR NUMBER[†]	$F(s)$	$f(t)$
(1)	$\dfrac{K}{s + a}$	$Ke^{-at}u(t)$
(2)	$\dfrac{K}{(s + a)^2}$	$Kte^{-at}u(t)$
(3)	$\dfrac{K}{s + \alpha - j\beta} + \dfrac{K^*}{s + \alpha + j\beta}$	$2|K|e^{-\alpha t}\cos(\beta t + \theta)u(t)$
(4)	$\dfrac{K}{(s + \alpha - j\beta)^2} + \dfrac{K^*}{(s + \alpha + j\beta)^2}$	$2t|K|e^{-\alpha t}\cos(\beta t + \theta)u(t)$

[†]*Note:* In pairs (1) and (2), K is a real quantity, whereas in pairs (3) and (4), K is the complex quantity $|K|\underline{/\theta}$.

DRILL EXERCISES

15.13 Find $f(t)$ if $F(s) = 40/(s^2 + 4s + 5)^2$.   **ANSWER:** $f(t) = (-20te^{-2t}\cos t + 20e^{-2t}\sin t)u(t)$.

PARTIAL FRACTION EXPANSION: IMPROPER RATIONAL FUNCTIONS

We conclude the discussion of partial fraction expansions by returning to an observation made at the beginning of this section, namely, that improper rational functions pose no serious problem in finding inverse transforms. An improper rational function can always be expanded into a polynomial plus a proper rational function. The polynomial portion of the expansion inverse-transforms into impulse functions and derivatives of impulse functions. The proper rational function portion of the expansion is inverse-transformed by the techniques outlined in this section. To illustrate the procedure, we use the function

$$F(s) = \frac{s^4 + 13s^3 + 66s^2 + 200s + 300}{s^2 + 9s + 20}. \qquad \text{(15.88)}$$

Dividing the denominator into the numerator until the remainder is a proper rational function gives

$$F(s) = s^2 + 4s + 10 + \frac{30s + 100}{s^2 + 9s + 20}. \qquad \text{(15.89)}$$

Next we expand the proper rational function into a sum of partial fractions:

$$\frac{30s + 100}{s^2 + 9s + 20} = \frac{30s + 100}{(s + 4)(s + 5)} = \frac{-20}{s + 4} + \frac{50}{s + 5}. \qquad \textbf{(15.90)}$$

Substituting Eq. (15.90) into Eq. (15.89) yields

$$F(s) = s^2 + 4s + 10 - \frac{20}{s + 4} + \frac{50}{s + 5}. \qquad \textbf{(15.91)}$$

Now we can inverse-transform Eq. (15.91) by inspection. Hence

$$f(t) = \frac{d^2\delta}{dt^2} + 4\frac{d\delta}{dt} + 10\delta - (20e^{-4t} - 50e^{-5t})u(t). \qquad \textbf{(15.92)}$$

DRILL EXERCISES

15.14 Find $f(t)$ if $F(s) = s^2/[(s + 1)(s + 2)]$.

ANSWER: $f(t) = \delta(t) + (e^{-t} - 4e^{-2t})u(t)$.

15.15 Find $f(t)$ if

$$F(s) = (2s^3 + 8s^2 + 2s - 4)/(s^2 + 5s + 4).$$

ANSWER: $f(t) = 2\dfrac{d\delta}{dt} - 2\delta + 4e^{-4t}u(t)$.

15.8 POLES AND ZEROS OF $F(s)$

The rational function of Eq. (15.45) also may be expressed as the ratio of two factored polynomials. That is, we may write $F(s)$ as

$$F(s) = \frac{K(s + z_1)(s + z_2) \cdots (s + z_n)}{(s + p_1)(s + p_2) \cdots (s + p_m)}, \qquad \textbf{(15.93)}$$

where K is the constant a_n/b_m. For example, we may also write the function

$$F(s) = \frac{8s^2 + 120s + 400}{2s^4 + 20s^3 + 70s^2 + 100s + 48}$$

as

$$F(s) = \frac{8(s^2 + 15s + 50)}{2(s^4 + 10s^3 + 35s^2 + 50s + 24)}$$

$$= \frac{4(s + 5)(s + 10)}{(s + 1)(s + 2)(s + 3)(s + 4)}. \qquad \textbf{(15.94)}$$

The roots of the denominator polynomial—that is, $-p_1$, $-p_2$, $-p_3$, . . . , $-p_m$—are called the *poles of* $F(s)$, because at these values of s, $F(s)$ becomes infinitely large. In the function described by Eq. (15.94), the poles of $F(s)$ are -1, -2, -3, and -4.

The roots of the numerator polynomial—that is, $-z_1$, $-z_2$, $-z_3$, . . . , $-z_n$—are called the *zeros* of $F(s)$, because at these values of s, $F(s)$ becomes zero. In the function described by Eq. (15.94), the zeros of $F(s)$ are -5 and -10.

In what follows you may find that being able to visualize the poles and zeros of $F(s)$ as points on a complex s plane sometimes is convenient. A complex plane is needed because the roots of the polynomials may be complex. In the complex s plane we use the horizontal axis to plot the real values of s and the vertical axis to plot the imaginary values of s.

As an example of plotting the poles and zeros of $F(s)$, consider the function

$$F(s) = \frac{10(s + 5)(s + 3 - j4)(s + 3 + j4)}{s(s + 10)(s + 6 - j8)(s + 6 + j8)}. \quad \textbf{(15.95)}$$

The poles of $F(s)$ are at 0, -10, $-6 + j8$, and $-6 - j8$. The zeros are at -5, $-3 + j4$, and $-3 - j4$. Figure 15.17 shows the poles and zeros plotted on the s plane, where X's represent poles and O's represent zeros.

Note that the poles and zeros designated by $-p_1$, $-p_2$, . . . , $-p_m$ and $-z_1$, $-z_2$, . . . , $-z_n$, respectively, are located in the finite s plane. Also, $F(s)$ can have either an rth-order pole or an rth-order zero at infinity. For example, the function described by Eq. (15.94) has a second-order zero at infinity, because for large values of s the function reduces to $4/s^2$, and $F(s) = 0$ when $s = \infty$. Here we are interested in the poles and zeros located in the finite s plane. Therefore, when we refer to the poles and zeros of a rational function of s, we are referring to the finite poles and zeros.

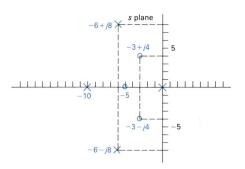

FIGURE 15.17 Plotting poles and zeros on the s plane.

15.9 INITIAL- AND FINAL-VALUE THEOREMS

The initial- and final-value theorems are useful because they enable us to determine from $F(s)$ the behavior of $f(t)$ at 0 and ∞. Hence we can check the initial and final values of $f(t)$ before actually finding the inverse transform of $F(s)$.

The initial-value theorem states that

$$\lim_{t \to 0^+} f(t) = \lim_{s \to \infty} sF(s), \quad \textbf{(15.96)}$$

and the final-value theorem states that

$$\lim_{t \to \infty} f(t) = \lim_{s \to 0} sF(s). \qquad (15.97)$$

The initial-value theorem is based on the assumption that $f(t)$ contains no impulse functions. In Eq. (15.97), we must add the restriction that the theorem is valid only if the poles of $F(s)$—except for a first-order pole at the origin—lie in the left half of the s plane.

To prove Eq. (15.96), we start with the operational transform of the first derivative:

$$\mathcal{L}\left\{\frac{df}{dt}\right\} = sF(s) - f(0^-) = \int_{0^-}^{\infty} \frac{df}{dt} e^{-st} dt. \qquad (15.98)$$

Now we take the limit as $s \to \infty$:

$$\lim_{s \to \infty} [sF(s) - f(0^-)] = \lim_{s \to \infty} \int_{0^-}^{\infty} \frac{df}{dt} e^{-st} dt. \qquad (15.99)$$

Observe that the right-hand side of Eq. (15.99) may be written as

$$\lim_{s \to \infty} \left(\int_{0^-}^{0^+} \frac{df}{dt} e^0 dt + \int_{0^+}^{\infty} \frac{df}{dt} e^{-st} dt \right).$$

As $s \to \infty$, $(df/dt)e^{-st} \to 0$; hence the second integral vanishes in the limit. The first integral reduces to $f(0^+) - f(0^-)$, which is independent of s. Thus the right-hand side of Eq. (15.99) becomes

$$\lim_{s \to \infty} \int_{0^-}^{\infty} \frac{df}{dt} e^{-st} dt = f(0^+) - f(0^-). \qquad (15.100)$$

As $f(0^-)$ is independent of s, the left-hand side of Eq. (15.99) may be written

$$\lim_{s \to \infty} [sF(s) - f(0^-)] = \lim_{s \to \infty} [sF(s)] - f(0^-). \qquad (15.101)$$

From Eqs. (15.100) and (15.101),

$$\lim_{s \to \infty} sF(s) = f(0^+) = \lim_{t \to 0^+} f(t),$$

which completes the proof of the initial-value theorem.

The proof of the final-value theorem also starts with Eq. (15.98). Here we take the limit as $s \to 0$:

$$\lim_{s \to 0} [sF(s) - f(0^-)] = \lim_{s \to 0} \left(\int_{0^-}^{\infty} \frac{df}{dt} e^{-st} dt \right). \qquad (15.102)$$

The integration is with respect to t and the limit operation is with respect to s, so the right-hand side of Eq. (15.102) reduces to

$$\lim_{s \to 0} \left(\int_{0^-}^{\infty} \frac{df}{dt} e^{-st} dt \right) = \int_{0^-}^{\infty} \frac{df}{dt} dt. \qquad (15.103)$$

Because the upper limit on the integral is infinite, this integral also may be written as a limit process:

$$\int_{0^-}^{\infty} \frac{df}{dt}\, dt = \lim_{t \to \infty} \int_{0^-}^{t} \frac{df}{dy}\, dy, \qquad \textbf{(15.104)}$$

where we use y as the symbol of integration to avoid confusion with the upper limit on the integral. Carrying out the integration process yields

$$\lim_{t \to \infty} [f(t) - f(0^-)] = \lim_{t \to \infty} [f(t)] - f(0^-). \qquad \textbf{(15.105)}$$

Substituting Eq. (15.105) into Eq. (15.102) gives

$$\lim_{s \to 0} [sF(s)] - f(0^-) = \lim_{t \to \infty} [f(t)] - f(0^-). \qquad \textbf{(15.106)}$$

Because $f(0^-)$ cancels, Eq. (15.106) reduces to the final-value theorem, namely,

$$\lim_{s \to 0} sF(s) = \lim_{t \to \infty} f(t).$$

The final-value theorem is useful only if $f(\infty)$ exists. This condition is true only if all the poles of $F(s)$—except for a simple pole at the origin—lie in the left half of the s plane.

APPLICATION OF INITIAL- AND FINAL-VALUE THEOREMS

To illustrate the application of the initial- and final-value theorems, we apply them to some of the functions we used to illustrate partial fraction expansions. As we already have the time-domain expressions, we can easily test the theorems. We begin by applying the initial-value theorem to the transform pair of Eq. (15.53):

$$\lim_{s \to \infty} sF(s) = \lim_{s \to \infty} \frac{96(s + 5)(s + 12)}{(s + 8)(s + 6)}$$

$$= \lim_{s \to \infty} \frac{96[1 + (5/s)][1 + (12/s)]}{[1 + (8/s)][1 + (6/s)]} = 96;$$

$$\lim_{t \to 0^+} f(t) = \lim_{t \to 0^+} (120 + 48e^{-6t} - 73e^{-8t})u(t)$$

$$= (120 + 48 - 72)(1) = 96.$$

In applying the final-value theorem, we recognize that all the poles of $sF(s)$ lie in the left half of the s plane. Thus

$$\lim_{s \to o} sF(s) = \frac{(96)(5)(12)}{(8)(6)} = 120;$$

$$\lim_{t \to \infty} f(t) = (120 + 0 - 0)(1) = 120.$$

As a second example, consider the transform pair given by Eq. (15.63). The initial-value theorem gives

$$\lim_{s \to \infty} sF(s) = \lim_{s \to \infty} \frac{100s^2[1 + (3/s)]}{s^3[1 + (6/s)][1 + (6/s) + (25/s^2)]} = 0;$$

$$\lim_{t \to 0^+} f(t) = [-12 + 20 \cos(-53.13°)](1) = -12 + 12 = 0.$$

The final-value theorem gives

$$\lim_{s \to 0} sF(s) = \lim_{s \to 0} \frac{100s(s + 3)}{(s + 6)(s^2 + 6s + 25)} = 0;$$

$$\lim_{t \to \infty} f(t) = \lim_{t \to \infty} [-12e^{-6t} + 20e^{-3t} \cos(4t - 53.13°)]u(t) = 0.$$

As a final example of illustrating these theorems, consider the expression for $V(s)$ given by Eq. (15.43). Although we cannot calculate $v(t)$ until the circuit parameters are specified, we can check to see if $V(s)$ predicts the correct values of $v(0^+)$ and $v(\infty)$. We know from the statement of the problem that generated $V(s)$ that $v(0^+)$ is zero. We also know that $v(\infty)$ must be zero because the ideal inductor is a perfect short circuit across the dc current source. Finally, we know that the poles of $V(s)$ must lie in the left half of the s plane because R, L, and C are positive constants. Hence the poles of $sV(s)$ also lie in the left half of the s plane.

Applying the initial-value theorem yields

$$\lim_{s \to \infty} sV(s) = \lim_{s \to \infty} \frac{s(I_{dc}/C)}{s^2[1 + 1/(RCs) + 1/(LCs^2)]} = 0.$$

Applying the final-value theorem gives

$$\lim_{s \to \infty} sV(s) = \lim_{s \to \infty} \frac{s(I_{dc}/C)}{s^2 + \dfrac{1}{RC}s + \dfrac{1}{LC}} = 0.$$

The derived expression for $V(s)$ correctly predicts the initial and final values of $v(t)$. The initial- and final-value theorems provide one method of testing the s-domain expressions for the unknown variables before you work out the inverse transform.

DRILL EXERCISES

15.16 Use the initial- and final-value theorems to find the initial and final values of $f(t)$ in Drill Exercises 15.9, 15.12, and 15.13.   **ANSWER:** 0, 0; 4, 1; and 0, 0.

15.17 a) Use the initial-value theorem to find the initial value of v in Drill Exercise 15.7.

   b) Can the final-value theorem be used to find the steady-state value of v? Why?

ANSWER: (a) 0; (b) no, because V has a pair of poles on the imaginary axis.

SUMMARY

The Laplace transform is an important tool for analyzing linear, lumped-parameter circuits because it

- facilitates the analysis of circuits that can be described only by two or more simultaneous differential equations;

- facilitates the use of various signal sources (impulse, step, ramp, and so on); and

- strengthens the connection between the time-domain and frequency-domain behavior of a circuit.

Application of the Laplace transform method requires an understanding of

- functional transforms, or Laplace transforms of a time-domain function (see Table 15.1);

- operational transforms, or transforms that relate mathematical operations in the time domain to their corresponding operation in the frequency domain (see Table 15.2);

- inverse transform, which involves finding the time-domain expression that corresponds to a frequency-domain expression (see Table 15.1);

- rational function, a ratio of two polynomials in s where all the exponents of s are integers, which may be either

  a proper rational function in which the order of the polynomial in the denominator is *greater* than the order of the polynomial in the numerator or

  an improper rational function in which the order of the polynomial in the denominator is *less* than the order of the polynomial in the numerator;

- partial fractions, which involves expansion of a proper rational function into a sum of partial fractions;

- poles, or the roots of the denominator polynomial;

- zeros, or the roots of the numerator polynomial;

- initial-value theorem, which involves finding the initial value of a time-domain function from its Laplace transform; and

- final-value theorem, which involves finding the final value of a time-domain function from its Laplace transform.

The Laplace transform method of circuit analysis is divided into five steps:

- writing the integrodifferential time-domain equations that describe the circuit;

- transforming the integrodifferential equations into s-domain algebraic equations;

- solving the s-domain algebraic equations for the unknowns of interest;

- inverse-transforming the s-domain variables back to the time domain; and

- testing the derived time-domain expressions to verify that they make sense in terms of the known physical behavior of the circuit.

PROBLEMS

15.1 Use step functions to write the expression for each of the functions shown in Fig. P15.1.

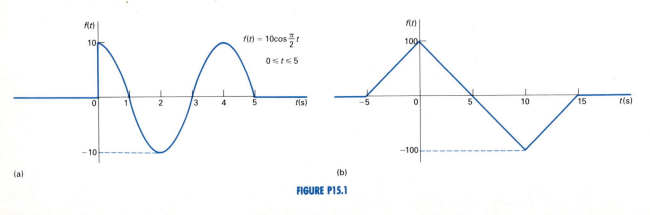

(a)

(b)

FIGURE P15.1

15.2 Step functions can be used to define a *window* function. Thus $u(t) - u(t - 5)$ defines a window one unit high and five units wide, located on the time axis between 0 and 5. Use the concept of the window function to write an expression for the function shown in Fig. P15.2.

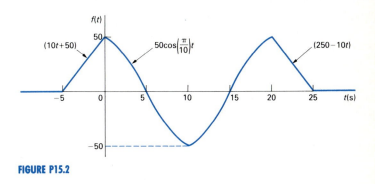

FIGURE P15.2

15.3 Explain why the following function generates an impulse function as $\epsilon \to 0$:

$$f(t) = \frac{\epsilon/\pi}{\epsilon^2 + t^2}, \qquad -\infty \leq t \leq \infty.$$

15.4 In Section 15.3, we used the sifting property of the impulse function to show that $\mathcal{L}\{\delta(t)\} = 1$. Show that we can obtain the same result by finding the Laplace transform of the rectangular pulse that exists between $\pm\epsilon$ in Fig. 15.9 and then finding the limit of this transform as $\epsilon \to 0$.

15.5 The triangular pulses shown in Fig. P15.5 are equivalent to the rectangular pulses in Fig. 15.12(b) because they both enclose the same area $(1/\epsilon)$ and they both approach infinity proportional to $1/\epsilon^2$ as $\epsilon \to 0$. Use this triangular-pulse representation for $\delta'(t)$ to find the Laplace transform of $\delta''(t)$.

FIGURE P15.5

15.6 a) Show that

$$\int_{-\infty}^{\infty} f(t)\delta'(t - a)dt = -f'(a).$$

(*Hint:* Integrate by parts.)

b) Use the formula in part (a) to show that

$$\mathcal{L}\{\delta'(t)\} = s.$$

15.7 Find the Laplace transform of each of the following functions:

a) $f(t) = te^{-at}$;

b) $f(t) = \cos \omega t$;

c) $f(t) = \cos (\omega t + \theta)$;

d) $f(t) = \sinh t$.

15.8 Find the Laplace transform, when $\epsilon \to 0$, of the derivative of the exponential function illustrated in Fig. 15.8 by each of the following two methods.

   a) First differentiate the function and then find the transform of the resulting function.

   b) Use the operational transform given by Eq. (15.26).

15.9 Show that
$$\mathcal{L}\{\delta^{(n)}(t)\} = s^n.$$

15.10 Find each of the following:

   a) $\mathcal{L}\left\{\dfrac{d}{dt}\sin \omega t\right\}$;

   b) $\mathcal{L}\left\{\dfrac{d}{dt}\cos \omega t\right\}$;

   c) $\mathcal{L}\left\{\dfrac{d^3}{dt^3}t^2\right\}$.

   d) Check the results of parts (a), (b), and (c) by first differentiating and then transforming.

15.11 Find each of the following:

   a) $\mathcal{L}\left\{\displaystyle\int_{0^-}^{t} e^{-ax}\,dx\right\}$;

   b) $\mathcal{L}\left\{\displaystyle\int_{0^-}^{t} y\,dy\right\}$.

   c) Check the results of parts (a) and (b) by first integrating and then transforming.

15.12 Show that
$$\mathcal{L}\{e^{-at}f(t)\} = F(s + a).$$

15.13 a) Sketch each of the following functions:
$$f(t) = -20e^{-5(t-2)}u(t - 2);$$
$$f(t) = (8t - 8)[u(t - 1) - u(t - 2)]$$
$$+(24 - 8t)[u(t - 2) - u(t - 4)]$$
$$+ (8t - 40)[u(t - 4) - u(t - 5)].$$

   b) Find the Laplace transform of each function.

15.14 a) Find the Laplace transform of the function illustrated in Fig. P15.14.

   b) Find the Laplace transform of the first derivative of the function illustrated in Fig. P15.14.

   c) Find the Laplace transform of the second derivative of the function illustrated in Fig. P15.14.

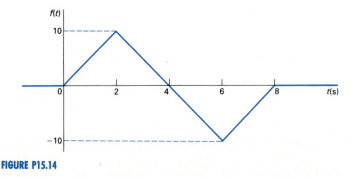

FIGURE P15.14

15.15 Find the Laplace transform of each of the following functions:

a) $f(t) = \dfrac{d}{dt}(e^{-at}\sin \omega t)$;

b) $f(t) = \displaystyle\int_{0^-}^{t} e^{-ax}\cos \omega x\, dx$.

c) Verify the results obtained in parts (a) and (b) by first carrying out the indicated mathematical operation and then finding the Laplace transform.

15.16 a) Show that

$$\mathscr{L}\{f(at)\} = \frac{1}{a}F\left(\frac{s}{a}\right).$$

b) Use the result of part (a) along with the answer derived in Problem 15.7(d) to find

$$\mathscr{L}\{\sinh \beta t\}.$$

15.17 a) Given that $F(s) = \mathscr{L}\{f(t)\}$, show that

$$-\frac{dF(s)}{ds} = \mathscr{L}\{tf(t)\}.$$

b) Show that

$$(-1)^n\frac{d^n F(s)}{ds^n} = \mathscr{L}\{t^n f(t)\}.$$

c) Use the result of part (b) to find $\mathscr{L}\{t^5\}$, $\mathscr{L}\{t \sin \beta t\}$, and $\mathscr{L}\{te^{-t}\cosh t\}$.

15.18 a) Show that if $F(s) = \mathscr{L}\{f(t)\}$ and $[f(t)/t]$ is Laplace-transformable, then

$$\int_{s}^{\infty} F(u)\, du = \mathscr{L}\left\{\frac{f(t)}{t}\right\}.$$

(*Hint:* Use the defining integral to write

$$\int_{s}^{\infty} F(u)\, du = \int_{s}^{\infty}\left(\int_{0^-}^{\infty} f(t)e^{-ut}\, dt\right) du$$

and then reverse the order of integration.)

b) Start with the result obtained in Problem 15.17(c) for $\mathscr{L}\{t \sin \beta t\}$ and use the operational transform given in part (a) of this problem to find $\mathscr{L}\{\sin \beta t\}$.

15.19 There is no energy stored in the circuit shown in Fig. P15.19 at the time the switch is closed.

a) Derive the integrodifferential equation that governs the behavior of the current i_o.

b) Show that

$$I_o(s) = \frac{V_{dc}/L}{s^2 + (R/L)s + (1/LC)}.$$

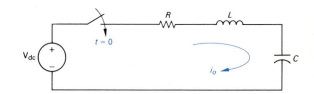

FIGURE P15.19

15.20 The circuit parameters in the circuit seen in Fig. P15.19 have the following values: $R = $ 600 Ω, $L = 1$ H, $C = 4$ μF, and $V_{dc} = $ 40 V. Find $i_o(t)$ for $t \geq 0$.

15.21 The switch in the circuit in Fig. P15.21 has
been in position a for a long time. At $t = 0$
the switch moves instantaneously to position b.

a) Derive the integrodifferential equation that
governs the behavior of the current i_o for
$t \geq 0^+$.

b) Show that

$$I_o(s) = \frac{I_{dc}[s + (1/RC)]}{[s^2 + (1/RC)s + (1/LC)]}$$

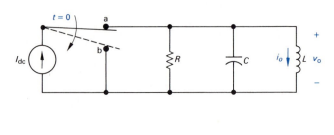

FIGURE P15.21

15.22 The circuit parameters in the circuit in Fig.
P15.21 are $R = 160\,\Omega$, $L = 0.10$ H, and
$C = 0.625\ \mu$F. If $I_{dc} = 300$ mA, find $v_o(t)$ for
$t \geq 0$.

15.23 Find $f(t)$ for each of the following functions:

a) $F(s) = \dfrac{8s^2 + 37s + 32}{(s + 1)\ (s + 2)(s + 4)}$

b) $F(s) = \dfrac{8s^3 + 89s^2 + 311s + 300}{s(s + 2)(s^2 + 8s + 15)}$

c) $F(s) = \dfrac{22s^2 + 60s + 58}{(s + 1)(s^2 + 4s + 5)}$

d) $F(s) = \dfrac{250(s + 7)(s + 14)}{s(s^2 + 14s + 50)}$

15.24 Find $f(t)$ for each of the following functions:

a) $F(s) = \dfrac{300}{s(s + 10)^2}$

b) $F(s) = \dfrac{500(s + 4)}{s^2(s^2 + 2s + 10)}$

c) $F(s) = \dfrac{400}{s(s + 2)^3}$

d) $F(s) = \dfrac{750(s + 10)^2}{s(s + 5)^4}$

e) $F(s) = \dfrac{200}{s(s^2 + 6s + 10)^2}$

15.25 Find $f(t)$ for each of the following functions:

a) $F(s) = \dfrac{5s^2 + 38s + 80}{s^2 + 6s + 8}$

b) $F(s) = \dfrac{10s^2 + 512s + 7186}{s^2 + 48s + 625}$

c) $F(s) = \dfrac{s^3 + 5s^2 - 50s - 100}{s^2 + 15s + 50}$

15.26 There is no energy stored in the circuit shown in Fig. P15.26 at the time the switch is closed.

a) Derive the integrodifferential equations that govern the behavior of the mesh currents i_1 and i_2.

b) Show that

$$I_1(s) = \frac{V_g[s^2 + (1/LC)]}{R[s^2 + (1/RC)s + (1/LC)]}$$

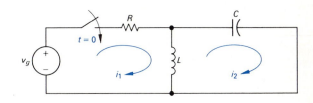

FIGURE P15.26

15.27 The values of the circuit parameters in the circuit shown in Fig. P15.26 are $R = 100\ \Omega$, $L = 6.25$ mH, and $C = 0.1\ \mu$F. If $v_g = 300\ u(t)$ V, find $i_1(t)$ for $t \geq 0$.

15.28 Derive the transform pair given by Eq. (15.67).

15.29 a) Derive the transform pair given by Eq. (15.86).

b) Derive the transform pair given by Eq. (15.87).

15.30 Apply the initial- and final-value theorems to each transform pair in Problem 15.23.

15.31 Apply the initial- and final-value theorems to each transform pair in Problem 15.24.

15.32 Use the initial- and final-value theorems to check the initial and final values of the current in Problem 15.19.

15.33 Use the initial- and final-value theorems to check the initial and final values of the current in Problem 15.21.

15.34 Use the initial- and final-value theorems to check the initial and final values of the currents i_1 and i_2 in Problem 15.27.

15.35 a) Write the two simultaneous differential equations that describe the circuit shown in Fig. P15.35 in terms of the mesh currents i_1 and i_2.

b) Laplace-transform the equations derived in part (a). Assume that initial energy stored in the circuit is zero.

c) Solve the equations in part (b) for $I_1(s)$ and $I_2(s)$.

d) Find $i_1(t)$ and $i_2(t)$.

e) Find $i_1(\infty)$ and $i_2(\infty)$.

f) Do the solutions for i_1 and i_2 make sense? Explain.

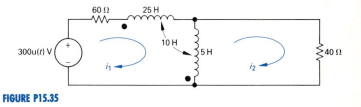

FIGURE P15.35

THE LAPLACE TRANSFORM IN CIRCUIT ANALYSIS

CHAPTER 16

The Laplace transform has two characteristics that make it an attractive tool in circuit analysis. First, it transforms a set of linear, constant-coefficient, differential equations into a set of linear polynomial equations. Second, it automatically introduces into the polynomial equations the initial values of the current and voltage variables. Thus initial conditions are an inherent part of the transform process. This property contrasts with the classical approach to the solution of a differential equation in which initial conditions are considered when the unknown coefficients are evaluated.

In this chapter we develop a systematic method for using the Laplace transform to find the transient behavior of a circuit. The five-step procedure we presented in Section 15.6 is the basis of our current discussion. The first step in utilizing the transform method efficiently is to eliminate the need to write the integro-differential equations that describe the circuit. We do so by developing s-domain equivalent circuits for the circuit elements. This approach allows construction of the circuit to be analyzed directly in the s domain. After formulating the s-domain circuit, we use the analytical tools already developed (such as node voltages, mesh currents, and circuit reductions) to describe the circuit by algebraic equations. The solution of these s-domain equations yield the unknown currents and voltages as rational functions of s, which we then inverse-transform by partial fraction expansions. Finally, we test the time-domain expressions to ensure that the solutions make sense in terms of the initial conditions and final values.

In Section 16.1 we develop the s-domain equivalent circuits for the circuit elements. The dimension of a transformed voltage

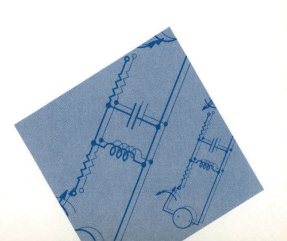

is volt-seconds, and the dimension of a transformed current is ampere-seconds. A voltage-to-current ratio in the s domain carries the dimension of volts per amperes. Hence an impedance in the s domain is measured in ohms and an admittance is measured in siemens, or mhos.

16.1 CIRCUIT ELEMENTS IN THE s DOMAIN

The procedure for developing the s-domain equivalent circuit for each circuit element is simple. First, we write the time-domain equation that relates the terminal voltage to the terminal current. Next, we take the Laplace transform of the time-domain equation. This step generates an algebraic relationship between the s-domain current and voltage. Finally, we construct a circuit model that satisfies the relationship between the s-domain current and voltage. We use the passive sign convention in all the derivations.

RESISTOR IN THE s DOMAIN

We begin with the resistance element. From Ohm's law,

$$v = Ri. \qquad (16.1)$$

As R is a constant, the Laplace transform of Eq. (16.1) is

$$V = RI, \qquad (16.2)$$

where

$$V = \mathcal{L}\{v\} \qquad \text{and} \qquad I = \mathcal{L}\{i\}.$$

Equation (16.2) states that the s-domain equivalent circuit of a resistor is simply a resistance of R ohms that carries a current of I ampere-seconds and has a terminal voltage of V volt-seconds.

Figure 16.1 shows the time- and frequency-domain circuits of the resistor. Note that going from the time domain to the frequency domain does not change the resistance element.

INDUCTOR IN THE s DOMAIN

Figure 16.2 shows an inductor carrying an initial current of I_0 amperes. The time-domain equation that relates the terminal voltage to the terminal current is

$$v = L\frac{di}{dt}. \qquad (16.3)$$

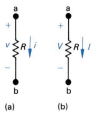

FIGURE 16.1 The resistance element: (a) time domain; (b) frequency domain.

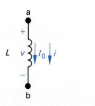

FIGURE 16.2 An inductor of L henrys carrying an initial current of I_0 amperes.

The Laplace transform of Eq. (16.3) gives

$$V = L[sI - i(0^-)] = sLI - LI_0. \qquad (16.4)$$

Two different circuit configurations satisfy Eq. (16.4). The first s-domain equivalent circuit consists of an impedance of sL ohms in series with an independent voltage source of LI_0 volt-seconds, as shown in Fig. 16.3. Note that the polarity marks on the voltage source LI_0 agree with the minus sign in Eq. (16.4). Note also that I_0 carries its own algebraic sign. That is, if the initial value of i is opposite to the reference direction for i, then I_0 has a negative value.

The second s-domain equivalent circuit that satisfies Eq. (16.4) consists of an impedance of sL ohms in parallel with an independent current source of I_0/s ampere-seconds, as shown in Fig. 16.4. We can derive the alternative equivalent circuit shown in Fig. 16.4 in several ways. One way is simply to solve Eq. (16.4) for the current I and then construct the circuit to satisfy the resulting equation. Thus

$$I = \frac{V + LI_0}{sL} = \frac{V}{sL} + \frac{I_0}{s}. \qquad (16.5)$$

Two other ways to derive the circuit shown in Fig. 16.4 are (1) to find the Norton equivalent of the circuit shown in Fig. 16.3 and (2) to start with the inductor current as a function of the inductor voltage and then find the Laplace transform of the resulting integral equation. We leave these two approaches to Problems 16.1 and 16.2.

If the initial energy stored in the inductor is zero, that is, if $I_0 = 0$, the s-domain equivalent circuit of the inductor reduces to an inductor with an impedance of sL ohms. Figure 16.5 shows this circuit.

CAPACITOR IN THE s DOMAIN

An initially charged capacitor also has two s-domain equivalent circuits. Figure 16.6 shows a capacitor initially charged to V_0 volts. The terminal current is

$$i = C\frac{dv}{dt}. \qquad (16.6)$$

Transforming Eq. (16.6) yields

$$I = C[sV - v(0^-)]$$

or

$$I = sCV - CV_0, \qquad (16.7)$$

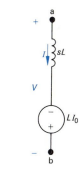

FIGURE 16.3 The series equivalent circuit for an inductor of L henrys carrying an initial current of I_0 amperes.

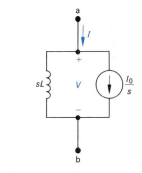

FIGURE 16.4 The parallel equivalent circuit for an inductor of L henrys carrying an initial current of I_0 amperes.

FIGURE 16.5 The s-domain circuit for an inductor when the initial current is zero.

FIGURE 16.6 A capacitor of C farads initially charged to V_0 volts.

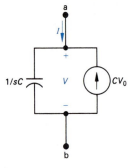

FIGURE 16.7 The parallel equivalent circuit for a capacitor initially charged to V_0 volts.

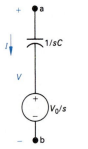

FIGURE 16.8 The series equivalent circuit for a capacitor initially charged to V_0 volts.

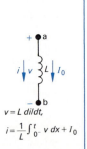

FIGURE 16.9 The s-domain circuit for a capacitor when the initial voltage is zero.

which indicates that the s-domain current I is the sum of two branch currents. One branch consists of an admittance of sC mhos and the second branch consists of an independent current source of CV_0 ampere-seconds. Figure 16.7 shows this equivalent circuit.

We derive the series equivalent circuit for the charged capacitor by solving Eq. (16.7) for V:

$$V = \left(\frac{1}{sC}\right) I + \frac{V_0}{s}. \tag{16.8}$$

Figure 16.8 shows the circuit that satisfies Eq. (16.8).

In the equivalent circuits shown in Figs. 16.7 and 16.8, V_0 carries its own algebraic sign. That is, if the polarity of V_0 is opposite to the reference polarity for v, V_0 is a negative quantity. If the initial voltage on the capacitor is zero, both equivalent circuits reduce to an impedance of $1/sC$ ohms, as shown in Fig. 16.9.

Table 16.1 summarizes the s-domain circuits developed in this section. We illustrate the application of these circuits in Section 16.3.

TABLE 16.1

SUMMARY OF THE s-DOMAIN EQUIVALENT CIRCUITS

Time Domain	Frequency Domain
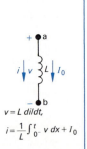	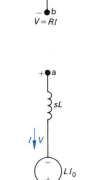 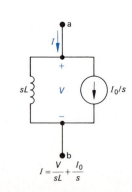

TABLE 16.1 (cont.)

SUMMARY OF THE s-DOMAIN EQUIVALENT CIRCUITS

Time Domain	Frequency Domain

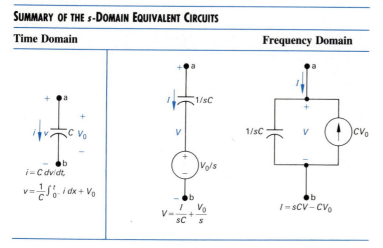

16.2 CIRCUIT ANALYSIS IN THE s DOMAIN

Before illustrating how to use the s-domain equivalent circuits in analysis, we need to lay some groundwork for all the work that follows.

First, we know that if no energy is stored in the inductor or capacitor, the relationship between the terminal voltage and current for each passive element takes the form:

$$V = ZI, \tag{16.9}$$

where Z refers to the s-domain impedance of the element. Thus a resistor has an impedance of R ohms, an inductor has an impedance of sL ohms, and a capacitor has an impedance of $1/sC$ ohms. The relationship contained in Eq. (16.9) also is contained in Figs. 16.1(b), 16.5, and 16.9. Equation (16.9) is sometimes referred to as *Ohm's law for the s domain*.

The reciprocal of the impedance is admittance. Therefore the s-domain admittance of a resistor is $1/R$ mhos, an inductor has an admittance of $1/sL$ mhos, and a capacitor has an admittance of sC mhos.

The rules for combining impedances and admittances in the s domain are the same as those for combining impedances and admittances in phasor-domain circuits. Thus series-parallel simplifications and Δ-to-Y conversions are applicable to s-domain analysis.

In addition, Kirchhoff's laws apply to s-domain currents and voltages. Their applicability stems from the operational transform, which states that the Laplace transform of a sum of time-domain functions is the sum of the transforms of the individual functions (see Table 15.2). Because the algebraic sum of the currents at a junction is zero in the time domain, the algebraic sum of the transformed currents also is zero. A similar statement holds for the algebraic sum of the transformed voltages around a closed path. The s-domain version of Kirchhoff's laws is

$$\text{alg } \Sigma I\text{'s} = 0; \tag{16.10}$$

$$\text{alg } \Sigma V\text{'s} = 0. \tag{16.11}$$

Because the voltage and current at the terminals of a passive element are related by an algebraic equation and because Kirchhoff's laws still hold, all the techniques of circuit analysis developed for pure resistive networks may be used in s-domain analysis. Thus node voltages, mesh currents, source transformations, and Thévenin–Norton equivalents all are valid techniques, even when energy is stored initially in the inductors and capacitors. Initially stored energy requires that we modify Eq. (16.9), but the modifications simply require the addition of independent sources either in series or parallel with the element impedances. The addition of these sources is governed by Kirchhoff's laws.

DRILL EXERCISES

16.1 A 125-Ω resistor, a 4-mH inductor, and a 0.1-μF capacitor are connected in parallel.

  a) Express the admittance of this parallel combination of elements as a rational function of s.

  b) Compute the numerical values of the zeros and poles.

ANSWER: (a) $10^{-7}(s^2 + 80{,}000s + 25 \times 10^8)/s$; (b) $-z_1 = -40{,}000 - j30{,}000$; $-z_2 = -40{,}000 + j30{,}000$, $p_1 = 0$.

16.2 The parallel circuit in Drill Exercise 16.1 is placed in series with a 500-Ω resistor.

  a) Express the impedance of this series combination as a rational function of s.

  b) Compute the numerical values of the zeros and poles.

ANSWER: (a) $500(s + 50{,}000)^2/(s^2 + 80{,}000s + 25 \times 10^8)$; (b) $-z_1 = -z_2 = -50{,}000$; $-p_1 = -40{,}000 - j30{,}000$, $-p_2 = -40{,}000 + j30{,}000$.

16.3 ILLUSTRATIVE EXAMPLES

We illustrate how to use the Laplace transform to determine the transient behavior of linear, lumped-parameter circuits with some circuit structures that we analyzed in Chapters 8 and 9. We start by analyzing familiar circuits to show that the Laplace transform approach yields the same results obtained in the earlier work.

NATURAL RESPONSE OF AN *RC* CIRCUIT

The first circuit that we analyze is the *RC* circuit shown in Fig. 16.10. The capacitor is initially charged to V_0 volts, and we are interested in the time-domain expressions for i and v. We first analyzed this circuit in Chapter 8, so you may want to review Section 8.2 before proceeding with this analysis.

We start by first finding i. In transferring the circuit in Fig. 16.10 to the s domain, we have a choice of two equivalent circuits for the charged capacitor. Because we are interested in the current, the series-equivalent circuit is more attractive; it results in a single-mesh circuit in the frequency domain. Thus we construct the s-domain circuit shown in Fig. 16.11.

Summing the voltages around the mesh in Fig. 16.11 generates the expression

$$\frac{V_0}{s} = \frac{1}{sC}I + RI. \tag{16.12}$$

Solving Eq. (16.12) for I yields

$$I = \frac{CV_0}{RCs + 1} = \frac{V_0/R}{s + (1/RC)}. \tag{16.13}$$

Note that the expression for I is a proper rational function of s and can be inverse-transformed by inspection:

$$i = \frac{V_0}{R}e^{-t/RC}u(t), \tag{16.14}$$

which is equivalent to the expression for the current derived by the classical methods discussed in Chapter 8. In Chapter 8, the current is given by Eq. (8.23) where τ is used in place of RC.

After we have found i, the easiest way to determine v is simply to apply Ohm's law; that is, from the circuit,

$$v = Ri = V_0e^{-t/RC}u(t). \tag{16.15}$$

We now illustrate a way to find v from the circuit without first finding i. In this alternative approach, we return to the original circuit of Fig. 16.10 and transfer it to the s domain using the parallel equivalent circuit for the charged capacitor. The parallel equivalent circuit is attractive now because its use makes possible describing the resulting circuit in terms of a single node voltage. Figure 16.12 shows the new s-domain equivalent circuit.

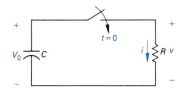

FIGURE 16.10 The capacitor discharge circuit.

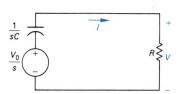

FIGURE 16.11 An s-domain equivalent circuit for the circuit shown in Fig. 16.10.

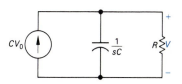

FIGURE 16.12 An s-domain equivalent circuit for the circuit shown in Fig. 16.10.

The node-voltage equation that describes the circuit in Fig. 16.12 is

$$\frac{V}{R} + sCV = CV_0. \qquad (16.16)$$

Solving Eq. (16.16) for V gives

$$V = \frac{V_0}{s + (1/RC)}. \qquad (16.17)$$

Inverse-transforming Eq. (16.17) leads to the same expression for v given by Eq. (16.15), namely,

$$v = V_0 e^{-t/RC} = V_0 e^{-t/\tau} u(t). \qquad (16.18)$$

Our purpose in deriving v by direct use of the transform method is to show that the choice of which s-domain equivalent circuit to use is influenced by which response signal is of interest.

DRILL EXERCISES

16.3 The switch in the circuit shown has been in position a for a long time. At $t = 0$, the switch is thrown to position b.

a) Find I, V_1, and V_2 as rational functions of s.

b) Find the time-domain expressions for i, v_1, and v_2.

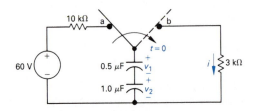

ANSWER: (a) $I = 0.02/(s + 1000)$, $V_1 = 40/s + 1000$, $V_2 = 20/(s + 1000)$; (b) $i = 20e^{-1000t} u(t)$ mA, $v_1 = 40e^{-1000t} u(t)$ V, $v_2 = 20e^{-1000t} u(t)$ V.

STEP RESPONSE OF A PARALLEL *RLC* CIRCUIT

Next we analyze the parallel *RLC* circuit that we first analyzed in Example 9.6, as shown in Fig. 16.13. The problem is to find the expression for i_L after the constant current source is switched across the parallel elements. The initial energy stored in the circuit is zero.

As before, we begin by constructing the s-domain equivalent circuit shown in Fig. 16.14. Note how easily an independent source can be transformed from the time domain to the frequency domain. We transfer the source to the s domain simply by determining the Laplace transform of the time-domain func-

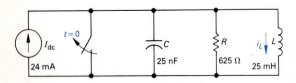

FIGURE 16.13 Step response of a parallel *RLC* circuit.

tion of the source. Here, opening the switch results in a step change in the current applied to the circuit. Therefore the s-domain current source is $\mathcal{L}\{I_{dc}u(t)\}$, or I_{dc}/s. To find I_L, we first solve for V and then use

$$I_L = \frac{V}{sL} \qquad (16.19)$$

to establish the s-domain expression for I_L. Summing the currents away from the top node in the circuit shown in Fig. 16.14 generates the expression

$$sCV + \frac{V}{R} + \frac{V}{sL} = \frac{I_{dc}}{s}. \qquad (16.20)$$

Solving Eq. (16.20) for V gives

$$V = \frac{I_{dc}/C}{s^2 + (1/RC)s + (1/LC)}. \qquad (16.21)$$

Substituting Eq. (16.21) into Eq. (16.19) gives

$$I_L = \frac{I_{dc}/LC}{s[s^2 + (1/RC)\, s + (1/LC)]}. \qquad (16.22)$$

Substituting the numerical values of R, L, C, and I_{dc} into Eq. (16.22) yields

$$I_L = \frac{384 \times 10^5}{s(s^2 + 64{,}000s + 16 \times 10^8)}. \qquad (16.23)$$

Before expanding Eq. (16.23) into a sum of partial fractions, we factor the quadratic term in the denominator:

$$I_L = \frac{384 \times 10^5}{s(s + 32{,}000 - j24{,}000)(s + 32{,}000 + j24{,}000)}. \qquad (16.24)$$

Now, we can test the s-domain expression for I_L by checking to see whether the final-value theorem predicts the correct value for i_L at $t = \infty$. All the poles of I_L, except for the first-order pole at the origin, lie in the left half of the s plane, so the theorem is applicable. We know from the behavior of the circuit that after the switch has been open for a long time the inductor will short circuit the current source. Therefore the final value of i_L must be 24 mA. The limit of sI_L as $s \to 0$ is

$$\lim_{s \to 0} sI_L = \frac{384 \times 10^5}{16 \times 10^8} = 24 \text{ mA}. \qquad (16.25)$$

(Currents in the s domain carry the dimension of ampere-second, so the dimension of sI_L will be amperes.)

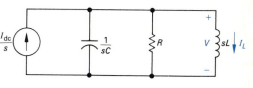

FIGURE 16.14 The s-domain equivalent circuit for the circuit shown in Fig. 16.13.

We now proceed with the partial fraction expansion of Eq. (16.24):

$$I_L = \frac{K_1}{s} + \frac{K_2}{s + 32{,}000 - j24{,}000}$$

$$+ \frac{K_2^*}{s + 32{,}000 + j24{,}000}. \qquad (16.26)$$

The partial fraction coefficients are

$$K_1 = \frac{384 \times 10^5}{16 \times 10^8} = 24 \times 10^{-3}; \qquad (16.27)$$

$$K_2 = \frac{384 \times 10^5}{(-32{,}000 + j24{,}000)(j48{,}000)}$$

$$= 20 \times 10^{-3}\underline{/126.87°}. \qquad (16.28)$$

Substituting the numerical values of K_1 and K_2 into Eq. (16.26) and inverse-transforming the resulting expression yields

$$i_L = [24 + 40e^{-32{,}000t} \cos (24{,}000t + 126.87°)] \times$$

$$u(t) \text{ mA}. \qquad (16.29)$$

The answer given by Eq. (16.29) is equivalent to the answer given for Example 9.7 because

$$40 \cos (24{,}000t + 126.87°) =$$

$$-24 \cos 24{,}000t - 32 \sin 24{,}000t.$$

If we weren't using a previous solution as a check, we would test Eq. (16.29) to make sure that $i_L(0)$ satisfied the given initial conditions and $i_L(\infty)$ satisfied the known behavior of the circuit.

DRILL EXERCISES

16.4 The energy stored in the circuit shown is zero at the time when the switch is closed.

a) Find the s-domain expression for I.

b) Find the time-domain expression for i when $t > 0$.

c) Find the s-domain expression for V.

d) Find the time-domain expression for v when $t > 0$.

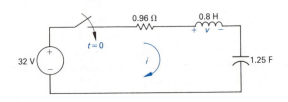

ANSWER: (a) $I = 40/(s^2 + 1.2s + 1)$; (b) $i = (50e^{-0.6t} \sin 0.8t)u(t)$ A; (c) $V = 32s/(s^2 + 1.2s + 1)$; (d) $v = [40e^{-0.6t} \cos (0.8t + 36.87°)]u(t)$ V.

TRANSIENT RESPONSE OF A PARALLEL *RLC* CIRCUIT

Another example of using the Laplace transform to find the transient behavior of a circuit arises from replacing the dc current source with a sinusoidal current source in the circuit shown in Fig. 16.13. The current source is

$$i_g = I_m \cos \omega t \text{ A}, \qquad (16.30)$$

where $I_m = 24$ mA and $\omega = 40,000$ rad/s. As before, we assume that the initial energy stored in the circuit is zero.

The s-domain expression for the source current is

$$I_g = \frac{sI_m}{s^2 + \omega^2}. \qquad (16.31)$$

The voltage across the parallel elements is

$$V = \frac{(I_g/C)s}{s^2 + (1/RC)s + (1/LC)}. \qquad (16.32)$$

Substituting Eq. (16.31) into Eq. (16.32) results in

$$V = \frac{(I_m/C)s^2}{(s^2 + \omega^2)[s^2 + (1/RC)s + (1/LC)]}, \qquad (16.33)$$

from which

$$I_L = \frac{V}{sL} = \frac{(I_m/LC)s}{(s^2 + \omega^2)[s^2 + (1/RC)s + (1/LC)]}. \qquad (16.34)$$

Substituting the numerical values of I_m, ω, R, L, and C into Eq. (16.34) gives

$$I_L = \frac{384 \times 10^5 s}{(s^2 + 16 \times 10^8)(s^2 + 64,000s + 16 \times 10^8)}. \qquad (16.35)$$

We now write the denominator in factored form:

$$I_L = \frac{384 \times 10^5 s}{(s - ja)(s + ja)(s + \alpha - j\beta)(s + \alpha + j\beta)}, \qquad (16.36)$$

where $a = 40,000$, $\alpha = 32,000$, and $\beta = 24,000$.

We can't test the final value of i_L with the final-value theorem, because I_L has a pair of poles on the imaginary axis, that is, poles at $\pm j4 \times 10^4$. Thus we must first find i_L and then check the validity of the expression from known circuit behavior.

If we expand Eq. (16.36) into a sum of partial fractions, we generate the equation:

$$I_L = \frac{K_1}{s - j40,000} + \frac{K_1^*}{s + j40,000} + \frac{K_2}{s + 32,000 - j24,000}$$

$$+ \frac{K_2^*}{s + 32,000 + j24,000}. \qquad (16.37)$$

The numerical values of the coefficients K_1 and K_2 are

$$K_1 = \frac{384 \times 10^5(j40,000)}{(j80,000)(32,000 + j16,000)(32,000 + j64,000)}$$

$$= 7.5 \times 10^{-3}\underline{/-90°}; \quad \text{(16.38)}$$

$$K_2 = \frac{384 \times 10^5(-32,000 + j24,000)}{(-32,000 - j16,000)(-32,000 + j64,000)j48,000}$$

$$= 12.5 \times 10^{-3}\underline{/90°}. \quad \text{(16.39)}$$

Substituting the numerical values from Eqs. (16.38) and (16.39) into Eq. (16.37) and inverse-transforming the resulting expression yields

$$i_L = [15 \cos (40,000t - 90°)$$
$$+ 25e^{-32,000t} \cos (24,000t + 90°)] \text{ mA}$$
$$= (15 \sin 40,000t - 25e^{-32,000t} \sin 24,000t)u(t) \text{ mA}. \quad \text{(16.40)}$$

We now test Eq. (16.40) to see whether it makes sense in terms of the given initial conditions and the known circuit behavior after the switch has been open for a long time. For $t = 0$, Eq. (16.40) predicts zero initial current, which agrees with the initial energy of zero in the circuit. Equation (16.40) predicts a steady-state current of

$$i_{L_{ss}} = 15 \sin 40,000t \text{ mA}, \quad \text{(16.41)}$$

which can be verified by the phasor method (Chapter 10).

THE STEP RESPONSE OF A MULTIPLE-MESH CIRCUIT

Refer to the circuit shown in Fig. 16.15. The problem is to find the branch currents i_1 and i_2 that arise when the 336-V dc voltage source is applied suddenly to the circuit. The initial energy stored in the circuit is zero. We were not able to solve this type of transient response by the analytic technique presented in Chapter 8 because it involves the solution of two simultaneous first-order differential equations. Although there are techniques for solving simultaneous differential equations in the time domain, we chose not to introduce them in this text. However, because the Laplace transform method changes the problem to the solution of algebraic equations, we can now analyze multiple-mesh and node circuits.

Figure 16.16 shows the s-domain equivalent circuit of the circuit shown in Fig. 16.15. The two mesh-current equations are

$$\frac{336}{s} = (42 + 8.4s)I_1 - 42I_2; \quad \text{(16.42)}$$

$$0 = -42I_1 + (90 + 10s)I_2. \quad \text{(16.43)}$$

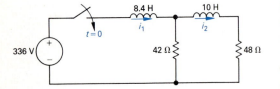

FIGURE 16.15 A multiple-mesh RL circuit.

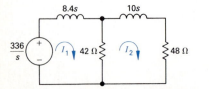

FIGURE 16.16 The s-domain equivalent circuit for the circuit shown in Fig. 16.15.

Using Cramer's method to solve for I_1 and I_2, we obtain

$$\Delta = \begin{vmatrix} 42 + 8.4s & -42 \\ -42 & 90 + 10s \end{vmatrix}$$

$$= 84(s^2 + 14s + 24)$$

$$= 84(s + 2)(s + 12); \qquad (16.44)$$

$$N_1 = \begin{vmatrix} \dfrac{336}{s} & -42 \\ 0 & 90 + 10s \end{vmatrix}$$

$$= \frac{3360(s + 9)}{s}; \qquad (16.45)$$

$$N_2 = \begin{vmatrix} 8.4s + 42 & \dfrac{336}{s} \\ -42 & 0 \end{vmatrix}$$

$$= \frac{14,112}{s}. \qquad (16.46)$$

Based on Eqs. (16.44)–(16.46),

$$I_1 = \frac{N_1}{\Delta} = \frac{40(s + 9)}{s(s + 2)(s + 12)}; \qquad (16.47)$$

$$I_2 = \frac{N_2}{\Delta} = \frac{168}{s(s + 2)(s + 12)}. \qquad (16.48)$$

Expanding I_1 and I_2 into a sum of partial fractions gives

$$I_1 = \frac{15}{s} - \frac{14}{s + 2} - \frac{1}{s + 12}; \qquad (16.49)$$

$$I_2 = \frac{7}{s} - \frac{8.4}{s + 2} + \frac{1.4}{s + 12}. \qquad (16.50)$$

We obtain the expressions for i_1 and i_2 by inverse-transforming Eqs. (16.49) and (16.50), respectively:

$$i_1 = (15 - 14e^{-2t} - e^{-12t})u(t) \text{ A}; \qquad (16.51)$$

$$i_2 = (7 - 8.4e^{-2t} + 1.4e^{-12t})u(t) \text{ A}. \qquad (16.52)$$

Next we test the solutions to see whether they make sense in terms of the circuit. Because no energy is stored in the circuit at the instant when the switch is closed, both $i_1(0^-)$ and $i_2(0^-)$ must be zero. The solutions agree with these initial values. After the switch has been closed for a long time, the two inductors appear as short circuits. Therefore the final values of i_1 and i_2 are

$$i_1(\infty) = \frac{336(90)}{42(48)} = 15 \text{ A}, \qquad (16.53)$$

and

$$i_2(\infty) = \frac{15(42)}{90} = 7 \text{ A}. \qquad (16.54)$$

The solutions also agree with these final values.

One final test involves the numerical values of the exponents and calculating the voltage drop across the 42-Ω resistor by three different methods. From the circuit, the voltage across the 42-Ω resistor (positive at the top of the resistor) is

$$v = 42(i_1 - i_2) = 336 - 8.4\frac{di_1}{dt} = 48i_2 + 10\frac{di_2}{dt}. \qquad (16.55)$$

We leave to you verification that regardless of which form of Eq. (16.55) is used, the voltage is

$$v = (336 - 235.2e^{-2t} - 100.80e^{-12t})u(t) \text{ V}.$$

We are confident that the solutions for i_1 and i_2 are correct.

DRILL EXERCISES

16.5 The dc current and voltage sources are applied simultaneously to the circuit shown. No energy is stored in the circuit at the instant of application.

a) Derive the s-domain expressions for V_1 and V_2.

b) For $t > 0$, derive the time-domain expressions for v_1 and v_2.

c) Calculate $v_1(0^+)$ and $v_2(0^+)$.

d) Compute the steady-state values of v_1 and v_2.

ANSWER: (a) $V_1 = [5(s + 3)]/[s(s + 0.5)(s + 2)]$, $V_2 = [2.5(s^2 + 6)]/[s(s + 0.5)(s + 2)]$;
(b) $v_1 = (15 - \frac{50}{3}e^{-0.5t} + \frac{5}{3}e^{-2t})u(t)$ V,
$v_2 = (15 - \frac{125}{6}e^{-0.5t} + \frac{25}{3}e^{-2t})u(t)$ V; (c) $v_1(0^+) = 0$,
$v_2(0^+) = 2.5$ V; (d) $v_1 = v_2 = 15$ V.

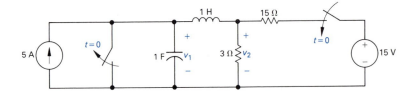

USE OF THÉVENIN'S EQUIVALENT

Here, we show how to use Thévenin's equivalent in the s domain. Figure 16.17 illustrates the circuit to be analyzed. The problem is to find the capacitor current that results from closing the switch. The energy stored in the circuit prior to closing the switch is zero.

To find i_C we first construct the s-domain equivalent circuit of the circuit shown in Fig. 16.17 and then find the Thévenin equivalent of the s-domain circuit with respect to the terminals of the capacitor. Figure 16.18 shows the s-domain circuit.

The Thévenin voltage is the open-circuit voltage at terminals a and b. Under open-circuit conditions, there is no voltage across the 60-Ω resistor. Hence

$$V_{Th} = \frac{(480/s)(0.002s)}{20 + 0.002s} = \frac{480}{s + 10^4}. \qquad (16.56)$$

The Thévenin impedance seen from terminals a and b equals the 60-Ω resistor in series with the parallel combination of the 20-Ω resistor and the 2-mH inductor. Thus

$$Z_{Th} = 60 + \frac{0.002s(20)}{20 + 0.002s} = \frac{80(s + 7500)}{s + 10^4}. \qquad (16.57)$$

Using the Thévenin equivalent, we reduce the circuit shown in Fig. 16.18 to that shown in Fig. 16.19. It indicates that the capacitor current I_C equals the Thévenin voltage divided by the total series impedance. Thus

$$I_C = \frac{480/(s + 10^4)}{[80(s + 7500)/(s + 10^4)] + [(2 \times 10^5)/s]}. \qquad (16.58)$$

We simplify Eq. (16.58) to

$$I_C = \frac{6s}{s^2 + 10,000s + 25 \times 10^6}$$

$$= \frac{6s}{(s + 5000)^2}. \qquad (16.59)$$

A partial fraction expansion of Eq. (16.59) generates

$$I_C = \frac{-30,000}{(s + 5000)^2} + \frac{6}{(s + 5000)}, \qquad (16.60)$$

the inverse transform of which is

$$i_C = (-30,000te^{-5000t} + 6e^{-5000t})u(t) \text{ A.} \qquad (16.61)$$

We now test Eq. (16.61) to see whether it makes sense in terms of known circuit behavior. From Eq. (16.61),

$$i_C(0) = 6 \text{ A.} \qquad (16.62)$$

This result agrees with the initial current in the capacitor as calculated from the circuit in Fig. 16.17. The initial inductor current is zero and the initial capacitor voltage is zero, so the initial capacitor current is 480/80, or 6 A. The final value of the current is zero, which also agrees with Eq. (16.61).

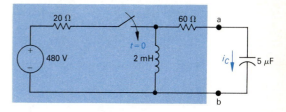

FIGURE 16.17 A circuit showing the use of Thévenin's equivalent in the s domain.

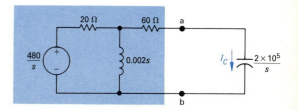

FIGURE 16.18 The s-domain model of the circuit shown in Fig. 16.17.

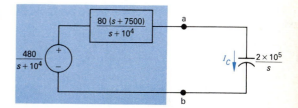

FIGURE 16.19 A simplified version of the circuit shown in Fig. 16.18 using a Thévenin equivalent.

Note also from Eq. (16.61) that the current reverses sign when t exceeds 6/30,000, or 200 μs. The fact that i_C reverses sign makes sense because, when the switch first closes, the capacitor begins to charge. Eventually this charge is reduced to zero because the inductor is a short circuit at $t = \infty$. The sign reversal of i_C reflects the charging and discharging of the capacitor.

Let's assume that the voltage drop across the capacitor v_{ab} also is of interest. Once we know i_C, we find v_{ab} by integration in the time domain; that is,

$$v_C = 2 \times 10^5 \int_{0^-}^{t} (6 - 30{,}000x)e^{-5000x}\,dx. \qquad \textbf{(16.63)}$$

Although the integration called for in Eq. (16.63) is not difficult, we may avoid it altogether by first finding the s-domain expression for V_C and then finding v_C by an inverse transform. Thus

$$V_C = \frac{1}{sC}I_C = \frac{2 \times 10^5}{s}\frac{6s}{(s + 5000)^2}$$

$$= \frac{12 \times 10^5}{(s + 5000)^2}, \qquad \textbf{(16.64)}$$

from which

$$v_C = 12 \times 10^5 te^{-5000t}u(t). \qquad \textbf{(16.65)}$$

We leave to you verification that Eq. (16.65) is consistent with Eq. (16.61) and that it also supports the observations made with regard to the behavior of i_C. (See Problem 16.35.)

DRILL EXERCISES

16.6 The initial charge on the capacitor in the circuit shown is zero.

a) Find the s-domain Thévenin equivalent circuit with respect to terminals a and b.

b) Find the s-domain expression for the current that the circuit delivers to a load consisting of a 0.4-H inductor in series with a 1-Ω resistor.

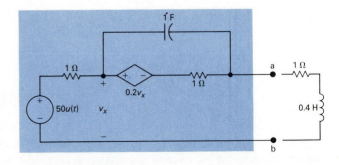

ANSWER: (a) $V_{Th} = V_{ab} = [50(s + 0.8)]/[s(s + 1)]$, $Z_{Th} = (s + 1.8)/(s + 1)$; (b) $I_{ab} = [125(s + 0.8)]/[s(s^2 + 6s + 7)]$.

CIRCUIT WITH MUTUAL INDUCTANCE

This last introductory example illustrates how to analyze by the Laplace transform the transient response of a circuit that contains mutual inductance. Figure 16.20 shows the circuit. The make-before-break switch has been in position a for a long time. At $t = 0$, the switch moves instantaneously to position b. The problem is to derive the time-domain expression for i_2.

We begin by redrawing the circuit in Fig. 16.20, with the switch in position b and the magnetically coupled coils replaced with a T-equivalent circuit. Figure 16.21 shows the circuit.

We now transfer the circuit shown in Fig. 16.21 to the s domain. In so doing, we note that

$$i_1(0^-) = \frac{60}{12} = 5 \text{ A}; \qquad (16.66)$$

$$i_2(0^-) = 0. \qquad (16.67)$$

Because we plan to use mesh analysis in the s domain, we use the series equivalent circuit for an inductor carrying an initial current. Figure 16.22 shows the s-domain circuit. Note that only one independent voltage source appears in the circuit. This source appears in the vertical leg of the tee to account for the initial value of the current in the 2-H inductor of $i_1(0^-) + i_2(0^-)$, or 5 A. The branch carrying i_1 has no voltage source because $L_1 - M = 0$.

The two s-domain mesh equations that describe the circuit in Fig. 16.22 are

$$(3 + 2s)I_1 + 2sI_2 = 10 \qquad (16.68)$$

and

$$2sI_1 + (12 + 8s)I_2 = 10. \qquad (16.69)$$

Solving for I_2 yields

$$I_2 = \frac{2.5}{(s + 1)(s + 3)}. \qquad (16.70)$$

Expanding Eq. (16.70) into a sum of partial fractions generates

$$I_2 = \frac{1.25}{s + 1} - \frac{1.25}{s + 3}. \qquad (16.71)$$

Then,

$$i_2 = (1.25e^{-t} - 1.25e^{-3t})u(t) \text{ A}. \qquad (16.72)$$

Equation (16.72) reveals that i_2 increases from zero to a peak value of 481.13 mA 549.31 ms after the switch is moved to position b. Thereafter, i_2 decreases exponentially toward zero. Figure 16.23 shows a plot of i_2 versus t. This response makes sense

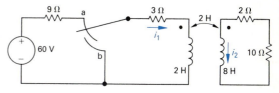

FIGURE 16.20 A circuit containing magnetically coupled coils.

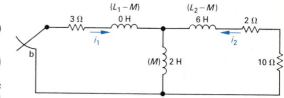

FIGURE 16.21 The circuit shown in Fig. 16.20 with the magnetically coupled coils replaced by a T-equivalent circuit.

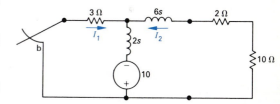

FIGURE 16.22 The s-domain equivalent circuit for the circuit shown in Fig. 16.21.

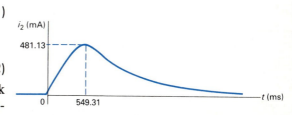

FIGURE 16.23 The plot of i_2 versus t for the circuit shown in Fig. 16.20.

in terms of the known physical behavior of the magnetically coupled coils. A current can exist in the L_2 inductor only if there is a time-varying current in the L_1 inductor. As i_1 decreases from its initial value of 5 A, i_2 increases from zero and then approaches zero as i_1 approaches zero.

DRILL EXERCISES

16.7 a) Verify from Eq. (16.72) that i_2 reaches a peak value of 481.13 mA at $t = 549.31$ ms.

b) Find i_1, for $t > 0$, for the circuit shown in Fig. 16.20.

c) Compute di_1/dt when i_2 is at its peak value.

d) Express i_2 as a function of di_1/dt when i_2 is at its peak value.

e) Use the results obtained in (c) and (d) to calculate the peak value of i_2.

ANSWER: (a) $di_2/dt = 0$ when $t = \frac{1}{2} \ln 3$ (s);
(b) $i_1 = 2.5(e^{-t} + e^{-3t})u(t)$ A; (c) -2.89A/s;
(d) $i_2 = -(Mdi_1/dt)/12$; (e) 481.13 mA.

16.4 THE IMPULSE FUNCTION IN CIRCUIT ANALYSIS

Impulse functions occur in circuit analysis either because of a switching operation or because the circuit is excited by an impulsive source. We begin our discussion by showing how to create an impulse function with a switching operation.

SWITCHING OPERATIONS

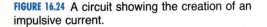

FIGURE 16.24 A circuit showing the creation of an impulsive current.

CAPACITOR CIRCUIT In the circuit shown in Fig. 16.24, the capacitor C_1 is charged to an initial voltage of V_0 at the time the switch is closed. The initial charge on C_2 is zero. The problem is to find the expression for $i(t)$ as $R \to 0$. Figure 16.25 shows the s-domain equivalent circuit.

From Fig. 16.25,

$$I = \frac{V_0/s}{R + (1/sC_1) + (1/sC_2)}$$

$$= \frac{V_0/R}{s + (1/RC_e)}, \qquad \textbf{(16.73)}$$

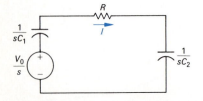

FIGURE 16.25 The s-domain equivalent circuit for the circuit shown in Fig. 16.24.

where we introduced C_e for the equivalent capacitance $C_1 C_2/(C_1 + C_2)$.

We inverse-transform Eq. (16.73) by inspection to obtain

$$i = \left(\frac{V_0}{R}e^{-t/RC_e}\right)u(t), \qquad \textbf{(16.74)}$$

which indicates that as R decreases the initial current (V_0/R) increases and the time constant (RC_e) decreases. Thus as R gets smaller, the current starts from a larger initial value and then drops off more rapidly. Figure 16.26 shows these characteristics of i.

Apparently, i is approaching an impulse function as R approaches zero, because the initial value of i is approaching infinity and the duration of i is approaching zero. We still have to determine whether the area under the current function is independent of R. Physically the total area under the i versus t curve represents the total charge transferred to C_2 after the switch is closed. Thus

$$\text{Area} = q = \int_{0^-}^{\infty} \frac{V_0}{R}e^{-t/RC_e}dt = V_0C_e, \qquad \textbf{(16.75)}$$

which says that the total charge transferred to C_2 is independent of R and equals V_0C_e coulombs. Thus, as R approaches zero, the current approaches an impulse strength V_0C_e; that is,

$$i \rightarrow V_0C_e\delta(t). \qquad \textbf{(16.76)}$$

The physical interpretation of Eq. (16.76) is that when $R = 0$ a finite amount of charge is transferred to C_2 instantaneously. Making R zero in the circuit shown in Fig. 16.24 shows why we get an instantaneous transfer of charge. With $R = 0$, we create a contradiction when we close the switch. That is, we apply a voltage across a capacitor that has a zero initial voltage. The only way to have an instantaneous change in capacitor voltage is to have an instantaneous transfer of charge. When the switch is closed, the voltage across C_2 does not jump to V_0 but to its final value of

$$v_2 = \frac{C_1 V_0}{C_1 + C_2}. \qquad \textbf{(16.77)}$$

We leave the derivation of Eq. (16.77) to you. (See Problem 16.42.)

If we set R equal to zero at the outset, the Laplace transform analysis will predict the impulsive current response. Thus

$$I = \frac{V_0/s}{(1/sC_1) + (1/sC_2)} = \frac{C_1 C_2 V_0}{C_1 + C_2} = C_e V_0. \qquad \textbf{(16.78)}$$

In writing Eq. (16.78), we use the capacitor voltages at $t = 0^-$.

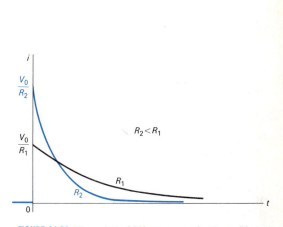

FIGURE 16.26 The plot of $i(t)$ versus t for two different values of R.

PSpice Use pulsed sources in PSpice to simulate an impulse voltage or current source: Chapter 16, Example 20

The inverse transform of a constant is the constant times the impulse function; therefore, from Eq. (16.78),

$$i = C_e V_0 \delta(t). \qquad (16.79)$$

The ability of the Laplace transform to predict correctly the occurrence of an impulsive response is one reason why the transform is widely used to analyze the transient behavior of linear, lumped-parameter, time-invariant circuits.

SERIES INDUCTOR CIRCUIT We illustrate a second switching operation that produces an impulsive response with the circuit shown in Fig. 16.27. The problem is to find the time-domain expression for v_o after the switch has been opened. Note that opening the switch forces an instantaneous change in the current of L_2, which causes v_o to contain an impulsive component.

Figure 16.28 shows the s-domain equivalent with the switch open. In deriving this circuit we recognized that the current in the 3-H inductor at $t = 0^-$ is 10 A and the current in the 2-H inductor at 0^- is zero. Using the initial conditions at $t = 0^-$ is a direct consequence of our using 0^- as the lower limit on the defining integral of the Laplace transform.

We derive the expression for V_o from a single node-voltage equation. Summing the currents away from the node between the 15-Ω resistor and the 30-V source gives

$$\frac{V_o}{2s + 15} + \frac{V_o - [(100/s) + 30]}{3s + 10} = 0. \qquad (16.80)$$

Solving for V_o yields

$$V_o = \frac{40(s + 7.5)}{s(s + 5)} + \frac{12(s + 7.5)}{s + 5}. \qquad (16.81)$$

We anticipate that v_o will contain an impulse term because the second term on the right-hand side of Eq. (16.81) is an improper rational function. We can express this improper fraction as a constant plus a rational function by simply dividing the denominator into the numerator; that is,

$$\frac{12(s + 7.5)}{s + 5} = 12 + \frac{30}{s + 5}. \qquad (16.82)$$

Combining Eq. (16.82) with the partial fraction expansion of the first term on the right-hand side of Eq. (16.81) gives

$$V_o = \frac{60}{s} - \frac{20}{s + 5} + 12 + \frac{30}{s + 5}$$

$$= 12 + \frac{60}{s} + \frac{10}{s + 5}, \qquad (16.83)$$

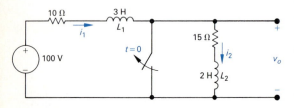

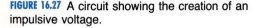

FIGURE 16.27 A circuit showing the creation of an impulsive voltage.

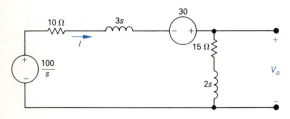

FIGURE 16.28 The s-domain equivalent circuit for the circuit shown in Fig. 16.27.

from which

$$v_o = 12\delta(t) + (60 + 10e^{-5t})u(t) \text{ V.} \qquad \textbf{(16.84)}$$

Does this solution make sense? Before answering that question, let's first derive the expression for the current when $t > 0^-$. After the switch has been opened, the current in L_1 is the same as the current in L_2. If we reference the current clockwise around the mesh, the s-domain expression is

$$I = \frac{(100/s) + 30}{5s + 25} = \frac{20}{s(s + 5)} + \frac{6}{s + 5}$$

$$= \frac{4}{s} - \frac{4}{s + 5} + \frac{6}{s + 5}$$

$$= \frac{4}{s} + \frac{2}{s + 5}. \qquad \textbf{(16.85)}$$

Inverse-transforming Eq. (16.85) gives

$$i = (4 + 2e^{-5t})u(t) \text{ A.} \qquad \textbf{(16.86)}$$

We now make the following observations: (1) Before the switch is opened, the current in L_1 is 10 A and the current in L_2 is 0 A; and (2) from Eq. (16.86) we know that at $t = 0^+$ the current in L_1 and L_2 is 6 A. Then, the current in L_1 changes instantaneously from 10 to 6 A, while the current in L_2 changes instantaneously from 0 to 6 A. From this value of 6 A, the current decreases exponentially to a final value of 4 A. This final value is easily verified from the circuit; that is, it should equal 100/25, or 4 A. Figure 16.29 shows these characteristics of i_1 and i_2 graphically.

How can we verify that these instantaneous jumps in the inductor current make sense in terms of the physical behavior of the circuit? First, we note that the switching operation places the two inductors in series. Any impulsive voltage appearing across the 3-H inductor must be exactly balanced by an impulsive voltage across the 2-H inductor, because the sum of the impulsive voltages around a closed path must equal zero. Faraday's law states that the induced voltage is proportional to the change in flux linkage ($v = d\lambda/dt$). Therefore in this series circuit the *change* in flux linkage must sum to zero. In other words, the total flux linkage immediately after switching is the same as that before switching. For the circuit here, the flux linkage before switching is

$$\lambda = L_1i_1 + L_2i_2 = 3(10) + 2(0) = 30 \text{ Wb-turns} \quad \textbf{(16.87)}$$

Immediately after switching, it is

$$\lambda = (L_1 + L_2)i(0^+) = 5i(0^+). \qquad \textbf{(16.88)}$$

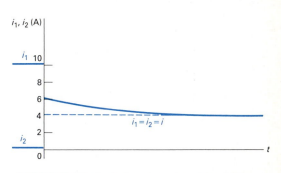

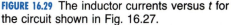

FIGURE 16.29 The inductor currents versus t for the circuit shown in Fig. 16.27.

Combining Eqs. (16.87) and (16.88) gives

$$i(0^+) = 30/5 = 6 \text{ A}. \qquad (16.89)$$

Thus the solution for i (Eq. 16.86) agrees with the principle of the conservation of flux linkage.

We now test the validity of Eq. (16.84). First we check the impulsive term $12\delta(t)$. The instantaneous jump of i_2 from 0 to 6 A at $t = 0$ gives rise to an impulse of strength $6\delta(t)$ in the derivative of i_2. This impulse in the derivative of i_2 gives rise to the $12\delta(t)$ in the voltage across the 2-H inductor. For $t > 0^+$, di_2/dt is $-10e^{-5t}$ A/s; therefore the voltage v_o is

$$v_o = 15(4 + 2e^{-5t}) + 2(-10e^{-5t})$$
$$= (60 + 10e^{-5t})u(t) \text{ V}. \qquad (16.90)$$

Equation (16.90) agrees with the last two terms on the right-hand side of Eq. (16.84); thus we have confirmed that Eq. (16.84) does make sense in terms of known circuit behavior.

We can also check the instantaneous drop from 10 to 6 A in the current i_1. This drop gives rise to an impulse of $-4\delta(t)$ in the derivative of i_1. Therefore the voltage across L_1 contains an impulse of $-12\delta(t)$ at the origin. This impulse exactly balances the impulse across L_2; that is, the sum of the impulsive voltages around a closed path equals zero.

PARALLEL INDUCTOR CIRCUIT We illustrate a third impulsive response that results from a switching operation with the circuit shown in Fig. 16.30. The switch has been closed for a long time prior to opening. The initial currents in the 2-H and 1-H inductors are 6 and 24 A, as shown in Fig. 16.30. The first step in solving for v, i_1, and i_2 is to construct the s-domain equivalent circuit, shown in Fig. 16.31.

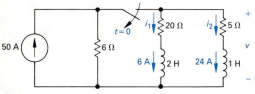

FIGURE 16.30 A circuit showing an impulsive response.

We find V by writing a single node-voltage equation:

$$\frac{V + 12}{2s + 20} + \frac{V + 24}{s + 5} = 0. \qquad (16.91)$$

Solving Eq. (16.91) for V gives

$$V = -\frac{20(s + 9)}{s + (25/3)}. \qquad (16.92)$$

Equation (16.92) is an improper rational function, so we divide the denominator into the numerator to obtain

$$V = -20 - \frac{40/3}{s + (25/3)}. \qquad (16.93)$$

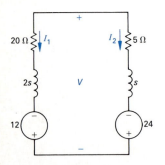

FIGURE 16.31 The s-domain equivalent circuit for the circuit shown in Fig. 16.30.

We now inverse-transform Eq. (16.93):

$$v(t) = -20\delta(t) - \frac{40}{3}e^{-(25/3)t}u(t). \qquad \textbf{(16.94)}$$

The solutions for I_1 and I_2 are

$$I_1 = \frac{V + 12}{2s + 20} = \frac{-10(s + 9)}{(s + 10)[s + (25/3)]} + \frac{6}{s + 10}$$

$$= -\frac{4}{s + (25/3)} \qquad \textbf{(16.95)}$$

and

$$I_2 = \frac{V + 24}{s + 5} = \frac{-20(s + 9)}{(s + 5)[s + (25/3)]} + \frac{24}{s + 5}$$

$$= \frac{4}{s + (25/3)}. \qquad \textbf{(16.96)}$$

The expressions for i_1 and i_2 are

$$i_1 = -4e^{-(25/3)t}u(t) \text{ A}; \qquad \textbf{(16.97)}$$

$$i_2 = 4e^{-(25/3)t}u(t) \text{ A}. \qquad \textbf{(16.98)}$$

Now, we must check to see whether the solutions make sense. We begin by noting that $i_1 = -i_2$ for $t > 0$. This result agrees with Kirchhoff's current law. Next, we note that when the switch is open, i_1 jumps from $+6$ to -4 A and i_2 jumps from $+24$ to $+4$ A. We verify that these abrupt changes are correct as follows. The two inductors are in parallel, so the change in flux linkage must be the same in each branch, or

$$\Delta i_1 L_1 = \Delta i_2 L_2. \qquad \textbf{(16.99)}$$

Kirchhoff's current law requires that

$$i_1(0^-) + \Delta i_1 + i_2(0^-) + \Delta i_2 = 0. \qquad \textbf{(16.100)}$$

When we substitute the numerical values $L_1 = 2$ H, $L_2 = 1$ H, $i_1(0^-) = 6$ A, and $i_2(0^-) = 24$ A into Eqs. (16.99) and (16.100), we find that the solutions for Δi_1 and Δi_2 are

$$\Delta i_1 = -10 \text{ A}; \qquad \textbf{(16.101)}$$

$$\Delta i_2 = -20 \text{ A}. \qquad \textbf{(16.102)}$$

These expressions agree with the initial jumps in i_1 and i_2, respectively.

Finally, the sudden drop in i_1 from $+6$ to -4 A generates an impulse in the derivative of i_1 equal to $-10\delta(t)$. Similarly, the abrupt change in i_2 from $+24$ to $+4$ A generates an impulse in the derivative of i_2 equal to $-20\delta(t)$. Thus

$$L_1\frac{di_1}{dt} = L_2\frac{di_2}{dt} = -20\delta(t) \qquad (16.103)$$

at the instant of switching. The result given by Eq. (16.103) agrees with the solution for v, namely, Eq. (16.94).

IMPULSIVE DRIVING SOURCES

The second type of circuit problem involving impulse functions is the use of impulsive driving sources. An impulsive source driving a circuit imparts a finite amount of energy into the system instantaneously. A mechanical analogy is striking a bell with an impulsive clapper blow. After the energy has been transferred to the bell, the natural response of the bell determines the metallic tone emitted and the tone's duration.

In the circuit shown in Fig. 16.32, an impulsive voltage source having a strength of V_0 volt-seconds is applied to a series connection of a resistor and an inductor. When the voltage source is applied, the initial energy in the inductor is zero; therefore the initial current is zero. Thus there is no voltage drop across R, so the impulsive voltage source appears directly across L. An impulsive voltage at the terminals of an inductor establishes an instantaneous current. The current is

$$i = \frac{1}{L}\int_{0^-}^{t} V_0\,\delta(x)\,dx. \qquad (16.104)$$

Using the sifting property of the impulse function, we find that Eq. (16.104) yields

$$i(0^+) = \frac{V_0}{L}\ \text{A}. \qquad (16.105)$$

Thus, in an infinitesimal moment, the impulsive voltage source has stored

$$w = \frac{1}{2}L\left(\frac{V_0}{L}\right)^2 = \frac{1}{2}\frac{V_0^2}{L}\ \text{J} \qquad (16.106)$$

in the inductor.

The current V_0/L now decays to zero in accordance with the natural response of the circuit that is,

$$i = \frac{V_0}{L}e^{-t/\tau}u(t), \qquad (16.107)$$

where $\tau = L/R$. When a circuit is driven by only an impulsive source, the response is always the natural response of the circuit.

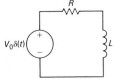

FIGURE 16.32 An *RL* circuit excited by an impulsive voltage source.

We may also obtain Eq. (16.107) by direct application of the Laplace transform method. Figure 16.33 shows the s-domain equivalent of the circuit in Fig. 16.32. Hence

$$I = \frac{V_0}{R + sL} = \frac{V_0/L}{s + (R/L)}. \qquad (16.108)$$

and

$$i = \frac{V_0}{L}e^{-(R/L)t} = \frac{V_0}{L}e^{-t/\tau}u(t). \qquad (16.109)$$

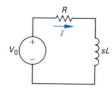

FIGURE 16.33 The s-domain equivalent circuit for the circuit shown in Fig. 16.32.

Thus the Laplace transform method gives the correct solution for $t \geq 0^+$.

Finally, we consider the case in which internally generated impulses and externally applied impulses occur simultaneously. The Laplace transform approach automatically ensures the correct solution for $t > 0^+$ if inductor currents and capacitor voltages at $t = 0^-$ are used in constructing the s-domain equivalent circuit and externally applied impulses are represented by their transforms. To illustrate, we add an impulsive voltage source of $50\delta(t)$ in series with the 100-V source to the circuit shown in Fig. 16.27. Figure 16.34 shows the new arrangement.

At $t = 0^-$, $i_1(0^-) = 10$ A and $i_2(0^-) = 0$ A. The Laplace transform of $50\delta(t) = 50$. If we use these values, the s-domain equivalent circuit is as shown in Fig. 16.35.

The expression for I is

$$I = \frac{50 + (100/s) + 30}{25 + 5s} = \frac{16}{s + 5} + \frac{20}{s(s + 5)}$$

$$= \frac{16}{s + 5} + \frac{4}{s} - \frac{4}{s + 5}$$

$$= \frac{12}{s + 5} + \frac{4}{s}, \qquad (16.110)$$

from which

$$i(t) = (12e^{-5t} + 4)u(t) \text{ A}. \qquad (16.111)$$

The expression for V_o is

$$V_o = (15 + 2s)I = \frac{32(s + 7.5)}{s + 5} + \frac{40(s + 7.5)}{s(s + 5)}$$

$$= 32\left(1 + \frac{2.5}{s + 5}\right) + \frac{60}{s} - \frac{20}{s + 5}$$

$$= 32 + \frac{60}{s + 5} + \frac{60}{s}, \qquad (16.112)$$

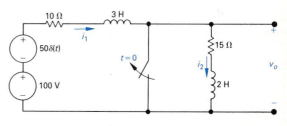

FIGURE 16.34 The circuit shown in Fig. 16.27 with an impulsive voltage source added in series with the 100-V source.

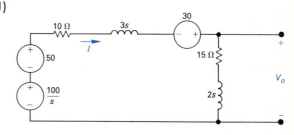

FIGURE 16.35 The s-domain equivalent circuit for the circuit shown in Fig. 16.34.

from which

$$v_o = 32\delta(t) + (60e^{-5t} + 60)u(t) \text{ V}. \qquad \textbf{(16.113)}$$

Now we test the results to see whether they make sense. From Eq. (16.111), we see that the current in L_1 and L_2 is 16 A at $t = 0^+$. As in the previous case, the switch operation causes i_1 to decrease instantaneously from 10 to 6 A, and at the same time, causes I_2 to increase from 0 to 6 A. Superimposed on these changes is the establishment of 10 A in L_1 and L_2 by the impulsive voltage source. That is,

$$i = \frac{1}{3+2}\int_{0^-}^{t} 50\delta(x)\, dx = 10 \text{ A}. \qquad \textbf{(16.114)}$$

Therefore i_1 increases suddenly from 10 to 16 A, while i_2 increases suddenly from 0 to 16 A. The final value of i is 4 A. Figure 16.36 shows i_1, i_2, and i graphically.

We may also find the abrupt changes in i_1 and i_2 without using superposition. In this circuit, the sum of the impulsive voltages across L_1 (3 H) and L_2 (2 H) equals $50\delta(t)$. Thus the change in flux linkage must sum to 50; that is,

$$\Delta\lambda_1 + \Delta\lambda_2 = 50. \qquad \textbf{(16.115)}$$

Because $\lambda = Li$, we express Eq. (16.115) as

$$3\,\Delta i_1 + 2\,\Delta i_2 = 50. \qquad \textbf{(16.116)}$$

But as i_1 and i_2 must be equal after the switching takes place,

$$i_1(0^-) + \Delta i_1 = i_2(0^-) + \Delta i_2. \qquad \textbf{(16.117)}$$

Then,

$$10 + \Delta i_1 = 0 + \Delta i_2. \qquad \textbf{(16.118)}$$

Solving Eqs. (16.116) and (16.118) for Δi_1 and Δi_2 yields

$$\Delta i_1 = 6 \text{ A}; \qquad \textbf{(16.119)}$$

$$\Delta i_2 = 16 \text{ A}. \qquad \textbf{(16.120)}$$

These expressions agree with the previous check.

Figure 16.36 also indicates that the derivatives of i_1 and i_2 will contain an impulse at $t = 0$. Specifically, the derivative of i_1 will have an impulse of $6\delta(t)$, and the derivative of i_2 will have an impulse of $16\delta(t)$. Figure 16.37(a) and (b), respectively, illustrate the derivatives of i_1 and i_2.

Now let's turn to Eq. (16.113). The impulsive component $32\delta(t)$ agrees with the impulse of $16\delta(t)$ of di_2/dt at the origin. The terms $60e^{-5t} + 60$ agree with the fact that for $t > 0^+$,

$$v_o = 15i + 2\frac{di}{dt}.$$

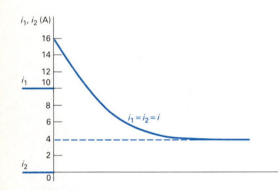

FIGURE 16.36 The inductor currents versus t for the circuit shown in Fig. 16.34.

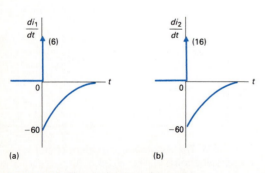

(a)   (b)

FIGURE 16.37 The derivative of i_1 and i_2.

We test the impulsive component of di_1/dt by noting that it produces an impulsive voltage of $(3)6\delta(t)$, or $18\delta(t)$, across L_1. This voltage, along with $32\delta(t)$ across L_2, adds to $50\delta(t)$. Thus the algebraic sum of the impulsive voltages around the mesh adds to zero.

We summarize our discussion of impulse functions in circuit analysis as follows. The Laplace transform will correctly predict the creation of impulsive currents and voltages that arise from switching. However, the s-domain equivalent circuits must be based on initial conditions at $t = 0^-$, that is, on the initial conditions that exist prior to the disturbance caused by the switching. The Laplace transform will correctly predict the response to impulsive driving sources by simply representing these sources in the s domain by their correct transforms.

DRILL EXERCISES

16.8 The switch in the circuit shown has been in position a for a long time. At $t = 0$ the switch moves to position b. Compute (a) $v_1(0^-)$; (b) $v_2(0^-)$; (c) $v_3(0^-)$; (d) $i(t)$; (e) $v_1(0^+)$; (f) $v_2(0^+)$; and (g) $v_3(0^+)$.

ANSWER: (a) 80 V; (b) 20 V; (c) 0 V; (d) $32\delta(t)$ μA; (e) 16 V; (f) 4 V; (g) 20 V.

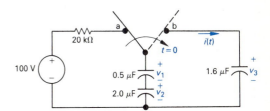

16.9 The switch in the circuit shown has been closed for a long time. The switch opens at $t = 0$. Compute (a) $i_1(0^-)$; (b) $i_1(0^+)$; (c) $i_2(0^-)$; (d) $i_2(0^+)$; (e) $i_1(t)$; (f) $i_2(t)$; and (g) $v(t)$.

ANSWER: (a) 0.8 A; (b) 0.6 A; (c) 0.2 A; (d) -0.6 A; (e) $0.6e^{-2\times10^6 t}u(t)$ A; (f) $-0.6e^{-2\times10^6 t}u(t)$ A; (g) $-1.6 \times 10^{-3}\,\delta(t) - 7200e^{-2\times10^6 t}u(t)$ V.

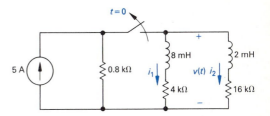

SUMMARY

Effective use of the Laplace transform to find the transient response of linear, lumped-parameter circuits depends on the following techniques.

- s-domain equivalent circuits: Each circuit element has an s-domain model as summarized in Table 16.1.

- Initial conditions: Inductor currents and capacitor voltages at $t = 0^-$ show up as independent sources in the s-domain circuit element model. (See Table 16.1.)

- s-domain circuit: The s-domain model of a circuit is constructed using the models of the circuit elements.

- Kirchhoff's laws: Kirchhoff's laws hold for s-domain currents and voltages.

- Terminal relationships: When initial conditions are zero, the relationship between the terminal current and voltage reduces to an s-domain version of Ohm's law, or $V = ZI$. When the initial conditions are not zero, the terminal behavior of an element is described by combining the s-domain version of Ohm's law with the initial-condition sources.

- s-domain analysis: All the techniques of circuit analysis introduced in Chapter 4 may be used to find the s-domain currents and voltages.

- Time-domain expressions: The time-domain expressions for currents and voltages are found using the partial fraction expansion technique discussed in Chapter 15.

The Laplace transform method predicts the existence of impulsive currents and voltages caused by switching and also predicts the circuit's response to impulsive sources.

PROBLEMS

16.1 Find the Norton equivalent of the circuit shown in Fig. 16.3.

16.2 Derive the s-domain equivalent circuit shown in Fig. 16.4 by expressing the inductor current i as a function of the terminal voltage v and then finding the Laplace transform of this time-domain integral equation.

16.3 Find the Thévenin equivalent of the circuit shown in Fig. 16.7.

16.4 A 2-kΩ resistor, a 6.25-H inductor, and a 250-nF capacitor are in parallel.

  a) Express the s-domain impedance of this parallel combination as a rational function.

  b) Give the numerical values of the poles and zeros of the impedance.

16.5 A 5000-Ω resistor is in series with a 10-H inductor. This series combination is in parallel with a 2.5-μF capacitor.

a) Express the equivalent s-domain admit-

tance of these paralleled branches as a rational function.

b) Determine the numerical values of the poles and zeros.

16.6 Find the poles and zeros of the impedance seen looking into the terminals a, b of the circuit shown in Fig. P16.6.

FIGURE P16.6

16.7 Find the poles and zeros of the impedance seen looking into the terminals a, b of the circuit shown in Fig. P16.7.

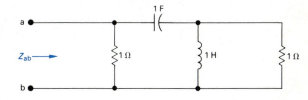

FIGURE P16.7

16.8 The switch in the circuit shown in Fig. P16.8 has been in position x for a long time. At $t = 0$ the switch moves instantaneously to position y.

a) Construct an s-domain circuit for $t > 0$.

b) Find V_a.

c) Find v_a.

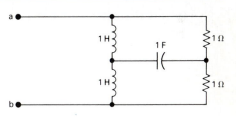

FIGURE P16.8

16.9 The switch in the circuit in Fig. P16.9 has been in position a for a long time. At $t = 0$ the switch moves instantaneously to position b.

a) Construct the s-domain circuit for $t > 0$.

b) Find V_o.

c) Find I_L.

d) Find v_o for $t > 0$.

e) Find i_L for $t > 0$.

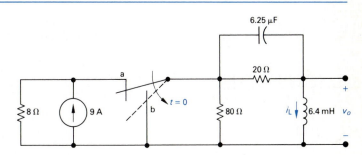

FIGURE P16.9

16.10 a) Find the s-domain expression for V_o in the circuit in Fig. P16.10.

b) Use the s-domain expression derived in part (a) to predict the initial and final values of v_o.

c) Find the time-domain expression for v_o.

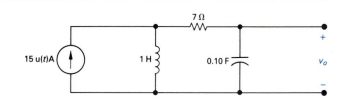

FIGURE P16.10

16.11 Find the time-domain expression for the current in the inductor in the circuit in Fig.

P16.10. Assume the reference direction for i_L is down.

16.12 Find V_o and v_o in the circuit shown in Fig. P16.12 if the initial energy is zero and the switch is closed at $t = 0$.

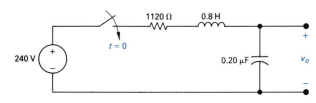

FIGURE P16.12

16.13 Repeat Problem 16.12 if the initial voltage on the capacitor is 72 V positive at the lower terminal.

16.14 The switch in the circuit in Fig. P16.14 has been in position a for a long time. At $t = 0$ it moves instantaneously from a to b.

a) Construct the s-domain circuit for $t > 0$.

b) Find $I_o(s)$.

c) Find $i_o(t)$ for $t \geq 0$.

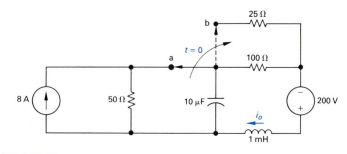

FIGURE P16.14

16.15 There is no energy stored in the circuit in Fig. P16.15 at the time the current source is energized.

a) Find V_o, I_o, and I_L.

b) Find v_o, i_o, and i_L for $t \geq 0$.

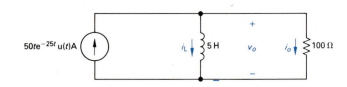

FIGURE P16.15

16.16 The switch in the circuit in Fig. P16.16 has been closed for a long time. At $t = 0$ the switch is opened. Find $v_o(t)$ for $t \geq 0$.

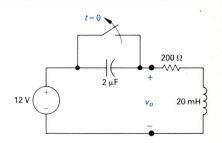

FIGURE P16.16

16.17 The switch in the circuit in Fig. P16.17 has been closed for a long time before opening at $t = 0$.

a) Construct the s-domain equivalent circuit for $t > 0$.

b) Find I_o.

c) Find i_o for $t \geq 0$.

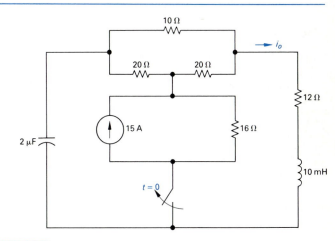

FIGURE P16.17

16.18 The switch in the circuit seen in Fig. P16.18 has been in position a for a long time. At $t = 0$ it moves instantaneously to position b.

a) Find V_o.

b) Find v_o.

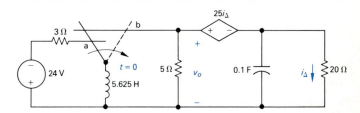

FIGURE P16.18

16.19 Find v_o in the circuit shown in Fig. P16.19 if $i_g = 15\, u(t)$ A. There is no energy stored in the circuit at $t = 0$.

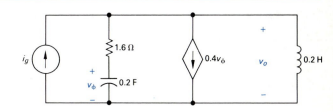

FIGURE P16.19

16.20 The switch in the circuit in Fig. P16.20 has been closed for a long time before opening at $t = 0$. Find v_o for $t \geq 0$.

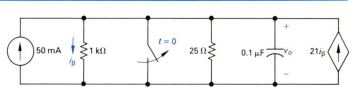

FIGURE P16.20

16.21 There is no energy stored in the circuit in Fig. P16.21 at the time the switch is closed.

a) Find v_o for $t \geq 0$.

b) Does your solution make sense in terms of known circuit behavior? Explain.

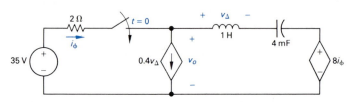

FIGURE P16.21

16.22 There is no energy stored in the capacitors in the circuit in Fig. P16.22 at the time the switch is closed.

a) Construct the s-domain circuit for $t > 0$.

b) Find I_1, V_1, and V_2.

c) Find i_1, v_1, and v_2.

d) Do your answers for i_1, v_1, and v_2 make sense in terms of known circuit behavior? Explain.

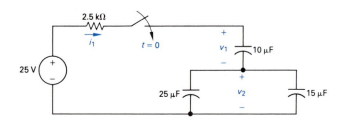

FIGURE P16.22

16.23 The make-before-break switch in the circuit seen in Fig. P16.23 has been in position a for a long time before moving instantaneously to position b at $t = 0$.

a) Construct the s-domain equivalent circuit for $t > 0$.

b) Find V_L and v_L.

c) Find V_C and v_C.

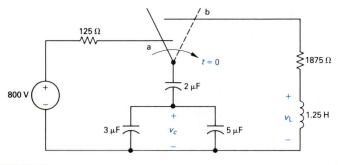

FIGURE P16.23

16.24 The make-before-break switch in the circuit in Fig. P16.24 has been in position a for a long time. At $t = 0$ it moves instantaneously to position b. Find i_o for $t \geq 0$.

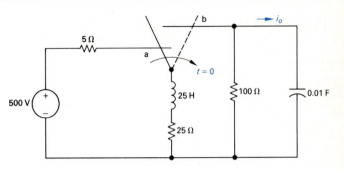

FIGURE P16.24

16.25 There is no energy stored in the circuit in Fig. P16.25 at the time the sources are energized.

a) Find $I_1(s)$ and $I_2(s)$.

b) Use the initial- and final-value theorems to check the initial and final values of $i_1(t)$ and $i_2(t)$.

c) Find $i_1(t)$ and $i_2(t)$ for $t \geq 0$.

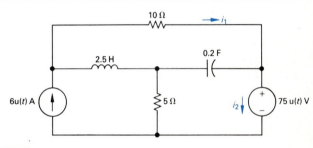

FIGURE P16.25

16.26 There is no energy stored in the circuit in Fig. P16.26 at the time the voltage source is turned on. If $v_g = 54\, u(t)$ V find:

a) V_o and I_o;

b) v_o and i_o;

c) Do the solutions for v_o and i_o make sense in terms of known circuit behavior? Explain.

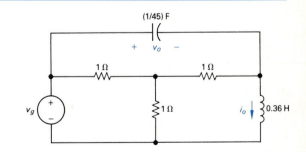

FIGURE P16.26

16.27 There is no energy stored in the circuit in Fig. P16.27 at the time the current source is energized.

a) Find I_a and I_b.

b) Find i_a and i_b.

c) Find V_a, V_b, and V_c.

d) Find v_a, v_b and v_c.

e) Assume a capacitor will break down whenever its terminal voltage is 1000 V. How long after the current source turns on will one of the capacitors break down?

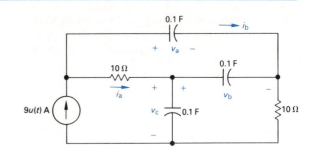

FIGURE P16.27

16.28 There is no energy stored in the circuit in Fig. P16.28 at $t = 0^-$.

a) Find V_o.

b) Find v_o.

c) Does your solution for v_o make sense in terms of known circuit behavior? Explain.

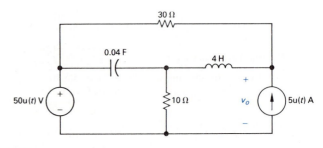

FIGURE P16.28

16.29 The initial energy in the circuit in Fig. P16.29 is zero. The ideal voltage source is $120\, u(t)$ V.

a) Find $I_o(s)$.

b) Use the initial- and final-value theorems to find $i_o(0^+)$ and $i_o(\infty)$.

c) Do the values obtained in part (b) agree with known circuit behavior? Explain.

d) Find $i_o(t)$.

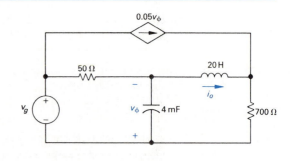

FIGURE P16.29

16.30 Scale the circuit in Problem 16.29 so that the 50-Ω resistor is increased to 5kΩ and the frequency of the current response is increased by a factor of 5000. Find $i_o(t)$.

16.31 There is no energy stored in the circuit in Fig. P16.31 at the time the current source turns on. Given that $i_g = 50\,u(t)$ A:

a) find $V_o(s)$;

b) use the initial- and final-value theorems to find $v_o(0^+)$ and $v_o(\infty)$;

c) do the results obtained in part (b) agree with known circuit behavior?; and

d) find $v_o(t)$.

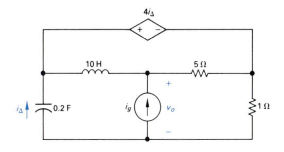

FIGURE P16.31

16.32 Scale the circuit in Problem 16.31 so that the inductor is reduced to 250 mH, and the capacitor is reduced to 8 μF.

a) Specify the scaling factors k_m and k_f.

b) Construct the s-domain equivalent circuit of the scaled circuit.

c) Find $v_o(t)$ for the scaled circuit.

16.33 The switch in the circuit shown in Fig. P16.33 has been open for a long time. The voltage of the sinusoidal source is $v_g = V_m \sin(\omega t + \phi)$. The switch closes at $t = 0$. Note that the angle ϕ in the voltage expression determines the value of the voltage at the moment when the switch closes, that is, $v_g(0) = V_m \sin \phi$.

a) Use the Laplace transform method to find i for $t > 0$.

b) Using the expression derived in part (a), write the expression for the current after the switch has been closed for a long time.

c) Using the expression derived in part (a), write the expression for the transient component of i.

d) Find the steady-state expression for i using the phasor method. Verify that your expression is equivalent to that obtained in part (b).

e) Specify the value of ϕ so that the circuit passes directly into steady-state operation when the switch is closed.

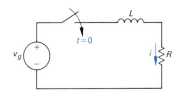

FIGURE P16.33

16.34 The two switches in the circuit shown in Fig. P16.34 operate simultaneously. There is no energy stored in the circuit at the instant the switches close. Find $i(t)$ for $t \geq 0^+$ by first finding the s-domain Thévenin equivalent of the circuit to the left of the terminals a, b.

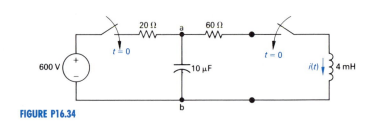

FIGURE P16.34

16.35 Beginning with Eq. (16.65), show that the capacitor current in the circuit in Fig. 16.17 is positive for $0 < t < 200$ μs and negative for $t > 200$ μs. Also show that at 200 μs the current is zero and this corresponds to when dv_C/dt is zero.

16.36 The switch in the circuit seen in Fig. P16.36 has been closed for a long time before opening at $t = 0$. Use the Laplace transform method of analysis to find v_o.

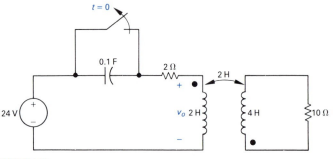

FIGURE P16.36

16.37 The circuit in Fig. P16.36 is scaled so that the capacitor changes to 0.10 μF and the 4-H inductance changes to 40 mH. Write the expression for v_o.

16.38 There is no energy stored in the circuit in Fig. P16.38 at the time the switch is closed.

a) Find V_o.

b) Use the initial- and final-value theorems to find $v_o(0^+)$ and $v_o(\infty)$.

c) Find v_o.

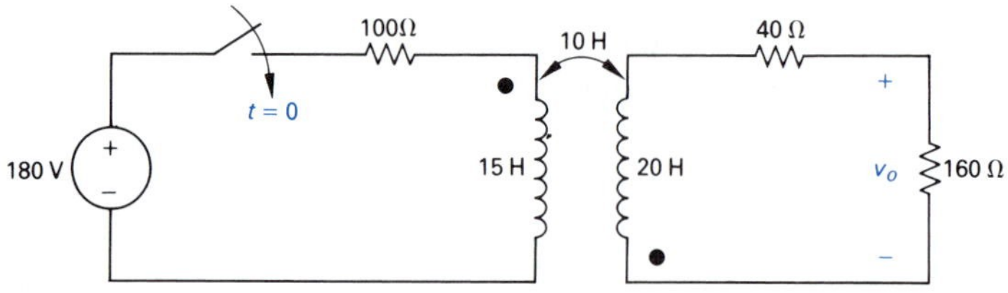

FIGURE P16.38

16.39 Find v_o in the circuit in Fig. P16.38 if the polarity dot on the 20-H coil is at the top.

16.40 The make-before-break switch in the circuit seen in Fig. P16.40 has been in position a for a long time. At $t = 0$ it moves instantaneously to position b. Find i_o for $t \geq 0$.

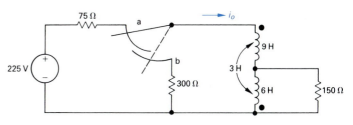

FIGURE P16.40

16.41 In the circuit in Fig. P16.41 switch 1 closes at $t = 0$ and the make-before-break switch moves instantaneously from position a to position b.

a) Construct the s-domain equivalent circuit for $t > 0$.

b) Find I_1.

c) Use the initial- and final-value theorems to check the initial and final values of i_1.

d) Find i_1 for $t \geq 0^+$.

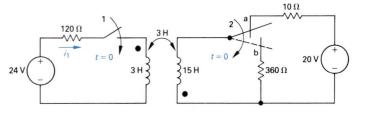

FIGURE P16.41

16.42 Show that after $V_0 C_e$ coulombs are transferred from C_1 to C_2 in the circuit shown in Fig. 16.24, the voltage across each capacitor is $C_1 V_0/(C_1 + C_2)$. (*Hint:* Use the conservation-of-charge principle.)

16.43 The inductor L_1 in the circuit shown in Fig. P16.43 is carrying an initial current of ρ amperes at the instant the switch opens. Find (a) $v(t)$; (b) $i_1(t)$; (c) $i_2(t)$; and (d) $\lambda(t)$, where $\lambda(t)$ is the total flux linkage in the circuit.

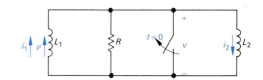

FIGURE P16.43

16.44 a) Let $R \to \infty$ in the circuit shown in Fig. P16.43 and use the solutions derived in Problem 16.43 to find $v(t)$, $i_1(t)$, and $i_2(t)$.

b) Let $R = \infty$ in the circuit shown in Fig. P16.43 and use the Laplace transform method to find $v(t)$, $i_1(t)$, and $i_2(t)$.

16.45 The switch in the circuit shown in Fig. P16.45 has been open for a long time before closing at $t = 0$.

a) Find v_o and i_o for $t \geq 0$.

b) Test your solutions and make sure they are in agreement with known circuit behavior.

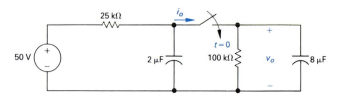

FIGURE P16.45

16.46 The parallel combination of R_2 and C_2 in the circuit shown in Fig. P16.46 represents the input circuit to a cathode-ray oscilloscope. The parallel combination of R_1 and C_1 is a circuit model of a compensating lead that is used to connect the CRO to the source. There is no energy stored in C_1 or C_2 at the time when the 5-V source is connected to the CRO via the compensating lead. The circuit values are $C_1 = 2$ pF, $C_2 = 8$ pF, $R_1 = 2.5$ MΩ, and $R_2 = 10$ MΩ.

a) Find v_o.

b) Find i_o.

c) Repeat parts (a) and (b) given C_1 is changed to 32 pF.

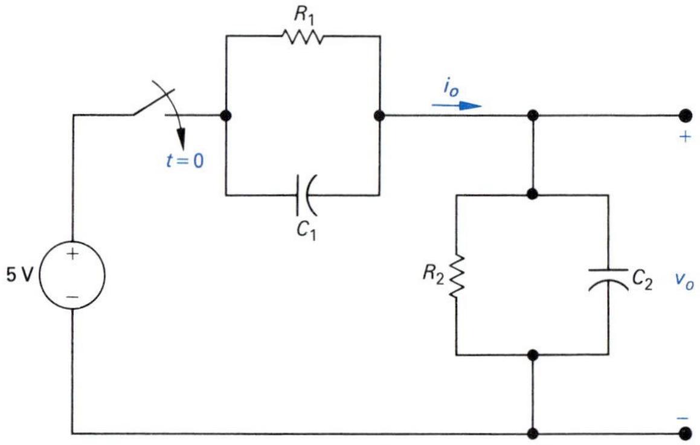

FIGURE P16.46

16.47 Show that if $R_1 C_1 = R_2 C_2$ in the circuit shown in Fig. P16.46, v_o will be a scaled replica of the source voltage.

16.48 There is no energy stored in the circuit in Fig. P16.48 at the time the impulsive current is applied.

a) Find v_o for $t \geq 0^+$.

b) Does your solution make sense in terms of known circuit behavior? Explain.

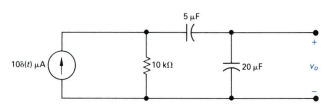

FIGURE P16.48

16.49 There is no energy stored in the circuit in Fig. P16.49 at the time the impulsive voltage is applied.

a) Find $v_o(t)$ for $t \geq 0$.

b) Does your solution make sense in terms of known circuit behavior? Explain.

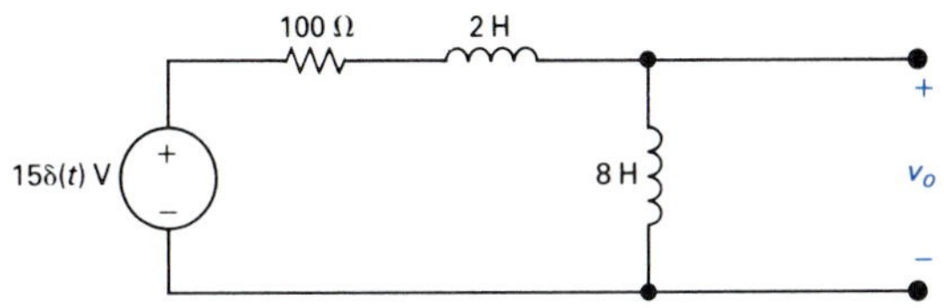

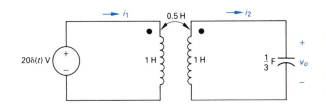

FIGURE P16.49

16.50 There is no energy stored in the circuit in Fig. P16.50 at the time the impulse voltage is applied.

a) Find i_1 for $t \geq 0^+$.

b) Find i_2 for $t \geq 0^+$.

c) Find v_o for $t \geq 0^+$.

d) Do your solutions for i_1, i_2, and v_o make sense in terms of known circuit behavior? Explain.

FIGURE P16.50

16.51 The operational amplifier in the circuit shown in Fig. P16.51 is ideal. There is no energy stored in the circuit at the time it is energized. If $v_g = 6000tu(t)$ V, find (a) V_o, (b) v_o, (c) how long it takes to saturate the operational amplifier, and (d) how small the rate of increase in v_g must be to prevent saturation.

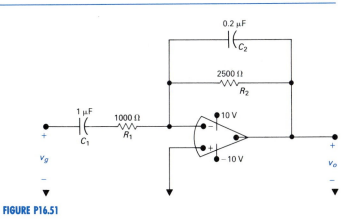

FIGURE P16.51

16.52 The operational amplifier in the circuit seen in Fig. P16.52 is ideal. There is no energy stored in the capacitors at the time the circuit is energized. Determine (a) V_o, (b) v_o, and (c) how long it takes to saturate the operational amplifier.

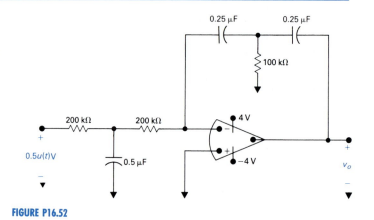

FIGURE P16.52

16.53 Find $v_o(t)$ in the circuit shown in Fig. P16.53 if the ideal op amp operates within its linear range and $v_g = 40u(t)$ mV.

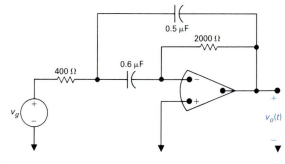

FIGURE P16.53

16.54 The operational amplifier in the circuit shown in Fig. P16.54 is ideal. There is no energy stored in the capacitors at the instant the circuit is energized.

a) Find v_o if $v_{g1} = 20u(t)$ V and $v_{g2} = 10u(t)$ V.

b) How many milliseconds after the two voltage sources are turned on does the op amp saturate?

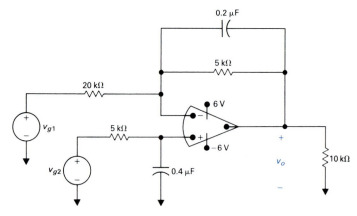

FIGURE P16.54

16.55 The magnetically coupled coils in the circuit seen in Fig. P16.55 carry initial currents of 15 and 10 A, as shown.

a) Find the initial energy stored in the circuit.

b) Find I_1 and I_2.

c) Find i_1 and i_2.

d) Find the total energy dissipated in the 120- and the 270-Ω resistors.

e) Repeat parts (a) through (d), with the dot on the 18-H inductor at the lower terminal.

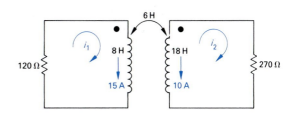

FIGURE P16.55

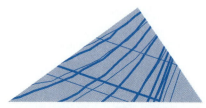

THE TRANSFER FUNCTION

CHAPTER 17

In this chapter we introduce the concept of the transfer function and show how to use it in circuit analysis. After defining the transfer function and discussing its properties, we divide our discussion of its applications into three parts. First we discuss the use of the transfer function in a partial fraction expansion. Next, we show how to use the transfer function in conjunction with the convolution integral to find the response of a circuit. Finally, we show how to use the transfer function to find the sinusoidal steady-state response of a circuit.

We use the transfer function to find the response of a circuit to externally applied excitation sources. The contribution of initially stored energy to the response is not included in the transfer function solution. Therefore we must first illustrate how to use the principle of superposition to divide the response function into two parts: one owing to excitation sources and one owing to initial conditions.

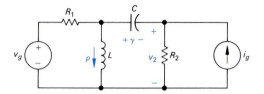

FIGURE 17.1 A circuit showing the use of superposition in *s*-domain analysis.

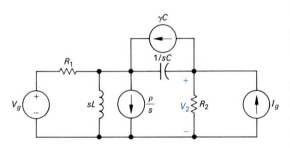

FIGURE 17.2 The *s*-domain equivalent for the circuit of Fig. 17.1.

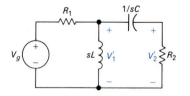

FIGURE 17.3 The circuit shown in Fig. 17.2 with V_g acting alone.

17.1 AN *S*-DOMAIN APPLICATION OF SUPERPOSITION

Because we are analyzing linear, lumped-parameter circuits, we can use superposition to divide the response into components that can be identified with particular driving sources and initial conditions. Here, we simply confirm this observation with a specific illustration so that, as you study the following material, you may keep the concept of the transfer function in proper perspective.

Figure 17.1 shows our illustrative circuit. We assume that at the instant when the two sources are applied to the circuit the inductor is carrying an initial current of ρ amperes and that the capacitor is carrying an initial voltage of γ volts. The desired response of the circuit is the voltage across the resistor R_2, labeled v_2 in Fig. 17.1.

Figure 17.2 shows the *s*-domain equivalent circuit. We opted for the parallel equivalents for L and C because we anticipated solving for V_2 using the node-voltage method.

To find V_2 by superposition, we calculate the component of V_2 resulting from each source acting alone and then sum the components. We begin with V_g acting alone. Opening each source deactivates the three current sources. Figure 17.3 shows the resulting circuit. We added the node voltage V_1' in Fig. 17.3 to aid the analysis. The primes on V_1 and V_2 indicate that they are the components of V_1 and V_2 because of V_g acting alone. The two equations that describe the circuit in Fig. 17.3 are

$$\left(\frac{1}{R_1} + \frac{1}{sL} + sC\right)V_1' - sCV_2' = \frac{V_g}{R_1}; \qquad (17.1)$$

$$-sCV_1' + \left(\frac{1}{R_2} + sC\right)V_2' = 0. \qquad (17.2)$$

For convenience we introduce the notation:

$$Y_{11} = \frac{1}{R_1} + \frac{1}{sL} + sC; \qquad (17.3)$$

$$Y_{12} = -sC; \qquad (17.4)$$

$$Y_{22} = \frac{1}{R_2} + sC. \qquad (17.5)$$

Substituting Eqs. (17.3), (17.4), and (17.5) into Eqs. (17.1) and (17.2) gives

$$Y_{11}V_1' + Y_{12}V_2' = V_g/R_1; \qquad (17.6)$$

$$Y_{12}V_1' + Y_{22}V_2' = 0. \qquad (17.7)$$

Solving Eqs. (17.6) and (17.7) for V_2' gives

$$V_2' = \frac{-Y_{12}/R_1}{Y_{11} Y_{22} - Y_{12}^2} V_g. \qquad \textbf{(17.8)}$$

With the current source I_g acting alone, the circuit shown in Fig. 17.2 reduces to that shown in Fig. 17.4. Here V_1'' and V_2'' are the components of V_1 and V_2 resulting from I_g. If we use the notation introduced in Eqs. (17.3)–(17.5), the two node-voltage equations that describe the circuit in Fig. 17.4 are

$$Y_{11} V_1'' + Y_{12} V_2'' = 0; \qquad \textbf{(17.9)}$$

$$Y_{12} V_1'' + Y_{22} V_2'' = I_g. \qquad \textbf{(17.10)}$$

Solving Eqs. (17.9) and (17.10) for V_2'' yields

$$V_2'' = \frac{Y_{11}}{Y_{11} Y_{22} - Y_{12}^2} I_g. \qquad \textbf{(17.11)}$$

To find the component of V_2 resulting from the initial energy stored in the inductor, V_2''', we must solve the circuit shown in Fig. 17.5, where

$$Y_{11} V_1''' + Y_{12} V_2''' = -\rho/s; \qquad \textbf{(17.12)}$$

$$Y_{12} V_1''' + Y_{22} V_2''' = 0. \qquad \textbf{(17.13)}$$

Thus

$$V_2''' = \frac{Y_{12}/s}{Y_{11} Y_{22} - Y_{12}^2} \rho. \qquad \textbf{(17.14)}$$

We find the component of $V_2(V_2'''')$ resulting from the initial energy stored in the capacitor from the circuit shown in Fig. 17.6. The node-voltage equations describing this circuit are

$$Y_{11} V_1'''' + Y_{12} V_2'''' = \gamma C; \qquad \textbf{(17.15)}$$

$$Y_{12} V_1'''' + Y_{22} V_2'''' = -\gamma C. \qquad \textbf{(17.16)}$$

Solving for V_2'''' yields

$$V_2'''' = \frac{-(Y_{11} + Y_{12})C}{Y_{11} Y_{22} - Y_{12}^2} \gamma. \qquad \textbf{(17.17)}$$

The expression for V_2 is

$$V_2 = V_2' + V_2'' + V_2''' + V_2''''$$

$$= \frac{-(Y_{12}/R_1)}{Y_{11} Y_{22} - Y_{12}^2} V_g + \frac{Y_{11}}{Y_{11} Y_{22} - Y_{12}^2} I_g$$

$$+ \frac{Y_{12}/s}{Y_{11} Y_{22} - Y_{12}^2} \rho + \frac{-C(Y_{11} + Y_{12})}{Y_{11} Y_{22} - Y_{12}^2} \gamma. \qquad \textbf{(17.18)}$$

We can find V_2 without using superposition by solving the two

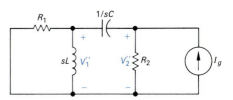

FIGURE 17.4 The circuit shown in Fig. 17.2 with I_g acting alone.

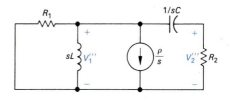

FIGURE 17.5 The circuit shown in Fig. 17.2 with the energized inductor acting alone.

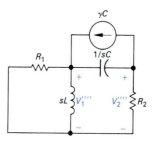

FIGURE 17.6 The circuit shown in Fig. 17.2 with the energized capacitor acting alone.

node-voltage equations that describe the circuit shown in Fig. 17.2. Thus

$$Y_{11} V_1 + Y_{12} V_2 = \frac{V_g}{R_1} + \gamma C - \frac{\rho}{s}; \qquad (17.19)$$

$$Y_{12} V_1 + Y_{22} V_2 = I_g - \gamma C. \qquad (17.20)$$

We leave to you (Problem 17.2) verification that the solution of Eqs. (17.19) and (17.20) for V_2 gives the same result as Eq. (17.18).

The fact that we can subdivide the response of a linear circuit into components that can be associated with specific sources of energy enables us to develop a technique that focuses on the relationship between an output and a source. This relationship is known as the *transfer function*, which we introduce next.

DRILL EXERCISES

17.1 The energy stored in the circuit shown is zero at the instant the two sources are turned on.

a) Find the component of v for $t > 0$ owing to the voltage source.

b) Find the component of v for $t > 0$ owing to the current source.

c) Find the expression for v when $t > 0$.

ANSWER: (a) $25e^{-t} - 25e^{-4t}$ V; (b) $5e^{-t} - 5e^{-4t}$ V; (c) $30e^{-t} - 30^{-4t}$ V.

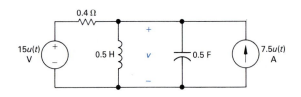

17.2 THE TRANSFER FUNCTION

The *transfer function* is defined as the s-domain ratio of the output (response) to the input (source). In finding this ratio, we set all initial conditions equal to zero. Furthermore, the ratio applies to a single source. If more than one source exists in the circuit, we define a transfer function for each source and then use superposition to find the total response. Here, we focus on circuits in which the initial energy stored is zero and the excitation is from a single source.

The transfer function is

$$H(s) = \frac{Y(s)}{X(s)}, \qquad \textbf{(17.21)}$$

where $Y(s)$ is the Laplace transform of the output signal and $X(s)$ is the Laplace transform of the input signal. Note that the transfer function depends on what is defined as the output signal. Thus a circuit can generate many transfer functions. Consider, for example, the series circuit shown in Fig. 17.7. If the current is defined as the response signal of the circuit,

$$H(s) = \frac{I}{V_g} = \frac{1}{R + sL + 1/sC} = \frac{sC}{s^2LC + RCs + 1}. \qquad \textbf{(17.22)}$$

In deriving Eq. (17.22), we recognized that I corresponds to the output $Y(s)$ and V_g corresponds to the input $X(s)$.

If the voltage across the capacitor is defined as the output signal of the circuit shown in Fig. 17.7, the transfer function is

$$H(s) = \frac{V}{V_g} = \frac{1/sC}{R + sL + (1/sC)} = \frac{1}{s^2LC + RCs + 1}. \qquad \textbf{(17.23)}$$

Example 17.1 illustrates the computation of a transfer function for known numerical values of R, L, and C.

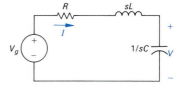

FIGURE 17.7 A series RLC circuit.

E X A M P L E   17.1

The voltage source v_g drives the circuit shown in Fig. 17.8. The response signal is the voltage across the capacitor, v_o.

a) Calculate the numerical expression for the transfer function.
b) Calculate the numerical values for the poles and zeros of the transfer function.

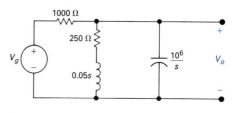

FIGURE 17.8 The circuit for Example 17.1.

S O L U T I O N

a) The first step in finding the transfer function is to construct the s-domain equivalent circuit, as shown in Fig. 17.9. By definition, the transfer function is the ratio of V_o/V_g, which can be computed from a single node-voltage equation. Summing the currents away from the upper node generates

$$\frac{V_o - V_g}{1000} + \frac{V_o}{250 + 0.05s} + \frac{V_o s}{10^6} = 0.$$

Solving for V_o yields

$$V_o = \frac{1000(s + 5000)V_g}{s^2 + 6000s + 25 \times 10^6}.$$

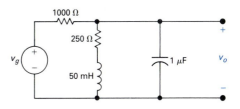

FIGURE 17.9 The s-domain equivalent circuit for the circuit shown in Fig. 17.8.

Hence the transfer function is

$$H(s) = \frac{V_o}{V_g} = \frac{1000(s + 5000)}{s^2 + 6000s + 25 \times 10^6}.$$

b) The poles of $H(s)$ are the roots of the denominator polynomial. Therefore

$$-p_1 = -3000 - j4000;$$

$$-p_2 = -3000 + j4000.$$

The zeros of $H(s)$ are the roots of the numerator polynomial; thus $H(s)$ has a zero at

$$-z_1 = -5000.$$

LOCATION OF POLES AND ZEROS OF $H(s)$

For linear, lumped-parameter circuits, $H(s)$ is always a rational function of s. Complex poles and zeros always appear in conjugate pairs. The poles of $H(s)$ must lie in the left half of the s plane if the response to a bounded driving source is to be bounded. The zeros of $H(s)$ may lie in either the right half or the left half of the s plane. If the zeros of $H(s)$ lie only in the left half of the s plane, $H(s)$ is said to be *minimum phase*.

With these general characteristics of $H(s)$ in mind, we next discuss the role that $H(s)$ plays in determining the response function. We begin with the partial fraction expansion technique for finding $y(t)$.

DRILL EXERCISES

17.2 a) Derive the numerical expression for the transfer function V_o/I_g for the circuit shown.

b) Give the numerical value of each pole and zero of $H(s)$.

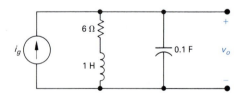

ANSWER: (a) $H(s) = 10(s + 6)/(s^2 + 6s + 10)$;
(b) $-p_1 = -3 + j1$, $-p_2 = -3 - i1$. $-z = -6$.

17.3 THE TRANSFER FUNCTION IN PARTIAL FRACTION EXPANSIONS

From Eq. (17.21) we can write the circuit output as the product of the transfer function and the driving function:

$$Y(s) = H(s)X(s). \qquad (17.24)$$

We have already noted that $H(s)$ is a rational function of s. Reference to Table 15.1 will show that $X(s)$ also is a rational function of s for the excitation functions of most interest in circuit analysis.

Expanding the right-hand side of Eq. (17.24) into a sum of partial fractions produces a term for each pole of $H(s)$ and a term for each pole of $X(s)$. The terms generated by the poles of $H(s)$ give rise to the transient component of the total response, whereas the terms generated by the poles of $X(s)$ give rise to the steady-state component of the response. By steady-state response, we mean the response that exists after the transient components have become negligible. Example 17.2 illustrates these general observations.

EXAMPLE 17.2

The circuit in Example 17.1 (Fig. 17.8) is driven by a voltage source whose voltage increases linearly with time, namely, $v_g = 50tu(t)$.

a) Use the transfer function to find v_o.

b) Identify the transient component of the response.

c) Identify the steady-state component of the response.

d) Sketch v_o versus t for $0 \leq t \leq 1.5$ ms.

SOLUTION

a) From Example 17.1,

$$H(s) = \frac{1000(s + 5000)}{s^2 + 6000s + 25 \times 10^6}.$$

The transform of the driving voltage is $50/s^2$; therefore the s-domain expression for the output voltage is

$$V_o = \frac{1000(s + 5000)}{(s^2 + 6000s + 25 \times 10^6)} \frac{50}{s^2}.$$

The partial fraction expansion of V_o is

$$V_o = \frac{K_1}{s + 3000 - j4000} + \frac{K_1^*}{s + 3000 + j4000} + \frac{K_2}{s^2} + \frac{K_3}{s}.$$

We evaluate the coefficients K_1, K_2, and K_3 by using the techniques described in Section 15.7:

$$K_1 = 5\sqrt{5} \times 10^{-4}\underline{/79.70°};$$
$$K_1^* = 5\sqrt{5} \times 10^{-4}\underline{/-79.70°};$$
$$K_2 = 10;$$
$$K_3 = -4 \times 10^{-4}.$$

The time-domain expression for v_o is

$$v_o = [10\sqrt{5} \times 10^{-4}e^{-3000t}\cos(4000t + 79.70°)$$
$$+ 10t - 4 \times 10^{-4}]u(t)\ V.$$

b) The transient component of v_o is

$$10\sqrt{5} \times 10^{-4}e^{-3000t}\cos(4000t + 79.70°).$$

Note that this term is generated by the poles $(-3000 + j4000)$ and $(-3000 - j4000)$ of the transfer function.

c) The steady-state component of the response is

$$(10t - 4 \times 10^{-4})u(t).$$

These two terms are generated by the second-order pole (at the origin) of the driving voltage.

d) Figure 17.10 shows a sketch of v_o versus t. Note that the deviation from the steady-state solution $10,000t - 0.4$ mV is imperceptible after approximately 1 ms.

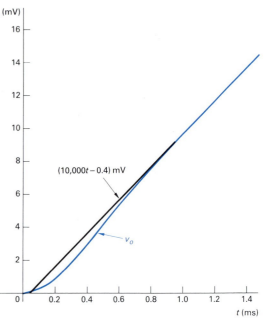

FIGURE 17.10 The graph of v_o versus t for Example 17.2.

OBSERVATIONS ON THE USE OF H(S) IN CIRCUIT ANALYSIS

Example 17.2 clearly shows how the transfer function $H(s)$ relates to the response of a circuit through a partial fraction expansion. However, the example does raise questions about the practicality of driving a circuit with an ever increasing voltage that generates a response that also increases without limit. Eventually the circuit components will fail under the stress of excessive voltage, and when that happens our linear model no longer is valid. The ramp response is of interest in practical applications where the ramp function increases to a maximum value over a finite time interval. If the runup time is long compared to the time constants of the circuit, the solution assuming an unbounded ramp is valid for this runup time interval.

We make two additional observations regarding Eq. (17.24).

First, if a unit impulse source drives the circuit, the response of the circuit equals the inverse transform of the transfer function. Thus if

$$x(t) = \delta(t), \quad \text{then} \quad X(s) = 1$$

and

$$Y(s) = H(s). \qquad (17.25)$$

Hence, from Eq. (17.25),

$$y(t) = h(t), \qquad (17.26)$$

where the inverse transform of the transfer function equals the unit impulse response of the circuit. Note that this is also the natural response of the circuit because the application of an impulsive source is equivalent to instantaneously storing energy in the circuit. The subsequent release of this stored energy gives rise to the natural response. (See Problem 17.4.)

The second observation relates to the response of the circuit due to a delayed input. If the input is delayed by a seconds,

$$\mathcal{L}\{x(t-a)u(t-a)\} = e^{-as}X(s),$$

and, from Eq. (17.24), the response becomes

$$Y(s) = H(s)X(s)e^{-as}. \qquad (17.27)$$

If $y(t) = \mathcal{L}^{-1}\{H(s)X(s)\}$, then, from Eq. (17.27),

$$y(t-a)u(t-a) = \mathcal{L}^{-1}\{H(s)X(s)e^{-as}\}. \qquad (17.28)$$

Therefore delaying the input by a seconds simply delays the response function by a seconds. A circuit that exhibits this characteristic is said to be *time-invariant*.

We now discuss how to use the transfer function in conjunction with the convolution integral.

DRILL EXERCISES

17.3 Find (a) the unit step and (b) the unit impulse response of the circuit shown in Drill Exercise 17.2.

ANSWER: (a) $[6 + 10e^{-3t} \cos(t + 126.87°)]u(t)$ V; (b) $31.62e^{-3t} \cos(t - 71.57°)u(t)$ V.

17.4 The unit impulse response of a circuit is

$$v_o(t) = 10{,}000e^{-70t} \cos(240t + \theta) \text{ V},$$

where $\tan \theta = \frac{7}{24}$.

a) Find the transfer function of the circuit.

b) Find the unit step response of the circuit.

ANSWER: (a) $9600s/(s^2 + 140s + 62{,}500)$; (b) $40e^{-70t} \sin 240t$ V.

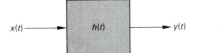

FIGURE 17.11 Block diagram of a general circuit.

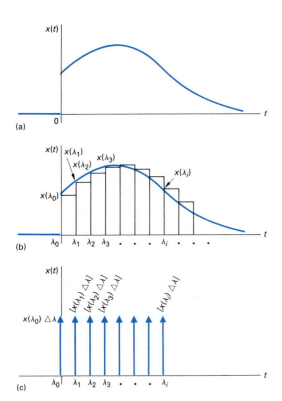

FIGURE 17.12 The excitation signal of $x(t)$: (a) a general excitation signal; (b) approximating $x(t)$ with a series of pulses; (c) approximating $x(t)$ with a series of impulses.

17.4 THE TRANSFER FUNCTION AND THE CONVOLUTION INTEGRAL

The convolution integral relates the output $y(t)$ of a linear, time-invariant circuit to the input $x(t)$ of the circuit and the circuit's impulse response $h(t)$. The integral relationship can be expressed in two ways:

$$y(t) = \int_{-\infty}^{\infty} h(\lambda)x(t - \lambda)\, d\lambda = \int_{-\infty}^{\infty} h(t - \lambda)x(\lambda)\, d\lambda. \quad (17.29)$$

We are interested in the convolution integral for several reasons. First, it allows us to work entirely in the time domain in situations where $x(t)$ and $h(t)$ may be known only through experimental data. Where experimental data are the bases for computations, the transform method may be awkward or even impossible. Second, the convolution integral introduces the concepts of memory and weighting function into analysis. We show how the concept of memory enables us to look at the impulse response (or weighting function) $h(t)$ and predict, to some degree, how closely the output waveform replicates the input waveform. Finally, the convolution integral provides a formal procedure for finding the inverse transform of products of Laplace transforms.

We based the derivation of Eq. (17.29) on the assumption that the circuit is linear and time-invariant. Because the circuit is linear, the principle of superposition is valid, and because the circuit is time-invariant, the response to a delayed input is delayed by exactly the amount of the input delay. Now consider Fig. 17.11, in which the block containing $h(t)$ represents any linear, time-invariant circuit whose impulse response is known, $x(t)$ represents the excitation signal, and $y(t)$ represents the desired output signal.

We assume that $x(t)$ is the general excitation signal shown in Fig. 17.12(a). For convenience we also assume that $x(t) = 0$ for $t < 0^-$. [Once you see the derivation of the convolution integral assuming $x(t) = 0$ for $t < 0^-$, the extension of the integral to include excitation functions that exist over all time becomes apparent.] Note that we permit a discontinuity in $x(t)$ at the origin, that is, a jump between 0^- and 0^+.

Now we approximate $x(t)$ by a series of rectangular pulses of uniform width $\Delta\lambda$, as shown in Fig. 17.12(b). Thus

$$x(t) = x_0(t) + x_1(t) + \cdots + x_i(t) + \cdots, \quad (17.30)$$

where $x_i(t)$ is a rectangular pulse that equals $x(\lambda_i)$ between λ_i and λ_{i+1} and is zero elsewhere. Note that the ith pulse can be expressed in terms of step functions; that is

$$x_i(t) = x(\lambda_i)\{u(t - \lambda_i) - u[t - (\lambda_i + \Delta\lambda)]\}.$$

The next step in the approximation of $x(t)$ is to make $\Delta\lambda$ small enough that the ith component can be approximated by an impulse function of strength $x(\lambda_i)\, \Delta\lambda$. Figure 17.12(c) shows the impulse representation, with the strength of each impulse shown in brackets beside each arrow. The impulse representation of $x(t)$ is

$$x(t) = x(\lambda_0)\, \Delta\lambda\delta(t - \lambda_0) + x(\lambda_1)\, \Delta\lambda\delta(t - \lambda_1) + \cdots$$
$$+ x(\lambda_i)\, \Delta\lambda\delta(t - \lambda_i) + \cdots. \qquad \textbf{(17.31)}$$

Now when $x(t)$ is represented by a series of impulse functions (which occur at equally spaced intervals of time, that is, at λ_0, $\lambda_1, \lambda_2, \ldots$), the response function $y(t)$ consists of the sum of a series of uniformly delayed impulse responses. The strength of each impulse response depends on the strength of the impulse driving the circuit. For example, let's assume that the unit impulse response of the circuit contained in the box in Fig. 17.11 is the exponential decay function shown in Fig. 17.13(a). Then the approximation of $y(t)$ is the sum of the impulse responses shown in Fig. 17.13(b).

Analytically the expression for $y(t)$ is

$$y(t) = x(\lambda_0)\, \Delta\lambda h(t - \lambda_0) + x(\lambda_1)\, \Delta\lambda h(t - \lambda_1)$$
$$+ x(\lambda_2)\, \Delta\lambda h(t - \lambda_2) + \cdots$$
$$+ x(\lambda_i)\, \Delta\lambda h(t - \lambda_i) + \cdots. \qquad \textbf{(17.32)}$$

As $\Delta\lambda \to 0$, the summation in Eq. (17.32) approaches a continuous integration, or

$$\sum_{i=0}^{\infty} x(\lambda_i)h(t - \lambda_i)\, \Delta\lambda \;\; \to \;\; \int_0^{\infty} x(\lambda)h(t - \lambda)\, d\lambda. \qquad \textbf{(17.33)}$$

Therefore

$$y(t) = \int_0^{\infty} x(\lambda)h(t - \lambda)\, d\lambda. \qquad \textbf{(17.34)}$$

If $x(t)$ exists over all time, then the lower limit on Eq. (17.34) becomes $-\infty$; thus, in general,

$$y(t) = \int_{-\infty}^{\infty} x(\lambda)h(t - \lambda)\, d\lambda, \qquad \textbf{(17.35)}$$

which is the second form of the convolution integral given in Eq. (17.29). We derive the first form of the integral from Eq. (17.35) by making a change in the variable of integration. We let

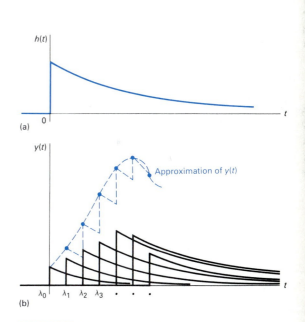

FIGURE 17.13 The approximation of $y(t)$: (a) impulse response of the box shown in Fig. 17.11; (b) summing the impulse responses.

$u = t - \lambda$ and then note that $du = -d\lambda$, $u = -\infty$ when $\lambda = \infty$, and $u = +\infty$ when $\lambda = -\infty$. Then we can write Eq. (17.35) as

$$y(t) = \int_{\infty}^{-\infty} x(t - u)h(u)(-du)$$

or

$$y(t) = \int_{-\infty}^{\infty} x(t - u)h(u) \, du. \qquad (17.36)$$

But as u is just a symbol of integration, Eq. (17.36) is equivalent to the first form of the convolution integral, Eq. (17.29).

The integral relationship between $y(t)$, $h(t)$, and $x(t)$, expressed in Eq. (17.29), often is written in a shorthand notation:

$$y(t) = h(t) * x(t) = x(t) * h(t), \qquad (17.37)$$

where the asterisk signifies the integral relationship between $h(t)$ and $x(t)$. Thus $h(t) * x(t)$ is read as "$h(t)$ is convolved with $x(t)$" and implies that

$$h(t) * x(t) = \int_{-\infty}^{\infty} h(\lambda)x(t - \lambda) \, d\lambda,$$

whereas $x(t) * h(t)$ is read as "$x(t)$ is convolved with $h(t)$" and implies that

$$x(t) * h(t) = \int_{-\infty}^{\infty} x(\lambda)h(t - \lambda) \, d\lambda.$$

The integrals in Eq. (17.29) give the most general relationship for the convolution of two functions. However, in our applications of the convolution integral we can change the lower limit to zero and the upper limit to t. Then we can write Eq. (17.29) as

$$y(t) = \int_0^t h(\lambda)x(t - \lambda) \, d\lambda = \int_0^t x(\lambda)h(t - \lambda) \, d\lambda. \qquad (17.38)$$

We change the limits for two reasons. First, for physically realizable circuits $h(t)$ is zero for $t < 0$. In other words, there can be no impulse response before an impulse is applied. Second, we start measuring time at the instant the excitation $x(t)$ is turned on; therefore $x(t) = 0$ for $t < 0^-$.

A graphic interpretation of the convolution integrals contained in Eq. (17.38) is important in the use of the integral as a computational tool. We begin with an interpretation of the first integral. For purposes of discussion, we assume that the impulse response of our circuit is the exponential decay function shown in Fig. 17.14(a) and that the excitation function has the waveform

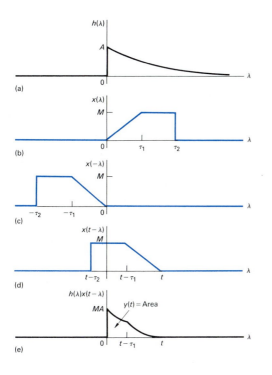

(a)

(b)

(c)

(d)

(e)

FIGURE 17.14 Graphic interpretation of the convolution integral $\int_0^t h(\lambda)x(t - \lambda)\,d\lambda$: (a) impulse response; (b) excitation function; (c) folded excitation function; (d) folded excitation function displaced t units; (e) product $h(\lambda)x(t - \lambda)$.

shown in Fig. 17.14(b). In each of these plots we replaced t with λ, the symbol of integration. Figure 17.14(c) and (d) illustrate the graphic interpretation of $x(t - \lambda)$. Replacing λ with $-\lambda$ simply folds the excitation function over the vertical axis, and replacing $-\lambda$ with $t - \lambda$ slides the "folded" function to the right. This folding operation gives rise to the term *convolution*. At any specified value of t, the response function $y(t)$ is the area under the product function $h(\lambda)x(t - \lambda)$, as shown in Fig. 17.14(e). It should be apparent from this plot why the lower limit on the convolution integral is zero and the upper limit is t. For $\lambda < 0$ the product $h(\lambda)x(t - \lambda)$ is zero because $h(\lambda)$ is zero. For $\lambda > t$ the product $h(\lambda)x(t - \lambda)$ is zero because $x(t - \lambda)$ is zero.

Figure 17.15 shows the graphic interpretation of the second form of the convolution integral. Note that the product function

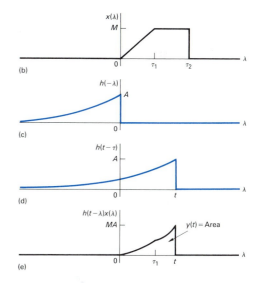

(b)

(c)

(d)

(e)

FIGURE 17.15 Graphic representation of the convolution integral $\int_0^t h(t - \lambda)x(\lambda)\,d\lambda$: (a) impulse response; (b) excitation function; (c) folded impulse response; (d) folded impulse response displaced t units; (e) product $h(t - \lambda)x(\lambda)$.

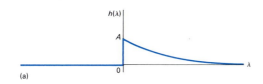

(a)

in Fig. 17.15(e) confirms the use of zero for the lower limit and t for the upper limit.

Example 17.3 illustrates how to use the convolution integral, in conjunction with the unit impulse response, to find the response of a circuit.

E X A M P L E 17.3

The excitation voltage v_i for the circuit shown in Fig. 17.16(a) is shown in Fig. 17.16(b).

a) Use the convolution integral to find $v_o(t)$.

b) Plot $v_o(t)$ over the range of $0 \le t \le 15$ s.

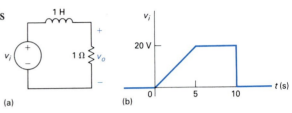

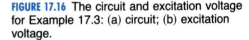

(a)   (b)

FIGURE 17.16 The circuit and excitation voltage for Example 17.3: (a) circuit; (b) excitation voltage.

S O L U T I O N

a) The first step in using the convolution integral is to find the unit impulse response of the circuit. We obtain the expression for V_o from the s-domain equivalent of the circuit in Fig. 17.16(a):

$$V_o = \frac{V_i}{s + 1} \quad (1).$$

When v_i is a unit impulse function $\delta(t)$,

$$v_o = h(t) = e^{-t}u(t),$$

from which

$$h(\lambda) = e^{-\lambda}u(\lambda).$$

Using the first form of the convolution integral in Eq. (17.38), we construct the impulse response and folded excitation function shown in Fig. 17.17, which are helpful in selecting the limits on the convolution integral. Sliding the folded excitation function to the right requires breaking the integration into three intervals: $0 \le t \le 5$, $5 \le t \le 10$, and $10 \le t \le \infty$. The breaks in the excitation function at 0, 5, and 10 s dictate these break points. Figure 17.18 shows the positioning of the folded excitation for each of these intervals. The analytical expression for v_i in the time interval $0 \le t \le 5$ is

$$v_i = 4t \quad (0 \le t \le 5 \text{ s}).$$

Hence the analytic expression for the folded excitation func-

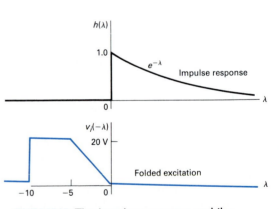

FIGURE 17.17 The impulse response and the folded excitation function for Example 17.3.

tion in the interval $t - 5 \le \lambda \le t$ is

$$v_i(t - \lambda) = 4(t - \lambda) \qquad (t - 5 \le \lambda \le t).$$

We can now set up the three integral expressions for $v_o(t)$.

For $0 \le t \le 5$ s:

$$v_o = \int_0^t 4(t - \lambda)e^{-\lambda}\, d\lambda = 4(e^{-t} + t - 1) \text{ V.}$$

For $5 \le t \le 10$ s:

$$v_o = \int_0^{t-5} 20e^{-\lambda}\, d\lambda + \int_{t-5}^t 4(t - \lambda)e^{-\lambda}\, d\lambda$$

$$= 4(5 + e^{-t} - e^{-(t-5)}) \text{ V.}$$

For $10 \text{ s} \le t \le \infty$:

$$v_o = \int_{t-10}^{t-5} 20e^{-\lambda}\, d\lambda + \int_{t-5}^t 4(t - \lambda)e^{-\lambda}\, d\lambda$$

$$= 4(e^{-t} - e^{-(t-5)} + 5e^{-(t-10)}) \text{ V.}$$

b) We have computed $v_o(t)$ for 1-s intervals of time using the appropriate equation. We tabulated the results in Table 17.1 and show them graphically in Fig. 17.19.

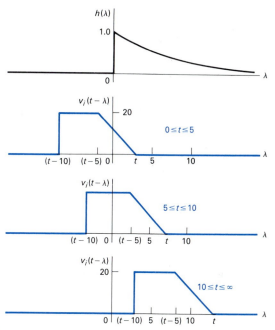

FIGURE 17.18 The displacement of $v_i(t - \lambda)$ for three different time intervals.

TABLE 17.1

NUMERICAL VALUES OF $v_o(t)$

t	v_o	t	v_o	t	v_o
1	1.47	6	18.54	11	7.35
2	4.54	7	19.56	12	2.70
3	8.20	8	19.80	13	0.99
4	12.07	9	19.93	14	0.37
5	16.03	10	19.97	15	0.13

FIGURE 17.19 The voltage response versus time for Example 17.3.

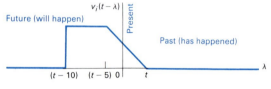

FIGURE 17.20 The past, present, and future values of the excitation function.

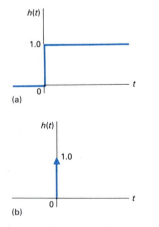

FIGURE 17.21 Weighting functions: (a) perfect memory; (b) no memory.

CONCEPTS OF MEMORY AND WEIGHTING FUNCTION

We mentioned at the beginning of this section that the convolution integral introduces the concepts of memory and weighting function into circuit analysis. Graphic interpretation of the convolution integral is the easiest way to introduce these concepts. The concepts of memory and weighting function become evident when you view the folding and sliding of the excitation function on a time scale characterized as past, present, and future. On such a time scale, the vertical axis, over which the excitation function $x(t)$ is folded, represents the present value, past values of $x(t)$ lie to the right of the vertical axis, and future values of $x(t)$ lie to the left of the vertical axis. Figure 17.20 shows this description of $x(t)$. For illustrative purposes we used the excitation function from Example 17.3.

When we combine the past, present, and future views of $x(t - \tau)$ with the impulse response of the circuit, we see that the impulse response weights $x(t)$ according to present and past values. For example, Fig. 17.18 shows that the impulse response in Example 17.3 gives less weight to past values of $x(t)$ than to the present value of $x(t)$. In other words, the circuit retains less and less about past input values. For example, in Fig. 17.19 v_o quickly approaches zero when the present value of the input is zero (that is, when $t > 10$ s). In other words, because the present value of the input receives more weight than the past values of the input, the output quickly approaches the present value of the input.

Multiplication of $x(t - \lambda)$ by $h(\lambda)$ gives rise to referring to the impulse response as the circuit *weighting* function. At the same time the weighting function is described in terms of how much memory it has. For example, if the impulse response, or weighting function, is flat, as shown in Fig. 17.21(a), it gives equal weight to all values of $x(t)$, past and present. Such a circuit has a perfect memory. However, if the impulse response is an impulse function, as shown in Fig. 17.21(b), it gives no weight to past values of $x(t)$. Such a circuit has no memory. Thus the more memory a circuit has, the more distortion there is between the waveform of the excitation function and the waveform of the response function. We can show this relation by assuming that the circuit has no memory—that is, $h(t) = A\delta(t)$—and then noting from the convolution integral that

$$y(t) = \int_0^t h(\lambda)x(t - \lambda)\, d\lambda$$

$$= \int_0^t A\delta(\lambda)x(t - \lambda)\, d\lambda$$

$$= Ax(t). \tag{17.39}$$

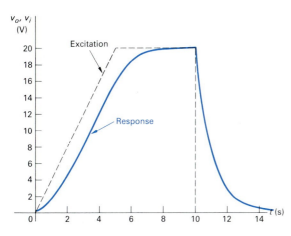

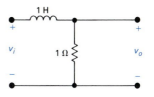

FIGURE 17.22 The input and output waveforms for Example 17.3.

Equation (17.39) shows that, if the circuit has no memory, the output is a scaled replica of the input.

The circuit shown in Example 17.3 illustrates the distortion between input and output for a circuit that has some memory. The distortion is clear when we plot the input and output waveforms on the same graph, as in Fig. 17.22.

DRILL EXERCISES

17.5 A rectangular voltage pulse $v_i = [u(t) - u(t - 1)]$ V is applied to the circuit shown. Use the convolution integral to find v_o.

ANSWER: $v_o = 1 - e^{-t}$ V, $\qquad 0 \le t \le 1$;
$v_o = (e - 1)e^{-t}$ V, $\qquad 1 \le t \le \infty$.

17.6 Interchange the inductor and resistor in Drill Exercise 17.5 and again use the convolution integral to find v_o.

ANSWER: $v_o = e^{-t}$ V, $\qquad 0 < t < 1$;
$v_o = (1 - e)e^{-t}$ V, $\qquad 1 < t \le \infty$.

17.5 THE TRANSFER FUNCTION AND THE STEADY-STATE SINUSOIDAL RESPONSE

We can use the transfer function to find the steady-state response of a circuit driven by a sinusoidal source. To show how the transfer function relates the steady-state response to the excitation source, we assume that

$$x(t) = A \cos (\omega t + \phi) \qquad \textbf{(17.40)}$$

and then use Eq. (17.24) to find the steady-state solution of $y(t)$. To find the Laplace transform of $x(t)$ we first write $x(t)$ as

$$x(t) = A \cos \omega t \cos \phi - A \sin \omega t \sin \phi, \qquad \textbf{(17.41)}$$

from which

$$X(s) = \frac{(A \cos \phi)s}{s^2 + \omega^2} - \frac{(A \sin \phi)\omega}{s^2 + \omega^2} = \frac{A(s \cos \phi - \omega \sin \phi)}{s^2 + \omega^2}.$$
$$\textbf{(17.42)}$$

Substituting Eq. (17.42) into Eq. (17.24) gives the s-domain expression for the response:

$$Y(s) = H(s)\frac{A(s \cos \phi - \omega \sin \phi)}{s^2 + \omega^2}. \qquad \textbf{(17.43)}$$

We now visualize the partial fraction expansion of Eq. (17.43). The number of terms in the expansion depends on the number of poles of $H(s)$. Because $H(s)$ is not specified beyond being the transfer function of a physically realizable circuit, the expansion of Eq. (17.43) is

$$Y(s) = \frac{K_1}{s - j\omega} + \frac{K_1^*}{s + j\omega}$$

$$+ \sum \text{ terms generated by the poles of } H(s). \qquad \textbf{(17.44)}$$

In Eq. (17.44) the first two terms result from the complex conjugate poles of the driving source; that is, $s^2 + \omega^2 = (s - j\omega)(s + j\omega)$. However, the terms generated by the poles of $H(s)$ do not contribute to the steady-state response of $y(t)$. The reason is that all the poles of $H(s)$ lie in the left half of the s plane, and consequently the corresponding time-domain terms approach zero as t increases. Thus the first two terms on the right-hand side of Eq. (17.44) determine the steady-state response. The problem is reduced to finding the partial fraction coefficient K_1:

$$K_1 = \left. \frac{H(s)A(s \cos \phi - \omega \sin \phi)}{s + j\omega} \right|_{s=j\omega}$$

$$= \frac{H(j\omega)A(j\omega \cos \phi - \omega \sin \phi)}{2j\omega}$$

$$= \frac{H(j\omega)A(\cos \phi + j \sin \phi)}{2} = \frac{1}{2}H(j\omega)Ae^{j\phi}. \qquad \textbf{(17.45)}$$

In general, $H(j\omega)$ is a complex quantity, which we recognize by writing it in polar form; thus

$$H(j\omega) = |H(j\omega)|e^{j\theta(\omega)}. \qquad (17.46)$$

Note from Eq. (17.46) that both the magnitude $|H(j\omega)|$ and phase angle $[\theta(\omega)]$ of the transfer function vary with the frequency ω. When we substitute Eq. (17.46) into Eq. (17.45), the expression for K_1 becomes

$$K_1 = \frac{A}{2}|H(j\omega)|e^{j[\theta(\omega)+\phi]}. \qquad (17.47)$$

We obtain the steady-state solution for $y(t)$ by inverse-transforming Eq. (17.44) and, in the process, ignoring the terms generated by the poles of $H(s)$. Thus

$$y_{ss}(t) = A|H(j\omega)| \cos [\omega t + \phi + \theta(\omega)], \qquad (17.48)$$

which indicates how to use the transfer function to find the steady-state sinusoidal response of a circuit. The amplitude of the response equals the amplitude of the source, A, times the magnitude of the transfer function $|H(j\omega)|$. The phase angle of the response $[\phi + \theta(\omega)]$ equals the phase angle of the source (ϕ) plus the phase angle of the transfer function $[\theta(\omega)]$. We evaluate both $|H(j\omega)|$ and $\theta(\omega)$ at the frequency of the source (ω).

Example 17.4 illustrates how to use the transfer function to find the steady-state sinusoidal response of a circuit.

EXAMPLE 17.4

A sinusoidal voltage source drives the circuit shown in Fig. 17.23. The source voltage is 120 cos (5000t + 30°) V. Find the steady-state expression for v_o.

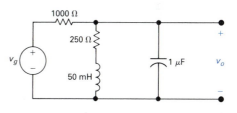

FIGURE 17.23 The circuit for Example 17.4.

SOLUTION

From Example 17.1,

$$H(s) = \frac{1000(s + 5000)}{s^2 + 6000s + 25 \times 10^6}.$$

The frequency of the voltage source is 5000 rad/s; hence we evaluate $H(s)$ at $H(j5000)$:

$$H(j5000) = \frac{1000(5000 + j5000)}{-25 \times 10^6 + j5000(6000) + 25 \times 10^6}$$

$$= \frac{1 + j1}{j6} = \frac{1 - j1}{6} = \frac{\sqrt{2}}{6}\angle{-45°}.$$

Then, from Eq. (17.48),

$$v_{o_{ss}} = \frac{(120)\sqrt{2}}{6} \cos{(5000t + 30° - 45°)}$$

$$= 20\sqrt{2} \cos{(5000t - 15°)} \text{ V.}$$

The ability to use the transfer function to calculate the steady-state sinusoidal response of a circuit is important for two primary reasons. First, we can use the transfer function to find the steady-state frequency response of the circuit. That is, because the amplitude and phase of the response depend on the magnitude and phase of $H(j\omega)$, we simply evaluate the magnitude and phase of $H(j\omega)$ over the frequency range of interest. The variation of $|H(j\omega)|$ and $\theta(\omega)$ with ω indicates exactly how the amplitude and phase of the steady-state response signal vary with ω. Second, if we know $H(j\omega)$, we also know $H(s)$, at least theoretically. In other words, we can reverse the process. Instead of using $H(s)$ to find $H(j\omega)$, we use $H(j\omega)$ to find $H(s)$. Once we know $H(s)$, we can find the response to other excitation sources. In this application, we determine $H(j\omega)$ experimentally and then construct $H(s)$ from the experimental data. Practically, this experimental approach is not always possible; however, in some cases it does provide a useful method for deriving $H(s)$. Theoretically, the relationship between $H(s)$ and $H(j\omega)$ provides a link between the time domain and the frequency domain.

DRILL EXERCISES

17.7 The current source in the circuit shown is delivering $3\sqrt{2} \cos{2t}$ A. Use the transfer function to compute the steady-state expression for v_o.

ANSWER: $20 \cos{(2t - 45°)}$ V.

17.8 a) For the circuit shown, find the steady-state expression for v_o when $v_g = 10 \cos{50,000t}$ V.

   b) Replace the 50-kΩ resistor with a variable resistor and compute the value of resistance necessary to cause v_o to lead v_g by 120°.

ANSWER: (a) $10 \cos{(50,000t + 90°)}$ V;
(b) 28,867.51 Ω.

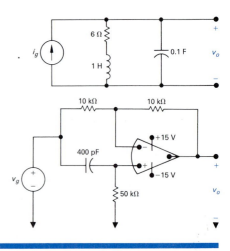

17.6 BODE DIAGRAMS

The behavior of the transfer function as the frequency of the sinusoidal source varies is an important characteristic of a circuit.
One effective way to describe how the amplitude and phase angle
of $H(j\omega)$ vary with frequency is graphically. The most efficient
method for generating and plotting the amplitude and phase data
is the digital computer. We can rely on the digital computer to
give us accurate numerical plots of $|H(j\omega)|$ and $\theta(\omega)$ versus ω.
However, in some situations some preliminary sketches using
Bode diagrams can greatly aid intelligent use of the computer. A
Bode diagram, or plot, is a graphic technique that gives a "feel"
for the frequency response of a circuit. Such plots are most useful in circuits where the poles and zeros of $H(s)$ are reasonably
well separated.

PSpice PROBE can create Bode plots once
PSpice has performed sinusoidal steady-state simulation: Section 15.3

A Bode diagram consists of two separate plots: One shows
how the amplitude of $H(j\omega)$ varies with frequency, and the second shows how the phase angle of $H(j\omega)$ varies with frequency.
Both plots are made on semilog graph paper. In both the amplitude and phase plots, the frequency is plotted on the horizontal
log scale and the amplitude and phase angle are plotted on the
linear vertical scale. These amplitude and phase diagrams are
called *Bode plots* in recognition of the pioneering work done in
this area by H. W. Bode.[†]

BODE DIAGRAMS: REAL, FIRST-ORDER POLES AND ZEROS

In order to simplify development of Bode diagrams, we assume
to start with that all the poles and zeros of $H(s)$ are real and first
order. For our purposes having a specific expression for $H(s)$ is
helpful. Hence we base the discussion on:

$$H(s) = \frac{K(s + z_1)}{s(s + p_1)},\qquad (17.49)$$

from which

$$H(j\omega) = \frac{K(j\omega + z_1)}{j\omega(j\omega + p_1)}.\qquad (17.50)$$

The first step in making Bode diagrams is to put the expression
for $H(j\omega)$ in a *standard form*, which we derive simply by dividing out the poles and zeros:

$$H(j\omega) = \frac{Kz_1(1 + j\omega/z_1)}{p_1(j\omega)(1 + j\omega/p_1)}.\qquad (17.51)$$

[†] See H. W. Bode, *Network Analysis and Feedback Design* (New York: Van
Nostrand, 1945).

Next we let K_o represent the constant quantity Kz_1/p_1 and at the same time express $H(j\omega)$ in polar form:

$$H(j\omega) = \frac{K_o|1 + j\omega/z_1|\underline{/\psi_1}}{|\omega|\underline{/90°}|1 + j\omega/p_1|\underline{/\beta_1}}$$

$$= \frac{K_o|1 + j\omega/z_1|}{\omega|1 + j\omega/p_1|}\underline{/\psi_1 - 90° - \beta_1}. \quad \textbf{(17.52)}$$

From Eq. (17.52),

$$|H(j\omega)| = \frac{K_o|1 + j\omega/z_1|}{\omega|1 + j\omega/p_1|}; \quad \textbf{(17.53)}$$

$$\theta(\omega) = \psi_1 - 90° - \beta_1. \quad \textbf{(17.54)}$$

By definition the phase angles ψ_1 and β_1 are

$$\psi_1 = \tan^{-1} \omega/z_1; \quad \textbf{(17.55)}$$

$$\beta_1 = \tan^{-1} \omega/p_1. \quad \textbf{(17.56)}$$

The Bode diagrams consist of plotting Eq. (17.53) (amplitude) and Eq. (17.54) (phase) as functions of ω.

STRAIGHT-LINE AMPLITUDE PLOT

The amplitude plot involves the multiplication and division of factors associated with the poles and zeros of $H(s)$. We reduce this multiplication and division to addition and subtraction by expressing the amplitude of $H(j\omega)$ in terms of a logarithmic value: the decibel (dB). The amplitude of $H(j\omega)$ in decibels is

$$A_{dB} = 20 \log_{10} |H(j\omega)|. \quad \textbf{(17.57)}$$

Expressing Eq. (17.53) in terms of decibels gives

$$A_{dB} = 20 \log_{10} \frac{K_o|1 + j\omega/z_1|}{\omega|1 + j\omega/p_1|}$$

$$= 20 \log_{10} K_o + 20 \log_{10} |1 + j\omega/z_1| - 20 \log_{10} \omega$$

$$- 20 \log_{10} |1 + j\omega/p_1|. \quad \textbf{(17.58)}$$

The key to plotting Eq. (17.58) is to plot each term in the equation separately and then combine the separate plots graphically. The individual factors are easy to plot because they can be approximated in all cases by straight lines.

The plot of $20 \log_{10} K_o$ is a horizontal straight line because K_o is not a function of frequency. The value of this term is positive for $K_o > 1$, zero for $K_o = 1$, and negative for $K_o < 1$.

Two straight lines approximate the plot of $20 \log_{10} |1 + j\omega/z_1|$. For small values of ω the magnitude of

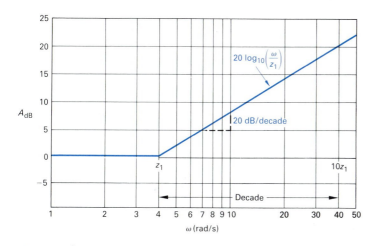

FIGURE 17.24 A straight-line approximation of the amplitude plot of a first-order zero.

$|1 + j\omega/z_1|$ is approximately 1, and therefore

$$20 \log_{10} |1 + j\omega/z_1| \to 0, \qquad \text{as } \omega \to 0. \quad \textbf{(17.59)}$$

For large values of ω the magnitude of $|1 + j\omega/z_1|$ is approximately ω/z_1, and therefore

$$20 \log_{10} |1 + j\omega/z_1| \to 20 \log_{10}(\omega/z_1), \qquad \text{as } \omega \to \infty. \quad \textbf{(17.60)}$$

On a log frequency scale, $20 \log_{10} (\omega/z_1)$ is a straight line with a slope of 20 dB/decade (a decade is a 10-to-1 change in frequency). This straight line intersects the 0-dB axis at $\omega = z_1$. This value of ω is called the *corner frequency*. Thus, on the basis of Eqs. (17.59) and (17.60), two straight lines can approximate the amplitude plot of a first-order zero, as shown in Fig. 17.24.

The plot of $-20 \log_{10} \omega$ is a straight line having a slope of -20 dB/decade that intersects the 0-dB axis at $\omega = 1$. Two straight lines approximate the plot of $-20 \log_{10} |1 + j\omega/p_1|$. Here the two straight lines intersect on the 0-dB axis at $\omega = p_1$. For large values of ω the straight line $-20 \log_{10} (\omega/p_1)$ has a slope of -20 dB/decade. Figure 17.25 shows the straight-line approximation of the amplitude plot of a first-order pole.

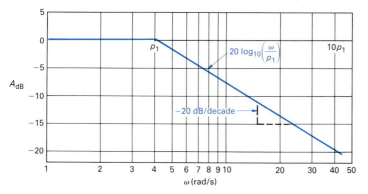

FIGURE 17.25 A straight-line approximation of the amplitude plot of a first-order pole.

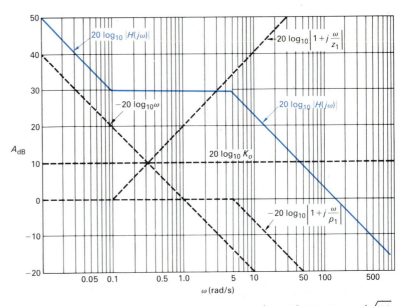

FIGURE 17.26 A straight-line approximation of the amplitude plot for Eq. (17.58).

Figure 17.26 shows a plot of Eq. (17.58) for $K_o = \sqrt{10}$, $z_1 = 0.1$ rad/s, and $p_1 = 5$ rad/s. Each term in Eq. (17.58) is labeled on Fig. 17.26, so you can verify that the individual terms sum to create the resultant plot, labeled $20 \log_{10} |H(j\omega)|$ in Fig. 17.26.

Example 17.5 also illustrates the construction of a straight-line amplitude plot for a transfer function characterized by first-order poles and zeros.

E X A M P L E   17.5

The numerical expression for a transfer function is

$$H(s) = \frac{10^4(s + 1)}{(s + 10)(s + 100)}.$$

a) Construct a straight-line approximation of the Bode amplitude plot.

b) Calculate $20 \log_{10} |H(j\omega)|$ at $\omega = 50$ rad/s and $\omega = 1000$ rad/s.

c) Plot the values computed in (b) on the straight-line graph.

S O L U T I O N

a) We begin by writing $H(j\omega)$ in standard form:

$$H(j\omega) = \frac{10(1 + j\omega)}{[1 + j(\omega/10)][1 + j(\omega/100)]}.$$

The expression for the amplitude of $H(j\omega)$ in decibels is

$$A_{dB} = 20 \log_{10} |H(j\omega)|$$
$$= 20 \log_{10} 10 + 20 \log_{10} |1 + j\omega|$$
$$-20 \log_{10} \left|1 + j\frac{\omega}{10}\right| - 20 \log_{10} \left|1 + j\frac{\omega}{100}\right|.$$

Figure 17.27 shows the straight-line plot. Each term contributing to the overall amplitude is identified.

b) We have

$$H(j50) = \frac{10(1 + j50)}{(1 + j5)(1 + j0.5)}$$
$$= 87.72\underline{/-16.40°};$$
$$20 \log_{10} |H(j50)| = 20 \log_{10} 87.72$$
$$= 38.86 \text{ dB};$$
$$H(j1000) = \frac{10(1 + j1000)}{(1 + j100)(1 + j10)}$$
$$= 9.95\underline{/-83.77°};$$
$$20 \log_{10} |9.95| = 19.96 \text{ dB}.$$

c) See Fig. 17.27.

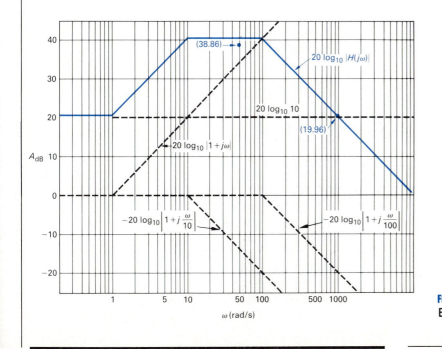

FIGURE 17.27 The amplitude plot for Example 17.5.

MORE ACCURATE AMPLITUDE PLOTS

We can make the straight-line plots for first-order poles and zeros more accurate by correcting the amplitude values at the corner frequency, one-half the corner frequency, and twice the corner frequency. At the corner frequency, the actual value in decibels is

$$
\begin{aligned}
A_{\mathrm{dB}_c} &= \pm 20 \log_{10} |1 + j1| \\
&= \pm 20 \log_{10} \sqrt{2} \\
&\approx \pm 3 \text{ dB.} \tag{17.61}
\end{aligned}
$$

The actual value of the amplitude in decibels at one-half of the corner frequency is

$$
\begin{aligned}
A_{\mathrm{dB}_{c/2}} &= \pm 20 \log_{10} \left| 1 + j\frac{1}{2} \right| \\
&= \pm 20 \log_{10} \sqrt{5/4} \\
&\approx \pm 1 \text{ dB.} \tag{17.62}
\end{aligned}
$$

At twice the corner frequency the actual value in decibels is

$$
\begin{aligned}
A_{\mathrm{dB}_{2c}} &= \pm 20 \log_{10} |1 + j2| \\
&= \pm 20 \log_{10} \sqrt{5} \\
&\approx \pm 7 \text{ dB.} \tag{17.63}
\end{aligned}
$$

In Eqs. (17.61), (17.62), and (17.63), the plus sign applies to a first-order zero and the minus sign applies to a first-order pole. The straight-line approximation of the amplitude plot gives 0 dB at the corner and one-half the corner frequencies and ± 6 dB at twice the corner frequency. Hence the corrections are ± 3 dB at the corner frequency and ± 1 dB at both one-half the corner frequency and twice the corner frequency. Figure 17.28 summarizes these corrections.

A 2-to-1 change in frequency is called an *octave*. A slope of 20 dB/decade is equivalent to 6.02 dB/octave which, for graphic purposes, is equivalent to 6 dB/octave. Thus the corrections enumerated correspond to one octave below and one octave above the corner frequency.

If the poles and zeros of $H(s)$ are well separated, inserting these corrections into the overall amplitude plot and achieving a reasonably accurate curve is relatively easy. However, if the poles and zeros are close together, the overlapping corrections are difficult to evaluate, and you're better off using the straight-line plot as a first estimate of the amplitude characteristic. Then use the computer to refine the calculations in the frequency range of interest.

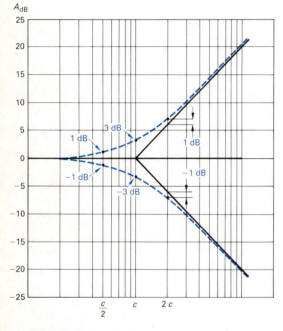

FIGURE 17.28 Corrected amplitude plots for a first-order zero and pole.

STRAIGHT-LINE PHASE-ANGLE PLOTS

We can also make phase-angle plots by using straight-line approximations. The phase angle associated with the constant K_o is zero, and the phase angle associated with a first-order zero or pole at the origin is a constant $\pm 90°$. For a first-order zero or pole not at the origin, the straight-line approximations are as follows. For frequencies less than one-tenth the corner frequency, the phase angle is assumed to be zero. For frequencies greater than 10 times the corner frequency, the phase angle is assumed to be $\pm 90°$. Between one-tenth the corner frequency and 10 times the corner frequency, the phase-angle plot is a straight line that goes through $0°$ at one-tenth the corner frequency, $\pm 45°$ at the corner frequency, and $\pm 90°$ at 10 times the corner frequency. In all these cases the plus sign applies to the first-order zero and the minus sign to the first-order pole. Figure 17.29 depicts the straight-line approximation for a first-order zero and pole. The dashed curves show the exact variation of the phase angle as the frequency varies. Note how closely the straight-line plot approximates the actual variation in phase angle. The maximum deviation between the straight-line plot and the actual plot is approximately $6°$.

Figure 17.30 depicts the straight-line approximation of the phase angle of the transfer function given by Eq. (17.49). Equation (17.54) gives the equation for the phase angle, the plot corresponds to $z_1 = 0.1$ rad/s and $p_1 = 5$ rad/s. An illustration of a phase-angle plot using the straight-line approximations is given in Example 17.6.

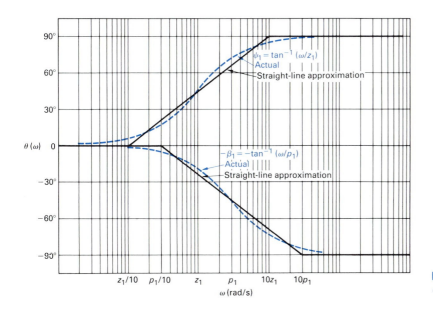

FIGURE 17.29 Phase-angle plots for a first-order zero and pole.

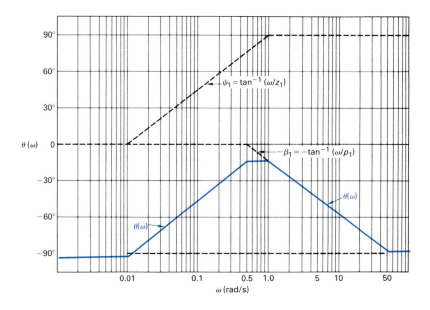

FIGURE 17.30 A straight-line approximation of the phase-angle plot for Eq. (17.54).

E X A M P L E   17.6

a) Make a straight-line phase-angle plot for the transfer function in Example 17.5.

b) Compute the phase angle $\theta(\omega)$ at $\omega = 50$, 500, and 1000 rad/s.

c) Plot the values of (b) on the diagram of (a).

S O L U T I O N

a) From Example 17.5,

$$H(j\omega) = \frac{10(1 + j\omega)}{[1 + j(\omega/10)][1 + j(\omega/100)]}$$

$$= \frac{10|1 + j\omega|}{|1 + j(\omega/10)||1 + j(\omega/100)|} \underline{/\psi_1 - \beta_1 - \beta_2}.$$

Therefore

$$\theta(\omega) = \psi_1 - \beta_1 - \beta_2,$$

where

$$\psi_1 = \tan^{-1}\omega, \qquad \beta_1 = \tan^{-1}(\omega/10), \qquad \text{and}$$

$$\beta_2 = \tan^{-1}(\omega/100).$$

Figure 17.31 depicts the straight-line approximation of $\theta(\omega)$.

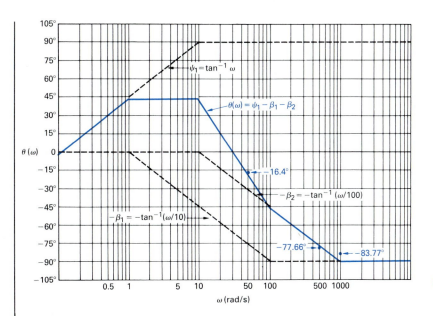

FIGURE 17.31 A straight-line approximation of $\theta(\omega)$ for Example 17.6.

b) We have

$$H(j50) = 87.72\underline{/-16.40°};$$

$$H(j500) = 19.61\underline{/-77.66°};$$

$$H(j1000) = 9.95\underline{/-83.77°}.$$

Thus

$$\theta(50) = -16.40°, \qquad \theta(500) = -77.66°,$$

and

$$\theta(1000) = -83.77°.$$

c) See Fig. 17.31.

DRILL EXERCISES

17.9 The numerical expression for a transfer function is

$$H(s) = \frac{10^5(s + 5)}{(s + 100)(s + 5000)}.$$

On the basis of a straight-line approximation of $|H(j\omega)|$ versus ω, estimate (a) the maximum amplitude of $H(j\omega)$ in decibels and (b) the value of $\omega > 0$ where the amplitude of $H(j\omega)$ equals unity.

ANSWER: (a) 26 dB; (b) 98 krad/s.

17.10 Approximate the phase angle of the transfer function in Drill Exercise 17.9 by means of a straight-line plot.

a) Use the straight-line plot to predict the phase angle of $H(s)$ at frequencies of 30, 50, 100, and 5000 rad/s.

b) Calculate the actual value of the phase angle at 30, 50, 100, and 5000 rad/s.

ANSWER: (a) 58.5°, 58.5°, 45°, −45°; (b) 63.49°, 57.15°, 40.99°, −43.91°.

17.7 BODE DIAGRAMS: COMPLEX POLES AND ZEROS

Complex poles and zeros in the expression for $H(s)$ require special attention when you make amplitude and phase-angle plots. Let's focus on the contribution that a pair of complex poles make to the amplitude and phase-angle plots. Once you understand the rules for handling complex poles, their application to a pair of complex zeros becomes apparent.

The complex poles and zeros of $H(s)$ always appear in conjugate pairs. The first step in making either an amplitude or a phase-angle plot of a transfer function that contains complex poles is to combine the conjugate pair into a single quadratic term. Thus for

$$H(s) = \frac{K}{(s + \alpha - j\beta)(s + \alpha + j\beta)}, \quad (17.64)$$

we first rewrite the product $(s + \alpha - j\beta)(s + \alpha + j\beta)$ as

$$(s + \alpha)^2 + \beta^2 = s^2 + 2\alpha s + \alpha^2 + \beta^2.$$

When making Bode diagrams, we write the quadratic term in the more convenient form:

$$s^2 + 2\alpha s + \alpha^2 + \beta^2 = s^2 + 2\zeta\omega_n s + \omega_n^2. \quad (17.65)$$

A direct comparison of the two forms shows that

$$\omega_n^2 = \alpha^2 + \beta^2 \quad (17.66)$$

and

$$\zeta\omega_n = \alpha. \quad (17.67)$$

The term ω_n is the corner frequency of the quadratic factor, and ζ is the damping coefficient of the quadratic term. The critical value of ζ is 1. If $\zeta < 1$, the roots of the quadratic factor are complex and we use Eq. (17.65) to represent the complex poles. If $\zeta \geq 1$, we factor the quadratic factor into $(s + p_1)(s + p_2)$ and plot amplitude and phase in accordance with the discussion in Section 17.6. Assuming that $\zeta < 1$, we rewrite Eq. (17.64) as

$$H(s) = \frac{K}{s^2 + 2\zeta\omega_n s + \omega_n^2}. \qquad (17.68)$$

We then write Eq. (17.68) in standard form by dividing through by the poles and zeros. For the quadratic term we divide through by ω_n^2, so

$$H(s) = \frac{K}{\omega_n^2} \frac{1}{1 + (s/\omega_n)^2 + 2\zeta(s/\omega_n)}, \qquad (17.69)$$

from which

$$H(j\omega) = \frac{K_o}{1 - (\omega^2/\omega_n^2) + j(2\zeta\omega/\omega_n)}, \qquad (17.70)$$

where

$$K_o = \frac{K}{\omega_n^2}.$$

Before discussing the amplitude and phase-angle diagrams associated with Eq. (17.70), we replace the ratio ω/ω_n by a new variable u for convenience. Then

$$H(j\omega) = \frac{K_o}{1 - u^2 + j2\zeta u}. \qquad (17.71)$$

Now we write $H(j\omega)$ in polar form:

$$H(j\omega) = \frac{K_o}{|(1 - u^2) + j2\zeta u|\underline{/\beta_1}}, \qquad (17.72)$$

from which

$$\begin{aligned} A_{\text{dB}} &= 20 \log_{10} |H(j\omega)| \\ &= 20 \log_{10} K_o - 20 \log_{10} |(1 - u^2) + j2\zeta u| \qquad (17.73) \end{aligned}$$

and

$$\theta(\omega) = -\beta_1 = -\tan^{-1}\frac{2\zeta u}{1 - u^2}. \qquad (17.74)$$

AMPLITUDE PLOT

The quadratic factor contributes to the amplitude of $H(j\omega)$ by means of the term $-20 \log_{10} |1 - u^2 + j2\zeta u|$. Because

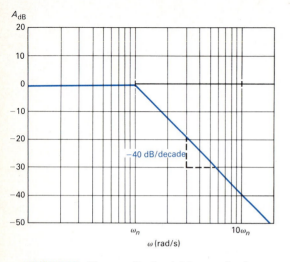

FIGURE 17.32 The amplitude plot for a pair of complex poles.

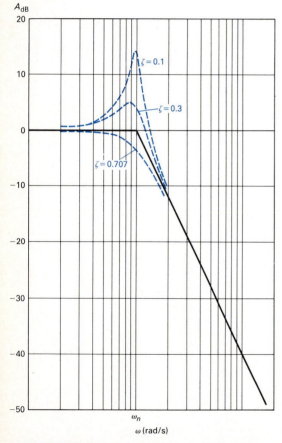

FIGURE 17.33 Effect of ζ on the amplitude plot.

$u = \omega/\omega_n$, $u \to 0$ as $\omega \to 0$, and $u \to \infty$ as $\omega \to \infty$. To see how the term behaves as ω ranges from 0 to ∞, we note that

$$-20 \log_{10} |(1 - u^2) + j2\zeta u| = -20 \log_{10} \sqrt{(1 - u^2)^2 + 4\zeta^2 u^2}$$
$$= -10 \log_{10} [u^4 + 2u^2(2\zeta^2 - 1) + 1]; \quad \textbf{(17.75)}$$

as $u \to 0$,

$$-10 \log_{10} [u^4 + 2u^2(2\zeta^2 - 1) + 1] \to 0; \quad \textbf{(17.76)}$$

and as $u \to \infty$,

$$-10 \log_{10} [u^4 + 2u^2(2\zeta^2 - 1) + 1] \to -40 \log_{10} u. \quad \textbf{(17.77)}$$

From Eqs. (17.76) and (17.77), we conclude that the approximate amplitude plot consists of two straight lines. For $\omega < \omega_n$ the straight line lies along the 0-dB axis, and for $\omega > \omega_n$ the straight line has a slope of -40 dB/decade. These two straight lines join on the 0-dB axis at $u = 1$ or $\omega = \omega_n$. Figure 17.32 shows the straight-line approximation for a quadratic factor with $\zeta < 1$.

CORRECTING THE STRAIGHT-LINE AMPLITUDE PLOT

Correcting the straight-line amplitude plot for a pair of complex poles is not as easy as correcting a first-order real pole, because the corrections depend on the damping coefficient ζ. Figure 17.33 shows the effect of ζ on the amplitude plot. Note that as ζ becomes very small, a large peak in the amplitude occurs in the neighborhood of the corner frequency $\omega_n(u = 1)$. When $\zeta \geq 1/\sqrt{2}$, the corrected amplitude plot lies entirely below the straight-line approximation. For sketching purposes, the straight-line amplitude plot can be corrected by locating four points on the actual curve. These four points correspond to one-half the corner frequency, the frequency at which the amplitude reaches its peak value, the corner frequency, and the frequency at which the amplitude is zero. Figure 17.34 shows these four points.

At one-half the corner frequency (point 1), the actual amplitude is

$$A_{dB}(\omega_n/2) = -10 \log_{10} (\zeta^2 + 0.5625). \quad \textbf{(17.78)}$$

The amplitude peaks (point 2) at a frequency of

$$\omega_p = \omega_n\sqrt{1 - 2\zeta^2} \quad \textbf{(17.79)}$$

and has a peak amplitude of

$$A_{dB}(\omega_p) = -10 \log_{10} [4\zeta^2(1 - \zeta^2)]. \quad \textbf{(17.80)}$$

At the corner frequency (point 3), the actual amplitude is

$$A_{dB}(\omega_n) = -20 \log_{10} 2\zeta. \quad \textbf{(17.81)}$$

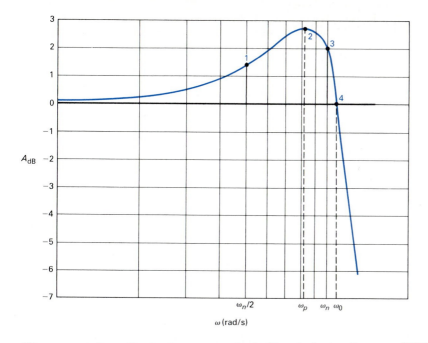

The corrected amplitude plot crosses the 0-dB axis (point 4) at

$$\omega_0 = \omega_n \sqrt{2(1 - 2\zeta^2)} = \sqrt{2}\, \omega_p. \qquad (17.82)$$

The derivations of Eqs. (17.78), (17.81), and (17.82) follow from Eq. (17.75). Evaluating Eq. (17.75) at $u = 0.5$ and $u = 1.0$, respectively, yields Eqs. (17.78) and (17.81). Equation (17.82) corresponds to finding the value of u that makes $u^4 + 2u^2(2\zeta^2 - 1) + 1 = 1$. The derivation of Eq. (17.79) requires differentiating Eq. (17.75) with respect to u and finding the value of u where the derivative is zero. Equation (17.80) is the evaluation of Eq. (17.75) at the value of u found in Eq. (17.79).

Example 17.7 illustrates the amplitude plot for a transfer function with a pair of complex poles.

E X A M P L E   17.7

The transfer function for a linear circuit is

$$H(s) = \frac{2500}{s^2 + 20s + 2500}.$$

a) What is the value of the corner frequency in radians per second?

b) What is the value of K_o?

c) What is the value of the damping coefficient?

d) Make a straight-line amplitude plot ranging from 10 to 500 rad/s.

e) Calculate and sketch the actual amplitude in decibels at $\omega_n/2$, ω_p, ω_n, and ω_0.

S O L U T I O N

a) From the expression for $H(s)$, $\omega_n^2 = 2500$; therefore $\omega_n = 50$ rad/s.

b) By definition K_o is $2500/\omega_n^2$, or 1.

c) The coefficient of s equals $2\zeta\omega_n$; therefore

$$\zeta = \frac{20}{2\omega_n} = 0.20.$$

d) See Fig. 17.35.

e) The actual amplitudes are

$$A_{dB}(\omega_n/2) = -10 \log_{10} (0.6025) = 2.2 \text{ dB};$$

$$\omega_p = 50\sqrt{0.92} = 47.96 \text{ rad/s};$$

$$A_{dB}(\omega_p) = -10 \log_{10} (0.16)(0.96) = 8.14 \text{ dB};$$

$$A_{dB}(\omega_n) = -20 \log_{10} (0.4) = 7.96 \text{ dB};$$

$$\omega_0 = \sqrt{2}\,\omega_p = 67.82 \text{ rad/s};$$

$$A_{dB}(\omega_0) = 0 \text{ dB}.$$

Figure 17.35 shows the corrected plot.

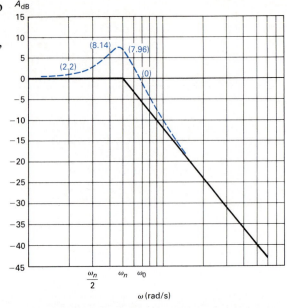

FIGURE 17.35 The amplitude plot for Example 17.7.

PHASE-ANGLE PLOTS

The phase-angle plot for a pair of complex poles is a plot of Eq. (17.74). The phase angle is zero at zero frequency, is $-90°$ at the corner frequency, and approaches $-180°$ as $\omega(u)$ becomes large. As in the case of the amplitude plot, ζ is important in determining the exact shape of the phase-angle plot. For small values of ζ the phase angle changes rapidly in the vicinity of the corner frequency. Figure 17.36 shows the effect of ζ on the phase-angle plot.

We can also make a straight-line approximation of the phase-angle plot for a pair of complex poles. We do so by drawing a line tangent to the phase-angle curve at the corner frequency and extending this line until it intersects with the $0°$ and $-180°$ lines. The line tangent to the phase-angle curve at $-90°$ has a slope of $-2.3/\zeta$ rad/decade $(-132/\zeta$ degrees/decade) and intersects the $0°$ and $-180°$ lines at $u_1 = 4.81^{-\zeta}$ and $u_2 = 4.81^{\zeta}$, respectively.

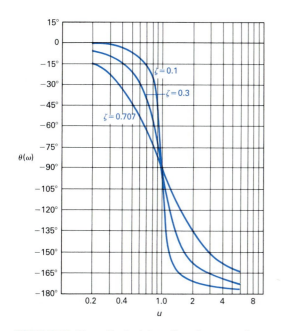

FIGURE 17.36 The effect of ζ on the phase-angle plot.

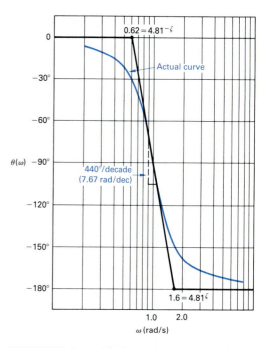

FIGURE 17.37 A straight-line approximation of the phase angle for a pair of complex poles.

Figure 17.37 depicts the straight-line approximation for $\zeta = 0.3$ and shows the actual phase-angle plot. Comparing the straight-line approximation to the actual curve indicates that the approximation is reasonable in the vicinity of the corner frequency. However, in the neighborhood of u_1 and u_2 the error is quite large.

In Example 17.8 we summarize our discussion of Bode diagrams.

E X A M P L E   17.8

The transfer function for a linear circuit is

$$H(s) = \frac{100(s + 1)}{s^2 + 8s + 100}.$$

a) Make a straight-line amplitude plot of $20 \log_{10} |H(j\omega)|$.

b) Estimate the maximum amplitude of $H(j\omega)$ in decibels using the straight-line plot adjusted for corrections.

c) Use the straight-line amplitude plot to estimate the frequency, greater than zero, where the amplitude of $H(j\omega)$ equals 1.

d) What is the actual amplitude at the frequency obtained in (c)?

e) Make a straight-line phase-angle plot of $H(j\omega)$.

f) What is the maximum value of $\theta(\omega)$ predicted by the straight-line plot?

g) What is the actual value of $\theta(\omega)$ at the frequency corresponding to the prediction of part (f)?

SOLUTION

a) The first step in making Bode diagrams is to put $H(j\omega)$ in standard form. Because $H(s)$ contains a quadratic factor, we first check the value of ζ. We find that $\zeta = 0.4$ and $\omega_n = 10$, so

$$H(s) = \frac{s + 1}{1 + (s/10)^2 + 0.8(s/10)},$$

from which

$$H(j\omega) = \frac{|1 + j\omega|\underline{/\psi_1}}{|1 - (\omega/10)^2 + j0.8(\omega/10)|\underline{/\beta_1}}.$$

Note that for the quadratic factor, $u = \omega/10$. The amplitude of $H(j\omega)$ in decibels is

$$A_{dB} = 20 \log_{10} |1 + j\omega|$$
$$- 20 \log_{10} \left[\left| 1 - \left(\frac{\omega}{10}\right)^2 + j0.8\left(\frac{\omega}{10}\right) \right| \right]$$

and the phase angle is

$$\theta(\omega) = \psi_1 - \beta_1,$$

where

$$\psi_1 = \tan^{-1} \omega \qquad \text{and} \qquad \beta_1 = \tan^{-1} \frac{0.8(\omega/10)}{1 - (\omega/10)^2}.$$

Figure 17.38 shows the amplitude plot.

b) From Fig. 17.38, the straight-line approximation of A_{dB} predicts a maximum amplitude of 20 dB at 10 rad/s. The correction owing to the first-order zero is negligible at this frequency. For $\zeta = 0.4$, the correction at 10 rad/s is $-20 \log_{10} (2\zeta)$, or 1.94 dB. The peak correction is $-10 \log_{10} [4\zeta^2(1 - \zeta^2)]$, or 2.70 dB, at $\omega_n\sqrt{1 - 2\zeta^2}$, or 8.25 rad/s. Therefore at the corner frequency the estimated value of A_{dB} is $20 + 1.94$, or 21.94 dB. At ω_p the estimated value of A_{dB} is $18.2 + 2.7$, or 20.9 dB. Thus the peak amplitude is estimated at 21.94 dB, which occurs at a frequency of 10 rad/s. We leave to you verification that the actual value of $20 \log_{10} |H(j10)|$ is 21.98 dB.

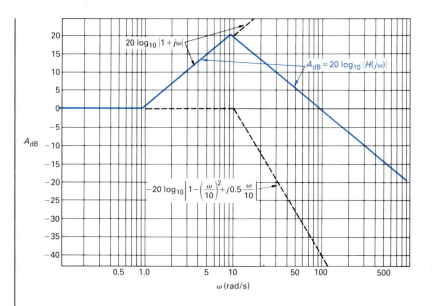

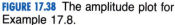

FIGURE 17.38 The amplitude plot for Example 17.8.

c) From Fig. 17.38, A_{dB} is zero at $\omega = 100$ rad/s. Zero decibels corresponds to $|H(j\omega)|$ equals unity. Therefore the straight-line approximation plot predicts that $|H(j\omega)| = 1.0$ at 100 rad/s.

d) The actual amplitude is

$$H(j100) = \frac{|1 + j100|}{|-99 + j8|} = 1.007.$$

Figure 17.39 shows the phase-angle plot. The straight-line

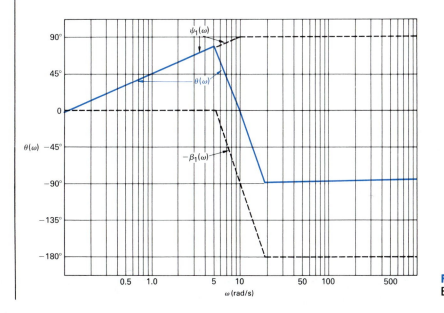

FIGURE 17.39 The phase-angle plot for Example 17.8.

approximation of the angle associated with the quadratic factor intercepts the 0° axis at $10(4.81^{-0.4})$, or 18.74 rad/s. [Note that the straight-line segment of $\theta(\omega)$ between 5.34 and 10 rad/s does not have the same slope as the segment between 10 and 18.74 rad/s.]

f) $\theta(\omega)_{max} \cong 70°$ at $\omega = 5.34$ rad/s.

g) We have $\theta(5.34) = 79.38 - 30.82 = 48.56°$. Note that the large error in the predicted angle is due to the fact that $\theta(\omega)_{max}$ occurs close to the intercept at u_1. Phase-angle estimates from straight-line plots, in general, are not very satisfactory. The straight-line plot does show the general behavior of the phase-angle variation. In this case, it predicts that the phase angle starts from 0°, rises to a maximum value, and eventually approaches $-90°$.

DRILL EXERCISES

17.11 The numerical expression for a current transfer function is

$$H(s) = \frac{I_o}{I_i} = \frac{25 \times 10^8}{s^2 + 20{,}000s + 25 \times 10^8}.$$

Compute the

a) corner frequency;

b) damping coefficient;

c) frequencies when $H(j\omega)$ is unity;

d) peak amplitude of $H(j\omega)$ in decibels;

e) frequency at which the peak occurs;

f) amplitude of $H(j\omega)$ at one-half the corner frequency.

ANSWER: (a) 50 krad/s; (b) 0.2; (c) 0, 67.82 krad/s; (d) 8.14 dB; (e) 47.96 krad/s; (f) 2.20 dB.

17.12 The numerical expression for a voltage transfer function is

$$H(s) = \frac{V_o}{V_g} = \frac{32 \times 10^5}{s^2 + 400s + 64 \times 10^4}.$$

a) Use a straight-line amplitude plot to find the frequency when the amplitude of $H(j\omega)$ equals unity.

b) What is the actual amplitude at the frequency found in (a)?

c) What is the peak amplitude of $H(j\omega)$ in decibels?

d) At what frequency does the output voltage reach its peak amplitude?

e) If the amplitude of the source voltage is 10 V, what is the peak amplitude of the output voltage in V?

ANSWER: (a) 1800 rad/s; (b) 1.19; (c) 20.28 dB; (d) 748.33 rad/s; (e) 103.28 V.

17.8 THE DECIBEL

We defined the amplitude of $H(j\omega)$ in decibels in Eq. (17.57). However, the original definition of the decibel involves power ratios. Tracing back to this definition is worthwhile because it is still a widely accepted use of the term. Telephone engineers who were concerned with the power loss across cascaded circuits used to transmit telephone signals introduced the decibel. Figure 17.40 defines the problem. There p_i is the power input to the system, p_1 is the power output of circuit A, p_2 is the power output of circuit B, and p_o is the power output of the system. The power gain of each circuit is the ratio of the power out to the power in. Thus

$$\sigma_A = \frac{p_1}{p_i}, \qquad \sigma_B = \frac{p_2}{p_1}, \qquad \text{and} \qquad \sigma_C = \frac{p_o}{p_2}.$$

The overall power gain of the system is simply the product of the individual gains, or

$$\frac{p_o}{p_i} = \frac{p_1}{p_i}\frac{p_2}{p_1}\frac{p_o}{p_2} = \sigma_A \sigma_B \sigma_C. \qquad \textbf{(17.83)}$$

The multiplication of power ratios is converted to addition by means of the logarithm; that is,

$$\log_{10}\frac{p_o}{p_i} = \log_{10}\sigma_A + \log_{10}\sigma_B + \log_{10}\sigma_C. \qquad \textbf{(17.84)}$$

This log ratio of the powers was named the *bel* in honor of Alexander Graham Bell. Thus we calculate the overall power gain, in bels, simply by summing the power gains, in bels, of each segment of the transmission system. In practice, the bel was an inconveniently large quantity. One-tenth of a bel was a more useful measure of power gain, hence the *decibel*. The number of decibels equals 10 times the number of bels, so

$$\text{Number of decibels} = 10\log_{10}\frac{p_o}{p_i}. \qquad \textbf{(17.85)}$$

When we use the decibel as a measure of power ratios, in some situations the resistance seen looking into the circuit equals the resistance loading the circuits, as illustrated in Fig. 17.41. When the input resistance equals the load resistance, we can convert the power ratio to either a voltage ratio or a current ratio:

$$\frac{p_o}{p_i} = \frac{v_{out}^2/R_L}{v_{in}^2/R_{in}} = \left(\frac{v_{out}}{v_{in}}\right)^2 \qquad \textbf{(17.86)}$$

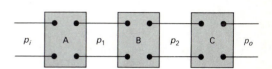

FIGURE 17.40 Three cascaded circuits.

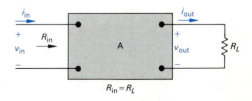

FIGURE 17.41 A circuit in which the input resistance equals the load resistance.

or

$$\frac{p_o}{p_i} = \frac{i_{\text{out}}^2 R_{\text{L}}}{i_{\text{in}}^2 R_{\text{L}}} = \left(\frac{i_{\text{out}}}{i_{\text{in}}}\right)^2. \qquad (17.87)$$

In this case Eqs. (17.86) and (17.87) show that the number of decibels becomes

$$\text{Number of decibels} = 20 \log_{10} \frac{v_{\text{out}}}{v_{\text{in}}}$$

$$= 20 \log_{10} \frac{i_{\text{out}}}{i_{\text{in}}}. \qquad (17.88)$$

The definition of the decibel used in Bode diagrams, that is Eq. (17.57), is borrowed from the results expressed by Eq. (17.88). Equation (17.57) goes beyond Eq. (17.88) because it applies to any transfer function involving a voltage ratio, a current ratio, a voltage-to-current ratio, or a current-to-voltage ratio. You should keep the original definition of the decibel firmly in mind because it is of fundamental importance in many engineering applications.

When you are working with transfer function amplitudes expressed in decibels, having a table that translates the decibel value to the actual value of the output/input ratio is helpful. Table 17.2 gives some useful pairs. The ratio corresponding to a negative decibel value is the reciprocal of the positive ratio. For example, -3 dB corresponds to an output/input ratio of $1/1.41$, or 0.707. Interestingly, -3 dB corresponds to the half-power frequencies of the parallel- and series-resonant circuits discussed in Chapter 14.

The decibel is also used as a unit of power by expressing in decibels the ratio of a known power to a reference power. Usually the reference power is 1 milliwatt and the power unit is called dBm which stands for decibels relative to 1 milliwatt. For example a power of 20 mW corresponds to ± 13 dBm.

AC voltmeters commonly provide dBm readings that assume not only a 1-mW reference power but also a 600-Ω reference resistance (a value commonly used in telephone systems). Since a power of 1 milliwatt in 600 Ω corresponds to 0.7746 V rms, that voltage is read as 0 dBm on the meter. For analog meters there usually is exactly a 10-dB difference between adjacent ranges. Although the scales may be marked 0.1, 0.3, 1, 3, 10, and so on, in fact 3.16 V on the 3-V scale lines up with 1 V on the 1-V scale.

Some voltmeters provide a switch to choose a reference resistance (50, 135, 600, or 900 ohms) or to select dBm or dBV (decibels relative to 1 volt).

TABLE 17.2

A TABLE OF dB-RATIO PAIRS

dB	RATIO	dB	RATIO
0	1.00	30	31.62
3	1.41	40	100.00
6	2.00	60	10^3
10	3.16	80	10^4
15	5.62	100	10^5
20	10.00	120	10^6

17.13 Assume the power delivered to a load is 80% of the power developed by the source. Express the loss in decibels.

ANSWER: −0.97 dB.

17.14 The insertion of an operational amplifier in the circuit in Problem 6.20 allows the source to control 25 times as much power in the load. Express the power gain in decibels.

17.15 An AC voltmeter reads 18.61 on its dBm scale. Assuming a reference resistance of 600 Ω what is the rms value of the voltage corresponding to this reading?

ANSWER: 6.6 V.

17.16 If the reference resistance is 600 Ω explain why 3.16 V on the 3-V scale lines up with 1 V on the 1-V scale when there is exactly a 10-dB difference between scales.

ANSWER: Because 1 V corresponds to 2.22 dBm and 12.22 dBm corresponds to 3.16 V.

SUMMARY

The transfer function plays an important role in the analysis of linear, lumped-parameter circuits. Its importance is based on the following observations.

- The transfer function relates the output signal to a single input signal in the s-domain. Specifically $H(s) = Y(s)/X(s)$ where $Y(s)$ is the Laplace transform of the output signal and $X(s)$ is the Laplace transform of the input signal. (If there is more than one input signal or if there is initial energy stored in the circuit, multiple transfer functions are defined and the output is found by invoking superposition.)

- The transfer function is a rational function of s. Its poles predict the nature of the transient terms in the output signal.

- The inverse transform of the transfer function [h(t)] is the unit impulse response of the circuit. The unit impulse response can be convolved with the input signal to find the output. (The convolution integral introduces the concepts of memory and weighting function. These concepts can be used to predict the fidelity of the transmission between the input signal and the output signal.)

- When the input signal is sinusoidal the steady-state sinusoidal output signal can be derived by replacing s by jω in the expression for the transfer function.

• The transfer function $H(j\omega)$ can be used to analyze the frequency response of the circuit because the amplitude and phase angle of $H(j\omega)$ determine the amplitude and phase angle of the output signal. (If the poles and zeros of $H(s)$ are spread out, Bode plots can be used to give a graphic picture of the amplitude and phase angle characteristics.)

The amplitude of $H(j\omega)$ is expressed in decibels, a unit first introduced by engineers concerned with the power loss in telephone transmission circuits. Decibels relative to 1 milliwatt (dBm) was also discussed.

PROBLEMS

17.1 There is no energy stored in the circuit seen in Fig. P17.1 at the time the two sources are energized.

  a) Use the principle of superposition to find V_o.

  b) Find v_o for $t > 0$.

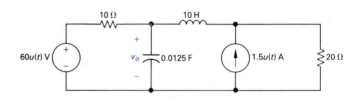

FIGURE P17.1

17.2 Verify that the solution of Eqs. (17.19) and (17.20) for V_2 yields the same expression as that given by Eq. (17.18).

17.3 Find the numerical expression for the transfer function (V_o/V_i) of each circuit in Fig. P17.3 and give the numerical value of the poles and zeros of each transfer function.

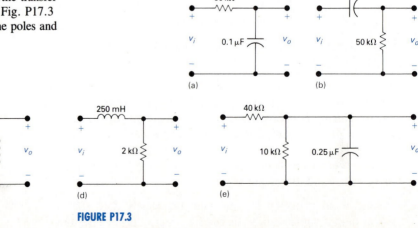

FIGURE P17.3

17.4 The voltage source in the circuit in Example 17.1 is changed to a unit impulse, that is, $v_g = \delta(t)$.

a) How much energy does the impulsive voltage source store in the capacitor?

b) How much energy does it store in the inductor?

c) Use the transfer function to find $v_o(t)$.

d) Show that the response found in part (c) is identical with the response generated by first charging the capacitor to 1000 V and then releasing the charge to the circuit, as shown in Fig. P17.4.

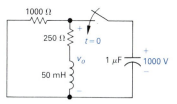

FIGURE P17.4

17.5 a) Find the numerical expression for the transfer function $H(s) = V_o/V_i$ for the circuit in Fig. P17.5.

b) Give the numerical value of each pole and zero of $H(s)$.

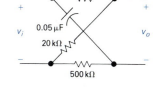

FIGURE P17.5

17.6 a) Derive the numerical expression of the transfer function $H(s) = V_o/V_g$ for the circuit in Fig. P17.6.

b) Give the numerical value of each pole and zero of $H(s)$.

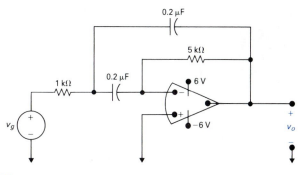

FIGURE P17.6

17.7 There is no energy stored in the circuit in Fig. P17.7 at the time the switch is opened. The sinusoidal current source is generating the signal $100 \cos 10^4 t$ mA. The response signal is the current i_o.

a) Find the transfer function I_o/I_g.

b) Find $I_o(s)$.

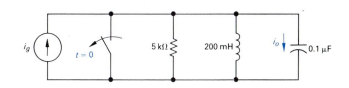

FIGURE P17.7

(continued)

c) Describe the nature of the transient component of $i_o(t)$ without solving for $i_o(t)$.

d) Describe the nature of the steady-state component of $i_o(t)$ without solving for $i_o(t)$.

e) Verify the observations made in parts (c) and (d) by finding $i_o(t)$.

17.8 a) Find the transfer function I_o/I_g as a function of μ for the circuit seen in Fig. P17.8.

b) Find the largest value of μ that will produce a bounded output signal for a bounded input signal.

c) Find i_o for $\mu = -0.5$, 0, 1.0, 1.5, and 2.00 if $i_g = 10u(t)$ A.

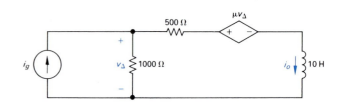

FIGURE P17.8

17.9 In the circuit of Fig. P17.9 v_o is the output signal and v_g is the input signal. Find the poles and zeros of the transfer function.

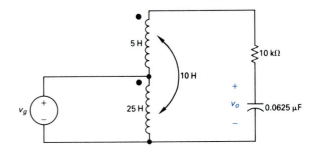

FIGURE P17.9

17.10 a) Find $h(t) * x(t)$ when $h(t)$ and $x(t)$ are the rectangular pulses shown in Fig. P17.10(a).

b) Repeat part (a) when $x(t)$ changes to the rectangular pulse shown in Fig. P17.10(b).

c) Repeat part (a) when $h(t)$ changes to the rectangular pulse shown in Fig. P17.10(c).

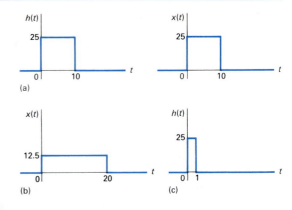

FIGURE P17.10

17.11 The voltage impulse response of a circuit is shown in Fig. P17.11(a). The input signal to this circuit is the triangular voltage pulse shown in Fig. P17.11(b).

a) Use the convolution integral to derive the expressions for the output voltage.

b) Sketch the output voltage over the interval 0 to 25 s.

c) Repeat parts (a) and (b) if the voltage impulse response of a second circuit is as shown in Fig. P17.11(c).

d) Compare the two output voltages relative to their being scaled replicas of the input voltage.

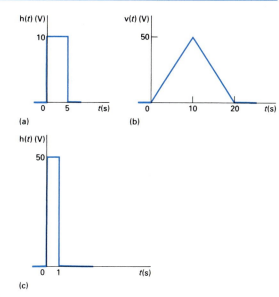

FIGURE P17.11

17.12 The voltage impulse response of a circuit is shown in Fig. P17.12(a). The input signal to the circuit is the rectangular voltage pulse shown in Fig. P17.12(b).

a) Derive the equations for the output voltage. Note the range of time for which each equation is applicable.

b) Sketch v_o for $-2 \le t \le 35$ s.

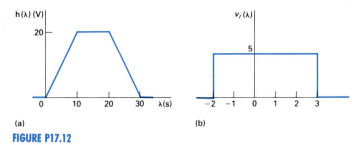

FIGURE P17.12

17.13 a) Use the convolution integral to find the output voltage of the circuit in Fig. P17.3(a) if the input voltage is the rectangular pulse shown in Fig. P17.13.

b) Sketch $v_o(t)$ versus t for the time interval $0 \le t \le 15$ ms.

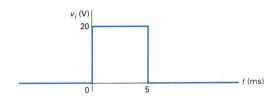

FIGURE P17.13

17.14 a) Repeat Problem 17.13, given that the resistor in the circuit in Fig. P17.3(a) is reduced to 5 kΩ.

b) Does decreasing the resistor increase or decrease the "memory" of the circuit?

c) Which circuit comes closest to transmitting a replica of the input voltage?

17.15 The input voltage in the circuit seen in Fig. P17.15 is

$$v_i = 5[u(t) - u(t - 0.5)] \text{ V}.$$

a) Use the convolution integral to find v_o.

b) Sketch v_o for $0 \le t \le 1$s.

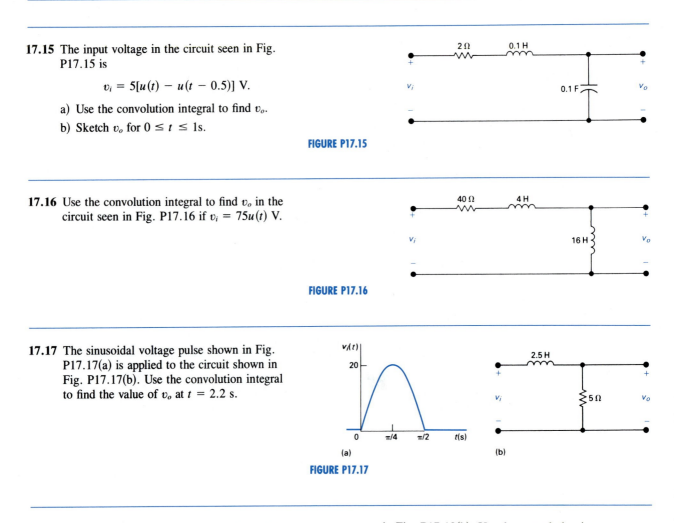

FIGURE P17.15

17.16 Use the convolution integral to find v_o in the circuit seen in Fig. P17.16 if $v_i = 75u(t)$ V.

FIGURE P17.16

17.17 The sinusoidal voltage pulse shown in Fig. P17.17(a) is applied to the circuit shown in Fig. P17.17(b). Use the convolution integral to find the value of v_o at $t = 2.2$ s.

FIGURE P17.17

17.18 The current source in the circuit shown in Fig. P17.18(a) is generating the waveform shown in Fig. P17.18(b). Use the convolution integral to find v_o at $t = 5$ ms.

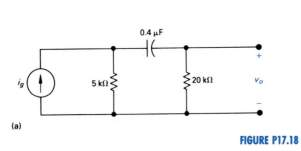

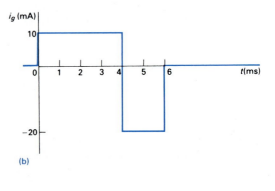

FIGURE P17.18

17.19 a) Use the convolution integral to find v_o in the circuit in Fig. P17.19(a) if i_g is the pulse shown in Fig. P17.19(b).

b) Use the convolution integral to find i_o.

c) Show that your solutions for v_o and i_o are consistent by calculating i_o at 100^- ms, 100^+ ms, 200^- ms, and 200^+ ms.

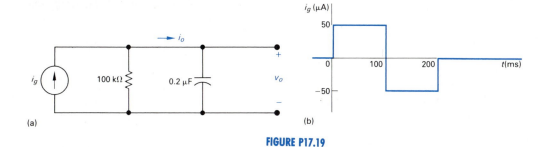

(a)

(b)

FIGURE P17.19

17.20 a) Find the impulse response of the circuit shown in Fig. P17.20(a) if v_g is the input signal and v_o is the output signal.

b) Given that v_g has the waveform shown in Fig. P17.20(b), use the convolution integral to find v_o.

c) Does v_o have the same waveform as v_g? Why?

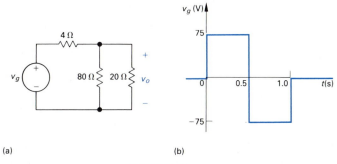

(a)

(b)

FIGURE P17.20

17.21 a) Find the impulse response of the circuit seen in Fig. P17.21 if v_g is the input signal and v_o is the output signal.

b) Assume that the voltage source has the waveform shown in Fig. P17.20(b). Use the convolution integral to find v_o.

c) Sketch v_o for $0 \leq t \leq 2$ s.

d) Does v_o have the same waveform as v_g? Why?

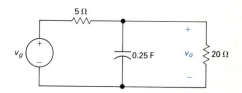

FIGURE P17.21

17.22 a) Show that if $y(t) = h(t) * x(t)$ then
$Y(s) = H(s)X(s)$.

b) Use the result given in part (a) to find $f(t)$ if

$$F(s) = \frac{a}{s(s + a)^2}.$$

17.23 The transfer function for a linear time-invariant circuit is

$$H(s) = \frac{V_o}{V_g} = \frac{4(s + 3)}{s^2 + 8s + 41}.$$

If $v_g = 40 \cos 3t$ V, what is the steady-state expression for v_o?

17.24 The operational amplifier in the circuit seen in Fig. P17.24 is ideal and is operating within its linear region.

a) Calculate the transfer function V_o/V_g.

b) If $v_g = 800 \cos 100t$ mV, what is the steady-state expression for v_o?

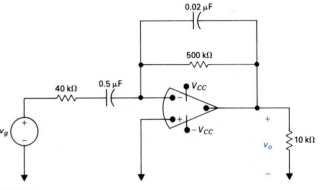

FIGURE P17.24

17.25 The operational amplifier in the circuit seen in Fig. P17.25 is ideal.

a) Find the transfer function V_o/V_g.

b) Find v_o if $v_g = 600u(t)$ mV.

c) Find the steady-state expression for v_o if $v_g = 2 \cos 10,000t$ V.

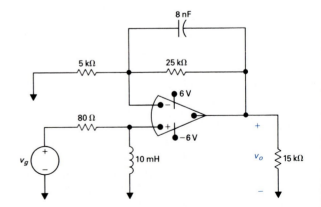

FIGURE P17.25

17.26 When an input voltage of $50u(t)$ V is applied to a circuit, the response is known to be

$$v_o = (20 - 500te^{-25t} - 20e^{-25t})u(t) \text{ V.}$$

What will the steady-state response be if $v_g = 100 \cos 50t$ V?

17.27 Make straight-line (uncorrected) amplitude and phase-angle plots for each of the transfer functions derived in Problem 17.3.

17.28 Make straight-line amplitude and phase-angle plots for the voltage transfer function derived in Problem 17.5.

17.29 a) Derive the numerical expression of the transfer function I_o/I_g for the circuit in Fig. P17.29.

  b) Make a corrected amplitude plot for the transfer function derived in part (a).

  c) At what frequency is the amplitude maximum?

  d) What is the maximum amplitude in decibels?

  e) At what frequencies is the amplitude down 3 dB from the maximum?

  f) What is the bandwidth of the circuit?

  g) Check your graphical results by calculating the actual amplitude in decibels at the frequencies read from the plot.

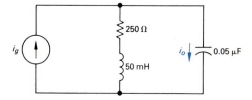

FIGURE P17.29

17.30 Use Bode diagrams to describe the behavior of the circuit in Problem 10.60 as R_x is varied from zero to infinity.

17.31 Given the following current transfer function:

$$H(s) = \frac{V_o}{V_i} = \frac{10^8}{s^2 + 3000s + 10^8}.$$

  a) At what frequencies (in radians per second) is the ratio of V_o/V_i equal to unity?

  b) At what frequency is the ratio maximum?

  c) What is the maximum value of the ratio?

17.32 The circuit shown in Fig. P17.32 resembles the interstage coupling network of an amplifier.

a) Show that

$$H(s) = \frac{V_o}{V_i} = \frac{(1/R_1C_2)s}{s^2 + [(1/R_1C_1) + (1/R_2C_2) + (1/R_1C_2)]s + (1/R_1C_1R_2C_2)}.$$

b) Find the numerical expression for $H(s)$ if $R_1 = 40\ k\Omega$, $C_1 = 0.1\ \mu F$, $R_2 = 10\ k\Omega$, and $C_2 = 250\ pF$.

c) Give the numerical values of the poles and zeros of $H(s)$.

d) Give an approximate numerical expression for $H(s)$ for values of s much less than the highest corner frequency, that is, for values of s close to the lowest corner frequency.

e) Give an approximate numerical expression for $H(s)$ for values of s much larger than the lowest corner frequency.

f) Show that the expression derived in part (d) is equivalent to the transfer function of the circuit in Fig. P17.32 when C_2 is neglected at low frequencies.

g) Show that the expression derived in part (e) is equivalent to the transfer function of the circuit in Fig. P17.32 when C_1 is neglected at high frequencies.

This problem illustrates that when the numerical values of the circuit parameters are known, it is sometimes possible to use different circuit models in different frequency ranges. Quite frequently electronic-amplifier equivalent circuits can be divided into models that apply to low-, mid-, and high-frequency ranges.

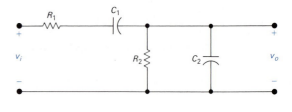

FIGURE P17.32

17.33 a) Find the resistance seen looking into the terminals a, b of the circuit in Fig. P17.33.

b) Find the power loss through the network, in decibels, when the output power is the power delivered to the 25-Ω resistor.

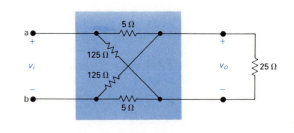

FIGURE P17.33

17.34 The amplitude plot of a transfer function is shown in Fig. P17.34. What is the numerical expression for $H(s)$?

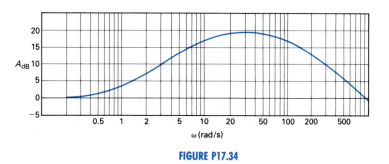

FIGURE P17.34

FOURIER SERIES

CHAPTER 18

In the preceding chapters, we devoted a considerable amount of discussion to steady-state sinusoidal analysis. One reason for this interest in the sinusoidal excitation function is that it allows finding the steady-state response to nonsinusoidal, but periodic, excitations. A periodic function is a function that repeats itself every T seconds. For example, the triangular wave illustrated in Fig. 18.1 is a nonsinusoidal, but periodic, waveform.

A periodic function is one that satisfies the relationship

$$f(t) = f(t \pm nT), \qquad (18.1)$$

where n is an integer $(1, 2, 3, \ldots)$ and T is the period. The function shown in Fig. 18.1 is periodic because

$$f(t_0) = f(t_0 - T) = f(t_0 + T) = f(t_0 + 2T) = \cdots$$

for any arbitrarily chosen value of t_0. Note that T is the smallest time interval that a periodic function may be shifted (in either direction) to produce a function that is identical with itself.

Why the interest in periodic functions? One reason is that many electrical sources of practical value generate periodic waveforms. For example, nonfiltered electronic rectifiers driven from a sinusoidal source produce rectified sine waves that are nonsinusoidal but periodic. Figure 18.2(a) and (b) show the waveforms of the full-wave and half-wave sinusoidal rectifiers, respectively.

The sweep generator used to control the electron beam of a cathode-ray oscilloscope produces a periodic triangular wave like that shown in Fig. 18.3.

Electronic oscillators, which are useful in laboratory testing of equipment, are designed to produce nonsinusoidal periodic waveforms. Function generators, which are capable of producing square-wave, triangular-wave, and rectangular-pulse waveforms,

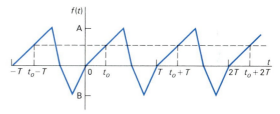

FIGURE 18.1 A periodic waveform.

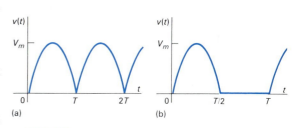

FIGURE 18.2 Output waveforms of a nonfiltered sinusoidal rectifier: (a) full-wave rectification; (b) half-wave rectification.

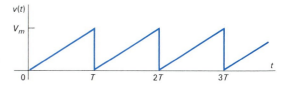

FIGURE 18.3 The triangular waveform of a cathode-ray oscilloscope sweep generator.

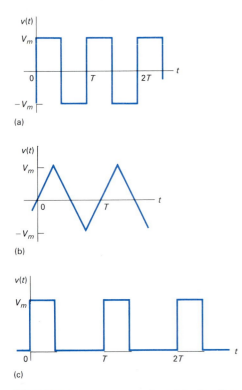

FIGURE 18.4 Waveforms produced by function generators used in laboratory testing: (a) square wave; (b) triangular wave; (c) rectangular-pulse waveform.

are found in most testing laboratories. Figure 18.4 illustrates typical waveforms.

Another practical problem that stimulates interest in periodic functions is that power generators, although designed to produce a sinusoidal waveform, cannot in practice be made to produce a pure sine wave. The distorted sinusoidal wave, however, is periodic. Engineers naturally are interested in ascertaining the con-. sequences of exciting power systems with a slightly distorted sinusoidal voltage.

Interest in periodic functions also stems from the general observation that any nonlinearity in an otherwise linear circuit creates a nonsinusoidal periodic function. The rectifier circuit alluded to earlier is one example of this phenomenon. Magnetic saturation, which occurs in both machines and transformers, is another example of a nonlinearity that generates a nonsinusoidal periodic function. An electronic clipping circuit, which uses transistor saturation, is yet another illustration of using nonlinearity to produce a periodic function.

Moreover, nonsinusoidal periodic functions are important in the analysis of nonelectrical systems. Problems involving mechanical vibration, fluid flow, and heat flow all make use of periodic functions. In fact, the study and analysis of heat flow in a metal rod led the French mathematician Jean Baptiste Joseph Fourier (1768–1830) to the trigonometric series representation of a periodic function. This series bears his name and is the starting point for finding the steady-state response to periodic excitations of electrical circuits.

18.1 FOURIER SERIES ANALYSIS: AN OVERVIEW

What Fourier discovered in investigating heat-flow problems is that a periodic function can be represented by an infinite sum of sine or cosine functions that are harmonically related. That is, the period of any trigonometric term in the infinite series is an integral multiple, or harmonic, of the fundamental period T of the periodic function. Thus for periodic $f(t)$ Fourier showed that $f(t)$ can be expressed as

$$f(t) = a_v + \sum_{n=1}^{\infty} a_n \cos n\omega_0 t + b_n \sin n\omega_0 t, \qquad (18.2)$$

where n is the integer sequence 1, 2, 3, . . .

In Eq. (18.2), a_v, a_n, and b_n are known as the *Fourier coefficients* and are calculated from $f(t)$. The term $\omega_0(2\pi/T)$ represents the fundamental frequency of the periodic function $f(t)$. The integral multiples of ω_0—that is, $2\omega_0$, $3\omega_0$, $4\omega_0$, and so on—are known as the *harmonic frequencies of $f(t)$*. Thus $2\omega_0$ is the second harmonic, $3\omega_0$ is the third harmonic, and $n\omega_0$ is the nth harmonic of $f(t)$.

We discuss determination of the Fourier coefficients in Section 18.2. Before pursuing the analytic details of using a Fourier series in circuit analysis, we first need to look at the problem in general terms. From an applications point of view, we can express all the periodic functions of interest in terms of a Fourier series. Mathematically, the conditions on a periodic function $f(t)$ that ensure expressing $f(t)$ as a convergent Fourier series (known as *Dirichlet's conditions*) are that

1. $f(t)$ be single-valued,

2. $f(t)$ have a finite number of discontinuities in the periodic interval,

3. $f(t)$ have a finite number of maxima and minima in the periodic interval, and

4. the integral $\int_{t_0}^{t_0+T} |f(t)|\,dt$ exists.

Any periodic function generated by a physically realizable source satisfies Dirichlet's conditions. These are *sufficient* conditions, not *necessary* conditions. Thus if $f(t)$ meets these requirements, we know that we can express it as a Fourier series. However, if $f(t)$ does not meet these requirements, we still may be able to express it as a Fourier series. The *necessary* conditions on $f(t)$ are not known.

After we have determined $f(t)$ and calculated the Fourier coefficients (a_v, a_n, b_n), we resolve the periodic source into a dc source (a_v) plus a sum of sinusoidal sources (a_n, b_n). Because the periodic source is driving a linear circuit, we may use the principle of superposition to find the steady-state response. In particular, we first calculate the response to each source generated by the Fourier series representation of $f(t)$ and then add the individual responses to obtain the total response. The steady-state response owing to a specific sinusoidal source is most easily found with the phasor method of analysis.

The procedure for finding the steady-state response to a periodic excitation is straightforward and involves no new techniques of circuit analysis. This procedure does produce the Fourier series representation of the steady-state response; consequently, the actual shape of the response is unknown. Furthermore, the

PSpice Circuits with pulsed sources can be simulated by PSpice: Chapter 16

response waveform can be estimated only by adding a sufficient number of terms together. Even though the Fourier series approach to finding the steady-state response does have some drawbacks, it introduces a way of thinking about a problem that is as important as introducing a way of getting quantitative results. In fact, the conceptual picture is even more important in some respects than the quantitative one.

18.2 THE FOURIER COEFFICIENTS

After defining a periodic function over its fundamental period, we determine the Fourier coefficients from the relationships:

$$a_v = \frac{1}{T} \int_{t_0}^{t_0+T} f(t)\ dt, \tag{18.3}$$

$$a_k = \frac{2}{T} \int_{t_0}^{t_0+T} f(t)\ \cos k\omega_0 t\ dt, \tag{18.4}$$

and

$$b_k = \frac{2}{T} \int_{t_0}^{t_0+T} f(t)\ \sin k\omega_0 t\ dt. \tag{18.5}$$

In Eqs. (18.4) and (18.5) the subscript k indicates the kth coefficient in the integer sequence 1, 2, 3, ... Note that a_v is the average value of $f(t)$, a_k is twice the average value of $f(t)$ $\cos k\omega_0 t$, and b_k is twice the average value of $f(t) \sin k\omega_0 t$.

We easily derive Eqs. (18.3), (18.4), and (18.5) from Eq. (18.2) by recalling the following integral relationships, which hold when m and n are integers:

$$\int_{t_0}^{t_0+T} \sin m\omega_0 t\ dt = 0, \qquad \text{for all } m; \tag{18.6}$$

$$\int_{t_0}^{t_0+T} \cos m\omega_0 t\ dt = 0, \qquad \text{for all } m; \tag{18.7}$$

$$\int_{t_0}^{t_0+T} \cos m\omega_0 t\ \sin n\omega_0 t\ dt = 0, \qquad \text{for all } m, n; \tag{18.8}$$

$$\int_{t_0}^{t_0+T} \sin m\omega_0 t\ \sin n\omega_0 t\ dt = 0, \qquad \text{for } m \neq n;$$

$$= \frac{T}{2} \qquad \text{for } m = n; \tag{18.9}$$

and

$$\int_{t_0}^{t_0+T} \cos m\omega_0 t \cos n\omega_0 t \; dt = 0 \qquad \text{for } m \neq n;$$

$$= \frac{T}{2} \qquad \text{for } m = n. \qquad \textbf{(18.10)}$$

We leave verification of Eqs. (18.6)–(18.10) to you as Problem 18.1.

To derive Eq. (18.3), we simply integrate both sides of Eq. (18.2) over one period:

$$\int_{t_0}^{t_0+T} f(t) \; dt = \int_{t_0}^{t_0+T} \left(a_v + \sum_{n=1}^{\infty} a_n \cos n\omega_0 t + b_n \sin n\omega_0 t \right) dt$$

$$= \int_{t_0}^{t_0+T} a_v \; dt + \sum_{n=1}^{\infty} \int_{t_0}^{t_0+T} (a_n \cos n\omega_0 t + b_n \sin n\omega_0 t) \; dt$$

$$= a_v T + 0. \qquad \textbf{(18.11)}$$

Equation (18.3) follows directly from Eq. (18.11).

To derive the expression for the kth value of a_n, we first multiply Eq. (18.2) by $\cos k\omega_0 t$ and then integrate both sides over one period of $f(t)$:

$$\int_{t_0}^{t_0+T} f(t) \cos k\omega_0 t \; dt = \int_{t_0}^{t_0+T} a_v \cos k\omega_0 t \; dt$$

$$+ \sum_{n=1}^{\infty} \int_{t_0}^{t_0+T} (a_n \cos n\omega_0 t \cos k\omega_0 t + b_n \sin n\omega_0 t \cos k\omega_0 t) \; dt$$

$$= 0 + a_k \left(\frac{T}{2} \right) + 0. \qquad \textbf{(18.12)}$$

Solving Eq. (18.12) for a_k yields the expression in Eq. (18.4).

We obtain the expression for the kth value of b_n by first multiplying both sides of Eq. (18.2) by $\sin k\omega_0 t$ and then integrating each side over one period of $f(t)$.

Example 18.1 shows how to use Eqs. (18.3)–(18.5) to find the Fourier coefficients for a specific periodic function.

PSpice PSpice can compute the Fourier coefficients of a periodic voltage or current: Chapter 17

E X A M P L E   18.1

Find the Fourier series for the periodic voltage shown in Fig. 18.5.

S O L U T I O N

When using Eqs. (18.3)–(18.5) to find a_v, a_k, and b_k, we may choose the value of t_0. For the periodic voltage of Fig. 18.5, the

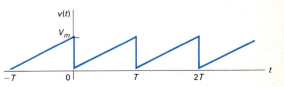

FIGURE 18.5 The periodic voltage for Example 18.1.

best choice for t_0 is zero. Any other choice makes the required integrations more cumbersome. The expression for $v(t)$ between 0 and T is

$$v(t) = \left(\frac{V_m}{T}\right) t.$$

The equation for a_v is

$$a_v = \frac{1}{T} \int_0^T \left(\frac{V_m}{T}\right) t \, dt = \frac{1}{2} V_m.$$

The equation for the kth value of a_n is

$$a_k = \frac{2}{T} \int_0^T \left(\frac{V_m}{T}\right) t \cos k\omega_0 t \, dt$$

$$= \frac{2V_m}{T} \left(\frac{1}{k^2\omega_0^2} \cos k\omega_0 t + \frac{t}{k\omega_0} \sin k\omega_0 t \, \Big|_0^T\right)$$

$$= \frac{2V_m}{T^2} \left[\frac{1}{k^2\omega_0^2} (\cos 2\pi k - 1)\right] = 0 \qquad \text{for all } k.$$

The equation for the kth value of b_n is

$$b_k = \frac{2}{T} \int_0^T \left(\frac{V_m}{T}\right) t \sin k\omega_0 t \, dt$$

$$= \frac{2V_m}{T^2} \left(\frac{1}{k^2\omega_0^2} \sin k\omega_0 t - \frac{t}{k\omega_0} \cos k\omega_0 t \, \Big|_0^T\right)$$

$$= \frac{2V_m}{T^2} \left(0 - \frac{T}{k\omega_0} \cos 2\pi k\right)$$

$$= \frac{-V_m}{\pi k}.$$

The Fourier series for $v(t)$ is

$$v(t) = \frac{V_m}{2} - \frac{V_m}{\pi} \sum_{n=1}^{\infty} \frac{1}{n} \sin n\omega_0 t$$

$$= \frac{V_m}{2} - \frac{V_m}{\pi} \sin \omega_0 t - \frac{V_m}{2\pi} \sin 2\omega_0 t - \frac{V_m}{3\pi} \sin 3\omega_0 t - \cdots .$$

Finding the Fourier coefficients, in general, is tedious. Therefore anything that simplifies the task is beneficial. Fortunately, a periodic function that possesses certain types of symmetry greatly reduces the amount of work involved in finding the coefficients. In Section 18.3 we discuss how symmetry affects the coefficients in a Fourier series.

DRILL EXERCISES

18.1 Derive the expressions for a_v, a_k, and b_k for the periodic voltage function shown if $V_m = 60\pi$ V.

ANSWER: $a_v = 37.5\pi$ V, $a_k = \dfrac{30}{k} \sin \dfrac{k\pi}{2}$ V,

$b_k = \dfrac{30}{k} \left(1 - \cos \dfrac{k\pi}{2} \right)$ V.

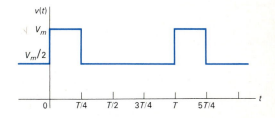

18.2 Refer to Drill Exercise 18.1.

a) What is the average value of the periodic voltage?

b) Compute the numerical values of a_1–a_5 and b_1–b_5.

c) If $T = 628.32$ ms, what is the fundamental frequency in radians per second?

d) What is the frequency of the fifth harmonic in hertz?

e) Write the Fourier series up to and including the fifth harmonic.

ANSWER: (a) 117.81 V; (b) 30 V, 0 V, −10 V, 0 V, and 6 V; 30 V, 30 V, 10 V, 0 V, and 6 V; (c) 10 rad/s; (d) 7.96 Hz; (e) $v(t) = 117.81 + 30 \cos 10t + 30 \sin 10t + 30 \sin 20t - 10 \cos 30t + 10 \sin 30t + 6 \cos 50t + 6 \sin 50t$ V.

18.3 THE EFFECT OF SYMMETRY ON THE FOURIER COEFFICIENTS

Four types of symmetry may be used to simplify the task of evaluating the Fourier coefficients:

1. even-function symmetry;

2. odd-function symmetry;

3. half-wave symmetry; and

4. quarter-wave symmetry.

The effect of each type of symmetry on the Fourier coefficients follows.

EVEN-FUNCTION SYMMETRY

A function is defined as even if

$$f(t) = f(-t). \qquad \textbf{(18.13)}$$

Functions that satisfy Eq. (18.13) are said to be even because polynomial functions with only even exponents possess this characteristic. For even periodic functions, the equations for the

Fourier coefficients reduce to

$$a_v = \frac{2}{T} \int_0^{T/2} f(t)\ dt;$$ (18.14)

$$a_k = \frac{4}{T} \int_0^{T/2} f(t) \cos k\omega_0 t\ dt;$$ (18.15)

$$b_k = 0 \qquad \text{for all } k.$$ (18.16)

Note that all the b coefficients are zero if the periodic function is even. Figure 18.6 illustrates an even periodic function. The derivations of Eqs. (18.14), (18.15), and (18.16) follow directly from Eqs. (18.3), (18.4), and (18.5). In each derivation, we select $t_0 = -T/2$ and then break the interval of integration into the range from $-T/2$ to 0 and 0 to $T/2$, or

$$a_v = \frac{1}{T} \int_{-T/2}^{T/2} f(t)\ dt$$
$$= \frac{1}{T} \int_{-T/2}^{0} f(t)\ dt + \frac{1}{T} \int_{0}^{T/2} f(t)\ dt.$$ (18.17)

Now we change the variable of integration in the first integral on the right-hand side of Eq. (18.17). Specifically, we let $t = -x$ and note that $f(t) = f(-x) = f(x)$ because the function is even. We also observe that $x = T/2$ when $t = -T/2$ and $dt = -dx$. Then

$$\int_{-T/2}^{0} f(t)\ dt = \int_{T/2}^{0} f(x)(-dx) = \int_{0}^{T/2} f(x)\ dx,$$ (18.18)

which shows that the integration from $-T/2$ to 0 is identical to the integration from 0 to $T/2$; therefore Eq. (18.17) is the same as Eq. (18.14). The derivation of Eq. (18.15) proceeds along similar lines. Here,

$$a_k = \frac{2}{T} \int_{-T/2}^{0} f(t) \cos k\omega_0 t\ dt + \frac{2}{T} \int_{0}^{T/2} f(t) \cos k\omega_0 t\ dt.$$ (18.19)

But

$$\int_{-T/2}^{0} f(t) \cos k\omega_0 t\ dt = \int_{T/2}^{0} f(x) \cos (-k\omega_0 x)(-dx)$$
$$= \int_{0}^{T/2} f(x) \cos k\omega_0 x\ dx.$$ (18.20)

As before, the integration from $-T/2$ to 0 is identical to that from 0 to $T/2$. Combining Eq. (18.20) with Eq. (18.19) yields Eq. (18.15).

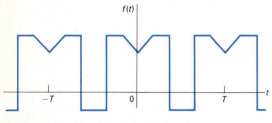

FIGURE 18.6 An even periodic function $f(t) = f(-t)$.

All the b coefficients are zero when $f(t)$ is an even periodic function because the integration from $-T/2$ to 0 is the exact negative of the integration from 0 to $T/2$. That is,

$$\int_{-T/2}^{0} f(t) \sin k\omega_0 t \, dt = \int_{T/2}^{0} f(x) \sin (-k\omega_0 x)(-dx)$$

$$= -\int_{0}^{T/2} f(x) \sin k\omega_0 x \, dx. \quad \textbf{(18.21)}$$

When we use Eqs. (18.14) and (18.15) to find the Fourier coefficients, the interval of integration must be between 0 and $T/2$.

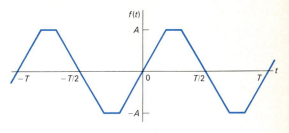

FIGURE 18.7 An odd periodic function $f(t) = -f(-t)$.

ODD-FUNCTION SYMMETRY

A function is defined as odd if

$$f(t) = -f(-t). \quad \textbf{(18.22)}$$

Functions that satisfy Eq. (18.22) are said to be odd because polynomial functions with only odd exponents have this characteristic. The expressions for the Fourier coefficients are

$$a_v = 0; \quad \textbf{(18.23)}$$

$$a_k = 0 \qquad \text{for all } k; \quad \textbf{(18.24)}$$

$$b_k = \frac{4}{T} \int_{0}^{T/2} f(t) \sin k\omega_0 t \, dt. \quad \textbf{(18.25)}$$

Note that all the a coefficients are zero if the periodic function is odd. Figure 18.7 shows an odd periodic function.

We use the same thought process to derive Eqs. (18.23), (18.24), and (18.25) that we used to derive Eqs. (18.14), (18.15), and (18.16). We leave the derivations to you in Problem 18.4.

The evenness, or oddness, of a periodic function can be destroyed by shifting the function along the time axis. In other words, the judicious choice of where $t = 0$ may give a periodic function even or odd symmetry. For example, the triangular function shown in Fig. 18.8(a) is neither even nor odd. However, we can make the function even, as shown in Fig. 18.8(b), or odd, as shown in Fig. 18.8(c).

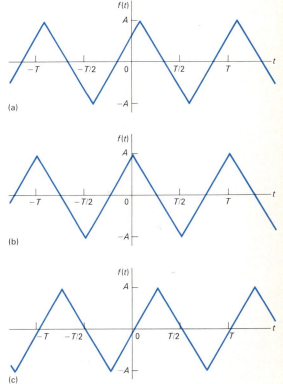

FIGURE 18.8 How the choice of where $t = 0$ can make a periodic function even, odd, or neither even nor odd: (a) a periodic triangular wave that is neither even nor odd; (b) the triangular wave of (a) made even by shifting the function along the t axis; (c) the triangular wave of (a) made odd by shifting the function along the t axis.

HALF-WAVE SYMMETRY

A periodic function possesses half-wave symmetry if it satisfies the constraint

$$f(t) = -f(t - T/2). \quad \textbf{(18.26)}$$

Equation (18.26) states that a periodic function has half-wave symmetry if, after it is shifted one-half period and inverted, it is identical with the original function. For example, the functions shown in Figs. 18.7 and 18.8 have half-wave symmetry, whereas those shown in Figs. 18.5 and 18.6 do not. Note that half-wave symmetry is not a function of where $t = 0$.

If a periodic function has a half-wave symmetry, both a_k and b_k are zero for even values of k. Moreover, a_v also is zero because the average value of a function with half-wave symmetry is zero. The expressions for the Fourier coefficients are

$$a_v = 0; \tag{18.27}$$

$$a_k = 0, \qquad \text{for } k \text{ even;} \tag{18.28}$$

$$a_k = \frac{4}{T} \int_0^{T/2} f(t) \cos k\omega_0 t \, dt, \qquad \text{for } k \text{ odd;} \tag{18.29}$$

$$b_k = 0, \qquad \text{for } k \text{ even;} \tag{18.30}$$

$$b_k = \frac{4}{T} \int_0^{T/2} f(t) \sin k\omega_0 t \, dt, \qquad \text{for } k \text{ odd.} \tag{18.31}$$

We derive Eqs. (18.27)–(18.31) by starting with Eqs. (18.3), (18.4), and (18.5) and choosing the interval of integration as $-T/2$ to $T/2$. We then divide this range into the intervals $-T/2$ to 0 and 0 to $T/2$. For example, the derivation for a_k is

$$a_k = \frac{2}{T} \int_{t_0}^{t_0+T} f(t) \cos k\omega_0 t \, dt$$

$$= \frac{2}{T} \int_{-T/2}^{T/2} f(t) \cos k\omega_0 t \, dt$$

$$= \frac{2}{T} \int_{-T/2}^{0} f(t) \cos k\omega_0 t \, dt + \frac{2}{T} \int_{0}^{T/2} f(t) \cos k\omega_0 t \, dt. \tag{18.32}$$

Now we change a variable in the first integral on the right-hand side of Eq. (18.32). Specifically, we let

$$t = x - T/2;$$

then

$$x = T/2, \qquad \text{when } t = 0;$$

$$x = 0, \qquad \text{when } t = -T/2;$$

and

$$dt = dx.$$

We rewrite the first integral as

$$\int_{-T/2}^{0} f(t) \cos k\omega_0 t \, dt = \int_{0}^{T/2} f(x - T/2) \cos k\omega_0(x - T/2) \, dx.$$

$$(18.33)$$

Note that

$$\cos k\omega_0(x - T/2) = \cos(k\omega_0 x - k\pi) = \cos k\pi \cos k\omega_0 x$$

and, by hypothesis,

$$f(x - T/2) = -f(x).$$

Therefore Eq. (18.33) becomes

$$\int_{-T/2}^{0} f(t) \cos k\omega_0 t \, dt = \int_{0}^{T/2} [-f(x)] \cos k\pi \cos k\omega_0 x \, dx.$$

$$(18.34)$$

Incorporating Eq. (18.34) into Eq. (18.32) gives

$$a_k = \frac{2}{T}(1 - \cos k\pi)\int_{0}^{T/2} f(t) \cos k\omega_0 t \, dt. \quad (18.35)$$

But $\cos k\pi$ is 1 when k is even and -1 when k is odd. Therefore Eq. (18.35) generates Eqs. (18.28) and (18.29).

We leave it to you to verify that this same thought process can be used to derive Eqs. (18.30) and (18.31). (See Problem 18.5.)

We summarize our observations about a periodic function that possesses half-wave symmetry by noting that the Fourier series representation of such a function has zero average, or dc, value and contains only odd harmonics.

QUARTER-WAVE SYMMETRY

The term *quarter-wave symmetry* describes a periodic function that has half-wave symmetry and, in addition, has symmetry about the midpoint of the positive and negative half-cycles. The function illustrated in Fig. 18.9(a) has quarter-wave symmetry about the midpoint of the positive and negative half-cycles. The function shown in Fig. 18.9(b) does not have quarter-wave symmetry, although it does have half-wave symmetry.

A periodic function that has quarter-wave symmetry can always be made either even or odd by the proper choice of the point where t equals zero. For example, the function shown in Fig. 18.9(a) is odd and can be made even by shifting the function $T/4$ units either right or left along the t axis. However, the function in Fig. 18.9(b) can never be made either even or odd. To take advantage of quarter-wave symmetry in the calculation

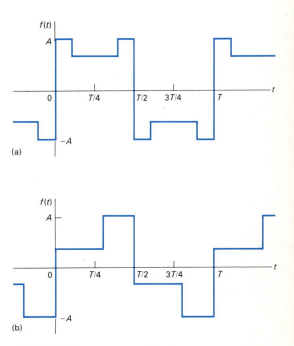

FIGURE 18.9 (a) A function that has quarter-wave symmetry; (b) a function that does not have quarter-wave symmetry.

of the Fourier coefficients, you must choose the point where t equals zero to make the function either even or odd.

If the function is made even, then

$a_v = 0,$ because of half-wave symmetry;

$a_k = 0,$ for k even, because of half-wave symmetry;

$$a_k = \frac{8}{T}\int_0^{T/4} f(t) \cos k\omega_0 t \, dt, \qquad \text{for } k \text{ odd;} \qquad \textbf{(18.36)}$$

$b_k = 0,$ for all k, because the function is even.

Equations (18.36) result from the function's quarter-wave symmetry in addition to its being even. Recall that quarter-wave symmetry is superimposed on half-wave symmetry, so we can eliminate a_v and a_k for k even. Comparing Eq. (18.36) with Eq. (18.29), shows that combining quarter-wave symmetry with evenness allows shortening of the range of integration from 0 to $T/2$ to 0 to $T/4$. We leave the derivation of Eq. (18.36) to you in Problem 18.6.

If the quarter-wave symmetric function is made odd,

$a_v = 0,$ because the function is odd;

$a_k = 0,$ for all k, because the function is odd;

$b_k = 0,$ for k even, because of half-wave symmetry;   **(18.37)**

$$b_k = \frac{8}{T}\int_0^{T/4} f(t) \sin k\omega_0 t \, dt, \qquad \text{for } k \text{ odd.}$$

Equations (18.37) are a direct consequence of quarter-wave symmetry and oddness. Again quarter-wave symmetry allows shortening the interval of integration from 0 to $T/2$ to 0 to $T/4$. We leave the derivation of Eq. (18.37) to you in Problem 18.7.

Example 18.2 shows how to use symmetry to simplify the task of finding the Fourier coefficients.

E X A M P L E   18.2

Find the Fourier series representation for the current waveform shown in Fig. 18.10.

S O L U T I O N

We begin by looking for degrees of symmetry in the waveform. We find that the function is odd and, in addition, has half-wave and quarter-wave symmetry. Because the function is odd, all the a coefficients are zero; that is, $a_v = 0$ and $a_k = 0$ for all k. Because the function has half-wave symmetry, $b_k = 0$ for even values of k. Because the function has quarter-wave symmetry, the expression for b_k for odd values of k is

$$b_k = \frac{8}{T} \int_0^{T/4} i(t) \sin k\omega_0 t \, dt.$$

In the interval $0 \le t \le T/4$, the expression for $i(t)$ is

$$i(t) = \frac{4I_m}{T} t.$$

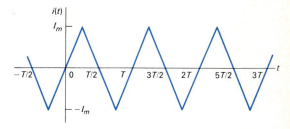

FIGURE 18.10 The periodic waveform for Example 18.2.

Thus

$$b_k = \frac{8}{T} \int_0^{T/4} \frac{4I_m}{T} t \sin k\omega_0 t \, dt.$$

$$= \frac{32I_m}{T^2} \left(\frac{\sin k\omega_0 t}{k^2 \omega_0^2} - \frac{t \cos k\omega_0 t}{k\omega_0} \Big|_0^{T/4} \right)$$

$$= \frac{8I_m}{\pi^2 k^2} \sin \frac{k\pi}{2} \qquad (k \text{ is odd}).$$

The Fourier series representation of $i(t)$ is

$$i(t) = \frac{8I_m}{\pi^2} \sum_{n=1,3,5,\dots}^{\infty} \frac{1}{n^2} \sin \frac{n\pi}{2} \sin n\omega_0 t = \frac{8I_m}{\pi^2} \times$$

$$\left(\sin \omega_0 t - \frac{1}{9} \sin 3\omega_0 t + \frac{1}{25} \sin 5\omega_0 t - \frac{1}{49} \sin 7\omega_0 t + \cdots \right).$$

DRILL EXERCISES

18.3 Derive the Fourier series for the periodic voltage shown.

ANSWER: $v_g(t) = \frac{12V_m}{\pi^2} \sum_{n=1,3,5,\dots}^{\infty} \frac{\sin(n\pi/3)}{n^2} \sin n\omega_0 t.$

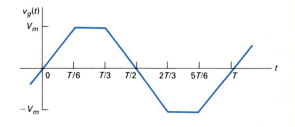

18.4 An Alternative Trigonometric Form of the Fourier Series

In circuit applications of the Fourier series, we combine the cosine and sine terms in the series into a single term for convenience. Doing so allows representation of each harmonic of $v(t)$, or $i(t)$, as a single phasor quantity. The cosine and sine terms may be merged in either a cosine expression or a sine expression. Because we chose the cosine format in the phasor method of analysis, we choose the cosine expression for the alternative form of the series. Thus we write the Fourier series in Eq. (18.2) as

$$f(t) = a_v + \sum_{n=1}^{\infty} A_n \cos(n\omega_0 t - \theta_n), \qquad (18.38)$$

where A_n and θ_n are defined by the complex quantity

$$a_n - jb_n = \sqrt{a_n^2 + b_n^2}\underline{/-\theta_n} = A_n\underline{/-\theta_n}. \qquad (18.39)$$

We derive Eqs. (18.38) and (18.39) with the phasor method to add the cosine and sine terms in Eq. (18.2). We begin by expressing the sine functions as cosine functions; that is, we rewrite Eq. (18.2) as

$$f(t) = a_v + \sum_{n=1}^{\infty} a_n \cos n\omega_0 t + b_n \cos(n\omega_0 t - 90°). \qquad (18.40)$$

Adding the terms under the summation sign using phasors gives

$$\mathcal{P}\{a_n \cos n\omega_0 t\} = a_n\underline{/0°} \qquad (18.41)$$

and

$$\mathcal{P}\{b_n \cos(n\omega_0 t - 90°)\} = b_n\underline{/-90°} = -jb_n. \qquad (18.42)$$

Then

$$\mathcal{P}\{a_n \cos n\omega_0 t + b_n \cos(n\omega_0 t - 90°)\} = a_n - jb_n$$
$$= \sqrt{a_n^2 + b_n^2}\underline{/-\theta_n}$$
$$= A_n\underline{/-\theta_n}. \qquad (18.43)$$

When we inverse-transform Eq. (18.43), we get

$$a_n \cos n\omega_0 t + b_n \cos(n\omega_0 t - 90°) = \mathcal{P}^{-1}\{A_n\underline{/-\theta_n}\}$$
$$= A_n \cos(n\omega_0 t - \theta_n).$$
$$(18.44)$$

Substituting Eq. (18.44) into Eq. (18.40) yields Eq. (18.38). Equation (18.43) corresponds to Eq. (18.39).

If the periodic function is either even or odd, A_n reduces to either a_n (even) or b_n (odd), and θ_n is either $0°$ (even) or $90°$ (odd).

The derivation of the alternate form of the Fourier series for a given periodic function is illustrated in Ex. 19.3.

E X A M P L E   18.3

a) Derive the expressions for a_k and b_k for the periodic function shown in Fig. 18.11.

b) Write the first four terms of the Fourier series representation of $v(t)$ using the format of Eq. (18.38).

S O L U T I O N

a) The voltage $v(t)$ is neither even nor odd, nor does it have half-wave symmetry. Therefore we use Eqs. (18.4) and (18.5) to find a_k and b_k. Choosing t_0 as zero, we obtain

$$a_k = \frac{2}{T}\left[\int_0^{T/4} V_m \cos k\omega_0 t \, dt + \int_{T/4}^T (0) \cos k\omega_0 t \, dt\right]$$

$$= \frac{2V_m}{T} \frac{\sin k\omega_0 t}{k\omega_0}\bigg|_0^{T/4} = \frac{V_m}{k\pi} \sin \frac{k\pi}{2}$$

and

$$b_k = \frac{2}{T}\int_0^{T/4} V_m \sin k\omega_0 t \, dt$$

$$= \frac{2V_m}{T}\left(\frac{-\cos k\omega_0 t}{k\omega_0}\bigg|_0^{T/4}\right)$$

$$= \frac{V_m}{k\pi}\left(1 - \cos \frac{k\pi}{2}\right).$$

b) The average value of $v(t)$ is

$$a_v = \frac{V_m(T/4)}{T} = \frac{V_m}{4}.$$

The values of $a_k - jb_k$ for $k = 1, 2,$ and 3 are

$$a_1 - jb_1 = \frac{V_m}{\pi} - j\frac{V_m}{\pi} = \frac{\sqrt{2}V_m}{\pi}\underline{/-45°};$$

$$a_2 - jb_2 = 0 - j\frac{V_m}{\pi} = \frac{V_m}{\pi}\underline{/-90°};$$

$$a_3 - jb_3 = \frac{-V_m}{3\pi} - j\frac{V_m}{3\pi} = \frac{\sqrt{2}V_m}{3\pi}\underline{/-135°}.$$

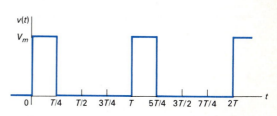

FIGURE 18.11 The periodic function for Example 18.3.

Thus the first four terms in the Fourier series representation of $v(t)$ are

$$v(t) = \frac{V_m}{4} + \frac{\sqrt{2}V_m}{\pi} \cos{(\omega_0 t - 45°)} + \frac{V_m}{\pi} \cos{(2\omega_0 t - 90°)}$$

$$+ \frac{\sqrt{2}V_m}{3\pi} \cos{(3\omega_0 t - 135°)} + \cdots.$$

Next we illustrate how to use a Fourier series representation of a periodic excitation function to find the steady-state response of a linear circuit.

DRILL EXERCISES

18.4 a) Compute A_1-A_5 and $\theta_1-\theta_5$ for the periodic function shown if $V_m = \pi V$.

b) Using the format of Eq. (18.38), write the Fourier series for $v(t)$ up to and including the fifth harmonic.

ANSWER: (a) 42.43, 30, 14.14, 0, 8.49 V, and +45°, +90°, +135°, not defined, +45°; (b) $v(t) = 117.81 + 42.43 \cos{(10t - 45°)} + 30 \cos{(20t - 90°)} + 14.14 \cos{(30t - 135°)} + 8.49 \cos{(50t - 45°)} + \ldots$ V.

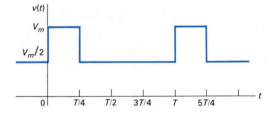

18.5 AN ILLUSTRATIVE APPLICATION

The *RC* circuit shown in Fig. 18.12(a) illustrates how to use a Fourier series in circuit analysis. The circuit is energized with the periodic square-wave voltage shown in Fig. 18.12(b). The voltage across the capacitor is the desired response, or output, signal.

The first step in finding the steady-state response is to represent the periodic excitation source with its Fourier series. After noting that the source has odd, half-wave, and quarter-wave symmetry, we know that the Fourier coefficients reduce to b_k, with k restricted to odd integer values, or

$$b_k = \frac{8}{T} \int_0^{T/4} V_m \sin k\omega_0 t \ dt$$

$$= \frac{4V_m}{\pi k} \qquad (k \text{ is odd}). \qquad \textbf{(18.45)}$$

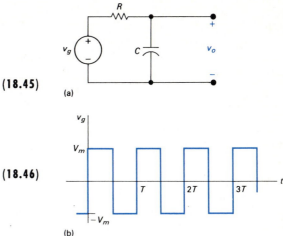

FIGURE 18.12 An *RC* circuit excited by a periodic voltage: (a) the *RC* series circuit; (b) the square-wave voltage.

Then the Fourier series representation of v_g is

$$v_g = \frac{4V_m}{\pi} \sum_{n=1,3,5,\ldots}^{\infty} \frac{1}{n} \sin n\omega_0 t. \qquad \textbf{(18.46)}$$

Writing the series in expanded form, we have

$$v_g = \frac{4V_m}{\pi} \sin \omega_0 t + \frac{4V_m}{3\pi} \sin 3\omega_0 t$$

$$+ \frac{4V_m}{5\pi} \sin 5\omega_0 t + \frac{4V_m}{7\pi} \sin 7\omega_0 t + \cdots \qquad \textbf{(18.47)}$$

The voltage source expressed by Eq. (18.47) is the equivalent of infinitely many series-connected sinusoidal sources, each source having its own amplitude and frequency. To find the contribution of each source to the output voltage, we use the principle of superposition.

For any one of the sinusoidal sources, the phasor-domain expression for the output voltage is

$$\mathbf{V}_o = \frac{\mathbf{V}_g}{1 + j\omega RC}. \qquad \textbf{(18.48)}$$

All the voltage sources are expressed as sine functions, we interpret a phasor in terms of the sine instead of the cosine, as we have done in the past. That is, when we go from the phasor back to the time domain, we simply write the time-domain expressions as $\sin(\omega t + \theta)$ instead of $\cos(\omega t + \theta)$.

The phasor output voltage owing to the fundamental frequency sinusoidal source is

$$\mathbf{V}_{01} = \frac{(4V_m/\pi)\underline{/0°}}{1 + j\omega_0 RC}. \qquad \textbf{(18.49)}$$

Writing $\mathbf{V}_{o1}$ in polar form gives

$$\mathbf{V}_{o1} = \frac{(4V_m)\underline{/-\beta_1}}{\pi\sqrt{1 + \omega_0^2 R^2 C^2}}, \qquad \textbf{(18.50)}$$

where

$$\beta_1 = \tan^{-1}\omega_0 RC. \qquad \textbf{(18.51)}$$

From Eq. (18.50), the time-domain expression for the fundamental frequency component of v_o is

$$v_{o1} = \frac{4V_m}{\pi\sqrt{1 + \omega_0^2 R^2 C^2}} \sin(\omega_0 t - \beta_1). \qquad (18.52)$$

We derive the third-harmonic component of the output voltage in a similar manner. The third-harmonic phasor voltage is

$$\mathbf{V}_{o3} = \frac{(4V_m/3\pi)\underline{/0°}}{1 + j3\omega_0 RC}$$

$$= \frac{4V_m}{3\pi\sqrt{1 + 9\omega_0^2 R^2 C^2}}\underline{/-\beta_3}, \qquad (18.53)$$

where

$$\beta_3 = \tan^{-1} 3\omega_0 RC. \qquad (18.54)$$

The time-domain expression for the third-harmonic output voltage is

$$v_{o3} = \frac{4V_m}{3\pi\sqrt{1 + 9\omega_0^2 R^2 C^2}} \sin(3\omega_0 t - \beta_3). \qquad (18.55)$$

Hence the expression for the kth-harmonic component of the output voltage is

$$v_{ok} = \frac{4V_m}{k\pi\sqrt{1 + k^2\omega_0^2 R^2 C^2}} \sin(k\omega_0 t - \beta_k) \ (k \text{ is odd}), \quad (18.56)$$

where

$$\beta_k = \tan^{-1} k\omega_0 RC \qquad (k \text{ is odd}). \qquad (18.57)$$

We now write down the Fourier series representation of the output voltage:

$$v_o(t) = \frac{4V_m}{\pi} \sum_{n=1,3,5,\ldots}^{\infty} \frac{\sin(n\omega_0 t - \beta_n)}{n\sqrt{1 + (n\omega_0 RC)^2}}. \qquad (18.58)$$

Derivation of Eq. (18.58) was not difficult. But, although we have an analytic expression for the steady-state output, what $v_o(t)$ looks like is not immediately apparent from Eq. (18.58). As we mentioned earlier, this shortcoming is a problem with the Fourier series approach. Equation (18.58) is not useless because it gives some feel for the steady-state waveform of $v_o(t)$, if we focus on the frequency response of the circuit. For example, if C is large, $1/n\omega_0 C$ is small for the higher-order harmonics. Thus the capacitor short circuits the high-frequency components of the input waveform. Thus the higher-order harmonics in Eq. (18.58) are negligible compared to the lower-order harmonics. Equation

(18.58) reflects this condition in that, for large C,

$$v_o \cong \frac{4V_m}{\pi\omega_0 RC} \sum_{n=1,3,5,\ldots}^{\infty} \frac{1}{n^2} \sin(n\omega_0 t - 90°)$$

$$\cong \frac{-4V_m}{\pi\omega_0 RC} \sum_{n=1,3,5,\ldots}^{\infty} \frac{1}{n^2} \cos n\omega_0 t. \qquad (18.59)$$

Equation (18.59) shows that the amplitude of the harmonic in the output is decreasing by $1/n^2$ compared to $1/n$ for the input harmonics. If C is so large that only the fundamental component is significant, then to a first approximation

$$v_o(t) \cong \frac{-4V_m}{\pi\omega_0 RC} \cos \omega_0 t, \qquad (18.60)$$

and Fourier analysis tells us that the square-wave input is deformed into a sinusoidal output.

Now let's see what happens as $C \to 0$. The circuit shows that v_o and v_g are the same when $C = 0$, because the capacitive branch looks like an open circuit at all frequencies. Equation (18.58) predicts the same result because, as $C \to 0$,

$$v_o = \frac{4V_m}{\pi} \sum_{n=1,3,5,\ldots}^{n} \frac{1}{n} \sin n\omega_0 t. \qquad (18.61)$$

But Eq. (18.61) is identical to Eq. (18.46), and therefore $v_o \to v_g$ as $C \to 0$.

Thus Eq. (18.58) has proved to be useful because it enabled us to predict that the output will be a highly distorted replica of the input waveform if C is large and a reasonable replica of the input waveform if C is small. In Chapter 17, we looked at distortion between the input and output in terms of how much memory the system weighting function had. In the frequency domain, we look at distortion between the steady-state input and output in terms of how the amplitude and phase of the harmonics are altered as they are transmitted through the circuit. When the network significantly alters the amplitude and phase relationship among the harmonics at the output relative to their relationship at the input, the output is a distorted version of the input. Thus in the frequency domain we speak of amplitude distortion and phase distortion.

For the circuit here, amplitude distortion is present because the amplitudes of the input harmonics decrease as $1/n$, whereas the amplitudes of the output harmonics decrease as

$$\frac{1}{n}\frac{1}{\sqrt{1 + (n\omega_0 RC)^2}}.$$

This circuit also exhibits phase distortion because the phase angle of each input harmonic is zero, whereas the phase angle of the nth harmonic in the output signal is $-\tan^{-1} n\omega_0 RC$.

DIRECT APPROACH TO STEADY-STATE RESPONSE: AN EXAMPLE

For the simple RC circuit shown in Fig. 18.12(a), we can derive the expression for the steady-state response without resorting to the Fourier series representation of the excitation function. We do this extra analysis here because to do so adds to an understanding of the Fourier series approach to obtaining a steady-state solution.

To find the steady-state expression for v_o by straightforward circuit analysis, we reason as follows. The square-wave excitation function alternates between charging the capacitor toward $+V_m$ and $-V_m$. After the circuit reaches steady-state operation, this alternate charging becomes periodic. We also know from the analysis of the single-time constant RC circuit (Chapter 8) that the response to abrupt changes in the driving voltage is exponential. Thus the steady-state waveform of the voltage across the capacitor in the circuit shown in Fig. 18.12(a) is as shown in Fig. 18.13.

The analytic expressions for $v_o(t)$ in the time intervals $0 \leq t \leq T/2$ and $T/2 \leq t \leq T$ are

$$v_o = V_m + (V_1 - V_m)e^{-t/RC}, \qquad 0 \leq t \leq T/2; \quad \textbf{(18.62)}$$

$$v_o = -V_m + (V_2 + V_m)e^{-[t-(T/2)]/RC}, \qquad T/2 \leq t \leq T. \quad \textbf{(18.63)}$$

We derive Eqs. (18.62) and (18.63) by using the methods of Chapter 8, as summarized by Eq. (8.57). We obtain the values of V_1 and V_2 by noting from Eq. (18.62) that

$$V_2 = V_m + (V_1 - V_m)e^{-T/2RC} \qquad \textbf{(18.64)}$$

and from Eq. (18.63) that

$$V_1 = -V_m + (V_2 + V_m)e^{-T/2RC}. \qquad \textbf{(18.65)}$$

Solving Eqs. (18.64) and (18.65) for V_1 and V_2 yields

$$V_2 = -V_1 = \frac{V_m(1 - e^{-T/2RC})}{1 + e^{-T/2RC}}. \qquad \textbf{(18.66)}$$

Substituting Eq. (18.66) into Eqs. (18.62) and (18.63) gives

$$v_o = V_m - \frac{2V_m}{1 + e^{-T/2RC}} e^{-t/RC}, \qquad 0 \leq t \leq T/2 \quad \textbf{(18.67)}$$

and

$$v_o = -V_m + \frac{2V_m}{1 + e^{-T/2RC}} e^{-[t-(T/2)]/RC}, \quad T/2 \leq t \leq T. \quad \textbf{(18.68)}$$

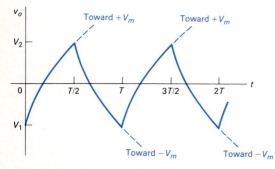

FIGURE 18.13 The steady-state waveform of v_o for the circuit of Fig. 18.12(a).

Equations (18.67) and (18.68) indicate that $v_o(t)$ has half-wave symmetry and that therefore the average value of v_o is zero. This result agrees with the Fourier series solution for the steady-state response—namely, that as the excitation function has no zero frequency component, the response can have no zero frequency component. Equations (18.67) and (18.68) also show the effect of changing the size of the capacitor. If C is small, the exponential functions quickly vanish, and $v_o = V_m$ between 0 and $T/2$ and $-V_m$ between $T/2$ and T. In other words, $v_o \to v_g$ as $C \to 0$. If C is large, the output waveform becomes triangular in shape, as Fig. 18.14 shows. Note that for large C we may approximate the exponential terms $e^{-t/RC}$ and $e^{-[t-(T/2)]/RC}$ by the linear terms $1 - (t/RC)$ and $1 - \{[t - (T/2)]/RC\}$, respectively. Equation (18.59) gives the Fourier series of this triangular waveform. Figure 18.14 summarizes the results. The dashed line in Fig. 18.14 is the input voltage, the solid blue line depicts the output voltage when C is small, and the solid black line depicts the output voltage when C is large.

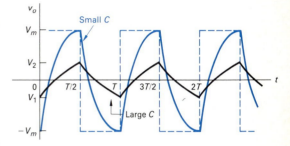

FIGURE 18.14 The effect of capacitor size on the steady-state response.

DRILL EXERCISES

18.5 a) Show that for large values of C Eq. (18.67) can be approximated by the expression

$$v_o(t) \cong \frac{-V_m T}{4RC} + \frac{V_m}{RC} t.$$

Note that this expression is the equation of the triangular wave for $0 \le t \le T/2$. [*Hints:* (1) Let $e^{-t/RC} \cong 1 - (t/RC)$ and $e^{-T/2RC} \cong 1 - (T/2RC)$; (2) put the resulting expression over the common denominator $2 - (T/2RC)$; (3) simplify the numerator; and (4) for large C assume that $T/2RC$ is much less than 2.]

b) Substitute the peak value of the triangular wave into the solution for Problem 18.9 (see Fig. P18.9b) and show that the result is Eq. (18.59).

ANSWER: (a) Derivation; (b) from Problem 18.9, $a_k = -8V_p/\pi^2 k^2$, where $k = 1, 3, 5, \ldots$, and V_p is the peak value of the triangular wave. Therefore for the triangular wave described in (a),

$$a_k = \frac{-2V_m T}{\pi^2 k^2 RC} = \frac{-4V_m}{\pi \omega_0 RC k^2}.$$

Finally, we verify that the steady-state response of Eqs. (18.67) and (18.68) is equivalent to the Fourier series solution in Eq. (18.58). To do so we simply derive the Fourier series representation of the periodic function described by Eqs. (18.67) and (18.68). We have already noted that the periodic voltage response has half-wave symmetry. Therefore the Fourier series contains only odd harmonics. For k odd,

$$a_k = \frac{4}{T} \int_0^{T/2} \left(V_m - \frac{2V_m e^{-t/RC}}{1 + e^{-T/2RC}} \right) \cos k\omega_0 t \, dt$$

$$= \frac{-8RCV_m}{T[1 + (k\omega_0 RC)^2]} \qquad (k \text{ is odd}); \qquad \textbf{(18.69)}$$

$$b_k = \frac{4}{T} \int_0^{T/2} \left(V_m - \frac{2V_m e^{-t/RC}}{1 + e^{-T/2RC}} \right) \sin k\omega_0 t \, dt$$

$$= \frac{4V_m}{k\pi} - \frac{8k\omega_0 V_m R^2 C^2}{T[1 + (k\omega_0 RC)^2]} \qquad (k \text{ is odd}). \qquad \textbf{(18.70)}$$

To show that the results obtained from Eqs. (18.69) and (18.70) are consistent with Eq. (18.58), we must prove that

$$\sqrt{a_k^2 + b_k^2} = \frac{4V_m}{k\pi} \frac{1}{\sqrt{1 + (k\omega_0 RC)^2}} \qquad \textbf{(18.71)}$$

and

$$\frac{a_k}{b_k} = -k\omega_0 RC. \qquad \textbf{(18.72)}$$

We leave the verification of Eqs. (18.69)–(18.72) to you. (See Problems 18.19 and 18.20.) Equations (18.71) and (18.72) are used with Eqs. (18.38) and (18.39) to derive the Fourier series expression in Eq. (18.58). We leave the details to you. (See Problem 18.21.)

Using this illustrative circuit, we showed how to use the Fourier series in conjunction with the principle of superposition to obtain the steady-state response to a periodic driving function. Again, the principal shortcoming of the Fourier series approach is the difficulty of ascertaining the waveform of the response. However, we also noted that, by thinking in terms of a circuit's frequency response, we can deduce a reasonable approximation of the steady-state response by using a finite number of appropriate terms in the Fourier series representation. (See Problems 18.25 and 18.27.)

18.6 AVERAGE-POWER CALCULATIONS WITH PERIODIC FUNCTIONS

If we have the Fourier series representation of the voltage and current at a pair of terminals in a linear, lumped-parameter circuit, we can easily express the average power at the terminals as a function of the harmonic voltages and currents. Using the trigonometric form of the Fourier series expressed in Eq. (18.38), we write the periodic voltage and current at the terminals of a network as

$$v = V_{\mathrm{dc}} + \sum_{n=1}^{\infty} V_n \cos(n\omega_0 t - \theta_{vn}) \qquad \textbf{(18.73)}$$

and

$$i = I_{\mathrm{dc}} + \sum_{n=1}^{\infty} I_n \cos(n\omega_0 t - \theta_{in}). \qquad \textbf{(18.74)}$$

The notation used in Eqs. (18.73) and (18.74) is defined as follows:

V_{dc} = the amplitude of the dc-voltage component;

V_n = the amplitude of the nth-harmonic voltage;

θ_{vn} = the phase angle of the nth-harmonic voltage;

I_{dc} = the amplitude of the dc-current component;

I_n = the amplitude of the nth-harmonic current;

θ_{in} = the phase angle of the nth-harmonic current.

We assume that the current reference is in the direction of the reference voltage drop across the terminals (passive sign convention) so that the instantaneous power at the terminals is vi. The average power is

$$P = \frac{1}{T} \int_{t_0}^{t_0+T} p \; dt = \frac{1}{T} \int_{t_0}^{t_0+T} vi \; dt. \qquad \textbf{(18.75)}$$

To find the expression for the average power, we substitute Eqs. (18.73) and (18.74) into Eq. (18.75) and integrate. At first glance this appears to be a formidable task since the product vi requires multiplying two infinite series! However, the only terms to survive integration are products of voltage and current at the same frequency. A review of Eqs. (18.8), (18.9), and (18.10) should convince you of the validity of this observation. Therefore Eq. (18.75) reduces to

$$P = \frac{1}{T} V_{\mathrm{dc}} I_{\mathrm{dc}} t \Big|_{t_0}^{t_0+T} + \sum_{n=1}^{\infty} \frac{1}{T} \int_{t_0}^{t_0+T} V_n I_n \cos(n\omega_0 t - \theta_{vn}) \times$$
$$\cos(n\omega_0 t - \theta_{in}) \; dt. \qquad \textbf{(18.76)}$$

Now, using the trigonometric identity

$$\cos \alpha \cos \beta = \tfrac{1}{2} \cos(\alpha - \beta) + \tfrac{1}{2} \cos(\alpha + \beta),$$

we simplify Eq. (18.76) to

$$P = V_{dc}I_{dc} + \frac{1}{T}\sum_{n=1}^{\infty}\frac{V_nI_n}{2}\int_{t_0}^{t_0+T}[\cos(\theta_{vn} - \theta_{in}) + \cos(2n\omega_0 t$$

$$- \theta_{vn} - \theta_{in})]\,dt. \qquad (18.77)$$

The second term under the integral sign integrates to zero, so

$$P = V_{dc}I_{dc} + \sum_{n=1}^{\infty}\frac{V_nI_n}{2}\cos(\theta_{vn} - \theta_{in}). \qquad (18.78)$$

Equation (18.78) is particularly important because it states that in the case of an interaction between a periodic voltage and the corresponding periodic current, the total average power is the sum of the average powers obtained from the interaction of currents and voltages of the same frequency. Currents and voltages of different frequencies do not interact to produce average power. Therefore in average-power calculations involving periodic functions, the total average power is the superposition of the average powers associated with each harmonic voltage and current. Example 18.4 illustrates the computation of average power involving a periodic voltage.

E X A M P L E   18.4

Assume that the periodic square-wave voltage in Example 18.3 is applied across the terminals of a 15-Ω resistor. The value of V_m is 60 V and that of T is 5 ms.

a) Write the first five nonzero terms of the Fourier series representation of $v(t)$. Use the trigonometric form given in Eq. (18.38).

b) Calculate the average power associated with each term in (a).

c) Calculate the total average power delivered to the 15-Ω resistor.

d) What percentage of the total power is delivered by the first five terms of the Fourier series?

S O L U T I O N

a) The dc component of $v(t)$ is

$$a_v = \frac{(60)(T/4)}{T} = 15 \text{ V}.$$

From Example 18.3 we have

$$A_1 = \sqrt{2}\, 60/\pi = 27.01 \text{ V};$$

$$\theta_1 = 45°;$$

$$A_2 = 60/\pi = 19.10 \text{ V};$$

$$\theta_2 = 90°;$$

$$A_3 = 20\sqrt{2}/\pi = 9.00 \text{ V};$$

$$\theta_3 = 135°;$$

$$A_4 = 0;$$

$$A_5 = 5.40 \text{ V};$$

$$\theta_5 = 45°;$$

$$\omega_0 = \frac{2\pi}{T} = \frac{2\pi(1000)}{5} = 400\pi \text{ rad/s.}$$

Thus the first five nonzero terms of the Fourier series are

$$v(t) = 15 + 27.01 \cos (400\pi t - 45°)$$

$$+ 19.10 \cos (800\pi t - 90°)$$

$$+ 9.00 \cos (1200\pi t - 135°)$$

$$+ 5.40 \cos (2000\pi t - 45°) + \cdots \text{ V.}$$

b) The voltage is applied to the terminals of a resistor, so we can find the power associated with each term as follows:

$$P_{dc} = \frac{15^2}{15} = 15 \text{ W};$$

$$P_1 = \frac{1}{2}\frac{27.01^2}{15} = 24.32 \text{ W};$$

$$P_2 = \frac{1}{2}\frac{19.10^2}{15} = 12.16 \text{ W};$$

$$P_3 = \frac{1}{2}\frac{9^2}{15} = 2.70 \text{ W};$$

$$P_5 = \frac{1}{2}\frac{5.4^2}{15} = 0.97 \text{ W.}$$

c) To obtain the total average power delivered to the 15-Ω resistor, we first calculate the rms value of $v(t)$:

$$V_{rms} = \sqrt{\frac{(60)^2(T/4)}{T}} = \sqrt{900} = 30 \text{ V.}$$

The total average power delivered to the 15-Ω resistor is

$$P_T = \frac{30^2}{15} = 60 \text{ W}.$$

d) The total power delivered by the first five nonzero terms is

$$P = P_{dc} + P_1 + P_2 + P_3 + P_5 = 55.15 \text{ W}.$$

This is $(55.15/60)(100)$, or 91.92 percent of the total.

DRILL EXERCISES

18.6 The trapezoidal voltage function in Drill Exercise 18.3 is applied to the circuit shown. If $12V_m = 986.96$ V and $T = 6283.19$ ms, estimate the average power delivered to the 1-Ω resistor.

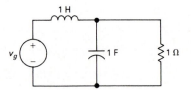

ANSWER: 3750 W.

18.7 THE RMS VALUE OF A PERIODIC FUNCTION

The rms value of a periodic function can be expressed in terms of the Fourier coefficients; by definition,

$$F_{rms} = \sqrt{\frac{1}{T} \int_{t_0}^{t_0+T} f(t)^2 \, dt}. \qquad (18.79)$$

Representing $f(t)$ by its Fourier series yields

$$F_{rms} = \sqrt{\frac{1}{T} \int_{t_0}^{t_0+T} \left[a_v + \sum_{n=1}^{\infty} A_n \cos(n\omega_0 t - \theta_n) \right]^2 dt}. \qquad (18.80)$$

The integral of the squared time function simplifies because the only terms to survive integration over a period are the product of the dc term and the harmonic products of the same frequency. All other products integrate to zero. Therefore Eq. (18.80) reduces to

$$F_{rms} = \sqrt{\frac{1}{T}\left(a_v^2 T + \sum_{n=1}^{\infty} \frac{T}{2} A_n^2\right)}$$

$$= \sqrt{a_v^2 + \sum_{n=1}^{\infty} \frac{A_n^2}{2}}$$

$$= \sqrt{a_v^2 + \sum_{n=1}^{\infty} \left(\frac{A_n}{\sqrt{2}}\right)^2}. \qquad \textbf{(18.81)}$$

Equation (18.81) states that the rms value of a periodic function is the square root of the sum obtained by adding the square of the rms value of each harmonic to the square of the dc value. For example, let's assume that a periodic voltage is represented by the finite series

$$v = 10 + 30 \cos (\omega_0 t - \theta_1) + 20 \cos (2\omega_0 t - \theta_2)$$

$$+ 5 \cos (3\omega_0 t - \theta_3) + 2 \cos (5\omega_0 t - \theta_5).$$

The rms value of this voltage is

$$V = \sqrt{10^2 + (30/\sqrt{2})^2 + (20/\sqrt{2})^2 + (5/\sqrt{2})^2 + (2/\sqrt{2})^2}$$

$$= \sqrt{764.5} = 27.65 \text{ V}.$$

In the usual case infinitely many terms are required to represent a periodic function by a Fourier series, and therefore Eq. (18.81) yields an estimate of the true rms value. We illustrate this result in Example 18.5.

E X A M P L E   18.5

Use Eq. (18.81) to estimate the rms value of the voltage in Example 18.4.

S O L U T I O N

From Example 18.4,

$V_{dc} = 15$ V,

$V_1 = 27.01/\sqrt{2}$ V,   the rms value of the fundamental,

$V_2 = 19.10/\sqrt{2}$ V,   the rms value of the second harmonic,

$V_3 = 9.00/\sqrt{2}$ V,   the rms value of the third harmonic,

$V_5 = 5.40/\sqrt{2}$ V,   the rms value of the fifth harmonic.

Therefore

$$V_{\text{rms}} = \sqrt{15^2 + \left(\frac{27.01}{\sqrt{2}}\right)^2 + \left(\frac{19.10}{\sqrt{2}}\right)^2 + \left(\frac{9.00}{\sqrt{2}}\right)^2 + \left(\frac{5.40}{\sqrt{2}}\right)^2}$$

$$= 28.76 \text{ V}.$$

From Example 18.4, the true rms value is 30 V. We approach this value by including more and more harmonics in Eq. (18.81). For example, if we include the harmonics through $k = 9$, Eq. (18.31) yields a value of 29.32 V.

DRILL EXERCISES

18.7 a) Find the rms value of the voltage shown for $V_m = 100$ V.

b) Estimate the rms value of the voltage using the first three terms in the Fourier series representation of $v_g(t)$.

ANSWER: (a) 74.5356 V; (b) 74.5306 V.

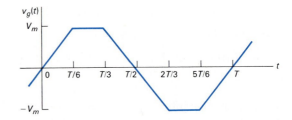

18.8   THE EXPONENTIAL FORM OF THE FOURIER SERIES

The exponential form of the Fourier series is of interest because it allows us to express the series concisely. The exponential form of the series is

$$f(t) = \sum_{n=-\infty}^{\infty} C_n e^{jn\omega_0 t}, \qquad (18.82)$$

where

$$C_n = \frac{1}{T}\int_{t_0}^{t_0+T} f(t)e^{-jn\omega_0 t}\, dt. \qquad (18.83)$$

To derive Eqs. (18.82) and (18.83), we return to Eq. (18.2) and replace the cosine and sine functions with their exponential equivalents:

$$\cos n\omega_0 t = \frac{e^{jn\omega_0 t} + e^{-jn\omega_0 t}}{2}; \qquad (18.84)$$

$$\sin n\omega_0 t = \frac{e^{jn\omega_0 t} - e^{-jn\omega_0 t}}{2j}. \qquad (18.85)$$

Substituting Eqs. (18.84) and (18.85) into Eq. (18.2) gives

$$f(t) = a_v + \sum_{n=1}^{\infty} \frac{a_n}{2}(e^{jn\omega_0 t} + e^{-jn\omega_0 t}) + \frac{b_n}{2j}(e^{jn\omega_0 t} - e^{-jn\omega_0 t})$$

$$= a_v + \sum_{n=1}^{\infty} \left(\frac{a_n - jb_n}{2}\right)e^{jn\omega_0 t} + \left(\frac{a_n + jb_n}{2}\right)e^{-jn\omega_0 t}. \qquad (18.86)$$

Now we define C_n as

$$C_n = \frac{1}{2}(a_n - jb_n) = \frac{A_n}{2}\underline{/-\theta_n}, \; n = 1, 2, 3, \cdots. \qquad (18.87)$$

From the definition of C_n,

$$C_n = \frac{1}{2}\left[\frac{2}{T}\int_{t_0}^{t_0+T} f(t)\cos n\omega_0 t \, dt - j\frac{2}{T}\int_{t_0}^{t_0+T} f(t)\sin n\omega_0 t \, dt\right]$$

$$= \frac{1}{T}\int_{t_0}^{t_0+T} f(t)(\cos n\omega_0 t - j\sin n\omega_0 t) \, dt$$

$$= \frac{1}{T}\int_{t_0}^{t_0+T} f(t)e^{-jn\omega_0 t} \, dt, \qquad (18.88)$$

which completes the derivation of Eq. (18.83). To complete the derivation of Eq. (18.82), we first observe from Eq. (18.88) that

$$C_0 = \frac{1}{T}\int_{t_0}^{t_0+T} f(t) \, dt = a_v. \qquad (18.89)$$

Next we note that

$$C_{-n} = \frac{1}{T}\int_{t_0}^{t_0+T} f(t)e^{jn\omega_0 t} \, dt = C_n^* = \frac{1}{2}(a_n + jb_n). \qquad (18.90)$$

Substituting Eqs. (18.87), (18.89), and (18.90) into Eq. (18.86) yields

$$f(t) = C_0 + \sum_{n=1}^{\infty} (C_n e^{jn\omega_0 t} + C_n^* e^{-jn\omega_0 t})$$

$$= \sum_{n=0}^{\infty} C_n e^{jn\omega_0 t} + \sum_{n=1}^{\infty} C_n^* e^{-jn\omega_0 t}. \qquad (18.91)$$

Note that the second summation on the right-hand side of Eq.

(18.91) is equivalent to summing $C_n e^{jn\omega_0 t}$ from -1 to $-\infty$; that is,

$$\sum_{n=1}^{\infty} C_n^* e^{-jn\omega_0 t} = \sum_{n=-1}^{-\infty} C_n e^{jn\omega_0 t}. \qquad (18.92)$$

As the summation from -1 to $-\infty$ is the same as the summation from $-\infty$ to -1, we use Eq. (18.92) to rewrite Eq. (18.91):

$$f(t) = \sum_{n=0}^{\infty} C_n e^{jn\omega_0 t} + \sum_{-\infty}^{-1} C_n e^{jn\omega_0 t}$$

$$= \sum_{-\infty}^{\infty} C_n e^{jn\omega_0 t}, \qquad (18.93)$$

which completes the derivation of Eq. (18.82).

Example 18.6 illustrates the process of finding the exponential Fourier series representation of a periodic function.

E X A M P L E   18.6

Find the exponential Fourier series for the periodic voltage shown in Fig. 18.15.

S O L U T I O N

Using $-\tau/2$ as the starting point for the integration, we have, from Eq. (18.83),

$$C_n = \frac{1}{T} \int_{-\tau/2}^{\tau/2} V_m e^{-jn\omega_0 t} \, dt$$

$$= \frac{V_m}{T} \left(\frac{e^{-jn\omega_0 t}}{-jn\omega_0} \right)_{-\tau/2}^{\tau/2}$$

$$= \frac{jV_m}{n\omega_0 T} (e^{jn\omega_0 \tau/2} - e^{jn\omega_0 \tau/2})$$

$$= \frac{2V_m}{n\omega_0 T} \sin n\omega_0 \tau/2.$$

Here, because $v(t)$ has even symmetry, $b_n = 0$ for all n and hence we expect C_n to be real. Moreover, the amplitude of C_n follows a $(\sin x)/x$ distribution, as indicated when we rewrite C_n:

$$C_n = \frac{V_m \tau}{T} \frac{\sin (n\omega_0 \tau/2)}{n\omega_0 \tau/2}.$$

We say more about this subject in Section 18.9. The exponential

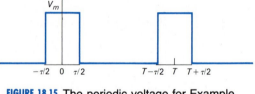

FIGURE 18.15 The periodic voltage for Example 18.6.

series representation of $v(t)$ is

$$v(t) = \sum_{n=-\infty}^{n=\infty} \left(\frac{V_m \tau}{T}\right) \frac{\sin (n\omega_0 \tau/2)}{n\omega_0 \tau/2} e^{jn\omega_0 t}$$

$$= \left(\frac{V_m \tau}{T}\right) \sum_{n=-\infty}^{n=\infty} \frac{\sin (n\omega_0 \tau/2)}{n\omega_0 \tau/2} e^{jn\omega_0 t}.$$

We may also express the rms value of a periodic function in terms of the complex Fourier coefficients. From Eqs. (18.81), (18.87), and (18.89),

$$F_{\text{rms}} = \sqrt{a_v^2 + \sum_{n=1}^{\infty} \frac{a_n^2 + b_n^2}{2}}; \qquad (18.94)$$

$$|C_n| = \frac{\sqrt{a_n^2 + b_n^2}}{2}; \qquad (18.95)$$

$$C_0^2 = a_v^2. \qquad (18.96)$$

Substituting Eqs. (18.95) and (18.96) into Eq. (18.94) yields the desired expression:

$$F_{\text{ms}} = \sqrt{C_0^2 + 2\sum_{n=1}^{\infty} |C_n|^2}. \qquad (18.97)$$

DRILL EXERCISES

18.8 Derive the expression for the Fourier coefficients C_n for the periodic function shown. [*Hint:* Take advantage of symmetry by using the fact that $C_n n = (a_n - jb_n)/2$.].

ANSWER: $C_n = -j\dfrac{10}{\pi n}\left(1 + \cos\dfrac{n\pi}{3}\right),$ n odd.

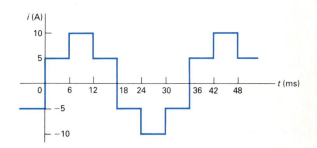

18.9 a) Calculate the rms value of the periodic current in Drill Exercise 18.8.

b) Using $C_1 - C_{11}$, estimate the rms value.

c) What is the percentage error in the value obtained in (b) based on the true value found in (a)?

d) For this periodic function, how many terms must be used to estimate the rms value before the error is less than 1%?

ANSWER: (a) $\sqrt{50}$ A; (b) 6.980 A; (c) -1.28%; (d) $n = 17$; therefore, the first six nonzero harmonic terms of the series are required?

18.9 AMPLITUDE AND PHASE SPECTRA

A periodic time function is defined by its Fourier coefficients and its period. That is, when we know a_v, a_n, b_n, and T, we can construct $f(t)$, at least theoretically. When we know a_n and b_n, we also know the amplitude (A_n) and phase angle ($-\theta_n$) of each harmonic. Again, we cannot, in general, visualize what the periodic function looks like in the time domain from a description of the coefficients and phase angles; nevertheless we recognize that these quantities characterized the periodic function completely. Thus, with sufficient computing time, we can synthesize the time-domain waveform from the amplitude and phase-angle data. Also, when a periodic driving function is exciting a circuit that is highly frequency selective, the Fourier series of the steady-state response is dominated by just a few terms. Thus the description of the response in terms of amplitude and phase may provide an understanding of the output waveform.

We can present graphically the description of a periodic function in terms of the amplitude and phase angle of each term in the Fourier series of $f(t)$. The plot of the amplitude of each term versus the frequency is called the *amplitude spectrum* of $f(t)$, and the plot of the phase angle versus the frequency is called the *phase spectrum of $f(t)$*. Because the amplitude and phase-angle data occur at discrete values of the frequency (that is, at ω_0, $2\omega_0$, $3\omega_0$, . . .), these plots also are referred to as *line spectra*.

PSpice ➤ PROBE can create plots of the frequency spectrum contained in any time domain waveform: Chapter 17

AN EXAMPLE

Amplitude- and phase-spectra plots are based on either Eq. (18.38) (A_n and $-\theta_n$) or Eq. (18.82) (C_n). We focus on Eq. (18.82) and leave the plots based on Eq. (18.38) to Problem 18.43. To illustrate the amplitude and phase-angle spectra, which are based on the exponential form of the Fourier series, we use the periodic voltage of Example 18.6. To aid the discussion, we assume that $V_m = 5$ V and $\tau = T/5$. From Example 18.6,

$$C_n = \frac{V_m \tau}{T} \frac{\sin(n\omega_0 \tau/2)}{n\omega_0 \tau/2}, \qquad (18.98)$$

which for the assumed values of V_m and τ reduces to

$$C_n = 1\frac{\sin(n\pi/5)}{n\pi/5}. \qquad (18.99)$$

Figure 18.16 shows the plot of the magnitude of C_n from Eq. (18.99) for values of n ranging from -10 to $+10$. Figure 18.16 clearly shows that the amplitude spectrum is bounded by the en-

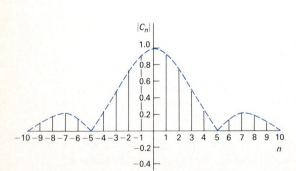

FIGURE 18.16 The plot of $|C_n|$ versus n when $\tau = T/5$ for Example 18.6.

velope of the $|(\sin x)/x|$ function. We used the order of the harmonic as the frequency scale because the numerical value of T is not specified. When we know T, we also know ω_0 and the frequency corresponding to each harmonic.

Figure 18.17 shows the plot of $|(\sin x)/x|$ versus x, where x is in radians. Figure 18.17 shows that the function goes through zero whenever x is an integral multiple of π. From Eq. (18.98),

$$n\omega_0\left(\frac{\tau}{2}\right) = \frac{n\pi\tau}{T} = \frac{n\pi}{T/\tau}. \qquad (18.100)$$

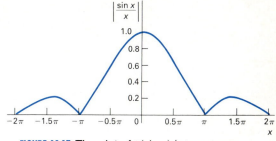

FIGURE 18.17 The plot of $|(\sin x)/x|$ versus x.

From Eq. (18.100) we deduce that the amplitude spectrum goes through zero whenever $n\tau/T$ is an integer. For example, in the plot, τ/T is $\frac{1}{5}$, and therefore the envelope goes through zero at $n = 5, 10, 15$, and so on. In other words, the fifth, tenth, fifteenth, ..., harmonics all are zero. As the reciprocal of τ/T becomes an increasingly larger and larger integer, the number of harmonics between every π radians increases. If $n\tau/T$ is not an integer, the amplitude spectrum still follows the $|(\sin x)/x|$ envelope. However, the envelope is not zero at an integral multiple of ω_0.

As C_n is real for all n, the phase angle associated with C_n is either zero or 180°, depending on the algebraic sign of $(\sin n\pi/5)/(n\pi/5)$. For example, the phase angle is zero for $n = 0, \pm 1, \pm 2, \pm 3$, and ± 4. The phase angle is not defined at $n = \pm 5$, because $C_{\pm 5}$ is zero. The phase angle is 180° at $n = \pm 6, \pm 7, \pm 8$, and ± 9 and is not defined at ± 10. This pattern repeats itself as n takes on larger integer values. Figure 18.18 shows the phase angle of C_n given by Eq. (18.89).

Now, what happens to the amplitude and phase spectra if $f(t)$ is shifted along the time axis? To find out we shift the periodic voltage in Example 18.6 t_0 units to the right. By hypothesis,

$$v(t) = \sum_{n=-\infty}^{\infty} C_n e^{jn\omega_0 t}; \qquad (18.101)$$

therefore

$$v(t - t_0) = \sum_{n=-\infty}^{\infty} C_n e^{jn\omega_0(t-t_0)} = \sum_{n=-\infty}^{\infty} C_n e^{-jn\omega_0 t_0} e^{jn\omega_0 t}, \qquad (18.102)$$

which indicates that shifting the origin has no effect on the am-

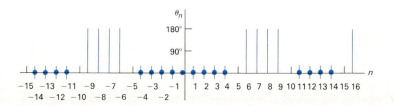

FIGURE 18.18 The phase angle of C_n.

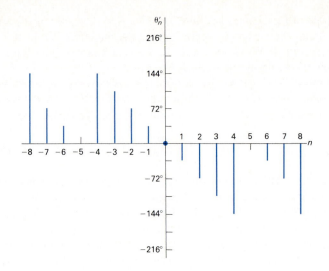

FIGURE 18.19 The plot of θ'_n versus n for Eq. (18.104).

plitude spectrum, because

$$|C_n| = |C_n e^{-jn\omega_0 t_0}|. \tag{18.103}$$

However, reference to Eq. (18.87) reveals the phase spectrum has changed to $-(\theta_n + n\omega_0 t_0)$ radians.

For example, let's shift the periodic voltage in Example 18.6 $\tau/2$ units to the right. As before, we assume that $\tau = T/5$; then the new phase angle θ'_n is

$$\theta'_n = -(\theta_n + n\pi/5). \tag{18.104}$$

We plotted Eq. (18.104) in Fig. 18.19 for n ranging from -8 to $+8$. Note that no phase angle is associated with a zero amplitude coefficient.

You may wonder why we have devoted so much attention to the amplitude spectrum of the periodic pulse in Example 18.6. The reason is that this particular periodic waveform provides an excellent way to illustrate the transition from the Fourier series representation of a periodic function to the Fourier transform representation of a nonperiodic function. We discuss the Fourier transform in Chapter 19.

DRILL EXERCISES

18.10 The Drill Exercise 18.8 function is shifted along the time axis 9 ms to the left. Write the exponential Fourier series for the periodic current.

ANSWER:

$$i(t) = \frac{10}{\pi} \sum_{n=-\infty(\text{odd})}^{n=\infty} \frac{1}{n}\left(1 + \cos\frac{n\pi}{3}\right)e^{(j\pi/2)(n-1)}e^{jn\omega_0 t} \text{ A}.$$

18.10 MEAN-SQUARE ERROR

In practical applications, using a Fourier series to represent a periodic function forces us to truncate the series to a finite number of turns. Thus the periodic function is actually being represented by a partial sum; that is,

$$f(t) \cong S_N(t) = \sum_{n=-N}^{N} C_n e^{jn\omega_0 t}. \qquad (18.105)$$

The error in this approximation to $f(t)$ is the difference between $f(t)$ and its representation by the partial sum; that is

$$\epsilon(t) = f(t) - S_N(t). \qquad (18.106)$$

This error function is a direct consequence of having to work with a finite number of terms in the series representation of $f(t)$.

When we focus on the error function $\epsilon(t)$, we begin to raise questions about its behavior, which in turn raises questions about the series being used to represent $f(t)$. Are there other trigonometric series that can be used to approximate $f(t)$? [A trigonometric series is a Fourier series only if the coefficients are found in accordance with Eqs. (18.3), (18.4), and (18.5) or Eq. (18.83)]. The answer to the question is *yes*, but in this text we do not investigate these other types of trigonometric series.[†] We bring this aspect of series analysis to your attention simply to point out the nature of the error function when $f(t)$ is represented by a Fourier series.

The phrase *mean-square error* designates the mean (average) value of the error squared. If we let $\overline{\epsilon^2}$ denote the mean-square error, then by definition

$$\overline{\epsilon^2} = \frac{1}{T} \int_{t_0}^{t_0+T} \epsilon^2(t) \, dt. \qquad (18.107)$$

The *mean-square error is minimum if the coefficients in the partial-sum approximation to $f(t)$ are Fourier coefficients*.

In order to prove that the Fourier coefficients minimize the integral in Eq. (18.107), we proceed as follows. First, we postulate that $f(t)$ is approximated by a finite trigonometric series and the coefficients in the series are unspecified; that is, we let

$$S_N(t) = \sum_{n=-N}^{N} D_n e^{jn\omega_0 t}, \qquad (18.108)$$

[†] See James B. Ley, Samuel G. Lutz, and Charles F. Rehberg, *Linear Circuit Analysis*, Chapter 6 (New York: McGraw-Hill, 1959); E. A. Gullemin, *The Mathematics of Circuit Analysis*, Chapter 7 (New York: John Wiley & Sons, 1949).

where the coefficients D_n are not necessarily the Fourier coefficients. The second step is to solve for the values of D_n in Eq. (18.108) that minimize the integral in Eq. (18.107).

We express the mean-square error as a function of the partial sum given by Eq. (18.108):

$$\overline{\epsilon^2} = \frac{1}{T}\int_{-T/2}^{T/2}\left[f(t) - \sum_{n=-N}^{N} D_n e^{jn\omega_0 t}\right]^2 dt. \quad \textbf{(18.109)}$$

In writing Eq. (18.109), for convenience we chose $t_o = -T/2$.

If we let D_k represent the kth value of D_n, the problem is reduced to finding the expression for D_k such that

$$\frac{d\overline{\epsilon^2}}{dD_k} = 0 \quad \textbf{(18.110)}$$

for all k.

From Eq. (18.109),

$$\frac{d\overline{\epsilon^2}}{dD_k} = \frac{d}{dD_k}\left\{\frac{1}{T}\int_{-T/2}^{T/2}\left[f(t) - \sum_{n=-N}^{N} D_n e^{jn\omega_0 t}\right]^2 dt\right\}. \quad \textbf{(18.111)}$$

The integration is with respect to time, and therefore we can move the differentiation inside the integral. Thus Eq. (18.111) leads to

$$\frac{d\overline{\epsilon^2}}{dD_k} = \frac{1}{T}\int_{-T/2}^{T/2}\frac{d}{dD_k}\left[f(t) - \sum_{n=-N}^{N} D_n e^{jn\omega_0 t}\right]^2 dt$$

$$= \frac{1}{T}\int_{-T/2}^{T/2} 2\left[f(t) - \sum_{n=-N}^{N} D_n e^{jn\omega_0 t}\right](-e^{jk\omega_0 t})\, dt$$

$$= -\frac{2}{T}\int_{-T/2}^{T/2}\left[f(t) - \sum_{n=-N}^{N} D_n e^{jn\omega_0 t}\right]e^{jk\omega_0 t}\, dt. \quad \textbf{(18.112)}$$

In order for the right-hand side of Eq. (18.112) to equal zero, the integral must be zero; hence

$$\int_{-T/2}^{T/2} f(t)e^{jk\omega_0 t}\, dt = \int_{-T/2}^{T/2} e^{jk\omega_0 t} \sum_{n=-N}^{N} D_n e^{jn\omega_0 t}\, dt. \quad \textbf{(18.113)}$$

The next step in the derivation of the expression for D_n is to evaluate the right-hand side of Eq. (18.113). We begin by noting that the integration is with respect to time and the summation is with respect to N; therefore the operations can be interchanged. Letting I represent the integral of the right-hand side, we have

$$I = \sum_{n=-N}^{N} D_n \int_{-T/2}^{T/2} e^{j(k+n)\omega_0 t}\, dt = \sum_{n=-N}^{N} D_n \left.\frac{e^{j(k+n)\omega_0 t}}{j(k+n)\omega_0}\right|_{-T/2}^{T/2}$$

$$= \sum_{n=-N}^{N} D_n T\frac{\sin(k+n)\pi}{(k+n)\pi}. \quad \textbf{(18.114)}$$

Now we make the observation that

$$\frac{\sin (k + n)\pi}{(k + n)\pi} = 0,$$

except when $n = -k$. At $n = -k$ the ratio is 1. Therefore Eq. (18.114) reduces to

$$I = TD_{-k}. \qquad (18.115)$$

Because Eq. (18.115) equals the left-hand side of Eq. (18.113), we have

$$D_{-k} = \frac{1}{T}\int_{-T/2}^{T/2} f(t)e^{jk\omega_0 t}\, dt, \qquad (18.116)$$

which is equivalent to

$$D_k = \frac{1}{T}\int_{-T/2}^{T/2} f(t)e^{-jk\omega_0 t}\, dt. \qquad (18.117)$$

Equation (18.117) is the formula for the Fourier coefficients. Therefore we have shown that the mean-square error is minimized when $f(t)$ is approximated by a partial sum of trigonometric terms with Fourier coefficients.

DRILL EXERCISES

18.11 a) Calculate the mean-square error if the periodic current shown is approximated by the first term in its Fourier series and $I_m = 10$ A.

b) Repeat (a) for the periodic current approximated by $10 \sin \omega_0 t$. (*Hint:* Note that because of symmetry the computation of $\overline{\epsilon^2}$ can be made by integrating from 0 to $T/4$ and then dividing this result by $T/4$.)

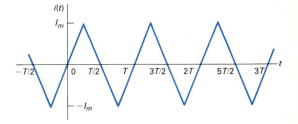

ANSWER: (a) 0.4822 A^2; (b) 2.2764 A^2.

SUMMARY

Periodic waveforms occur in many engineering systems. A Fourier series is used to predict the steady-state response of a system when the system is excited by a periodic signal. The following observations apply to use of the series as an analytic tool in circuit analysis and design.

- A periodic function is a function that repeats itself every T seconds.

- A period is the smallest time interval (T) that a periodic function can be shifted to produce a function identical to itself.

- The Fourier series is an infinite series used to represent a periodic function. The series consists of a constant term and infinitely many harmonically related cosine and sine terms.

- The fundamental frequency is the frequency determined by the fundamental period ($f_0 = 1/T$ or $\omega_0 = 2\pi f_0$).

- The harmonic frequency is an integer multiple of the fundamental frequency.

- Fourier coefficients are the constant term and the coefficient of each cosine and sine term in the series. [See Eqs. (18.3)–(18.5).]

- Five types of symmetry are used to simplify the computation of the Fourier coefficients (see Section 18.3):

  even, in which all sine terms in the series are zero;

  odd, in which all cosine terms and the constant term are zero;

  half-wave, in which all even harmonics are zero;

  quarter-wave, half-wave, even, in which the series contains only odd harmonic cosine terms; and

  quarter-wave, half-wave, odd, in which the series contains only odd harmonic sine terms.

- In the alternative form, each harmonic represented by the sum of a cosine and sine term is combined into a single term of the form $A_n \cos(n\omega_0 t - \theta_n)$.

- For steady-state response, the Fourier series of the response signal is determined by first finding the response to each component of the input signal. The individual responses are added (superimposed) to form the Fourier series of the response signal. The response to the individual terms in the input series is found by either phasor-domain or s-domain analysis.

- The waveform of the response signal is difficult to obtain without the aid of a computer. Sometimes the frequency-response (or filtering) characteristics of the circuit can be used to ascertain how closely the output waveform matches the input waveform.

- Only harmonics of the same frequency interact to produce average power. The total average power is the sum of the average power associated with each frequency.

- The root mean square (rms) value of a periodic function can be estimated from the Fourier coefficients. [See Eqs. (18.81), (18.94), and (18.97).]

- The Fourier series also may be written in exponential form by using Euler's identity to replace the cosine and sine terms with their exponential equivalents.

The Fourier series enables finding the steady-state response to a periodic excitation by transferring the analysis to the frequency domain.

PROBLEMS

18.1 a) Verify Eqs. (18.6) and (18.7).

    b) Verify Eq. (18.8). [*Hint:* Use the trigonometric identity $\cos \alpha \sin \beta = \frac{1}{2} \sin (\alpha + \beta) - \frac{1}{2} \sin (\alpha - \beta)$.]

    c) Verify Eq. (18.9). [*Hint:* Use the trigonometric identity $\sin \alpha \sin \beta = \frac{1}{2} \cos (\alpha - \beta) - \frac{1}{2} \cos (\alpha + \beta)$.]

    d) Verify Eq. (18.10). [*Hint:* Use the trigonometric identity $\cos \alpha \cos \beta = \frac{1}{2} \cos (\alpha - \beta) + \frac{1}{2} \cos (\alpha + \beta)$.]

18.2 Derive Eq. (18.5).

18.3 Find the Fourier series expressions for the periodic voltage functions shown in Fig. P18.3. Note that Fig. P18.3(a) illustrates the square wave; Fig. P18.3(b) illustrates the full-wave rectified sine wave, where $v(t) = V_m \sin (\pi/T)t$, $0 \le t \le T$; and Fig. P18.3(c) illustrates the half-wave rectified sine wave, where $v(t) = V_m \sin (2\pi/T)t$, $0 \le t \le T/2$.

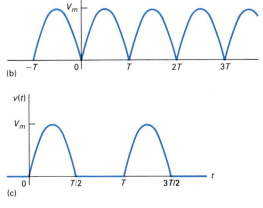

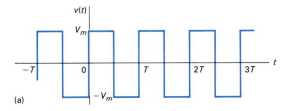

FIGURE P18.3

18.4 Derive the expressions for the Fourier coefficients of an odd periodic function. (*Hint:* Use the same technique as the one used in the text in deriving Eqs. 18.14, 18.15, and 18.16.)

18.5 Show that if $f(t) = -f(t - T/2)$, the Fourier coefficients b_k are given by the expressions:

$$b_k = 0, \qquad \text{for } k \text{ even};$$

$$b_k = \frac{4}{T} \int_0^{T/2} f(t) \sin k\omega_0 t\, dt, \qquad \text{for } k \text{ odd}.$$

(*Hint:* Use the same technique as the one used in the text to derive Eqs. 18.28 and 18.29.)

18.6 Derive Eq. (18.36). [*Hint:* Start with Eq. (18.29) and divide the interval of integration into 0 to $T/4$ and $T/4$ to $T/2$. Note that because of evenness and quarter-wave symmetry, $f(t) = -f(T/2 - t)$ in the interval $T/4 \le t \le T/2$. Let $x = T/2 - t$ in the second interval and combine the resulting integral with the integration between 0 and $T/4$.]

18.7 Derive Eq. (18.37). Follow the hint given in Problem 18.6 except that because of oddness and quarter-wave symmetry, $f(t) = f(T/2 - t)$ in the interval $T/4 \le t \le T/2$.

18.8 For each of the periodic functions in Fig. P18.8, specify

a) ω_0 in radians per second;

b) f_0 in hertz;

c) the value of a_v;

d) the equations for a_k and b_k.

e) For each function, express $v(t)$ as a Fourier series.

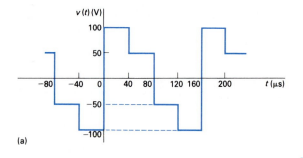

(a)

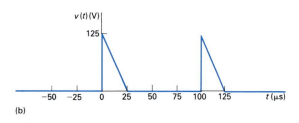

(b)

FIGURE P18.8

18.9 Find the Fourier series of each periodic function shown in Fig. P18.9.

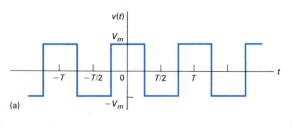

(a)

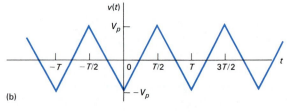

(b)

FIGURE P18.9

18.10 Derive the Fourier series for the periodic voltage shown in Fig. P18.10, given that

$$v(t) = 80 \sin \frac{2\pi}{T} t, \qquad 0 \le t \le T/2;$$

$$v(t) = 200 \sin \frac{2\pi}{T} \left(t - \frac{T}{2} \right), \qquad T/2 \le t \le T.$$

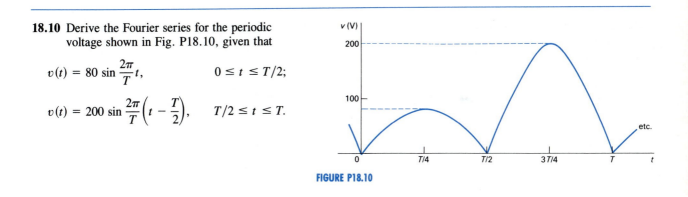

FIGURE P18.10

18.11 a) Derive the Fourier series for the periodic current function shown in Fig. P18.11.

  b) Repeat part (a) if the vertical reference axis is shifted $T/2$ units to the right.

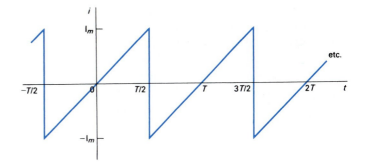

FIGURE P18.11

18.12 It is given that $f(t) = 4t^2$ over the interval $-2 < t < 2$ s.

  a) Construct a periodic function that satisfies this $f(t)$ between -2 and $+2$ s, has a period of 8 s, and has half-wave symmetry.

  b) Is the function even or odd?

  c) Does the function have quarter-wave symmetry?

  d) Derive the Fourier series for $f(t)$.

18.13 Repeat Problem 18.12 given that $f(t) = t^3$ over the interval $-2 < t < 2$ s.

18.14 It is given that $v(t) = 20t \cos 0.2\pi t$ V over the interval $-5 \le t \le 5$ s. The function then repeats itself.

  a) What is the fundamental frequency in radians per second?

  b) Is the function even?

  c) Is the function odd?

  d) Does the function have half-wave symmetry?

18.15 One period of a periodic function is described by the following equations:

$$v(t) = 10 \text{ V}, \qquad\qquad -5 \text{ ms} \le t \le 5 \text{ ms};$$

$$v(t) = (20 - 2000t) \text{ V}, \quad 5 \text{ ms} \le t \le 15 \text{ ms};$$

$$v(t) = -10 \text{ V}, \qquad\qquad 15 \text{ ms} \le t \le 25 \text{ ms};$$

$$v(t) = (2000t - 60) \text{ V}, \quad 25 \text{ ms} \le t \le 35 \text{ ms}.$$

a) What is the fundamental frequency in hertz?

b) Is the function even?

c) Is the function odd?

d) Does the function have half-wave symmetry?

e) Does the function have quarter-wave symmetry?

f) Give the numerical expressions for a_v, a_k, and b_k.

18.16 The periodic function shown in Fig. P18.16 is odd and has both half-wave and quarter-wave symmetry.

a) Sketch one full cycle of the function over the interval $-T/4 \le t \le 3T/4$.

b) Derive the expression for the Fourier coefficients b_k.

c) Write the first five nonzero terms in the Fourier expansion of $f(t)$.

d) Use the first five nonzero terms to estimate $f(T/4)$.

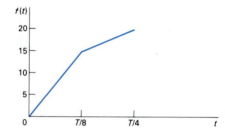

FIGURE P18.16

18.17 It is sometimes possible to use symmetry to find the Fourier coefficients even though the original function is not symmetrical! With this thought in mind, consider the function in Drill Exercise 18.1. Observe that $v(t)$ can be divided into the two functions illustrated in Fig. P18.17(a) and (b). Furthermore, we can make $v_2(t)$ an even function by shifting it $T/8$ units to the left. This is illustrated in Fig. P18.17(c).

At this point we note that $v(t) = v_1(t) + v_2(t)$ and that the Fourier series of $v_1(t)$ is a single-term series consisting of $V_m/2$. To find the Fourier series of $v_2(t)$ we first find the Fourier series of $v_2(t + T/8)$ and then shift this series $T/8$ units to the right. Use the technique outlined above to verify the Fourier series given as the answer to Drill Exercise 18.2(e).

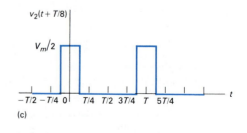

FIGURE P18.17

18.18 a) Derive the Fourier series for the periodic
function shown in Fig. P18.18 when
$I_m = 5\pi^2$ A. Write the series in the form of
Eq. (18.38).

b) Use the first five nonzero terms to estimate
$i(T/4)$.

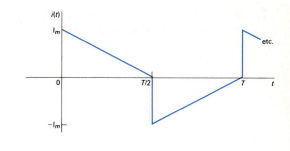

FIGURE P18.18

18.19 Derive Eqs. (18.69) and (18.70).

18.20 a) Derive Eq. (18.71). [*Hint:* Note that
$b_k = (4V_m/\pi k) + k\omega_0 RCa_k$. Use this ex-
pression for b_k to find $a_k^2 + b_k^2$ in terms of
a_k. Now use the expression for a_k to derive
Eq. (18.71).]

b) Derive Eq. (18.72).

18.21 Show that when we combine Eqs. (18.71) and
(18.72) with Eqs. (18.38) and (18.39), the result is Eq. (18.58). [*Hint:* Note from the
definition of β_k that

$$\frac{a_k}{b_k} = -\tan \beta_k$$

and from the definition of θ_k that

$$\tan \theta_k = -\cot \beta_k.$$

Now use the trigonometric identity

$$\tan x = \cot (90 - x)$$

to show that $\theta_k = (90 + \beta_k)$.]

18.22 The square-wave voltage shown in Fig.
P18.22(a) is applied to the circuit shown in
Fig. P18.22(b).

a) Find the Fourier series representation of
the steady-state current i.

b) Find the steady-state expression for i by
straightforward circuit analysis.

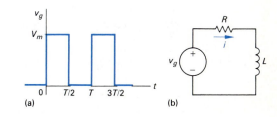

FIGURE P18.22

18.23 The periodic square-wave current seen in Fig.
P18.23(a) is applied to the circuit shown in

Fig. P18.23(b). Derive the first four nonzero
terms in the Fourier series that represents the
steady-state current i_o if $I_m = 105\pi$ mA.

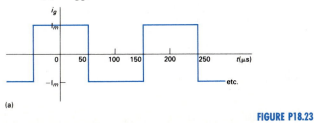

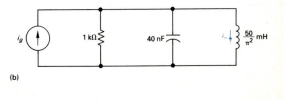

FIGURE P18.23

18.24 The periodic voltage described in Problem 18.15 is applied to the circuit in Fig. P18.24. Derive the first three nonzero terms in the Fourier series that represent the steady-state current i_o.

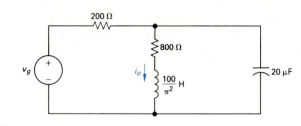

FIGURE P18.24

18.25 It is claimed that the circuit shown in Fig. P18.25(a) will convert the square-wave voltage shown in Fig. P18.25(b) to an al-most sinusoidal voltage with a frequency of 25 Mrad/s. Do you agree? Explain your answer.

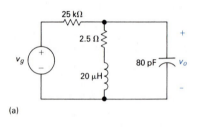

(a)

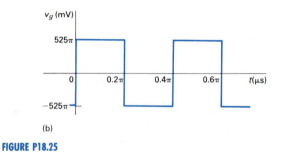

(b)

FIGURE P18.25

18.26 The periodic current shown in Fig. P18.26(a) is applied to the circuit shown in Fig. P18.26(b). Derive the first four nonzero terms in the Fourier series representation of the steady-state output current i_o when $I_m = 1260 \sqrt{2}\pi$ mA and $T = 80\pi$ μs.

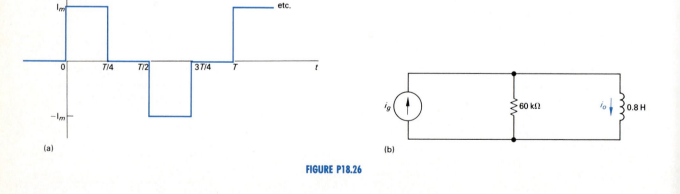

(a)

(b)

FIGURE P18.26

18.27 The full-wave rectified sine-wave voltage shown in Fig. P18.27(a) is applied to the circuit shown in Fig. P18.27(b).

a) Find the first four nonzero terms in the Fourier series representation of i_o.

b) What is a good approximation for the steady-state expression of i_o?

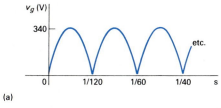

(a)

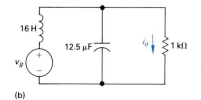

(b)

FIGURE P18.27

18.28 The periodic current shown in Fig. P18.28(a) is used to energize the circuit shown in Fig. P18.28(b). Write the time-domain expression for the fifth-harmonic current in the expression for i_o.

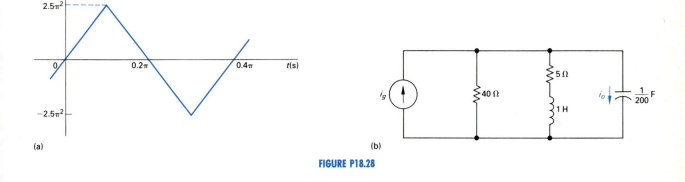

(a)

(b)

FIGURE P18.28

18.29 A periodic voltage having a period of 0.2π seconds is given by the following Fourier series:

$$v_g = 50 \sum_{n=1,3,5,}^{\infty} \frac{\pi^2 n^2 - 8}{n^3} \sin \frac{n\pi}{2} \cos n\omega_0 t \text{ V.}$$

This periodic voltage is applied to the circuit shown in Fig. P18.29. Find the amplitude and phase angle of the component of v_o that has a frequency of 250 rad/s.

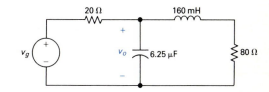

FIGURE P18.29

18.30 a) Derive the expressions for the Fourier coefficients for the periodic current shown in Fig. P18.30.

b) Write the first four nonzero terms of the series using the alternative trigonometric form given by Eq. 18.38.

c) Use the first four nonzero terms of the expression derived in part (b) to estimate the rms value of i_g when $I_m = 100$ mA.

d) Find the exact rms value of i_g.

e) Calculate the percent error in the estimated rms value.

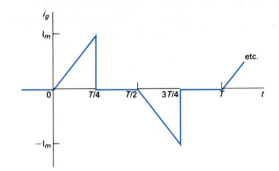

FIGURE P18.30

18.31 a) Use the first four nonzero terms in the Fourier series approximation of the periodic voltage shown in Fig. P18.31 to estimate its rms value.

b) Calculate the true rms value of the voltage.

c) Calculate the percent error in the estimated value.

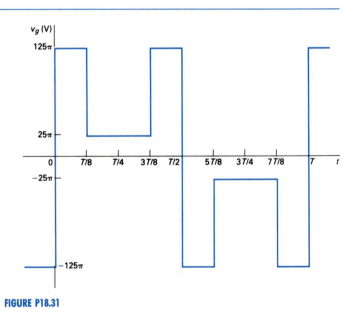

FIGURE P18.31

18.32 The voltage and current at the terminals of a network are

$$v = 20 + 250 \cos(300t + 30°) + 100 \sin(900t + 45°) \text{ V}$$

$$i = 2.5 + 8 \sin(300t + 75°) + 4 \cos(900t + 15°) \text{ A}.$$

The current is in the direction of the voltage drop across the terminals.

a) What is the average power at the terminals?

b) What is the rms value of the voltage?

c) What is the rms value of the current?

18.33 A half-wave rectified sinusoidal voltage is applied to the terminals of a 20-Ω resistor. The peak amplitude of the sinusoidal voltage is 170 V.

a) Find the total power delivered to the 20-Ω resistor.

b) Find the total power delivered to the 20-Ω resistor if the half-wave rectified sinusoidal voltage is approximated by the first three nonzero terms of its Fourier series.

18.34 A periodic current (i_o) as shown in Fig. P18.34 exists in a 12-Ω resistor.

a) Write the first four terms of the Fourier series for i_o using the format specified in Eq. 18.38.

b) Estimate the power dissipated in the 12-Ω resistor using the Fourier series spproximation derived in part (a).

c) Calculate the actual power dissipated in the 12-Ω resistor.

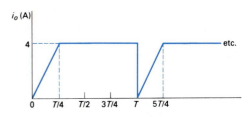

FIGURE P18.34

18.35 The triangular-wave voltage source is applied to the circuit shown in Fig. P18.35(a). The triangular-wave voltage is shown in Fig. P18.35(b). Estimate the average power delivered to the 90-Ω resistor when the circuit is in steady-state operation.

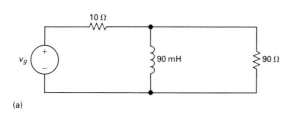

(a)

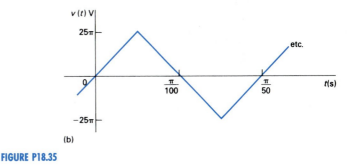

(b)

FIGURE P18.35

18.36 Assume the periodic function described in Problem 18.16 is a voltage with a peak amplitude of 20 V.

a) Find the rms value of the voltage.

b) If this voltage is applied to a 15-Ω resistor what is the average power dissipated in the resistor?

c) If v_g is approximated by using just the fundamental frequency term of its Fourier series what is the average power delivered to the 15-Ω resistor?

d) What is the percent error in the estimation of the power dissipated?

18.37 Derive the expression for C_n in the exponential Fourier series for the periodic waveform shown in Fig. P18.37.

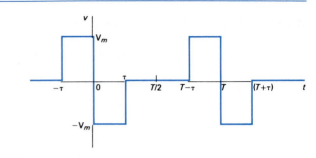

FIGURE P18.37

18.38 Derive the expression for the complex Fourier coefficients for the periodic voltage shown in Fig. P18.38.

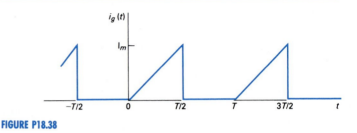

FIGURE P18.38

18.39 a) The periodic current in Problem 18.38 exists in a 30-Ω resistor. If $I_m = 8$ A what is the average power delivered to the resistor?

b) Assume $i_g(t)$ is approximated by a truncated exponential form of the Fourier series consisting of the first 5 nonzero terms, i.e.

$n = 0, 1, 2, 3,$ and 4. What is the rms value of the current using this approximation.

c) If the approximation in part (b) is used to represent i_g what is the percent error in the calculated power?

18.40 The periodic current source in the circuit shown in Fig. P18.40(a) has the waveform shown in Fig. P18.40(b).

a) Derive the expression for C_n.

b) Write the exponential form of the Fourier series for i_g up to and including the fourth harmonic.

c) Repeat part (b) for v_o when $T = 10\pi$ μs.

d) Use the truncated series of part (c) to estimate the average power delivered by the periodic current source.

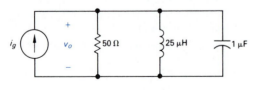

(a)

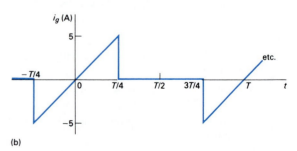

(b)

FIGURE P18.40

18.41 a) Find the rms value of the periodic current in Fig. P18.40(b).

b) Use the exponential series derived in Problem 18.40(b) to estimate the rms value of i_g.

c) What is the percent error in the estimated rms value of i_g?

18.42 Assume the capacitor in Fig. P18.40(a) is disconnected from the circuit. Estimate the average power dissipated in the 50-Ω resistor.

18.43 a) Make an amplitude and phase plot, based on Eq. (18.38), for the periodic voltage in Example 18.3. Assume V_m is 40 V. Plot both amplitude and phase versus $n\omega_0$ where $n = 0, 1, 2, 3, \ldots$.

b) Repeat part (a), but base the plots on Eq. (18.82).

18.44 A periodic function is represented by a Fourier series that has a finite number of terms. The amplitude and phase spectra are shown in Fig. P18.44(a) and (b), respectively.

a) Write the expression for the periodic current using the form given by Eq. (18.38).

b) Is the current an even or odd function of t?

c) Does the current have half-wave symmetry?

d) Calculate the rms value of the current in milliamperes.

e) Write the exponential form of the Fourier series.

f) Make the amplitude- and phase-spectra plots on the basis of the exponential series.

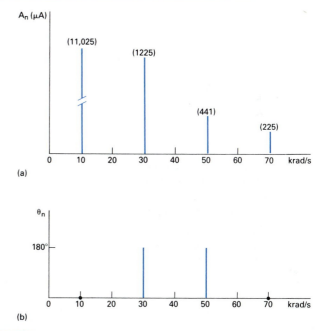

(a)

(b)

FIGURE P18.44

18.45 The signal voltage in the circuit of Fig. P18.45(a) is the periodic square wave shown in Fig. P18.45(b).

a) Assume $v_o(0) = 4$ V. Sketch v_{o1} versus t for $t \geq 0$.

b) Write the Fourier series for v_{o1}.

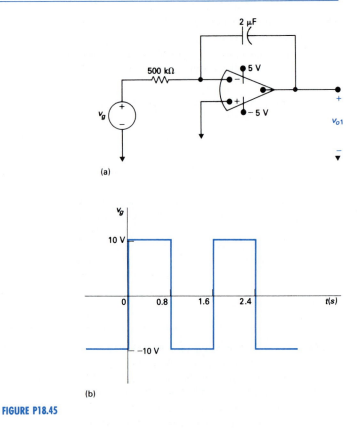

(a)

(b)

FIGURE P18.45

18.46 The input signal voltage in the circuit in Fig. P18.46 is the output voltage in the circuit in Fig. P18.45(a). Use the same time origin as specified in Problem 18.45.

a) Sketch v_{o2} versus t if $v_{o2}(0) = 0$.

b) Write the Fourier series for v_{o2}.

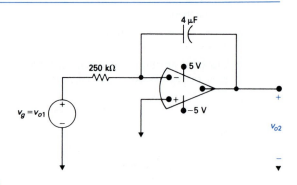

FIGURE P18.46

18.47 The output voltage in the circuit in Fig. P18.45(a) is used to drive the circuit in Fig. P18.46. This cascade connection of the two integrators is shown in Fig. P18.47.

a) Find the transfer function V_{o2}/V_g.

b) Assume v_g is the square-wave voltage shown in Fig. P18.45(b). Calculate the am-plitude of the fundamental- and the third-harmonic components of v_g.

c) Use the transfer function derived in part (a) to calculate the fundamental- and the third-harmonic components of v_{o2}.

d) Compare the terms obtained in part (c) with the terms generated by the Fourier series derived in Problem 18.46.

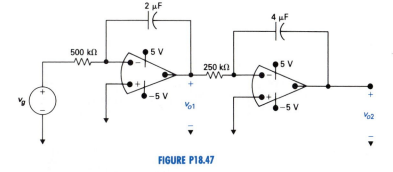

FIGURE P18.47

18.48 a) Approximate the periodic square-wave voltage shown in Fig. P18.48 with the first term in its Fourier series representation. Using this partial-sum approximation to $v(t)$, sketch $\epsilon(t)$ versus t over the range $-T/2 < t < T/2$.

b) Sketch $\epsilon^2(t)$ versus t over the range $-T/2 < t < T/2$.

c) Calculate $\overline{\epsilon^2}$.

d) Approximate $v(t)$ with the single term $10 \sin \omega_0 t$. Calculate $\overline{\epsilon^2}$.

e) Compare the value of $\overline{\epsilon^2}$ in part (c) with that obtained in part (d).

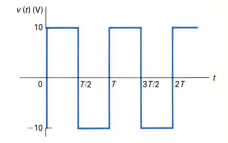

FIGURE P18.48

THE FOURIER TRANSFORM

CHAPTER 19

In Chapter 18 we discussed the representation of a periodic function by means of a Fourier series. This series representation enables us to describe the periodic function in terms of the frequency-domain attributes of amplitude and phase. The Fourier transform enables us to extend this frequency-domain description to functions that are not periodic. Through the Laplace transform we already introduced the idea of transforming an aperiodic function from the time domain to the frequency domain. You may wonder, then, why yet another type of transformation is necessary. Strictly speaking, the Fourier transform is not a new transform. It is a special case of the *bilateral* Laplace transform, with the real part of the complex frequency set equal to zero. However, in terms of physical interpretation the Fourier transform is more satisfyingly viewed as a limiting case of a Fourier series. We present this point of view in Section 19.1 when we derive the Fourier transform equations.

The Fourier transform is more useful than the Laplace transform in certain communications theory and signal processing situations. Although we cannot pursue the Fourier transform in depth, its introduction seems appropriate when the ideas underlying the Laplace transform and the Fourier series are still fresh in your mind.

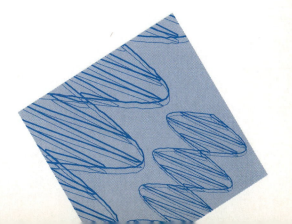

19.1 DERIVATION OF THE FOURIER TRANSFORM

We begin derivation of the Fourier transform as a limiting case of a Fourier series with the exponential form of the series:

$$f(t) = \sum_{n=-\infty}^{\infty} C_n e^{jn\omega_0 t}, \tag{19.1}$$

where

$$C_n = \frac{1}{T} \int_{-T/2}^{T/2} f(t) e^{-jn\omega_0 t} \, dt. \tag{19.2}$$

In Eq. (19.2), we elected to start the integration at $t_0 = -T/2$.

Allowing the fundamental period T to increase without limit accomplishes the transition from a periodic to an aperiodic function. That is, if T becomes infinite, the function never repeats itself and hence is aperiodic. As T increases, the separation between adjacent harmonic frequencies becomes smaller and smaller. In particular,

$$\Delta\omega = (n + 1)\omega_0 - n\omega_0 = \omega_0 = \frac{2\pi}{T}, \tag{19.3}$$

and as T gets larger and larger the incremental separation $\Delta\omega$ approaches a differential separation $d\omega$. From Eq. (19.3),

$$\frac{1}{T} \to \frac{d\omega}{2\pi}, \qquad \text{as } T \to \infty. \tag{19.4}$$

As the period increases, the frequency moves from being a discrete variable to becoming a continuous variable, or

$$n\omega_0 \to \omega, \qquad \text{as } T \to \infty. \tag{19.5}$$

In terms of Eq. (19.2), as the period increases, the Fourier coefficients C_n get smaller. In the limit $C_n \to 0$ as $T \to \infty$. This result makes sense because we expect the Fourier coefficients to vanish as the function loses its periodicity. Note, however, the limiting value of the product $C_n T$; that is

$$C_n T \to \int_{-\infty}^{\infty} f(t) e^{-j\omega t} \, dt, \qquad \text{as } T \to \infty. \tag{19.6}$$

In writing Eq. (19.6) we took advantage of the relationship in Eq. (19.5). The integral in Eq. (19.6) is the Fourier transform of $f(t)$ and is denoted

$$F(\omega) = \mathcal{F}\{f(t)\} = \int_{-\infty}^{\infty} f(t) e^{-j\omega t} \, dt. \tag{19.7}$$

We obtain an explicit expression for the inverse Fourier transform by investigating the limiting form of Eq. (19.1) as $T \to \infty$.

We begin by multiplying and dividing by T:

$$f(t) = \sum_{n=-\infty}^{\infty} (C_n T) e^{jn\omega_0 t} \left(\frac{1}{T}\right). \tag{19.8}$$

As $T \to \infty$, the summation approaches integration, $C_n T \to F(\omega)$, $n\omega_0 \to \omega$, and $1/T \to d\omega/2\pi$. Thus in the limit Eq. (19.8) becomes

$$f(t) = \frac{1}{2\pi} \int_{-\infty}^{\infty} F(\omega) e^{j\omega t}\, d\omega. \tag{19.9}$$

Equations (19.7) and (19.9) define the Fourier transform. Equation (19.7) transforms the time-domain expression $f(t)$ into its corresponding frequency-domain expression $F(\omega)$. Equation (19.9) defines the inverse operation of transforming $F(\omega)$ into $f(t)$.

Let's now derive the Fourier transform of the single voltage pulse shown in Fig. 19.1. Note that this single voltage pulse corresponds to the periodic voltage in Example 19.6 if we let $T \to \infty$. The Fourier transform of $v(t)$ comes directly from Eq. (19.7):

$$F(\omega) = \int_{-\tau/2}^{\tau/2} V_m e^{-j\omega t}\, dt$$

$$= V_m \frac{e^{-j\omega t}}{(-j\omega)} \Bigg|_{-\tau/2}^{\tau/2}$$

$$= \frac{V_m}{-j\omega} \left(-2j \sin \frac{\omega\tau}{2}\right), \tag{19.10}$$

which can be put in the form of $(\sin x)/x$ by multiplying the numerator and denominator by τ. Then,

$$F(\omega) = V_m \tau \frac{\sin \omega\tau/2}{\omega\tau/2}. \tag{19.11}$$

For the periodic train of voltage pulses in Example 18.6, the expression for the Fourier coefficients is

$$C_n = \frac{V_m \tau}{T} \frac{\sin n\omega_0 \tau/2}{n\omega_0 \tau/2}. \tag{19.12}$$

Comparing Eqs. (19.11) and (19.12) clearly shows that as the time-domain function goes from periodic to aperiodic, the amplitude spectrum goes from a discrete line spectrum to a continuous spectrum. Furthermore, the envelope of the line spectrum has the same shape as the continuous spectrum. Thus as T increases, the spectrum of lines gets denser and the amplitudes get smaller, but the envelope doesn't change shape. Thus we have the physical interpretation of the Fourier transform $F(\omega)$ as be-

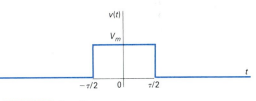

FIGURE 19.1 A voltage pulse.

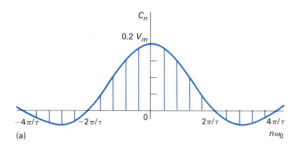

(a)

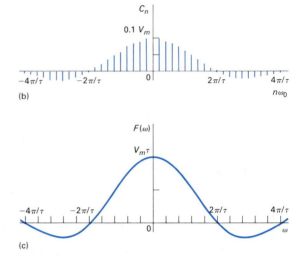

(b)

(c)

FIGURE 19.2 Transition of the amplitude spectrum as $f(t)$ goes from periodic to aperiodic: (a) C_n versus $n\omega_0$, $T/\tau = 5$; (b) C_n versus $n\omega_0$, $T/\tau = 10$; (c) $F(\omega)$ versus ω.

ing a measure of the frequency content of $f(t)$. Figure 19.2 illustrates these observations graphically. The amplitude spectrum plot is based on the assumption that τ is constant and T is increasing.

19.2 CONVERGENCE OF THE FOURIER INTEGRAL

A function of time $f(t)$ has a Fourier transform if the integral in Eq. (19.7) converges. If $f(t)$ is a "well-behaved" function that differs from zero over a *finite* interval of time, convergence is no problem. "Well-behaved" implies that $f(t)$ is single-valued and encloses a finite area over the range of integration. In practical terms all pulses of finite duration in which we are interested are well-behaved functions. The evaluation of the Fourier transform of the rectangular pulse discussed in Section 19.1 illustrates this point.

If $f(t)$ is different from zero over an infinite interval, the convergence of the Fourier integral depends on the behavior of $f(t)$ as $t \rightarrow \infty$. A single-valued function that is nonzero over an infinite interval has a Fourier transform if the integral

$$\int_{-\infty}^{\infty} |f(t)| \, dt$$

exists and if any discontinuities in $f(t)$ are finite. An example of this type of function is the exponential decaying function illustrated in Fig. 19.3.

The Fourier transform of $f(t)$ is

$$F(\omega) = \int_{-\infty}^{\infty} f(t)e^{-j\omega t}\, dt = \int_{0}^{\infty} Ke^{-at}e^{j\omega t}\, dt$$

$$= \frac{Ke^{-(a+j\omega)t}}{-(a+j\omega)}\bigg|_{0}^{\infty} = \frac{K}{-(a+j\omega)}(0-1)$$

$$= \frac{K}{a+j\omega}. \tag{19.13}$$

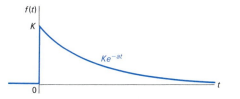

FIGURE 19.3 The exponential function $Ke^{-at}u(t)$.

A third important group of functions have great practical interest but do not, in a strict sense, have a Fourier transform. For example, the integral in Eq. (19.7) doesn't converge if $f(t)$ is a constant. The same can be said if $f(t)$ is a sinusoidal function $(\cos \omega_o t)$ or a step function $[Ku(t)]$. These functions are of great interest in circuit analysis, but, in order to include them in Fourier analysis, we must resort to some mathematical subterfuge. We create a function in the time domain that has a Fourier transform and at the same time can be made arbitrarily close to the function of interest. Next we find the Fourier transform of the approximating function and then evaluate the limiting value of $F(\omega)$ as the approximating function approaches $f(t)$. We define the limiting value of $F(\omega)$ as the *Fourier transform of $f(t)$*. We demonstrate this technique by finding the Fourier transform of a constant.

We can approximate a constant with the exponential function

$$f(t) = Ae^{-\epsilon|t|}, \quad \epsilon > 0. \tag{19.14}$$

As $\epsilon \to 0$, $f(t) \to A$. Figure 19.4 shows the approximation graphically.

The Fourier transform of $f(t)$ is

$$F(\omega) = \int_{-\infty}^{0} Ae^{\epsilon t}e^{-j\omega t}\, dt + \int_{0}^{\infty} Ae^{-\epsilon t}e^{-j\omega t}\, dt. \tag{19.15}$$

Carrying out the integration called for in Eq. (19.15) yields

$$F(\omega) = \frac{A}{\epsilon - j\omega} + \frac{A}{\epsilon + j\omega} = \frac{2\epsilon A}{\epsilon^2 + \omega^2}. \tag{19.16}$$

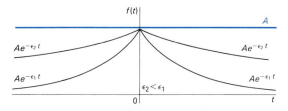

FIGURE 19.4 Approximation of a constant with an exponential function.

The function given by Eq. (19.16) generates an impulse function at $\omega = 0$ as $\epsilon \to 0$. You can verify this result by showing that $F(\omega)$ approaches infinity at $\omega = 0$ as $\epsilon \to 0$, that the duration of $F(\omega)$ approaches zero as $\epsilon \to 0$, and that the area under $F(\omega)$ is independent of ϵ. The area under $F(\omega)$ is the strength of the impulse and is

$$\int_{-\infty}^{\infty} \frac{2\epsilon A}{\epsilon^2 + \omega^2}\, d\omega = 4\epsilon A \int_{0}^{\infty} \frac{d\omega}{\epsilon^2 + \omega^2} = 2\pi A. \qquad (19.17)$$

In the limit $f(t)$ approaches a constant A, and $F(\omega)$ approaches an impuse function $2\pi A\delta(\omega)$. Therefore the Fourier transform of a constant A is defined as $2\pi A\delta(\omega)$, or

$$\mathcal{F}\{A\} = 2\pi A\delta(\omega). \qquad (19.18)$$

In Section 19.4 we say more about Fourier transforms that are defined through a limit process. Before doing so, in Section 19.3 we show how to take advantage of the Laplace transform to find the Fourier transform of functions for which the Fourier integral converges.

DRILL EXERCISES

19.1 Use the defining integral to find the Fourier transform of the following functions:

a) $f(t) = -A, \qquad -\tau/2 \le t < 0;$
$\quad\, f(t) = A, \qquad\quad 0 < t \le \tau/2;$
$\quad\, f(t) = 0, \qquad\quad$ elsewhere.

b) $f(t) = 0, \qquad\qquad t < 0;$
$\quad\, f(t) = te^{-at}, \qquad t \ge 0, a > 0.$

ANSWER: (a) $-j\left(\dfrac{2A}{\omega}\right)\left(1 - \cos\dfrac{\omega\tau}{2}\right)$; (b) $\dfrac{1}{(a + j\omega)^2}$.

19.2 The Fourier transform of $f(t)$ is given by

$$F(\omega) = 0, \qquad -\infty \le \omega < -2;$$
$$F(\omega) = 2, \qquad -2 < \omega < -1;$$
$$F(\omega) = 1, \qquad -1 < \omega < 1;$$
$$F(\omega) = 2, \qquad\; 1 < \omega < 2;$$
$$F(\omega) = 0, \qquad\; 2 < \omega \le \infty.$$

Find $f(t)$.

ANSWER: $f(t) = \dfrac{1}{\pi t}(2\sin 2t - \sin t).$

19.3 USING LAPLACE TRANSFORMS TO FIND FOURIER TRANSFORMS

We can use a table of unilateral, or one-sided, Laplace transform pairs to find the Fourier transform of functions *for which the Fourier integral converges*. The Fourier integral converges when all the poles of $F(s)$ lie in the left half of the s plane. [If $F(s)$ has poles in the right half of the s plane or along the imaginary axis,

$f(t)$ does not satisfy the constraint that $\int_{-\infty}^{\infty} |f(t)| \, dt$ exists.] The following rules apply to the use of Laplace transforms to find the Fourier transforms of such functions.

1. If $f(t)$ is zero for $t \leq 0^-$, we obtain the Fourier transform of $f(t)$ from the Laplace transform of $f(t)$ simply by replacing s by $j\omega$. Thus

$$\mathcal{F}\{f(t)\} = \mathcal{L}\{f(t)\}_{s=j\omega}. \qquad (19.19)$$

For example, if

$$f(t) = 0, \qquad\qquad t \leq 0^-;$$

$$f(t) = e^{-at} \cos \omega_0 t, \qquad t \geq 0^+;$$

then

$$\mathcal{F}\{f(t)\} = \frac{s + a}{(s + a)^2 + \omega_0^2}\bigg|_{s=j\omega}$$

$$= \frac{j\omega + a}{(j\omega + a)^2 + \omega_0^2}.$$

2. Because the range of integration on the Fourier integral goes from $-\infty$ to $+\infty$, the Fourier transform of a negative-time function exists. A *negative-time function* is a function that is nonzero for negative values of time and zero for positive values of time. To find the Fourier transform of a negative-time function, we proceed as follows. First, we reflect the negative-time function over to the positive-time domain and then find its one-sided Laplace transform. We obtain the Fourier transform of the original time function by replacing s with $-j\omega$. Therefore, when $f(t) = 0$ for $t \geq 0^+$,

$$\mathcal{F}\{f(t)\} = \mathcal{L}\{f(-t)\}_{s=-j\omega}. \qquad (19.20)$$

For example, if

$$f(t) = 0 \qquad (\text{for } t \geq 0^+)$$

and

$$f(t) = e^{at} \cos \omega_0 t \qquad (\text{for } t \leq 0^-),$$

then

$$f(-t) = 0 \qquad (\text{for } t \leq 0^-)$$

and

$$f(-t) = e^{-at} \cos \omega_0 t \qquad (\text{for } t \geq 0^+).$$

Both $f(t)$ and its mirror image are plotted in Fig. 19.5.

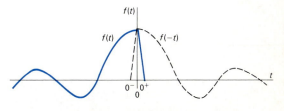

FIGURE 19.5 Reflection of a negative-time function over to the positive-time domain.

The Fourier transform of $f(t)$ is

$$\mathcal{F}\{f(t)\} = \mathcal{L}\{f(-t)\}_{s=-j\omega} = \left.\frac{s + a}{(s + a)^2 + \omega_0^2}\right|_{s=-j\omega}$$

$$= \frac{-j\omega + a}{(-j\omega + a)^2 + \omega_0^2}.$$

3. Functions that are nonzero over all time can be resolved into positive- and negative-time functions. We use Eqs. (19.19) and (19.20) to find the Fourier transform of the positive- and negative-time functions, respectively. The Fourier transform of the original function is the sum of the two transforms. Thus if we let

$$f^+(t) = f(t) \qquad (\text{for } t > 0)$$

and

$$f^-(t) = f(t) \qquad (\text{for } t < 0),$$

then

$$f(t) = f^+(t) + f^-(t)$$

and

$$\mathcal{F}\{f(t)\} = \mathcal{F}\{f^+(t)\} + \mathcal{F}\{f^-(t)\}$$

$$= \mathcal{L}\{f^+(t)\}_{s=j\omega} + \mathcal{L}\{f^-(-t)\}_{s=-j\omega}. \qquad \text{(19.21)}$$

An example of using Eq. (19.21) involves finding the Fourier transform of $e^{-a|t|}$. For the original function the positive- and negative-time functions are

$$f^+(t) = e^{-at} \qquad \text{and} \qquad f^-(t) = e^{at}.$$

Then,

$$\mathcal{L}\{f^+(t)\} = \frac{1}{s + a}$$

and

$$\mathcal{L}\{f^-(-t)\} = \frac{1}{s + a}.$$

Therefore, from Eq. (19.21),

$$\mathcal{F}\{e^{-a|t|}\} = \left.\frac{1}{s + a}\right|_{s=j\omega} + \left.\frac{1}{s + a}\right|_{s=-j\omega}$$

$$= \frac{1}{j\omega + a} + \frac{1}{-j\omega + a}$$

$$= \frac{2a}{\omega^2 + a^2}.$$

If $f(t)$ is even, Eq. (19.21) reduces to

$$\mathcal{F}\{f(t)\} = \mathcal{L}\{f(t)\}_{s=j\omega} + \mathcal{L}\{f(t)\}_{s=-j\omega}. \qquad \textbf{(19.22)}$$

If $f(t)$ is odd, then Eq. (19.21) becomes

$$\mathcal{F}\{f(t)\} = \mathcal{L}\{f(t)\}_{s=j\omega} - \mathcal{L}\{f(t)\}_{s=-j\omega}. \qquad \textbf{(19.23)}$$

DRILL EXERCISES

19.3 Find the Fourier transform of each function. In each case a is a positive real constant.

a) $f(t) = 0,$ $\quad\quad\quad\quad t < 0;$
$\quad f(t) = e^{-at}\sin \omega_0 t,$ $\quad t \geq 0.$

b) $f(t) = 0,$ $\quad\quad\quad\quad t > 0;$
$\quad f(t) = -te^{at},$ $\quad\quad t \leq 0.$

c) $f(t) = te^{-at},$ $\quad\quad\quad t \geq 0;$
$\quad f(t) = te^{at},$ $\quad\quad\quad t \leq 0.$

ANSWER: (a) $\dfrac{\omega_0}{(a + j\omega)^2 + \omega_0^2}$; (b) $\dfrac{1}{(a - j\omega)^2}$;

(c) $\dfrac{-j4a\omega}{(a^2 + \omega^2)^2}.$

19.4 FOURIER TRANSFORMS IN THE LIMIT

As we pointed out in Section 19.2, the Fourier transforms of several practical functions must be defined by a limit process. We now return to these types of functions and develop their transforms.

FOURIER TRANSFORM OF A SIGNUM FUNCTION

We showed that the Fourier transform of a constant A is $2\pi A\delta(\omega)$ in Eq. (19.18). The next function of interest is the *signum function*, defined as $+1$ for $t > 0$ and -1 for $t < 0$. The signum function is denoted sgn (t) and can be expressed in terms of unit step functions, or

$$\text{sgn } (t) = u(t) - u(-t). \qquad \textbf{(19.24)}$$

Figure 19.6 shows the function graphically.

To find the Fourier transform of the signum function, we first create a function that approaches the signum function in the limit:

$$\text{sgn } (t) = \lim_{\epsilon \to 0} [e^{-\epsilon t}u(t) - e^{\epsilon t}u(-t)]. \qquad \textbf{(19.25)}$$

The function inside the brackets, plotted in Fig. 19.7, has a Fourier transform because the Fourier integral converges. Be-

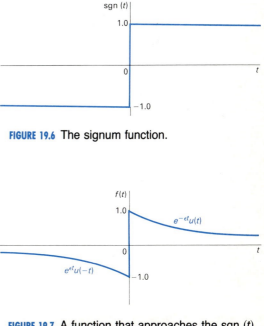

FIGURE 19.6 The signum function.

FIGURE 19.7 A function that approaches the sgn (t) as ϵ approaches zero.

cause $f(t)$ is an odd function, we use Eq. (19.23) to find its Fourier transform:

$$\mathcal{F}\{f(t)\} = \frac{1}{s + \epsilon}\bigg|_{s=j\omega} - \frac{1}{s + \epsilon}\bigg|_{s=-j\omega}$$

$$= \frac{1}{j\omega + \epsilon} - \frac{1}{-j\omega + \epsilon}$$

$$= \frac{-2j\omega}{\omega^2 + \epsilon^2}, \tag{19.26}$$

as $\epsilon \to 0$, $f(t) \to$ sgn (t), and $\mathcal{F}\{f(t)\} \to 2/j\omega$. Therefore

$$\mathcal{F}\{\text{sgn }(t)\} = \frac{2}{j\omega}. \tag{19.27}$$

FOURIER TRANSFORM OF A UNIT-STEP FUNCTION

To find the Fourier transform of the unit-step function, we use Eqs. (19.18) and (19.27). We do so by recognizing that the unit-step function can be expressed as

$$u(t) = \frac{1}{2} + \frac{1}{2}\text{sgn }(t). \tag{19.28}$$

Thus

$$\mathcal{F}\{u(t)\} = \mathcal{F}\left\{\frac{1}{2}\right\} + \mathcal{F}\left\{\frac{1}{2}\text{sgn }(t)\right\}$$

$$= \pi\delta(\omega) + \frac{1}{j\omega}. \tag{19.29}$$

FOURIER TRANSFORM OF A COSINE FUNCTION

To find the Fourier transform of the cos $\omega_0 t$, we return to the inverse-transform integral of Eq. (19.9) and observe that if

$$F(\omega) = 2\pi\delta(\omega - \omega_0), \tag{19.30}$$

then

$$f(t) = \frac{1}{2\pi}\int_{-\infty}^{\infty} [2\pi\delta(\omega - \omega_0)]e^{j\omega t}\, d\omega. \tag{19.31}$$

Using the sifting property of the impulse function, we reduce Eq. (19.31) to

$$f(t) = e^{j\omega_0 t}. \tag{19.32}$$

Then, from Eqs. (19.30) and (19.32),

$$\mathcal{F}\{e^{j\omega_0 t}\} = 2\pi\delta(\omega - \omega_0). \tag{19.33}$$

TABLE 19.1

FOURIER TRANSFORMS OF ELEMENTARY FUNCTIONS

$f(t)$	$F(\omega)$
$\delta(t)$ (impulse)	1
A (constant)	$2\pi A\delta(\omega)$
$\text{sgn}(t)$ (signum)	$2/j\omega$
$u(t)$ (step)	$\pi\delta(\omega) + 1/j\omega$
$e^{-at}u(t)$ (positive-time exponential)	$1/(a + j\omega)$
$e^{at}u(-t)$ (negative-time exponential)	$1/(a - j\omega)$
$e^{-a\lvert t\rvert}$ (positive- and negative-time exponential)	$2a/(a^2 + \omega^2)$
$e^{j\omega_0 t}$ (complex exponential)	$2\pi\delta(\omega - \omega_0)$
$\cos \omega_0 t$ (cosine)	$\pi[\delta(\omega + \omega_0) + \delta(\omega - \omega_0)]$
$\sin \omega_0 t$ (sine)	$j\pi[\delta(\omega + \omega_0) - \delta(\omega - \omega_0)]$

We now use Eq. (19.33) to find the Fourier transform of the $\cos \omega_0 t$, because

$$\cos \omega_0 t = \frac{e^{j\omega_0 t} + e^{-j\omega_0 t}}{2}. \qquad (19.34)$$

Thus

$$\begin{aligned}
\mathscr{F}\{\cos \omega_0 t\} &= \tfrac{1}{2}(\mathscr{F}\{e^{j\omega_0 t}\} + \mathscr{F}\{e^{-j\omega_0 t}\}) \\
&= \tfrac{1}{2}[2\pi\delta(\omega - \omega_0) + 2\pi\delta(\omega + \omega_0)] \\
&= \pi\delta(\omega - \omega_0) + \pi\delta(\omega + \omega_0). \qquad (19.35)
\end{aligned}$$

The Fourier transform of $\sin \omega_0 t$ involves similar manipulation, which we leave for Drill Exercise 19.4. Table 19.1 presents a summary of the transform pairs of the important elementary functions.

This completes development of the Fourier transforms of the elementary functions. We now turn to the properties of the transform that enhance our ability to describe aperiodic time-domain behavior in terms of frequency-domain behavior.

DRILL EXERCISES

19.4 Find $\mathscr{F}\{\sin \omega_0 t\}$.

ANSWER: $j\pi[\delta(\omega + \omega_0) - \delta(\omega - \omega_0)]$.

19.5 SOME MATHEMATICAL PROPERTIES

The first mathematical property that we call to your attention is that $F(\omega)$ is a complex quantity and can be expressed in either rectangular or polar form. Thus from the defining integral,

$$F(\omega) = \int_{-\infty}^{\infty} f(t)e^{-j\omega t}\, dt$$

$$= \int_{-\infty}^{\infty} f(t)(\cos \omega t - j \sin \omega t)\, dt$$

$$= \int_{-\infty}^{\infty} f(t) \cos \omega t\, dt - j \int_{-\infty}^{\infty} f(t) \sin \omega t\, dt. \qquad (19.36)$$

Now we let

$$A(\omega) = \int_{-\infty}^{\infty} f(t) \cos \omega t\, dt \qquad (19.37)$$

and

$$B(\omega) = -\int_{-\infty}^{\infty} f(t) \sin \omega t\, dt. \qquad (19.38)$$

Thus, using the definitions given by Eqs. (19.37) and (19.38) in Eq. (19.36) we get

$$F(\omega) = A(\omega) + jB(\omega) = |F(\omega)|e^{j\theta(\omega)}. \qquad (19.39)$$

The following observations about $F(\omega)$ are pertinent.

1. The real part of $F(\omega)$, that is, $A(\omega)$, is an even function of ω; in other words, $A(\omega) = A(-\omega)$.

2. The imaginary part of $F(\omega)$, that is, $B(\omega)$, is an odd function of ω; in other words, $B(\omega) = -B(-\omega)$.

3. The magnitude of $F(\omega)$, that is, $\sqrt{A^2(\omega) + B^2(\omega)}$, is an even function of ω.

4. The phase angle of $F(\omega)$, that is,

$$\theta(\omega) = \tan^{-1} B(\omega)/A(\omega),$$

   is an odd function of ω.

5. Replacing ω by $-\omega$ generates the conjugate of $F(\omega)$; in other words, $F(-\omega) = F^*(\omega)$.

Hence, if $f(t)$ is an even function, $F(\omega)$ is real, and if $f(t)$ is an odd function, $F(\omega)$ is imaginary. If $f(t)$ is even, from Eqs. (19.37) and (19.38),

$$A(\omega) = 2 \int_{0}^{\infty} f(t) \cos \omega t\, dt \qquad (19.40)$$

and

$$B(\omega) = 0. \qquad (19.41)$$

If $f(t)$ is an odd function,

$$A(\omega) = 0; \qquad (19.42)$$

$$B(\omega) = -2 \int_0^\infty f(t) \sin \omega t \, dt. \qquad (19.43)$$

We leave the derivations of Eqs. (19.40)–(19.43) for you as Problems 19.5 and 19.6.

If $f(t)$ is an even function, its Fourier transform is an even function, and if $f(t)$ is an odd function, its Fourier transform is an odd function. Moreover, if $f(t)$ is an even function, from the inverse Fourier integral,

$$f(t) = \frac{1}{2\pi} \int_{-\infty}^\infty F(\omega)e^{j\omega t} \, d\omega = \frac{1}{2\pi} \int_{-\infty}^\infty A(\omega)e^{j\omega t} \, d\omega$$

$$= \frac{1}{2\pi} \int_{-\infty}^\infty A(\omega)(\cos \omega t + j \sin \omega t) \, d\omega$$

$$= \frac{1}{2\pi} \int_{-\infty}^\infty A(\omega) \cos \omega t \, d\omega + 0$$

$$= \frac{2}{2\pi} \int_0^\infty A(\omega) \cos \omega t \, d\omega. \qquad (19.44)$$

Now compare Eq. (19.44) with Eq. (19.40). Note that, except for a factor of $1/2\pi$, these two equations have the same form. Thus the waveforms of $A(\omega)$ and $f(t)$ become interchangeable if $f(t)$ is an even function. For example, we have already observed that a rectangular pulse in the time domain produces a frequency spectrum of the form $\sin \omega/\omega$. Specifically, Eq. (19.11) expresses the Fourier transform of the voltage pulse shown in Fig. 19.1. Hence a rectangular pulse in the frequency domain must be generated by a time-domain function of the form $\sin t/t$. We can illustrate this requirement by finding the time-domain function $f(t)$ corresponding to the frequency spectrum shown in Fig. 19.8. From Eq. (19.44),

$$f(t) = \frac{2}{2\pi} \int_0^{\omega_0/2} M \cos \omega t \, d\omega = \frac{2M}{2\pi} \left(\frac{\sin \omega t}{t}\right) \Big|_0^{\omega_0/2}$$

$$= \frac{1}{2\pi} \left(2M \frac{\sin \omega_0 t/2}{t/2}\right)$$

$$= \frac{1}{2\pi} \left(M\omega_0 \frac{\sin \omega_0 t/2}{\omega_0 t/2}\right). \qquad (19.45)$$

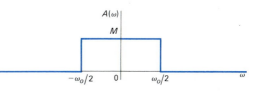

FIGURE 19.8 A rectangular frequency spectrum.

We say more about the frequency spectrum of a rectangular pulse in the time domain versus the rectangular frequency spectrum of $(\sin t)/t$ after we introduce Parseval's theorem.

DRILL EXERCISES

19.5 If $f(t)$ is a real function of t, show that the inversion integral reduces to

$$f(t) = \frac{1}{2\pi} \int_{-\infty}^{\infty} [A(\omega) \cos \omega t - B(\omega) \sin \omega t]\, d\omega.$$

ANSWER: Derivation.

19.6 If $f(t)$ is a real, odd function of t, show that the inversion integral reduces to

$$f(t) = -\frac{1}{2\pi} \int_{-\infty}^{\infty} B(\omega) \sin \omega t\, d\omega.$$

ANSWER: Derivation.

19.6 OPERATIONAL TRANSFORMS

Fourier transforms, like Laplace transforms, can be classified as functional and operational. So far, we have concentrated on the functional transforms. We now discuss some of the important operational transforms. With regard to the Laplace transform, these operational transforms are similar to those discussed in Chapter 15. Hence we leave their proofs to you as Problems 19.7–19.14.

1. Multiplication by a Constant   If $\mathscr{F}\{f(t)\} = F(\omega)$,

$$\mathscr{F}\{Kf(t)\} = KF(\omega). \tag{19.46}$$

2. Addition (Subtraction)   If

$$\mathscr{F}\{f_1(t)\} = F_1(\omega),$$

$$\mathscr{F}\{f_2(t)\} = F_2(\omega),$$

and

$$\mathscr{F}\{f_3(t)\} = F_3(\omega),$$

then

$$\mathscr{F}\{f_1(t) - f_2(t) + f_3(t)\} = F_1(\omega) - F_2(\omega) + F_3(\omega). \tag{19.47}$$

3. Differentiation   The Fourier transform of the first derivative of $f(t)$ is

$$\mathscr{F}\left\{\frac{df(t)}{dt}\right\} = j\omega F(\omega). \tag{19.48}$$

The nth derivative of $f(t)$ is

$$\mathcal{F}\left\{\frac{d^n f(t)}{dt^n}\right\} = (j\omega)^n F(\omega). \qquad \textbf{(19.49)}$$

Equations (19.48) and (19.49) are valid if $f(t)$ is zero at $\pm\infty$.

4. Integration   If $g(t) = \int_{-\infty}^t f(x)\,dx$,

$$\mathcal{F}\{g(t)\} = \frac{F(\omega)}{j\omega}. \qquad \textbf{(19.50)}$$

Equation (19.50) is valid if

$$\int_{-\infty}^{\infty} f(x)\,dx = 0.$$

5. Scale Change   Dimensionally, time and frequency are reciprocals. Therefore when time is stretched out, frequency is compressed and vice versa, as reflected in the functional transform

$$\mathcal{F}\{f(at)\} = \frac{1}{a}F\left(\frac{\omega}{a}\right), \qquad a > 0. \qquad \textbf{(19.51)}$$

Note that when $0 < a < 1.0$ time is stretched out, whereas when $a > 1.0$ time is compressed.

6. Translation in the Time Domain   The effect of translating a function in the time domain is to alter the phase spectrum and leave the amplitude spectrum untouched. Thus

$$\mathcal{F}\{f(t - a)\} = e^{-j\omega a} F(\omega). \qquad \textbf{(19.52)}$$

If a is positive in Eq. (19.52), the time function is delayed, and if a is negative, the time function is advanced.

7. Translation in the Frequency Domain   Translation in the frequency domain corresponds to multiplication by the complex exponential in the time domain:

$$\mathcal{F}\{e^{j\omega_0 t} f(t)\} = F(\omega - \omega_0). \qquad \textbf{(19.53)}$$

8. Modulation   Amplitude modulation is the process of varying the amplitude of a sinusoidal carrier. If the modulating signal is denoted $f(t)$, the modulated carrier becomes $f(t)\cos\omega_0 t$. The amplitude spectrum of the modulated carrier is one-half the amplitude spectrum $f(t)$ centered at $\pm\omega_0$, that is,

$$\mathcal{F}\{f(t)\cos\omega_0 t\} = \tfrac{1}{2}F(\omega - \omega_0) + \tfrac{1}{2}F(\omega + \omega_0). \qquad \textbf{(19.54)}$$

9. Convolution in the Time Domain   Convolution in the time domain corresponds to multiplication in the frequency domain.

That is,

$$y(t) = \int_{-\infty}^{\infty} x(\lambda)h(t - \lambda)\, d\lambda,$$

becomes

$$\mathscr{F}\{y(t)\} = Y(\omega) = X(\omega)H(\omega). \qquad \textbf{(19.55)}$$

Equation (19.55) is important in applications of the Fourier transform because it states that the transform of the response function $Y(\omega)$ is the product of the input transform $X(\omega)$ and the system function $H(\omega)$. We say more about this relationship in Section 19.7.

10. Convolution in the Frequency Domain   Convolution in the frequency domain corresponds to finding the Fourier transform of the product of two time functions. Thus if

$$f(t) = f_1(t)f_2(t),$$

then

$$F(\omega) = \frac{1}{2\pi} \int_{-\infty}^{\infty} F_1(u)F_2(\omega - u)\, du. \qquad \textbf{(19.56)}$$

Table 19.2 summarizes these 10 operational transforms and another operational transform that we introduce in Problem 19.14.

TABLE 19.2

OPERATIONAL TRANSFORMS

$f(t)$	$F(\omega)$
$Kf(t)$	$KF(\omega)$
$f_1(t) - f_2(t) + f_3(t)$	$F_1(\omega) - F_2(\omega) + F_3(\omega)$
$d^n f(t)/dt^n$	$(j\omega)^n F(\omega)$
$\displaystyle\int_{-\infty}^{t} f(x)\, dx$	$F(\omega)/j\omega$
$f(at)$	$\dfrac{1}{a} F\!\left(\dfrac{\omega}{a}\right), \quad a > 0$
$f(t - a)$	$e^{-j\omega a} F(\omega)$
$e^{j\omega_0 t} f(t)$	$F(\omega - \omega_0)$
$f(t) \cos \omega_0 t$	$\frac{1}{2} F(\omega - \omega_0) + \frac{1}{2} F(\omega + \omega_0)$
$\displaystyle\int_{-\infty}^{\infty} x(\lambda)h(t - \lambda)\, d\lambda$	$X(\omega)H(\omega)$
$f_1(t)f_2(t)$	$\dfrac{1}{2\pi} \displaystyle\int_{-\infty}^{\infty} F_1(u)F_2(\omega - u)\, du$
$t^n f(t)$	$(j)^n \dfrac{d^n F(\omega)}{d\omega^n}$

DRILL EXERCISES

19.7 a) Find the second derivative of the function described in Problem 19.1(b) on p. 819.

b) Find the Fourier transform of the second derivative.

c) Use the result obtained in (b) to find the Fourier transform of the function in (a). (*Hint:* Use the operational transform of differentiation.)

ANSWER:

(a) $\dfrac{d^2 f}{dt^2} = \dfrac{2A}{\tau}\delta\left(t + \dfrac{\tau}{2}\right) - \dfrac{4A}{\tau}\delta(t) + \dfrac{2A}{\tau}\left(t - \dfrac{\tau}{2}\right);$

(b) $\dfrac{4A}{\tau}\left(\cos\dfrac{\omega\tau}{2} - 1\right);$ (c) $\dfrac{4A}{\omega^2\tau}\left(1 - \cos\dfrac{\omega\tau}{2}\right).$

19.8 The rectangular pulse shown can be expressed as the difference between two step voltages; that is,

$$v(t) = V_m u\left(t + \dfrac{\tau}{2}\right) - V_m u\left(t - \dfrac{\tau}{2}\right) \text{ V}.$$

Use the operational transform for translation in the time domain to find the Fourier transform of $v(t)$.

ANSWER: $V(\omega) = V_m \tau \dfrac{\sin(\omega\tau/2)}{(\omega\tau/2)}.$

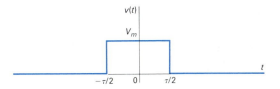

19.7 CIRCUIT APPLICATIONS

The Laplace transform is used more widely to find the response of a circuit than is the Fourier transform for two reasons. First, the Laplace transform integral converges for a wider range of driving functions, and second, it accommodates initial conditions. Despite the advantages of the Laplace transform in solving this type of problem, we can use the Fourier transform to find the response. The fundamental relationship underlying the use of the Fourier transform in transient analysis is Eq. (19.55), which relates the transform of the response $Y(\omega)$ to the transform of the input $X(\omega)$ and the transfer function $H(\omega)$ of the circuit. Note that $H(\omega)$ is the familiar $H(s)$ with s replaced by $j\omega$.

Example 19.1 illustrates how to use the Fourier transform to find the response of a circuit.

EXAMPLE 19.1

Use the Fourier transform to find $i_o(t)$ in the circuit shown in Fig. 19.9. The current source $i_g(t)$ is the signum function 20 sgn t A.

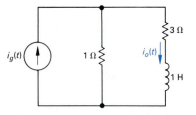

FIGURE 19.9 The circuit for Example 19.1.

SOLUTION

The Fourier transform of the driving source is

$$I_g(\omega) = \mathcal{F}\{20 \text{ sgn } t\} = 20\left(\frac{2}{j\omega}\right) = \frac{40}{j\omega}.$$

The transfer function of the circuit is the ratio of $\mathbf{I}_o$ to $\mathbf{I}_g$; so

$$H(\omega) = \frac{\mathbf{I}_o}{\mathbf{I}_g} = \frac{1}{4 + j\omega}.$$

The Fourier transform of $i_o(t)$ is

$$I_o(\omega) = I_g(\omega)H(\omega) = \frac{40}{j\omega(4 + j\omega)}.$$

Expanding $I_o(\omega)$ into a sum of partial fractions yields

$$I_o(\omega) = \frac{K_1}{j\omega} + \frac{K_2}{4 + j\omega}.$$

Evaluating K_1 and K_2 gives

$$K_1 = \frac{40}{4} = 10 \quad \text{and} \quad K_2 = \frac{40}{-4} = -10.$$

Therefore

$$I_o(\omega) = \frac{10}{j\omega} - \frac{10}{4 + j\omega}.$$

The response is

$$i_o(t) = \mathcal{F}^{-1}[I_o(\omega)] = 5 \text{ sgn } t - 10e^{-4t}u(t).$$

Figure 19.10 shows the response. Does the solution make sense in terms of known circuit behavior? The answer is *yes*, for the following reasons. The current source delivers −20 A to the circuit between −∞ and 0 seconds. The resistance in each branch governs how the −20 A divides between the two branches. In particular, one-fourth of the −20 A appears in the i_o branch; therefore i_o is −5 for $t < 0$. When the current source jumps from −20 A to +20 A at $t = 0$, i_o approaches its final value of +5 A exponentially with a time constant of $\frac{1}{4}$ s.

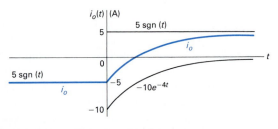

FIGURE 19.10 The plot of $i_o(t)$ versus t.

An important characteristic of the Fourier transform is that it directly yields the steady-state response to a sinusoidal driving function. The reason is that the Fourier transform of cos $\omega_0 t$ is based on the assumption that the function exists over all time. Example 19.2 illustrates this feature.

E X A M P L E   19.2

The current source in the circuit in Example 19.1 (Fig. 19.9) is changed to a sinusoidal source. The expression for the current is

$$i_g(t) = 50 \cos 3t \text{ A}.$$

Use the Fourier transform method to find $i_o(t)$.

S O L U T I O N

The transform of the driving function is

$$I_g(\omega) = 50\pi [\delta(\omega - 3) + \delta(\omega + 3)].$$

As before, the transfer function of the circuit is

$$H(\omega) = \frac{1}{4 + j\omega}.$$

The transform of the current response then is

$$I_o(\omega) = 50\pi \frac{\delta(\omega - 3) + \delta(\omega + 3)}{4 + j\omega}.$$

Because of the sifting property of the impulse function, the easiest way to find the inverse transform of $I_o(\omega)$ is by the inversion integral:

$$i_o(t) = \mathcal{F}^{-1}\{I_o(\omega)\} = \frac{50\pi}{2\pi} \int_{-\infty}^{\infty} \left[\frac{\delta(\omega - 3) + \delta(\omega + 3)}{4 + j\omega} \right] e^{j\omega t} \, d\omega$$

$$= 25\left(\frac{e^{j3t}}{4 + j3} + \frac{e^{-j3t}}{4 - j3} \right)$$

$$= 25\left(\frac{e^{j3t} e^{-j36.87°}}{5} + \frac{e^{-j3t} e^{j36.87°}}{5} \right)$$

$$= 5[2 \cos(3t - 36.87°)]$$

$$= 10 \cos(3t - 36.87°).$$

We leave to you verification that the solution for $i_o(t)$ is identical to that obtained by phasor analysis.

DRILL EXERCISES

19.9 The current source in the circuit shown delivers a current of 10 sgn t A. The response is the voltage across the 1-H inductor. Compute (a) $I_g(\omega)$; (b) $H(j\omega)$; (c) $V_o(\omega)$; (d) $v_o(t)$; (e) $i_1(0^-)$; (f) $i_1(0^+)$; (g) $i_2(0^-)$; (h) $i_2(0^+)$; (i) $v_o(0^-)$; and (j) $v_o(0^+)$.

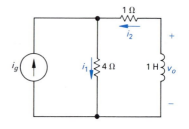

ANSWER: (a) $20/j\omega$; (b) $4j\omega/(5 + j\omega)$; (c) $80/(5 + j\omega)$; (d) $80e^{-5t}u(t)$ V; (e) -2 A; (f) 18 A; (g) 8 A; (h) 8 A; (i) 0 V; (j) 80 V.

19.10 The voltage source in the circuit shown is generating the voltage

$$v_g = e^t u(-t) + u(t) \text{ V}.$$

a) Use the Fourier transform method to find v_a.

b) Compute $v_a(0^-)$, $v_a(0^+)$, and $v_a(\infty)$.

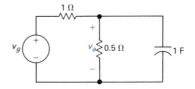

ANSWER: (a) $v_a = \frac{1}{4}e^t u(-t) - \frac{1}{12}e^{-3t}u(t) + \frac{1}{6} + \frac{1}{6}$ sgn t V; (b) $\frac{1}{4}$ V, $\frac{1}{4}$ V, $\frac{1}{3}$ V.

19.8 PARSEVAL'S THEOREM

Parseval's theorem relates the energy associated with a time-domain function of finite energy content to the Fourier transform of the function. Imagine that the time-domain function $f(t)$ is either the voltage across or the current in a 1-Ω resistor. The energy associated with this function then is

$$W_{1\Omega} = \int_{-\infty}^{\infty} f^2(t)\ dt. \qquad (19.57)$$

Parseval's theorem holds that this same energy can be calculated by an integration in the frequency domain, or specifically,

$$\int_{-\infty}^{\infty} f^2(t)\ dt = \frac{1}{2\pi} \int_{-\infty}^{\infty} |F(\omega)|^2\ d\omega. \qquad (19.58)$$

Therefore the 1-Ω energy associated with $f(t)$ can be calculated either by integrating the square of $f(t)$ over all time or by integrating $1/2\pi$ times the square of the Fourier transform of $f(t)$

over all frequency. Parseval's theorem is valid if both integrals exist.

The average power associated with time-domain signals of finite energy is zero when averaged over all time. Therefore, when comparing signals of this type, we resort to the energy content of the signal. Using a 1-Ω resistor as the base for the energy calculation is convenient for comparing the energy content of voltage and current signals.

We begin the derivation of Eq. (19.58) by rewriting the kernel of the integral on the left-hand side as $f(t)$ times itself and then expressing one $f(t)$ in terms of the inversion integral:

$$\int_{-\infty}^{\infty} f^2(t) \, dt = \int_{-\infty}^{\infty} f(t)f(t) \, dt$$

$$= \int_{-\infty}^{\infty} f(t)\left[\frac{1}{2\pi}\int_{-\infty}^{\infty} F(\omega)e^{j\omega t} \, d\omega\right] dt. \qquad \textbf{(19.59)}$$

We move $f(t)$ inside the interior integral because the integration is with respect to ω and then factor the constant $1/2\pi$ outside both integrations. Thus Eq. (19.59) becomes

$$\int_{-\infty}^{\infty} f^2(t) \, dt = \frac{1}{2\pi}\int_{-\infty}^{\infty}\left[\int_{-\infty}^{\infty} F(\omega)f(t)e^{j\omega t} \, d\omega\right] dt. \qquad \textbf{(19.60)}$$

We reverse the order of integration and in so doing recognize that $F(\omega)$ can be factored out of the integration with respect to t. Thus

$$\int_{-\infty}^{\infty} f^2(t) \, dt = \frac{1}{2\pi}\int_{-\infty}^{\infty} F(\omega)\left[\int_{-\infty}^{\infty} f(t)e^{j\omega t} \, dt\right] d\omega. \qquad \textbf{(19.61)}$$

The interior integral is $F(-\omega)$, so Eq. (19.61) reduces to

$$\int_{-\infty}^{\infty} f^2(t) \, dt = \frac{1}{2\pi}\int_{-\infty}^{\infty} F(\omega)F(-\omega) \, d\omega. \qquad \textbf{(19.62)}$$

In Section 19.6, we noted that $F(-\omega) = F^*(\omega)$. Thus the product $F(\omega)F(-\omega)$ is simply the magnitude of $F(\omega)$ squared, and Eq. (19.62) is equivalent to Eq. (19.58). We also noted that $|F(\omega)|$ is an even function of ω. Therefore we can also write Eq. (19.58) as

$$\int_{-\infty}^{\infty} f^2(t) \, dt = \frac{1}{\pi}\int_{0}^{\infty} |F(\omega)|^2 \, d\omega. \qquad \textbf{(19.63)}$$

EXAMPLE ILLUSTRATING PARSEVAL'S THEOREM

We can best demonstrate the validity of Eq. (19.63) with a specific example. If

$$f(t) = e^{-a|t|},$$

the left-hand side of Eq. (19.63) becomes

$$\int_{-\infty}^{\infty} e^{-2a|t|} \, dt = \int_{-\infty}^{0} e^{2at} \, dt + \int_{0}^{\infty} e^{-2at} \, dt$$

$$= \frac{e^{2at}}{2a}\bigg|_{-\infty}^{0} + \frac{e^{-2at}}{-2a}\bigg|_{0}^{\infty}$$

$$= \frac{1}{2a} + \frac{1}{2a} = \frac{1}{a}. \qquad (19.64)$$

The Fourier transform of $f(t)$ is

$$F(\omega) = \frac{2a}{a^2 + \omega^2},$$

and therefore the right-hand side of Eq. (19.63) becomes

$$\frac{1}{\pi}\int_{0}^{\infty} \frac{4a^2}{(a^2 + \omega^2)^2} \, d\omega = \frac{4a^2}{\pi}\frac{1}{2a^2}\left(\frac{\omega}{\omega^2 + a^2} + \frac{1}{a}\tan^{-1}\frac{\omega}{a}\right)\bigg|_{0}^{\infty}$$

$$= \frac{2}{\pi}\left(0 + \frac{\pi}{2a} - 0 - 0\right)$$

$$= \frac{1}{a}. \qquad (19.65)$$

Note that the result given by Eq. (19.65) is the same as that given by Eq. (19.64).

INTERPRETATION OF PARSEVAL'S THEOREM

Parseval's theorem gives a physical interpretation that the magnitude of the Fourier transform squared $[|F(\omega)|^2]$ is an energy density (in joules per hertz). To see it we write the right-hand side of Eq. (19.63) as

$$\frac{1}{\pi}\int_{0}^{\infty} |F(2\pi f)|^2 2\pi \, df = 2\int_{0}^{\infty} |F(2\pi f)|^2 \, df, \quad (19.66)$$

where $|F(2\pi f)|^2 \, df$ is the energy in an infinitesimal band of frequencies (df) and the total 1-Ω energy associated with $f(t)$ is the summation (integration) of $|F(2\pi f)|^2 \, df$ over all frequency. We can associate a portion of the total energy with a specified band of frequencies. That is, the 1-Ω energy in the frequency band from ω_1 to ω_2 is

$$W_{1\Omega} = \frac{1}{\pi}\int_{\omega_1}^{\omega_2} |F(\omega)|^2 \, d\omega. \qquad (19.67)$$

Note that expressing the integration in the frequency domain as

$$\frac{1}{2\pi} \int_{-\infty}^{\infty} |F(\omega)|^2 \, d\omega$$

instead of

$$\frac{1}{\pi} \int_{0}^{\infty} |F(\omega)|^2 \, d\omega,$$

allows Eq. (19.67) to be written in the form:

$$W_{1\Omega} = \frac{1}{2\pi} \int_{-\omega_2}^{-\omega_1} |F(\omega)|^2 \, d\omega + \frac{1}{2\pi} \int_{\omega_1}^{\omega_2} |F(\omega)|^2 \, d\omega. \qquad \textbf{(19.68)}$$

Figure 19.11 shows the graphic interpretation of Eq. (19.68).

Examples 19.3, 19.4, and 19.5 illustrate calculations involving Parseval's theorem.

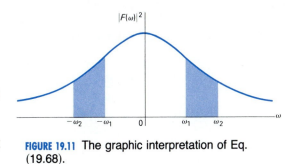

FIGURE 19.11 The graphic interpretation of Eq. (19.68).

E X A M P L E   19.3

The current in a 40-Ω resistor is

$$i = 20e^{-2t}u(t) \text{ A.}$$

What percentage of the total energy dissipated in the resistor can be associated with the frequency band $0 \le \omega \le 2\sqrt{3}$ rad/s?

S O L U T I O N

The total energy dissipated in the 40-Ω resistor is

$$W_{40\Omega} = 40 \int_{0}^{\infty} 400e^{-4t} \, dt$$

$$= 16,000 \left. \frac{e^{-4t}}{-4} \right|_{0}^{\infty}$$

$$= 4000 \text{ J.}$$

We can check this total energy calculation with Parseval's theorem:

$$F(\omega) = \frac{20}{2 + j\omega}.$$

Therefore

$$|F(\omega)| = \frac{20}{\sqrt{4 + \omega^2}}$$

and

$$
W_{40\Omega} = \frac{40}{\pi} \int_0^{\infty} \frac{400 \, d\omega}{4 + \omega^2} = \frac{16,000}{\pi} \left(\frac{1}{2} \tan^{-1} \frac{\omega}{2} \Big|_0^{\infty} \right)
$$

$$
= \frac{8000}{\pi} \left(\frac{\pi}{2} \right) = 4000 \text{ J.}
$$

The energy associated with the frequency band $0 \leq \omega \leq 2\sqrt{3}$ rad/s is

$$
W_{40\Omega} = \frac{40}{\pi} \int_0^{2\sqrt{3}} \frac{400 \, d\omega}{4 + \omega^2} = \frac{16,000}{\pi} \left(\frac{1}{2} \tan^{-1} \frac{\omega}{2} \Big|_0^{2\sqrt{3}} \right)
$$

$$
= \frac{8000}{\pi} \frac{\pi}{3} = \frac{8000}{3} \text{ J.}
$$

Hence the percentage of the total energy associated with this range of frequencies is

$$
\eta = \frac{8000/3}{4000} \times 100 = 66.67\%.
$$

E X A M P L E   19.4

The input voltage to an ideal bandpass filter is

$$
v(t) = 120e^{-24t} u(t) \text{ V.}
$$

The ideal filter passes all frequencies that lie between 24 and 48 rad/s, without attenuation, and completely rejects all frequencies outside this passband.

a) Sketch $|V(\omega)|^2$ for the filter input voltage.
b) Sketch $|V_o(\omega)|^2$ for the filter output voltage.
c) What percentage of the total 1-Ω energy content of the signal at the input of the filter is available in the signal at the output of the filter?

S O L U T I O N

a) The Fourier transform of the filter input voltage is

$$
V(\omega) = \frac{120}{24 + j\omega}.
$$

Therefore

$$|V(\omega)|^2 = \frac{14,400}{576 + \omega^2}.$$

Figure 19.12 shows the sketch of $|V(\omega)|^2$ versus ω.

b) The ideal bandpass filter rejects all frequencies outside the passband, so the plot of $|V_o(\omega)|^2$ versus ω appears as shown in Fig. 19.13.

c) The total 1-Ω energy available at the input to the filter is

$$W_i = \frac{1}{\pi} \int_0^\infty \frac{14,400}{576 + \omega^2} d\omega = \frac{14,400}{\pi} \left(\frac{1}{24} \tan^{-1} \frac{\omega}{24} \bigg|_0^\infty \right)$$

$$= \frac{600\,\pi}{\pi\,2} = 300 \text{ J}.$$

The total 1-Ω energy available at the output of the filter is

$$W_0 = \frac{1}{\pi} \int_{24}^{48} \frac{14,400}{576 + \omega^2} d\omega = \frac{600}{\pi} \tan^{-1} \frac{\omega}{24} \bigg|_{24}^{48}$$

$$= \frac{600}{\pi}(\tan^{-1} 2 - \tan^{-1} 1) = \frac{600}{\pi}\left(\frac{\pi}{2.84} - \frac{\pi}{4}\right)$$

$$= 61.45 \text{ J}.$$

The percentage of the input energy available at the output is

$$\eta = \frac{61.45}{300} \times 100 = 20.48\%.$$

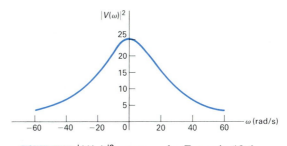

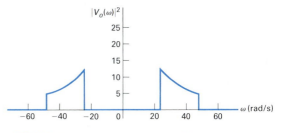

FIGURE 19.13 $|V_o(\omega)|^2$ versus ω for Example 19.4.

Parseval's theorem makes possible calculation of the energy available at the output of the filter without our knowing the time-domain expression for $v_o(t)$.

E X A M P L E   19.5

The input voltage to the low-pass RC filter circuit shown in Fig. 19.14 is

$$v_i(t) = 15e^{-5t}u(t) \text{ V}.$$

a) What percentage of the 1-Ω energy available in the input signal is available in the output signal?

b) What percentage of the output energy is associated with the frequency range $0 \le \omega \le 10$ rad/s?

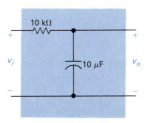

FIGURE 19.14 The low-pass RC filter for Example 19.5.

S O L U T I O N

a) The 1-Ω energy in the input signal to the filter is

$$W_i = \int_0^\infty (15e^{-5t})^2 \, dt = 225 \frac{e^{-10t}}{-10}\bigg|_0^\infty = 22.5 \text{ J}.$$

The Fourier transform of the output voltage is

$$V_o(\omega) = V_i(\omega)H(\omega),$$

where

$$V_i(\omega) = \frac{15}{5 + j\omega}$$

and

$$H(\omega) = \frac{1/RC}{1/RC + j\omega} = \frac{10}{10 + j\omega}.$$

Hence

$$V_o(\omega) = \frac{150}{(5 + j\omega)(10 + j\omega)}$$

and

$$|V_o(\omega)|^2 = \frac{22{,}500}{(25 + \omega^2)(100 + \omega^2)}.$$

The 1-Ω energy available in the output signal of the filter is

$$W_o = \frac{1}{\pi} \int_0^\infty \frac{22{,}500 \, d\omega}{(25 + \omega^2)(100 + \omega^2)}.$$

We can easily evaluate the integral by expanding the kernel into a sum of partial fractions:

$$\frac{22{,}500}{(25 + \omega^2)(100 + \omega^2)} = \frac{300}{25 + \omega^2} - \frac{300}{100 + \omega^2}.$$

Then,

$$W_o = \frac{300}{\pi}\left\{\int_0^\infty \frac{d\omega}{25 + \omega^2} - \int_0^\infty \frac{d\omega}{100 + \omega^2}\right\}$$

$$= \frac{300}{\pi}\left[\frac{1}{5}\left(\frac{\pi}{2}\right) - \frac{1}{10}\left(\frac{\pi}{2}\right)\right] = 15 \text{ J}.$$

The energy available in the output signal therefore is 66.67 percent of the energy available in the input signal; that is,

$$\eta = \frac{15}{22.5}(100) = 66.67\%.$$

b) The output energy associated with the frequency range $0 \leq \omega \leq 10$ rad/s is

$$W_0' = \frac{300}{\pi} \left(\int_0^{10} \frac{d\omega}{25 + \omega^2} - \int_0^{10} \frac{d\omega}{100 + \omega^2} \right)$$

$$= \frac{300}{\pi} \left(\frac{1}{5} \tan^{-1} \frac{10}{5} - \frac{1}{10} \tan^{-1} \frac{10}{10} \right) = \frac{30}{\pi} \left(\frac{2\pi}{2.84} - \frac{\pi}{4} \right)$$

$$= 13.64 \text{ J}.$$

The total 1-Ω energy in the output signal is 15 J, so the percentage associated with the frequency range 0 to 10 rad/s is 90.97 percent.

ENERGY CONTAINED IN A RECTANGULAR VOLTAGE PULSE

We conclude the discussion of Parseval's theorem by calculating the energy associated with the rectangular voltage pulse. In Section 19.1 we found the Fourier transform of the voltage pulse to be

$$V(\omega) = V_m \tau \frac{\sin \omega\tau/2}{\omega\tau/2}. \qquad (19.69)$$

To aid the discussion, we have redrawn the voltage pulse and its Fourier transform in Fig. 19.15(a) and (b), respectively. It shows that, as the width of the voltage pulse (τ) becomes smaller, the dominant portion of the amplitude spectrum (that is, the spectrum from $-2\pi/\tau$ to $2\pi/\tau$) spreads out over a wider range of frequencies. This result agrees with our earlier comments regarding the operational transform involving a scale change—in other words, when time is compressed, frequency is stretched out and vice versa. In order to transmit a single rectangular pulse with reasonable fidelity, the bandwidth of the system must be at least wide enough to accommodate the dominant portion of the amplitude spectrum. Thus the cutoff frequency should be at least $2\pi/\tau$ rad/s, or $1/\tau$ Hz.

We can use Parseval's theorem to calculate the fraction of the total energy associated with $v(t)$ that lies in the frequency range $0 \leq \omega \leq 2\pi/\tau$. From Eq. (19.69),

$$W = \frac{1}{\pi} \int_0^{2\pi/\tau} V_m^2 \tau^2 \frac{\sin^2 \omega\tau/2}{(\omega\tau/2)^2} d\omega. \qquad (19.70)$$

To carry out the integration called for in Eq. (19.70), we let

$$x = \frac{\omega\tau}{2}, \qquad (19.71)$$

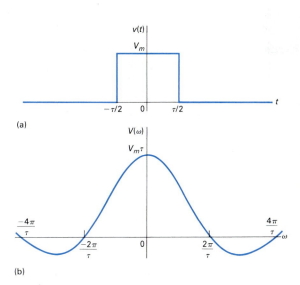

FIGURE 19.15 The rectangular voltage pulse and its Fourier transform: (a) the rectangular voltage pulse; (b) the Fourier transform of $v(t)$.

noting that

$$dx = \frac{\tau}{2} d\omega \qquad (19.72)$$

and

$$x = \pi, \quad \text{when } \omega = 2\pi/\tau. \qquad (19.73)$$

If we use the results given by Eqs. (19.71)–(19.73), Eq. (19.70) becomes

$$W = \frac{2V_m^2 \tau}{\pi} \int_0^\pi \frac{\sin^2 x}{x^2} dx. \qquad (19.74)$$

We can integrate the integral in Eq. (19.74) by parts. If we let

$$u = \sin^2 x \qquad (19.75)$$

and

$$dv = \frac{dx}{x^2}, \qquad (19.76)$$

then

$$du = 2 \sin x \cos x \, dx = \sin 2x \, dx \qquad (19.77)$$

and

$$v = -\frac{1}{x}. \qquad (19.78)$$

Hence

$$\int_0^\pi \frac{\sin^2 x}{x^2} dx = -\frac{\sin^2 x}{x} \Big|_0^\pi - \int_0^\pi -\frac{1}{x} \sin 2x \, dx$$

$$= 0 + \int_0^\pi \frac{\sin 2x \, dx}{x}. \qquad (19.79)$$

Substituting Eq. (19.79) into Eq. (19.74) yields

$$W = \frac{4V_m^2 \tau}{\pi} \int_0^\pi \frac{\sin 2x}{2x} dx. \qquad (19.80)$$

To evaluate the integral in Eq. (19.80), we must first put it in the form of $\sin y/y$. We do so by letting $y = 2x$ and noting that $dy = 2 \, dx$ and $y = 2\pi$ when $x = \pi$. Thus Eq. (19.80) becomes

$$W = \frac{2V_m^2 \tau}{\pi} \int_0^{2\pi} \frac{\sin y}{y} dy. \qquad (19.81)$$

The value of the integral in Eq. (19.81) can be found in a table of sine integrals.[†] Its value is 1.41815, so

$$W = \frac{2V_m^2 \tau}{\pi}(1.41815). \qquad \textbf{(19.82)}$$

The total 1-Ω energy associated with $v(t)$ can be calculated either from the time-domain integration or the evaluation of Eq. (19.81) with the upper limit equal to infinity. In either case, the total energy is

$$W_t = V_m^2 \tau. \qquad \textbf{(19.83)}$$

The fraction of the total energy associated with the band of frequencies between 0 and $2\pi/\tau$ is

$$\begin{aligned}
\eta &= \frac{W}{W_t} \\
&= \frac{2V_m^2 \tau (1.41815)}{\pi (V_m^2 \tau)} \\
&= 0.9028. \qquad \textbf{(19.84)}
\end{aligned}$$

Therefore approximately 90 percent of the energy associated with $v(t)$ is contained in the dominant portion of the amplitude spectrum.

DRILL EXERCISES

19.11 The voltage across a 50-Ω resistor is

$$v = 4te^{-t}u(t) \text{ V}.$$

What percentage of the total energy dissipated in the resistor can be associated with the frequency band $0 \le \omega \le \sqrt{3}$ rad/s?

ANSWER: 94.23%.

19.12 Assume that the magnitude of the Fourier transform of $v(t)$ is as shown. This voltage is applied to a 6-kΩ resistor. Calculate the total energy delivered to the resistor.

ANSWER: 4 J.

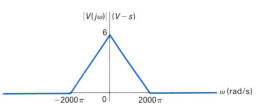

[†]M. Abramowitz and I. Stegun, *Handbook of Mathematical Functions*, p. 244 (New York: Dover, 1965).

SUMMARY

The Fourier transform is of interest because it gives a frequency-domain description of an aperiodic time-domain function. The ability to transform aperiodic time-domain functions into the frequency domain is particularly useful in the analysis and design of communication circuits. The following is a summary of the important characteristics of the Fourier transform as it applies to circuit analysis.

- Depending on the nature of the time-domain signal, one of three approaches to finding its Fourier transform may be utilized.

  1. If the time-domain signal is a well-behaved pulse of finite duration, the integral that defines the Fourier transform is used.

  2. If the one-sided Laplace transform of $f(t)$ exists and all the poles of $F(s)$ lie in the left half of the s plane, $F(s)$ may be used to find $F(\omega)$.

  3. If $f(t)$ is a constant, a signum function, a step function, or a sinusoidal function, the Fourier transform is found by using a limit process.

- Functional and operational Fourier transforms that are useful in circuit analysis are tabulated in Tables 19.1 and 19.2.

- The Fourier transform of a response signal $y(t)$ is

$$Y(\omega) = X(\omega)H(\omega),$$

  where $X(\omega)$ is the Fourier transform of the input signal $x(t)$ and $H(\omega)$ is the transfer function $H(s)$ evaluated at $s = j\omega$.

- The Fourier transform accommodates both negative-time functions and positive-time functions and therefore is suited to problems that are described in terms of events that start at $t = -\infty$. However, the unilateral Laplace transform is suited to problems that are described in terms of initial conditions and events that occur for $t > 0$.

- The magnitude of the Fourier transform squared is a measure of the energy density (joules per hertz) in the frequency domain (Parseval's theorem). Thus the Fourier transform permits associating a fraction of the total energy contained in $f(t)$ with a specified band of frequencies.

PROBLEMS

19.1 Use the defining integral to find the Fourier transform of the following functions:

a) $f(t) = A \sin \frac{\pi}{2} t,$ $\qquad -2 \le t \le 2,$

$\quad f(t) = 0,$ $\qquad\qquad$ elsewhere;

b) $f(t) = \frac{2A}{\tau} t + A,$ $\qquad -\frac{\tau}{2} \le t \le 0,$

$\quad f(t) = -\frac{2A}{\tau} t + A,$ $\qquad 0 \le t \le \tau/2,$

$\quad f(t) = 0,$ $\qquad\qquad$ elsewhere.

19.2 Find the Fourier transform of each of the following functions. In all of the functions, a is a positive real constant and $-\infty \le t \le \infty$.

a) $f(t) = |t| e^{-a|t|};$

b) $f(t) = t^3 e^{-a|t|};$

c) $f(t) = e^{-a|t|} \cos \omega_0 t;$

d) $f(t) = e^{-a|t|} \sin \omega_0 t;$

e) $f(t) = \delta(t - t_0).$

19.3 Use the inversion integral (Eq. 19.9) to show that $\mathcal{F}^{-1}\{2/j\omega\} = \text{sgn } t.$ (*Hint:* Use Drill Exercise 19.6.)

19.4 Find $\mathcal{F}\{\cos \omega_0 t\}$ by using the approximating function

$$f(t) = e^{-\epsilon|t|} \cos \omega_0 t,$$

where ϵ is a positive real constant.

19.5 Show that if $f(t)$ is an even function,

$$A(\omega) = 2 \int_0^\infty f(t) \cos \omega t \, dt$$

and

$$B(\omega) = 0.$$

19.6 Show that if $f(t)$ is an odd function,

$$A(\omega) = 0$$

and

$$B(\omega) = -2 \int_0^\infty f(t) \sin \omega t \, dt.$$

19.7 a) Show that $\mathcal{F}\{df(t)/dt\} = j\omega F(\omega)$, where $F(\omega) = \mathcal{F}\{f(t)\}$. (*Hint:* Use the defining integral and integrate by parts.)

b) What is the restriction on $f(t)$ if the result given in part (a) is valid?

c) Show that $\mathcal{F}\{d^n f(t)/dt^n\} = (j\omega)^n F(\omega)$, where $F(\omega) = \mathcal{F}\{f(t)\}$.

19.8 a) Show that

$$\mathcal{F}\left\{\int_{-\infty}^t f(x)\, dx\right\} = \frac{F(\omega)}{j\omega},$$

where $F(\omega) = \mathcal{F}\{f(x)\}$. (*Hint:* Use the defining integral and integrate by parts.)

b) What is the restriction on $f(x)$ if the result given in part (a) is valid?

c) If $f(x) = e^{-ax} u(x)$, can the operational transform in part (a) be used? Explain.

19.9 a) Show that

$$\mathscr{F}\{f(at)\} = \frac{1}{a}F\left(\frac{\omega}{a}\right), \qquad a > 0.$$

b) Given $f(at) = e^{-a|t|}$ for $a > 0$, sketch $F(\omega) = \mathscr{F}\{f(at)\}$ for $a = 0.5$, 1.0, and 2.0. Do your sketches reflect the observation that "compression" in the time domain corresponds to "stretching" in the frequency domain?

19.10 Derive each of the following operational transforms:

a) $\mathscr{F}\{f(t - a)\} = e^{-j\omega a}F(\omega)$;

b) $\mathscr{F}\{e^{j\omega_0 t}f(t)\} = F(\omega - \omega_0)$;

c) $\mathscr{F}\{f(t)\cos\omega_0 t\} = \frac{1}{2}F(\omega - \omega_0) + \frac{1}{2}F(\omega + \omega_0)$.

19.11 Given $y(t) = \int_{-\infty}^{\infty}x(\lambda)h(t - \lambda)\,d\lambda$, show that $Y(\omega) = \mathscr{F}\{y(t)\} = X(\omega)H(\omega)$, where $X(\omega) = \mathscr{F}\{x(t)\}$ and $H(\omega) = \mathscr{F}\{h(t)\}$. (*Hint:* Use the defining integral to write $\mathscr{F}\{y(t)\} = \int_{-\infty}^{\infty}[\int_{-\infty}^{\infty}x(\lambda)h(t - \lambda)d\lambda]e^{-j\omega t}\,dt$. Next, reverse the order of integration and then make a change in the variable of integration, that is, let $u = t - \lambda$.)

19.12 Given $f(t) = f_1(t)f_2(t)$, show that $F(\omega) = (1/2\pi)\int_{-\infty}^{\infty}F_1(u)F_2(\omega - u)\,du$. [*Hint:* First, use the defining integral to express $F(\omega)$ as

$$F(\omega) = \int_{-\infty}^{\infty}f_1(t)f_2(t)e^{-j\omega t}\,dt.$$

Second, use the inversion integral to write

$$f_1(t) = \frac{1}{2\pi}\int_{-\infty}^{\infty}F_1(u)e^{j\omega t}\,du.$$

Third, substitute the expression for $f_1(t)$ into the defining integral and then interchange the order of integration.]

19.13 It is given that $f(t) = f_1(t)f_2(t)$, where

$$f_1(t) = \cos\omega_0 t \qquad \text{and}$$

$$f_2(t) = 1, \qquad -\tau/2 < t < \tau/2,$$

$$f_2(t) = 0, \qquad \text{elsewhere.}$$

a) Use convolution in the frequency domain to find $F(\omega)$.

b) What happens to $F(\omega)$ as the width of $f_2(t)$ increases so that $f(t)$ includes more and more cycles of $f_1(t)$?

19.14 a) Show that

$$(j)^n\left[\frac{d^n F(\omega)}{d\omega^n}\right] = \mathscr{F}\{t^n f(t)\}.$$

b) Use the result of part (a) to find each of the following Fourier transforms:

(i) $\mathscr{F}\{te^{-at}u(t)\}$;

(ii) $\mathscr{F}\{|t|e^{-a|t|}\}$;

(iii) $\mathscr{F}\{te^{-a|t|}\}$.

19.15 a) Use the Fourier transform method to find $i_o(t)$ in the circuit shown in Fig. P19.15. The initial value of $i_o(t)$ is zero and the source voltage is $200u(t)$ V.

b) Sketch $i_o(t)$ versus t.

FIGURE P19.15

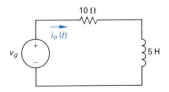

19.16 a) Use the Fourier transform method to find $i_o(t)$ in the circuit shown in Fig. P19.16 if $i_g = 200 \text{ sgn } (t) \; \mu\text{A}$.

b) Does your solution make sense in terms of known circuit behavior? Explain.

FIGURE P19.16

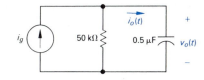

19.17 Repeat Problem 19.16 except replace $i_o(t)$ with $v_o(t)$.

19.18 The voltage source in the circuit in Fig. P19.18 is given by the expression

$$v_g = 30e^{-5|t|} \text{ V}.$$

a) Find $v_o(t)$.

b) What is the value of $v_o(0^-)$?

c) What is the value of $v_o(0^+)$?

d) Use the Laplace transform method to find $v_o(t)$ for $t > 0^+$.

e) Does the solution obtained in part (d) agree with $v_o(t)$ for $t > 0^+$ from part (a)?

FIGURE P19.18

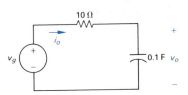

19.19 Repeat Problem 19.18 except replace $v_o(t)$ with $i_o(t)$.

19.20 Use the Fourier transform method to find v_o in the circuit in Fig. P19.20 if $v_g = 2.5 \; \delta(t)$ V. There is no energy stored in the circuit at the instant the signal voltage is applied.

FIGURE P19.20

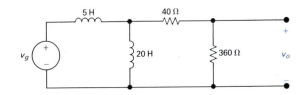

19.21 a) Repeat Problem 19.20 if v_g changes to

$$v_g = -75 \; e^{50t} \; u(-t) + 75 \; e^{-50t} \; u(t) \text{ V}.$$

b) Let i_1 be the current in the 20-H inductor. Assume the reference direction for i_1 is down. Find i_1.

19.22 a) Use the Fourier transform method to find v_o in the circuit seen in Fig. P19.22 when

$$i_g = 4 \text{ sgn } (t) \text{ A.}$$

b) Find $v_o(0^-)$ and $v_o(0^+)$.

c) Find $v_o(\infty)$.

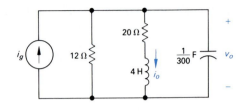

FIGURE P19.22

19.23 Repeat Problem 19.22 with v_o replaced by i_o.

19.24 a) Use the Fourier transform method to find i_o in the circuit shown in Fig. P19.24. The current source generates the current

$$i_g = 15 \text{ sgn } (t) \text{ mA.}$$

b) Calculate $i_o(0^-)$ and $i_o(0^+)$.

c) Do the results in part (b) make sense in terms of known circuit behavior? Explain.

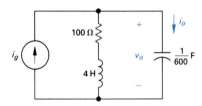

FIGURE P19.24

19.25 Repeat Problem 19.24 with i_o replaced by v_o.

19.26 The voltage source in the circuit in Fig. P19.26 is generating the signal

$$v_g = 5 \text{ sgn } (t) - 5 + 30e^{-5t} u(t) \text{ V.}$$

a) Find $v_o(0^-)$ and $v_o(0^+)$.

b) Find $i_o(0^-)$ and $i_o(0^+)$.

c) Find v_o.

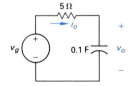

FIGURE P19.26

19.27 a) Use the Fourier transform method to find v_o in the circuit in Fig. P19.27 when

$$i_g = 18\, e^{10t}\, u(-t) - 18\, e^{-10t}\, u(t) \text{ A}.$$

  b) Find $v_o(0^-)$.

  c) Find $v_o(0^+)$.

  d) Do the answers obtained in parts (b) and (c) make sense in terms of known circuit behavior? Explain.

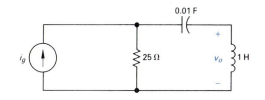

FIGURE P19.27

19.28 The unit voltage impulse response for the system shown in Fig. P19.28 is

$$h(t) = (50e^{-40t} - 25e^{-10t})u(t) \text{ V}.$$

The input voltage is

$$v_i = 100\, \text{sgn}\,(t) \text{ V}.$$

Use the Fourier transform to find v_o.

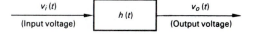

FIGURE P19.28

19.29 It is given that $F(\omega) = e^{\omega}u(-\omega) + e^{-\omega}u(\omega)$.

  a) Find $f(t)$.

  b) Find the 1-Ω energy associated with $f(t)$ via time-domain integration.

  c) Repeat part (b) using frequency-domain integration.

  d) Find the value of ω_1 if $f(t)$ has 90% of the energy in the frequency band $0 \leq \omega \leq \omega_1$.

19.30 The input current signal in the circuit seen in Fig. P19.30 is

$$i_g = 30e^{-2t}\ \mu\text{A} \quad t \geq 0^+$$

What percentage of the total 1-Ω energy content in the output current signal lies in the frequency range 0 to 4 rad/s?

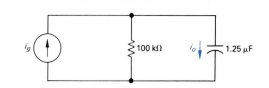

FIGURE P19.30

19.31 The input voltage in the circuit seen in Fig. P19.31 is

$$v_g = 30e^{-|t|} \text{ V.}$$

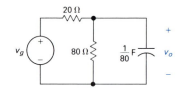

a) Calculate the 1-Ω energy content of the input signal.

b) Calculate the 1-Ω energy content of the output signal v_o.

FIGURE P19.31

19.32 a) Calculate the 1-Ω energy content of the output signal obtained in Problem 19.21(a) using the time domain expression for v_o.

b) Repeat part (a) using the frequency domain expression for $v_o(\omega)$.

19.33 The input voltage in the circuit shown in Fig. P19.33 is $v_i = 9.6e^{-400t} u(t)$ V. What percentage of the total energy content in the output voltage lies in the frequency range 0 to 400 rad/s?

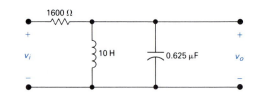

FIGURE P19.33

19.34 The circuit shown in Fig. P19.34 is driven by the voltage

$$v_1 = 60e^{-5t}u(t) \text{ V.}$$

What percentage of the total 1-Ω energy content in the output voltage v_2 lies in the frequency range $0 \le \omega \le 10$ rad/s?

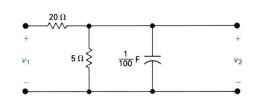

FIGURE P19.34

19.35 The amplitude spectrum of the input voltage to the high-pass RC filter in Fig. P19.35 is

$$V_i(\omega) = \frac{200}{|\omega|}, \qquad 100 \le |\omega| \le 200 \text{ rad/s,}$$

$$V_i(\omega) = 0, \qquad \text{elsewhere.}$$

a) Sketch $|V_i(\omega)|^2$ for $-300 \le \omega \le 300$ rad/s.

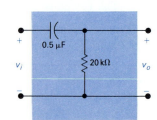

FIGURE P19.35

b) Sketch $|V_o(\omega)|^2$ for $-300 \le \omega \le 300$ rad/s.

c) Calculate the 1-Ω energy in the signal at the input of the filter.

d) Calculate the 1-Ω energy in the signal at the output of the filter.

19.36 The input voltage to the high-pass *RC* filter circuit in Fig. P19.36 is

$$v_i(t) = Ae^{-at}u(t).$$

Let α denote the corner frequency of the filter, that is, $\alpha = 1/RC$.

a) What percentage of the energy in the signal at the output of the filter is associated with the frequency band $0 \le \omega \le \alpha$ if $\alpha = a$?

b) Repeat part (a), given that $\alpha = \sqrt{3}\,a$.

c) Repeat part (a), given that $\alpha = a/\sqrt{3}$.

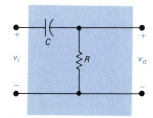

FIGURE P19.36

Two-Port Circuits

CHAPTER 20

So far, we frequently have focused on the behavior of a circuit at a specified pair of terminals. Recall that we introduced the Thévenin and Norton equivalent circuits solely for the purpose of simplifying the analysis of a circuit relative to a pair of terminals. In analyzing some electrical systems, focusing on two pairs of terminals also is convenient. In particular this point of view is convenient when a signal is fed into one pair of terminals and then, after being processed by the system, is extracted at a second pair of terminals. Because the terminal pairs represent the points in the system where signals are either fed in or extracted, they are referred to as the *ports* of the system. In this chapter, we limit the discussion to circuits that have one input and one output port.

Figure 20.1 illustrates the basic two-port building block. Use of this building block is subject to several restrictions. First, there can be no energy stored within the circuit. Second, there can be no independent sources within the circuit; dependent sources are permissible. Third, the current into the port must equal the current out of the port; that is, $i_1 = i_1'$ and $i_2 = i_2'$. Fourth, all external connections must be made to either the input port or the output port; no external connections between ports are allowed, that is, between terminals a and c, a and d, b and c, or b and d. These restrictions simply limit the range of circuit problems to which the two-port formulation is applicable.

The fundamental principle underlying two-port modeling of a system is that only the terminal variables, that is, i_1, v_1, i_2, and v_2, are of interest. We have no interest in calculating the currents and voltages inside the circuit. We have already stressed terminal behavior in the analysis of operational amplifier circuits. In this chapter we formalize that approach by introducing the two-port parameters.

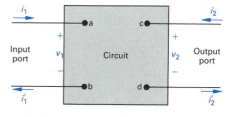

FIGURE 20.1 The two-port building block.

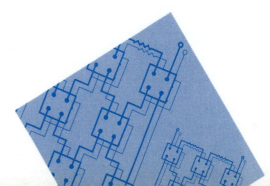

20.1 THE TERMINAL EQUATIONS

In viewing a circuit as a two-port network, we are interested in relating the current and voltage at one port to the current and voltage at the second port. Figure 20.1 shows the reference polarities of the terminal voltages and the reference directions of the terminal currents. The references at each port are symmetric with respect to each other; that is, at each port the current is directed into the upper terminal and each port voltage is a rise from the lower to the upper terminal. This symmetry makes generalizing the analysis of a network easier in terms of two-port blocks and is the reason for its universal use in the literature.

The most general description of the two-port network is carried out in the s domain. For purely resistive networks the analysis reduces to solving resistive circuits. Sinusoidal steady-state problems can be solved either by first finding the appropriate s-domain expressions and then replacing s by $j\omega$ or by direct analysis in the phasor domain. Here, we write all equations in the s domain; resistive networks and sinusoidal steady-state solutions become special cases. Figure 20.2 shows the basic building block in terms of the s-domain variables I_1, V_1, I_2, and V_2.

Of the four terminal variables I_1, V_1, I_2, and V_2, only two are independent. Thus for any circuit, once we specify two of the variables we can find the two remaining unknowns. For example, knowing V_1 and V_2 and the circuit within the box, we can determine I_1 and I_2. Thus we can describe a two-port network with just two simultaneous equations. However, there are six different ways in which to combine the four variables:

$$V_1 = z_{11}I_1 + z_{12}I_2,$$
$$V_2 = z_{21}I_1 + z_{22}I_2; \tag{20.1}$$

$$I_1 = y_{11}V_1 + y_{12}V_2,$$
$$I_2 = y_{21}V_1 + y_{22}V_2; \tag{20.2}$$

$$V_1 = a_{11}V_2 - a_{12}I_2,$$
$$I_1 = a_{21}V_2 - a_{22}I_2; \tag{20.3}$$

$$V_2 = b_{11}V_1 - b_{12}I_1,$$
$$I_2 = b_{21}V_1 - b_{22}I_1; \tag{20.4}$$

$$V_1 = h_{11}I_1 + h_{12}V_2,$$
$$I_2 = h_{21}I_1 + h_{22}V_2; \tag{20.5}$$

$$I_1 = g_{11}V_1 + g_{12}I_2,$$
$$V_2 = g_{21}V_1 + g_{22}I_2. \tag{20.6}$$

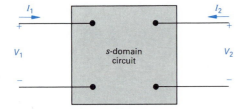

FIGURE 20.2 The s-domain two-port basic building block.

These six sets of equations may also be considered as three pairs of mutually inverse relations. The first set (Eqs. 20.1) gives the input and output voltages as functions of the input and output currents. The second set (Eqs. 20.2) gives the inverse relationship, that is, the input and output currents as functions of the input and output voltages. Equations (20.3) and (20.4) are inverse relations, as are Eqs. (20.5) and (20.6).

The coefficients of the current and/or voltage variables on the right-hand side of Eqs. (20.1)–(20.6) are called the *parameters* of the two-port circuit. Thus when using Eqs. (20.1), we refer to the z parameters of the circuit. Similarly, we refer to the y parameters, the a parameters, the b parameters, the h parameters, and the g parameters of the network.

20.2 THE TWO-PORT PARAMETERS

We can determine the parameters for any circuit by computation or measurement. The computation or measurement to be made come directly from the parameter equations. For example, suppose that the problem is to find the z parameters for a circuit. From Eqs. (20.1),

$$z_{11} = \left.\frac{V_1}{I_1}\right|_{I_2=0}, \tag{20.7}$$

$$z_{12} = \left.\frac{V_1}{I_2}\right|_{I_1=0}, \tag{20.8}$$

$$z_{21} = \left.\frac{V_2}{I_1}\right|_{I_2=0}, \tag{20.9}$$

and

$$z_{22} = \left.\frac{V_2}{I_2}\right|_{I_1=0}. \tag{20.10}$$

Equations (20.7)–(20.10) reveal that the four z parameters can be described as follows:

1. z_{11} is the impedance seen looking into port 1 when port 2 is open.

2. z_{12} is a transfer impedance. It is the ratio of the port-1 voltage to the port-2 current when port 1 is open.

3. z_{21} is a transfer impedance. It is the ratio of the port-2 voltage to the port-1 current when port 2 is open.

4. z_{22} is the impedance seen looking into port 2 when port 1 is open.

Therefore the impedance parameters may be either calculated or measured by first opening port 2 and determining the ratios V_1/I_1 and V_2/I_1 and then opening port 1 and determining the ratios V_1/I_2 and V_2/I_2. Example 20.1 illustrates determination of the z parameters for a resistive circuit.

E X A M P L E   20.1

Find the z parameters for the circuit shown in Fig. 20.3.

S O L U T I O N

The circuit is purely resistive, so the s-domain circuit is also purely resistive. With port 2 open, that is, $I_2 = 0$, the resistance seen looking into port 1 is the 20-Ω resistor in parallel with the series combination of the 5- and 15-Ω resistors. Therefore

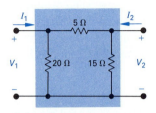

FIGURE 20.3 The circuit for Example 20.1.

$$z_{11} = \left.\frac{V_1}{I_1}\right|_{I_2=0} = \frac{(20)(20)}{40} = 10 \ \Omega.$$

When I_2 is zero, V_2 is

$$V_2 = \frac{V_1}{15 + 5}(15) = 0.75 \ V_1,$$

and therefore

$$z_{21} = \left.\frac{V_2}{I_1}\right|_{I_2=0} = \frac{0.75 \ V_1}{V_1/10} = 7.5 \ \Omega.$$

When I_1 is zero, the resistance seen looking into port 2 is the 15-Ω resistor in parallel with the series combination of the 5- and 20-Ω resistors. Therefore

$$z_{22} = \left.\frac{V_2}{I_2}\right|_{I_1=0} = \frac{(15)(25)}{40} = 9.375 \ \Omega.$$

When port 1 is open, I_1 is zero and the voltage V_1 is

$$V_1 = \frac{V_2}{5 + 20}(20) = 0.8 \ V_2.$$

With port 1 open, the current into port 2 is

$$I_2 = \frac{V_2}{9.375}.$$

Hence

$$z_{12} = \left.\frac{V_1}{I_2}\right|_{I_1=0} = \frac{0.8 \ V_2}{V_2/9.375} = 7.5 \ \Omega.$$

Equations (20.7)–(20.10) and Example 20.1 show why the parameters in Eqs. (20.1) are called the z parameters. Each parameter is the ratio of a voltage to a current and therefore is an impedance with the dimension of ohms.

We use the same process to determine the remaining port parameters, which are either calculated or measured. A port parameter is obtained by either opening or shorting a port. Moreover, a port parameter is an impedance, an admittance, or a dimensionless ratio. The dimensionless ratio is either the ratio of two voltages or the ratio of two currents. Equations (20.11)–(20.15) summarize these observations.

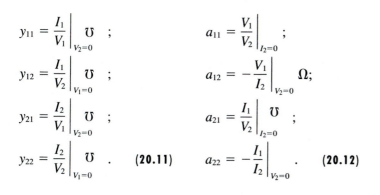

$$y_{11} = \left.\frac{I_1}{V_1}\right|_{V_2=0} \text{℧} \; ; \qquad a_{11} = \left.\frac{V_1}{V_2}\right|_{I_2=0} ;$$

$$y_{12} = \left.\frac{I_1}{V_2}\right|_{V_1=0} \text{℧} \; ; \qquad a_{12} = -\left.\frac{V_1}{I_2}\right|_{V_2=0} \Omega;$$

$$y_{21} = \left.\frac{I_2}{V_1}\right|_{V_2=0} \text{℧} \; ; \qquad a_{21} = \left.\frac{I_1}{V_2}\right|_{I_2=0} \text{℧} \; ;$$

$$y_{22} = \left.\frac{I_2}{V_2}\right|_{V_1=0} \text{℧} \; . \quad (20.11) \qquad a_{22} = -\left.\frac{I_1}{I_2}\right|_{V_2=0} . \quad (20.12)$$

$$b_{11} = \left.\frac{V_2}{V_1}\right|_{I_1=0} ; \qquad h_{11} = \left.\frac{V_1}{I_1}\right|_{V_2=0} \Omega;$$

$$b_{12} = -\left.\frac{V_2}{I_1}\right|_{V_1=0} \Omega; \qquad h_{12} = \left.\frac{V_1}{V_2}\right|_{I_1=0} ;$$

$$b_{21} = \left.\frac{I_2}{V_1}\right|_{I_1=0} \text{℧} \; ; \qquad h_{21} = \left.\frac{I_2}{I_1}\right|_{V_2=0} ;$$

$$b_{22} = -\left.\frac{I_2}{I_1}\right|_{V_1=0} . \quad (20.13) \qquad h_{22} = \left.\frac{I_2}{V_2}\right|_{I_1=0} \text{℧} \; . \quad (20.14)$$

$$g_{11} = \left.\frac{I_1}{V_1}\right|_{I_2=0} \text{℧} \; ;$$

$$g_{12} = \left.\frac{I_1}{I_2}\right|_{V_1=0} ;$$

$$g_{21} = \left.\frac{V_2}{V_1}\right|_{I_2=0} ;$$

$$g_{22} = \left.\frac{V_2}{I_2}\right|_{V_1=0} \Omega. \quad (20.15)$$

DRILL EXERCISES

20.1 Find the y parameters for the circuit in Fig. 20.3.

ANSWER: $y_{11} = 0.25\ \mho$; $y_{12} = y_{21} = -0.2\ \mho$; $y_{22} = \frac{4}{15}\ \mho$.

20.2 Find the a and b parameters for the circuit in Fig. 20.3.

ANSWER: $a_{11} = \frac{4}{3}$; $a_{12} = 5\ \Omega$; $a_{21} = \frac{2}{15}\ \mho$; $a_{22} = 1.25$; $b_{11} = 1.25$; $b_{12} = 5\ \Omega$; $b_{21} = \frac{2}{15}\ \mho$; $b_{22} = \frac{4}{3}$.

The two-port parameters are also described in relation to the reciprocal sets of equations. The impedance and admittance parameters are grouped into the immittance parameters. The term *immitance* denotes a quantity that is either an impedance or an admittance. The a and b parameters are called the *transmission* parameters because they describe the voltage and current at one end of the two-port in terms of the voltage and current at the other end. The immittance and transmission parameters are the natural choices for relating the port variables. That is, they relate either voltage variables to current variables or input variables to output variables. The h and g parameters relate cross-variables, that is, an input voltage and output current to an output voltage and input current. Therefore the h and g parameters are called *hybrid* parameters. Example 20.2 illustrates how a set of measurements made at the terminals of a two-port circuit can be used to calculate the a parameters.

EXAMPLE 20.2

The following measurements pertain to a two-port circuit operating in the sinusoidal steady-state. With port 2 open, a voltage equal to 150 cos 4000t V is applied to port 1. The current into port 1 is 25 cos (4000t − 45°) A and the port-2 voltage is 100 cos (4000t + 15°) V. With port 2 short-circuited, a voltage equal to 30 cos 4000t V is applied to port 1. The current into port 1 is 1.5 cos (4000t + 30°) A and the current into port 2 is 0.25 cos (4000t + 150°) A. Find the a parameters that can describe the sinusoidal steady-state behavior of the circuit.

SOLUTION

The first set of measurements give

$$\mathbf{V}_1 = 150\underline{/0°}\ \text{V}; \qquad \mathbf{I}_1 = 25\underline{/-45°}\ \text{A};$$

$$\mathbf{V}_2 = 100\underline{/15°}\ \text{V}; \qquad \mathbf{I}_2 = 0\ \text{A}.$$

From Eqs. (20.12),

$$a_{11} = \left.\frac{\mathbf{V}_1}{\mathbf{V}_2}\right|_{I_2=0} = \frac{150\underline{/0°}}{100\underline{/15°}} = 1.5\underline{/-15°}$$

and

$$a_{21} = \left.\frac{\mathbf{I}_1}{\mathbf{V}_2}\right|_{I_2=0} = \frac{25\underline{/-45°}}{100\underline{/15°}} = 0.25\underline{/-60°}\ \mho.$$

The second set of measurements give

$$\mathbf{V}_1 = 30\underline{/0°}\ \text{V}; \qquad \mathbf{I}_1 = 1.5\underline{/30°}\ \text{A};$$
$$\mathbf{V}_2 = 0; \qquad \mathbf{I}_2 = 0.25\underline{/150°}\ \text{A}.$$

Therefore

$$a_{12} = \left.-\frac{\mathbf{V}_1}{\mathbf{I}_2}\right|_{V_2=0} = \frac{-30\underline{/0°}}{0.25\underline{/150°}} = 120\underline{/30°}\ \Omega$$

and

$$a_{22} = \left.-\frac{\mathbf{I}_1}{\mathbf{I}_2}\right|_{V_2=0} = \frac{-1.5\underline{/30°}}{0.25\underline{/150°}} = 6\underline{/60°}.$$

DRILL EXERCISES

20.3 The following measurements were made on a two-port resistive circuit. With 10 mV applied to port 2 and port 1 open, the current into port 2 is 0.25 μA and the voltage across port 1 is 5 μV. With port 2 short-circuited and 50 mV applied to port 1, the current into port 1 is 50 μA and the current into port 2 is 2 mA. Find the h parameters of the network.

ANSWER: $h_{11} = 1000\ \Omega$; $h_{12} = 5 \times 10^{-4}$; $h_{21} = 40$; $h_{22} = 25\ \mu\mho$.

RELATIONSHIPS AMONG THE TWO-PORT PARAMETERS

Because the six sets of equations relate to the same variables, the parameters associated with any pair of equations must be related to the parameters of all the other pairs of equations. In other words, if we know one set of parameters, we can derive all the other sets of parameters from the known set. Because of the amount of algebra involved in deriving all the interrelationships, we merely list the results in Table 20.1.

Although we do not derive all the relationships listed in Table 20.1 here, we do derive the relationships between the z parame-

TABLE 20.1

PARAMETER CONVERSION TABLE

$$z_{11} = \frac{y_{22}}{\Delta y} = \frac{a_{11}}{a_{21}} = \frac{b_{22}}{b_{21}} = \frac{\Delta h}{h_{22}} = \frac{1}{g_{11}}$$

$$z_{12} = -\frac{y_{12}}{\Delta y} = \frac{\Delta a}{a_{21}} = \frac{1}{b_{21}} = \frac{h_{12}}{h_{22}} = -\frac{g_{12}}{g_{11}}$$

$$z_{21} = -\frac{y_{21}}{\Delta y} = \frac{1}{a_{21}} = \frac{\Delta b}{b_{21}} = -\frac{h_{21}}{h_{22}} = \frac{g_{21}}{g_{11}}$$

$$z_{22} = \frac{y_{11}}{\Delta y} = \frac{a_{22}}{a_{21}} = \frac{b_{11}}{b_{21}} = \frac{1}{h_{22}} = \frac{\Delta g}{g_{11}}$$

$$y_{11} = \frac{z_{22}}{\Delta z} = \frac{a_{22}}{a_{12}} = \frac{b_{11}}{b_{12}} = \frac{1}{h_{11}} = \frac{\Delta g}{g_{22}}$$

$$y_{12} = -\frac{z_{12}}{\Delta z} = -\frac{\Delta a}{a_{12}} = -\frac{1}{b_{12}} = -\frac{h_{12}}{h_{11}} = \frac{g_{12}}{g_{22}}$$

$$y_{21} = -\frac{z_{21}}{\Delta z} = -\frac{1}{a_{12}} = -\frac{\Delta b}{b_{12}} = \frac{h_{21}}{h_{11}} = -\frac{g_{21}}{g_{22}}$$

$$y_{22} = \frac{z_{11}}{\Delta z} = \frac{a_{11}}{a_{12}} = \frac{b_{22}}{b_{12}} = \frac{\Delta h}{h_{11}} = \frac{1}{g_{22}}$$

$$a_{11} = \frac{z_{11}}{z_{21}} = -\frac{y_{22}}{y_{21}} = \frac{b_{22}}{\Delta b} = -\frac{\Delta h}{h_{21}} = \frac{1}{g_{21}}$$

$$a_{12} = \frac{\Delta z}{z_{21}} = -\frac{1}{y_{21}} = \frac{b_{12}}{\Delta b} = \frac{-h_{11}}{h_{21}} = \frac{g_{22}}{g_{21}}$$

$$a_{21} = \frac{1}{z_{21}} = -\frac{\Delta y}{y_{21}} = \frac{b_{21}}{\Delta b} = -\frac{h_{22}}{h_{21}} = \frac{g_{11}}{g_{21}}$$

$$a_{22} = \frac{z_{22}}{z_{21}} = -\frac{y_{11}}{y_{21}} = \frac{b_{11}}{\Delta b} = -\frac{1}{h_{21}} = \frac{\Delta g}{g_{21}}$$

$$b_{11} = \frac{z_{22}}{z_{12}} = -\frac{y_{11}}{y_{12}} = \frac{a_{22}}{\Delta a} = \frac{1}{h_{12}} = -\frac{\Delta g}{g_{12}}$$

$$b_{12} = \frac{\Delta z}{z_{12}} = -\frac{1}{y_{12}} = \frac{a_{12}}{\Delta a} = \frac{h_{11}}{h_{12}} = -\frac{g_{22}}{g_{12}}$$

$$b_{21} = \frac{1}{z_{12}} = -\frac{\Delta y}{y_{12}} = \frac{a_{21}}{\Delta a} = \frac{h_{22}}{h_{12}} = -\frac{g_{11}}{g_{12}}$$

$$b_{22} = \frac{z_{11}}{z_{12}} = \frac{y_{22}}{y_{12}} = \frac{a_{11}}{\Delta a} = \frac{\Delta h}{h_{12}} = -\frac{1}{g_{12}}$$

$$h_{11} = \frac{\Delta z}{z_{22}} = \frac{1}{y_{11}} = \frac{a_{12}}{a_{22}} = \frac{b_{12}}{b_{11}} = \frac{g_{22}}{\Delta g}.$$

$$h_{12} = \frac{z_{12}}{z_{22}} = -\frac{y_{12}}{y_{11}} = \frac{\Delta a}{a_{22}} = \frac{1}{b_{11}} = -\frac{g_{12}}{\Delta g}$$

$$h_{21} = -\frac{z_{21}}{z_{22}} = \frac{y_{21}}{y_{11}} = -\frac{1}{a_{22}} = -\frac{\Delta b}{b_{11}} = -\frac{g_{21}}{\Delta g}$$

$$h_{22} = \frac{1}{z_{22}} = \frac{\Delta y}{y_{11}} = \frac{a_{21}}{a_{22}} = \frac{b_{21}}{b_{11}} = \frac{g_{11}}{\Delta g}$$

$$g_{11} = \frac{1}{z_{11}} = \frac{\Delta y}{y_{22}} = \frac{a_{21}}{a_{11}} = \frac{b_{21}}{b_{22}} = \frac{h_{22}}{\Delta h}$$

$$g_{12} = -\frac{z_{12}}{z_{11}} = \frac{y_{12}}{y_{22}} = -\frac{\Delta a}{a_{11}} = -\frac{1}{b_{22}} = -\frac{h_{12}}{\Delta h}$$

$$g_{21} = \frac{z_{21}}{z_{11}} = -\frac{y_{21}}{y_{22}} = \frac{1}{a_{11}} = \frac{\Delta b}{b_{22}} = -\frac{h_{21}}{\Delta h}$$

$$g_{22} = \frac{\Delta z}{z_{11}} = \frac{1}{y_{22}} = \frac{a_{12}}{a_{11}} = \frac{b_{12}}{b_{22}} = \frac{h_{11}}{\Delta h}$$

$$\Delta z = z_{11}z_{22} - z_{12}z_{21}$$

$$\Delta y = y_{11}y_{22} - y_{12}y_{21}$$

$$\Delta a = a_{11}a_{22} - a_{12}a_{21}$$

$$\Delta b = b_{11}b_{22} - b_{12}b_{21}$$

$$\Delta h = h_{11}h_{22} - h_{12}h_{21}$$

$$\Delta g = g_{11}g_{22} - g_{12}g_{21}$$

ters and y parameters and the relationships between the z parameters and the a parameters. These derivations illustrate the general thought process involved in relating one set of parameters to the other. To find the z parameters as functions of the y parameters we first solve Eqs. (20.2) for V_1 and V_2. We then compare the coefficients of I_1 and I_2 in the resulting expressions to the coefficients of I_1 and I_2 in Eqs. (20.1). From Eqs. (20.2),

$$V_1 = \frac{\begin{vmatrix} I_1 & y_{12} \\ I_2 & y_{22} \end{vmatrix}}{\begin{vmatrix} y_{11} & y_{12} \\ y_{21} & y_{22} \end{vmatrix}} = \frac{y_{22}}{\Delta y}I_1 - \frac{y_{12}}{\Delta y}I_2 \qquad \textbf{(20.16)}$$

and

$$V_2 = \frac{\begin{vmatrix} y_{11} & I_1 \\ y_{21} & I_2 \end{vmatrix}}{\Delta y} = -\frac{y_{21}I_1}{\Delta y} + \frac{y_{11}}{\Delta y}I_2. \qquad \textbf{(20.17)}$$

Comparing Eqs. (20.16) and (20.17) with Eqs. (20.1) shows

$$z_{11} = \frac{y_{22}}{\Delta y}, \qquad \textbf{(20.18)}$$

$$z_{12} = -\frac{y_{12}}{\Delta y}, \qquad \textbf{(20.19)}$$

$$z_{21} = -\frac{y_{21}}{\Delta y}, \qquad \textbf{(20.20)}$$

and

$$z_{22} = \frac{y_{11}}{\Delta y}. \qquad \textbf{(20.21)}$$

To find the z parameters as functions of the a parameters, we rearrange Eqs. (20.3) in the form of Eqs. (20.1) and then compare coefficients. From the second equation in Eqs. (20.3),

$$V_2 = \frac{1}{a_{21}}I_1 + \frac{a_{22}}{a_{21}}I_2. \qquad \textbf{(20.22)}$$

Therefore substituting Eq. (20.22) into the first equation of Eqs. (20.3) yields

$$V_1 = \frac{a_{11}}{a_{21}}I_1 + \left(\frac{a_{11}a_{22}}{a_{21}} - a_{12} \right) I_2. \qquad \textbf{(20.23)}$$

From Eq. (20.23),

$$z_{11} = \frac{a_{11}}{a_{21}}; \qquad \textbf{(20.24)}$$

$$z_{12} = \frac{\Delta a}{a_{21}}. \qquad \textbf{(20.25)}$$

From Eq. (20.22),

$$z_{21} = \frac{1}{a_{21}}; \qquad \textbf{(20.26)}$$

$$z_{22} = \frac{a_{22}}{a_{21}}. \qquad \textbf{(20.27)}$$

Example 20.3 illustrates the usefulness of the parameter conversion table.

E X A M P L E   20.3

Two sets of measurements are made on a two-port resistive circuit. The first set of measurements is made with port 2 open, and the second set of measurements is made with port 2 short-circuited. The results are

Port 2 Open	**Port 2 Short-Circuited**
$V_1 = 10$ mV,	$V_1 = 24$ mV,
$I_1 = 10$ μA,	$I_1 = 20$ μA,
$V_2 = -40$ V;	$I_2 = 1$ mA.

Find the h parameters of the circuit.

S O L U T I O N

We can find h_{11} and h_{21} directly from the short-circuit test:

$$h_{11} = \frac{V_1}{I_1}\bigg|_{V_2=0} = \frac{24 \times 10^{-3}}{20 \times 10^{-6}} = 1.2 \ k\Omega$$

and

$$h_{21} = \frac{I_2}{I_1}\bigg|_{V_2=0} = \frac{10^{-3}}{20 \times 10^{-6}} = 50.$$

The parameters h_{12} and h_{22} cannot be obtained directly from the open-circuit test. However, a check of Eqs. (20.7)–(20.15) indicates that the four a parameters can be derived from the test data. Therefore h_{12} and h_{22} can be obtained through the conversion table. Specifically,

$$h_{12} = \frac{\Delta a}{a_{22}} \quad \text{and} \quad h_{22} = \frac{a_{21}}{a_{22}}.$$

The a parameters are

$$a_{11} = \frac{V_1}{V_2}\bigg|_{I_2=0} = \frac{10 \times 10^{-3}}{-40} = -0.25 \times 10^{-3};$$

$$a_{21} = \frac{I_1}{V_2}\bigg|_{I_2=0} = \frac{10 \times 10^{-6}}{-40} = -0.25 \times 10^{-6} \mho;$$

$$a_{12} = -\frac{V_1}{I_2}\bigg|_{V_2=0} = -\frac{24 \times 10^{-3}}{10^{-3}} = -24 \ \Omega;$$

$$a_{22} = -\frac{I_1}{I_2}\bigg|_{V_2=0} = -\frac{20 \times 10^{-6}}{10^{-3}} = -20 \times 10^{-3}.$$

The numerical value of Δa is

$$\Delta a = a_{11}a_{22} - a_{12}a_{21}$$

$$= 5 \times 10^{-6} - 6 \times 10^{-6} = -10^{-6}.$$

Thus

$$h_{12} = \frac{\Delta a}{a_{22}} = \frac{-10^{-6}}{-20 \times 10^{-3}} = 5 \times 10^{-5}$$

and

$$h_{22} = \frac{a_{21}}{a_{22}} = \frac{-0.25 \times 10^{-6}}{-20 \times 10^{-3}} = 12.5 \ \mu\mho.$$

DRILL EXERCISES

20.4 The following measurements were made on a two-port resistive circuit: With port 1 open, $V_2 = 15$ V, $V_1 = 10$ V, and $I_2 = 30$ A; with port 1 short-circuited, $V_2 = 10$ V, $I_2 = 4$ A, and $I_1 = -5$ A. Calculate the y parameters.

ANSWER: $y_{11} = 0.75 \ \mho$; $y_{12} = -0.5 \ \mho$; $y_{21} = 2.4 \ \mho$; $y_{22} = 0.4 \ \mho$.

RECIPROCAL TWO-PORT CIRCUIT

If a two-port circuit is reciprocal, the following relationships exist among the port parameters:

$$z_{12} = z_{21}; \qquad \textbf{(20.28)}$$

$$y_{12} = y_{21}; \qquad \textbf{(20.29)}$$

$$a_{11}a_{22} - a_{12}a_{21} = \Delta a = 1; \qquad \textbf{(20.30)}$$

$$b_{11}b_{22} - b_{12}b_{21} = \Delta b = 1; \qquad \textbf{(20.31)}$$

$$h_{12} = -h_{21}; \qquad \textbf{(20.32)}$$

$$g_{12} = -g_{21}. \qquad \textbf{(20.33)}$$

A two-port circuit is reciprocal if the interchange of an ideal voltage source at one port with an ideal ammeter at the second port produces the same ammeter reading. Consider, for example, the resistive circuit shown in Fig. 20.4. When a voltage source of 15 V is applied to the port ad, it produces a current of 1.75 A in the ammeter at port cd. The ammeter current is easily

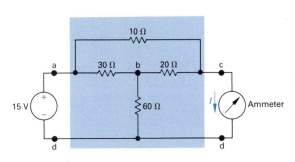

FIGURE 20.4 A reciprocal two-port circuit.

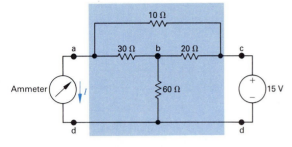

FIGURE 20.5 The circuit shown in Fig. 20.4 with the voltage source and ammeter interchanged.

determined once we know the voltage V_{bd}. Thus

$$\frac{V_{bd}}{60} + \frac{V_{bd} - 15}{30} + \frac{V_{bd}}{20} = 0, \quad \textbf{(20.34)}$$

and $V_{bd} = 5$ V. Therefore

$$I = \frac{5}{20} + \frac{15}{10} = 1.75 \text{ A}. \quad \textbf{(20.35)}$$

If the voltage source and ammeter are interchanged, the ammeter will still read 1.75 A. We verify this result by solving the circuit shown in Fig. 20.5:

$$\frac{V_{bd}}{60} + \frac{V_{bd}}{30} + \frac{V_{bd} - 15}{20} = 0. \quad \textbf{(20.36)}$$

From Eq. (20.36), $V_{bd} = 7.5$ V. The current I_{ad} equals

$$I_{ad} = \frac{7.5}{30} + \frac{15}{10} = 1.75 \text{ A}. \quad \textbf{(20.37)}$$

A two-port circuit also is reciprocal if the interchange of an ideal current source at one port with an ideal voltmeter at the second port produces the same reading of the voltmeter.

DRILL EXERCISES

20.5 a) Calculate the reading of the ideal voltmeter in the circuit shown.

b) Interchange the voltmeter and the ideal current source. Calculate the voltmeter reading.

ANSWER: (a) 32 V; (b) 32 V.

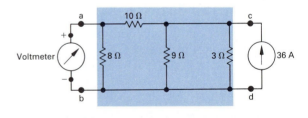

For a reciprocal two-port circuit only three calculations or measurements are needed to determine a set of parameters.

A reciprocal two-port circuit is symmetric if its ports can be interchanged without disturbing the values of the terminal currents and voltages. Figure 20.6 shows four examples of symmetric two-port circuits. If a reciprocal two-port circuit is symmetric, the following additional relationships exist among the port

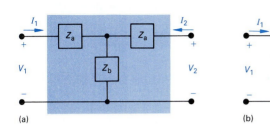

(a)   (b)   (c)

parameters:

$$z_{11} = z_{22};\qquad\qquad (20.38)$$

$$y_{11} = y_{22};\qquad\qquad (20.39)$$

$$a_{11} = a_{22};\qquad\qquad (20.40)$$

$$b_{11} = b_{22};\qquad\qquad (20.41)$$

$$h_{11}h_{22} - h_{12}h_{21} = \Delta h = 1;\qquad (20.42)$$

$$g_{11}g_{22} - g_{12}g_{21} = \Delta g = 1.\qquad (20.43)$$

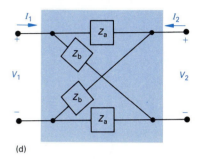

(d)

FIGURE 20.6 Four examples of symmetric two-port circuits: (a) symmetric tee; (b) symmetric pi; (c) symmetric bridged tee; (d) symmetric lattice.

For a symmetric reciprocal network, only two calculations or measurements are necessary to determine all the two-port parameters.

DRILL EXERCISES

20.6 The following measurements were made on a symmetric, reciprocal, resistive, two-port network: With port 2 open, $V_1 = 95$ V and $I_1 = 5$ A; with a short circuit across port 2, $V_1 = 11.52$ V and $I_2 = -2.72$ A. Calculate the z parameters of the two-port network.

ANSWER: $z_{11} = z_{22} = 19\ \Omega$, $z_{12} = z_{21} = 17\ \Omega$.

20.3 ANALYSIS OF THE TERMINATED TWO-PORT CIRCUIT

In the typical application of a two-port model of a network, the circuit is driven at port 1 and loaded at port 2. Figure 20.7 shows the s-domain circuit diagram for the typically terminated two-port model. Here, Z_g represents the internal impedance of the source, V_g the internal voltage of the source, and Z_L the load

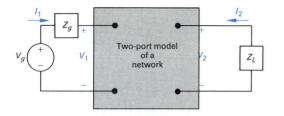

FIGURE 20.7 A terminated two-port model.

impedance. Analysis of the circuit shown in Fig. 20.7 involves expressing the terminal currents and voltages as functions of the two-port parameters, V_g, Z_g, and Z_L.

Six characteristics of the terminated two-port circuit define its terminal behavior:

1. the input impedance $Z_{in} = V_1/I_1$, or admittance $Y_{in} = I_1/V_1$;

2. the output current I_2;

3. the Thévenin voltage and impedance (V_{Th}, Z_{Th}) with respect to port 2;

4. the current gain I_2/I_1;

5. the voltage gain V_2/V_1; and

6. the voltage gain V_2/V_g.

SIX CHARACTERISTICS IN TERMS OF THE z PARAMETERS

To illustrate how these six characteristics are derived, we develop the expressions using the z parameters to model the two-port portion of the circuit. Table 20.2 summarizes the expressions involving the y, a, b, h, and g parameters. The derivation of any one of the desired expressions involves the algebraic manipulation of the two-port equations along with the two constraint equations imposed by the terminations. If we use the z-parameter equations, the four that describe the circuit in Fig. 20.7 are

$$V_1 = z_{11}I_1 + z_{12}I_2; \tag{20.44}$$

$$V_2 = z_{21}I_1 + z_{22}I_2; \tag{20.45}$$

$$V_1 = V_g - I_1 Z_g; \tag{20.46}$$

$$V_2 = -I_2 Z_L. \tag{20.47}$$

Equations (20.46) and (20.47) describe the constraints imposed by the terminations.

To find the impedance seen looking into port 1, that is, $Z_{in} = V_1/I_1$, we proceed as follows. In Eq. (20.45) we replace V_2 with $-I_2 Z_L$ and solve the resulting expression for I_2:

$$I_2 = \frac{-z_{21}I_1}{z_L + z_{22}}, \tag{20.48}$$

which we then substitute into Eq. (20.44) and solve for Z_{in}:

$$Z_{in} = z_{11} - \frac{z_{12}z_{21}}{z_{22} + Z_L}. \tag{20.49}$$

TABLE 20.2

TERMINATED TWO-PORT EQUATIONS

z **PARAMETERS**

$$Z_{in} = z_{11} - \frac{z_{12}z_{21}}{z_{22} + Z_L}$$

$$I_2 = \frac{-z_{21}V_g}{(z_{11} + Z_g)(z_{22} + Z_L) - z_{12}z_{21}}$$

$$V_{Th} = \frac{z_{21}}{z_{11} + Z_g}V_g$$

$$Z_{Th} = z_{22} - \frac{z_{12}z_{21}}{z_{11} + Z_g}$$

$$\frac{I_2}{I_1} = \frac{-z_{21}}{z_{22} + Z_L}$$

$$\frac{V_2}{V_1} = \frac{z_{21}Z_L}{z_{11}Z_L + \Delta z}$$

$$\frac{V_2}{V_g} = \frac{z_{21}Z_L}{(z_{11} + Z_g)(z_{22} + Z_L) - z_{12}z_{21}}$$

y **PARAMETERS**

$$Y_{in} = y_{11} - \frac{y_{12}y_{21}Z_L}{1 + y_{22}Z_L}$$

$$I_2 = \frac{y_{21}V_g}{1 + y_{22}Z_L + y_{11}Z_g + \Delta y Z_g Z_L}$$

$$V_{Th} = \frac{-y_{21}V_g}{y_{22} + \Delta y Z_g}$$

$$Z_{Th} = \frac{1 + y_{11}Z_g}{y_{22} + \Delta y Z_g}$$

$$\frac{I_2}{I_1} = \frac{y_{21}}{y_{11} + \Delta y Z_L}$$

$$\frac{V_2}{V_1} = \frac{-y_{21}Z_L}{1 + y_{22}Z_L}$$

$$\frac{V_2}{V_g} = \frac{y_{21}Z_L}{y_{12}y_{21}Z_g Z_L - (1 + y_{11}Z_g)(1 + y_{22}Z_L)}$$

a **PARAMETERS**

$$Z_{in} = \frac{a_{11}Z_L + a_{12}}{a_{21}Z_L + a_{22}}$$

$$I_2 = \frac{-V_g}{a_{11}Z_L + a_{12} + a_{21}Z_g Z_L + a_{22}Z_g}$$

$$V_{Th} = \frac{V_g}{a_{11} + a_{21}Z_g}$$

$$Z_{Th} = \frac{a_{12} + a_{22}Z_g}{a_{11} + a_{21}Z_g}$$

$$\frac{I_2}{I_1} = \frac{-1}{a_{21}Z_L + a_{22}}$$

$$\frac{V_2}{V_1} = \frac{Z_L}{a_{11}Z_L + a_{12}}$$

$$\frac{V_2}{V_g} = \frac{Z_L}{(a_{11} + a_{21}Z_g)Z_L + a_{12} + a_{22}Z_g}$$

b **PARAMETERS**

$$Z_{in} = \frac{b_{22}Z_L + b_{12}}{b_{21}Z_L + b_{11}}$$

$$I_2 = \frac{-V_g \Delta b}{b_{11}Z_g + b_{21}Z_g Z_L + b_{22}Z_L + b_{12}}$$

$$V_{Th} = \frac{V_g \Delta b}{b_{22} + b_{21}Z_g}$$

$$Z_{Th} = \frac{b_{11}Z_g + b_{12}}{b_{21}Z_g + b_{22}}$$

$$\frac{I_2}{I_1} = \frac{-\Delta b}{b_{11} + b_{21}Z_L}$$

$$\frac{V_2}{V_1} = \frac{\Delta b Z_L}{b_{12} + b_{22}Z_L}$$

$$\frac{V_2}{V_g} = \frac{\Delta b Z_L}{b_{12} + b_{11}Z_g + b_{22}Z_L + b_{21}Z_g Z_L}$$

h **PARAMETERS**

$$Z_{in} = h_{11} - \frac{h_{12}h_{21}Z_L}{1 + h_{22}Z_L}$$

$$I_2 = \frac{h_{21}V_g}{(1 + h_{22}Z_L)(h_{11} + Z_g) - h_{12}h_{21}Z_L}$$

$$V_{Th} = \frac{-h_{21}V_g}{h_{22}Z_g + \Delta h}$$

$$Z_{Th} = \frac{Z_g + h_{11}}{h_{22}Z_g + \Delta h}$$

$$\frac{I_2}{I_1} = \frac{h_{21}}{1 + h_{22}Z_L}$$

$$\frac{V_2}{V_1} = \frac{-h_{21}Z_L}{\Delta h Z_L + h_{11}}$$

$$\frac{V_2}{V_g} = \frac{-h_{21}Z_L}{(h_{11} + Z_g)(1 + h_{22}Z_L) - h_{12}h_{21}Z_L}$$

(continued)

TABLE 20.2 (*continued*)

g **PARAMETERS**

$$Y_{in} = g_{11} - \frac{g_{12} g_{21}}{g_{22} + Z_L}$$

$$I_2 = \frac{-g_{21} V_g}{(1 + g_{11} Z_g)(g_{22} + Z_L) - g_{12} g_{21} Z_g}$$

$$V_{Th} = \frac{g_{21} V_g}{1 + g_{11} Z_g}$$

$$Z_{Th} = g_{22} - \frac{g_{12} g_{21} Z_g}{1 + g_{11} Z_g}$$

$$\frac{I_2}{I_1} = \frac{-g_{21}}{g_{11} Z_L + \Delta g}$$

$$\frac{V_2}{V_1} = \frac{g_{21} Z_L}{g_{22} + Z_L}$$

$$\frac{V_2}{V_g} = \frac{g_{21} Z_L}{(1 + g_{11} Z_g)(g_{22} + Z_L) - g_{12} g_{21} Z_g}$$

To find the terminal current I_2, we first solve Eq. (20.44) for I_1 after replacing V_1 with the right-hand side of Eq. (20.46). The result is

$$I_1 = \frac{V_g - z_{12} I_2}{z_{11} + Z_g}. \tag{20.50}$$

We now substitute Eq. (20.50) into Eq. (20.48) and solve the resulting equation for I_2:

$$I_2 = \frac{-z_{21} V_g}{(z_{11} + Z_g)(z_{22} + Z_L) - z_{12} z_{21}}. \tag{20.51}$$

The Thévenin voltage with respect to port 2 equals V_2 when $I_2 = 0$. With $I_2 = 0$, Eqs. (20.44) and (20.45) combine to yield

$$V_2 \bigg|_{I_2=0} = z_{21} I_1 = z_{21} \frac{V_1}{z_{11}}. \tag{20.52}$$

But $V_1 = V_g - I_1 Z_g$ and $I_1 = V_g/(Z_g + z_{11})$; therefore substituting the results into Eq. (20.52) yields the open-circuit value of V_2:

$$V_2 \bigg|_{I_2=0} = V_{Th} = \frac{z_{21}}{Z_g + z_{11}} V_g. \tag{20.53}$$

The Thévenin, or output, impedance is the ratio V_2/I_2 when V_g is replaced by a short circuit. When V_g is zero, Eq. (20.46) reduces to

$$V_1 = -I_1 Z_g. \tag{20.54}$$

Substituting Eq. (20.54) into Eq. (20.44) gives

$$I_1 = \frac{-z_{12} I_2}{z_{11} + Z_g}. \tag{20.55}$$

We now use Eq. (20.55) to replace I_1 in Eq. (20.45), with the result that

$$\frac{V_2}{I_2}\bigg|_{V_g=0} = Z_{\text{Th}} = z_{22} - \frac{z_{12}z_{21}}{z_{11} + Z_g}. \qquad (20.56)$$

The current gain I_2/I_1 comes directly from Eq. (20.48):

$$\frac{I_2}{I_1} = \frac{-z_{21}}{Z_L + z_{22}}. \qquad (20.57)$$

To derive the expression for the voltage gain V_2/V_1 we start by replacing I_2 in Eq. (20.45) with its value from Eq. (20.47); thus

$$V_2 = z_{21}I_1 + z_{22}\left(\frac{-V_2}{Z_L}\right). \qquad (20.58)$$

Next we solve Eq. (20.44) for I_1 as a function of V_1 and V_2:

$$z_{11}I_1 = V_1 - z_{12}\left(\frac{-V_2}{Z_L}\right)$$

or

$$I_1 = \frac{V_1}{z_{11}} + \frac{z_{12}V_2}{z_{11}Z_L}. \qquad (20.59)$$

We now replace I_1 in Eq. (20.58) with Eq. (20.59) and solve the resulting expression for V_2/V_1:

$$\frac{V_2}{V_1} = \frac{z_{21}Z_L}{z_{11}Z_L + z_{11}z_{22} - z_{12}z_{21}}$$

$$= \frac{z_{21}Z_L}{z_{11}Z_L + \Delta z}. \qquad (20.60)$$

To derive the voltage ratio V_2/V_g we first combine Eqs. (20.44), (20.46), and (20.47) to find I_1 as a function of V_2 and V_g:

$$I_1 = \frac{z_{12}V_2}{Z_L(z_{11} + Z_g)} + \frac{V_g}{z_{11} + Z_g}. \qquad (20.61)$$

We now use Eqs. (20.61) and (20.47) in conjunction with Eq. (20.45) to derive an expression involving only V_2 and V_g; that is,

$$V_2 = \frac{z_{21}z_{12}V_2}{Z_L(z_{11} + Z_g)} + \frac{z_{21}V_g}{z_{11} + Z_g} - \frac{z_{22}}{Z_L}V_2, \qquad (20.62)$$

which we can manipulate to get the desired voltage ratio:

$$\frac{V_2}{V_g} = \frac{z_{21}Z_L}{(z_{11} + Z_g)(z_{22} + Z_L) - z_{12}z_{21}}. \qquad (20.63)$$

The first entries in Table 20.2 summarize the expressions for these six attributes of the terminated two-port circuit. Also listed in Table 20.2 are the corresponding expressions in terms of the y, a, b, h, and g parameters.

Example 20.4 illustrates the usefulness of the relationships listed in Table 20.2.

E X A M P L E   20.4

The two-port circuit shown in Fig. 20.8 is described in terms of its b parameters, the values of which are

$b_{11} = -20, \quad b_{12} = -3000 \ \Omega, \quad b_{21} = -2 \ \text{mU}, \quad b_{22} = -0.2.$

a) Find the phasor voltage $\mathbf{V}_2$.

b) Find the average power delivered to the 5-kΩ load.

c) Find the average power delivered to the input port.

d) Find the load impedance for maximum average-power transfer.

e) Find the maximum average power delivered to the load in (d).

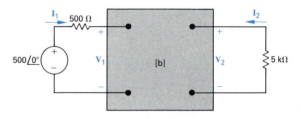

FIGURE 20.8 The circuit for Example 20.4.

S O L U T I O N

a) To find $\mathbf{V}_2$ we have two choices from the entries in Table 20.2. We may choose to find $\mathbf{I}_2$ and then find $\mathbf{V}_2$ from the relationship $\mathbf{V}_2 = -\mathbf{I}_2 Z_L$, or we may find the voltage gain $\mathbf{V}_2/\mathbf{V}_g$ and calculate $\mathbf{V}_2$ from the gain. We use the latter approach. For the b-parameter values given we have

$$\Delta b = (-20)(-0.2) - (-3000)(-2 \times 10^{-3})$$

$$= 4 - 6$$

$$= -2.$$

From Table 20.2,

$$\frac{\mathbf{V}_2}{\mathbf{V}_g} = \frac{\Delta b Z_L}{b_{12} + b_{11} Z_g + b_{22} Z_L + b_{21} Z_g Z_L}$$

$$= \frac{(-2)(5000)}{-3000 + (-20)500 + (-0.2)5000 + [-2 \times 10^{-3}(500)(5000)]}$$

$$= \frac{10}{19}.$$

Then,

$$\mathbf{V}_2 = \left(\frac{10}{19}\right)500 = 263.16\underline{/0°} \ \text{V}.$$

b) The average power delivered to the 5000-Ω load is

$$P_2 = \frac{263.16^2}{2(5000)} = 6.93 \text{ W.}$$

c) To find the average power delivered to the input port, we first find the input impedance Z_{in}. From Table 20.2,

$$Z_{in} = \frac{b_{22}Z_L + b_{12}}{b_{21}Z_L + b_{11}}$$

$$= \frac{(-0.2)(5000) - 3000}{-2 \times 10^{-3}(5000) - 20}$$

$$= \frac{400}{3} = 133.33 \ \Omega.$$

Now $\mathbf{I}_1$ follows directly:

$$\mathbf{I}_1 = \frac{500}{500 + 133.33} = 789.47 \text{ mA.}$$

The average power delivered to the input port is

$$P_1 = \frac{0.78947^2}{2}(133.33)$$

$$= 41.55 \text{ W.}$$

d) The load impedance for maximum power transfer equals the conjugate of the Thévenin impedance seen from looking into port 2. From Table 20.2,

$$Z_{Th} = \frac{b_{11}Z_g + b_{12}}{b_{21}Z_g + b_{22}}$$

$$= \frac{(-20)(500) - 3000}{(-2 \times 10^{-3})(500) - 0.2}$$

$$= \frac{13,000}{1.2} = 10,833.33 \ \Omega.$$

Therefore $Z_L = Z_{Th}^* = 10,833.33 \ \Omega.$

e) To find the maximum average power delivered by Z_L we first find $\mathbf{V}_2$ from the voltage-gain expression $\mathbf{V}_2/\mathbf{V}_g$. When Z_L is 10,833.33 Ω, this gain is

$$\frac{\mathbf{V}_2}{\mathbf{V}_g} = 0.8333.$$

Thus

$$\mathbf{V}_2 = (0.8333)(500) = 416.67 \text{ V}$$

and

$$P_2(\text{maximum}) = \frac{1}{2}\frac{416.67^2}{10,833.33}$$

$$= 8.01 \text{ W.}$$

DRILL EXERCISES

20.7 The b parameters of the two-port network shown are $b_{11} = 2000/3$, $b_{12} = \frac{2}{3}$ MΩ, $b_{21} = \frac{1}{15}$ ℧, and $b_{22} = -100/3$. The network is driven by a sinusoidal current source having a maximum amplitude of 100 μA and an internal impedance of $1000 + j0$ Ω. The network is terminated in a resistive load of 10 kΩ.

a) Calculate the average power delivered to the load resistor.

b) Calculate the load resistance for maximum average power.

c) Calculate the maximum average power delivered to the resistor in (b).

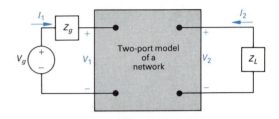

ANSWER: (a) 80 mW; (b) 40 kΩ; (c) 125 mW.

20.4 INTERCONNECTED TWO-PORT CIRCUITS

In the design of a large, complex system, synthesizing the system is usually made easier by first designing subsections of the system. Interconnecting these simpler, easier-to-design, smaller units then completes the system. If the subsections are modeled by two-port circuits, synthesis of the complete system involves the analysis of interconnected two-port circuits.

Two-port circuits may be interconnected five ways: (1) in cascade, (2) in series, (3) in parallel, (4) in series–parallel, and (5) in parallel–series. Figure 20.9 depicts these five basic interconnections.

We analyze and illustrate only the cascade connection in this section. However, if the four latter connections meet certain requirements, we can obtain the parameters that describe the inter-

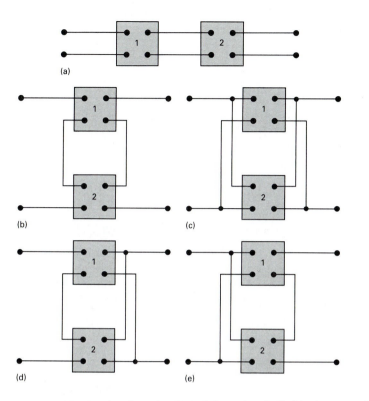

(a)

(b)

(c)

(d)

(e)

FIGURE 20.9 The five basic interconnections of two-port circuits: (a) cascade; (b) series; (c) parallel; (d) series–parallel; (e) parallel–series.

connected circuits by simply adding the individual network parameters. In particular, the z parameters describe the series connection, the y parameters the parallel connection, the h parameters the series–parallel connection, and the g parameters the parallel–series connection.[†]

The cascade connection is important because of its frequent occurrence in the modeling of large systems. Unlike the other four basic interconnections, there are no restrictions on using the parameters of the individual two-port circuits to obtain the parameters of the interconnected circuits. The a parameters are best suited for describing the cascade connection. We analyze the cascade connection using the circuit shown in Fig. 20.10, where we use a single prime to denote the a parameters of the first circuit and a double prime to denote the a parameters of the second circuit. The output voltage and current of the first circuit are labeled V_2' and I_2', and the input voltage and current of the second circuit are labeled V_1' and I_1'. The problem is to derive the a-parameter equations that relate V_2 and I_2 to V_1 and I_1. That

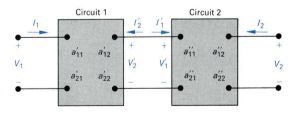

FIGURE 20.10 The cascade connection.

[†] A detailed discussion of these four interconnections is presented in Henry Ruston and Joseph Bordogna, *Electric Networks: Functions, Filters, Analysis,* Chapter 4 (New York: McGraw-Hill, 1966).

is, we seek the pair of equations

$$V_1 = a_{11} V_2 - a_{12} I_2 \qquad (20.64)$$

and

$$I_1 = a_{21} V_2 - a_{22} I_2, \qquad (20.65)$$

where the a parameters are given explicitly in terms of the a parameters of the individual circuits.

We begin the derivation by noting from Fig. 20.10 that

$$V_1 = a'_{11} V'_2 - a'_{12} I'_2; \qquad (20.66)$$

$$I_1 = a'_{21} V'_2 - a'_{22} I'_2. \qquad (20.67)$$

The interconnection means that $V'_2 = V'_1$ and $I'_2 = -I'_1$. Substituting these constraints into Eqs. (20.66) and (20.67) yields

$$V_1 = a'_{11} V'_1 + a'_{12} I'_1; \qquad (20.68)$$

$$I_1 = a'_{21} V'_1 + a'_{22} I'_1. \qquad (20.69)$$

The voltage V'_1 and the current I'_1 are related to V_2 and I_2 through the a parameters of the second circuit:

$$V'_1 = a''_{11} V_2 - a''_{12} I_2; \qquad (20.70)$$

$$I'_1 = a''_{21} V_2 - a''_{22} I_2. \qquad (20.71)$$

We substitute Eqs. (20.70) and (20.71) into Eqs. (20.68) and (20.69) to generate the relationships between V_1, I_1 and V_2, I_2:

$$V_1 = (a'_{11} a''_{11} + a'_{12} a''_{21}) V_2 - (a'_{11} a''_{12} + a'_{12} a''_{22}) I_2 \qquad (20.72)$$

$$I_1 = (a'_{21} a''_{11} + a'_{22} a''_{21}) V_2 - (a'_{21} a''_{12} + a'_{22} a''_{22}) I_2. \qquad (20.73)$$

By comparing Eqs. (20.72) and (20.73) to Eqs. (20.64) and (20.65), we get the desired expressions for the a parameters of the interconnected networks, namely,

$$a_{11} = a'_{11} a''_{11} + a'_{12} a''_{21}; \qquad (20.74)$$

$$a_{12} = a'_{11} a''_{12} + a'_{12} a''_{22}: \qquad (20.75)$$

$$a_{21} = a'_{21} a''_{11} + a'_{22} a''_{21}; \qquad (20.76)$$

$$a_{22} = a'_{21} a''_{12} + a'_{22} a''_{22}. \qquad (20.77)$$

Example 20.5 illustrates how to use Eqs. (20.74)–(20.77) to analyze the cascade connection with two amplifier circuits.

E X A M P L E   20.5

Two identical amplifiers are connected in cascade, as shown in Fig. 20.11. Each amplifier is described in terms of its h parameters. The values are $h_{11} = 1000 \, \Omega$, $h_{12} = 0.0015$, $h_{21} = 100$, and $h_{22} = 100 \, \mu\mho$. Find the voltage gain V_2/V_g.

SOLUTION

The first step in finding the overall voltage gain V_2/V_g is to convert from h parameters to a parameters. The amplifiers are identical, so one set of a parameters describes the amplifiers:

$$a'_{11} = \frac{-\Delta h}{h_{21}} = \frac{+0.05}{100} = 5 \times 10^{-4},$$

$$a'_{12} = \frac{-h_{11}}{h_{21}} = \frac{-1000}{100} = -10 \ \Omega,$$

$$a'_{21} = \frac{-h_{22}}{h_{21}} = \frac{-100 \times 10^{-6}}{100} = -10^{-6} \ \mho,$$

$$a'_{22} = \frac{-1}{h_{21}} = \frac{-1}{100} = -10^{-2}.$$

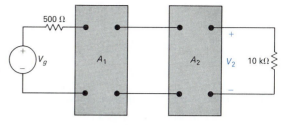

FIGURE 20.11 The circuit for Example 20.5.

Next we use Eqs. (20.74)–(20.77) to compute the a parameters of the cascaded amplifiers:

$$a_{11} = a'_{11}a'_{11} + a'_{12}a'_{21}$$
$$= 25 \times 10^{-8} + (-10)(-10^{-6}) = 10.25 \times 10^{-6};$$

$$a_{12} = a'_{11}a'_{12} + a'_{12}a'_{22}$$
$$= (5 \times 10^{-4})(-10) + (-10)(-10^{-2}) = 0.095 \ \Omega;$$

$$a_{21} = a'_{21}a'_{11} + a'_{22}a'_{21}$$
$$= (-10^{-6})(5 \times 10^{-4}) + (-10^{-6})(-0.01)$$
$$= 0.0095 \times 10^{-6} \ \mho;$$

$$a_{22} = a'_{21}a'_{12} + a'_{22}a'_{22}$$
$$= (-10)(-10^{-6}) + (-10^{-2})^2 = 1.1 \times 10^{-4}.$$

From Table 20.2,

$$\frac{V_2}{V_g} = \frac{Z_L}{(a_{11} + a_{21}Z_g)Z_L + a_{12} + a_{22}Z_g}$$

$$= \frac{10^4}{[10.25 \times 10^{-6} + 0.0095 \times 10^{-6}(500)]10^4 + 0.095 + 1.1 \times 10^{-4}(500)}$$

$$= \frac{10^4}{0.15 + 0.095 + 0.055} = \frac{10^5}{3} = 33,333.33.$$

Thus an input signal of 150 μV is amplified to an output signal of 5 V. For an alternative approach to finding the voltage gain V_2/V_g see Problem 20.41.

If more than two units are connected in cascade, the *a* parameters of the equivalent two-port circuit can be found by successively reducing the original set of two-port circuits one pair at a time.

DRILL EXERCISES

20.8 Each element in the symmetric bridged-tee circuit shown is a 15-Ω resistor. Two of these bridged tees are connected in cascade between a dc voltage source and a resistive load. The dc voltage source has a no-load voltage of 100 V and an internal resistance of 8 Ω. The load resistor is adjusted until maximum power is delivered to the load. Calculate (a) the load resistance; (b) the load voltage; and (c) the load power.

ANSWER: (a) 14.44 Ω; (b) 16 V; (c) 17.73 W.

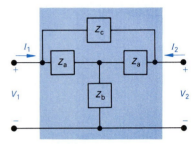

SUMMARY

The two-port model is used to describe the performance of a circuit in terms of the voltage and current at its input and output ports. Application of the two-port model is based on the following conditions.

- The model is limited to circuits in which

  no independent sources are inside the circuit between the ports;

  no energy is stored inside the circuit between the ports;

  the current into the port is equal to the current out of the port; and

  no external connections exist between the input and output port.

- Two of the four terminal variables (V_1, I_1, V_2, I_2) are independent; therefore only two simultaneous equations involving the four variables are needed to describe the circuit.

- The six possible sets of simultaneous equations involving the four terminal variables are called the z-, y-, a-, b-, h-, and g-parameter equations. [See Eqs. (20.1)–(20.6).]

• The parameter equations are written in the s domain. The dc values of the parameters are obtained by setting $s = 0$, and the sinusoidal steady-state values are obtained by setting $s = j\omega$.

• The relationships among the six sets of parameters are given in Table 20.1.

• Any set of parameters may be calculated or measured by invoking appropriate short-circuit and open-circuit conditions at the input and output ports. [See Eqs. (20.7)–(20.15).]

• A two-port circuit is reciprocal if the interchange of an ideal voltage source at one port with an ideal ammeter at the second port produces the same ammeter reading. The effect of reciprocity on the two-port parameters is given by Eqs. (20.28)–(20.33).

• A reciprocal two-port circuit is symmetrical if its ports can be interchanged without disturbing the values of the terminal currents and voltages. The added effect of symmetry on the two-port parameters is given by Eqs. (20.38)–(20.43).

• The performance of a two-port circuit connected to a Thévenin equivalent source and a load is summarized by the relationships given in Table 20.2.

• Large networks can be divided into subnetworks by means of interconnected two-port models. The cascade connection was used to illustrate analysis of interconnected two-ports.

PROBLEMS

20.1 Find the h and g parameters for the circuit in Example 20.1.

20.2 Find the y parameters for the circuit shown in Fig. P20.2.

FIGURE P20.2

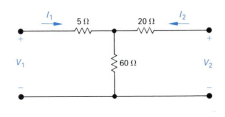

20.3 Find the *a* parameters for the circuit shown in Fig. P20.3.

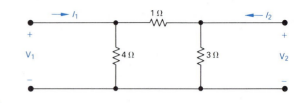

FIGURE P20.3

20.4 Find the *h* parameters of the circuit in Fig. P20.4.

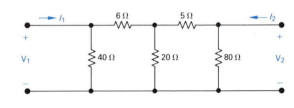

FIGURE P20.4

20.5 Find the *b* parameters for the circuit in Fig. P20.5.

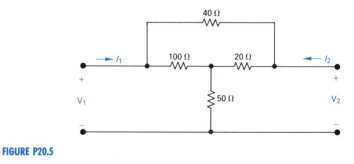

FIGURE P20.5

20.6 Use the results obtained in Problem 20.5 to calculate the *g* parameters of the circuit in Fig. P20.5.

20.7 Find the *z* parameters for the circuit in Fig. P20.7.

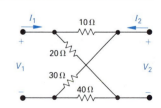

FIGURE P20.7

20.8 Use the results obtained in Problem 20.7 to calculate the y parameters for the circuit in Fig. P20.7.

20.9 Select the values of R_1, R_2, and R_3 in the circuit in Fig. P20.9 so that $a_{11} = 1.2$, $a_{12} = 34\ \Omega$, $a_{21} = 20$ mmho, and $a_{22} = 1.4$.

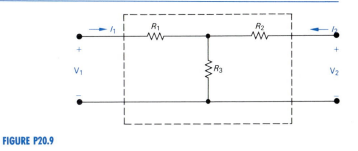

FIGURE P20.9

20.10 Find the h parameters of the two port circuit shown in Fig. P20.10.

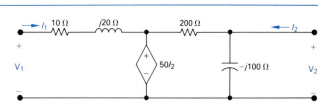

FIGURE P20.10

20.11 The following direct-current measurements were made on the two-port network shown in Fig. P20.11.

Port 2 Open	Port 2 Short-Circuited
$V_1 = 20$ mV	$V_1 = 10$ V
$I_1 = 0.25\ \mu A$	$I_1 = 200\ \mu A$
$V_2 = -5$ V	$I_2 = 50\ \mu A$

Calculate the g parameters for the network.

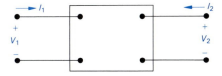

FIGURE P20.11

20.12 a) Use the measurements given in Problem 20.11 to find the y parameters for the network.

b) Check your calculations by finding the y parameters directly from the g parameters found in Problem 20.11.

20.13 Find the phasor-domain values of the y parameters for the two-port circuit shown in Fig. P20.13.

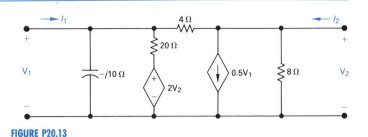

FIGURE P20.13

20.14 Find the a parameters for the two-port circuit shown in Fig. P20.13.

20.15 Find the h parameters for the operational-amplifier circuit shown in Fig. P20.15.

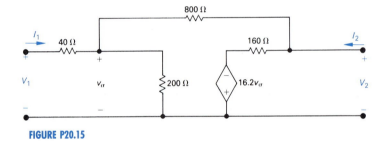

FIGURE P20.15

20.16 Derive the expressions for the h parameters as functions of the a parameters.

20.17 Derive the expressions for the y parameters as functions of the b parameters.

20.18 Derive the expressions for the g parameters as functions of the z parameters.

20.19 The operational amplifier in the circuit shown in Fig. P20.19 is ideal. Find the g parameters of the circuit.

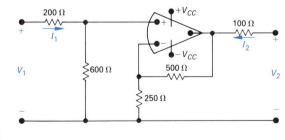

FIGURE P20.19

20.20 Find the s-domain expressions for the a parameters of the two-port circuit shown in Fig. P20.20.

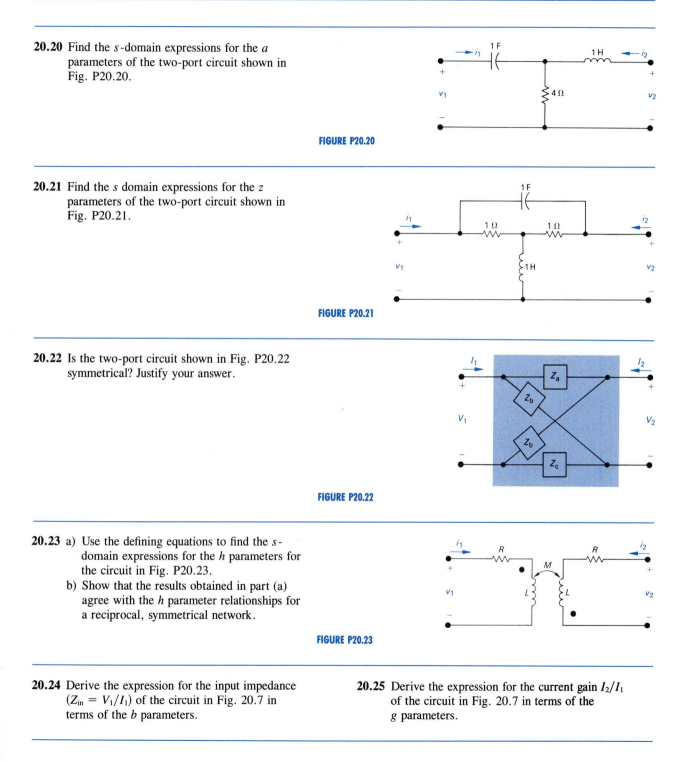

FIGURE P20.20

20.21 Find the s domain expressions for the z parameters of the two-port circuit shown in Fig. P20.21.

FIGURE P20.21

20.22 Is the two-port circuit shown in Fig. P20.22 symmetrical? Justify your answer.

FIGURE P20.22

20.23 a) Use the defining equations to find the s-domain expressions for the h parameters for the circuit in Fig. P20.23.
   b) Show that the results obtained in part (a) agree with the h parameter relationships for a reciprocal, symmetrical network.

FIGURE P20.23

20.24 Derive the expression for the input impedance ($Z_{\text{in}} = V_1/I_1$) of the circuit in Fig. 20.7 in terms of the b parameters.

20.25 Derive the expression for the current gain I_2/I_1 of the circuit in Fig. 20.7 in terms of the g parameters.

20.26 Derive the expression for the voltage gain V_2/V_1 of the circuit in Fig. 20.7 in terms of the y parameters.

20.27 Derive the expression for the voltage gain V_2/V_g of the circuit in Fig. 20.7 in terms of the h parameters.

20.28 Find the Thévenin equivalent circuit with respect to port 2 of the circuit in Fig. 20.7 in terms of the z parameters.

20.29 The linear transformer in the circuit shown in Fig. P20.29 has a coefficient of coupling of 0.75. The transformer is driven by a sinusoidal voltage source whose internal voltage is $v_g = 260 \cos 4000t$ V. The internal impedance of the source is $25 + j0\ \Omega$.

a) Find the phasor-domain a parameters of the linear transformer.

b) Use the a parameters to derive the Thévenin equivalent circuit with respect to the terminals of the 1000-Ω load.

c) Derive the steady-state time-domain expression for v_2.

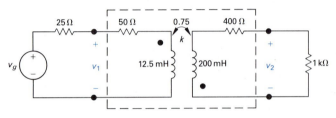

FIGURE P20.29

20.30 The b parameters of the amplifier in the circuit shown in Fig. P20.30 are

$$b_{11} = 25, \quad b_{12} = 1000\ \Omega, \quad b_{21} = -1.25\ \mho,$$

$$b_{22} = -40$$

Find the ratio of the output power to that supplied by the ideal voltage source.

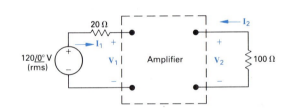

FIGURE P20.30

20.31 The g parameters for the two-port circuit in Fig. P20.31 are

$$g_{11} = \frac{1}{6} - j\frac{1}{6}\ \mho; \quad g_{12} = -\frac{1}{2} + j\frac{1}{2};$$

$$g_{21} = \frac{1}{2} - j\frac{1}{2}; \quad \text{and} \quad g_{22} = 1.5 + j2.5\ \Omega.$$

The load impedance Z_L is adjusted for maxi-

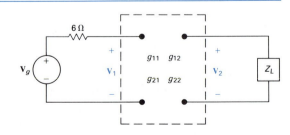

FIGURE P20.31

mum average power transfer to Z_L. The ideal
voltage source is generating a sinusoidal voltage
of

$$v_g = 42\sqrt{2} \cos 5000\, t \text{ V.}$$

a) Find the rms value of V_2.

b) Find the average power delivered to Z_L.

c) What percentage of the average power developed by the ideal voltage source is delivered by Z_L?

20.32 The y parameters for the two-port power amplifier circuit in Fig. P20.32 are:

$$y_{11} = 2 \text{ mmho;} \quad y_{12} = -2\,\mu\text{mho;}$$

$$y_{21} = 100 \text{ mmho;} \quad \text{and} \quad y_{22} = -50\,\mu\text{mho.}$$

The internal impedance of the source is
$2500 + j\,0\ \Omega$ and the load impedance is
$70{,}000 + j\,0\ \Omega$. The ideal voltage source is
generating a voltage

$$v_g = 80\sqrt{2} \cos 4000\, t \text{ mV.}$$

a) Find the rms value of V_2.

b) Find the average power delivered to Z_L.

c) Find the average power developed by the ideal voltage source.

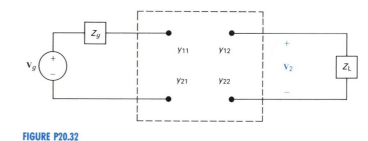

FIGURE P20.32

20.33 For the terminated two-port amplifier circuit in
Fig. P20.32 find:

a) the value of Z_L for maximum average power
transfer to Z_L;

b) the maximum average power delivered to Z_L;
and

c) the average power developed by the ideal
voltage source when maximum power is delivered to Z_L.

20.34 a) Find the s-domain expressions for the
h parameters of the circuit in Fig. P20.34.

b) Port 2 in Fig. P20.34 is terminated in a resistance of $400\ \Omega$ and port 1 is driven by a
step voltage source $v_1(t) = 30u(t)$ V. Find
$v_2(t)$ for $t > 0$ if $C = 0.2\,\mu\text{F}$ and $L = 200$ mH.

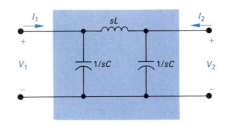

FIGURE P20.34

20.35 a) Find the *a* parameters for the two-port network in Fig. P20.35.

b) Assume $Z_L = s + 2$ Ω, $Z_g = 1$ Ω, and $v_g = 15\, u(t)$ V. Find $i_2(t)$ for $t > 0$.

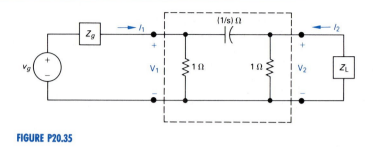

FIGURE P20.35

20.36 The following measurements were made on a resistive two-port network. With port 2 open and 100 V applied to port 1 the port 1 current is 1.125 A and the port 2 voltage is 104 V. With port 1 open and 24 V applied to port 2 the port 2 current is 0.25 A and the port 1 voltage is 20 V. All measurements are based on the polarity reference system given in Fig. 20.2. Find the maximum power that this two-port circuit can deliver to a resistive load at port 2 when port 1 is driven by an ideal voltage source of 160 V dc.

20.37 The following dc measurements were made on the resistive network shown in Fig. P20.37.

Measurement 1	Measurement 2
$V_1 = 25$ V	$V_1 = 41$ V
$I_1 = 1$ A	$I_1 = 1$ A
$V_2 = 0$ V	$V_2 = 20$ V
$I_2 = -0.5$ A	$I_2 = 0$ A

A variable resistor R_o is connected across port 2 and adjusted for maximum power transfer to R_o. Find the maximum power.

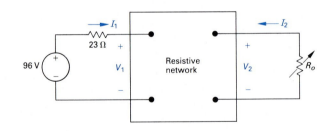

FIGURE P20.37

20.38 The two networks shown in Fig. P20.38 are reciprocal and symmetric. The load resistor R_L is adjusted for maximum power transfer to R_L. The networks are driven from a dc voltage source that generates 135 V. If $h_{11} = 9\ \Omega$, $h_{12} = 0.80$, $b_{11} = (7/6)$, and $b_{12} = 3\ \Omega$ calculate the maximum power delivered to R_L in watts.

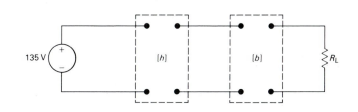

FIGURE P20.38

20.39 The g parameters and h parameters for the resistive two-ports shown in Fig. P20.39 are

$$g_{11} = \frac{3}{35}\ \mho, \qquad h_{11} = 5000\Omega$$

$$g_{12} = \frac{20}{7}, \qquad h_{12} = -0.20$$

$$g_{21} = \frac{800}{7}, \qquad h_{21} = -4.0$$

$$g_{22} = \frac{50}{7}\ k\Omega; \qquad h_{22} = 200\mu\mho.$$

Find I_1 if $v_g = 30$ V.

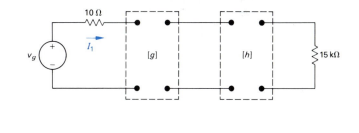

FIGURE P20.39

20.40 The networks D and E in the circuit in Fig. P20.40 are reciprocal and symmetrical. For network D it is known that $a'_{11} = 5$ and $a'_{12} = 24\ \Omega$. The impedance Z_o is adjusted for maximum average power transfer to Z_o. Find Z_o if $Z_g = (5 + j0)\ \Omega$.

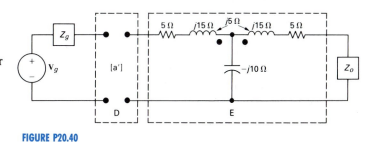

FIGURE P20.40

20.41 a) Show that the circuit in Fig. P20.41 is an equivalent circuit that is satisfied by the *h* parameter equations.

b) Use the *h* parameter equivalent circuit of part (a) to find the voltage gain V_2/V_g in the circuit in Fig. 20.11.

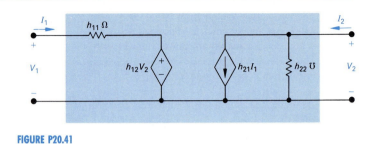

FIGURE P20.41

20.42 a) Show that the circuit in Fig. P20.42 is an equivalent circuit that is satisfied by the *z* parameter equations.

b) Assume that the equivalent circuit in Fig. P20.42 is driven by a voltage source having an internal impedance of Z_g ohms. Calculate the Thévenin equivalent circuit with respect to port 2. Check your results against the appropriate entries in Table 20.2.

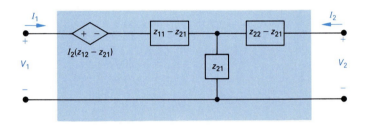

FIGURE P20.42

20.43 a) Show that the circuit in Fig. P20.43 is also an equivalent circuit that is satisfied by the *z* parameter equations.

b) Assume that the equivalent circuit in Fig. P20.43 is terminated in an impedance of Z_L ohms at port 2. Find the input impedance V_1/I_1. Check your result against the appropriate entry in Table 20.2.

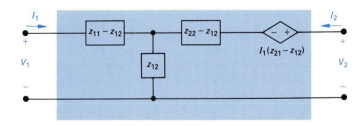

FIGURE P20.43

20.44 a) Derive two equivalent circuits that are satisfied by the y parameter equations. (*Hint:* Start with Eqs. 20.2. Add and subtract $y_{21} V_2$ to the first equation of the set. Construct a circuit that satisfies the resulting set of equations, by thinking in terms of node voltages. Derive an alternative equivalent circuit by first altering the second equation in Eqs. 20.2.)

b) Assume that port 1 is driven by a voltage source having an internal impedance Z_g and port 2 is loaded with an impedance Z_L. Find the current gain I_2/I_1. Check your result against the appropriate entry in Table 20.2.

20.45 Derive the equivalent circuit that is satisfied by the g parameter equations.

Appendix A

Solution of Linear Simultaneous Equations

Circuit analysis frequently involves the solution of linear simultaneous equations. Our purpose here is to review the use of determinants to solve such a set of equations. The theory of determinants (with applications) can be found in most intermediate-level algebra texts. (A particularly good reference for engineering students is Chapter 1 of *The Mathematics of Circuit Analysis* by E. A. Guillemin, Wiley, New York, 1949). In our review here we shall limit our discussion to the mechanics of solving simultaneous equations with determinants.

A.1 PRELIMINARY STEPS

The first step in solving a set of simultaneous equations by determinants is to write the equations in a rectangular (square) format. That is, the equations are arranged in a vertical stack such that each variable occupies the same horizontal position in every equation. For example, in Eqs. (A.1), the variables i_1, i_2, and i_3 occupy the first, second, and third position, respectively, on the left-hand side of each equation:

$$21i_1 - 9i_2 - 12i_3 = -33,$$
$$-3i_1 + 6i_2 - 2i_3 = 3, \tag{A.1}$$
$$-8i_1 - 4i_2 + 22i_3 = 50.$$

Alternatively one can describe this set of equations by saying that i_1 occupies the first column in the array, i_2 the second column, and i_3 the third column.

If one or more variables are missing from a given equation, they can be inserted by simply making their coefficient zero. Thus Eqs. (A.2) can be "squared up" as shown by Eqs. (A.3):

$$2v_1 - v_2 = 4,$$
$$4v_2 + 3v_3 = 16, \tag{A.2}$$
$$7v_1 + 2v_3 = 5;$$

$$2v_1 - v_2 + 0v_3 = 4,$$
$$0v_1 + 4v_2 + 3v_3 = 16, \tag{A.3}$$
$$7v_1 + 0v_2 + 2v_3 = 5.$$

A.2 CRAMER'S METHOD

The value of each unknown variable in the set of equations is expressed as the ratio of two determinants. If we let N, with an appropriate subscript, represent the numerator determinant and Δ represent the denominator determinant, then the kth unknown x_k is

$$x_k = \frac{N_k}{\Delta}. \tag{A.4}$$

The denominator determinant Δ is the same for every unknown variable and is called the *characteristic determinant* of the set of equations. The numerator determinant N_k varies with each unknown. Equation (A.4) is referred to as *Cramer's method* for solving simultaneous equations.

A.3 THE CHARACTERISTIC DETERMINANT

Once we have organized the set of simultaneous equations into an ordered array, as illustrated by Eqs. (A.1) and (A.3), it is a simple matter to form the characteristic determinant. The characteristic determinant is the square array made up from the coefficients of the unknown variables. For example, the characteristic determinants of Eqs. (A.1) and (A.3) are

$$\Delta = \begin{vmatrix} 21 & -9 & -12 \\ -3 & 6 & -2 \\ -8 & -4 & 22 \end{vmatrix} \tag{A.5}$$

and

$$\Delta = \begin{vmatrix} 2 & -1 & 0 \\ 0 & 4 & 3 \\ 7 & 0 & 2 \end{vmatrix}, \tag{A.6}$$

respectively.

A.4 THE NUMERATOR DETERMINANT

The numerator determinant N_k is formed from the characteristic determinant by replacing the kth column in the characteristic determinant by the column of values appearing on the right-hand side of the equations. For example, the numerator determinants for evaluating i_1, i_2, and i_3 in Eqs. (A.1) are

$$N_1 = \begin{vmatrix} -33 & -9 & -12 \\ 3 & 6 & -2 \\ 50 & -4 & 22 \end{vmatrix}, \tag{A.7}$$

$$N_2 = \begin{vmatrix} 21 & -33 & -12 \\ -3 & 3 & -2 \\ -8 & 50 & 22 \end{vmatrix}, \tag{A.8}$$

and

$$N_3 = \begin{vmatrix} 21 & -9 & -33 \\ -3 & 6 & 3 \\ -8 & -4 & 50 \end{vmatrix}. \tag{A.9}$$

The numerator determinants for the evaluation of v_1, v_2, and v_3 in Eqs. (A.3) are

$$N_1 = \begin{vmatrix} 4 & -1 & 0 \\ 16 & 4 & 3 \\ 5 & 0 & 2 \end{vmatrix}, \tag{A.10}$$

$$N_2 = \begin{vmatrix} 2 & 4 & 0 \\ 0 & 16 & 3 \\ 7 & 5 & 2 \end{vmatrix}, \tag{A.11}$$

and

$$N_3 = \begin{vmatrix} 2 & -1 & 4 \\ 0 & 4 & 16 \\ 7 & 0 & 5 \end{vmatrix}. \tag{A.12}$$

A.5 EVALUATION OF A DETERMINANT

The value of a determinant is found by expanding it in terms of its minors. The *minor* of any element in a determinant is the determinant that remains after the row and column occupied by the element have been deleted. For example, the minor of the element 6 in Eq. (A.7) is

$$\begin{vmatrix} -33 & -12 \\ 50 & 22 \end{vmatrix},$$

while the minor of the element 22 in Eq. (A.7) is

$$\begin{vmatrix} -33 & -9 \\ 3 & 6 \end{vmatrix}.$$

The *cofactor* of an element is its minor multiplied by the sign-controlling factor

$$-1^{(i+j)},$$

where i and j denote the row and column, respectively, occupied by the element. Thus the cofactor of the element 6 in Eq. (A.7) is

$$-1^{(2+2)} \begin{vmatrix} -33 & -12 \\ 50 & 22 \end{vmatrix}$$

and the cofactor of the element 22 is

$$-1^{(3+3)} \begin{vmatrix} -33 & -9 \\ 3 & 6 \end{vmatrix}.$$

The cofactor of an element is also referred to as its *signed minor*.

The sign-controlling factor $-1^{(i+j)}$ will equal plus or minus 1 depending on whether $i + j$ is an even or odd integer. Thus the algebraic sign of a cofactor alternates between ± 1 as we move along a row or column. For a 3-×-3 determinant, the + and − signs form the checkerboard pattern illustrated below:

$$\begin{vmatrix} + & - & + \\ - & + & - \\ + & - & + \end{vmatrix}$$

A determinant can be expanded along any row or column. Thus the first step in making an expansion is to select a row i or a column j. Once a row or column has been selected, each element in that row or column is multiplied by its signed minor, or cofactor. The value of the determinant is the sum of these products. As an example, let us evaluate the determinant in Eq. (A.5) by expanding it along its first column. Following the rules enumerated above, we write the expansion as

$$\Delta = 21(1) \begin{vmatrix} 6 & -2 \\ -4 & 22 \end{vmatrix} - 3(-1) \begin{vmatrix} -9 & -12 \\ -4 & 22 \end{vmatrix} - 8(1) \begin{vmatrix} -9 & -12 \\ 6 & -2 \end{vmatrix}.$$

(A.13)

The 2-×-2 determinants in Eq. (A.13) can also be expanded by minors. The minor of an element in a 2-×-2 determinant is a single element. It follows that the expansion reduces to multiplying the upper-left element by the lower-right element and then subtracting from this product the product of the lower-left element times the upper-right element. Using this observation, we evaluate Eq. (A.13) to

$$\Delta = 21(132 - 8) + 3(-198 - 48) - 8(18 + 72)$$
$$= 2604 - 738 - 720 = 1146.$$

(A.14)

Had we elected to expand the determinant along the second row of elements, we would have written

$$\Delta = -3(-1) \begin{vmatrix} -9 & -12 \\ -4 & 22 \end{vmatrix} + 6(+1) \begin{vmatrix} 21 & -12 \\ -8 & 22 \end{vmatrix} - 2(-1) \begin{vmatrix} 21 & -9 \\ -8 & -4 \end{vmatrix}$$
$$= 3(-198 - 48) + 6(462 - 96) + 2(-84 - 72)$$
$$= -738 + 2196 - 312 = 1146.$$

(A.15)

The numerical values of the determinants N_1, N_2, and N_3 given by Eqs. (A.7), (A.8), and (A.9) are

$$N_1 = 1146, \tag{A.16}$$

$$N_2 = 2292, \tag{A.17}$$

and

$$N_3 = 3438. \tag{A.18}$$

It follows from Eqs. (A.15) through (A.18) that the solutions for i_1, i_2, and i_3 in Eq. (A.1) are

$$i_1 = \frac{N_1}{\Delta} = 1 \text{ A},$$

$$i_2 = \frac{N_2}{\Delta} = 2 \text{ A},$$

and

$$i_3 = \frac{N_3}{\Delta} = 3 \text{ A}. \tag{A.19}$$

We will leave it to the reader to verify that the solutions for v_1, v_2, and v_3 in Eqs. (A.3) are

$$v_1 = \frac{49}{-5} = -9.8 \text{ V},$$

$$v_2 = \frac{118}{-5} = -23.6 \text{ V}, \tag{A.20}$$

and

$$v_3 = \frac{-184}{-5} = 36.8 \text{ V}.$$

A.6 MATRICES

A system of simultaneous linear equations can also be solved using matrices. In what follows we briefly review matrix notation, algebra, and terminology.[†]

[†]An excellent introductory-level text in matrix applications to circuit analysis is Lawrence P. Huelsman, *Circuits, Matrices, and Linear Vector Spaces* (New York: McGraw-Hill, 1963).

A *matrix* is by definition a rectangular array of elements; thus

$$
\mathbf{A} = \begin{bmatrix}
a_{11} & a_{12} & a_{13} & \cdots & a_{1n} \\
a_{21} & a_{22} & a_{23} & \cdots & a_{2n} \\
\cdot & \cdot & \cdot & \cdots & \cdot \\
a_{m1} & a_{m2} & a_{m3} & \cdots & a_{mn}
\end{bmatrix}
\tag{A.21}
$$

is a matrix with m rows and n columns. We describe $\mathbf{A}$ as being a matrix of order m by n, or $m \times n$, where m equals the number of rows and n the number of columns. We always specify the *rows first* and the *columns second*. The elements of the matrix—$a_{11}, a_{12}, a_{13}, \ldots$,—can be real numbers, complex numbers, or functions. We denote a matrix by a boldface capital letter.

The array in Eq. (A.21) is frequently abbreviated by writing

$$
\mathbf{A} = [a_{ij}]_{mn},
\tag{A.22}
$$

where a_{ij} is the element in the ith row and the jth column.

If $m = 1$, $\mathbf{A}$ is called a *row* matrix, i.e.,

$$
\mathbf{A} = [a_{11} \quad a_{12} \quad a_{13} \quad \cdots \quad a_{1n}].
\tag{A.23}
$$

If $n = 1$, $\mathbf{A}$ is called a *column* matrix, i.e.,

$$
\mathbf{A} = \begin{bmatrix}
a_{11} \\
a_{21} \\
a_{31} \\
\vdots \\
a_{m1}
\end{bmatrix}.
\tag{A.24}
$$

If $m = n$, $\mathbf{A}$ is called a *square* matrix. For example, if $m = n = 3$, the square 3-by-3 matrix is

$$
\mathbf{A} = \begin{bmatrix}
a_{11} & a_{12} & a_{13} \\
a_{21} & a_{22} & a_{23} \\
a_{31} & a_{32} & a_{33}
\end{bmatrix}.
\tag{A.25}
$$

We also point out that we use brackets $[\;]$ to denote a matrix, whereas we use vertical lines $|\;|$ to denote a determinant. It is important to know the difference. A *matrix* is a rectangular array of elements. A *determinant* is always a square array of elements *that is a function of the elements*. Thus if a matrix is square, we can define the determinant of $\mathbf{A}$. For example, if

$$
\mathbf{A} = \begin{bmatrix} 2 & 1 \\ 6 & 15 \end{bmatrix},
$$

then

$$
\det \mathbf{A} = \begin{vmatrix} 2 & 1 \\ 6 & 15 \end{vmatrix} = 30 - 6 = 24.
$$

A.7 MATRIX ALGEBRA

The equality, addition, and subtraction of matrices apply only to matrices of the same order. Two matrices are equal if, and only if, their corresponding elements are equal. That is $\mathbf{A} = \mathbf{B}$ if, and only if, $a_{ij} = b_{ij}$ for all i and j. For example, the two matrices in Eqs. (A.26) and (A.27) are equal because $a_{11} = b_{11}$, $a_{12} = b_{12}$, $a_{21} = b_{21}$, and $a_{22} = b_{22}$:

$$\mathbf{A} = \begin{bmatrix} 36 & -20 \\ 4 & 16 \end{bmatrix}, \tag{A.26}$$

$$\mathbf{B} = \begin{bmatrix} 36 & -20 \\ 4 & 16 \end{bmatrix}. \tag{A.27}$$

If $\mathbf{A}$ and $\mathbf{B}$ are of the same order, then

$$\mathbf{C} = \mathbf{A} + \mathbf{B} \tag{A.28}$$

implies

$$c_{ij} = a_{ij} + b_{ij}. \tag{A.29}$$

For example, if

$$\mathbf{A} = \begin{bmatrix} 4 & -6 & 10 \\ 8 & 12 & -4 \end{bmatrix} \tag{A.30}$$

and

$$\mathbf{B} = \begin{bmatrix} 16 & 10 & -30 \\ -20 & 8 & 15 \end{bmatrix}, \tag{A.31}$$

then

$$\mathbf{C} = \begin{bmatrix} 20 & 4 & -20 \\ -12 & 20 & 11 \end{bmatrix}. \tag{A.32}$$

The equation

$$\mathbf{D} = \mathbf{A} - \mathbf{B} \tag{A.33}$$

implies

$$d_{ij} = a_{ij} - b_{ij}. \tag{A.34}$$

For the matrices in Eqs. (A.30) and (A.31), we would have

$$\mathbf{D} = \begin{bmatrix} -12 & -16 & 40 \\ 28 & 4 & -19 \end{bmatrix}. \tag{A.35}$$

Matrices of the same order are said to be *conformable* for addition and subtraction.

Multiplying a matrix by a scalar k is equivalent to multiplying each element by the scalar. Thus $\mathbf{A} = k\mathbf{B}$ if, and only if, $a_{ij} = kb_{ij}$. It should be noted that k may be real or complex. As an example, we will multiply the matrix $\mathbf{D}$ in Eq. (A.35) by 5. The result is

$$5\mathbf{D} = \begin{bmatrix} -60 & -80 & 200 \\ 140 & 20 & -95 \end{bmatrix}. \qquad \textbf{(A.36)}$$

Matrix multiplication can be performed only if the number of columns in the first matrix is equal to the number of rows in the second matrix. In other words, the product $\mathbf{AB}$ requires the number of columns in $\mathbf{A}$ to equal the number of rows in $\mathbf{B}$. The order of the resulting matrix will be the number of rows in $\mathbf{A}$ by the number of columns in $\mathbf{B}$. Thus if $\mathbf{C} = \mathbf{AB}$, where $\mathbf{A}$ is of order $m \times p$ and $\mathbf{B}$ is of the order $p \times n$, then $\mathbf{C}$ will be a matrix of order $m \times n$. When the number of columns in $\mathbf{A}$ equals the number of rows in $\mathbf{B}$, we say $\mathbf{A}$ is conformable to $\mathbf{B}$ for multiplication.

An element in $\mathbf{C}$ is given by the formula

$$c_{ij} = \sum_{k=1}^{p} a_{ik} b_{kj}. \qquad \textbf{(A.37)}$$

The formula given by Eq. (A.37) is easy to use if one remembers that matrix multiplication is a row-by-column operation. Hence to get the ith, jth term in $\mathbf{C}$, each element in the ith *row* of $\mathbf{A}$ is multiplied by the corresponding element in the jth *column* of $\mathbf{B}$ and the resulting products are summed. The following example illustrates the procedure. We are asked to find the matrix $\mathbf{C}$ when

$$\mathbf{A} = \begin{bmatrix} 6 & 3 & 2 \\ 1 & 4 & 6 \end{bmatrix} \qquad \textbf{(A.38)}$$

and

$$\mathbf{B} = \begin{bmatrix} 4 & 2 \\ 0 & 3 \\ 1 & -2 \end{bmatrix}. \qquad \textbf{(A.39)}$$

First we note that $\mathbf{C}$ will be a 2-×-2 matrix and each element in $\mathbf{C}$ will require summing three products.

To find C_{11} we multiply the corresponding elements in row 1 of matrix $\mathbf{A}$ with the elements in column 1 of matrix $\mathbf{B}$ and then sum the products. We can visualize this multiplication and summing process by extracting the corresponding row and column from each matrix and then lining them up element by element.

So to find C_{11} we have

Row 1 of **A**	6	3	2
Column 1 of **B**	4	0	1

;

therefore

$$C_{11} = 6 \times 4 + 3 \times 0 + 2 \times 1 = 26.$$

To find C_{12} we visualize

Row 1 of **A**	6	3	2
Column 2 of **B**	2	3	-2

;

thus

$$C_{12} = 6 \times 2 + 3 \times 3 + 2 \times (-2) = 17.$$

For C_{21} we have

Row 2 of **A**	1	4	6
Column 1 of **B**	4	0	1

and

$$C_{21} = 1 \times 4 + 4 \times 0 + 6 \times 1 = 10.$$

Finally, for C_{22} we have

Row 2 of **A**	1	4	6
Column 2 of **B**	2	3	-2

;

from which

$$C_{22} = 1 \times 2 + 4 \times 3 + 6 \times (-2) = 2.$$

It follows that

$$\mathbf{C} = \mathbf{AB} = \begin{bmatrix} 26 & 17 \\ 10 & 2 \end{bmatrix}. \tag{A.40}$$

In general, matrix multiplication is not commutative, that is, $\mathbf{AB} \neq \mathbf{BA}.$ As an example, consider the product $\mathbf{BA}$ for the matrices in Eqs. (A.38) and (A.39). The matrix generated by this multiplication is of order 3×3, and each term in the resulting matrix requires adding two products. Therefore if $\mathbf{D} = \mathbf{BA},$ we have

$$\mathbf{D} = \begin{bmatrix} 26 & 20 & 20 \\ 3 & 12 & 18 \\ 4 & -5 & -10 \end{bmatrix}. \tag{A.41}$$

Obviously $\mathbf{C} \neq \mathbf{D}$. We will leave it to the reader to verify the elements in Eq. (A.41).

Matrix multiplication is associative and distributive. Thus

$$(\mathbf{AB})\mathbf{C} = \mathbf{A}(\mathbf{BC}), \tag{A.42}$$

$$\mathbf{A}(\mathbf{B} + \mathbf{C}) = \mathbf{AB} + \mathbf{AC}, \tag{A.43}$$

and

$$(\mathbf{A} + \mathbf{B})\mathbf{C} = \mathbf{AC} + \mathbf{BC}. \tag{A.44}$$

In Eqs. (A.42), (A.43), and (A.44) we assume that the matrices are conformable for addition and multiplication.

We have already noted that matrix multiplication is not commutative. There are two other properties of multiplication in scalar algebra that do not carry over to matrix algebra.

First, the matrix product $\mathbf{AB} = 0$ does not imply either $\mathbf{A} = 0$ or $\mathbf{B} = 0$. (*Note:* A matrix is equal to zero when all its elements are zero.) For example, if

$$\mathbf{A} = \begin{bmatrix} 1 & 0 \\ 2 & 0 \end{bmatrix} \quad \text{and} \quad \mathbf{B} = \begin{bmatrix} 0 & 0 \\ 4 & 8 \end{bmatrix},$$

then

$$\mathbf{AB} = \begin{bmatrix} 0 & 0 \\ 0 & 0 \end{bmatrix} = 0.$$

Hence the product is zero, but neither $\mathbf{A}$ nor $\mathbf{B}$ is zero.

Second, the matrix equation $\mathbf{AB} = \mathbf{AC}$ does not imply $\mathbf{B} = \mathbf{C}$. For example, if

$$\mathbf{A} = \begin{bmatrix} 1 & 0 \\ 2 & 0 \end{bmatrix}, \quad \mathbf{B} = \begin{bmatrix} 3 & 4 \\ 7 & 8 \end{bmatrix}, \quad \text{and} \quad \mathbf{C} = \begin{bmatrix} 3 & 4 \\ 5 & 6 \end{bmatrix},$$

then

$$\mathbf{AB} = \mathbf{AC} = \begin{bmatrix} 3 & 4 \\ 6 & 8 \end{bmatrix} \quad \text{but } \mathbf{B} \neq \mathbf{C}.$$

The *transpose* of a matrix is formed by interchanging the rows and columns. For example, if

$$\mathbf{A} = \begin{bmatrix} 1 & 2 & 3 \\ 4 & 5 & 6 \\ 7 & 8 & 9 \end{bmatrix}, \quad \text{then} \quad \mathbf{A}^T = \begin{bmatrix} 1 & 4 & 7 \\ 2 & 5 & 8 \\ 3 & 6 & 9 \end{bmatrix}.$$

The transpose of the sum of two matrices is equal to the sum of the transposes, i.e.,

$$(\mathbf{A} + \mathbf{B})^T = \mathbf{A}^T + \mathbf{B}^T. \tag{A.45}$$

The transpose of the product of two matrices is equal to the product of the transposes taken in reverse order. In other words,

$$[\mathbf{AB}]^T = \mathbf{B}^T \mathbf{A}^T. \tag{A.46}$$

Equation (A.46) can be extended to a product of any number of matrices. For example,

$$[\mathbf{ABCD}]^T = \mathbf{D}^T \mathbf{C}^T \mathbf{B}^T \mathbf{A}^T. \tag{A.47}$$

If $\mathbf{A} = \mathbf{A}^T$, the matrix is said to be *symmetric*. Only square matrices can be symmetric.

A.8 IDENTITY, ADJOINT, AND INVERSE MATRICES

The *identity matrix* is a square matrix where $a_{ij} = 0$ for $i \neq j$ and $a_{ij} = 1$ for $i = j$. In other words, all the elements in an identity matrix are zero except those along the main diagonal, where they are equal to 1. Thus

$$\begin{bmatrix} 1 & 0 \\ 0 & 1 \end{bmatrix}, \quad \begin{bmatrix} 1 & 0 & 0 \\ 0 & 1 & 0 \\ 0 & 0 & 1 \end{bmatrix}, \quad \begin{bmatrix} 1 & 0 & 0 & 0 \\ 0 & 1 & 0 & 0 \\ 0 & 0 & 1 & 0 \\ 0 & 0 & 0 & 1 \end{bmatrix}$$

are all identity matrices. Note that identity matrices are always square. We shall use the boldface $\mathbf{U}$ for the identity matrix.

The *adjoint* of a matrix $\mathbf{A}$ of order $n \times n$ is defined as

$$\text{adj } \mathbf{A} = [\Delta_{ji}]_{n \times n}, \tag{A.48}$$

where Δ_{ij} is the cofactor of a_{ij}. (See Section A.5 for the definition of a cofactor.) It follows from Eq. (A.48) that one can think of finding the adjoint of a square matrix as a two-step process. First construct a matrix made up of the cofactors of $\mathbf{A}$ and then transpose the matrix of cofactors. As an example we will find the adjoint of the 3-by-3 matrix

$$\mathbf{A} = \begin{bmatrix} 1 & 2 & 3 \\ 3 & 2 & 1 \\ -1 & 1 & 5 \end{bmatrix}.$$

The cofactors of the elements in **A** are

$$\Delta_{11} = 1(10 - 1) = 9,$$
$$\Delta_{12} = -1(15 + 1) = -16,$$
$$\Delta_{13} = 1(3 + 2) = 5,$$
$$\Delta_{21} = -1(10 - 3) = -7,$$
$$\Delta_{22} = 1(5 + 3) = 8,$$
$$\Delta_{23} = -1(1 + 2) = -3,$$
$$\Delta_{31} = 1(2 - 6) = -4,$$
$$\Delta_{32} = -1(1 - 9) = 8,$$
$$\Delta_{33} = 1(2 - 6) = -4.$$

The matrix of cofactors is

$$\mathbf{B} = \begin{bmatrix} 9 & -16 & 5 \\ -7 & 8 & -3 \\ -4 & 8 & -4 \end{bmatrix}.$$

It follows that the adjoint of **A** is

$$\text{adj } \mathbf{A} = \mathbf{B}^T = \begin{bmatrix} 9 & -7 & -4 \\ -16 & 8 & 8 \\ 5 & -3 & -4 \end{bmatrix}.$$

One can check the arithmetic of finding the adjoint of a matrix by using the theorem

$$\text{adj } \mathbf{A} \cdot \mathbf{A} = \det \mathbf{A} \cdot \mathbf{U}. \qquad \text{(A.49)}$$

Equation (A.49) tells us that the adjoint of **A** times **A** equals the determinant of **A** times the identity matrix, or for our example,

$$\det \mathbf{A} = 1(9) + 3(-7) - 1(-4) = -8.$$

If we let $\mathbf{C} = \text{adj } \mathbf{A} \cdot \mathbf{A}$ and use the technique illustrated in Section A.7, we find the elements of **C** to be

$$c_{11} = 9 - 21 + 4 = -8,$$
$$c_{12} = 18 - 14 - 4 = 0,$$
$$c_{13} = 27 - 7 - 20 = 0,$$
$$c_{21} = -16 + 24 - 8 = 0,$$
$$c_{22} = -32 + 16 + 8 = -8,$$
$$c_{23} = -48 + 8 + 40 = 0,$$
$$c_{31} = 5 - 9 + 4 = 0,$$
$$c_{32} = 10 - 6 - 4 = 0,$$
$$c_{33} = 15 - 3 - 20 = -8.$$

Therefore

$$\mathbf{C} = \begin{bmatrix} -8 & 0 & 0 \\ 0 & -8 & 0 \\ 0 & 0 & -8 \end{bmatrix} = -8 \begin{bmatrix} 1 & 0 & 0 \\ 0 & 1 & 0 \\ 0 & 0 & 1 \end{bmatrix}$$
$$= \det \mathbf{A} \cdot \mathbf{U}.$$

A square matrix $\mathbf{A}$ has an *inverse,* denoted as $\mathbf{A}^{-1}$, if

$$\mathbf{A}^{-1}\mathbf{A} = \mathbf{A}\mathbf{A}^{-1} = \mathbf{U}. \tag{A.50}$$

Equation (A.50) tells us that a matrix either premultiplied or postmultiplied by its inverse generates the identity matrix $\mathbf{U}$. For the inverse matrix to exist, it is necessary that the det $\mathbf{A}$ not equal zero. Only square matrices have inverses and the inverse is also square.

A formula for finding the inverse of a matrix is

$$\mathbf{A}^{-1} = \frac{\text{adj } \mathbf{A}}{\det \mathbf{A}}. \tag{A.51}$$

The formula in Eq. (A.51) becomes very cumbersome if $\mathbf{A}$ is of an order larger than 3 by 3.[†] Today the digital computer eliminates the drudgery of having to find the inverse of a matrix in numerical applications of matrix algebra.

It follows from Eq. (A.51) that the inverse of the matrix $\mathbf{A}$ in the previous example is

$$\mathbf{A}^{-1} = -1/8 \begin{bmatrix} 9 & -7 & -4 \\ -16 & 8 & 8 \\ 5 & -3 & -4 \end{bmatrix}$$
$$= \begin{bmatrix} -1.125 & 0.875 & 0.5 \\ 2 & -1 & -1 \\ -0.625 & 0.375 & 0.5 \end{bmatrix}.$$

We leave to you verification that $\mathbf{A}^{-1}\mathbf{A} = \mathbf{A}\mathbf{A}^{-1} = \mathbf{U}$.

A.9   PARTITIONED MATRICES

It is often convenient in matrix manipulations to partition a given matrix into submatrices. The original algebraic operations are then carried out in terms of the submatrices. In partitioning a

---

[†] The interested reader can learn alternate methods for finding the inverse in any introductory text on matrix theory. See, for example, Franz E. Hohn, *Elementary Matrix Algebra* (New York: Macmillan, 1973).

matrix, the placement of the partitions is completely arbitrary, with the one restriction that a partition must dissect the entire matrix. In selecting the partitions, it is also necessary to make sure the submatrices are conformable for the mathematical operations in which they are involved.

For example, consider using submatrices to find the product $\mathbf{C} = \mathbf{AB}$, where

$$\mathbf{A} = \begin{bmatrix} 1 & 2 & 3 & 4 & 5 \\ 5 & 4 & 3 & 2 & 1 \\ -1 & 0 & 2 & -3 & 1 \\ 0 & 1 & -1 & 0 & 1 \\ 0 & 2 & 1 & -2 & 0 \end{bmatrix}$$

and

$$\mathbf{B} = \begin{bmatrix} 2 \\ 0 \\ -1 \\ 3 \\ 0 \end{bmatrix} .$$

Assume that we decide to partition $\mathbf{B}$ into two submatrices $\mathbf{B}_{11}$ and $\mathbf{B}_{21}$; thus

$$\mathbf{B} = \begin{bmatrix} \mathbf{B}_{11} \\ \mathbf{B}_{21} \end{bmatrix} .$$

Now since $\mathbf{B}$ has been partitioned into a two-row column matrix, $\mathbf{A}$ must be partitioned into at least a two-column matrix; otherwise the multiplication cannot be performed. The location of the vertical partitions of the $\mathbf{A}$ matrix will depend on the definitions of $\mathbf{B}_{11}$ and $\mathbf{B}_{21}$. For example, if $\mathbf{B}_{11}$ is defined as

$\begin{bmatrix} 2 \\ 0 \\ -1 \end{bmatrix}$ and $\mathbf{B}_{21}$ as $\begin{bmatrix} 3 \\ 0 \end{bmatrix}$, then $\mathbf{A}_{11}$ must contain three columns and

$\mathbf{A}_{12}$ must contain two columns. Thus the partitioning shown in Eq. (A.52) would be acceptable for executing the product $\mathbf{AB}$:

$$\mathbf{C} = \begin{bmatrix} 1 & 2 & 3 & \vdots & 4 & 5 \\ 5 & 4 & 3 & \vdots & -2 & 1 \\ -1 & 0 & 2 & \vdots & 3 & 1 \\ 0 & 1 & -1 & \vdots & 0 & 1 \\ 0 & 2 & 1 & \vdots & -2 & 0 \end{bmatrix} \begin{bmatrix} 2 \\ 0 \\ -1 \\ \cdots \\ 3 \\ 0 \end{bmatrix} . \qquad \textbf{(A.52)}$$

If, on the other hand, we partition the $\mathbf{B}$ matrix so that $\mathbf{B}_{11}$ is defined as $\begin{bmatrix} 2 \\ 0 \end{bmatrix}$ and $\mathbf{B}_{21}$ as $\begin{bmatrix} -1 \\ 3 \\ 0 \end{bmatrix}$, then $\mathbf{A}_{11}$ must contain two columns and $\mathbf{A}_{12}$ must contain three columns. In this case the partitioning shown in Eq. (A.53) would be acceptable in executing the product $\mathbf{C} = \mathbf{AB}$:

$$
\mathbf{C} =
\begin{bmatrix}
1 & 2 & \vdots & 3 & 4 & 5 \\
5 & 4 & \vdots & 3 & 2 & 1 \\
-1 & 0 & \vdots & 2 & -3 & 1 \\
0 & 1 & \vdots & -1 & 0 & 1 \\
0 & 2 & \vdots & 1 & -2 & 0
\end{bmatrix}
\begin{bmatrix}
2 \\
0 \\
\cdots \\
-1 \\
3 \\
0
\end{bmatrix} . \tag{A.53}
$$

For purposes of discussion we will focus on the partitioning given in Eq. (A.52) and leave it to the reader to verify that the partitioning in Eq. (A.53) leads to the same result.

From Eq. (A.52) we can write

$$
\mathbf{C} = [\mathbf{A}_{11} \quad \mathbf{A}_{12}] \begin{bmatrix} \mathbf{B}_{11} \\ \mathbf{B}_{21} \end{bmatrix} = \mathbf{A}_{11}\mathbf{B}_{11} + \mathbf{A}_{12}\mathbf{B}_{21}. \tag{A.54}
$$

It follows from Eqs. (A.52) and (A.54) that

$$
\mathbf{A}_{11}\mathbf{B}_{11} =
\begin{bmatrix}
1 & 2 & 3 \\
5 & 4 & 3 \\
-1 & 0 & 2 \\
0 & 1 & -1 \\
0 & 2 & 1
\end{bmatrix}
\begin{bmatrix}
2 \\
0 \\
-1
\end{bmatrix}
=
\begin{bmatrix}
-1 \\
7 \\
-4 \\
1 \\
-1
\end{bmatrix} ,
$$

$$
\mathbf{A}_{12}\mathbf{B}_{21} =
\begin{bmatrix}
4 & 5 \\
2 & 1 \\
-3 & 1 \\
0 & 1 \\
-2 & 0
\end{bmatrix}
\begin{bmatrix}
3 \\
0
\end{bmatrix}
=
\begin{bmatrix}
12 \\
6 \\
-9 \\
0 \\
-6
\end{bmatrix} ,
$$

and

$$
\mathbf{C} =
\begin{bmatrix}
11 \\
13 \\
-13 \\
1 \\
-7
\end{bmatrix} .
$$

The $\mathbf{A}$ matrix could also be partitioned horizontally once the vertical partitioning is made consistent with the multiplication operation. In this simple multiplication problem, the horizontal partitions can be made at the discretion of the analyst. Therefore **C** could also be evaluated using the partitioning shown in Eq. (A.55):

$$C = \left[\begin{array}{ccc:cc} 1 & 2 & 3 & 4 & 5 \\ 5 & 4 & 3 & 2 & 1 \\ \hdashline -1 & 0 & 2 & -3 & 1 \\ 0 & 1 & -1 & 0 & 1 \\ 0 & 2 & 1 & -2 & 0 \end{array}\right] \left[\begin{array}{c} 2 \\ 0 \\ -1 \\ \hdashline \cdots \\ 3 \\ 0 \end{array}\right]. \qquad \textbf{(A.55)}$$

From Eq. (A.55) it follows that

$$C = \begin{bmatrix} A_{11} & A_{12} \\ A_{21} & A_{22} \end{bmatrix} \begin{bmatrix} B_{11} \\ B_{21} \end{bmatrix} = \begin{bmatrix} C_{11} \\ C_{21} \end{bmatrix}, \qquad \textbf{(A.56)}$$

where

$$C_{11} = A_{11}B_{11} + A_{12}B_{21},$$
$$C_{21} = A_{21}B_{11} + A_{22}B_{21}.$$

We shall leave it to the reader to verify that

$$C_{11} = \begin{bmatrix} 1 & 2 & 3 \\ 5 & 4 & 3 \end{bmatrix} \begin{bmatrix} 2 \\ 0 \\ -1 \end{bmatrix} + \begin{bmatrix} 4 & 5 \\ 2 & 1 \end{bmatrix} \begin{bmatrix} 3 \\ 0 \end{bmatrix}$$

$$= \begin{bmatrix} -1 \\ 7 \end{bmatrix} + \begin{bmatrix} 12 \\ 6 \end{bmatrix} = \begin{bmatrix} 11 \\ 13 \end{bmatrix},$$

$$C_{21} = \begin{bmatrix} -1 & 0 & 2 \\ 0 & 1 & -1 \\ 0 & 2 & 1 \end{bmatrix} \begin{bmatrix} 2 \\ 0 \\ -1 \end{bmatrix} + \begin{bmatrix} -3 & 1 \\ 0 & 1 \\ -2 & 0 \end{bmatrix} \begin{bmatrix} 3 \\ 0 \end{bmatrix}$$

$$= \begin{bmatrix} -4 \\ 1 \\ -1 \end{bmatrix} + \begin{bmatrix} -9 \\ 0 \\ -6 \end{bmatrix} = \begin{bmatrix} -13 \\ 1 \\ -7 \end{bmatrix},$$

and

$$C = \begin{bmatrix} 11 \\ 13 \\ -13 \\ 1 \\ -7 \end{bmatrix}.$$

We note in passing that the partitioning in Eq. (A.52) and Eq. (A.55) is conformable with respect to addition.

A.10 APPLICATIONS

The following examples demonstrate some applications of matrix algebra in circuit analysis.

E X A M P L E   A.1

Use the matrix method to solve for the node voltages v_1 and v_2 in Eqs. (4.5) and (4.6).

S O L U T I O N

The first step is to rewrite Eqs. (4.5) and (4.6) in matrix notation. Collecting the coefficients of v_1 and v_2 and at the same time shifting the constant terms to the right-hand side of the equations gives us

$$1.7v_1 - 0.5v_2 = 10$$
$$-0.5v_1 + 0.6v_2 = 2. \qquad \text{(A.57)}$$

It follows that in matrix notation Eq. (A.57) becomes

$$\begin{bmatrix} 1.7 & -0.5 \\ -0.5 & 0.6 \end{bmatrix}\begin{bmatrix} v_1 \\ v_2 \end{bmatrix} = \begin{bmatrix} 10 \\ 2 \end{bmatrix}, \qquad \text{(A.58)}$$

or

$$\mathbf{AV} = \mathbf{I} \qquad \text{(A.59)}$$

where

$$\mathbf{A} = \begin{bmatrix} 1.7 & -0.5 \\ -0.5 & 0.6 \end{bmatrix}$$

$$\mathbf{V} = \begin{bmatrix} v_1 \\ v_2 \end{bmatrix}$$

$$\mathbf{I} = \begin{bmatrix} 10 \\ 2 \end{bmatrix}.$$

To find the elements of the V matrix, we premultiply both sides of Eq. (A.59) by the inverse of A; thus

$$\mathbf{A}^{-1}\mathbf{AV} = \mathbf{A}^{-1}\mathbf{I}. \qquad \text{(A.60)}$$

Equation A.60 reduces to

$$\mathbf{UV} = \mathbf{A}^{-1}\mathbf{I}, \qquad \text{(A.61)}$$

or

$$V = A^{-1}I. \qquad \text{(A.62)}$$

It follows from Eq. (A.62) that the solutions for v_1 and v_2 are obtained by solving for the matrix product $A^{-1}I$.

To find the inverse of A, we first find the cofactors of A. Thus

$$\Delta_{11} = (-1)^2(0.6) = 0.6$$
$$\Delta_{12} = (-1)^3(-0.5) = 0.5$$
$$\Delta_{21} = (-1)^3(-0.5) = 0.5 \qquad \text{(A.63)}$$
$$\Delta_{22} = (-1)^4(1.7) = 1.7.$$

The matrix of cofactors is

$$B = \begin{bmatrix} 0.6 & 0.5 \\ 0.5 & 1.7 \end{bmatrix}. \qquad \text{(A.64)}$$

and the adjoint of A is

$$\text{adj } A = B^T = \begin{bmatrix} 0.6 & 0.5 \\ 0.5 & 1.7 \end{bmatrix}. \qquad \text{(A.65)}$$

The determinate of A is

$$\det A = \begin{bmatrix} 1.7 & -0.5 \\ -0.5 & 0.6 \end{bmatrix} = (1.7)(0.6) - (0.25) = 0.77. \qquad \text{(A.66)}$$

From Eqs. (A.65) and (A.66), we can write the inverse of the coefficient matrix, i.e.,

$$A^{-1} = \frac{1}{0.77}\begin{bmatrix} 0.6 & 0.5 \\ 0.5 & 1.7 \end{bmatrix}. \qquad \text{(A.67)}$$

Now the product $A^{-1}I$ is found, thus

$$A^{-1}I = \frac{100}{77}\begin{bmatrix} 0.6 & 0.5 \\ 0.5 & 1.7 \end{bmatrix}\begin{bmatrix} 10 \\ 2 \end{bmatrix}$$

$$= \frac{100}{77}\begin{bmatrix} 7 \\ 8.4 \end{bmatrix} = \begin{bmatrix} 9.09 \\ 10.91 \end{bmatrix}. \qquad \text{(A.68)}$$

It follows directly that

$$\begin{bmatrix} v_1 \\ v_2 \end{bmatrix} = \begin{bmatrix} 9.09 \\ 10.91 \end{bmatrix}, \qquad \text{(A.69)}$$

or $v_1 = 9.09$ V and $v_2 = 10.91$ V.

EXAMPLE   A.2

Use the matrix method to find the three mesh currents in the circuit in Fig. 4.24.

SOLUTION

The mesh current equations that describe the circuit in Fig. 4.24 are given in Eq. (4.34). The constraint equation imposed by the current controlled voltage source is given in Eq. (4.35). When Eq. (4.35) is substituted into Eq. (4.34) the following set of equations evolves:

$$25i_i - 5i_2 - 20i_3 = 50$$
$$-5i_i + 10i_2 - 4i_3 = 0 \qquad \textbf{(A.70)}$$
$$-5i_1 - 4i_2 + 9i_3 = 0.$$

In matrix notation Eqs. (A.70) reduces to

$$\mathbf{AI} = \mathbf{V} \qquad \textbf{(A.71)}$$

where

$$\mathbf{A} = \begin{bmatrix} 25 & -5 & -20 \\ -5 & 10 & -4 \\ -5 & -4 & +9 \end{bmatrix}$$

$$\mathbf{I} = \begin{bmatrix} i_1 \\ i_2 \\ i_3 \end{bmatrix}$$

$$\mathbf{V} = \begin{bmatrix} 50 \\ 0 \\ 0 \end{bmatrix}.$$

It follows from Eq. (A.71) that the solution for **I** is

$$I = \mathbf{A}^{-1}\mathbf{V}. \qquad \textbf{(A.72)}$$

We find the inverse of **A** using the relationship

$$\mathbf{A}^{-1} = \frac{\text{adj } \mathbf{A}}{\text{det } \mathbf{A}}. \qquad \textbf{(A.73)}$$

To find the adjoint of **A**, we first calculate the cofactors of **A**.

Thus

$$\Delta_{11} = (-1)^2(90 - 16) = 74$$

$$\Delta_{12} = (-1)^3(-45 - 20) = 65$$

$$\Delta_{13} = (-1)^4(20 + 50) = 70$$

$$\Delta_{21} = (-1)^3(-45 - 80) = 125$$

$$\Delta_{22} = (-1)^4(225 - 100) = 125$$

$$\Delta_{23} = (-1)^5(-100 - 25) = 125$$

$$\Delta_{31} = (-1)^4(20 + 200) = 220$$

$$\Delta_{32} = (-1)^5(-100 - 100) = 200$$

$$\Delta_{33} = (-1)^6(250 - 25) = 225.$$

The cofactor matrix is

$$\mathbf{B} = \begin{bmatrix} 74 & 65 & 70 \\ 125 & 125 & 125 \\ 220 & 200 & 225 \end{bmatrix}, \qquad \textbf{(A.74)}$$

from which we can write the adjoint of $\mathbf{A}$.

$$\text{adj } \mathbf{A} = \mathbf{B}^T = \begin{bmatrix} 74 & 125 & 220 \\ 65 & 125 & 220 \\ 70 & 125 & 225 \end{bmatrix} \qquad \textbf{(A.75)}$$

The determinate of $\mathbf{A}$ is

$$\det \mathbf{A} = \begin{bmatrix} 25 & -5 & -20 \\ -5 & 10 & -4 \\ -5 & -4 & 9 \end{bmatrix} =$$

$$25(90 - 16) + 5(-45 - 80) - 5(20 + 200) = 125.$$

It follows from Eq. (A.73) that

$$\mathbf{A}^{-1} = \frac{1}{125} \begin{bmatrix} 74 & 125 & 220 \\ 65 & 125 & 200 \\ 70 & 125 & 225 \end{bmatrix}. \qquad \textbf{(A.76)}$$

The solution for $\mathbf{I}$ is

$$\mathbf{I} = \frac{1}{125} \begin{bmatrix} 74 & 125 & 220 \\ 65 & 125 & 200 \\ 70 & 125 & 225 \end{bmatrix} \begin{bmatrix} 50 \\ 0 \\ 0 \end{bmatrix} = \begin{bmatrix} 29.60 \\ 26.00 \\ 28.00 \end{bmatrix}. \qquad \textbf{(A.77)}$$

The mesh currents follow directly from Eq. (A.77). Thus

$$\begin{bmatrix} i_i \\ i_2 \\ i_3 \end{bmatrix} = \begin{bmatrix} 29.6 \\ 26.0 \\ 28.0 \end{bmatrix}, \qquad \textbf{(A.78)}$$

or $i_1 = 29.6$ A, $i_2 = 26$ A, and $i_3 = 28$ A.

Example A.3 illustrates the application of the matrix method when the elements of the matrix are complex numbers.

E X A M P L E   A.3

Using the matrix method to find the phasor mesh currents $\mathbf{I}_1$ and $\mathbf{I}_2$ in the circuit in Fig. 10.36.

S O L U T I O N

Summing the voltages around mesh 1 generates the equation

$$(1 + j2)\mathbf{I}_1 + (12 - j16)(\mathbf{I}_1 - \mathbf{I}_2) = 150\underline{/0°}. \qquad \textbf{(A.79)}$$

Summing the voltages around mesh 2 produces the equation

$$(12 - j16)(\mathbf{I}_2 - \mathbf{I}_1) + (1 + j3)\mathbf{I}_2 + 39\mathbf{I}_x = 0. \qquad \textbf{(A.80)}$$

The current controlling the dependent voltage source is

$$\mathbf{I}_x = (\mathbf{I}_1 - \mathbf{I}_2). \qquad \textbf{(A.81)}$$

After substituting Eq. (A.81) into Eq. (A.80), the equations are put into a matrix format by first collecting, in each equation, the coefficients of $\mathbf{I}_1$ and $\mathbf{I}_2$; thus

$$\begin{aligned} (13 - j14)\mathbf{I}_1 - (12 - j16)\mathbf{I}_2 &= 150\underline{/0°} \\ (27 + j16)\mathbf{I}_1 - (26 + j13)\mathbf{I}_2 &= 0. \end{aligned} \qquad \textbf{(A.82)}$$

Now, using matrix notation, Eq. (A.82) is written

$$\mathbf{AI} = \mathbf{V}, \qquad \textbf{(A.83)}$$

where

$$\mathbf{A} = \begin{bmatrix} 13 - j14 & -(12 - j16) \\ 27 + j16 & -(26 + j13) \end{bmatrix},$$

$$\mathbf{I} = \begin{bmatrix} I_1 \\ I_2 \end{bmatrix} \quad \text{and} \quad \mathbf{V} = \begin{bmatrix} 150\underline{/0°} \\ 0 \end{bmatrix}.$$

It follows from Eq. (A.83) that

$$I = A^{-1}V. \tag{A.84}$$

The inverse of the coefficient matrix A is found using Eq. (A.73). In this case, the cofactors of A are

$$\Delta_{11} = (-1)^2(-26 - j13) = -26 - j13$$

$$\Delta_{12} = (-1)^3(27 + j16) = -27 - j16$$

$$\Delta_{21} = (-1)^3(-12 + j16) = 12 - j16$$

$$\Delta_{22} = (-1)^4(13 - j14) = 13 - j14.$$

The cofactor matrix B is

$$B = \begin{bmatrix} (-26 - j13) & (-27 - j16) \\ (12 - j16) & (13 - j14) \end{bmatrix}. \tag{A.85}$$

The adjoint of A is

$$\text{adj } A = B^T = \begin{bmatrix} (-26 - j13) & (12 - j16) \\ (-27 - j16) & (13 - j14) \end{bmatrix}. \tag{A.86}$$

The determinate of A is

$$\det A = \begin{vmatrix} (13 - j14) & -(12 - j16) \\ (27 + j16) & -(26 + j13) \end{vmatrix} \tag{A.87}$$

$$= -(13 - j14)(26 + j13) + (12 - j16)(27 + j16)$$

$$= 60 - j45.$$

The inverse of the coefficient matrix is

$$A^{-1} = \frac{\begin{bmatrix} (-26 - j13) & (12 - j16) \\ (-27 - j16) & (13 - j14) \end{bmatrix}}{(60 - j45)}. \tag{A.88}$$

Equation (A.88) can be simplified to

$$A^{-1} = \frac{60 + j45}{5625} \begin{bmatrix} (-26 - j13) & (12 - j16) \\ (-27 - j16) & (13 - j14) \end{bmatrix}$$

$$= \frac{1}{375} \begin{bmatrix} -65 - j130 & 96 - j28 \\ -60 - j145 & 94 - j17 \end{bmatrix}. \tag{A.89}$$

Substituting Eq. (A.89) into (A.84) gives us

$$\begin{bmatrix} I_1 \\ I_2 \end{bmatrix} = \frac{1}{375} \begin{bmatrix} (-65 - j130) & (96 - j28) \\ (-60 - j145) & (94 - j17) \end{bmatrix} \begin{bmatrix} 150\underline{/0^\circ} \\ 0 \end{bmatrix}$$

$$= \begin{bmatrix} (-26 - j52) \\ (-24 - j58) \end{bmatrix}. \tag{A.90}$$

It follows from Eq. (A.90) that

$$\mathbf{I_1} = (-26 - j52) = 58.14\underline{/-116.57°}\,\mathbf{A}$$
$$\mathbf{I_2} = (-24 - j58) = 62.77\underline{/-122.48°}\,\mathbf{A}.$$

(A.91)

In the first three examples, the matrix elements have been numbers—real numbers in Examples A.1 and A.2 and complex numbers in Example A.3. It is also possible for the elements of a matrix to be functions. Example A.4 illustrates the use of matrix algebra in a circuit problem where the elements in the coefficient matrix are functions.

E X A M P L E　A.4

Use the matrix method to derive expressions for the node voltages V_1 and V_2 in the circuit in Fig. A.1.

S O L U T I O N

Summing the currents away from nodes 1 and 2 generates the following set of equations:

$$\frac{V_1 - V_g}{R} + V_1 sC + (V_1 - V_2)sC = 0$$

(A.92)

$$\frac{V_2}{R} + (V_2 - V_1)sC + (V_2 - V_g)sC = 0.$$

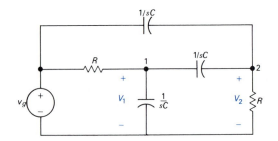

FIGURE A.1 Circuit for Example A.4.

Letting $G = 1/R$ and collecting the coefficients of V_1 and V_2 gives us

$$(G + 2sC)V_1 - sCV_2 = GV_g$$
$$-sCV_1 + (G + 2sC)V_2 = sCV_g.$$

(A.93)

Writing Eq. (A.93) in matrix notation yields

$$\mathbf{AV} = \mathbf{I}$$

(A.94)

where

$$\mathbf{A} = \begin{bmatrix} (G + 2sC) & -sC \\ -sC & G + 2sC \end{bmatrix}$$

$$\mathbf{V} = \begin{bmatrix} V_1 \\ V_2 \end{bmatrix}, \text{ and } \mathbf{I} = \begin{bmatrix} GV_g \\ sCV_g \end{bmatrix}.$$

It follows from Eq. (A.94) that

$$\mathbf{V} = \mathbf{A}^{-1}\mathbf{I}. \tag{A.95}$$

As before, we find the inverse of the coefficient matrix by first finding the adjoint of $\mathbf{A}$ and the determinate of $\mathbf{A}$.

The cofactors of $\mathbf{A}$ are

$$\Delta_{11} = (-1)^2[G + 2sC] = G + 2sC$$

$$\Delta_{12} = (-1)^3(-sC) = sC$$

$$\Delta_{21} = (-1)^3(-sC) = sC$$

$$\Delta_{22} = (-1)^4[G + 2sC] = G + 2sC.$$

The cofactor matrix is

$$\mathbf{B} = \begin{bmatrix} G + 2sC & sC \\ sC & G + 2sC \end{bmatrix}, \tag{A.96}$$

and therefore the adjoint of the coefficient matrix is

$$\text{adj } \mathbf{A} = \mathbf{B}^T = \begin{bmatrix} G + 2sC & sC \\ sC & G + 2sC \end{bmatrix}. \tag{A.97}$$

The determinate of A is

$$\det \mathbf{A} = \begin{vmatrix} G + 2sC & sC \\ sC & G + 2sC \end{vmatrix} = G^2 + 4sCG + 3s^2C^2. \tag{A.98}$$

The inverse of the coefficient matrix is

$$\mathbf{A}^{-1} = \frac{\begin{bmatrix} G + 2sC & sC \\ sC & G + 2sC \end{bmatrix}}{(G^2 + 4sCG + 3s^2C^2)}. \tag{A.99}$$

It follows from Eq. (A.95) that

$$\begin{bmatrix} V_1 \\ V_2 \end{bmatrix} = \frac{\begin{bmatrix} G + 2sC & sC \\ sC & G + 2sC \end{bmatrix}\begin{bmatrix} GVg \\ sCVg \end{bmatrix}}{(G^2 + 4sCG + 3s^2C^2)}. \tag{A.100}$$

Carrying out the matrix multiplication called for in Eq. (A.100) gives

$$\begin{bmatrix} V_1 \\ V_2 \end{bmatrix} = \frac{1}{(G^2 + 4sCG + 3s^2C^2)} \begin{bmatrix} (G^2 + 2sCG + s^2C^2)Vg \\ (2sCG + 2s^2C^2)Vg \end{bmatrix}. \tag{A.101}$$

Now the expressions for V_1 and V_2 can be written directly from

Eq. (A.101); thus

$$V_1 = \frac{(G^2 + 2sCG + s^2C^2)Vg}{(G^2 + 4sCG + 3s^2C^2)},$$ (A.102)

and

$$V_2 = \frac{2(sCG + s^2C^2)Vg}{(G^2 + 4sCG + 3s^2C^2)}.$$ (A.103)

The interested reader can test the validity of Eqs. (A.102) and (A.103) by using them to find $v_1(t)$ and $v_2(t)$ in Problem 16.27.

In our final example we will illustrate how matrix algebra can be used to analyze the cascade connection of two two-port circuits.

EXAMPLE A.5

Show by means of matrix algebra how the input variables V_1 and I_1 can be described as functions of the output variables V_2 and I_2 in the cascade connection shown in Fig. 20.10.

SOLUTION

We begin by expressing the relationship between the input and output variables of each two-port in matrix notation. Thus

$$\begin{bmatrix} V_1 \\ I_1 \end{bmatrix} = \begin{bmatrix} a'_{11} & -a'_{12} \\ a'_{21} & -a'_{22} \end{bmatrix}\begin{bmatrix} V'_2 \\ I'_2 \end{bmatrix},$$ (A.104)

and

$$\begin{bmatrix} V'_1 \\ I'_1 \end{bmatrix} = \begin{bmatrix} a''_{11} & -a''_{12} \\ a''_{21} & -a''_2 \end{bmatrix}\begin{bmatrix} V_2 \\ I_2 \end{bmatrix}.$$ (A.105)

Now the cascade connection imposes the constraints

$$V'_2 = V'_1$$

and (A.106)

$$I'_2 = -I'_1.$$

These constraint relationships are substituted into Eq. (A.104).

Thus

$$\begin{bmatrix} V_1 \\ I_1 \end{bmatrix} = \begin{bmatrix} a'_{11} & -a'_{12} \\ a'_{21} & -a'_{22} \end{bmatrix} \begin{bmatrix} V'_1 \\ -I'_1 \end{bmatrix}$$

$$= \begin{bmatrix} a'_{11} & a'_{12} \\ a'_{21} & a'_{22} \end{bmatrix} \begin{bmatrix} V'_1 \\ I'_1 \end{bmatrix}.$$

(A.107)

The relationship between the input variables (V_1, I_1) and the output variables (V_2, I_2) is obtained by substituting Eq. (A.105) into Eq. (A.107). The result is

$$\begin{bmatrix} V_1 \\ I_1 \end{bmatrix} = \begin{bmatrix} a'_{11} & a'_{12} \\ a'_{21} & a'_{22} \end{bmatrix} \begin{bmatrix} a''_{11} & -a''_{12} \\ a''_{21} & -a''_{22} \end{bmatrix} \begin{bmatrix} V_2 \\ I_2 \end{bmatrix}.$$

(A.108)

After multiplying the coefficient matrices, we have

$$\begin{bmatrix} V_1 \\ I_1 \end{bmatrix} = \begin{bmatrix} (a'_{11}a''_{11} + a'_{12}a''_{21}) & -(a'_{11}a''_{12} + a'_{12}a''_{22}) \\ (a'_{21}a''_{11} + a'_{22}a''_{21}) & -(a'_{21}a''_{12} + a'_{22}a''_{22}) \end{bmatrix} \begin{bmatrix} V_2 \\ I_2 \end{bmatrix}.$$

(A.109)

We call to the attention of the reader that Eq. (A.109) corresponds to writing Eqs. (20.72) and (20.73) in the text in matrix form.

APPENDIX B

COMPLEX NUMBERS

Complex numbers were invented and introduced into the number system to permit the extraction of square roots of negative numbers. The invention of complex numbers simplifies the solution of problems that would otherwise be very difficult. The equation $x^2 + 8x + 41 = 0$ has no solution in a number system that excludes complex numbers. We find the concept of a complex number, and the capability of manipulating these numbers algebraically, extremely useful in circuit analysis.

B.1 NOTATION

There are two ways to designate a complex number: the *cartesian*, or *rectangular*, form and the *polar*, or *trigonometric*, form. In the rectangular form, a complex number is written in terms of its real and imaginary components; hence

$$n = a + jb, \tag{B.1}$$

where a is the real component, b is the imaginary component, and j is by definition $\sqrt{-1}$.[†]

In polar form, a complex number is written in terms of its magnitude, or modulus, and angle, or argument; hence

$$n = ce^{j\theta}, \tag{B.2}$$

where c is the magnitude, θ is the angle, e is the base of the natural logarithm, and, as before, $j = \sqrt{-1}$. In the literature, the symbol $\underline{/\theta}$ is frequently used in place of $e^{j\theta}$; that is, the polar form is written

$$n = c\underline{/\theta}. \tag{B.3}$$

Although Eq. (B.3) is more convenient in printing text material, Eq. (B.2) is of primary importance in mathematical operations because the rules for manipulating an exponential quantity are well known. For example, since $(y^x)^n = y^{xn}$, then $(e^{j\theta})^n = e^{jn\theta}$; and since $y^{-x} = 1/y^x$, then $e^{-j\theta} = 1/e^{j\theta}$, and so forth.

Because there are two ways of expressing the same complex number, we need to relate one form to the other. The transition from the polar to the rectangular form makes use of Euler's identity

$$e^{\pm j\theta} = \cos\theta \pm j\sin\theta. \tag{B.4}$$

A complex number in polar form can be put in rectangular form by writing

$$ce^{j\theta} = c(\cos\theta + j\sin\theta)$$
$$= c\cos\theta + jc\sin\theta$$
$$= a + jb. \tag{B.5}$$

The transition from rectangular to polar form makes use of

[†] You may be more familiar with the notation i $= \sqrt{-1}$. In electrical engineering, i is used as the symbol for current and hence in electrical engineering literature, j is used to denote $\sqrt{-1}$.

the geometry of the right triangle, namely,

$$a + jb = (\sqrt{a^2 + b^2})\,e^{j\theta}$$

$$= ce^{j\theta}, \qquad\qquad \textbf{(B.6)}$$

where

$$\tan\theta = b/a. \qquad\qquad \textbf{(B.7)}$$

It is not obvious from Eq. (B.7) in which quadrant the angle θ lies. The ambiguity can be resolved by a graphical representation of the complex number.

B.2 GRAPHICAL REPRESENTATION OF A COMPLEX NUMBER

A complex number is represented graphically on a complex-number plane, which uses the horizontal axis for plotting the real component and the vertical axis for plotting the imaginary component. The angle of the complex number is measured *counterclockwise* from the *positive real axis*. The graphical plot of the complex number $n = a + jb = c/\underline{\theta}$, if we assume that a and b are both positive, is shown in Fig. B.1. The graphical representation of a complex number makes very clear the relationship between the rectangular and polar forms. Any point in the complex-number plane is uniquely defined by giving either its distance from each axis (that is, a and b) or its radial distance from the origin (c) and the angle of the radial measurement θ.

It follows from Fig. B.1 that θ is in the first quadrant when a and b are both positive, in the second quadrant when a is negative and b is positive, in the third quadrant when a and b are both negative, and in the fourth quadrant when a is positive and b is negative. These observations are illustrated in Fig. B.2, where we have plotted $4 + j3$, $-4 + j3$, $-4 - j3$, and $4 - j3$.

It should be pointed out that we can also specify θ as a clockwise angle from the positive real axis. Thus in Fig. B.2(c) we could also designate $-4 - j3$ as $5/\underline{-143.13°}$. In Fig. B.2(d) we observe that $5/\underline{323.13°} = 5/\underline{-36.87°}$. It is customary to express θ in terms of negative values when θ lies in the third or fourth quadrant.

The graphical interpretation of a complex number also shows the relationship between a complex number and its conjugate. *The conjugate of a complex number is formed by reversing the*

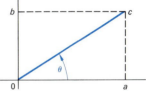

FIGURE B.1 The graphical representation of $a + jb$ when a and b are both positive.

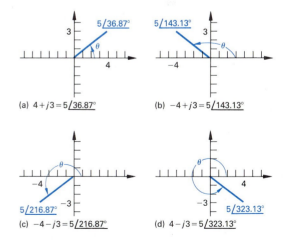

(a) $4 + j3 = 5/\underline{36.87°}$

(b) $-4 + j3 = 5/\underline{143.13°}$

(c) $-4 - j3 = 5/\underline{216.87°}$

(d) $4 - j3 = 5/\underline{323.13°}$

FIGURE B.2 The graphical representation of four complex numbers.

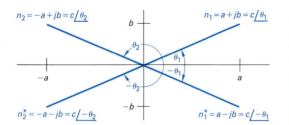

FIGURE B.3 The complex numbers n_1 and n_2 and conjugates n_1^* and n_2^*.

sign of its imaginary component. Thus the conjugate of $a + jb$ is $a - jb$ and the conjugate of $-a + jb$ is $-a - jb$. When we write a complex number in polar form, we form its conjugate by simply reversing the sign of the angle θ. Therefore the conjugate of $c\underline{/\theta}$ is $c\underline{/-\theta}$. The conjugate of a complex number is designated with an asterisk. In other words, n^* is understood to be the conjugate of n. When the conjugate of a complex number is plotted on the complex-number plane, we see that conjugation simply reflects the complex number about the real axis. This is illustrated in Fig. B.3.

B.3 ARITHMETIC OPERATIONS

ADDITION (SUBTRACTION)

In order to add or subtract complex numbers, we must express the numbers in rectangular form. Complex numbers are added by adding the real parts to form the real part of the sum and by adding the imaginary parts to form the imaginary part of the sum. Thus if we are given

$$n_1 = 8 + j16$$

and

$$n_2 = 12 - j3,$$

then

$$n_1 + n_2 = (8 + 12) + j(16 - 3) = 20 + j13.$$

Subtraction follows the same rule. Thus

$$n_2 - n_1 = (12 - 8) + j(-3 - 16) = 4 - j19.$$

If the numbers to be added or subtracted are given in polar form, they are first converted to rectangular form. For example, if

$$n_1 = 10\underline{/53.13°}$$

and

$$n_2 = 5\underline{/-135°},$$

then

$$n_1 + n_2 = 6 + j8 - 3.535 - j3.535$$

$$= (6 - 3.535) + j(8 - 3.535)$$

$$= 2.465 + j4.465 = 5.10\underline{/61.10°}$$

and

$$n_1 - n_2 = 6 + j8 - (-3.535 - j3.535)$$
$$= 9.535 + j11.535$$
$$= 14.966\underline{/50.42°}.$$

MULTIPLICATION (DIVISION)

Multiplication or division of complex numbers can be carried out with the numbers written in either rectangular or polar form. However, in most cases, the polar form is more convenient. As an example, let us find the product n_1n_2 when $n_1 = 8 + j10$ and $n_2 = 5 - j4$. Using the rectangular form, we have

$$n_1n_2 = (8 + j10)(5 - j4) = 40 - j32 + j50 + 40$$
$$= 80 + j18$$
$$= 82\underline{/12.68°}.$$

If we use the polar form, the multiplication n_1n_2 becomes

$$n_1n_2 = (12.81\underline{/51.34°})(6.40\underline{/-38.66°})$$
$$= 82\underline{/12.68°}$$
$$= 80 + j18.$$

The first step in dividing two complex numbers that are to be divided in rectangular form is to multiply the numerator and denominator by the *conjugate of the denominator*. This will reduce the denominator to a real number. We then divide the real number into the new numerator. As an example we will find the value of n_1/n_2, where $n_1 = 6 + j3$ and $n_2 = 3 - j1$. We have

$$\frac{n_1}{n_2} = \frac{6 + j3}{3 - j1} = \frac{(6 + j3)(3 + j1)}{(3 - j1)(3 + j1)}$$
$$= \frac{18 + j6 + j9 - 3}{9 + 1}$$
$$= \frac{15 + j15}{10} = 1.5 + j1.5$$
$$= 2.12\underline{/45°}.$$

In polar form, the division of n_1 by n_2 is

$$\frac{n_1}{n_2} = \frac{6.71\underline{/26.57°}}{3.16\underline{/-18.43°}} = 2.12\underline{/45°}$$
$$= 1.5 + j1.5.$$

B.4 USEFUL IDENTITIES

In working with complex numbers and quantities, the following
identities are very useful:

$$\pm j^2 = \mp 1, \tag{B.8}$$

$$(-j)(j) = 1, \tag{B.9}$$

$$j = \frac{1}{-j}, \tag{B.10}$$

$$e^{\pm j\pi} = -1, \tag{B.11}$$

$$e^{\pm j\pi/2} = \pm j. \tag{B.12}$$

Given that $n = a + jb = c\underline{/\theta}$, it follows that

$$nn^* = a^2 + b^2 = c^2, \tag{B.13}$$

$$n + n^* = 2a, \tag{B.14}$$

$$n - n^* = j2b, \tag{B.15}$$

$$n/n^* = 1\underline{/2\theta}. \tag{B.16}$$

B.5 INTEGER POWER OF A COMPLEX NUMBER

To raise a complex number to an integer power k, it is easier to
first write the complex number in polar form. Thus

$$n^k = (a + jb)^k$$
$$= (ce^{j\theta})^k = c^k e^{jk\theta}$$
$$= c^k(\cos k\theta + j\sin k\theta).$$

For example,

$$(2e^{j12°})^5 = 2^5 e^{j60°} = 32 e^{j60°}$$
$$= 16 + j27.71$$

and

$$(3 + j4)^4 = (5e^{j53.13°})^4 = 5^4 e^{j212.52°}$$
$$= 625 e^{j212.52°}$$
$$= -527 - j336.$$

B.6 ROOTS OF A COMPLEX NUMBER

To find the kth root of a complex number, we must recognize that we are solving the equation

$$x^k - ce^{j\theta} = 0, \qquad \textbf{(B.17)}$$

which is an equation of the kth degree and therefore has k roots.

To find the k roots, we first note that

$$ce^{j\theta} = ce^{j(\theta+2\pi)} = ce^{j(\theta+4\pi)} = \cdots. \qquad \textbf{(B.18)}$$

It follows from Eqs. (B.17) and (B.18) that

$$x_1 = (ce^{j\theta})^{1/k} = c^{1/k}e^{j\theta/k}, \qquad \textbf{(B.19)}$$

$$x_2 = [ce^{j(\theta+2\pi)}]^{1/k} = c^{1/k}e^{j(\theta+2\pi)/k}, \qquad \textbf{(B.20)}$$

$$x_3 = [ce^{j(\theta+4\pi)}]^{1/k} = c^{1/k}e^{j(\theta+4\pi)/k}, \qquad \textbf{(B.21)}$$

$$\vdots$$

We continue the process outlined by Eqs. (B.19), (B.20), and (B.21) until the roots start repeating. This will happen when the multiple of π is equal to $2k$. For example, let us find the four roots of $81e^{j60°}$. We have

$$x_1 = 81^{1/4}e^{j60/4} = 3e^{j15°},$$

$$x_2 = 81^{1/4}e^{j(60+360)/4} = 3e^{j105°},$$

$$x_3 = 81^{1/4}e^{j(60+720)/4} = 3e^{j195°},$$

$$x_4 = 81^{1/4}e^{j(60+1080)/4} = 3e^{j285°},$$

$$x_5 = 81^{1/4}e^{j(60+1440)/4} = 3e^{j375°} = 3e^{j15°}.$$

But x_5 is the same as x_1, so the roots have started to repeat. Therefore we know the four roots of $81e^{j60°}$ are the values given by x_1, x_2, x_3, and x_4.

It is worth noting that the roots of a complex number lie on a circle in the complex-number plane. The radius of the circle is $c^{1/k}$. The roots are uniformly distributed around the circle, the angle between adjacent roots being equal to $2\pi/k$ radians, or $360/k$ degrees. The four roots of $81e^{j60°}$ are shown plotted in Fig. B.4.

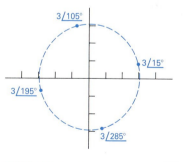

FIGURE B.4 The four roots of $81e^{j60°}$.

Appendix C

An Abbreviated Table of Trigonometric Identities

1. $\sin(\alpha \pm \beta) = \sin\alpha\cos\beta \pm \cos\alpha\sin\beta$

2. $\cos(\alpha \pm \beta) = \cos\alpha\cos\beta \mp \sin\alpha\sin\beta$

3. $\sin\alpha + \sin\beta = 2\sin\dfrac{\alpha+\beta}{2}\cos\dfrac{\alpha+\beta}{2}$

4. $\sin\alpha - \sin\beta = 2\cos\left(\dfrac{\alpha+\beta}{2}\right)\sin\left(\dfrac{\alpha-\beta}{2}\right)$

5. $\cos\alpha + \cos\beta = 2\cos\left(\dfrac{\alpha+\beta}{2}\right)\cos\left(\dfrac{\alpha-\beta}{2}\right)$

6. $\cos\alpha - \cos\beta = -2\sin\left(\dfrac{\alpha+\beta}{2}\right)\sin\left(\dfrac{\alpha-\beta}{2}\right)$

7. $2\sin\alpha\sin\beta = \cos(\alpha-\beta) - \cos(\alpha+\beta)$

8. $2\cos\alpha\cos\beta = \cos(\alpha-\beta) + \cos(\alpha+\beta)$

9. $2\sin\alpha\cos\beta = \sin(\alpha+\beta) + \sin(\alpha-\beta)$

10. $\sin 2\alpha = 2\sin\alpha\cos\alpha$

11. $\cos 2\alpha = 2\cos^2\alpha - 1 = 1 - 2\sin^2\alpha$

12. $\cos^2\alpha = \frac{1}{2} + \frac{1}{2}\cos 2\alpha$

13. $\sin^2\alpha = \frac{1}{2} - \frac{1}{2}\cos 2\alpha$

14. $\tan(\alpha \pm \beta) = \dfrac{\tan\alpha \pm \tan\beta}{1 \mp \tan\alpha\tan\beta}$

15. $\tan 2\alpha = \dfrac{2\tan\alpha}{1 - \tan^2\alpha}$

APPENDIX D

AN ABBREVIATED TABLE OF INTEGRALS

1. $\displaystyle\int xe^{ax}\,dx = \frac{e^{ax}}{a^2}(ax - 1)$

2. $\displaystyle\int x^2 e^{ax}\,dx = \frac{e^{ax}}{a^3}(a^2x^2 - 2ax + 2)$

3. $\displaystyle\int x \sin ax\,dx = \frac{1}{a^2}\sin ax - \frac{x}{a}\cos ax$

4. $\displaystyle\int x \cos ax\,dx = \frac{1}{a^2}\cos ax + \frac{x}{a}\sin ax$

5. $\displaystyle\int e^{ax}\sin bx\,dx = \frac{e^{ax}}{a^2 + b^2}(a \sin bx - b \cos bx)$

6. $\displaystyle\int e^{ax}\cos bx\,dx = \frac{e^{ax}}{a^2 + b^2}(a \cos bx + b \sin bx)$

7. $\displaystyle\int \frac{dx}{x^2 + a^2} = \frac{1}{a}\tan^{-1}\frac{x}{a}$

8. $\displaystyle\int \frac{dx}{(x^2 + a^2)^2} = \frac{1}{2a^2}\left(\frac{x}{x^2 + a^2} + \frac{1}{a}\tan^{-1}\frac{x}{a}\right)$

9. $\displaystyle\int \sin ax \sin bx\,dx =$

   $\dfrac{\sin (a - b)x}{2(a - b)} - \dfrac{\sin (a + b)x}{2(a + b)},\ a^2 \neq b^2$

10. $\displaystyle\int \cos ax \cos bx\,dx =$

$$\frac{\sin (a - b)x}{2(a - b)} + \frac{\sin (a + b)x}{2(a + b)}, \; a^2 \neq b^2$$

11. $\displaystyle\int \sin ax \cos bx\,dx =$

$$-\frac{\cos (a - b)x}{2(a - b)} - \frac{\cos (a + b)x}{2(a + b)}, \; a^2 \neq b^2$$

12. $\displaystyle\int \sin^2 ax\,dx = \frac{x}{2} - \frac{\sin 2ax}{4a}$

13. $\displaystyle\int \cos^2 ax\,dx = \frac{x}{2} + \frac{\sin 2ax}{4a}$

14. $\displaystyle\int_0^\infty \frac{a\,dx}{a^2 + x^2} = \frac{\pi}{2}, \; a > 0;$

$$= 0, \; a = 0;$$

$$= \frac{-\pi}{2}, \; a < 0$$

15. $\displaystyle\int_0^\infty \frac{\sin ax}{x}\,dx = \frac{\pi}{2}, \; a > 0$

$$= \frac{-\pi}{2}, \; a < 0$$

16. $\displaystyle\int x^2 \sin ax\,dx = \frac{2x}{a^2}\sin ax - \frac{a^2x^2 - 2}{a^3}\cos ax$

17. $\displaystyle\int x^2 \cos ax\,dx = \frac{2x}{a^2}\cos ax + \frac{a^2x^2 - 2}{a^3}\sin ax$

18. $\displaystyle\int e^{ax} \sin^2 bx\,dx =$

$$\frac{e^{ax}}{a^2 + 4b^2}\left[(a \sin bx - 2b \cos bx)\sin bx + \frac{2b^2}{a}\right]$$

19. $\displaystyle\int e^{ax} \cos^2 bx\,dx =$

$$\frac{e^{ax}}{a^2 + 4b^2}\left[(a \cos bx - 2b \sin bx)\cos bx + \frac{2b^2}{a}\right]$$

APPENDIX E

LIST OF EXAMPLES

CHAPTER 9

CHAPTER 10

ANSWERS TO SELECTED PROBLEMS

CHAPTER 1

1.1 1425.96 m/s

1.3 6.14 s

1.5 4 sin 5000t mC

1.7 156.04 μm/s

1.9 a) 100 W, absorbing
b) entering
c) lose

1.11 a) 1250 W   A to B
b) 1200 W   B to A
c) 5760 W   B to A
d) 16,500 W   A to B

1.13 a) 215,000 C
b) 2036.67 kJ

1.15 a) 250 μs
b) 13.53 mW
c) 6.25 μJ

1.17 3.02 W

1.20 a) Car A
b) 43.2 kJ

1.22 a) 223.80 W
b) 4 J

1.25 Yes, $p_{del} = p_{abs} = 40$ kW

CHAPTER 2

2.3 a) 3 A
b) 8 A
c) 960 W

2.5 NO, violates KCL

2.9 a) 2 A
b) $p_{4\Omega} = 100$ W, $p_{10\Omega} = 90$ W,
$p_{25\Omega} = 400$ W, $p_{50\Omega} = 50$ W
$p_{65\Omega} = 260$ W
c) $p_{diss} = p_{dev} = 900$ W

2.12 a) $p_{upper} = 4116$ W
$p_{lower} = 3087$ W
b) $p_{diss} = p_{dev} = 7203$ W

2.17 a) $p_{5\Omega} = 20$ W, $p_{2\Omega} = 175$ W,
$p_{15\Omega} = 135$ W, $p_{16\Omega} = 400$ W,
$p_{30\Omega} = 270$ W
b) 1000 W
c) $p_{diss} = p_{dev} = 1000$ W

2.22 $v_1 = -1.25$ V, $v_g = 1$ V

2.24 a) $v_1 = 25.6$ V, $v_2 = 24$ V
b) $p_{dev} = p_{diss} = 7818.24$ V

CHAPTER 3

3.3 a) 20 Ω
b) 15 Ω

3.6 −5.14 A

3.9 $i_o = 3.2$ A, $i_g = 14$ A

3.13 −200 V

3.17 a) $R_1 = 16.8$ kΩ, $R_2 = 5.6$ kΩ
b) 228.81 mW

3.22 $R_1 = 1875$ Ω, $R_2 = 3750$ Ω, $R_3 = 7500$ Ω,
$R_4 = 15,000$ Ω

3.27 12.5 μA

3.31 12.5 A

3.35 a) cannot be used
b) 375 V
c) 192 V, 128 V

3.39 a) 225 μA
b) 216 μA
c) -4%

3.43 6 mA, 600 mV

3.49 1.25 mA

3.51 10 Ω

3.56 a) 2.5 A
b) 8.25 A
c) 45 V
d) 7.5 kW

3.59 a) 80 Ω
b) 279 W

CHAPTER 4

4.2 -5 V

4.5 $v_1 = 100$ V, $v_2 = 20$ V

4.8 2430 W

4.11 a) $i_1 = 1$ mA, $i_2 = -20$ mA, $i_3 = 31$ mA
b) $p_{dev} = p_{diss} = 330$ mW

4.14 19.8 V

4.17 -23 V

4.20 5 V

4.24 602.5 W

4.26 a) $i_a = 6.4$ A, $i_b = 0.4$ A, $i_c = -6$ A
b) $i_a = 0$ A, $i_b = 2$ A, $i_c = 2$ A

4.29 153 W

4.33 2700 W

4.35 a) 2 mA
b) 304 mW
c) 900 μW

4.38 259.2 W

4.39 a) 25 W
b) 1.26%

4.44 a) $i_a = -7$ A, $i_b = -8$ A, $i_c = 8$ A, $i_d = 10$ A,
$i_e = 18$ A
b) $p_{dev} = p_{diss} = 8240$ W

4.47 1096 V

4.49 22.5 Ω, 2.5 Ω

4.52 a) -4.5 mA
b) 397.8 mW

4.55 $V_{Th} = 112$ V, $R_{Th} = 16$ Ω

4.58 $I_N = -1$ mA, $R_N = 3.75$ kΩ

4.61 3.50 μA

4.64 a) 32.5 kΩ
b) -1.1%

4.67 $V_{Th} = 0$ V, $R_{Th} = 12.5$ Ω

4.71 a) 5 kΩ
b) 957.03 μW

4.74 a) 2.5 Ω
b) 2250 W

4.77 7.13%

4.80 a) 0 Ω
b) 72 W

4.83 288 V

4.86 15 V

4.89 25 V

4.94 23.09 W

4.96 0 A

CHAPTER 5

5.2 bcd, bde, abe, ade, abc, cde, acd, ace

5.6 $v_1 = 100$ V, $v_2 = 45$ V

5.9 a) -6 V
b) 243 W

5.11 -4 V

5.13 $i_1 = 40$ A, $i_2 = 10$ A, $i_3 = -50$ A
$i_4 = 15$ A, $i_5 = 25$ A, $i_6 = 25$ A

5.16 45 W

5.19 5376.4 W

5.22 a) 310.65 W
b) It is not necessary to define a tree or the fundamental loops generated by the tree. It is not necessary to determine the identification and orientation of the link currents in each tree branch. Mesh currents can be identified by inspection. No branch carries more than two mesh currents.

CHAPTER 6

6.1 a) -12 V
b) -18 V
c) 10 V
d) -14 V

e) 18 V

f) $2 \leq v_b \leq 6$ V

6.4 -1 mA

6.7 a) -6.8 V $\leq v_o \leq -1.96$ V

b) 0.1039

6.10 a) 16.6 V

b) 9.32 V $\leq v_b \leq 17.32$ V

6.13 a) 108 kΩ

b) 270 μW

6.16 $0 \leq R_f \leq 75$ kΩ

6.18 a) 10.54 V

b) -4.55 V $\leq v_g \leq 4.55$ V

c) 181.76 kΩ

6.21 a) 7.37 V

b) $i_a = 10$ μA, $i_b = -10$ μA

c) 7.425 for v_a and 3.575 for v_b

6.25 $R_a = 20$ kΩ, $R_b = 200$ kΩ, $R_f = 47$ kΩ

6.26 a) -15.10 V

b) 34.3 kΩ

c) 250 kΩ

6.29 a) 16 V

b) -4.2 V $\leq v_c \leq 3.8$ V

6.32 a) 13.49

b) $v_1 = 999.45$ mV, $v_2 = 999.83$ mV

c) 387.78 μV

d) 692.47 pA

e) 13.5, 1 V, 1 V, 0 V, 0 A

6.35 $V_{Th} = 0$ V, $R_{Th} = -1000$ Ω

6.38 2.24 kΩ

6.40 $v_o = -4$V, $i_o = 2.8$ mA

6.44 $i_a = 14$ A, $i_b = 11.5$ A,

$i_c = 2.5$ A, $i_d = -3.5$ A, $i_e = -6$A

6.45 $i_1 = 3.5$ A, $i_2 = 1.5$ A

CHAPTER 7

7.2 a) $i = 0, t < 0$

$i = 80t$ A, $0 \leq t \leq 5$ ms

$i = 0.8 - 80t$ A, 5 ms $\leq t \leq 10$ ms

$i = 0$, 10 ms $\leq t \leq \infty$

b) $v = 0, t < 0$

$v = 2$ V, $0 < t < 5$ ms

$v = -2$ V, 5 ms $< t < 10$ ms

$v = 0$, 10 ms $< t < \infty$

$p = 0, t \leq 0$

$p = 160t$ W, $0 < t < 5$ ms

$p = -1.6 + 160t$ W, 5 ms $< t < 10$ ms

$p = 0$, 10 ms $\leq t \leq \infty$

$w = 0$, $t \leq 0$

$w = 80t^2$ J, $0 \leq t \leq 5$ ms

$w = 8 \times 10^{-3} - 1.6t + 80t^2$ J,

5 ms $\leq t \leq 10$ ms

$w = 0$, 10 ms $\leq t \leq \infty$

7.6 a) 2.77 ms

b) 64.27 V

7.7 a) $900e^{-10t}(1 - 10t)$ μV

b) -59.34 μW

c) delivering

d) 5.93 μJ

e) 10.96 μJ, 100 ms

7.9 339.57 W, delivering

7.12 19.2 V

7.14 a) $i = 0, t < 0$

b) $i = 80(\cos 1000t + \sin 1000t)e^{-2000t}$ mA,

$t > 0$

c) No

d) Yes, current jumps instantaneously from 0 to

80 mA at $t = 0$

e) 2000 μJ

7.17 a) 40 μJ

b) 360 μJ

7.20 8 H

7.23 8.05 ms

7.26 2 μF charged to 25 V, positive at terminal a

7.29 a) $-20e^{-25t}$ V

b) $-16e^{-25t} + 21$ V

c) $-4e^{-25t} - 21$ V

d) 320 μJ

e) 2525 μJ

f) 2205 μJ

g) Yes

CHAPTER 8

8.1 a) 25 Ω

b) 100 ms

c) 2.5 H

d) 51.2 J

e) 45.81 ms

8.4 $-37.5e^{-100t}$ V, $t \geq 0^+$

8.7 8.89%

8.10 52.68 Ω

8.13 a) 10 J

b) 12.5 J

8.16 33.33%

8.19 a) $9.9e^{-1000t}$ mA

b) 42.14%

8.22 a) 7.82 mJ

b) 27.73 ms

8.25 a) 250×10^{-6} A/V

b) $-1.8e^{-25t}$ V, $t \geq 0^+$

8.28 a) $-20e^{-50t}$ V, $t \geq 0$

b) 24.24%

c) $-(20/3)e^{-50t} + (50/3)$ V, $t \geq 0$

d) $-(40/3)e^{-50t} + (50/3)$ V, $t \geq 0$

e) 250 μJ

8.31 a) 100 V, 4 Ω, 10 mH

b) 1.73 ms

8.34 $2.5 + 7.5e^{-1250t}$ A, $t \geq 0$

$-150e^{-1250t}$ V, $t \geq 0^+$

8.37 $2.2 - 1.2e^{-1000t}$ A, $t \geq 0$

8.41 a) 0.25 mA, 800 kΩ, 250 nF, 2 ms

b) 1.83 ms

8.44 a) -7.2 mA

b) 0 A

c) 4 ms

d) $-7.2e^{-250t}$ mA, $t \geq 0^+$

e) $-36 + 12.96e^{-250t}$ V, $t \geq 0^+$

8.47 $-45 + 90e^{-800t}$ V, $t \geq 0$

8.50 a) $100 - 40e^{-1000t}$ V, $t \geq 0$

b) $-8e^{-1000t}$ mA, $t \geq 0^+$

c) $-(80/3)e^{-1000t} + (200/3)$ V, $t \geq 0$

d) $-(40/3)e^{-1000t} + (100/3)$ V, $t \geq 0$

e) 1 mJ

8.53 a) 6 A

b) 3.45 A

c) 1.0 A

d) -50.67 V

e) -44.06 V

8.56 17.62 ms

8.59 504.74 μA, left-to-right

8.64 a) $v_o(t) = 0$, $t < 0$

$v_o(t) = 50e^{-8000t}$ V, $0 < t < 75$ μs

$v_o(t) = -22.56e^{-8000(t - 75 \times 10^{-6})}$ V,

75 μs $< t < \infty$

b) $v_o(75^-$ μs$) = 27.44$ V,

$v_o(75^+$ μs$) = -22.56$ V

c) $i_o(75^-$ μs$) = i_o(75^+$ μs$) = 11.28$ mA

8.69 138.16 ms

8.73 12 kΩ

8.76 80 ms

8.80 $v_o = 500t$ V, $0 \leq t \leq 20$ ms

$v_o = -250t + 15$ V,

20 ms $\leq t \leq 140$ ms

8.84 a) 4.25 ms

b) 6.80 ms

CHAPTER 9

9.1 a) $s_1 = -100$ rad/s, $s_2 = -400$ rad/s

b) overdamped

c) 1562.5 Ω

d) $s_1 = -160 + j120$ rad/s

$s_2 = -160 - j120$ rad/s

e) 1250 Ω

9.4 -7.5 mA

9.7 a) 1000 Ω, 1 μF, 6000 V/s, 8 V

b) $-3000t\, e^{-500t} + 2e^{-500t}$ mA, $t \geq 0^+$

9.10 a) 31.25 kΩ

b) $(100 - 25 \times 10^5 t)e^{-5000t}$ V, $t \geq 0$

c) -150.6 V

d) 66.7%

9.13 $10(\sqrt{3} \sin 2000 \sqrt{3}\, t - \cos 2000 \sqrt{3}\, t)e^{-2000t}$ V,

$t \geq 0$

9.15 $(30 \cos 300t - 10 \sin 300t)e^{-100t}$ V, $t \geq 0$

9.19 $9 - 8e^{-40t} + 2e^{-160t}$ mA, $t \geq 0$

9.22 $20 + (5000t - 20)e^{-1000t}$ mA, $t \geq 0$

9.25 a) 0 V

b) 3000 V/s

c) $5e^{-800t} \sin 600t$ V, $t \geq 0$

9.28 3.2 Ω, 2 mH, 0 A, -10 A

9.31 a) 72 V

b) $-144,000$ V/s

c) $48e^{-1000t} + 24e^{-4000t}$ V, $t \geq 0^+$

9.34 a) $(10^4 t + 0.10)e^{-10^5 t}$ A, $t \geq 0$

b) $(25 \times 10^5 t + 50)e^{-10^5 t}$ V, $t \geq 0$

9.37 $400 - 7500t\, e^{-25t} - 300e^{-25t}$ V, $t \geq 0$

9.40 a) $2e^{-1000t}$ A, $t \geq 0$

b) $5 - 10e^{-1000t}$ V, $t \geq 0$

9.44 a) $500e^{-15,000t} \sin 20,000t$ V, $t \geq 0^+$

b) 46.36 μs

c) 199.54 V

d) $4007.22e^{-1500t} \sin 24{,}954.96t$ V, $t \geq 0^+$
60.54 μs
3652.77 V

9.47 a) $\dfrac{d^2v_o}{dt^2} = -\dfrac{v_g}{R^2C^2}$

b) $\dfrac{d^2v_o}{dt^2} = \dfrac{v_g}{R^2C^2}$, same except for a reversal in sign

c) two integrations of the input signal with one op-amp

CHAPTER 10

10.1 a) 100 V
b) 200 Hz
c) 1256.64 rad/s
d) 1.05 rad
e) 60°
f) 5 ms
g) (25/6) ms
h) $-100 \sin(4000\pi t)$ V
i) (5/6) ms
j) (35/12) ms

10.4 a) 62.83 krad/s
b) $10 \cos(2 \times 10^4 \pi t - 144°)$ A

10.8 169.71 V

10.10 a) 100π krad/s
b) 90°
c) $-15.92 \ \Omega$
d) 0.2 μF
e) $-j15.92 \ \Omega$

10.13 $17.68 \cos(50t - 135°)$ mA

10.16 $50\underline{/36.87°}$ m$\mho$ = $40 + j30$ m$\mho$

10.19 a) 2 H, 8 H
b) $0.25 \cos 500t$ A
$0.10 \cos 500t$ A

10.22 a) 954.93 Hz
b) $5 \cos 6000t$ mA

10.25 $252.98 \cos(40{,}000t - 55.31°)$ V

10.31 a) $25.30\underline{/55.3°}$ A, $50\underline{/36.87°}$ A,
$402.49\underline{/63.43°}$ V
b) $25.30 \cos(5000t + 55.30°)$ A
$50 \cos(5000t + 36.87°)$ A
$402.49 \cos(5000t + 63.43°)$ V

10.34 $-10 \cos(200t)$ mV

10.37 $80\underline{/90°}$ V = $j80$ V

10.40 $10 + j10$ A

10.43 $30 + j0$ A, $30 - j20$ A, $30 + j10$ A,
$-j30$A

10.46 $33.94 \cos(10^5 t - 98.13°)$ A
$32.20 \cos(10^5 t - 116.57°)$ A
$53.67 \cos(10^5 t + 26.57°)$ A

10.49 $\mathbf{V}_{ab} = \mathbf{V}_{Th} = 216 - j72$ V
$\mathbf{Z}_{Th} = 3.6 + j10.8 \ \Omega$

10.52 $\mathbf{V}_{ab} = \mathbf{V}_{Th} = 500 \underline{/-53.13°}$ V
$\mathbf{Z}_{Th} = 100 - j100 \ \Omega$

10.55 $30{,}104 \underline{/-4.76°} \ \Omega$

10.58 $38.65 \cos(25t + 165.07°) +$
$544.70 \cos(50t - 3.69°)$ V

10.61 a) $6.88 \cos(10^5 t - 35.54°)$ V
b) 4.36 V

10.64 $8.94 \cos(200t + 153.43°)$ V

CHAPTER 11

11.1 a) 409.58 W (abs)
286.79 VAR (abs)
b) 103.53 W (abs)
386.37 VAR (del)
c) 1000 W (del)
1732.05 VAR (del)
d) 250 W (del)
433.01 VAR (abs)

11.4 0.94 (lagging), 0.35; 0.35 leading, -0.94;
0.4 (leading), -0.92

11.7 a) 3 V
b) 4 W

11.10 56.25 mW, -70.31 m VAR,
$56.25 - j70.31$ mVA

11.15 3200 W

11.18 -0.8732

11.21 $g1$: 6200 W, 1490 VAR
$g2$: 5960 W, 970 VAR

11.24 a) 133.48 V
b) 256 W
c) 1788.59 μF
d) 126.83 V
e) 184.96 W

11.27 a) 36 W
b) $-250 \ \Omega$
c) 187.5 Ω
d) 23.58 W

11.30 a) 2.94 W
b) 200 Ω, 1 μF
c) 4.58 W, yes
d) 5 W
e) 250 Ω, 0.5 μF
f) yes

11.34 a) 20 Ω
b) 225 W

11.36 a) 4.47 Ω
b) 590.17 W
c) 625 W
d) 41.67%

11.39 5 mW

CHAPTER 12

12.1 a) balanced, positive
b) balanced, negative
c) balanced, positive
d) balanced, negative
e) unbalanced, unequal voltages
f) unbalanced, phase differences not equal to 120°

12.5 a) unbalanced, load impedances are not identical
b) $14.39\underline{/-107.06°}$ A

12.8 a) $\mathbf{I}_{aA} = 9.6\underline{/-16.26°}$ A
$\mathbf{I}_{bB} = 9.6\underline{/-136.26°}$ A
$\mathbf{I}_{cC} = 9.6\underline{/103.74°}$ A
b) $\mathbf{V}_{ab} = 415.69\underline{/30°}$ V
$\mathbf{V}_{bc} = 415.69\underline{/-90°}$ V
$\mathbf{V}_{ca} = 415.69\underline{/150°}$ V
c) $\mathbf{V}_{AN} = 234.29\underline{/-3.48°}$ V
$\mathbf{V}_{BN} = 234.29\underline{/-123.48°}$ V
$\mathbf{V}_{CN} = 234.29\underline{/116.52°}$ V
d) $\mathbf{V}_{AB} = 405.80\underline{/26.52°}$ V
$\mathbf{V}_{BC} = 405.80\underline{/-93.48°}$ V
$\mathbf{V}_{CA} = 405.80\underline{/146.52°}$ V

12.11 a) 24.76 A
b) 5.77 A
c) 15 A
d) 1350.23 V

12.13 a) $\mathbf{I}_{AB} = 9.6\underline{/-36.87°}$ A
$\mathbf{I}_{BC} = 9.6\underline{/-156.87°}$ A
$\mathbf{I}_{CA} = 9.6\underline{/83.13°}$ A
b) $\mathbf{I}_{aA} = 16.63\underline{/-66.87°}$ A
$\mathbf{I}_{bB} = 16.63\underline{/173.13°}$ A
$\mathbf{I}_{cC} = 16.63\underline{/53.13°}$ A

c) $\mathbf{I}_{ba} = 9.6\underline{/-36.87°}$ A
$\mathbf{I}_{cb} = 9.6\underline{/-156.87°}$ A
$\mathbf{I}_{ac} = 9.6\underline{/83.13°}$ A

12.18 a) $9.28\underline{/97.70°}$ A
b) 97.3%

12.21 a) $258,975 + j105,450$ VA
b) 98.47%

12.23 a) $0.68 + j3.05$ Ω
b) $R = 14.4$ Ω, $X = 3.20$ Ω

12.25 a) 4467.43 V
b) 97.83%

12.27 a) 9081.94 V
b) 8394.89 V
c) 97.88%
d) 99.22%
e) 15.35 μF

CHAPTER 13

13.1 a) 0.95
b) 24 mH
c) 2.5

13.4 upper terminal

13.8 a) $58.32 - j43.74$ Ω
b) $100 + j136.26$ Ω
c) 17.08 V
d) 21.87%

13.11 72 W

13.14 a) 5.20 W
b) 17.18%

13.17 0.50

13.20 $55.38 + j13.08$ Ω

13.25 156.25 W

13.28 a) $16 - j128$ Ω
b) $0.75 \underline{/180°}$ A, $96.75 \underline{/97.13°}$ V

13.31 $1000 \underline{/30°}$ Ω

13.35 a) $\omega(L_1 - M) = 40$ Ω, $\omega(L_2 - M) = 0$ Ω, $\omega M = 40$ Ω
b) $76.8 \cos(4000t - 53.13°)$ V
c) $76.8 \cos(4000t - 233.13°)$ V

13.36 a) 15 H, 40 H, 10 H
b) 0.4082

13.39 $86.10 \underline{/-41.99°}$ V

13.40 a) $4623.56 \underline{/-5.53°}$ V
b) $1334.71 \underline{/-35.53°}$ V

c) $88.98 \underline{/-35.53°}$ A

d) 356.29 kW

e) 368.16 kW

f) 96.77%

CHAPTER 14

14.1 a) 80 kΩ, 0.02 pF

b) 0.9692×10^{10} rad/s, 1.0317×10^{10} rad/s

14.4 a) 1591.55 Hz

b) 120 V

c) 1000 rad/s

d) 10

e) 9512.49 rad/s, 10,512.49 rad/s

f) 20%

14.7 20 kΩ, 2 mH, 0.32 nF, 12.5 mA

14.10 65.08 μH

14.13 $5.14 \underline{/-38.94°}$ Ω

14.16 a) 5.1 Mrad/s

b) 20.48 V

c) 5.14 Mrad/s

d) 20.58 V

e) 871.85 krad/s

f) 9.91

g) 5.85

14.19 a) 7 rad/s

b) 10 V

c) 5.6

d) 1.25 rad/s

e) 7.09 rad/s

f) 10.10 V

14.22 a) 37.69

b) 21.23 krad/s

c) 625 kΩ

d) 20.73 mH

e) 75.37 pF

14.26 a) 200 mV

b) 181.83 mV

c) 199.6 mV

14.31 a) 7000 rad/s

b) $10 \cos 7000t$ V

c) 0.75 V

d) 0 V

e) not necessarily since both the real and imaginary components of Z vary with frequency

f) $10.09 \cos(7270t - 7.54°)$ V

14.36 a) 40 mH, 25 F

b) 5 kΩ, 20 μH, 500 pF

14.39 1 MΩ, 1 kΩ, 263 mH, 1.05 nF

14.41 a) 10 rad/s

b) 5 V

c) 9.05 rad/s, 11.05 rad/s

d) 2 rad/s

e) 5

f) 4 kΩ, 5 nF

14.44 14.97 krad/s

CHAPTER 15

15.2 $f(t) = (10t + 50)[u(t + 5) - u(t)]$
$$+ 50 \cos\left(\frac{\pi}{10} t\right)[u(t) - u(t - 20)]$$
$$+ (250 - 10t)[u(t - 20) - u(t - 25)]$$

15.7 a) $1/(s + a)^2$

b) $\dfrac{s}{s^2 + \omega^2}$

c) $(s \cos\theta - \omega \sin\theta)/(s^2 + \omega^2)$

d) $1/(s^2 - 1)$

15.10 a) $s\omega/(s^2 + \omega^2)$

b) $s^2/(s^2 + \omega^2)$

c) 2

15.14 a) $\dfrac{5}{s^2}[1 - 2e^{-2s} + 2e^{-6s} - e^{-8s}]$

b) $\dfrac{5}{s}[1 - 2e^{-2s} + 2e^{-6s} - e^{-8s}]$

c) $5[1 - 2e^{-2s} + 2e^{-6s} - e^{-8s}]$

15.20 $[100e^{-300t} \sin 400t]u(t)$ mA

15.22 $80[e^{-8000t} - e^{-2000t}]u(t)$ V

15.25 a) $5\,\delta(t) + (12e^{-2t} - 4e^{-4t})u(t)$

b) $10\,\delta(t) + [40e^{-24t} \cos(7t - 36.87°)]u(t)$

c) $\delta'(t) - 10\,\delta(t) + [30e^{-5t} + 20e^{-10t}]u(t)$

15.27 $[3 - 5e^{-20,000t} + 5e^{-80,000t}]u(t)$ A

15.30 a) 8, 0

b) 8, 10

c) 22, 0

d) 250, 490

15.33 $i_o(0^+) = I_{dc}$; $i_o(\infty) = 0$

CHAPTER 16

16.4 a) $4 \times 10^6 s/(s^2 + 2000s + 640,000)$

b) $-p_1 = -400$ rad/s, $-p_2 = -1600$ rad/s, $-z_1 = 0$ rad/s

16.7 $-p_1 = -1$ rad/s, $-z_1 = 0$ rad/s

16.10 a) $150/(s^2 + 7s + 10)$

b) $v_o(0^+) = 0$, $v_o(\infty) = 0$

c) $50\left[e^{-2t} - e^{-5t}\right] u(t)$ V

16.13 $\dfrac{-72s^2 - 100{,}800s + 15 \times 10^8}{s(s^2 + 1400s + 6.25 \times 10^6)}$

$[240 + 325e^{-700t} \cos(2400t - 196.26°)] u(t)$ V

16.16 $(30{,}000te^{-5000t} + 12e^{-5000t}) u(t)$ V

16.19 $\left[-45te^{-5t} + 24e^{-5t}\right] u(t)$ V

16.22 b) $I_1 = 10 \times 10^{-3}/(s + 50)$

$V_1 = 1000/s(s + 50)$

$V_2 = 250/s(s + 50)$

c) $i_1 = 10e^{-50t} u(t)$ mA

$v_1 = [20 - 20e^{-50t}] u(t)$ V

$v_2 = [5 - 5e^{-50t}] u(t)$ V

16.24 $\left[-20e^{-t} \cos 2t\right] u(t)$ A

16.27 a) $I_a = 18/s(s + 3)$,

$I_b = \dfrac{9(s + 1)}{s(s + 3)}$

b) $i_a = 6(1 - e^{-3t}) u(t)$ A,

$i_b = (3 + 6e^{-3t}) u(t)$ A

c) $V_a = 90(s + 1)/s^2(s + 3)$

$V_b = -90(s - 1)/s^2(s + 3)$

$V_c = 90(s + 1)/s^2(s + 3)$

d) $v_a = (30t + 20 - 20e^{-3t}) u(t)$ V

$v_b = (30t - 40 + 40e^{-3t}) u(t)$ V

$v_c = (30t + 20 - 20e^{-3t}) u(t)$ V

e) 32.67 s

16.30 $[17.28 + 28.8e^{-10^5 t}$

$\cos(75{,}000t - 233.13°)] u(t)$ mA

16.34 $\left[7.5 - 75{,}000te^{-10{,}000t} - 7.5e^{-10{,}000t}\right] u(t)$ A

16.37 $\left[-60e^{-10^4 t} \sin 2 \times 10^4 t\right] u(t)$ V

16.40 $\left[e^{-20t} + 2e^{-50t}\right] u(t)$ A

16.45 a) $[40 - 30e^{-5t}] u(t)$ V

$80 \, \delta(t) + [400 + 900e^{-5t}] u(t)$ μA

16.48 a) $0.5e^{-25t} u(t)$ V

16.51 a) $-30 \times 10^6/s(s + 1000)(s + 2000)$

b) $[-15 + 30e^{-1000t} - 15e^{-2000t}] u(t)$ V

c) 1.7 ms

d) $\leq 4000 \, V/s$

16.54 a) $[7.5 - 15e^{-500t} + 7.5e^{-1000t}] u(t)$ V

b) 4.5 ms

CHAPTER 17

17.1 a) $V_o = \dfrac{480(s + 2)}{s(s + 4)(s + 6)} + \dfrac{240}{s(s + 4)(s + 6)}$

b) $(50 + 90e^{-4t} - 140e^{-6t}) u(t)$ V

17.4 a) 0.5 J

b) 0 J

c) $[1118.03e^{-3000t} \cos(4000t - 26.57°) \, u(t)$ V

17.8 a) $100(1 - \mu)/(s + 150 - 100\,\mu)$

b) $\mu < 1.5$

c) $7.5(1 - e^{-200t}) u(t)$ A $(\mu = -0.5)$

$\left(\dfrac{20}{3}\right) (1 - e^{-150t}) u(t)$ A $(\mu = 0)$

0 A $(\mu = 1.0)$

$-500t \, u(t)$ A $(\mu = 1.5)$

$20(1 - e^{50t}) u(t)$ A $(\mu = 2.0)$

17.10 a) $y(t) = 0$, $t \leq 0$

$y(t) = 625t$, $0 \leq t \leq 10$

$y(t) = 625(20 - t)$, $10 \leq t \leq 20$

$y(t) = 0$, $20 \leq t \leq \infty$

b) $y(t) = 0$, $t \leq 0$

$y(t) = 312.5t$, $0 \leq t \leq 10$

$y(t) = 3125$, $10 \leq t \leq 20$

$y(t) = 312.5(30 - t)$, $20 \leq t \leq 30$

$y(t) = 0$, $30 \leq t \leq \infty$

c) $y(t) = 0$, $t \leq 0$

$y(t) = 625t$, $0 \leq t \leq 1$

$y(t) = 625$, $1 \leq t \leq 10$

$y(t) = 625(11 - t)$, $10 \leq t \leq 11$

$y(t) = 0$, $11 \leq t \leq \infty$

17.13 a) $v_o = 0$, $t \leq 0$

$v_o = 20(1 - e^{-200t})$ V, $0 \leq t \leq 5$ ms

$v_o = 20(e - 1)e^{-200t}$ V, 5 ms $\leq t \leq \infty$

17.16 $60e^{-2t}u(t)$ V

17.19 a) $v_o = 0$, $t \leq 0$

$v_o = 5(1 - e^{-50t})$ V, $0 \leq t \leq 0.1$ s

$v_o = 5[2e^{-50(t-0.1)} - e^{-50t} - 1]$ V,

0.1 s $\leq t \leq 0.2$ s

$v_o = 5[2e^{-50(t-0.1)} - e^{-50(t-0.2)} - e^{-50t}]$ V,

0.2 s $\leq t \leq \infty$

b) $i_o = 50e^{-50t}$ μA, $0 < t < 0.1$ s

$i_o = 50[e^{-50t} - 2e^{-50(t-0.1)}]$ μA,

0.1 s $< t < 0.2$ s

$i_o = 50[e^{-50t} - 2e^{-50(t-0.1)} + e^{-50(t - 0.2)}]$ μA,

0.2 s $< t < \infty$

17.23 $12\sqrt{2}\cos(3t + 8.13°)$ V

17.26 $8\cos(50t - 126.87°)$ V

17.31 a) 0 rad/s, 13,820.27 rad/s
b) 9772.41 rad/s
c) 3.37

17.33 a) 25 Ω
b) -3.52 dB

CHAPTER 18

18.3 a) $\dfrac{4V_m}{\pi} \displaystyle\sum_{n=1,3,5}^{\infty} \dfrac{1}{n} \sin n\omega_o t$

b) $\dfrac{2V_m}{\pi} \left[1 + 2 \displaystyle\sum_{n=1}^{\infty} \dfrac{1}{(1 - 4n^2)} \cos n\omega_o t \right]$

c) $\dfrac{V_m}{\pi} + \dfrac{V_m}{2} \sin n\omega_o t + \dfrac{2V_m}{\pi} \displaystyle\sum_{n=2,4,6}^{\infty} \dfrac{1}{(1 - n^2)}$ $\cos n\omega_o t$

18.9 a) $\dfrac{4V_m}{\pi} \displaystyle\sum_{n=1,3,5}^{\infty} \left[\dfrac{1}{n} \sin \dfrac{n\pi}{2} \right] \cos n\omega_o t$

b) $-\dfrac{8V_p}{\pi^2} \displaystyle\sum_{n=1,3,5}^{\infty} \dfrac{1}{n^2} \cos n\omega_o t$

18.11 a) $-\dfrac{2I_m}{\pi} \displaystyle\sum_{n=1}^{\infty} \dfrac{1}{n} \cos n\pi \sin n\omega_o t$

b) $-\dfrac{2I_m}{\pi} \displaystyle\sum_{n=1}^{\infty} \dfrac{1}{n} \sin n\omega_o t$

18.14 a) 0.2π rad/s
b) no
c) yes
d) no

18.18 a) $10 \displaystyle\sum_{n=1,3,5}^{\infty} \left[\dfrac{\sqrt{(\pi n)^2 + 4}}{n^2} \right] \cos(n\omega_o t - \theta_n)$ A

A where $\tan \theta_n = \dfrac{\pi n}{2}$

b) 26.23 A

18.23 $514.91 \cos(\omega_o t - 11.25°) +$
$150.27 \cos(3\omega_o t + 30.83°) +$
$20.60 \cos(5\omega_o t - 168.75°) +$
$6.76 \cos(7\omega_o t + 7.22°) + \dots$ mA

18.26 $4781.36 \cos(\omega_o t - 63.43°) +$
$1187.94 \cos(3\omega_o t - 180°) +$
$518.61 \cos(5\omega_o t - 104.04°) +$
$283.62 \cos(7\omega_o t + 201.80°) + \dots$ mA

18.29 $16.40 \cos(250t + 3.28°)$ V

18.31 a) 265.26 V
b) 283.18V
c) -6.33%

18.35 92.35 W

18.37 $j\dfrac{V_m}{\pi n} (1 - \cos n\omega_o \tau)$

18.39 a) 320 W
b) 3.15 A
c) -6.76%

18.42 7.64 W

CHAPTER 19

19.2 a) $\dfrac{2(a^2 - \omega^2)}{(a^2 + \omega^2)^2}$

b) $\dfrac{-j48a\omega(a^2 - \omega^2)}{(a^2 + \omega^2)^4}$

c) $\dfrac{a}{a^2 + (\omega - \omega_o)^2} + \dfrac{a}{a^2 + (\omega + \omega_o)^2}$

d) $\dfrac{-ja}{a^2 + (\omega - \omega_o)^2} + \dfrac{ja}{a^2 + (\omega + \omega_o)^2}$

e) $e^{-j\omega t_o}$

19.13 a) $\dfrac{\tau}{2} \left\{ \dfrac{\sin[(\omega + \omega_o)\tau/2]}{(\omega + \omega_o)(\tau/2)} + \dfrac{\sin[(\omega - \omega_o)\tau/2]}{(\omega - \omega_o)(\tau/2)} \right\}$

b) $F(\omega) \rightarrow \pi[\delta(\omega - \omega_o) + \delta(\omega + \omega_o)]$

19.15 a) $20(1 - e^{-2t}) u(t)$ A

19.17 a) $10 \operatorname{sgn}(t) - 20e^{-40t} u(t)$ V

19.19 a) $2.5e^{5t}u(-t) + (-1.25e^{-t} + 3.75e^{-5t}) u(t)$ A
b) 2.5 A
c) 2.5 A
d) $(-1.25e^{-t} + 3.75e^{-5t}) u(t)$ A
e) yes

19.22 a) $30 \operatorname{sgn}(t) + (120e^{-10t} - 180e^{-20t}) u(t)$ V
b) $v_o(0^-) = -30$ V $= v_o(0^+)$
c) 30 V

19.26 a) $v_o(0^-) = v_o(0^+) = -10$ V
b) $i_o(0^-) = 0$ A, $i_o(0^+) = 8$ A
c) $5 \operatorname{sgn}(t) + (10e^{-2t} - 20e^{-5t}) u(t) -5$ V

19.30 15.86%

19.33 50%

19.36 a) 18.17%
b) 27.23%
c) 10.57%

CHAPTER 20

20.2 $y_{11} = 50$ m℧, $y_{12} = -37.5$ m℧
$y_{21} = -37.5$ m℧, $y_{22} = -40.625$ m℧

20.5 $b_{11} = 1.08$, $b_{12} = 32 \, \Omega$
$b_{21} = 16$ m℧, $b_{22} = 1.4$

20.7 $z_{11} = 24 \, \Omega$, $z_{12} = 2 \, \Omega$
$z_{21} = 2 \, \Omega$, $z_{22} = 21 \, \Omega$

20.10 $h_{11} = 10 + j20 \, \Omega$, $h_{12} = 0.2 + j0.4$
$h_{21} = 0$, $h_{22} = 4 + j8$ m℧

20.14 $a_{11} = -1.5$, $a_{12} = -4 \, \Omega$
$a_{21} = -(800 + j150)$ mΩ
$a_{22} = -1.2 - j0.4$

20.19 $g_{11} = 1.25$ m , $g_{12} = 0$
$g_{21} = 2.25$, $g_{22} = 100 \, \Omega$

20.29 a) $a_{11} = \dfrac{-1 + j1}{3}$, $a_{12} = \dfrac{-50}{3}(24 - j1) \, \Omega$

$a_{21} = \dfrac{j}{150}\text{℧}$, $a_{22} = \dfrac{-8}{3}(2 - j1)$

b) $\mathbf{V}_{Th} = 432.67 \,\underline{/-123.69°}$ V
$\mathbf{Z}_{Th} = 898.29 \,\underline{/47.43°} \, \Omega$

c) $248.88 \cos(4000t - 146.06°)$ V

20.31 a) 7.81 V

b) 21 W

c) 10%

20.34 a) $h_{11} = \dfrac{(1/C)s}{s^2 + (1/LC)}$

$h_{12} = \dfrac{(1/LC)}{s^2 + (1/LC)}$

$h_{21} = \dfrac{-(1/LC)}{s^2 + (1/LC)}$

$h_{22} = \dfrac{Cs[s^2 + (2/LC)]}{s^2 + (1/LC)}$

b) $[30 - 40e^{-2500t} + 10e^{-10,000t}]u(t)$ V

INDEX

PERIODIC FUNCTIONS

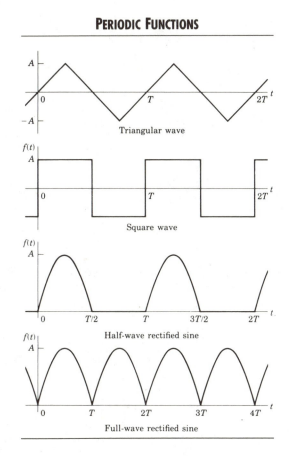

Triangular wave

Square wave

Half-wave rectified sine

Full-wave rectified sine

FOURIER SERIES

$$f(t) = \frac{8A}{\pi^2} \sum_{n=1,3,5,}^{\infty} \left[\frac{1}{n^2} \sin\left(\frac{n\pi}{2}\right) \right] \sin n\omega_0 t$$

Triangular wave

$$f(t) = \frac{A}{\pi} + \frac{A}{2} \sin \omega_0 t - \frac{2A}{\pi} \sum_{n=2,4,6,}^{\infty} \frac{\cos n\omega_0 t}{(n^2 - 1)}$$

Half-wave rectified sine

$$f(t) = \frac{4A}{\pi} \sum_{n=1,3,5,}^{\infty} \frac{1}{n} \sin n\omega_0 t$$

Square wave

$$f(t) = \frac{2A}{\pi} - \frac{4A}{\pi} \sum_{n=1}^{\infty} \frac{\cos n\omega_0 t}{(4n^2 - 1)}$$

Full-wave rectified sine